5th International Conference on Advanced Algorithms and Control Engineering (ICAACE 2022)

Journal of Physics: Conference Series Volume 2258

Sanya, China
20 – 22 January 2022

ISBN: 978-1-7138-5512-5
ISSN: 1742-6588

Printed from e-media with permission by:

Curran Associates, Inc.
57 Morehouse Lane
Red Hook, NY 12571

Some format issues inherent in the e-media version may also appear in this print version.

Institute of Physics
Dirac House, Temple Back
Bristol BS1 6BE UK

Phone: 44 1 17 929 7481
Fax: 44 1 17 920 0979

techtracking@iop.org

Additional copies of this publication are available from:

Curran Associates, Inc.
57 Morehouse Lane
Red Hook, NY 12571 USA
Phone: 845-758-0400
Fax: 845-758-2633
Email: curran@proceedings.com
Web: www.proceedings.com

TABLE OF CONTENTS

Journal of Physics: Conference Series 2258 (2022) 011001 doi:10.1088/1742-6596/2258/1/011001

Preface

Due to recent pandemic, 2022 5th International Conference on Advanced Algorithms and Control Engineering (ICAACE 2022) was held virtually online during January 20-22, 2022. The decision to hold the virtual conference was made in compliance with many restrictions and regulations that were imposed by countries around the globe. Such restrictions were made to minimize the risk of people contracting or spreading the COVID-19 through physical contact. There were 140 individuals who attended this on-line conference, represented many countries including Malaysia, India, Iran, and China.

ICAACE 2022 is to bring together innovative academics and industrial experts in the field of advanced algorithms and control engineering to a common forum. The primary goal of the conference is to promote research and developmental activities in advanced algorithms and control engineering. And another goal is to promote scientific information interchange between researchers, developers, engineers, students, and practitioners working all around the world.

During the conference, the conference model was divided into three sessions, including oral presentations, keynote speeches, and online Q&A discussion. In the first part, some scholars, whose submissions were selected as the excellent papers, were given about 5-10 minutes to perform their oral presentations one by one. Then in the second part, keynote speakers were each allocated 30-45 minutes to hold their speeches. In the second part, we invited two professors as our keynote speakers. Professor Jinming Wen, Jinan University, China. He is mainly engaged in the basic theoretical research and algorithm design of lattice cipher and compressed sensing. And then we had Assoc.Prof. Pavel Loskot, IEEE Senior Member, Zhejiang University-University of Illinois at Urbana-Champaign Institute (ZJUI), China. His current research interest focuses on problems involving statistical signal processing and importing methods from Telecommunication Engineering and Computer Science to other disciplines in order to improve the efficiency and the information power of system modeling and analysis. Their insightful speeches had triggered heated discussion in the third session of the conference. Every participant praised this conference for disseminating useful and insightful knowledge.

The proceedings are a compilation of the accepted papers and represent an interesting outcome of the conference. Topics include but are not limited to the following areas: Algorithms and Data Structures, Electrical systems, Machine Learning, Intelligent data analysis and Data Mining, Mathematical and computer modelling, Systems and automation, Process control and other related topics. All the papers have been through rigorous review and process to meet the requirements of international publication standard.

We would like to acknowledge all of those who supported ICAACE 2022. The help and contribution of each individual and institution was instrumental in the success of the conference. In particular, we would like to thank the organizing committee for its valuable inputs in shaping the conference program and reviewing the submitted papers.

We sincerely hope that the ICAACE 2022 turned out to be a forum for excellent discussions that enabled new ideas to come about, promoting collaborative research. We are sure that the proceedings will serve as an important research source of references and knowledge, which will lead to not only scientific and engineering findings but also new products and technologies.

The Committee of ICAACE 2022

Journal of Physics: Conference Series 2258 (2022) 011001 doi:10.1088/1742-6596/2258/1/011001

Committee member

General Conference Chair
Professor Jinming Wen, Jinan University, China

Academic Committees Member
Prof. Jun-Juh Yan, Shu-Te University, Taiwan, China
Prof. Zhongjun Ma, Guilin University of Electronic Technology, China
Prof. Johan Debayle, Ecole Nationale Supérieure des Mines (ENSM-SE), France
IR. MOHD SAFIRIN BIN KARIS, Universiti Teknikal Malaysia Melaka, Malaysia
Dr. Cihangir Kan, Xian Jiaotong Liverpool University, China
Dr. Shahzad Ashraf, Hohai University, China
Dr. Kishore Kumar Pedapenki, Vignan's Institute of Information Technology, India
Dr. Karan Aggarwal, Deemed to be University, India
Dr. TASHER ALI SHEIKH, Madanapalle Institute of Technology & Science (MITS), India
Dr. Mohammad Khishe, Iran University of Science and Technology, Iran
Dr. Ariffin, NOR HAPIZA, UNIVERSITI TEKNOLOGI MARA, MALAYSIA

International Technical Program Committees Member
Prof. Xiaodi Li, Shandong Normal University, China
Prof. D. Chimgee, National University of Mongolia, Mongolia
Prof. Ahrar Husain, Jamia Millia Islamia, India
Prof. Kuo-Chin Fan, National Central University, Taiwan, China
Prof. David Carabias Galindo, Universidad de Valladolid, Spain
Prof. George Carutasu, Romanian American University, Romania
Prof. Koh Kakusho, Kwansei Gakuin University, Sanda, Japan
Prof. Bin Zi, Hefei University of Technology, China
Lecturer Peng Chen, Southwest Jiaotong University, China
Dr. Filippo Gomez Paloma, University of Salerno, Italy
Dr. V. Rossano, University of Bari, Italy

Peer Review Statement

All papers published in this volume have been reviewed through processes administered by the Editors. Reviews were conducted by expert referees to the professional and scientific standards expected of a proceedings journal published by IOP Publishing.

- **Type of peer review:** Single Anonymous
- **Conference submission management system:** Morressier
- **Number of submissions received:** 151
- **Number of submissions sent for review:** 145
- **Number of submissions accepted:** 102
- **Acceptance Rate (Submissions Accepted / Submissions Received × 100):** 67.5
- **Average number of reviews per paper:** 2
- **Total number of reviewers involved:** 20
- **Contact person for queries:**
  Name: Xuexia Ye
  Email: xx.ye@keoaeic.org
  Affiliation: AEIC Academic Exchange Information Centre

Traffic classification algorithm using CNN and multi-head attention mechanism for representation learning

HaoLi[1a*]

[1] Xi'an Jiaotong University, Xi'an, China

[a*]lucashaoli121@gmail.com

Abstract—With the development of the times, the demand for privacy and security by people, enterprises and government units has become stronger and stronger, driven by this, encrypted traffic has shown a blowout growth, and the security of traffic has become an issue that cannot be ignored. To solve the traffic classification problem, this paper proposes a new traffic classification algorithm based on convolutional neural network and multi-head attention mechanism. In addition, this paper uses a feature engineering method based on representation learning and proposes a discard threshold to improve the quality of data sets obtained by feature engineering. The experimental results show that the algorithm model and the optimized feature engineering method proposed in this paper have good performance on the traffic classification task.

1. Introduction

In the field of network security, traffic classification refers to the classification of traffic into benign traffic and malicious traffic according to certain criteria. Recently, with the development of network security technology, more and more network traffic is encrypted via HTTPS to ensure the security of the transmission, but many malicious flows also use the same practice. This situation that poses a challenge to security analysts because Encrypted malicious traffic looks the same as normal traffic [1].

There are four common methods of classifying encrypted traffic [2]: port-based, deep packet inspection (DPI)-based, statistics-based and behaviour-based. For the port-based classification method, the prevalence of random and spoofed ports makes the method unreliable. DPI-based methods require decryption of encrypted traffic to be valid, but the cost of decrypting traffic is high. As a result, the main traffic classification method currently available is a statistical and behavioural based approach [3]. While this approach can solve problems that other traditional methods cannot, it also faces difficulties and challenges such as complex feature engineering, heavy reliance on expert experience, and the failure of older feature engineering as encryption protocols are updated.

In this paper, we use a feature engineering method based on representation learning [4] to transform raw traffic packets into traffic pictures, and on this basis, we propose an optimal discard threshold to optimise this feature engineering method. For traffic pictures, this paper proposes a classification model CMAFTC based on convolutional neural network and multi-head attention, which has better performance in the case of fourteen classifications.

2. Feature engineering based on representational learning

2.1 pcap package formatting

If an area is compromised by the same malicious traffic, we believe that it is likely to make the compromised machines communicate with the same IP address, so IP address information can be used as a feature to determine whether it is malicious traffic, so the IP header part should be reduced when formatting pcap packets. Through the analysis of IPv4 and IPv6 protocols [5], the header length of IPv4 is mostly 20bytes with a maximum of 40bytes, and the header length of IPv6 is 40bytes. Both types of header information contain key data such as the source IP address and destination IP address. Therefore, we intercepted 784bytes of pcap packets in byte stream format, which includes the IP header, TCP/UDP header and some load information. If the length of the pcap packet is less than 784bytes, 0x00 is used to make up the length. For packets that are too small in length, we consider that it does not provide valid information and therefore discard this part of the packet. The formatting method [4] is shown in Figure 1.

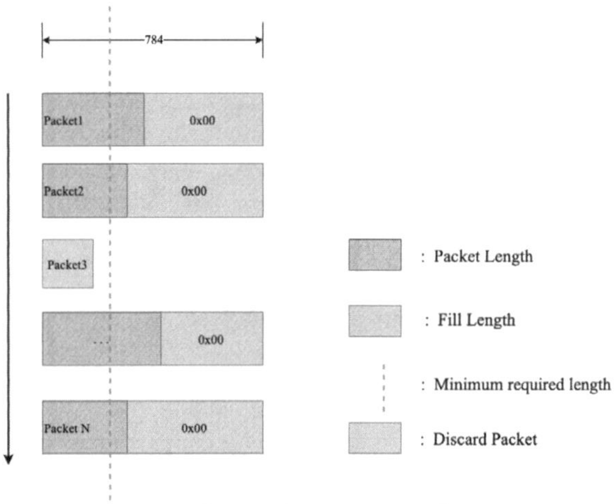

Fig.1 Package formatting method of a traffic flow

2.2 Conversion of pcap packages to images

According to the method of representation learning [4], the formatted pcap package is transformed into a 28*28bytes grayscale map, and the transformation process and results are shown in Figure 2.

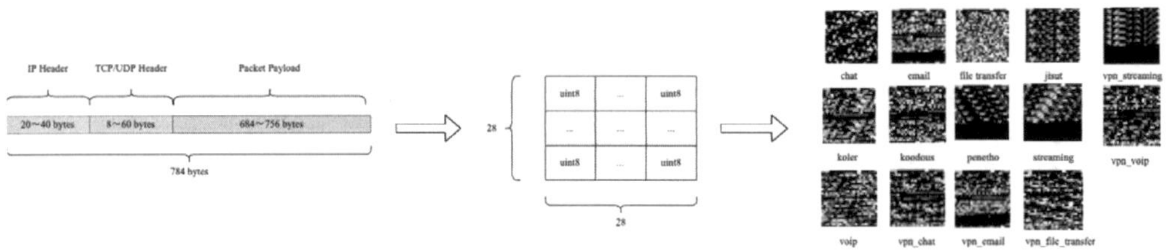

Fig. 2 Transformation process and results

2.3 Optimal discard threshold in pcap package formatting

With IP header lengths ranging from 20bytes to 40bytes in size and TCP/UDP headers ranging from 8bytes to 60bytes in size, we believe that some incomplete packets may not contain transport protocol header information or only contain a smaller amount of payload data. The content of the image data converted from such incomplete packets is almost always padded data, and the images appear to be padded black, as shown in Figure 3. There are a certain number of incomplete packets in each class of traffic, and the fill data are all the same, so these images will reduce the classification effect of the classification model.

Fig. 3 Image obtained from incomplete packet conversion

3. Classification models based on CNN and multi-head attention

3.1 Multi-head attention

The multi-head attention layer can acquire the global field of view without affecting the resolution of the image features, using Query-Key-Value to effectively capture the long-range dependencies in the image and add different weights to the different attributes. Multi-head attention uses different weight matrices W_q, W_k, W_v for Query, Key and Value to perform multiple linear transformations to obtain different vector attentions $head_i$, and finally merges the output global features, which can learn the key information in different subspaces very efficiently [6], which are calculated as follows:

$$Attentino(Q,K,V) = softmax(QK^T/\sqrt{d_k})V \quad (1)$$

$$\mathrm{MultiHead(Q,K,V) = Concat(head_1, head_2, \dots, head_h)W^o} \quad (2)$$

$$\mathrm{head_i = Attention(QW_i^Q, KW_i^K, VW_i^V)} \quad (3)$$

where d_k is the dimension of the K matrix; the Softmax function converts the result of the normalisation of Q and K after the dot product operation into a probability distribution, which is multiplied with the matrix V to obtain the result of the summation of the attention weights.

3.2 Model structure

In this paper, we propose CMAFTC (Convolutional Multi-head Attention For Traffic Classification), which is a cryptographic traffic classification model based on convolutional neural network and multi-head attention, and the model structure is shown in Figure 4.

pcap packets are firstly passed through data pre-processing to obtain traffic images, The image is randomly flipped horizontally and then the image is passed into the classification model. The image goes through a convolutional network layer for local feature extraction. First, a two-dimensional convolution with a convolution kernel size of 1 and a stride of 1 is used to expand the number of channels to 32, and then the image will pass through three groups of convolutional networks, each group of convolutional networks contains four convolution kernels with different sizes, after three sets of convolutional networks, the number of channels of the image will be expanded to 128. Then, the image matrix is positionally encoded and entered the multi-head attention layer for global feature calculation. For the multi-head attention layer, the number of input channels is 384, the number of multi-heads is 6, and the number of self-attention layers is 8. Then, the classification results are obtained through the fully connected layer and the Softmax activation function.

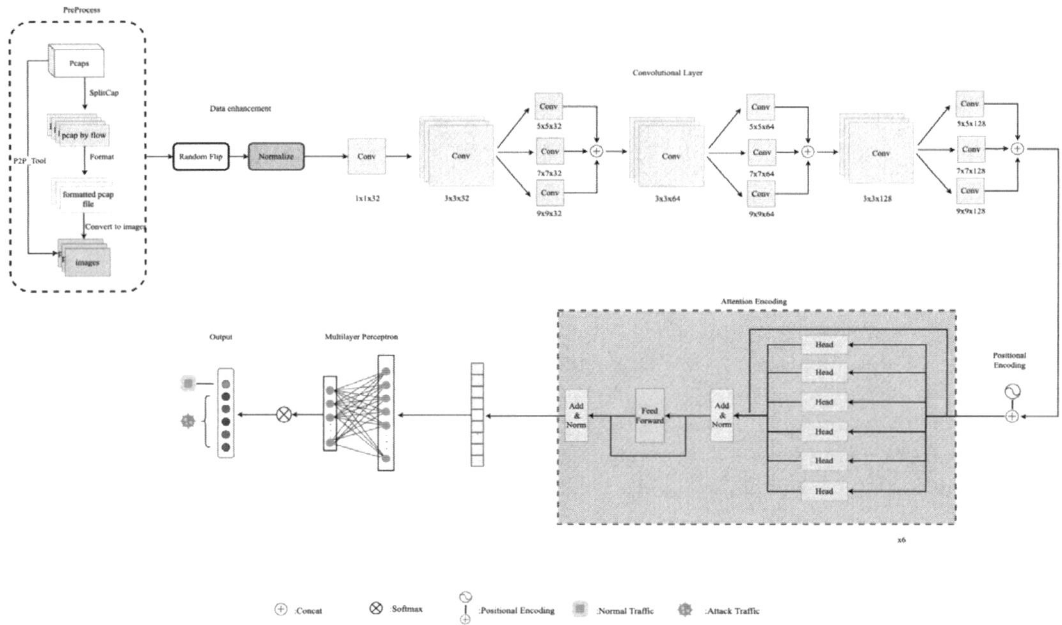

Fig. 4 CMAFTC algorithm model structure

4. Test Results and Discussions

4.1 Data set and experimental environment

The datasets for this experiment are classified as benign encrypted traffic dataset and malicious encrypted traffic dataset, where the benign traffic dataset uses ISCX2016 encrypted traffic dataset, which contains 12 types of regular network traffic based on Virtual Private Network (VPN) and encrypted network traffic based on encryption technology. The malicious encrypted traffic dataset is the 4 types of malicious encrypted traffic communicated with C&C (Command and Control) in the CICAndMal2017 dataset. The specific traffic types in the dataset and the final number of images transformed into each type of traffic are shown in Table 1.

Table 1. Experimental data set

Traffic type	Traffic name	Number of images
Benign traffic	NonVPN-Chat	9139
	NonVPN-Email	7060
	NonVPN-File Transfer	5717
	NonVPN-Streaming	1069
	NonVPN-VoIP	1837
	VPN-Chat	4243
	VPN-Email	432
	VPN-File Transfer	989
	VPN-Streaming	928
	VPN-VoIP	7286
Malicious (C&C communications) traffic	Koodous	24143
	Jisut	21139
	Penetho	20305
	Koler	25003

The hardware used for the experiments was a server with four 16-core CPUs, 256G of RAM and a GeForce RTX 3090 GPU, with Ubuntu 18.04.6, PyTorch 1.10.0, Python 3.8 and Cuda 11.3 installed.

4.2 Evaluation Metrics

To test the effectiveness of the CMAFTC model in classifying encrypted traffic from multiple aspects, we selected three evaluation metrics: precision, recall and F1 score, which are calculated as follows:

$$Precision = \frac{TP}{TP+FP} \tag{4}$$

$$Recall = \frac{TP}{TP+FN} \tag{5}$$

$$F1 - Score = \frac{2*Precision*Recall}{Precision+Recall} \tag{6}$$

where TP denotes the number of encrypted flows correctly classified as class a, FP denotes the number of encrypted flows incorrectly predicted as class a, TN denotes the number of encrypted flows correctly predicted as non-class a, and FN denotes the number of encrypted flows incorrectly predicted as non-class a.

4.3 Effect of discard threshold on classification effect

To explore the relationship between the discard threshold and the classification effect of the model, this paper set up several sets of experiments, these experiments were set to discard thresholds from 0bytes to 280bytes, the classification task was chosen to classify fourteen traffic categories, and the experimental model was chosen from the baseline paper as the 1D-CNN [4] model.

The experimental results are shown in Figures 5 and 6, where Figure 5 shows the number of images in the dataset after using different discard thresholds and Figure 6 shows the f1 score versus the discard threshold.

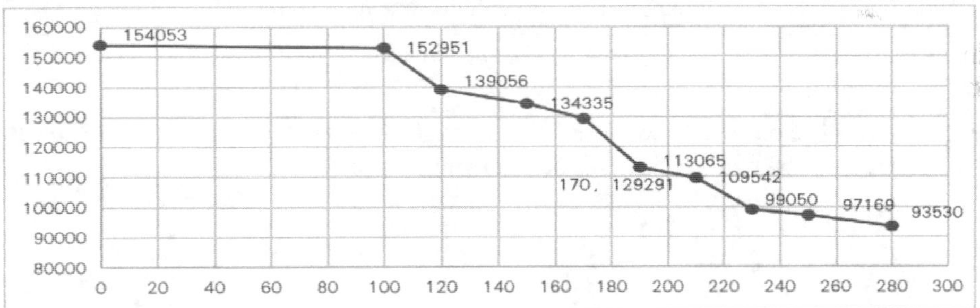

Fig 5. Relationship between the number of images and the discard threshold

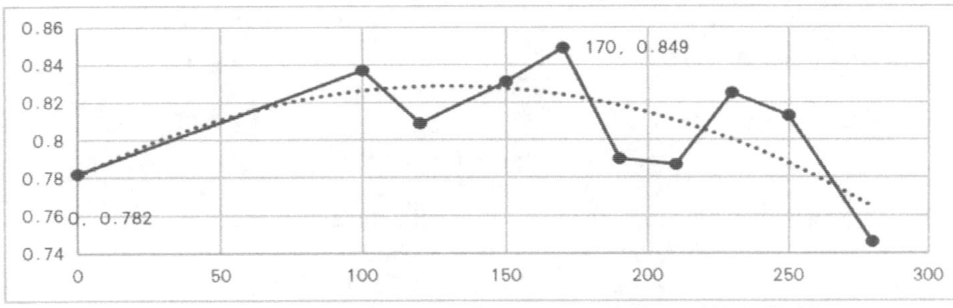

Fig 6. Relationship between f1 scores and discard threshold

From Figure 5, we can see that pcap packets with packet lengths between 0 and 100bytes are a very small fraction of the packets, which indicates that most of the pcap traffic packets contain header information of the transport protocol; when the rounding threshold is greater than 100, the number of rounded pcap traffic packets increases linearly with the increase of the rounding threshold. For each data set, the average result of three experiments is used as the result of the experiment. Figure 6 shows that the F1 score of the model classification increases and then decreases with the increase of the discard threshold, this is because as the discard threshold increases, the quality of the traffic images in the dataset is improved by discarding images with too little information, and the model can learn the difference between the features of the traffic better, so the classification of the model becomes better. This is because as the discard threshold increases, the total number of images in the dataset decreases, resulting in less data for the model to learn; in addition, as the discard threshold increases, traffic images containing more load information are retained, and this increased load information has a diminishing marginal benefit for the model to learn, so the classification effect decreases. In this experiment, 170bytes was the optimal discard threshold.

4.4 Classification experiments with different particle sizes

To verify the effectiveness of the CMAFTC model proposed in this paper, this section designs two experimental scenarios with different classification granularities, numbered 1 and 2. The first experiment is a triple-classification scenario, where traffic is classified into malicious traffic, VPN traffic, and non-VPN traffic based on protocols and usage properties; the second experiment is a fourteen-classification scenario, where traffic is classified into specific categories.

We use BCELoss as the loss function, set the batch size to 24 and set the learning rate to 1e-3. The learning rate decreases with epoch and becomes the original 0.978 every 2 epochs.

The baseline method was selected and implemented as 1D-CNN [4], C4.5. The specific experimental setup is shown in Table 2.

Table 2. Parameters of experiment

Number of Experiment	Classification granularity	Number of classes
1	Identification of protocol-encapsulated traffic	3
2	Encrypted traffic classification	14

For each group of experiments, we used the average of three experiments as the result, and the classification results of the two experiments are shown in Figure 7.

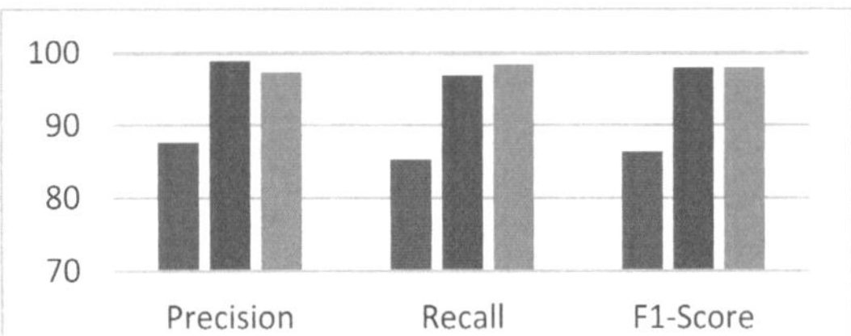

(a) Result of Experiment 1

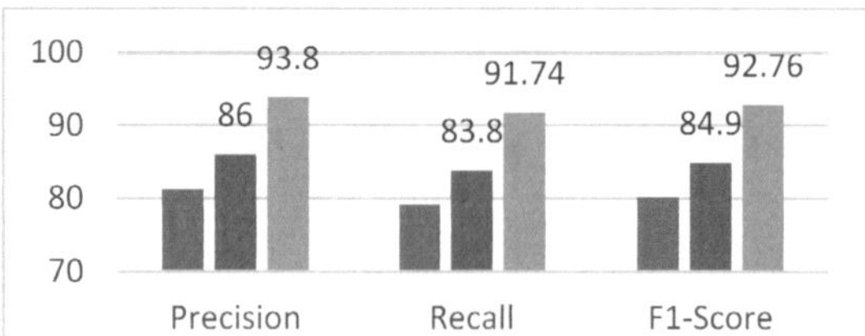

(b) Result of Experiment 2

Fig 7. Experimental results of flow classification

From the experimental results, it can be seen that in the three classification scenarios in Experiment 1, the 1D-CNN and CMAFTC models among the baseline algorithm models both performed well, with F1 scores above 90 for classification, while the C4.5 algorithm had slightly worse classification results; in the fourteen classification scenarios in Experiment 3, due to the small classification granularity, the classification results of the models all decreased, but it can be seen that the CMAFTC model significantly outperformed the other two baseline models, and compared to the suboptimal result 1D-CNN, the CMAFTC model improved 7.8% in precision, 7.94 in recall, and 7.86 in F1 score.

4.5 Discussion
The CMAFTC traffic classification model based on representation learning proposed in this paper can effectively improve the classification of traffic. However, the CMAFTC model still has shortcomings when facing specific scenarios. For example, when facing new niche traffic, the data samples that can be collected are very scarce, which makes it difficult for the model to learn the characteristics of the traffic, and due to the lack of data samples, the data is severely imbalanced, which causes the conventional classification prediction model to ignore the characteristics of these small samples that should really receive more attention, and therefore, these niche malicious traffic that should receive more attention is difficult to classify correctly in conventional classification models.

5. Conclusion
Based on the feature engineering method of representation learning, the optimal discard threshold is proposed and found, and the optimised feature engineering method can effectively improve the quality of the dataset. In addition, a new model of encrypted traffic classification and detection algorithm based on CNN convolutional neural network and multi-head attention mechanism is proposed. The algorithm extracts the local features of the image by CNN, encodes the position of the image and then obtains the global field of view of the image by multi-head attention and inputs it into the classifier to obtain the final classification result. The results show that the new feature engineering method outperforms the original feature engineering method, and the CMAFTC model proposed in this paper outperforms the baseline model in traffic classification. There are still many difficulties and challenges in the field of traffic classification, such as the problem of data scarcity. In future research, we will consider using transfer learning and adversarial generative networks to solve them.

Acknowledgments
This article could not have been completed without the help of my teachers and classmates.

References
[1] Z. Cao, G. Xiong, Y. Zhao, Z. Li and L. Guo, "A survey on encrypted traffic classification" in Applications and Techniques in Information Security, Springer, pp. 73-81, 2014.
[2] E. Biersack, C. Callegari and M. Matijasevic, Data traffic monitoring and analysis. Berlin: Springer, 2013.
[3] Y. Dhote, S. Agrawal. "A Survey on Feature Selection Techniques for Internet Traffic Classification". in Computational Intelligence and Communication Networks, Jabalpur, 2015, pp. 1375-1380.
[4] Wang, Wei, et al. "Malware traffic classification using convolutional neural network for representation learning." 2017 International Conference on Information Networking (ICOIN). IEEE, 2017.
[5] Ali, Amer Nizar Abu. "Comparison study between IPV4 & IPV6." International Journal of Computer Science Issues (IJCSI) 9.3 (2012): 314.
[6] Vaswani, Ashish, et al. "Attention is all you need." Advances in neural information processing systems. 2017.

Research on Point Cloud Registration Algorithm Based on Boundary Harris-3D Points and 4PCS

Zhiwei Xu[1a*], Di Wang[2b*]

Department of mechanical manufacturing and automation, Shanghai University, Shanghai, China

[a*]x188zw@163.com, [b*]diwang@mail.shu.edu.cn

Abstract—Point cloud registration based on the 4PCS+ICP algorithm will lead to many problems such as large registration error, low efficiency, and unstable running time, especially when it encounters large point cloud data, these problems will be more prominent, which greatly limits the application of the algorithm. Based on these shortcomings, we propose an optimized registration algorithm, called BH4PCS, to improve the applicability of the 4PCS algorithm. In our method, we first extract the boundary Harris-3D points through the edge detection algorithm and Harris-3D key point extraction method, then we use the 4PCS algorithm for point cloud coarse registration, and finally use KD-ICP to speed up the fine registration. After experimental verification and comparison with the existing artworks, our proposed algorithm registers the larger point cloud model by only using a small number of data points, optimizing the registration efficiency, accuracy, and operation stability of the 4PCS + ICP registration method, and has good robustness to outliers.

1. Introduction

Point cloud registration refers to the process of integrating and splicing multiple 3D point cloud data obtained from different angles, aligning corresponding areas, and finally obtaining a unified 3D point cloud in the same coordinate system[1]. At present, the process of point cloud registration is mainly divided into two stages: coarse registration and fine registration. Coarse registration has an important impact on fine registration, which can make the source point cloud roughly align the target point cloud with a certain accuracy, avoid the fine registration falling into a local minimum, and speed up the fine registration[2]. Iterative closest point(ICP) is the most commonly used fine registration algorithm due to its simple and intuitive calculation and good registration accuracy. However, its running speed and convergence to global optimization largely depend on the given initial transformation estimation and the establishment of a corresponding relationship[3]. To achieve good final results, many scholars have been studying various coarse registration methods to obtain a good initial alignment state of two point sets and then carry out fine registration through the ICP algorithm to achieve good registration accuracy.

Radu et al. proposed a method based on point feature histogram (PFH) for the corresponding point search problem. This method uses the high-dimensional (16D) hyperspace of the histogram to describe the local geometric features of points and searches the corresponding points pairs through this histogram[4]. Then, based on that method, they proposed a fast point feature histogram (FPFH) representation, which greatly reduced the computing time of PFH by caching the previously calculated values and simplifying the interconnection information between point clouds[5]. In addition, they also proposed an initial alignment algorithm based on sample consistency(SAC-IA) to quickly search the approximate corresponding points of two point clouds. Magnusson et al. proposed a 3D-NDT point cloud registration algorithm based on joint normal distribution[6]. This method uses an iterative subdivision to detect the model surface and establish the probability density distribution of spatial point cloud, which avoids the nearest neighbor search problem and improves the registration calculation speed, and in addition, it can store the scanned surface data in less memory space. Aiger et al. proposed a registration scheme using the 4PCS algorithm for coarse registration and ICP for fine registration. The 4PCS algorithm uses affine invariance to find the corresponding point pairs in the overlapping area between the two point clouds and calculates their initial transformation matrix to complete the initial registration[7]. After that, Pascal proposed a K4PCS method based on 4PCS and applied it to environmental point cloud model registration[8].

In our research, the coarse registration scheme of 4PCS will lead to problems such as large registration error, low efficiency, and unstable running time, especially when the point cloud data is large or the model is complex. Based on these shortcomings and to improve the applicability of this method, we propose an optimized coarse registration method based on Harris-3D points at the point cloud boundary and use KD-ICP for fine registration to improve the registration efficiency[9].

2. Point Cloud Registration

In our algorithm, as shown in Fig. 1, we first extract the boundary data of the point cloud model, then use the Harris-3D key point extraction algorithm to extract features points, and eliminate the wrong point pairs according to the median distance. In the next, the 4PCS method will be used for coarse registration, and then the KD-ICP algorithm is used to improve the efficiency of fine registration.

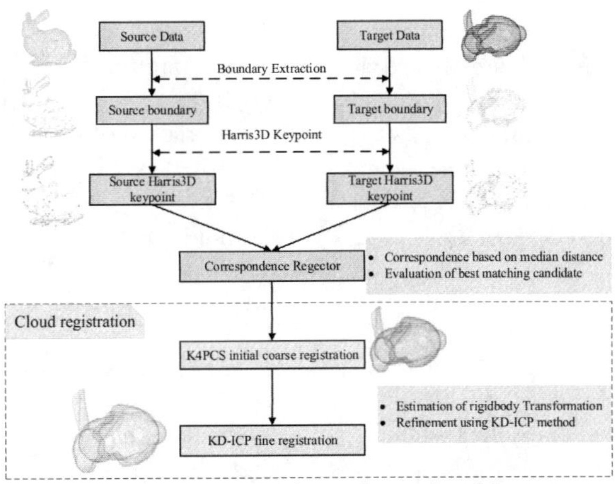

Fig. 1 Overview of our algorithm

2.1. Boundary extraction

The boundary points, as shown in Fig. 2, are mainly determined by the included angle of the vector projected in the tangent plane between the current point and other points in its neighborhood. The main process of this process includes the following steps:

(1) Construct The projection plane $\boldsymbol{S}$ by using the neighborhood relationship $N(p_i)$ of data point $\boldsymbol{p_i}$ and normal vector $\boldsymbol{n_i}$.

(2) Project the points in the set $N(p_i)$ into tangent plane S and define it as $N(p_i')$. Select a point p_j' in $N(p_i')$ and establish a local coordinate system with p_i as the origin, $\boldsymbol{n_i}$ as $\boldsymbol{w}$ axis, $\frac{\boldsymbol{p_i p_j'}}{|\boldsymbol{p_i p_j'}|}$ as $\boldsymbol{u}$ axis and $\boldsymbol{w}\times\boldsymbol{u}$ as $\boldsymbol{v}$ axis.

(3) Calculate the clockwise angle $\Psi = \{\delta_1, \dots \delta_{n-1}, \delta_n\}$ between the vector $\boldsymbol{p_i p_n'}$ and $\boldsymbol{u}$ axis, where p_n' represents other points except p_j' in $N(p_i')$.

(4) Make a difference between the adjacent included angles in $\boldsymbol{\Psi}$ to obtain the set $\boldsymbol{\Psi}' = \{\delta_1', \delta_2', \dots \delta_{n-1}'\}$. Find out the maximum angle δ_{max}' in $\boldsymbol{\Psi}'$ and compare it with the threshold. If δ_{max}' is greater than the threshold δ, it will be recognized as an edge point.

Note that the neighborhood points of the current point are obtained through the R-search method, that is, the sampling point is used as the ball center, and the ball composed of all points with radius R is used as the local neighborhood. In general, the number of boundary points extracted is about 10000 to 15000, which can meet the registration requirements.

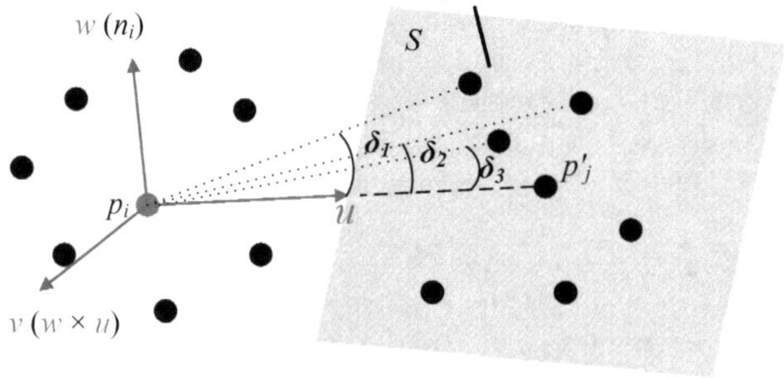

Fig. 2 Boundary extraction

2.2. Harris-3D Points

A Harris-3D algorithm is the spatial expansion of Harris-2D keypoint detection algorithm in a three-dimensional direction, which can directly extract the harris points of the point cloud in 3D space. Compared with Harris-2D, which uses the derivative value of the image to judge the corner, Harris-3D uses the normal of the discrete point cloud, that is, when the variation of the normal vector of the point exceeds the threshold, it will be considered as a Harris-3D corner[10]. The process of Harris-3D points extraction mainly includes the following steps:

(1) Calculate the normal vector $\boldsymbol{n_i}(n_x,\ n_y,\ n_z)^T$ of the point, and construct the normal covariance matrix N_{cov}, as in

$$N_{cov} = \frac{1}{N}\sum_{i\in N} \boldsymbol{n_i}\boldsymbol{n_i^T} = \frac{1}{N}\sum_{x,y,z} \begin{pmatrix} n_x^2 & n_x n_y & n_x n_z \\ n_y n_x & n_y^2 & n_y n_z \\ n_z n_x & n_z n_y & n_z^2 \end{pmatrix} \quad (1)$$

Where N is the number of points in the neighborhood of the current point.

(2) Calculate the corner response value of each point according to the corner response function F, i.e.,

$$F = Det(N_{cov}) - 0.04\big(tr(A_{cov})\big)^2 \quad (2)$$

Where Det is the determinant of the matrix N_{cov} and tr is the trace of A_{cov}.

(3) Determine whether the current point is a corner. If the value of F is greater than the given threshold ζ, and the current point is a local maximum point, we consider this point as a Harris-3D corner.

The number of Harris-3D points extracted is mainly determined by neighborhood radius R' and corner detection threshold τ. The former represents the sphere radius that is to be used for determining the nearest neighbors used for the key point detection, and the latter represents the threshold value for detecting corners, which can prevent key points from being extracted where there is no point cloud.

2.3. 4PCS

The main idea of 4PCS is to find out the corresponding four-point sets within a certain distance in the overlapping area of two-point clouds according to the affine invariance of transformation, and solve the transformation matrix according to the relationship between these point pairs.

Fig. 3 shows the principle of 4PCS. Given two input point clouds P and T, we first select an approximately coplanar four-point pair $M(a, b, c, d)$ in the point set P, where ac line and bd line intersect at point e. Define the intersection ratios ε_1 and ε_2 as follows

$$\varepsilon_1 = \frac{||a-e||}{||a-c||} \tag{3}$$

$$\varepsilon_2 = \frac{||d-e||}{||a-b||} \tag{4}$$

Note that ε_1 and ε_2 is invariant in rigid-body transformation. Four-point sets $N(m_1, m_2, n_1, n_2)$ are extracted from the target point cloud T, and after affine transformation, it can be consistent or approximately congruent with the point base M, which means

$$\frac{||m_1 - e_1||}{||m_1 - m_2||} \approx \varepsilon_1 \tag{5}$$

$$\frac{||n_1 - e_2||}{||n_1 - n_2||} \approx \varepsilon_2 \tag{6}$$

i.e., the intersection ratio of the two-point pairs $N'\{m_1, m_2\}$ and $N''\{n_1, n_2\}$ of the effective point set N needs to be approximate to that of M. Note that e_1 , e_2 follow

$$||e_1(N') - e_2(N'')|| < \zeta \tag{7}$$

$$||e_2(N') - e_1(N'')|| < \zeta \tag{8}$$

Where ζ is the distance threshold, which represents the maximum distance that e_1 and e_2 can be separated and to select the unique transformation matrix Tr_i, we set an evaluation score for it. Specifically, we select some points in P and transform them with Tr_i to see whether enough corresponding points in T are counted, then we score Tr_i according to the number of corresponding points, and finally, select Tr_i with the highest score as the best rigid transformation matrix.

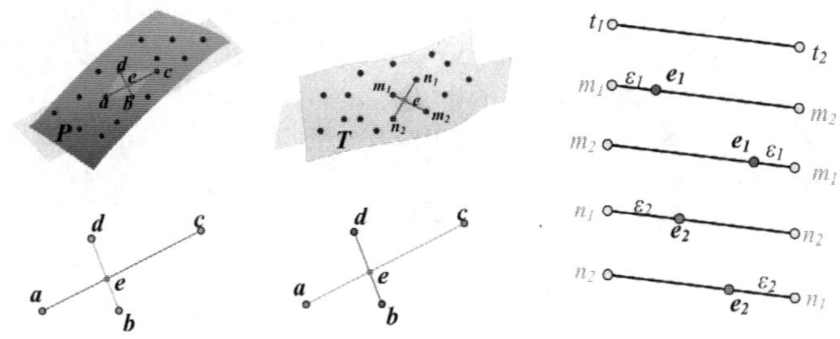

(a) (b)

Fig. 3 Principle of 4PCS

To calculate the optimal transformation matrix of the whole source point cloud P, L groups need to be tested of different basis sets M and when a constant number of random sampling points in P have enough corresponding points in T, the best transformation matrix TR for coarse registration is obtained.

3. Test Results and Discussions

All experiments are carried out on a standard 64-bit Windows 10 desktop computer with 16GB RAM and an Intel(R) Core(TM) i7-9700 3.00GHz CPU and we use PCL library and Microsoft Visual Studio as the program compilation and running environment[11].

As shown in Fig. 4, some Stanford models will be used as our experimental objects, such as Dragon, Lucy, Happy, Buddha, and Armadillo, and then two sets of data collected by the scanner will be used to test the actual effect.

Fig. 4 The experimental models

3.1. *Result and Analysis*

Results obtained by some advanced registration methods, such as FPFH +SAC-IA, NDT, and K4PCS are used to compare with our algorithm to see how well our algorithm does, and then the performance of these algorithms will be measured by geometric registration accuracy and running time.

The geometric registration accuracy between the target point cloud T and the transformed source point cloud P' is measured by the root-mean-square-error (*RMSE*), as shown in

$$RMSE = \sqrt{\frac{\sum_{i=1}^{n}(\widehat{D_\iota} - D_i)^2}{n}} \tag{9}$$

Where n is the number of corresponding point pairs, $\widehat{D_\iota}$ is the Euclidean distance between corresponding point pairs after registration, and D_i is the true value of Euclidean distance between corresponding point pairs. Ideally, the distance between corresponding points after complete registration is zero which means that the closer *RMSE* is to zero, the higher the registration accuracy is. In general, the 4-point base and its corresponding point base point cloud are randomly selected from two point clouds for registration, which means that the registration results obtained by 4PCS are not constant[12]. Therefore, to evaluate the final results, we need to run the program N_t ($N_t > 1$) times, and then the mean value is used as the evaluation index, i.e.,

$$\overline{RMSE} = \sqrt{\frac{\sum_{i=1}^{N_t}(RMSE_i)^2}{N_t}} \tag{10}$$

The value of $\overline{RMSE}$ is recorded in Table Ⅰ. According to that, we can see that our BH4PCS method has achieved very small error on each model and has high geometric accuracy, especially when the point cloud is a complete model, such as Armadillo and Dragon2, the accuracy of the other three methods are not satisfactory, which highlights the wider universality and applicability of our algorithm.

TABLE I. RMSE of registration algorithms

Model \ Algorithm(m)	NDT	FPFH+SAC-IA	K4PCS	BH4PCS
Dragon1	0.002	0.0011	0.0011	0.0011
Happy	0.0006	0.0003	0.0005	0.0003
Armadillo	0.04	0.6291	0.9027	0.0021
Dragon2	0.231	0.8378	0.8824	0.0001
Lucy	0.0044	0.0029	0.0032	0.002
Socket	0.0006	0.0005	0.0006	0.0004
Elephant	0.219	0.0214	0.4908	0.0702

The final registration results are shown in Table II and in the first column, the green point clouds represent the source point cloud and blue represent the target point cloud. According to this table, these algorithms, except the Dragon1 model and Elephant model based on NDT, have achieved good visual registration results.

TABLE II. Details of registration results

Original point clouds	Registration results			
	NDT	*FPFH+SAC-IA*	*K4PCS*	*BH4PCS*

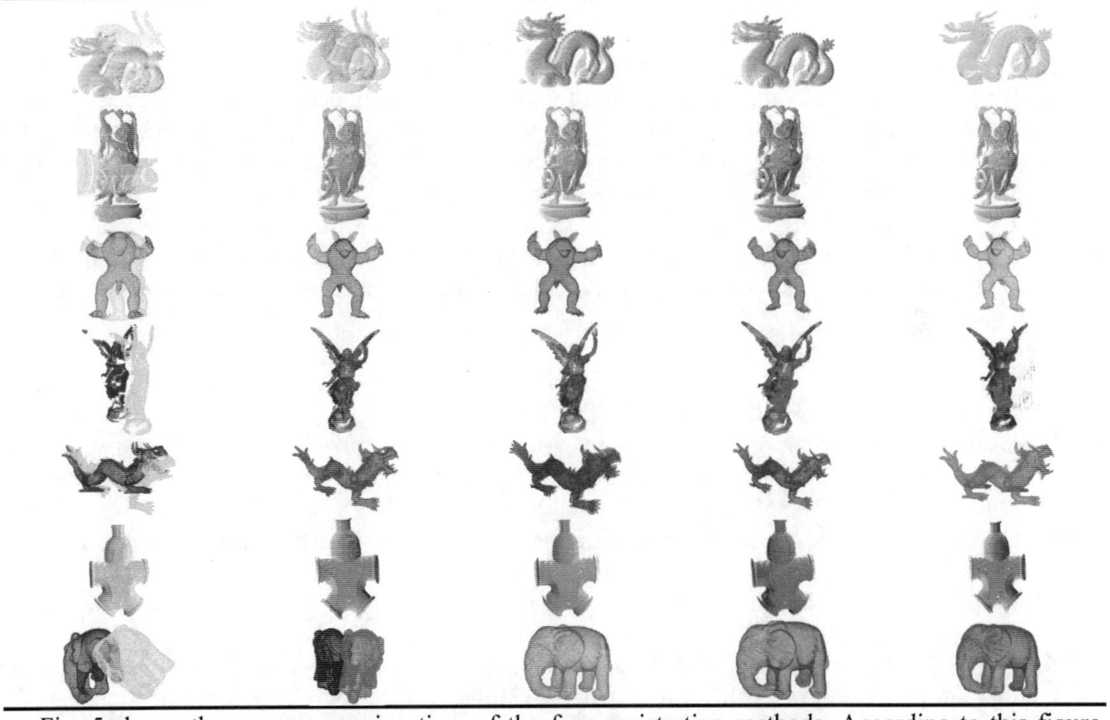

Fig. 5 shows the average running time of the four registration methods. According to this figure, K4PCS achieves the shortest running time, which means that this scheme has high registration efficiency. In practical experiments, however, it is found that the running time of K4PCS is unstable which means that the registration process of the same model may take a long time. The time of NDT is much higher than that of the other three schemes in the model of Armadillo and Dragon2, mainly because the algorithm spends a lot of time dividing the point cloud space and solving the probability density function of each point when aiming at point cloud model with a large number of point clouds and this algorithm, however, takes less time and ensures good registration accuracy when it comes to the flaking point clouds or incomplete models, indicating that it may be more suitable for the registration of such point cloud data.

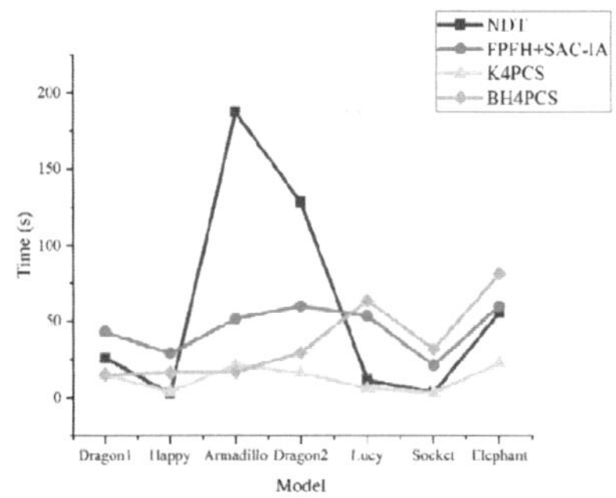

Fig. 5 Running time of four registration algorithms

According to the experimental results, our algorithm has no obvious advantage in registration time, and experiments show that BH4PCS consumes most of the time in boundary extraction and Harris-3D keypoints extraction, which may increase the whole time of the registration process. In general, our algorithm and FPFH+SAC-IA have achieved relatively consistent registration time and good operation stability. Moreover, it can be seen from Table Ⅰ that BH4PCS achieves greater geometric accuracy advantages than the other three algorithms.

4. Conclusion

In this paper, we focus on a new registration algorithm called BH4PCS, which is an optimization scheme for the 4PCS algorithm. The proposed algorithm first extracts the boundary 3D-Harris feature points of the point cloud, then uses these feature points to calculate the initial transformation matrix between the source point cloud and the target point cloud, and finally uses the KD-ICP algorithm for fine registration to complete the whole registration process. In addition, compared with some advanced registration methods, such as NDT, FPFH+SAC-IA, and KPCS, experiments show that our algorithm has achieved great advantages in geometric registration accuracy, improves the problem of operation instability of the 4PCS algorithm as well as wider universality and application scope.

References

[1] Luo, X., Zhong, Y., Li, R. Data registration in 3-D scanning systems [J]. Journal of Tsinghua University(Science and Technology), 2004, 44(8): 1104-6.

[2] Makovetskill, A., Voronin, S., Kober, V., et al. An algorithm for rough alignment of point clouds in three-dimensional space; proceedings of the 2020 International Conference on Information Technology and Nanotechnology (ITNT), F 26-29 May 2020, 2020 [C].

[3] Besl, P.J., Mckay, N.D. A method for registration of 3-D shapes [J]. IEEE Transactions on Pattern Analysis and Machine Intelligence, 1992, 14(2): 239-56.

[4] Rusu, R.B., Blodow, N., Marton, Z.C., et al. Aligning point cloud views using persistent feature histograms, F 2008]. IEEE.

[5] Rusu, R.B., Blodow, N., Beetz, M. Fast Point Feature Histograms (FPFH) for 3D registration; proceedings of the 2009 IEEE International Conference on Robotics and Automation, F 12-17 May 2009, 2009 [C].

[6] MAGNUSSON M, LILIENTHAL A, DUCKETT T. Scan registration for autonomous mining vehicles using 3D-NDT [J]. Journal of Field Robotics, 2007, 24(10): 803-27.

[7] Aiger, D., Mitra, N.J., Cohen-or, D. 4-points congruent sets for robust pairwise surface registration [J]. ACM Transactions on Graphics, 2008, 27(3): 1-10.

[8] Theiler, P.W., Wegner, J.D., Schindler, K. Keypoint-based 4-Points Congruent Sets – Automated marker-less registration of laser scans [J]. ISPRS Journal of Photogrammetry and Remote Sensing, 2014, 96.

[9] Li, S., Wang, J., Liang, Z., et al. Tree point clouds registration using an improved ICP algorithm based on kd-tree; proceedings of the 2016 IEEE International Geoscience and Remote Sensing

Symposium (IGARSS), F 10-15 July 2016, 2016 [C].
[10] Sipiran, I., Bustos, B. Harris 3D: a robust extension of the Harris operator for interest point detection on 3D meshes [J]. The Visual Computer, 2011, 27(11): 963-76.
[11] Rusu, R.B., Cousins, S. 3D is here: Point Cloud Library (PCL) [Z]. 2011 IEEE INTERNATIONAL CONFERENCE ON ROBOTICS AND AUTOMATION (ICRA). 2011
[12] Ge, X. Automatic markerless registration of point clouds with semantic-keypoint-based 4-points congruent sets [J]. ISPRS JOURNAL OF PHOTOGRAMMETRY AND REMOTE SENSING, 2017, 130: 344-57.

Interactive and Fast Low-Light Image Enhancement Algorithm and Application

Yangming Shi[1a], Xiaopo Wu[2b*], Ming Zhu[1c]

[1]Automation Department, University of Science and Technology of China, Hefei, China

[2] Department of Electronic Engineering, University of Science and Technology of China, Hefei, China

[a]ymshi@mail.ustc.edu.cn, [b*]wuxiaopo@ustc.edu.cn, [c]mzhu@ustc.edu.cn

Abstract— To obtain personalized outcomes for the low-light image enhancement, a novel interactive algorithm based on the well-designed Gamma Curve is proposed to enrich the enhancement techniques. Different from the previous works trying to enhance the image in solely brightness or naturalness by a specific designed deep network, the proposed method is capable of controlling the output results according to the user's preferences by the same framework with different parameters. There would be three main advantages brought by the proposed network: 1) Interactivity, which allows to generate enhancements results according to users' preferences in human-interactive manners; 2) Convenience, wherein the model only needs to train for once without using any reference images, and then can obtain results with different brightness during testing by adjusting the hyper-parameter. 3) Fastness, which results from the lightweight network and the excellent properties of the Gamma Curve to make the network operate in extraordinary high speed. Experiments demonstrate the superiority of our algorithm relative to the previous work. In addition, a multi-platform low-illumination enhancement software is explored to facilitate its application for the public.

1. Introduction

Low-light image enhancement tasks [1-2] aim to increase the image's brightness and recover hidden details (e.g., brilliant colors, natural structures, and high contrast) from low-light images to obtain better visual quality and boost the performance of downstream tasks. Over the past decades, different kinds of methods had been proposed to solve the low-light enhancement issue from multiple perspectives. These methods ranged from traditional algorithms to deep learning-based approaches. The traditional algorithms always enhanced the image brightness by using the prior knowledge (e.g., histogram equalization [3-6], Retinex theory [7-9]), while the deep learning-based ones [10-15] relied on the big data, deep models, and efficient optimization methods to achieve their goals. Zero-DCEs [10-11] were considered as the most representative type of methods in deep learning models. Inspired by professional image processing software, such as Photoshop of Adobe Inc., the Zero-DCEs tried to build a lightweight model to estimate the deep high-order curves for adjusting the brightness of images. Besides, the model was optimized by elaborate loss functions without requiring any reference images.

Although the previous works were dedicated to the enhancement of low-light image tasks for the most cases of one input, there still remained a challenge that was regardless of interactive enhancement for personalization results. In practice, it always depends on the preference of users to deal with the specific enhancement. For instance, some ones preferred to use images with high brightness, which contained more details but lower contrast as well as evident noise, while some others preferred those ones with standard brightness that were full of naturalness. The existing ways had tried either to train multi models to fit the various user's preferences which was regarded as an unreliable solution of time-consuming and labor-intensive, or to use professional software (e.g., Photoshop) which required specialized knowledge inconvenient for the public. In order to develop a general solution to satisfy the various requirements, the authors propose a novel interactive low-light enhancement algorithm based on a well-designed Gamma Curve, which is capable of controlling the output results according to the user's preferences to produce specific outputs. Besides, the creative method is convenient in that the model only needs to train once and adjust the hyper-parameters to obtain different luminance results during testing. Benefited from the lightweight network and Gamma Curve's insensitivity to the resolution of inputs, the proposed network runs very fast and therefore satisfies practical usages.

Main contributions of this paper can be summarized as:

1) To analyze how the higher-order curve work in Zero-DCEs and then illustrate the shortcomings of the curve for the interactive application;
2) To propose a novel Gamma Curve, which is fast, flexible. And four improved loss functions are introduced for the optimization of the Gamma Curve during training;
3) To develop a multi-platform low-light enhancement software that is easy to use by the general public based on the aforementioned algorithm.

The remaining of this paper is organized as follows. We try to introduce the related work in Section 2 and then propose the Gamma Curve in Section 3. Following that, we present the experimental results and the interactive software in Section 4, and finally conclude in Section 5.

2. Related work

2.1. Traditional and CNN-based low-light image enhancement

The mainstream low-light image enhancement methods can be divided into two categories, i.e., traditional and deep learning-based methods. The former tried to enhance images based on the prior knowledge (e.g., histogram equalization, Retinex theory), while the latter took advantage of the big data. LIME [7] was one of the most representative traditional low-light image enhancement algorithms, which proposed to enhance images by estimating corresponding luminance maps based on the Retinex theory. Subsequently, Wei et al. [13] proposed to use the convolutional neural networks (CNN) for the decomposition and enhancement of Retinex theory, while the Zhang et al. [15] further improved the performance by considering and dealing with the noise issue. Meanwhile, Lv et al. [14] proposed a multi-branch low-light image enhancement method (MBLLEN) that fused multiple immediate enhancement results and generated satisficed final ones. However, those methods could only generate single fixed results and run with a slow speed.

2.2. Analysis of Zero-DCEs

Different from most of the previous works to generate the enhancement results directly, the Zero-DCEs [10-11] calculated the outputs by high-order curve function:

$$LE\left(I_{low},a\right)=I_{low}+aI_{low}\left(1-I_{low}\right) \quad (1)$$

where I_{low} was the input low-light image, and a represented the pixel-wise curve parameters, which needed to estimate by a CNN network in advance.

Equation (1) has the following characteristics:

1) **Bounded**, it is obvious that the equation's output remains at [0,1] when $I_{low}\in\left[0,1\right]$ and $a\in\left[-1,1\right]$, the boundedness ensures that the information in the original image will be not lost after transformation.
2) **Monotonic**, the function increases monotonically with the input I_{low} and the parameter a, which preserves the structure information between adjacent pixels of the image.
3) **Differentiable**, this property is a must for the deep neural network training.

The Zero-DCE and Zero-DCE++ have proven the effectiveness of this high-order curve for low-light image enhancement task. Theoretically, any equation satisfying the above three conditions can potentially be a substitute for the higher-order curve.

A noteworthy phenomenon is that Zero-DCEs needed to repeat the above enhancement for eight or even more times to achieve satisfactory performance. We believe that the high-order curve contains only a low dynamic range which is not sufficient for enhancement, while it could be improved by the repeating operation. As can be seen from Fig.1 and Fig.2, it is obvious that repeating for more times increases the dynamic range of the output, which further makes the enhancement results more realistic.

3. Methodology

Inspired by Zero-DCEs, we propose a novel interactive Gamma Curve-based low-light image enhancement algorithm. In particular, it's found that the Gamma Curve possesses the essential properties of the deep curve and keeps flexible during the enhancement. On the one hand, the flexibility allows us to develop interactive software for personalized applications. On the other hand, smoother optimization methods can be designed to generate more photo-realistic (intermediate) results. The efficient Gamma Curve, the curve estimation network, the no-reference curve optimization method, and the interactive software will be detailed in the following.

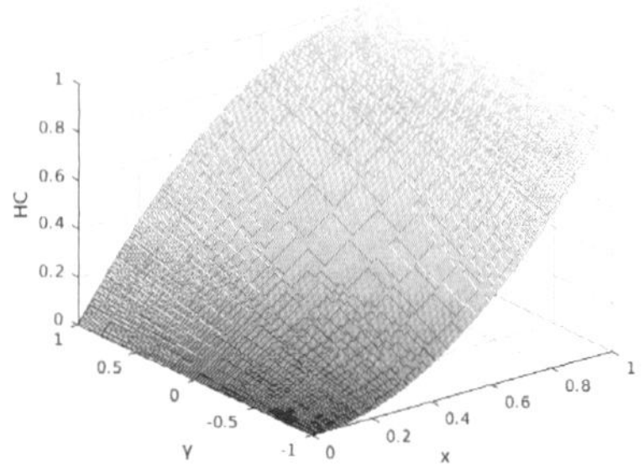

Fig. 1. 3D high-order curve (repeat once).

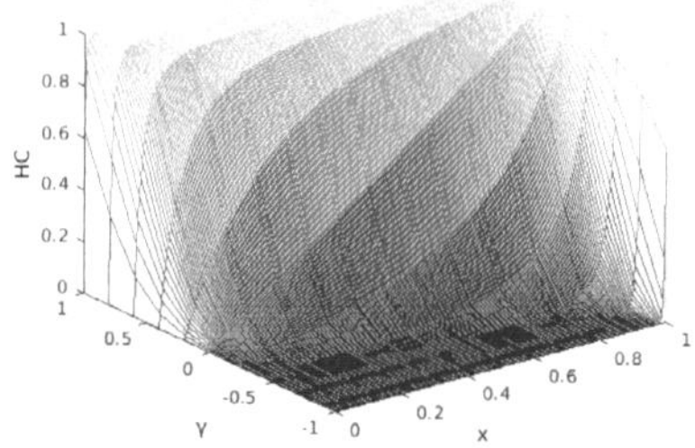

Fig. 2. 3D high-order curve (repeat 8 times).

3.1. Introduction of Gamma Curve

As the observation of Zero-DCEs in Section II, we comprehend that the depth curve controls the brightness and color of enhancement results. In addition to fulfilling the above three attributes, a good depth curve also needs to cover an extensive dynamic range to generate the expected results. To this end, this paper introduces the Gamma Curve as the deep enhancement curve based on the Gamma correction function [16], which is a commonly nonlinear system for luminance adjustment. Given a signal $x \in [0,1]$, the gamma correction is formulated by:

$$y = Ax^{\gamma} \tag{2}$$

where A is a constant and values 1 for most cases, while $\gamma \in [0,+\infty)$ for controlling the value of output y. Specifically, when $0<\gamma<1$, y is amplified with respect to x, and is opposite when $\gamma>1$. This function degenerates to a linear system when $\gamma=1$. As an evolution of gamma correction, the proposed Gamma Curve can be defined as:

$$GC(x,\alpha,c) = x^{(c|\alpha|+1)^{\mathrm{sign}(\alpha)}} \tag{3}$$

where $\mathrm{sign}(\bullet)$ represents the symbol function that defines as:

$$\mathrm{sign}(x) = \begin{cases} 1 & x \geq 0 \\ -1 & x < 0 \end{cases} \tag{4}$$

$c \in [1, +\infty)$ is an environment parameter to control the dynamic range of outputs (also called enhancement degree in low-light image enhancement tasks). Typically, the output increases with this parameter. And similar to (2), $\alpha \in [-1, 1]$ is the luminance adjustment parameter to be estimated. When $\alpha = 0$, the output keeps unchanged; When $\alpha > 0$, the output becomes relatively smaller compared with the input and is opposite when $\alpha < 0$. Fig. 3 and Fig. 4 presents the 2D and 3D curves clearly for $c = 8$.

Obviously, the proposed Gamma Curve satisfies the boundedness for the generation of reliable results, the monotonicity for preserving the image contrast, and the differentiability for neural network optimization. Besides, a major contribution of the Gamma Curve is that enhancement with different degrees can be easily implemented by adjusting the hyperparameter c. The input needs to be processed only once to obtain the expected results, whereas Zero-DCEs need to enhance for eight or more times to achieve similar results.

3.2. *Curve estimation network.*

We employ the estimation network in Zero-DCE++ (E) as the curve estimation network, comprising seven depthwise separable convolutional layers and the symmetric skip connection technique. It has proven the fastness of this lightweight network, which has only 57 KB size parameters. Given a low-light input I_{low}, the purpose of the network is to generate the corresponding pixel-wise gamma parameter α, which can be expressed as:

$$\alpha = \mathrm{E}(I_{low}) \qquad (5)$$

After that, the (intermediate) enhancement outputs can be obtained by (3), which is defined as:

$$I_{enh}^{c} = GC(I_{low}, \alpha, c) \qquad (6)$$

Therefore, results with different brightness will be enhanced for different c.

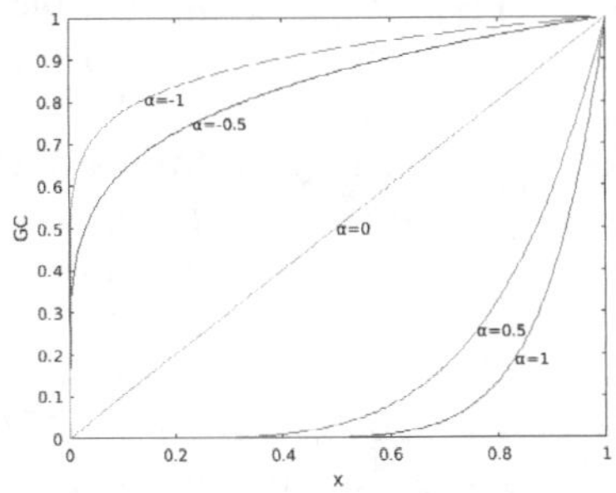

Fig. 3. 2D Gamma Curve when $c = 8$.

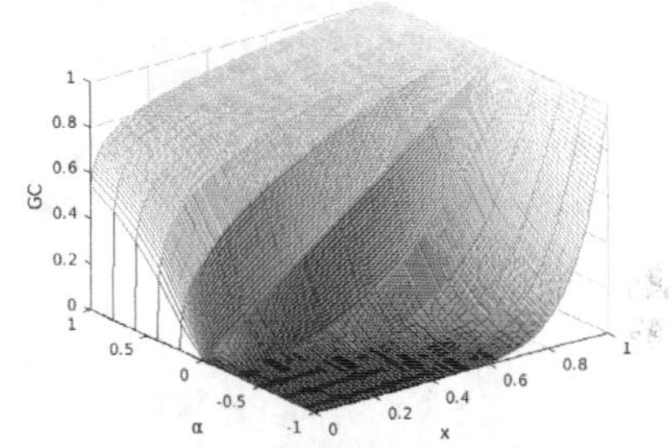

Fig. 4. 3D Gamma Curve when $c = 8$.

3.3. *No-reference curve estimation*

We redesign the multi-color consistency loss as well as the multi-spatial consistency loss, and employ the exposure control loss as well as the illumination smoothness loss for the optimization of the Gamma Curve estimation network.

1) Multi-color consistency loss: Different from the Zero-DCEs, which cares more about the final results, our method considers both the intermediate and final ones. Therefore, the color consistency of intermediate results should also be optimized. Referred to the color consistency loss in [3], our multi-color consistency loss can be defined as:

$$L_{mcol} = \frac{1}{8}\sum_{i=1}^{8}\sum_{\forall(p,q)\in\varepsilon}\left(I_{enh}^{i}(p) - I_{enh}^{i}(q)\right)^2, \qquad (7)$$
$$\varepsilon = \{(R,G),(R,B),(G,B)\},$$

that is the average of intermediate and final results.

2) Multi-spatial consistency loss: In Zero-DCEs, spatial consistency loss was only compared between the inputs and outputs that was not smooth enough because there usually exist enormous

differences between the inputs and outputs. To this end, we design a smoother multi-spatial consistency loss:

$$L_{mspa} = \frac{1}{8}\sum_{k=1}^{8}\sum_{i=1}^{K}\sum_{j\in\Omega(i)}\left(\left|I_{enh}^{k}(i) - I_{enh}^{k}(j)\right| - \left|I_{enh}^{k-1}(i) - I_{enh}^{k-1}(j)\right|\right)^2 \quad (8)$$

where K and $\Omega(i)$ are defined in [3]. The proposed loss calculates the average spatial consistency loss between the intermediate ones.

3) Exposure control loss: The enhancement outputs are expected to have uniform brightness as the normal-light images. Similar to the Zero-DCEs, we pull the mean value of the image's local patch close to a constant C, while $C = 0.5$ is the best choice for the Gamma Curve. Then the exposure loss is formulated by:

$$L_{\exp} = \frac{1}{M}\sum_{k=1}^{M}\left|\frac{1}{3}\sum_{h,w} I_{enh}^{8} - C\right| \quad (9)$$

where M is the patch number.

4) Illumination smoothness loss: our illumination smoothness loss is same as the Zero-DCEs without any change.

Finally, the total loss is the weighted average of the above four loss function.

$$L_{total} = \lambda_1 L_{mcol} + \lambda_2 L_{mspa} + \lambda_3 L_{\exp} + \lambda_4 L_{smooth} \quad (10)$$

where $\lambda_i (i = 1,2,3,4)$ are the weight to balance each component during training.

4. Experiments and application

4.1. Comparison with Zero-DCEs

To ensure a fair comparison, we used the same experimental settings as the Zero-DCEs, which trained on 2,002 multi exposure images, and tested on five public available datasets: LIME, VV, MEF, DICM, and NPE. In particular, we set the loss weights and the dynamic range adjustment parameter during training. The visual comparisons are presented in Fig.5. One could find that the LIME method outputs the over-enhanced result, while the MBLLEN and Zero-DCE++ generate the under-enhanced images. The result of Retinex-Net shows obvious artifacts, and the KinD fails to enhance in extremely low-light area. In addition, none of those methods can run in interactive manners, while our method allows the interactive enhancement. On the other hand, it can be seen from Table 1 that the proposed method outperforms the CNN-based methods and yields competitive natural image quality evaluator (NIQE) [17] metric with the Zero-DCE++, which also demonstrates the effectiveness of our method in recovering image's naturalness.

As the experiment results demonstrated, our well-designed Gamma Curve has two advantages:

1) **flexible**: the brightness can be controlled by adjusting the curve parameter c to match different users' preferences and to adapt for various tasks;
2) **fast**: As shown in Table 2, the estimation procedure runs with high speed by down-sampling first because the network is so lightweight and the Gamma Curve is not sensitive to the image's resolution. Hence, the algorithm can be easily integrated into software for public use.

Low-light input LIME Retinex-Net MBLLEN

Fig. 5. Enhancement results of the Zero-DCE++ and our method. As the c increases, the enhancement results are brightened and contain more details. Here, we only list the integer case for c , while c can also be a float number for more meticulous results.

4.2. Interactive software

Based on the advantages of the Gamma Curve, we build an interactive software that supports multiple platforms that is very convenient and does not have any difficulty in operation. In particular, we offer three options for users: 1) device: users can choose to run the program using the CPU or GPU; 2) down-sampling times: this option determines the running speed; 3) dynamic range adjustment parameter: it allows to adjust the image brightness by changing the dynamic range adjustment parameter. The application will be released on https://github.com/ymmshi/brighten_your_dark_image soon.

Table 1. The NIQE metric of the Zero-DCE++ and our method. The lower indicates the better, and the best are in bold.

Method	LIME	MEF	DICM	NPE	VV	ALL
LIME	4.347	3.827	3.634	3.701	2.328	3.568
Retinex-Net	4.908	4.905	4.315	4.350	2.612	4.218
MBLLEN	3.763	3.508	3.418	3.410	2.893	3.398
KinD++	4.586	3.476	**3.290**	3.534	2.180	3.413
Zero-DCE++	3.884	3.182	3.388	**3.270**	**2.049**	3.155
Our ($c=8$)	**3.563**	**3.175**	3.414	3.446	2.125	**3.125**

Table 2. Frame Per Second (FPS) is calculated on CPU (i7-8700K) and GPU (NVIDIA 1080Ti) when processing images with different resolutions (Down-sampling 16 times first). Our method achieves a fast speed.

Resolution	1080×720	2048×1024	4096×2160
FPS(CPU)	56.0	23.5	5.0
FPS(GPU)	1470.2	893.6	231.3

5. Conclusion

This work introduces a new interactive low-light enhancement model based on a well-designed Gamma Curve. Experiments demonstrate the superiority of our algorithm relative to previous work. In addition, thanks to the excellent proprieties: interactivity, convenience, and fastness of the proposed model, we further develop a multi-platform low-illumination enhancement software to facilitate its application for the public.

Acknowledgments
This work is supported partly by the National Natural Science Foundation of China under Grant 61901434 and the Anhui Provincial Natural Science Foundation under Grant 1908085QF25.

References
[1] Yang, Wenhan, et al. "Advancing image understanding in poor visibility environments: A collective benchmark study." IEEE Transactions on Image Processing 29 (2020): 5737-5752.
[2] Li, Chongyi, et al. "Low-Light Image and Video Enhancement Using Deep Learning: A Survey." IEEE Transactions on Pattern Analysis & Machine Intelligence 01 (2021): 1-1.
[3] Land, Edwin H. "The retinex theory of color vision." Scientific american 237.6 (1977): 108-129.
[4] Gonzalez, Rafael C., Richard E. Woods, and Barry R. Masters. "Digital image processing." (2009): 029901-029901.
[5] Pizer, Stephen M., et al. "Adaptive histogram equalization and its variations." Computer vision, graphics, and image processing 39.3 (1987): 355-368.
[6] Reza, Ali M. "Realization of the contrast limited adaptive histogram equalization (CLAHE) for real-time image enhancement." Journal of VLSI signal processing systems for signal, image and video technology 38.1 (2004): 35-44.
[7] Guo, Xiaojie, Yu Li, and Haibin Ling. "LIME: Low-light image enhancement via illumination map estimation." IEEE Transactions on image processing 26.2 (2016): 982-993.
[8] Rahman, Zia-ur, Daniel J. Jobson, and Glenn A. Woodell. "Multi-scale retinex for color image enhancement." Proceedings of 3rd IEEE International Conference on Image Processing. Vol. 3. IEEE, 1996.
[9] Fu, Xueyang, et al. "A weighted variational model for simultaneous reflectance and illumination estimation." Proceedings of the IEEE conference on computer vision and pattern recognition. 2016.
[10] Guo, Chunle, et al. "Zero-reference deep curve estimation for low-light image enhancement." Proceedings of the IEEE/CVF Conference on Computer Vision and Pattern Recognition. 2020.
[11] Li, Chongyi, et al. " Learning to Enhance Low-Light Image via Zero-Reference Deep Curve Estimation." IEEE Transactions on Pattern Analysis & Machine Intelligence 01 (2021): 1-1.
[12] Lore, Kin Gwn, Adedotun Akintayo, and Soumik Sarkar. "LLNet: A deep autoencoder approach to natural low-light image enhancement." Pattern Recognition 61 (2017): 650-662.
[13] Wei, Chen, et al. "Deep retinex decomposition for low-light enhancement." arXiv preprint arXiv:1808.04560 (2018).
[14] Lv, Feifan, et al. "MBLLEN: Low-Light Image/Video Enhancement Using CNNs." BMVC. 2018.
[15] Zhang, Yonghua, Jiawan Zhang, and Xiaojie Guo. "Kindling the darkness: A practical low-light image enhancer." Proceedings of the 27th ACM international conference on multimedia. 2019.
[16] Rahman, Shanto, et al. "An adaptive gamma correction for image enhancement." EURASIP Journal on Image and Video Processing 2016.1 (2016): 1-13.
[17] Mittal, Anish, Rajiv Soundararajan, and Alan C. Bovik. "Making a “completely blind” image quality analyzer." IEEE Signal processing letters 20.3 (2012): 209-212.

The weighted fusion prediction algorithm of acoustic interval optimized by PCA-PSO-BP and MLRM

Tao Hu[1], Xisheng Li[1,*] and Jia You[1]

[1] School of Automation and Electrical Engineering, University of Science and Technology Beijing, Beijing, 100083, China.

[*]Corresponding author's e-mail:lxs@ustb.edu.cn

Abstract. Acoustic interval is widely used in drilling operation, but the actual drilling operation is limited by well depth, geological conditions and economic cost, so it is difficult to directly measure the value of acoustic interval. A model between logging data and acoustic interval can be constructed to predict acoustic interval. In view of the shortcomings of traditional algorithms, this paper proposes a weighted fusion algorithm based on the combination optimization of the PCA, PSO, BP and MLRM, carries out the PCA and PSO combination optimization the BP prediction for non-linear logging data, the MLRM regression prediction for linear logging data, and then weights and fuses two groups of prediction results to obtain the final result. Through the test and comparison of several groups of logging data experimental results, the prediction effect of the fusion algorithm is better than the BP, and the prediction effect of the PCA and the PSO optimized the BP verifies the feasibility of the fusion algorithm proposed in this paper.

1. Introduction

Acoustic interval(AC) can be used for many purposes, such as pressure prediction[1], calculation of ultrasonic velocity or seam thickness and depth. AC is affected by many factors, such as depth, density and so on, but the measurement of AC[2] is limited by environment and economic cost, limitations of sensor accuracy[3]. Many scholars have proposed many methods to calculate AC. H J Yang[4] proposed a method based on the multiple regression model(MLRM), and constructed the linear relationship between gamma,density, three lateral resistivity and AC, which has a good application effect in sandy mudstone stratum. Hu[5] combined the principal component analysis(PCA) and the BP to reduce the dimension of input data, then the model is trained and AC is combined with other variables for reservoir lithology identification.The MLRM[6] is applicable to data with strong linear relationship, but the fitting error is large when the linear correlation is weak. The BP has good effect on the non-linear relationship with high accuracy.However, there are some problems such as local minimization and slow convergence speed.The particle swarm algorithm (PSO)can be used to optimize the weight and threshold of the BP instead of the steepest gradient descent method in the BP[7]. The PCA can remove the autocorrelation between independent variables, reduce the dimension of input data and remove noise.

2. Principle introduction

2.1. The BP network

The BP network uses the error back propagation learning algorithm to train the network, including input layer, hidden layer and output layer. The BP network structure is shown in figure 1.

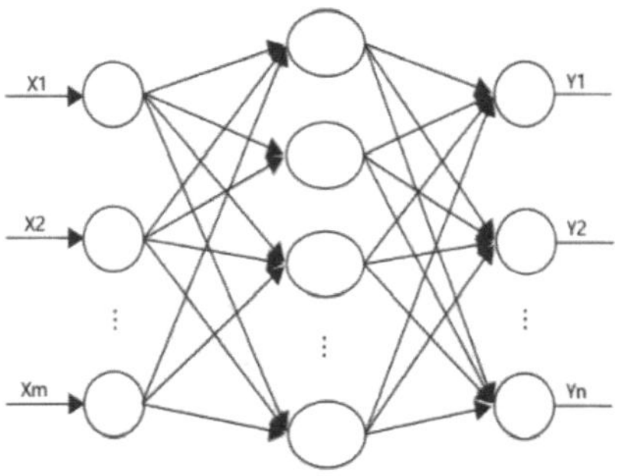

Figure 1. BP network structure.

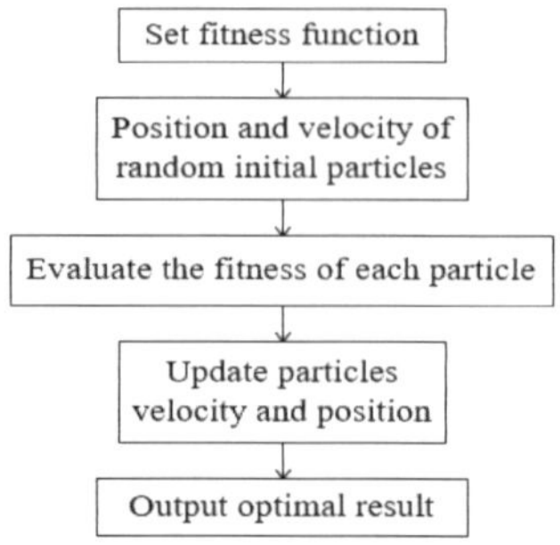

Figure 2. PSO algorithm steps.

X is the input data and Y is the output data. Set the number of the BP input and output layer nodes as m and n respectively, a is a random number between 1 and 10, the number of hidden layer nodes h and the common sigmoid activation functions of the BP network are defined in equation (1) and (2):

$$h = \sqrt{m+n} + a \tag{1}$$

$$f(x) = \frac{1}{1+e^{-x}} \tag{2}$$

2.2. The PSO algorithm

The PSO algorithm is a global optimization algorithm based on swarm intelligence, optimizing the weights and thresholds in the BP network. The PSO has the advantages of fast convergence speed and easy implementation. As shown in equation (3) and (4), V_i is the update formula of velocity, X_i and P_i are the vector and optimal position of the i-th particle respectively.

$$V_i(k+1) = wV_i(k) + c_1 r_1[P_i - X_i(k)] + c_2 r_2[P_j - X_i(k)] \tag{3}$$

$$X_i(k+1) = X_i(k) + V_i(k+1) \tag{4}$$

K is the number of iterations, w is the inertia weight, c_1 and c_2 are learning factors, r_1 and r_2 are rand-dom numbers between 0 and 1. The principle of the PSO algorithm is shown in figure 2.

2.3. The PCA algorithm

The PCA algorithm can map n-dimensional features to k-dimensional principal component orthogonal features, so as to reduce the dimension of data from n to k. According to the cumulative contribution rate C_i, the matrix is composed of the eigenvectors of the first k eigenvalues.

First, supposing that the original data of n demension is x, and x_{ij} is the data in row i and column j, as shown in equation(5) and (6), $\bar{x}$ *and* s_j are the mean and standard deviation of the data in column i of x,

$$\bar{x} = \frac{\sum_{i=1}^{n} x_i}{n} \tag{5}$$

$$s_j = \sqrt{\frac{\sum_{i=1}^{n}(x_i-\bar{x})^2}{n-1}} \tag{6}$$

And the standardized result x_{ij}^* is defined in equation (7):

$$x_{ij}^* = \frac{x_{ij}-\bar{x}}{s_j} \tag{7}$$

Then calculate the correlation coefficient matrix R of the sample data, r_{ij} is the correlation coefficien of data x_{ij}, x_{ki} and x_{kj} are the i-th row and k-th column, the i-th row and j-th column elements of x,respectively, r_{ij} is defined in equation (8):

$$r_{ij} = \frac{1}{n}\sum_{k=1}^{n}(x_{ki} - \bar{x}_i)(x_{kj} - \bar{x}_j) \tag{8}$$

Finally,calculate the eigenvalues λ_j and eigenvectors of the correlation coefficient matrix μ_j, and the contribution rate of each eigenvalue, then select the appropriate one according to the cumulative contribution rate C_i.

2.4. The MLRM algorithm

When there is a linear relationship between independent and dependent variables, The MLRM can be constructed[8]. The model between dependent and independent variables is as follows, where is an *n*-dimensional independent variable, y is the dependent variable, and $a_1, a_2, \dots a_n$ are the regression coefficients, c is a constant number, and ε is the residual, which is the absolute value of the difference between the actual value and the predicted value, the function is defined as equation(9).

$$y = a_1x_1 + a_2x_2 + \cdots a_nx_n + c + \varepsilon \tag{9}$$

2.5. Principle of fusion algorithm

In this paper,15 kinds of logging data are used. The BP has a good effect on non-linear data fitting,the MLRM has a better effect only when the linear correlation between input data and AC is high. The principle of this paper is shown in figure3. Firstly, the linear correlation test is carried out between the input and output data, the dimension of weak correlation data is reduced by the PCA, the BP is optimized by the PSO, so as to realize the combination optimization of the PCA-PSO-BP. The MLRM prediction is used for strong correlation data, finally fuse the two groups of prediction results.Assuming that the result predicted by the PCA-PSO-BP for non-linear data is y_1,and that predicted by the MLRM for linear data is y_2, the final weighted fusion prediction result y is defined as equation (10).

$$\mathbf{y} = \frac{y_1^2 + y_2^2}{y_1 + y_2} \tag{10}$$

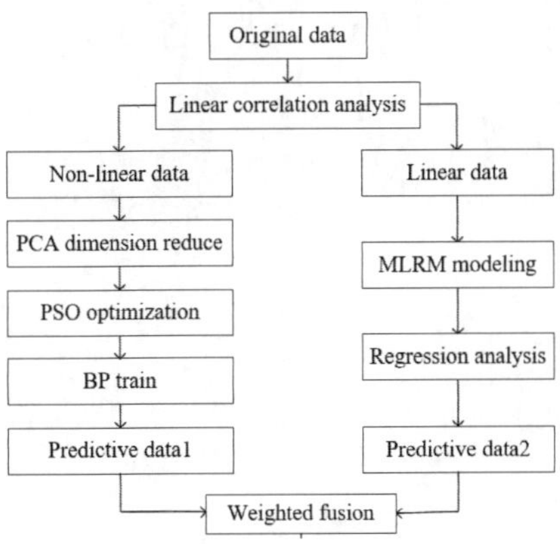

Figure 3. Algorithm principle of this paper.

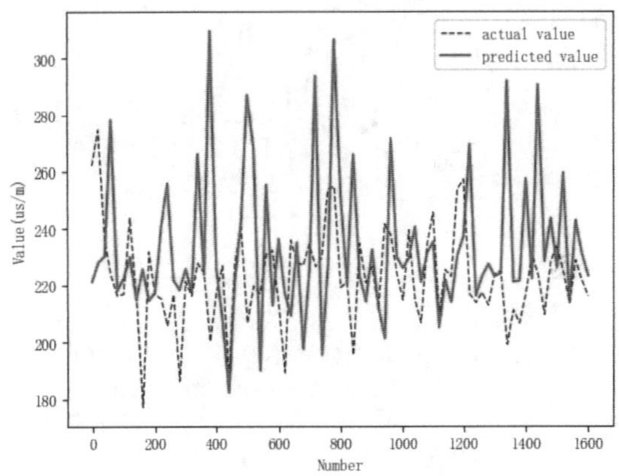

Figure 4. Comparison between BP predicted value and actual value of all data.

3.Experiments and results

3.1The Data preprocessing and BP

The depth of the logging data in this paper is within 1000 meters, including DEPTH (depth(m)), RML (micro lateral resistivity(Ω·m)), RMN (micro potential(Ω·m)), R4.0 (4m resistivity(Ω·m)), RILD (deep induced resistivity(Ω·m)), RILM (medium induced resistivity(Ω·m)), RLL8 (shallow eight side logging(Ω·m)), SP (natural potential(mv)), RT (true resistivity(Ω·m)), DEN(density(g/cm^3)), CNL (neutron(v/v)), GR (natural gamma(api)), CAL(well diameter(cm)), PORT(total porosity(%)), SH (shale content(g/cm^3)) and AC(acoustic interval(us/m)). The first 15 are independent variables and AC is dependent variable. After eliminating missing data and outlier points, 5390 groups of data are obtained.

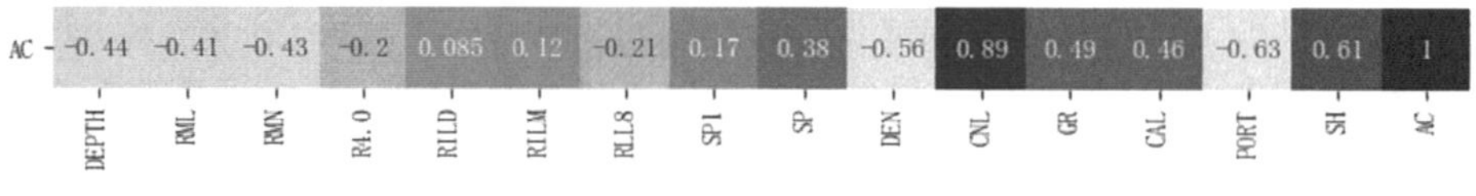

Figure 5. Linear correlation coefficient between independent variable and AC.

Usually when the absolute value of linear correlation coefficient is above 0.6, it can be considered strongly. Figure5 shows the linear correlation coefficient between the first 15 data and AC,then CNL, PORT and SH,which are called linear data, are fitted with AC by the MLRM. The PCA dimensionality reduction is used to remove the autocorrelation between the remaining 12 kinds of data,which are called non-linear data. In this paper, the linear data and the non-linear data are collectively named all data.

First, all of the data is used as the input of the BP, the first 70% of the data,which has a total of 3773 rows of data, is used for training the model, and they are named 'predicted value' in figure4, figure6, figure7, figure9 and figure10. The last 30% of the data, which has 1617 rows of data, is used for prediction, and they are named 'actual data' in figures, so there are 1617 data in x-axis of different figures. The BP prediction results of all data and non-linear data are shown in figure4 and figure6 respectively, the prediction standard deviations are 34.24 and 30.35 respectively. In all figures of this paper, the 'Number' of x-axis means the test dataset points, while the 'Value' of y-axis means the actual and predicted value,and the data unit of the y-axis is 'us/m'. It can be seen that the BP prediction results of the non-linear data have a lower error than the BP prediction results of all data from figures.

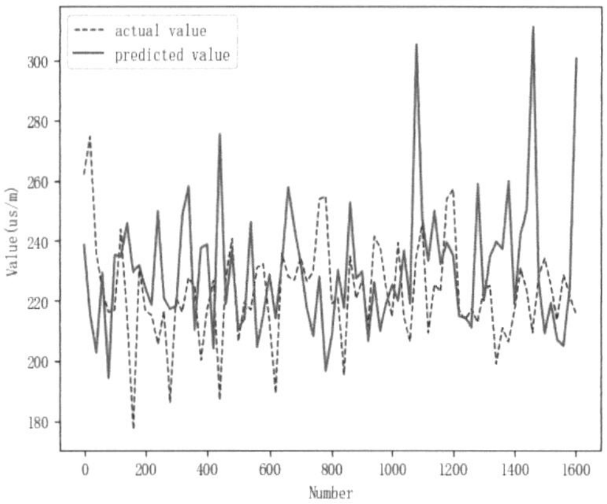

Figure 6. Comparison between BP predicted and actual value of non-linear data.

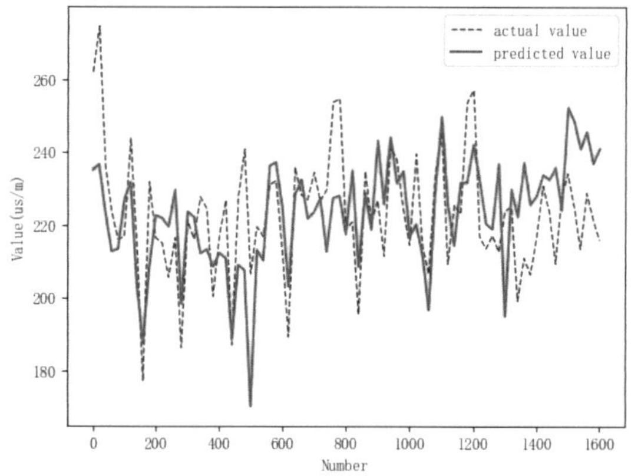

Figure7.Comparison between PCA-PSO-BP predicted and actual value of non-linear data.

3.2.The PSO-BP network

The number of the BP input, hidden and output nodes are 15, 6 and 1 respectively, the population particles is 100, the maximum and minimum value of inertia weight are 0.9 and 0.5, the number of iterations is 10000, the learning factors *c1* and *c2* are 2, and the minimum error is 0.001. Using 12 kinds of non-linear data as input , the standard deviation is 18.59. It can be seen that the prediction effect of the PSO-BP is better than that of the BP.

3.3.The PCA-BP network

12 variables with low linear correlation with AC, including DEPTH, RML, RMN, R4 0, GR, RILD, RILM, RLL8, SP, RT, DEN and CAL, the cumulative contribution rate is shown in table 1:

Table 1. The cumulative contribution rate of each component.

C_1	C_2	C_3	C_4	C_5	C_6	C_7	C_8	C_9	C_{10}	C_{11}	C_{12}
0.24	0.45	0.59	0.70	0.78	0.85	0.90	0.94	0.97	0.98	0.99	1.00

The cumulative contribution rate is often taken as 0.85, 0.90 or 0.95. After dimensionality reduction to 7, 8 and 9 dimensions respectively, it can be used as the BP input. The average standard deviations of the prediction results are 29.45, 31.41 and 26.86 respectively. It can be seen that the BP algorithm has the best prediction effect when the dimension is reduced from 12 to 9.

3.4. The PCA-PSO-BP network

12 kinds of low linear correlation data are called as non-linear data. The prediction results obtained by the PSO-PCA combined optimization the BP algorithm are shown in figure 7, with an error of 15.08. The prediction error of the BP algorithm is 30.35.The PCA-PSO-BP prediction results of non-linear data are better than the BP predicted results, and the error is greatly reduced, the prediction effect of the BP and improved algorithm on non-linear data is better than all data.

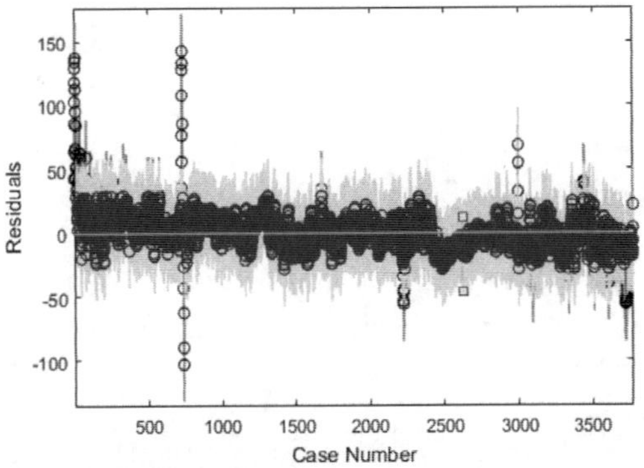

Figure 8. MLRM residual of linear data.

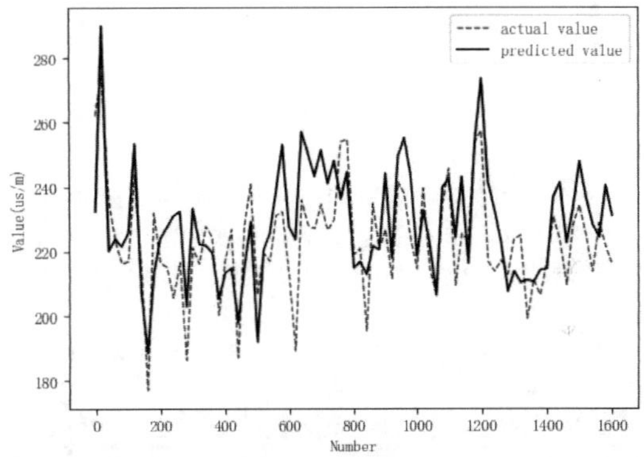

Figure 9. Comparison between MLRM predicted and actual value of linear data.

3.5. The MLRM prediction

Taking CNL, PORT and SH as linear independent variables, AC as dependent variables, and the first 70% data as input to obtain residual results of MLRM, as shown in figure 8,the result of MLRM is:

$$AC_{reg} = 3.744394 * CNL + 0.958333 * PORT + 0.903291 * SH + 71.299993 \quad (11)$$

The prediction result of the MLRM on linear data is shown in figure 9, the prediction standard deviation is 12.48. It can be seen that the prediction effect of MLRM on linear data is better than that of the BP algorithm on all data.

3.6. Weighted fusion prediction based on PCA-PSO-BP and MLRM combination optimization

In this paper, the combination optimization method of the PCA, PSO, BP and MLRM is adopted. The MLRM prediction is adopted for 3 kinds of linear data in logging data, and PCA-PSO-BP improved algorithm regression prediction is adopted for 12 kinds of non-linear data. Finally, the final prediction results, calculated by equation(10), are obtained by weighted fusion of the two groups of prediction results, as shown in figure10. The coincidence degree between the fused results and the original data is higher, the standard deviation between the improved fusion prediction result and the real data is 11.97. The prediction result of the fusion algorithm is better than that of BP, PCA-BP, PSO-BP and PCA-PSO-BP for all data and non-linear data, better than that of MLRM for the regression prediction of linear data.

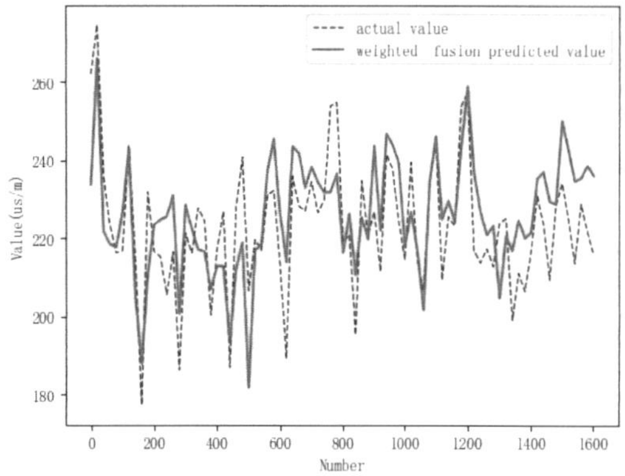

Figure 10. Comparison between predicted value by weighted fusion predicted and actual value of all data.

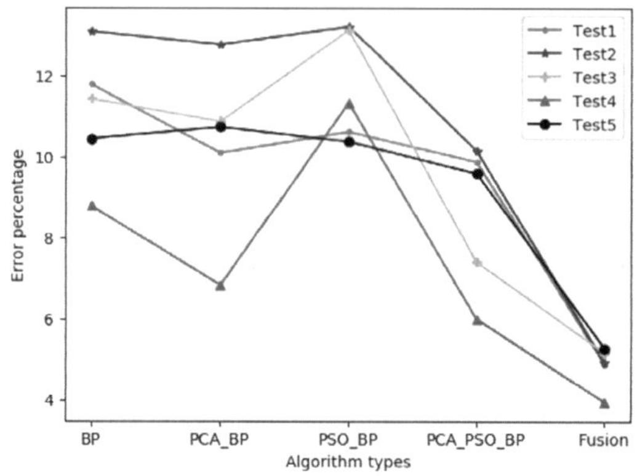

Figure 11. Comparison between MLRM predicted and actual value of all data.

To further verify the effect of the fusion prediction algorithm proposed in this paper, more groups of different logging data are added for experiments, the data of Test 1 is the data used in chapter 3.1.The newly added logging data are also located within 1000 meters. The data volume in Test2, Test3, Test4 and Test5 are 6880, 7110, 5370 and 5730 respectively, using different algorithms to obtain the percentage error results. The percentage error of different algorithms is shown in figure11.

Table 2 Standard deviation of predicted values of different algorithms for different groups of data.

Type	All data				Non-linear	Linear	All data
Test	BP	PCA-BP	PSO-BP	PCA-PSO-BP	PCA-PSO-BP	MLRM	Weighted Fusion
1	34.24	29.72	27.66	26.86	15.08	12.48	11.97
2	37.65	34.49	26.43	18.82	24.66	9.86	13.71
3	36.05	33.75	29.33	26.73	25.01	12.16	16.37
4	27.43	25.90	23.79	21.45	20.60	12.01	10.23
5	51.76	33.87	30.78	32.04	30.62	11.21	17.56

4.Conclusion

This paper analyzes the linear correlation between multidimensional logging data and AC, divides logging data into linear and non-linear data, reduces the dimension of non-linear data by PCA, optimizes the BP prediction by the PSO, uses the MLRM prediction for linear data, finally weights and fuses the two results to obtain the final prediction result. After experiments, the effect of the fusion prediction algorithm in this paper is better than that of the BP, and the BP optimized by the combination of the PCA and the PSO, which verifies the feasibility of the algorithm in this paper. The algorithm has good prediction effect, low complexity, easy to implement, and has a wide application prospect.

Acknowledgments

This project is funded by the National Key R&D Program (2019YFB2101901) and the Key Laboratory of Industrial Process Knowledge Automation of the Ministry of Education (FRF-BD-19-002A),and I would like to thank professor Li and associate professor You for their strong support for this paper.

References

[1] Xu,Y., Pang,X.Q., Hu,T., Xu,W., Zheng,X.W.(2018)Selection of Formation Pressure Prediction Methods Based on Mudstone Interval Transit Time [J]. Special Oil & Gas Reservoirs,25(04):46-50.

[2] Quan,X.S.,Lv,H.R.,Liu,C.C.,Wang,H.,Wu,D.B.,Chen,P.,Zhou,H.(2021)An investigation on bolt

stress ultrasonic measurement based on acoustic time difference algorithm with adaptive hybrid extended Kalman filter[J]. Measurement,186.

[3] Xi,P.F.,Yang,M.H.,Guo,W.H.,Shi,J.G.(2019)A Method for Identifying Abnormally Compacted Strata Based on the Fluctuation of Interval Transit Time Data [J]. Petroleum Drilling Techniques,47(06):111-115.

[4] Yang,H.J.,Qiao,B.Q.(2021)Research on reconstruction method of pseudo acoustic TDOA logging curve based on multiple regression model [J]. Uranium Geology,37(03):500-505.

[5] Hu,J.L.,Gao,Y.C.,Yu,J.F.,Lu,L.,Zhang,H.J.,Yang,Z.Q.(2016)Lithology Identification of Unconven-tional Reservoirs Based on PCA-BP Neural Network[J]. Journal of Shandong University of Science and Technology(Natural Science).35(05):9-16.

[6] Leng,J.F.,Gao,X.,Zhu,X.P. (2016)Application of multivariate linear regression statistical prediction model [J].(07):82-85.

[7] Lee,C.Y.,Cheng,Y.H.(2020) Motor Fault Detection Using Wavelet Transform and Improved PSO-BP Neural Network[J]. Processes,8(10).

[8] Yang,B.R.(2017)Passenger Vehicle Market Forecasting Model Based On Multivariate Linear Regression and BP Neural Network[D]. Huazhong University of Science and Technology.

Design of real-time target location algorithm for industrial scene

Rui Wang[1*], Di Li[2]

[1] School of Mechanical and Automotive Engineering, South China University of Technology, Guangzhou, Guangdong, 510641, China

[2] School of Mechanical and Automotive Engineering, South China University of Technology, Guangzhou, Guangdong, 510641, China

[*]Corresponding author's e-mail: 201920100379@mail.scut.edu.cn

Abstract: In order to meet the needs of real-time visual inspection of workpiece targets in industrial environment. In this paper an algorithm for locating workpiece based on edge feature is designed by using image information. At the same time, a model description method based on generalized Hough model and an acceleration strategy based on pyramid search are proposed to solve the problem of insufficient real-time performance of the algorithm. The algorithm is mainly divided into two stages. First, in the offline stage matching model is created. In the online stage, according to the model created in advance, the pyramid based accelerated search strategy is used to match, and the parallel computing method is adopted to calculate the similarity. Experimental results show that the algorithm has strong anti-occlusion and anti-interference ability, and can achieve real-time target positioning under any Angle and coordinates. At the same time, it is proved that the proposed acceleration strategy has better acceleration performance than the existing matching speed method.

1. Introduction

In today's industry in the field of vision, developed a variety of visual positioning method, among them because the target positioning method based on edge gradient feature has good stability and resistance to block ability strong characteristic, the field of vision in fixed scene has wide application, in the industrial scene, because of mechanical control, production efficiency and other project requirements, the calculation efficiency of the positioning algorithm has higher demand. Therefore, how to improve the efficiency of target positioning has always been a hot issue in the field of visual positioning.

In the target location algorithm based on edge gradient, since the calculation speed is mainly determined by the amount of calculation, how to reduce the amount of calculation becomes the key to accelerate. The amount of calculation in the algorithm is mainly composed of two parts: one is the description method of the model, that is, how to describe the model with less amount of data; The second is the calculation in the matching process, that is, how to remove invalid calculation in the matching process[1] .

In this paper, a fast template generation method and a pyramid-based target search acceleration strategy are designed to solve the above problems, and greedy value is introduced to speed up the similarity calculation. Finally, the location algorithm can obtain the accurate Angle and position information of the workpiece in real time. The algorithm overview is shown in Fig.1.

2. Propose Object Matching and Localization Method

2.1. Overview

The traditional template matching method based on gradient direction is mainly divided into offline stage and online stage[2], The main purpose of the offline stage is to calculate the template information that can be calculated in advance, so as to reduce the amount of calculation in the online stage, improve the matching speed in the online stage, achieve fast real-time template matching, and save the information extracted in the offline stage, which is the model required for online matching. The two-stage design can facilitate model switching and flexibly change the positioning target. The algorithm flow of the proposed method is shown in Fig.2.

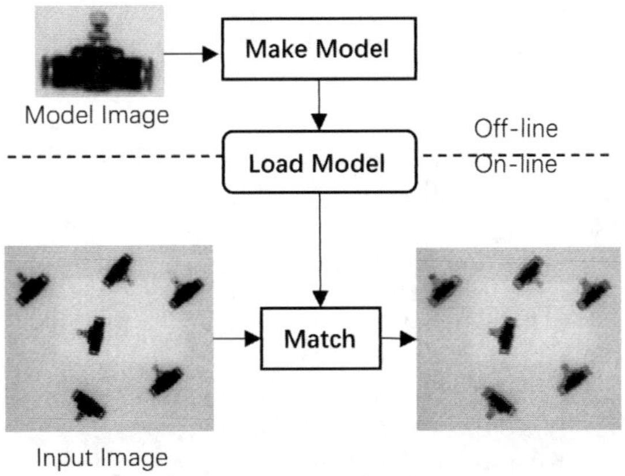

Fig.1 Overview of location algorithm

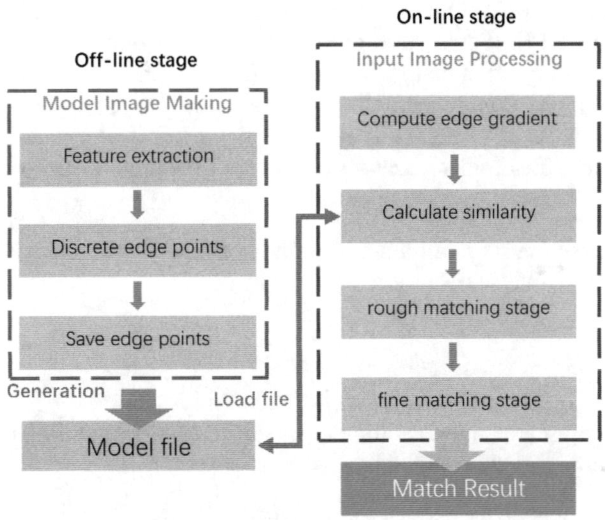

Fig.2 Algorithm flow of the proposed method

2.2 Offline model generation

In order to accelerate the matching speed, it is necessary to minimize the amount of calculation in the matching process. In the offline stage, the template image needs to be described by a specific data structure, and the simplified data structure can effectively reduce the amount of calculation. In this paper, a model description method based on generalized Hough Transform Algorithm (GHT) is used which can only use the template extracted edge information, thus greatly reducing the matching, as well as model production cost of calculation.

In generalized Hough transformation model, a central point is selected in the template(x_c, y_c), The model in this article builds the model from the ROI detected manually or automatically[3] , The model consists of a set of points $p_i = (x_i, y_i)^T$,and their corresponding direction vectors $d_i = (t_i, u_i)^T$

$$M = (p_i, d_i) \in R^2 \times R^2, for i = 1, \cdots, n \tag{1}$$

In this equation, $p_i = (x_i, y_i)^T$ is the relative coordinate of the template edge point to the template center point,

Multiple discrete parameter space templates need to be created in the offline phase. In order to simplify shape information and operation, discrete sampling method is adopted to reduce the number of edge points. The quantized edge point set P of its discrete strategy model is expressed as

$$P = (x, y) \in M, x mod n = 0, y mod n = 0 \tag{2}$$

In this equation, n is the discrete sampling step, which is determined by the complexity of the actual template image edge. the template of the discrete degree depends on the size of the template image, create multiple direction template, due to the aspect ratio of the template image is not fixed, the rotation on the template, the aspect ratio of the outer rectangle of the template will also be larger. Therefore, when creating the template, in this paper first expands the edge of the template image and creates a square area, called the model image, whose side length l is

$$l = min\left(2^i > max(r, c)\right), i \in (4, \cdots, n) \tag{3}$$

In this equation, r is the height of the template image, c Is the width of the template image. Rotation and movement formula of template image in model image[4]

$$T = \begin{bmatrix} a & b & (1-a) * c/2 - b * r/2 + p_x \\ -b & a & b * c/2 - (1-a) * r/2 + p_y \\ 0 & 0 & 1 \end{bmatrix} \tag{4}$$

In this equation

$$(a, b, p_x, p_y) = (s * cos(\theta), s * sin(\theta), (l-c)/2, (l-r)/2) \tag{5}$$

In this equation s is the scaling factor, θ is a rotation angle. After extracting the edge information of the current template image, in order to speed up the speed of other multiple direction template production, according to the generalized Hoff transform model, the current edge point set is used to transform the edge points using the transformation matrix T. The original image traversal operation is converted to the operation of the edge point set, thereby reducing the amount of calculation, the edge of the rotation.p' [5]

$$p' = T * p \tag{6}$$

At the same time, corresponding gradient vector value of the transformed edge point d' is

$$d' = (R^{-1})^T d_i, R = \begin{bmatrix} a & -b \\ b & a \end{bmatrix} \tag{7}$$

Since the inverse of the matrix is a time-consuming work, for the second order matrix, the formula can be derived to avoid the inverse operation of the matrix

$$(t', u') = \left(\frac{at + bu}{a^2 - b^2}, \frac{-bt + au}{a^2 - b^2}\right) \tag{8}$$

2.3 Similarity measures

In the process of matching search, the matching results need to be measured in the image to be searched according to the edge information provided by the model[6]

$$s_j = \frac{1}{n}\sum_{i=1}^{j} \frac{\langle d_i, e_{q+p'} \rangle}{\|d'_i\| \cdot \|e_{q+p'}\|} \tag{9}$$

In this equation, $e_{x,y} = (v_{x,y}, w_{x,y})^T$is gradient vector of $q = (x, y)$ in the image, $d_i = (t_i, u_i)^T$is gradient vector of template feature point $p_i = (x_i, y_i)^T$ in model image. In order to speed up the search, it is necessary to discard some calculations with low similarity in time, thus introducing the concept of greedy value.

$$S_m < MIN\left(\left(S^{min} - 1 + \frac{1 - g \cdot S^{min}}{1 - g} \cdot \frac{m}{n}\right), \left(S^{min} \cdot \frac{m}{n}\right)\right) \tag{10}$$

In this equation, g is greed value, n is the number of template edges, m is already calculated edge point, S^{min} is average similarity threshold, the calculation of similarity is a cumulative process. If the above expression is met in the cumulative process, the probability of matching is low and the calculation process can be terminated in time to reduce the calculation time.

2.4 Online Search strategy

After the model file is created, it is necessary to use the information of the model file to match the target image in the online matching stage. If the target image is traversed directly, a large number of invalid calculations will be generated. In order to solve the above problems, it is necessary to design an appropriate search strategy to speed up the matching speed.

Search strategy is to determine the search speed of the decisive factor, but match the robustness at the same time restricts the improvement of speed, so the search strategy formulation, the need to achieve the balance between, commonly used template search is through the pyramid zoom strategy to accelerate the matching speed, this paper will search strategy is mainly divided into coarse and fine match two stages. Multiple potential matching points are obtained in the upper pyramid, and matching points are screened in the lower pyramid. In the process of pre-matching, it is necessary to calculate the similarity of templates with multiple angles. If serial calculation takes a lot of time, in order to speed up the calculation process, OpenMP can be used for parallel calculation of multiple templates at the same time in the actual code writing.

In the coarse matching stage, the potential target points are found in the top pyramid according to the similarity strategy. In this stage, the traversal calculation is carried out in the top pyramid and a series of pre-matching points are obtained according to the similarity threshold. After get pre-matching goal, in the fine matching stage according to match the target, in the next layer pyramid searching, also note that the pyramid descends, the size of the image increases, in order to reduce extraneous calculations, you need to design a certain target area reduction strategy, so as to reduce the amount of calculation.

The target area reduction strategy is to determine the potential search area position in the upper pyramid, when calculating the similarity degree, since the template model is stored in the template center and the relative coordinates of the edge point, the area can be further reduced according to this feature. Inside the area, only the pre-match target point is used to travel within a certain range, according to the pre-matching target point information, the search area is divided in the process of matching, and the next layer is searched in a certain Angle and coordinate range, where the upper-left coordinate of the search area is (r_1, c_1), The lower right corner coordinates is (r_2, c_2).

$$(c_1, r_1) = (result_x - \frac{w}{2^{i+1}} - 2 \cdot (i+1), result_y - \frac{h}{2^{i+1}} - 2 \cdot (i+1)) \tag{11}$$

$$(c_2, r_2) = (result_x + \frac{w}{2^{i+1}} + 2 \cdot (i+1), result_y + \frac{h}{2^{i+1}} + 2 \cdot (i+1)) \tag{12}$$

In these equations, $result_x$, $result_y$ are the horizontal and vertical coordinate points of the upper level matching coordinate points, w is the width of the model image, h is the height of the model image, i is the number of pyramid layers the value of i ranges from 0 to 7, Where 0 is the lowest layer, which is the original image size.

The value of the angular search range within the search area.

$$(angle_{upper}, angle_{lower}) = (result_{angle} + 2 \cdot i, result_{angle} - 2 \cdot i) \tag{13}$$

In this equation, $result_{angle}$ is the value of pre-matching Angle. Coordinates in the search area, the upper-left coordinates of the search area is (x_1, y_1) The lower right corner coordinates is (x_2, y_2)

$$(x_1, y_1) = (result_x - c_1 - i, result_y - r_1 - i) \tag{14}$$

$$(x_2, y_2) = (x_1 + 2 \cdot i, y_1 + 2 \cdot i) \tag{15}$$

Since the upper pyramid is a discrete sampling of the original image, the edge information of the upper pyramid is not complete, so a larger step size can be used in the upper search, so the step size of the Angle search increases with the increase of the number of pyramid layers, the value of angle search step size e_a is

$$e_a = a_s \cdot 2^i \tag{16}$$

In this equation a_s is the step of rotation Angle when making templates. In order to speed up the calculation speed in the matching process, the parallel method is used to speed up the matching speed in the programming process. When the potential target point is obtained at the top of the pyramid, a separate thread is used for each potential target point, and the search strategy proposed in this paper is used for target search.

3. EXPERIMENTS AND RESULTS

3.1 Stability analysis of algorithm

Due to common influencing factors in the workpiece positioning mainly include light, noise, and occlusion in the actual industrial scene, so in the experimental process, in order to verify the stability of the algorithm, multiple images with occlusion and overlap were selected to verify the anti-occlusion ability of the algorithm and the matching ability in the actual scene with noise and uneven illumination, so as to verify the stability of the algorithm. Fig.3(b) are the image with overlap, and Fig.3(e) are the image with uneven illumination and noise in actual shooting

3.2 Efficiency analysis of the algorithm

Computational efficiency refers to the algorithm of the time required to complete the corresponding work, the experiment platform of the experiment for 6 core CPUs, dominant frequency of 2.6 GHz computer, in the platform test, the test set owner if through the actual collection of size of 785x555 picture, in the process of experiment validation, by randomly placing a plurality of angles of angles and identifies, and adding a plurality of items in the picture, the matching angle range is set to 0 to 360 degrees. In the experiment, five different types of templates were made for matching experiment, and the matching time and success rate were recorded.

Through experiments, it can be seen that the maximum matching speed of the algorithm can be stable at about 6ms under the condition that the contour edge features are clear and there is no occlusion as shown in Fig.3(a)(e). When the edge features are not clear and there are many model description points, the velocity is stable at about 20~50ms as shown in Fig.3(c)(d)(g)(h), and the recognition accuracy can reach more than 95%. In the case of stable photo effect and reasonable parameter selection, the recognition can reach 100% as shown in Fig.4. Experimental results show that this method is feasible and can meet the requirements of fast and accurate positioning and recognition of workpiece.

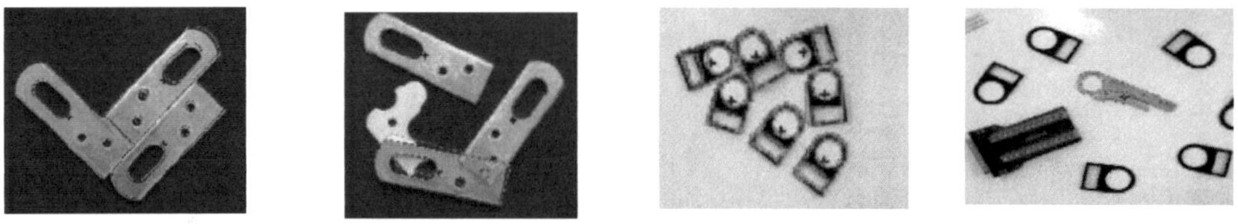

(a) image without overlap (b) image with overlap (c) Signage (d) wire cutter

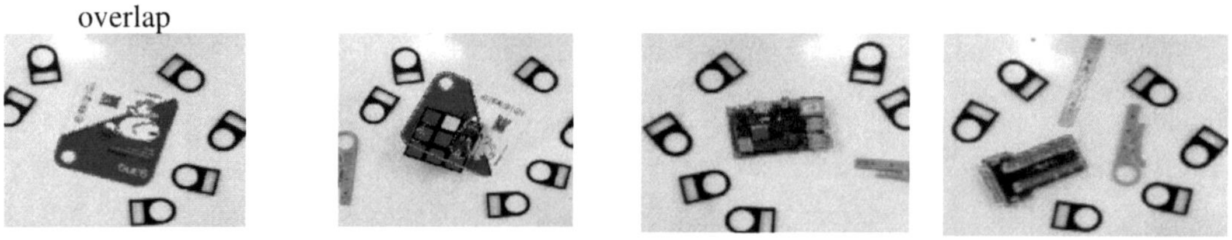

(e) image with uneven illumination (f) image with overlap and uneven illumination (g) Raspberry Pi (h)DB25-G

Fig.3 Matching results

In order to verify the acceleration effect of the search strategy proposed in this paper, industrial cameras were used to collect multiple images, and the algorithm proposed in this paper and the algorithm referred to in the algorithm design process were used to conduct the running speed test, and the statistical results of the algorithm duration were shown in Table1. As can be seen from Table 1, compared with other methods, this method has significant advantages in matching efficiency, and the comparison with the original algorithm proves that the target search acceleration strategy proposed in this paper can significantly improve the matching efficiency of the original algorithm.

Table 1. Calculation times (ms)

Method	Average Time
Unused acceleration[4]	120
Literature[8]	60
Literature[7]	520
Line-2D[1]	50
Proposed method	25.6

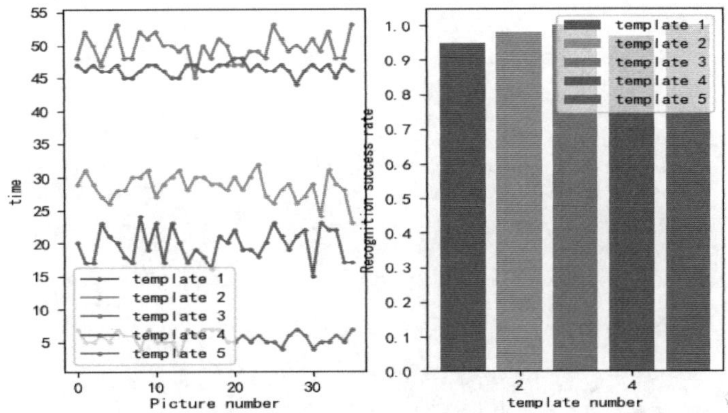

Fig.4 Matching time and success rate

4. CONCLUSION

In this paper, the position and Angle are identified based on the template matching algorithm of edge features. The speed optimization is designed by the data structure of the template model and the pyramid search strategy corresponding to it, so as to reduce the calculation time.

The above tests show that the algorithm designed in this paper has good stability, strong resistance to occlusion, noise and other disturbances, and can achieve matching at any Angle, and the matching time can meet the real-time matching requirements.

References

[1] Bottger, T., Ulrich, M. and Steger, C. (2018) 'Subpixel-precise Tracking of Rigid Objects in Real-time [arXiv]', arXiv, p. 12 pp.-12 pp.

[2] 'HARMONIC DEFORMATION MODEL FOR EDGE BASED TEMPLATE MATCHING': (2008) in Proceedings of the Third International Conference on Computer Vision Theory and Applications. International Conference on Computer Vision Theory and Applications, Funchal, Madeira, Portugal: SciTePress - Science and and Technology Publications, pp. 75–82.

[3] Hinterstoisser, S. et al. (2012) 'Gradient Response Maps for Real-Time Detection of Textureless Objects', IEEE Transactions on Pattern Analysis and Machine Intelligence, 34(5), pp. 876–888.

[4] Hofhauser, A., Steger, C. and Navab, N. (2008) 'Edge-Based Template Matching and Tracking for Perspectively Distorted Planar Objects', in Bebis, G. et al. (eds) Advances in Visual

Computing. Berlin, Heidelberg: Springer Berlin Heidelberg (Lecture Notes in Computer Science), pp. 35–44.
[5] Si, Y. et al. (2019) 'A Fast and Robust Template Matching Method with Rotated Gradient Features and Image Pyramid', in Yu, H. et al. (eds) Intelligent Robotics and Applications. Cham: Springer International Publishing (Lecture Notes in Computer Science), pp. 505–516.
[6] Ulrich, M. and Steger, C. (no date) 'Performance Evaluation of 2D Object Recognition Techniques', p. 31.
[7] Xu, X., van Beek, P. and Feng, X. (2014) 'High-speed object matching and localization using gradient orientation features', in Röning, J. and Casasent, D. (eds). IS&T/SPIE Electronic Imaging, San Francisco, California, USA, p. 902507.
[8] Yang, D. et al. (2020) 'Real-time matching strategy for rotary objects using digital image correlation', Applied Optics, 59(22), p. 6648.

A computational optimization research on ant colony optimization for the traveling salesman problem

Yufan Sheng

Institute of Logistics Science and Engineering, Shanghai Maritime University, Shanghai 201306, China

syf3708@163.com

Abstract—The traveling salesman problem (TSP) is one of typical combinatorial optimization problems. Ant colony optimization (ACO) is an effective method to solve the traveling salesman problem, but there are some non-negligible shortcomings hidden in the original algorithm. The primary objective of this research is to optimize the ACO to produce quality work throughout solving TSP. To this end, the hybrid SOS-MMAS algorithm is proposed. Concretely, apply the advanced Max-Min Ant System (MMAS) as the basic algorithm to raise task scheduling efficiency, meanwhile introduce symbiotic organisms search (SOS) into the MMAS to optimize the key parameters. Experiments were carried out on typical TSP instances of different scales, and the SOS-ACO and ACO algorithms were compared with SOS-MMAS, which proved the excellent performance of SOS-MMAS in solving TSP. Rationality of the algorithm design and high performance has been illuminated by experimentation. In addition, the model also could serve to suggest further research of TSP or other related areas.

1. Introduction

Traveling Salesman Problem (TSP) is one of the well-known problems in the field of optimization. The TSP will be solved if there exists a shortest route that visits each destination once and permits the salesman to return home, and this route is called a Hamiltonian Cycle (HC) [1]. TSP can be applied to many engineering fields, such as vehicle routing, logistics distribution, electronic circuit design, and network optimization, etc.[2] Short-term and high-precision are the dual challenges faced by ACO while solving the TSP, particularly for large-scale TSP[4].

As a heuristic algorithm, ACO's performance is subject to its global exploration and local exploitation capabilities[3]. Besides improving the ACO mechanism, reasonable allocation of algorithm parameters is also an effective way to improve algorithm performance [5].On the basis of ACO, Peker et al. used the Taguchi method to optimize related parameters, and experiments showed that the Taguchi method can play a positive role in finding better parameters[6]. Mahi et al. used particle swarm optimization (PSO) to optimize the parameters of ACO, and introduced 3-OPT algorithm to upgrade the ACO's performance[7]. The aforementioned methods for optimizing ACO parameters, contain their own parameters. These parameters which are generally given by designers will affect the performance of the hybrid algorithm. Unlike most heuristic algorithms, SOS has the advantage of requiring no parameters. At present, SOS has been used to solve various continuous and combinatorial optimization problems[8].

Max-Min Ant System (MMAS) is an advanced ant colony optimization algorithm, which can effectively avoid premature convergence during the searching procedure by limiting the pheromone strength, so as to solve the problem of early search stagnation that may occur in ACO, thereby increasing the search ability in the initial stage of the algorithm[9].

In order to enhance system performance and operating efficiency, a hybrid algorithm combining the SOS algorithm with the MMAS algorithm is put forward and verified through solving TSP instances.

2. Computational methods

2.1 Basic MMAS Methods

The benefits of using MMAS to solve TSP are mainly manifested in its processing of pheromone. (1) Only allow the iterative optimal ant(ant that constructs the shortest path in this iteration), or the current optimal ant releases pheromone; (2) The pheromone concentration on the path is limited to the range of $[\tau_{min}, \tau_{max}]$; (3) The initial value of the pheromone is set as the upper limit of its concentration; (4) Whenever the execution enters a standstill, the number of pheromones on all edges in the problem space will be reinitialized.

2.1.1 Initial definition

The number of ant colonies is defined as *m*, and the number of iterations is defined as *K*.

2.1.2 State transition probability

In the ant colony algorithm, ant *k* moves from the current city *i* to the next city *j* according to the state transition probability between *i* and *j*. When visiting all cities, the ant finds an HC. The state transition probability depends on the pheromone on the path and the heuristic information between *i* and *j*. The calculation formula is shown in Eq.(1)

$$P_{ij}^{k}(t) = \begin{cases} \dfrac{[\tau_{ij}(t)]^{\alpha}[\eta_{ij}(t)]^{\beta}}{\sum_{u\in allowed_k}[\tau_{iu}(t)]^{\alpha}[\eta_{iu}]^{\beta}} & j \in \mathrm{J}_k(i) \\ 0 & \text{Otherwise} \end{cases} \tag{1}$$

The variables in formula (1) are marked as follows:

$\tau_{ij}(t)$ is the pheromone concentration of the path between the target point *i* and the target point *j* during the *t*-th iteration. $\eta_{ij}(t)$ represents the heuristic factor between *i* and *j* in the *t*th iteration. The calculation formula of the heuristic factor is defined as Eq.(2). α represents the relative importance of pheromone, and β represents the relative importance of heuristic factors. $J_k(i)$ represents the set of cities that ant *k* is allowed to choose next at point *i*.

$$\eta_{ij} = \frac{1}{d_{ij}(t)} \tag{2}$$

$d_{ij}(t)$ is defined as the distance between city *i* and city *j* during the *t*-th iteration.

2.1.3 Roulette wheel to choose the next target point

After the transition probabilities of all optional target points are calculated, the roulette wheel method is used to make the final selection.

$$q_j = \sum_{j=1}^{m} p_{ij}^{k}(t) \tag{3}$$

In Eq.(3), q_j is equivalent to the span on the turntable. The larger the span, the easier it is to be selected, and *m* represents the number of alternative paths in the next step.

Then randomly generate a number *r* in the interval (0, 1) as shown in Eq.(4). Compare q_j and *r*, and select the smallest q_j greater than *r*. In Eq.(5), The index *j* corresponding to this q_j will become the target point of ant *k* in the next step.

$$r = \mathrm{rand}(0,1) \tag{4}$$

$$j = index\{\min[q_j, q_j > r]\} \tag{5}$$

2.1.4 Pheromone update

The MMAS pheromone update rules are as follows:

$$\tau_{ij}(t+1) = (1-\rho)\cdot\tau_{ij}(t) + \Delta\tau_{ij}^{best} \tag{6}$$

$$\Delta\tau_{ij}^{best} = \begin{cases} \frac{1}{L^{best}}, \textit{if the edge(i,j) belongs to the best tour} \\ 0, \textit{Otherwise} \end{cases} \quad (7)$$

In Eq.(6) and (7), $\rho \in[0,1]$ is the pheromone evaporation rate. $\Delta\tau_{ij}^{best}$ is the pheromone accumulated on the path between city i and j, which means that after each iteration, only the information on the path to which the optimal solution belongs is updated.

2.1.5 Iteration

Perform iterative operations according to the aforementioned MMAS algorithm process to find a current optimal or close to optimal path.

2.2 The SOS algorithm for optimizing MMAS parameters

Symbiotic Organisms Search (SOS) is a heuristic search algorithm based on biological phenomena, which is mainly used to solve the optimization problem of numerical generation of solutions in continuous space. The main process is divided into three stages: mutualism, commensalism and parasitism.

2.2.1 The mutualism phase.

In the biological mutualism calculus, X_i is set to represent the i-th member in the population, and X_j is a randomly chosen object to interact with X_i.The updates of X_i and X_j after interaction are represented by Eqs.(8)-(10).

$$X_{inew} = X_i + \text{rand}(0,1) \times (\text{X}_{\text{best}} - \text{Mutual\_Vector} \times BF_1) \quad (8)$$

$$X_{jnew} = X_j + \text{rand}(0,1) \times (\text{X}_{\text{best}} - \text{Mutual\_Vector} \times BF_2) \quad (9)$$

$$\text{Mutual\_Vector} = \frac{X_i + X_j}{2} \quad (10)$$

X_{best} represents the organism with the highest fitness value in the current iteration. BF1 and BF2 are mutualism benefit factors, set to 1 or 2. X_{best} represents the most adaptable organism in the current ecosystem. When the new fitness is better than the fitness of the previous organism, X_i is updated to X_{inew}, X_j is updated to X_{jnew}, otherwise it remains unchanged.

2.2.2 The commensalism phase

Different from the mutualism phase, only one of the two organisms benefits from the commensalism phase, while the other remains unchanged. X_j is randomly selected from the population to interact with X_i. In this mode, the update relationship between X_i and X_j is computed as Eq.(11).

$$X_{\text{inew}} = X_i + \text{rand}(-1,1) \times \left(X_{best} - X_j\right) \quad (11)$$

The term $\left(X_{best} - X_j\right)$ represents the benefit of improvingX_i's viability. With the assistance of X_j, X_i will be transformed into a new organism. Calculate the fitness value of the new candidate solution of X_i. If it is greater than the fitness value of the previous organism, then X_i will be updated to X_{inew}, otherwise it will remain unchanged. The goal of this step is to calculate promising organisms close to X_{best}.

2.2.3 The parasitism phase

In order to maintain the diversity of the population, *Parasite_vector* is specially set to avoid the algorithm from falling into premature convergence. The setting of artificial parasites is created by copying biological X_i in the search space, and the generation process is computed as Eq.(12)

$$Parasite\_vector = \text{rand}(0,1) * X_i \quad (12)$$

In Eq.(12), an organism X_i is randomly selected as the carrier of the parasite. When the *Parasite_vector* has a better fitness value, it will kill the organism X_i and replace its position in the population. On the contrary, the organism X_i has a better fitness value, from which it gains immunity, resulting in the *Parasite_vector* unable to survive in the system.

2.3 The hybrid SOS–MMAS

Since the performance of MMAS is largely affected by related parameters, while the SOS method itself does not contain parameters, a hybrid SOS-MMAS method is proposed, in order to use SOS to optimize the two key parameters α and β in MMAS.

Set the number of ants to m, the organism $Y = [\alpha, \beta]$ in SOS is composed of α and β. In the process, each ant uses the α and β values in Y to find its own HC, and HC^* is the length of the optimal HC. Therefore, the fitness value of Y is calculated according to Eq.(13).

$$Fitness(Y) = \frac{1}{HC^*} \tag{13}$$

As shown in Figure 1, for the hybrid SOS-MMAS, first, optimize the parameters through SOS. In the calculation process, only organisms with higher fitness values are retained and used as input parameters of MMAS. Then, the advanced MMAS algorithm is used to solve the TSP. Finally, when the entire program ends, the shortest HC will be output.

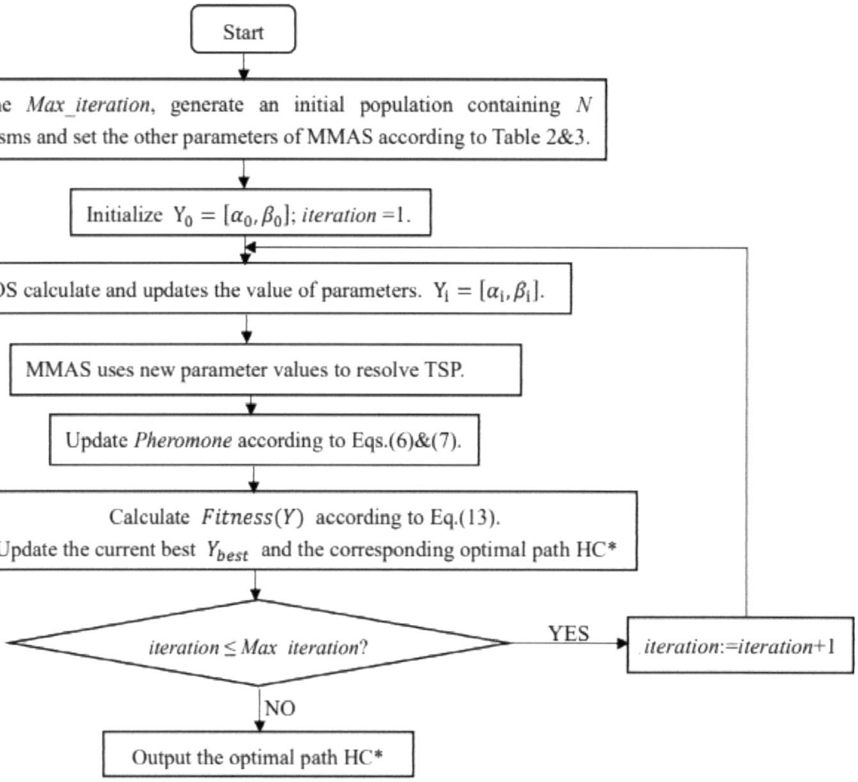

Figure 1. The flowchart of the hybrid SOS–MMAS

3. Experimental environment

TSPs of different scales are selected as shown in Table 1. The number of cities is specified in the range of 50-600, and 11 typical problems are selected from the TSPLIB[10] for calculation. These problems belong to the EUC_2D type, that is, weights are given by the 2-dimensional Euclidean distance.

Table 1. Basic information of selected TSP

Number	TSP Name	No. of Cities	Type	Bounds
1	Eil51	51	EUC_2D	426
2	Eil76	76	EUC_2D	538
3	lin105	105	EUC_2D	14379
4	KroA150	150	EUC_2D	26524

5	KroB200	200	EUC_2D	29437
6	Pr264	264	EUC_2D	49135
7	linph318	318	EUC_2D	41345
8	Rd400	400	EUC_2D	15278
9	pcb442	442	EUC_2D	50778
10	d493	493	EUC_2D	35002
11	u574	574	EUC_2D	36905

In addition, *Bounds* represents information about the new boundary or best solution of the problem, and can also be understood as representing the length of the most famous solution of the instance. SOS-MMAS is written in python language and runs on an Intel(R) Core(TM) i5-6200U computer, CPU 2.40 GHz, 8GB RAM.

Table 2 shows the value/value range of related parameters. N represents the population size of SOS, m represents the number of ants, ρ is the pheromone evaporation coefficient, and Q is the pheromone enhancement coefficient.

Table 2. Parameter setting

Parameter	α	β	N	m	ρ	Q
Value	[1,5]	[1,5]	10	20	0.1	1000

In Table 3, for TSPs with different city sizes, different maximum iteration times are set to save calculation time.

Table 3. Maximum iteration setting

No. of Cities	**Maximum iteration**
n<200	200.00
200≤n<400	300.00
n≥400	400.00

4. Results & Discussion

To verify the stability of the hybrid SOS-MMAS, select TSP instances of different scales in TSPLIB for verification. For each TSP, the SOS-MMAS algorithm is run 10 times. When MMAS and ACO are initialized, the values of α and β must be given. Through previous studies, it is found that when $\alpha = 1$ and $\beta = 5$, the ant colony algorithm achieves better results[11]. Therefore, the initial values of α and β in this study are set as 1 and 5 respectively, that is, initial $Y_0 = [1, 5]$.

4.1 Experimental results of SOS–MMAS

Table 4 shows the experimental results of SOS-MMAS. The *Best* column represents the optimal one of the corresponding TSP in 10 runs, and the *Average* column represents the average value of the 10 runs

Table 4. Experimental results

No. of cities	**Bounds**	**Best**	**Average**	**Ewa(%)**	**Ewb(%)**
51	426	426	428	0.49	0.49
76	538	538	542	0.69	0.69
105	14379	14371	14448	0.48	0.54
150	26524	26617	26675	0.57	0.22
200	29437	29471	29604	0.57	0.45
264	49135	49135	49253	0.24	0.24
318	41345	41820	42010	1.61	0.45
400	15278	15429	15530	1.65	0.66
442	50778	51291	51679	1.77	0.76
493	35002	35368	35614	1.75	0.70
574	36905	37453	37696	2.14	0.65

Eq.(14) and (15) are the error calculation formulas. *Ewa* represents the error of the *Average* relative to *Bounds*, which can be used to evaluate the calculation performance of SOS-MMAS; *Ewb* represents the error of the *Average* relative to *Best*, which can be used to judge the robustness of the algorithm.

$$Ewa(\%) = \frac{Average - BKS}{BKS} \times 100\% \tag{14}$$

$$Ewb(\%) = \frac{Average - Best}{Best} \times 100\% \tag{15}$$

The results show that for the smaller-scale TSP, the performance of SOS-MMAS is very eye-catching. At the same time, for the larger-scale problems, the solution is relatively close to the value of *Bounds*. For all instances, the *Maximum Ewa* is about 2%, and the *Average Ewa* is about 1%; the *maximum Ewb* is about 0.7%, and the *Average Ewb* is about 0.5%. The results show that the SOS-MMAS algorithm has strong performance and robustness to solve TSP.

4.2 Comparisons between SOS–MMAS and SOS–ACO, ACO

Furthermore, in order to compare the operating efficiency of SOS-MMAS, SOS-ACO and ACO, for each instance, each algorithm runs ten times. The results are shown in Table 5.

Table 5. Comparison results

	Bounds	**Best**	**Average**	**Ewa(%)**	**Ewb(%)**
SOS–MMAS	426	426	428	0.49	0.49
	538	538	542	0.69	0.69
	14379	14371	14448	0.48	0.54
	26524	26617	26675	0.57	0.22
	29437	29471	29604	0.57	0.45
	49135	49135	49253	0.24	0.24
	41345	41820	42010	1.61	0.45
	15278	15429	15530	1.65	0.66
	50778	51291	51679	1.77	0.76
	35002	35368	35614	1.75	0.70
	36905	37453	37696	2.14	0.65
	Average error(%)			1.09	0.53
SOS–ACO	426	430.00	436	2.44	1.49
	538	549.00	553	2.83	0.77
	14379	14782.00	14832	3.15	0.34
	26524	27407.81	27662	4.29	0.93
	29437	30365.23	30859	4.83	1.63
	49135	52062.00	52687	7.23	1.20
	41345	44242.23	44797	8.35	1.25
	15278	16430.00	16496	7.97	0.40
	50778	54057.85	55140	8.59	2.00
	35002	37572.66	38320	9.48	1.99
	36905	40243.65	40887	10.79	1.60
	Average error(%)			6.36	1.24
ACO	426	449.00	453	6.38	0.94
	538	570.00	574	6.77	0.77
	14379	15422.80	15555	8.18	0.86
	26524	29251.83	29709	12.01	1.56
	29437	31681.92	32073	8.96	1.24
	49135	54007.00	54545	11.01	1.00
	41345	46742.85	47210	14.18	1.00
	15278	17568.00	17803	16.53	1.34

	50778	58755.70	59434	17.05	1.15
	35002	41019.48	41834	19.52	1.99
	36905	44400.49	44933	21.75	1.20
	Average error(%)			13	1.18

From the results in Table 5, the solution of the typical TSPs calculated using SOS-MMAS is significantly better than the solutions obtained by the other 2 methods. At the same time, the parameter optimization benefits of SOS are also reflected in SOS-ACO, and the obtained solution is significantly better than the pure ACO algorithm.

Observe the comparison result of the average error, *Ewa (SOS-MMAS)<Ewa (SOS-ACO) <Ewa (ACO)*, indicating that SOS-MMAS does have outstanding performance in solving TSP; *Ewb(SOS-MMAS) <Ewb(SOS-ACO)≈Ewb(ACO)*, showing that SOS-MMAS has stronger robustness. The above results further indicate that SOS-MMAS algorithm has better solving effect in fewer iterations. Compared with the traditional ACO algorithm and SOS-ACO algorithm, this algorithm has the double advantages of higher accuracy and faster convergence.

5. Conclusions

This paper proposes a hybrid algorithm SOS-MMAS for the traveling salesman problem. On the one hand, the proposed hybrid algorithm makes full use of the advantages of MMAS in solving TSP, including fast convergence rate and avoiding premature convergence. On the other hand, considering that SOS has the benefits of simple operation and easy implementation, the hybrid algorithm uses it to optimize the two key parameters of MMAS.

Compared with SOS-ACO and ACO, SOS-MMAS can find a better solution in fewer iterations, accompanied by a smaller average error. In addition, the hybrid SOS-MMAS algorithm has superior optimization and stronger robustness to solve TSP. Simultaneously, the comparative experimental results of SOS-ACO prove that SOS method does have good adaptability in optimizing the key parameters of the ant colony system. In the future, researchers can consider applying SOS-MMAS to other combinatorial optimization problems, and can also add some effective local search strategies to enrich the algorithm, such as k-Opt, for further research.

References

[1] Corinne, B., (2013)The Travelling Salesman Problem. University of Pittsburgh Publishing, Pittsburgh.

[2] Baniasadi, P., Foumani, M., Smith-Miles, K., Ejov, V. (2020) A transformation technique for the clustered generalized traveling salesman problem with applications to logistics. European J. Oper. Res, 285(2):444–457.

[3] Ouaarab, A., Ahiod, B., Yang, X.S. (2015)Random-key cuckoo search for the travelling salesman problem. Soft Computing, 19:1099–1106.

[4] Muren, J., Wu, J., Zhou, L., Du, Z.P., Lv,Y. (2019)Mixed steepest descent algorithm for the traveling salesman problem and application in air logistics.Transp.Res. E, 126:87–102.

[5] Yang, X.S., Deb, S., Loomes, M., Karamanoglu, M. (2013) A framework for self-tuning optimization algorithm. Neural Comput, Appl. 23 (7–8):2051–2057.

[6] Peker, M., Sen, B., Kumru, P.Y. (2013) An efficient solving of the traveling salesman problem. Turk. J. Electr. Eng. Comput,21:2015–2036.

[7] Mahi, M.(2015) A new hybrid method based on particle swarm optimization, ant colony optimization and 3-opt algorithms for traveling salesman problem. Appl. Soft. Comput, 30:484–490.

[8] Ezugwu, E.S., Adewumi, A.O.(2017) Discrete symbiotic organisms search algorithm for traveling salesman problem. Expert Syst. Appl, 87:70–78.

[9] Stutzle, T., Hoos, H,H. (2000) Max-min ant systems. Future Generation Computer Systems, 16(19):889-891.

[10] Reinelt, G. (1991) TSPLIB- a traveling salesman problem library. ORSA J. Comput., 3:376–384.

[11] Dorigo, M., Caro, G.D., Gambardella, L.M. (1999) Ant algorithms for discrete optimization. Artif. Life 5 (2):137–172.

Research on Bi-LSTM machine reading comprehension algorithm based on attention mechanism

Tong Shi[1a*], Xiaohui Li[1], Zhonglin Liu[1] and Licai Wang[1]

[1]Department 5,North China Institute of Computing Technology ,Beijing, 10008,China

[a*]252682335@qq.com

Abstract—Machine reading comprehension is a very important research field in Natural Language Processing (NLP), but it is not sufficient for military research, and traditional machine reading comprehension algorithms do not fully explore every word in different contexts. In response to the above problems, this paper proposes a machine reading comprehension model that incorporates a multi-attention mechanism. It is brave to solve machine reading comprehension tasks in the military field. The model in this paper performs word embedding, part-of-speech embedding, and knowledge base embedding for each word in the article. Among them, knowledge embedding uses military knowledge bases to make up for the lack of information on military entities and relationships in military articles, and independently carries out part-of-speech embedding to solve the problem of part of speech judgment of polysemous words in specific contexts. The experimental results show that the machine reading comprehension model with multi-attention mechanism proposed in this paper can achieve better results on machine reading comprehension tasks, and the accuracy rate is improved by 1.6% compared with ROUGE-L.

1. Introduction

Machine reading comprehension (MRC) plays a very important role in natural language processing. Its goal is to enable the computer to interpret the articles we encounter in our daily life and find the answers we need from the articles according to the questions we ask. When we need to read a lot of text, machine reading comprehension can help us save a lot of energy and time, so machine reading comprehension has become one of the hot research fields in the field of artificial intelligence. Artificial intelligence technology has become more and more widely used in our life. Natural language processing (NLP) can fully mine the semantic information in articles through massive corpus, so that computers can understand the semantics and context of articles and replace human reading comprehension, so as to improve work efficiency [1][2][3].

The original machine reading comprehension model usually extracts text features through rules or manual methods. This feature extraction method relies on manual extraction, and the migration ability of each model is not good. At the same time, with the continuous increase of the amount of data in the data set, this old method has been gradually abandoned, The end-to-end reading comprehension model based on deep learning has become the main method of machine reading comprehension[4] .

2. Related work

Before the advent of the pre-training language model, for most end-to-end deep learning machine reading comprehension models, the overall architecture of "coding interaction and reasoning output" is usually used to deal with reading comprehension problems. The overall architecture is shown in Fig.1.

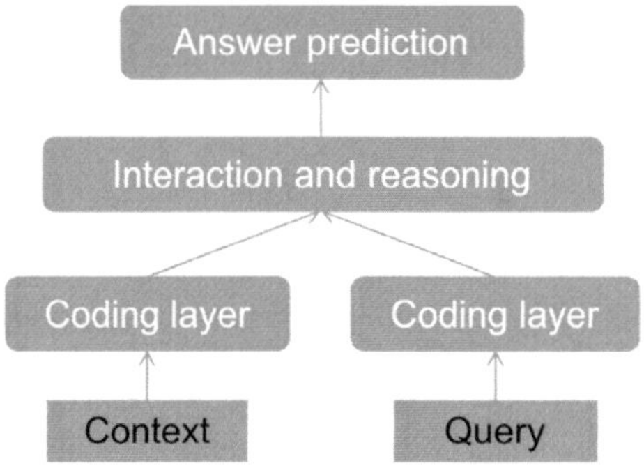

Fig.1 Common framework.

For the previous coders, the methods of word vector and cyclic neural network are often used to encode the text we want to read and the problems we want to understand [5][6][7].

Word vector (word2vec) is to transform words or words in the text into dense vectors. So that the computer can understand the meaning. Before the generation of word2vec, one hot was used in the early stage of natural language processing to convert words or words into discrete individual symbols , but one hot coding has a very serious problem. The size of word vector generated by one hot coding depends on the number of words or words in the article. If the number of words or words is too large, It will make the generated vectors very sparse, resulting in very high memory requirements for the computer. The word vector generated by word2vec is a dense vector, and there is also a mutual relationship between synonyms, which is no longer an independent word or word.

Recurrent neural network (RNN) [8], in terms of any input sequence, its output result at each time has a mutual relationship with the output result in front of this output result, and the specific expression is that RNN network will remember the output before each sequence and act on the current output, Connect each hidden layer in the network.

For the "interaction and reasoning" module, attention mechanism is often used to match the semantics of the read articles and questions.

Attention mechanism has become very popular in recent years. At first, attention mechanism was applied in machine translation tasks[9], and after that, attention mechanism began to be applied to more natural language processing tasks, such as text summarization tasks[10]. Attention mechanism [11] is a selection mechanism used to allocate limited information processing capacity.

Bidaf (bidirectional attention flow) [12] is a classic machine reading comprehension model. It uses a total of six network layers to deal with machine reading comprehension tasks, and uses a two-way attention flow structure to obtain the interactive information between articles and problems. Qanet (question answering network) [13] uses local convolution and global self attention mechanism to improve the training speed and inference speed of the model. Although the design of these models is very exquisite, because these models use the traditional word embedding method, it is difficult for these models to solve the problem of polysemy [14][15].

Based on the traditional model, this paper uses three different ways to model the text in order to mine the different meanings of each word in different contexts, so as to improve the accuracy of the model.

3. Model building

This paper establishes a machine reading comprehension model integrating multi attention mechanism. The model structure is shown in the figure below.

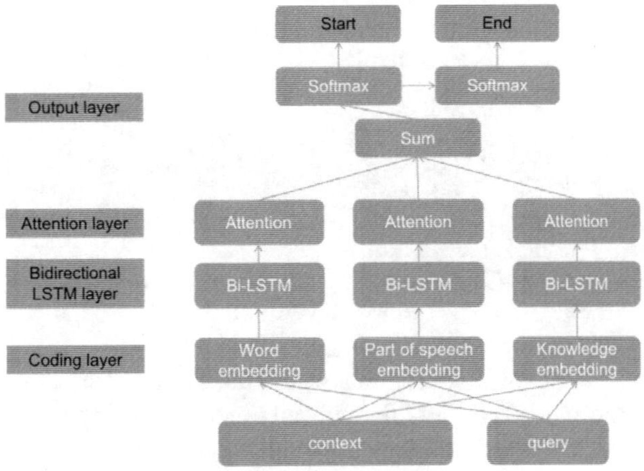

Fig.2 Model structure.

This model consists of three networks: word embedding network, part of speech attention network and knowledge embedded attention network. The article and problem extract features through three different feature extraction methods, then input three different bidirectional LSTM networks integrated with attention mechanism, and finally normalize the results. The model structure of this paper is introduced in detail below.

3.1. Coding layer

The function of the coding layer is to convert a single word or word in an article or problem into a high-dimensional word vector, so that each word or word can obtain its corresponding vector representation.

However, for a word, its part of speech also has a certain meaning in the article or problem, and considering that there are many entities such as characters and weapons in the articles and problems in the data set used this time, we consider extracting text features from three aspects for articles and problems.

For an article, it is composed of multiple paragraphs, and each paragraph is composed of multiple words. If the total number of paragraphs in an article is K and the total number of words in each article is P, it can be expressed as $\{p_1, p_2, \dots, p_m\}$. And a problem is also composed of multiple words. If the total number of words in a problem is N, it can be expressed as $\{q_1, q_2, \dots, q_n\}$.

(1)Word embedding

For word embedding, this paper uses distributed representation learning to embed each word p_i.

(2)Part of speech embedding

For part of speech embedding, this paper marks the part of speech of each word p_i in the article, and then represents and embeds the part of speech of each word.

(3)Knowledge embedding

For knowledge embedding, this paper uses named entity recognition to obtain the entity information in the article. If the word p_i can find the corresponding entity or relationship in the constructed military knowledge base, then take the entity corresponding to the word p_i in the knowledge base as the feature extraction vector. If the corresponding entity cannot be found in the constructed military knowledge base, then select the feature extraction vector of the previous entity of the entity as the supplementary vector to ensure that the length of the input vector is consistent.

The following describes the three coding methods used in this paper.

3.2. Word embedding

This paper uses the improved CBOW model to learn the distributed representation of words in text. The structure of CBOW model is shown in the figure below.

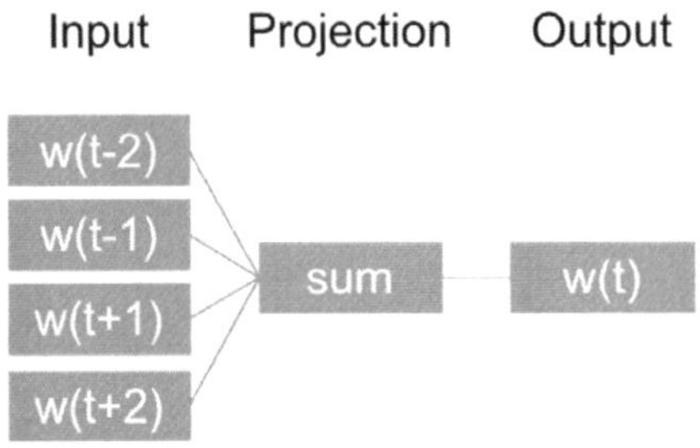

Fig.3 CBOW model

As shown in the figure, CBOW model is mainly divided into three levels: input layer, projection layer and output layer. The model predicts the current word according to the context information of each word. The context of each word is selected by sliding window method, that is, the words within the current word window are taken as the context information. However, there will be a problem in this process. Because the CBOW model will use the sliding window method to select the context information of the current word, the prediction of the current word is only determined by the window size, which has nothing to do with the full-text information of the text. Therefore, the generated word vector can not reflect the importance of the current word in the full-text.

To solve the above problems, this paper uses a weighted cbow model structure, namely tfidf-cbow model structure, so that the current word vector can reflect the importance of the current word in the document.

TFIDF is mainly composed of TF and IDF, and its formula is:

$$\mathrm{TFIDF} = \mathrm{TF} \times \mathrm{IDF} \tag{1}$$

TF represents the frequency of a word in the document, and IDF is an indicator to measure the importance of a word.

The IDF formula is as follows:

$$\mathrm{IDF} = \log(\mathrm{N}/\mathrm{n} + 1) \tag{2}$$

In the above formula, n represents the number of texts containing a word t in the text set, and N represents the total number of texts in the text set. When there are more texts containing the word t, the IDF value of the word t will be smaller and smaller, indicating that the word t is unimportant in the text.

Firstly, the cbow model in word2vec is used to train the corpus, and the word vector obtained is expressed as follows:

$$v_{w(i)} = CBOW(w(i)) \tag{3}$$

Then, based on the trained word vector, the weight TFIDF is introduced, and the formula is as follows:

$$v_{w(i)} = TFIDF \times CBOW(w(i)) \tag{4}$$

The combination of TFIDF model and cbow model can strengthen the semantic expression ability of the word vector trained by cbow model, so that each word can be context related.

3.3. Part of speech embedding

After obtaining the part of speech and entity category of words in the text, the coding vector of part of speech is obtained by random initialization method, and the feature vector will be used to establish the reading comprehension model.

In this paper, the bilstm attention CRF model is used to label the text data and obtain the part of speech of each word in the text. Firstly, the bilstm layer is used to bi directionally encode the input text, and the attention mechanism is used to increase the context related semantic information before decoding. Finally, the semantic vector containing the context information is input into the CRF layer for decoding. The CRF layer can output the tag sequence with the greatest probability, so as to obtain the category of each character, such as noun, verb, address Part of speech or entity category such as person's name.

3.4. Knowledge embedding

Based on the established military knowledge map, this paper represents and learns the structure of the knowledge map, and represents the nodes in the knowledge base and the relationship between nodes as feature vectors.

In this paper, compgcn [20] algorithm is used to represent and learn the knowledge base. Compgcn algorithm is a heterogeneous graph representation algorithm proposed for multiple relational graphs, which can represent and learn the relationship between each knowledge node and knowledge nodes in the knowledge base.

3.5. Bi-LSTM reading comprehension model with attention mechanism

In this paper, a bidirectional LSTM network based on attention mechanism is used to calculate the results of each modeling method separately.

Usually, we use bidirectional recurrent neural network to model text sequence, because text is a sequence in order. Assuming that the text to be converted is x, first convert the index number of the word into a real value vector through a word embedding layer. The formula is as follows:

$$e(x) = W_e * x \quad (5)$$

Then we will use a deformation of the cyclic neural network and use the bidirectional LSTM network for modeling to solve the dependence of the traditional cyclic neural network on long distance and the disappearance of gradient. The specific formula is as follows:

$$\overrightarrow{h_s(x)} = \mathrm{LSTM}(e(x)) \quad (6)$$

$$\overleftarrow{h_s(x)} = \mathrm{LSTM}(e(x)) \quad (7)$$

$$h_s(x) = [\overrightarrow{h_s(x)}; \overleftarrow{h_s(x)}] \quad (8)$$

In reading comprehension task, attention mechanism is an important technology to connect text and problems. Given the text h_{doc} and problem h_{query}, we can calculate the pairwise relationship between each word in them. Here we call $M(i, j)$ is matching matrix or similarity matrix. The formula is as follows:

$$M(i, j) = h_{doc}(i)^T * h_{query}(j) \quad (9)$$

On the basis of this matrix, the attention results of the text are obtained by averaging by column and normalizing.

3.6. Multi attention mechanism modeling and training

After obtaining the word embedding code, part of speech code and knowledge embedding code of each word in the problem and text, three reading comprehension models containing attention are established to complete the attention modeling from multiple dimensions, that is, using word embedding code alone, word embedding code splicing part of speech code, word embedding code splicing knowledge embedding code. Finally, the output of the three attention model networks is modeled by probability, and the probability that all words in the paragraph are the beginning or end of the answer range is predicted.

Paragraph P and question Q pass through three attention networks, the answer interval scores $S_{w(s|e)}(q, p_i)$, $S_{t(s|e)}(q, p_i)$ and $S_{k(s|e)}(q, p_i)$ of different features will be obtained according to the text modeling of the three aspects. For each answer interval score, it contains two scores, namely, the start score and the end score of the answer.

Then add the three scores to find the sum $S_{(s|e)}(q, p_i)$ of the scores of question Q in the answer interval on paragraph P:

$$S_{w(s|e)}(q, p_i) + S_{t(s|e)}(q, p_i) + S_{k(s|e)}(q, p_i) = S_{(s|e)}(q, p_i) \quad (10)$$

Because the output results of the three networks are unified at the end of this paper, the softmax function is used to normalize the summed answer interval score, and the final result probability is transformed into 0 to 1. The formula is as follows:

$$p_{(s|e)}(p_i|q) = e^{S_{(s|e)}(q,p_i)} / \sum_{k=1}^{m} e^{S_{(s|e)}(q,p_k)} \quad (11)$$

In this way, we get the final probability $p_{(s|e)}(p_i|q)$ that each word p_i is the starting position or ending position of the answer.

Using the above calculation method, we can obtain the probability $p_{(s|e)}(p_i|q)$ of the starting position and ending position of the answer range on a certain paragraph, but this answer is the optimal solution of a problem in a certain paragraph of this article. If we want to obtain the optimal solution in this article, we need to find a word range with the greatest probability in all paragraphs of the article, After normalization, the probabilities of the start position and end position of the answer add up to 1. If we want to find the word range with the greatest probability, we only need to maximize the probability product of the start position and end position of the answer, that is, $p_s(p_i|q) * p_e(p_i|q)$ is the largest. Finally, the answer probabilities generated by all paragraphs in the article are compared together, and the one with the greatest probability is the optimal solution we need.

4. Experimental results and analysis

4.1. Experimental data set

The data set used in this paper comes from the Les-totle military reading comprehension competition. Through processing the data set, a total of 96027 samples are obtained, including 90625 samples in the training set and 5402 samples in the test set. This data set is a reading comprehension data set for military texts. The specific data set examples are as follows:

Article:Israeli air force C-130J transport aircraft,China Aviation News Network News: according to the Jerusalem Post, the Israeli military announced...

Question:Which model of transport aircraft did the Israeli air force purchase from Lockheed Martin Space Systems Company

Answer：C-130J super Hercules

Answer start position：174

Answer end position：185

Each article in the dataset corresponds to multiple questions. Each question corresponds to an answer in the article, but the corresponding answer may be empty and does not exist in the article.

4.2. evaluation method

Considering the need to compare the similarity between the predicted answer and the real answer, this paper uses the standard Rouge (Recall-Oriented Understudy for Gisting Evaluation) to evaluate the effect of the model. Rouge is a performance index used to evaluate automatic summarization and machine translation. Rouge compares the abstracts and translations automatically generated by the machine with the artificially determined reference abstracts and translations, and calculates the score of each abstract or translation, so as to detect the similarity between the abstracts and texts automatically generated by the machine and the artificially determined reference abstracts and translations.

Rouge-l calculates the length of the longest common substring between two sequences, so as to obtain the similarity score between the two sequences. The calculation formula is as follows:

$$\mathrm{LCS} = \mathrm{lsc}(x, y) \quad (12)$$

$$R = LCS/len(x) \quad (13)$$

$$P = LCS/len(y) \quad (14)$$

$$Rate = len(x)/len(y) \quad (15)$$

$$F = [(1 + Rate^2) \times R \times P]/(R + Rate^2 \times P) \quad (16)$$

In the above formula: X is the real answer to our question, y is the answer predicted by the model, LCS is the length of the longest common substring of the real answer and the model predicted answer, R and P represent the recall rate and accuracy rate respectively, and F is the final score rouge-l.

4.3. Experimental results and analysis

Table 1 shows the performance of different models on the data sets used in this paper.

Table 1. Experimental result.

Model	ROUGE-L
BiDAF	77.32
Match-LSTM	76.81
QANet	88.43
R-net	89.92
the model of this paper	91.52

As can be seen from table 1, compared with the original basic model, the model in this paper has been improved in rouge-l by 1.6%.

It can be seen that by extracting the different attributes of each word in the article, the network can learn richer semantic information in the text, add the results and normalize the probability, which can improve the accuracy of the model.

5. Conclusion

This paper adopts a machine reading comprehension model integrating multi attention mechanism. For each word in the article, not only the traditional dictionary coding, but also the part of speech and knowledge embedding coding are used. Different attention mechanism networks are used to extract the text features of the article and the problem. Finally, the result probability is summed and normalized, which can effectively improve the accuracy of the model.

The knowledge base used by the existing model is based on the vertical domain knowledge base. In the future, we will consider expanding the used knowledge base in order to apply the model to a wider range.

References

[1] Karl Moritz Hermann,Tomás Kociský,Edward Grefenstette,Lasse Espeholt,Will Kay,Mustafa Suleyman,Phil Blunsom. Teaching Machines to Read and Comprehend.[J]. CoRR,2015,abs/1506.03340:

[2] CUI Y , LIU T , CHE W , et al. A Span-Extraction Dataset for Chinese Machine Reading Comprehension[J]. 2018.

[3] WANG Z , LIU J , XIAO X , et al. Joint Training of Candidate Extraction and Answer Selection for Reading Comprehension[C]// Meeting of the Association for Computational Linguistics. 2018.

[4] CUI Y , CHEN Z , WEI S , et al. Attention-over-Attention Neural Networks for Reading Comprehension[J]. 2016.

[5] WANG S , JIANG J . Machine Comprehension Using Match-LSTM and Answer Pointer[J]. 2016.

[6] KADLEC R , SCHMID M , BA J, et al. Text Understanding with the Attention Sum Reader Network[J]. 2016.
[7] GOLDBERG Y, LEVY O. word2vec Explained: deriving Mikolov et al. 's negative-sampling word-embedding method [J]. Eprint Arxiv, 2014.
[8] BAHDANAU D , CHO K , BENGIO Y . Neural Machine Translation by Jointly Learning to Align and Translate[J]. Computer Sc]ience, 2014.
[9] RUSH A M , CHOPRA S , WESTON J . A Neural Attention Model for Abstractive Sentence Summarization[J]. Computer Science, 2015.
[10] VASWANI A , SHAZZER N , PARMAR N , et al. Attention Is All You Need[J]. arXiv, 2017.
[11] SEO M , KEMBHAVI A , FARHADI A , et al. Bidirectional Attention Flow for Machine Comprehension[J]. 2016.
[12] YU A W , DOHAN D, LUONG M T , et al. QANet: Combining Local Convolution with Global Self-Attention for Reading Comprehension[J]. 2018.
[13] PENNINGTON J , SOCHER R , MANNING C . Glove: Global Vectors for Word Representation[C]// Conference on Empirical Methods in Natural Language Processing. 2014.
[14] LE Q V , Mikolov T . Distributed Representations of Sentences and Documents. JMLR.org, 2014.
[15] Vashishth S , Sanyal S , Nitin V , et al. Composition-based Multi-Relational Graph Convolutional Networks[J]. 2019.

An Contract Theory based Federated Learning Aggregation Algorithm in IoT Network

Mengmeng Tian[1*]

[1] Software College, Northeastern University, Shenyang, Liaoning,, CHXX0119, China

*Corresponding author's e-mail: 1901270@stu.neu.edu.cn

Abstract. Taking IoT devices as the edge nodes is one of the powerful way to offloading the federated task since IoT devices are closer to the data generation end. The aggregation efficiency of federated learning in the IoT environment is inefficiency since the server of federated learning can not know the data quality of heterogeneous IoT device. How to encourage IoT edge clients to participate in federated learning and maximize the aggregation effect of the global model is an important problem. This paper proposes a federated learning aggregation model based on contract theory incentive mechanism. Our experimental results show that the proposed algorithm effectively improves the aggregation efficiency of federated learning compared with FedAvg algorithm.

1. Introduction

Different from traditional machine learning, federated learning is a distributed machine learning framework. The central server can aggregate the models uploaded by the IoT edge nodes, which not only reduces the burden of the server but also effectively protects the data privacy of customers[1,2]. It allows client nodes to use their own local data for calculation without uploading local data. However, there are still some problems in federated learning with the IoT devices as an edge computing node since the the IoT devices has heterogeneous and limited resources. In 2017, the FedAvg[3] algorithm proposed by Google aggregates the model of each edge node on average, without considering the differences of edge computing nodes. Giving an average global aggregation weight to the model uploaded by low-quality edge computing nodes undoubtedly reduces the effect of the global model. Similarly, a global aggregation model based on data volume is proposed in[4]. The aggregation weight of the global model does not account for the proportion of the client data volume in the total data volume. However, the federated learning can not observe the amount of data on the client side of the edge computing node, so it is unreasonable to allocate aggregation weight according to the amount of data. What's more, the quantity of data can not represent the quality of data.

Based on the data quality of federated learning, an incentive mechanism and aggregation algorithm based on the Stackelberg game is proposed in [5,6]. However, the federated learning environment is asymmetric information, that is, the central server of federated learning cannot observe the data of the client, and they need to interact constantly to obtain the information of data quality. Thus, here is a part of information loss and the federated learning server can not achieve the optimal aggregation effect in the single global iteration.

Contract theory is a powerful tool to solve the problem of incentive mechanism construction in the environment of information asymmetry, such as data plan design[7], the design of advertising reward[8], and so on. In this paper, we will use the contract theory to construct the incentive mechanism and design the aggregation algorithm. Firstly, we estimate the data quality of IoT edge nodes. Then, we provide a corresponding contract-based incentive to reduce the loss caused by information asymmetry. Finally, the aggregation weight of the model uploaded by IoT edge client is given according to the contract selection. The experimental results show that our proposed contract based incentive mechanism effectively improves the aggregation efficiency of federated learning.

2. System Model

In this section, we introduce the system model. Firstly, we propose the basic process of the interaction between the federated learning server and the IoT edge nodes. Secondly, we give the data quality estimation method of the IoT edge nodes. Thirdly, we propose the optimal incentive mechanism based on contract theory. Finally, a new global aggregation algorithm is proposed.

2.1. Basic Flow

Figure 1 shows the architecture of federated learning assisted by IoT edge devices. The federated central server will hire IoT edge nodes to use their local edge data for model calculation. The federated learning central server gives a series of contracts, which specify the test accuracy and corresponding rewards required by the computing model of IoT edge nodes. The IoT client selects one of the contracts according to its own local data quality and train the model based on their data. The local IoT model will be uploaded and be tested by central server. Only the models that pass the accuracy test specified in the contract can get the corresponding reward.

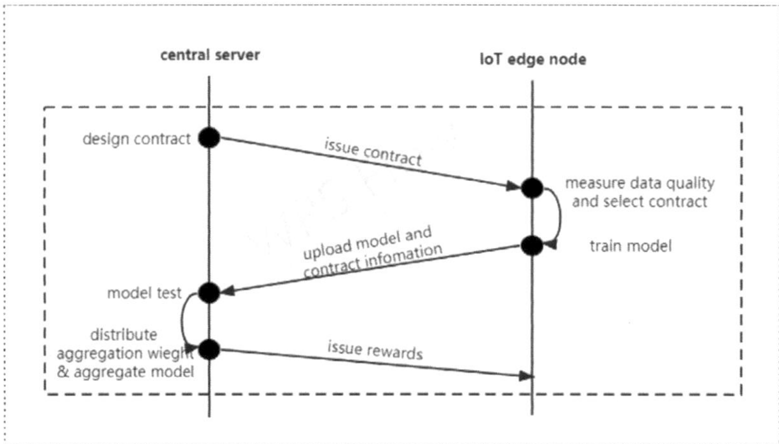

Figure 1. Interaction in one global aggregation.

2.2. Measurement of IoT Data Quality

The quality of data is reflected in the feature covered by the data[9]. The greater the feature coverage, the higher the quality of local data. The feature coverage of can be measured by the ratio of the number of features in the neighborhood with radius r to the number of features in the total square, marked by M(r). Therefore, the average feature coverage in the K dimension can be shown as

$$\alpha = \frac{1}{\sqrt{K}} \int_0^{\sqrt{K}} M(r)dr \qquad (1)$$

where a denotes the feature coverage in average. Thuswe can sort the data of the IoT edge client according to the quality. Suppose there is a total of N from low to high, which is defined as

$$\alpha_1 < \ldots < \alpha_i < \ldots < \alpha_N \qquad (2)$$

2.3. Incentive Design for Federated Learning

We link the data quality a_i and contract y_i of the IoT edge nodes through the restriction of economic benefits. Limited by economic incentives, only when IoT edge nodes choose a contract designed specifically for their own data quality types can they obtain benefits and maximize benefits. Rational IoT edge nodes will choose the contract y_i fitted for their data quality a_i. Thus, the selected contract can reflect the data quality of the IoT edge nodes.

2.3.1. Contract and Utility

For IoT edge nodes with different data quality a_i, we design a contract item $y_i=(R_i, A_i)$ for i in [1, N]. The contract is composed of R_i and A_i, in which A_i is the test accuracy that the IoT edge client needs with data quality a_i to meet, and R_i is the reward after completing the corresponding federated learning task. Assume that the data quality is a_i, the utility of IoT edge node with selecting a random contract $y_r=(R_r, A_r)$ show as equation (3), where v(R) is the value function about reward in contract which satisfies v'(R) >0 and v''(R) <0. A is the corresponding accuracy required by contract. c represents the cost per unit accuracy of IoT edge devices.

$$\mathrm{UIoT}_i = \alpha_i v(R_r) - cA_r \quad (3) \qquad USer = \sum_{r=1}^{N} \lambda_r (\gamma A_r - R_r) \quad (4)$$

The central server of federal learning can not accurately know the information about data quality of IoT edge devices, but has the priori distribution information l_i about IoT edge devices. Therefore, its economic utility of federated learning central server is shown as equation (4), where g represents the conversion coefficient from model accuracy A_r to revenue of the model, which represents the benefits obtained from the model uploaded by the edge client of the IoT. The utility is the economic benefit obtained by the uploading model minus the reward issued by the server.

The design of contract must meet two basic conditions. Firstly, it is profitable for the IoT edge nodes to select the contract corresponding to their data quality, that is, individual rationality(IR) constraint; Secondly, the IoT edge client can achieve the optimal benefit only when it is in the contract corresponding to its own data quality, called incentive compatibility(IC) constraint. Shown as

$$\alpha_i \mathrm{v}(R_i) - cA_i \geq 0, i \in [1, N] \quad (5) \qquad \alpha_i \mathrm{v}(R_i) - cA_i \geq \alpha_i v(R_j) - cA_j, \quad i, j \in [1, N] \quad (6)$$

While the IR constraint ensure that rational IoT edge clients can obtain benefits from federated learning tasks, the IC constraints ensures that rational clients will truly report their data quality. In addition, fairness requirements for federated learning task rewards require a IoT edge client with higher data quality will get more rewards. The monotonicity(Mon) constraint shows as follow.

$$0 \leq R_1 \leq \ldots \leq R_i \leq \ldots \leq R_N \quad (7)$$

Therefore, the goal of our contract design is to solve the following optimization problems.

$$\begin{aligned} &\max \sum_{i=1}^{N} \lambda_i (\gamma A_i - R_i) \\ &\mathrm{IR}: \alpha_i \mathrm{v}(R_i) - cA_i \geq 0, \\ &IC: \alpha_i v(R_i) - cA_i \geq \alpha_i v(R_j) - cA_j, \mathrm{i}, j \in [1, N] \\ &Mon: 0 \leq R_1 \leq \ldots \leq R_i \leq \ldots \leq R_N \end{aligned} \quad (8)$$

2.3.2. Contract Solution

There are N IR constraints and N(N-1) IC constraints in the optimization problem which are redundant and can be derived from some tight constraints. Next, we will reduce the IR and IC constraints on the contract, and then obtain the optimal solution of the contract according to all the tight constraints.

Combined with the data quality ranking of the IoT edge client in equation (1) and IC constraint, we have the following derivation

$$\mathrm{U}IoT_{\mathrm{i}} = \alpha_{\mathrm{i}}\mathrm{v}(R_{\mathrm{i}}) - cA_{\mathrm{i}} \geq \alpha_{\mathrm{i}}v(R_{\mathrm{i-1}}) - cA_{\mathrm{i-1}} > \alpha_{\mathrm{i-1}}v(R_{\mathrm{i-1}}) - cA_{\mathrm{i-1}} > \ldots > \alpha_1 v(R_1) - cA_1 = \mathrm{U}IoT_1 \geq 0 \quad (9)$$

Therefore, as long as the IoT edge client with the lowest data quality meets the IR condition, nodes with the other types of data quality will also meet the condition. Without losing generality, we set UIoT_1 to 0.

For IC constraints, we find that as long as the adjacent type of data quality of the client node's IC constraints, more IC constraints can be introduced. Consider IoT edge clients with three adjacent data quality types, which meet the conditions $a_{i-1} < a_i < a_{i+1}$. The three data quality clients are divided into two types: low data quality clients choose to fit the contract of higher data quality clients, i.e., equations (10) and (11); and high data quality clients choose to fit the contract of lower data quality clients, i.e., equations (12) and (13).

$$\alpha_{\mathrm{i-1}}\mathrm{v}(R_{\mathrm{i-1}}) - cA_{\mathrm{i-1}} \geq \alpha_{\mathrm{i-1}}v(R_{\mathrm{i}}) - cA_{\mathrm{i}} \quad (10) \qquad \alpha_{\mathrm{i}}\mathrm{v}(R_{\mathrm{i}}) - cA_{\mathrm{i}} \geq \alpha_{\mathrm{i}}v(R_{\mathrm{i+1}}) - cA_{\mathrm{i+1}} \quad (11)$$

$$\alpha_{\mathrm{i+1}}\mathrm{v}(R_{i+1}) - cA_{i+1} \geq \alpha_{\mathrm{i+1}}v(R_{\mathrm{i}}) - cA_{i} \quad (12) \qquad \alpha_{\mathrm{i}}v(R_{\mathrm{i}}) - cA_{\mathrm{i}} \geq \alpha_{\mathrm{i}}v(R_{\mathrm{i-1}}) - cA_{\mathrm{i-1}} \quad (13)$$

Taking equations (10) and (11) as examples, by appropriately transforming, we can obtain the IC constraint of client selection contract $y_{i+1}=(R_{i+1}, A_{i+1})$ with data quality a_{i-1}.

By shifting equation (11) and combining with the data quality ranking relationship of IoT edge clients, we have

$$cA_{i+1} - cA_{i} \geq \alpha_{\mathrm{i}}\left(\mathrm{v}(R_{\mathrm{i+1}}) - v(R_{i})\right) \geq \alpha_{\mathrm{i-1}}\left(\mathrm{v}(R_{\mathrm{i+1}}) - v(R_{i})\right) \Rightarrow \alpha_{\mathrm{i-1}}v(R_{i}) - cA_{i} \geq \alpha_{\mathrm{i-1}}v(R_{i+1}) - cA_{i+1} \ (14).$$

Combined (14) with inequality (10), we can get

$$\alpha_{\mathrm{i-1}}v(R_{i-1}) - cA_{i-1} \geq \alpha_{\mathrm{i-1}}v(R_{\mathrm{i}}) - cA_{i} \geq \alpha_{\mathrm{i-1}}v(R_{\mathrm{i+1}}) - cA_{i+1}$$

Thus, the following inequality can be derived according (10) and (11)

$$\alpha_{\mathrm{i-1}}\mathrm{v}(R_{i-1}) - cA_{i-1} \geq \alpha_{\mathrm{i-1}}v(R_{\mathrm{i+1}}) - cA_{i+1}$$

Similarly, we can also get the following relationship from inequality (12) and (13).

$$\alpha_{\mathrm{i+1}}\mathrm{v}(R_{\mathrm{i+1}}) - cA_{\mathrm{i+1}} \geq \alpha_{\mathrm{i+1}}v(R_{\mathrm{i-1}}) - cA_{\mathrm{i-1}}$$

Through the above push, we reduce the IC constraints of the IoT edge client to 2(N-1) like (11) and (13). Combined with our optimization conditions, we will further reduce the redundant IC constraints. Shifting inequality (11), we can obtain

$$A_{\mathrm{i+1}} \geq A_{\mathrm{i}} + \frac{\alpha_{\mathrm{i}}}{\mathrm{c}}\left(\mathrm{v}(R_{\mathrm{i+1}}) - \mathrm{v}(R_{\mathrm{i}})\right) \Rightarrow A_{\mathrm{i}} \geq A_{\mathrm{i-1}} + \frac{\alpha_{\mathrm{i-1}}}{\mathrm{c}}\left(\mathrm{v}(R_{i}) - \mathrm{v}(R_{\mathrm{i-1}})\right) \quad (15)$$

Shifting inequality (13), we can obtain

$$A_i \le \frac{\alpha_i}{c}\left(v(R_i) - v(R_{i-1})\right) + A_{i-1} \tag{16}$$

Inequalities (15) and (16) respectively represent the test accuracy range required by the federated learning central server for IoT edge nodes with the same data quality. The goal of this paper is to improve the global aggregation accuracy of the model as much as possible, so we take the upper bound of A_i as its optimal condition, shown in equation (17).

$$A_i = A_{i-1} + \frac{\alpha_i}{c}\left(v(R_i) - v(R_{i-1})\right) = A_1 + \frac{1}{c}\sum_{j=1}^{i} \alpha_i\left(v(R_i) - v(R_{i-1})\right) \tag{17}$$

That is, the tight constraint of the IC condition is that IoT edge client node with data quality a_i select the contract y_i, y_{i-1}. To sum up, the optimal contract is the solution of the following equation (17).

$$\begin{aligned} &\max \sum_{i=1}^{N} \lambda_i(\gamma A_i - R_i) \\ &\alpha_1 v(R_1) - cA_1 = 0, \\ &\alpha_i v(R_i) - cA_i = \alpha_i v(R_{i-1}) - cA_{i-1}, i \in [1, N] \\ &0 \le R_1 \le \ldots \le R_i \le \ldots \le R_N \end{aligned} \tag{18}$$

To solve the equation, we relax the monotonicity constraint firstly and find the solution of optimization problem as equation (18). Then, according to the algorithm in [10], the optimal contract satisfying monotonicity is obtained. The optimal value of R_i is

$$R_i = \arg\max \sum_{i=1}^{N} \lambda_i \left[\frac{\gamma}{c}\alpha_i v(R_i) - R_i + \frac{\gamma}{c}(\alpha_i - \alpha_{i+1}) v(R_i) \sum_{j=i+1}^{N} \lambda_j \right] \tag{19}$$

The optimal contract accuracy A_i can be obtained by bringing the equation (19) into equation (17).

2.4. Contract based Aggregation Algorithm

The federated server aggregate of the global model according to the matching of the optimal contract to optimize the aggregation accuracy. Since the choice of contract reflects the data quality of IoT edge nodes, thus, a model uploaded by a higher IoT edge clinet with higher contract selection will be given a higher weight. For an IoT edge node with contract y_i=(R_i, A_i) selected, when the uploaded model passes the weight accuracy test of the server, the weight assigned to its model when participating in the global model aggregation is i/N. Therefore, the global model aggregation algorithm of the central server of Federated learning shown in equation (20)

$$W_T^{Ser} = \frac{1}{\sum_{i=1}^{M} i} \sum_{i=1}^{M} i \times W_T^{IoT_i}, \ M \le N \tag{20}$$

where W_{T+1}^{Ser} is the T-th global aggregation model and $W_{T}^{IoT_i}$ is the model of IoT edge client with contract selection y_i. M is the number of IoT edge client who have passed the model test.

3. Numerous Results

In this section, we will show our experimental results. Our contract designs N = 70 kinds of data quality to fully cover the data quality of IoT edge client, and we assume that the prior distribution of the client is uniform, satisfies l_i=1/N. The value function is setting as v(R)=log(1+x), and g is 500 and c is 3.5, respectively. This paper mainly makes a comparative experiment with the FedAvg algorithm within public dataset MNIST. The total rewards of the two algorithms are consistent. FedAvg algorithm adopts uniform reward and aggregation since it has nothing about the data quality of IoT edge nodes. The optimal contract shown as Figure 2. Compared with FedAvg algorithm, our proposed algorithm shows better fairness. For IoT edge nodes with higher accuracy, we will give more rewards, which is conducive to encouraging high-quality work nodes to join.

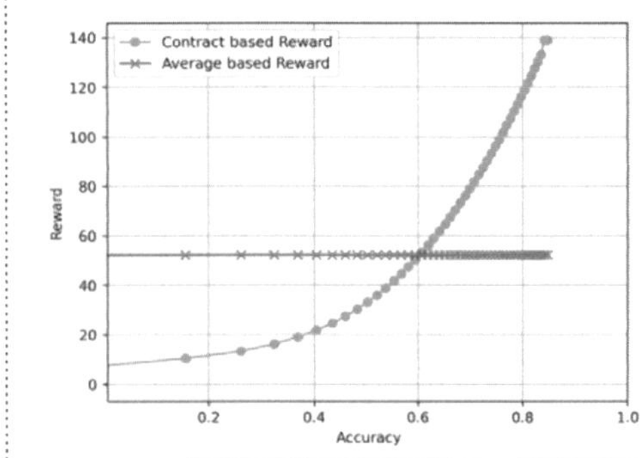

Figure 2. Incentive with reward and accuracy.

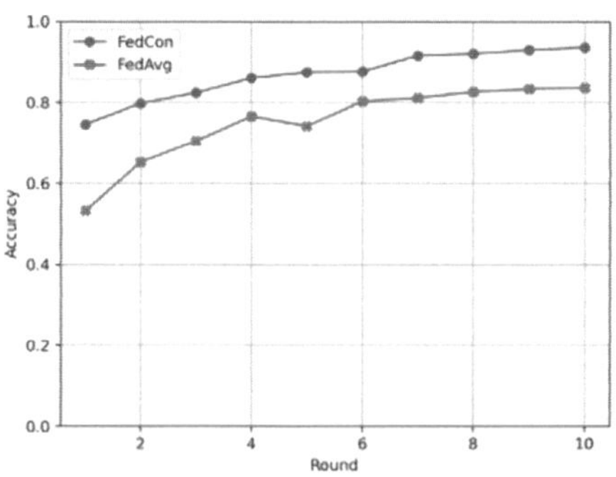

Figure 3. Aggregation effect.

The interaction between 10 IoT edge nodes and one federated server shown as Figure 1. We first issue the optimal contract shown in Figure 2. Then the client will select the contract of the object according to its own data quality and utility model, then train model and upload the model. Before each round of global iteration, the IoT edge nodes will select a certain number of pictures as their own data from the beginning of MNIST dataset. 600 to 15000 data are selected from these 10 nodes. Table 1 is an example(first iteration) about the data quality of MNIST according the data quality measurement method.

Table 1. MNIST data quality

volume	total datasets	500	1100	1500	2000	3000	4500	6500	10000	15000
data quality	0.8483	0.7814	0.7942	0.8005	0.8051	0.8158	0.8209	0.8256	0.8306	0.7351

Our simulation experiment uses the pySyft[11] framework. 10 rounds of global aggregation and iterations of federated learning is implemented. In each round of iterations, 10 client nodes participate in federated learning. Two more batch pictures based on the data volume of the previous iteration is used before each iteration to simulate the dynamic change of IoT edge nodes.

The training model structure of IoT node is 3x3x32 convolution layer, 2x2 max pooling layer, 3x3x32 convolution layer, 2x2 max pooling layer, dropout layer with probability 0.9, fully connected layers and softmax. By using CrossEntry loss function and SGD(Stochastic Gradient Descent) algorithm, these IoT nodes train the local model continuously with batch size 100.

The federated server tests the uploaded model and aggreagte thoese models. 10000 test data is used for contract accuracy test. In each round of global iteration, the federated learning server will use the test data to judge the accuracy of the model and check whether it meets the accuracy of the contract they choose. Models uploaded by clients that do not meet the contract rules will not be rewarded. In order to prevent failed models from affecting the overall aggregation accuracy, these models will be discarded.

We compare the accuracy of the global model after aggregation in 10 rounds based on FedCon(contract aggregation) algorithm and FedAvg algorithm. The test is also based on 10000 data from MNIST. Figure 3 shows the 10 rounds of iterative process with global aggregation. The aggregation efficiency of our proposed federated learning algorithm is significantly better than that of FedAvg algorithm. Since the contract selection information can reflect the data quality of IoT edge client which can be used to the global aggregation. A node with higher data quality will be given a higher weight during global aggregation. In contrast, FedAvg algorithm can not reflect the data quality of IoT edge nodes from the incentive mechanism, resulting in low aggregation efficiency. For the resource constrained environment of the Internet of Things, the efficient aggregation efficiency improves the utilization of resources and reduces the consumption of resources.

4. Conclusion

In this paper, we propose an algorithm of federated learning incentive mechanism based on contract theory. Since the IoT edge nodes are close to the data generation end, we use IoT nodes to offload federated learning tasks. Firstly, we estimate the data quality of Internet of things edge devices with heterogeneous resources. Secondly, the optimal contract is designed for the edge nodes of the Internet of things with different data quality. The restriction of economic benefits will make the edge nodes of the Internet of things actively expose their data quality. And then an federated learning algorithm is proposed according to the contract selection, so as to improve the accuracy of the global model in each round.

References

[1] A. Imteaj, U. Thakker, S. Wang, J. Li, M. Hadi Amini. (2022) A Survey on Federated Learning for Resource-Constrained IoT Devices. IEEE Internet Things J., 9: 1-24

[2] Z. Xiong, Z. Cai, D. Takabi, W. Li. (2022) Privacy Threat and Defense for Federated Learning With Non-i.i.d. Data in AIoT. IEEE Trans. Ind. Informatics, 18: 1310-1321

[3] B. McMahan, E. Moore, D. Ramage, S. Hampson, B. Agüera y Arcas. (2017) Communication-Efficient Intelligence and Statistics. Washington. pp. 1273-1282

[4] Y. Zhan, P. Li, K. Wang, S. Guo, and Y. Xia. (2020) Big Data Analytics by Crowdlearning: Architecture and Mechanism Design, IEEE Network, 34:143-147

[5] Y. Zhan, J. Zhang, P. Li, and Y. Xia. (2019) Crowdtraining: Architecture and Incentive Mechanism for Deep Learning Training in the Internet of Things, IEEE Network, 33: 89–95.

[6] Y. Zhan, P. Li, Z. Qu, D. Zeng, and S. Guo. (2020) A Learning-based Incentive Mechanism for Federated Learning, IEEE Internet Things J., 7:6360-6368

[7] Y. Wei, J. Yu, T. Lok, L. Gao. (2019) A Novel Mobile Data Contract Design with Time Flexibility. IEEE Trans. Mob. Comput. 18: 986-999

[8] Z. Xiong, J. Kang, D. Niyato, P. Wang, H. Vincent Poor, S. Xie. (2020) A Multi-Dimensional Contract Approach for Data Rewarding in Mobile Networks. IEEE Trans. Wirel. Commun. 19: 5779-5793

[9] P. Jin, L. Lu, Y. Tang, G. Em Karniadakis. (2020) Quantifying the Generalization Error in Deep Learning in terms of Data Distribution and Neural Network Smoothness. Neural Networks 130: 85-99

[10] L. Gao, X. Wang, Y. Xu, Q. Zhang. (2011) Spectrum Trading in Cognitive Radio Networks: A Contract-Theoretic Modeling Approach. IEEE J. Sel. Areas Commun. 29: 843-855

[11] T. Ryffel, A. Trask, M. Dahl, B. Wagner, J. Mancuso, D. Rueckert, J. Passerat-Palmbach. (2018) A Generic Framework for Privacy Preserving Deep Learning. CoRR abs/1811.04017

Traffic Sign Detection Algorithm Based On Improved YOLOv4

Xiru Wu[1, a], Haozhe Cao[2, b]

[1]College of Electronic Engineering and Automation, Guilin University of Electronic Technology, Guilin, 541004, Guangxi, China

[2]College of Electronic Engineering and Automation, Guilin University of Electronic Technology, Guilin, 541004, Guangxi, China

E-mail: [a]xiruwu_guet@163.com, [b]313682196@qq.com

ABSTRACT: This paper proposes a traffic sign detection algorithm using improved lightweight YOLOv4 in order to solve the problems of large parameters quantity, poor real-time performance, and low accuracy for the unmanned driving system. Firstly, we use the MobileNetV2 network to replace the original feature extraction network in YOLOv4 to improve the detection speed by reducing the number of model parameters. Secondly, we add an attention mechanism to the backbone network to effectively extract feature information. Trim the network layer of the prediction part and add residual structure, so that feature reuse enhancement can avoid the gradient disappearing when the network layer is deeper. The size of the improved network model is reduced by 79.5%. The experimental results show that the average accuracy of the network model is 2.5% lower than that of the YOLOV4 network, and the detection speed is 58.8% higher than that of the YOLOv4 network.

1. INTRODUCTION

In recent years, China's economic development, increase the national motor vehicle number, road safety is increasingly serious, according to concerning sectional statistic, in most of the traffic accident is caused by human factors, such as not obey the traffic rules, drink driving, speeding, etc., and vehicle and road accidents caused by less than 10% [1], [2]. In recent years, in order to avoid traffic accidents and solve traffic jams and other problems, the field of unmanned driving has become the current hot research direction, a complete unmanned driving system needs to sense the environment on the road, plan the car route, and reach the designated location. And detection of traffic signs in unmanned technology plays a vital role, through the computer vision technology for traffic sign detection and recognition, and will feedback the result to the driver or unmanned systems, with the vehicle's speed and direction for manual or automatic control, to avoid traffic accidents to the greatest extent. Therefore, it is of great significance to the detection of traffic signs [3], [4].

At present, there are many methods for the detection of traffic signs. Early scholars proposed that traffic signs can be detected according to shape or color [5], and methods such as scale-invariant feature transformation and support vector machine can also be combined to improve the detection rate [6]. However, the reality is very complicated. When traffic signs are similar to their background colors or extreme weather occurs, they are prone to misdetection or missing detection. In recent years, with the continuous rise of artificial intelligence, more and more scholars have applied deep learning to the field of traffic sign detection. Zhu et al. [7] used CNN convolutional neural network to identify traffic signs in the field. Zhang et al. [8] proposed a detection method for small traffic signs and traffic signs in severe weather, which achieved high accuracy and had good real-time performance and robustness on The Chinese Traffic Sign Dataset (CCTSDB) and German Traffic sign Detection Dataset (GTSDB). Tabernik et al. [9] proposed a traffic sign detection algorithm based on RCNN network, and the error rate could be reduced to less than 3%.

Deep learning has significant improvement in accuracy compared with the traditional method, but in practice, due to the amount of calculation model and the quantity are large, lead to detection speed too slow, and the high demand for hardware, the calculation of real vehicle system ability is limited, make traffic sign detection used in unmanned system has many problems, therefore, Detection of traffic signs has important research value [10], [11].

Based on the above investigation and research, this paper proposes an improved lightweight YOLOv4 traffic sign detection algorithm. The algorithm takes YOLOv4 as the basic network, replaces its trunk feature extraction network with MobileNetV2, and adds attention mechanism to ensure accuracy while reducing model parameters. Simplify and strengthen path aggregation network (PANet) in feature extraction network. The improved algorithm was verified on TT100K dataset and compared with other algorithms such as YOLOv3 and YOLOv4. It was proved that although the detection accuracy of the improved algorithm was reduced, the detection speed was greatly improved.

2. NETWORK STRUCTURE ANALYSIS

2.1. YOLOv4 Target Detection Algorithm

The network structure of YOLOv4 is shown in Fig 1. Its network structure consists of three parts, the main feature extraction network CSPDarknet-53, the enhanced feature extraction network and the prediction part. The enhanced feature extraction network includes SPP structure and PANet structure. The input images were used to obtain 13×13, 26×26 and 52×52 scale feature maps. In order to better learn the features in the images, the 52×52 feature images were pooled in three different sizes through SPP structure, and then they were fused with the 26×26 and 13×13 feature images by up-sampling and down-sampling. Finally, the prediction part is to transform the extracted feature map into prediction results, in which scales 1, 2 and 3 are respectively responsible for the prediction of large, medium and small objects.

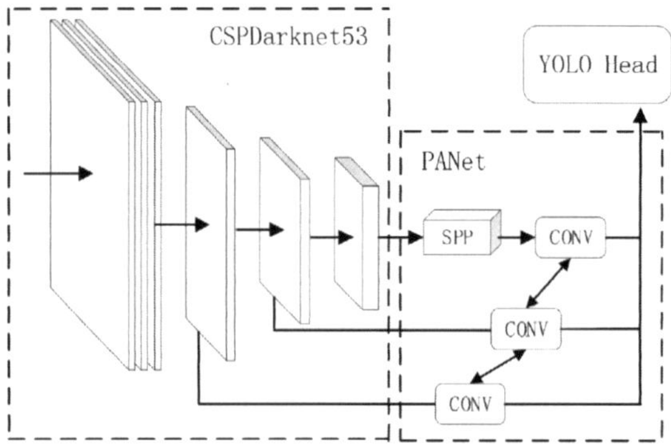

Figure 1 YOLOv4 network structure diagram

MobileNet is a lightweight convolutional neural network model, which uses depth-level separable convolution to replace standard convolution to solve the computational efficiency and parameter number problems of convolutional networks. When standard convolution calculation is used, the calculation of the corresponding region of all channels is taken into account in each step of calculation, which requires a large number of parameters and calculations in the convolution process. Depth-detachable convolution is composed of two parts, one-channel convolution one-sum convolution. The schematic diagram of depth-detachable convolution layer is shown in Fig. 2.

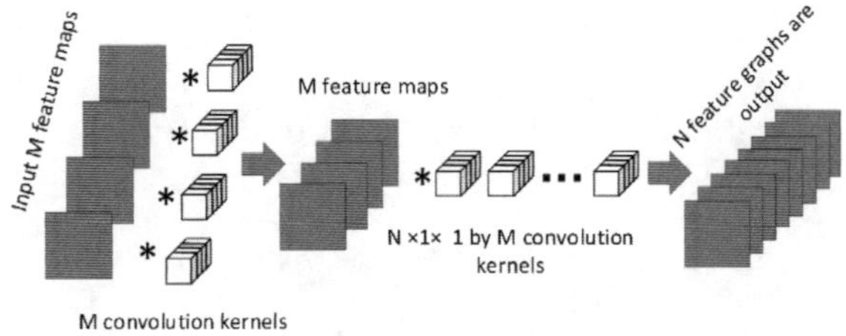

Figure. 2 Structure diagram of depth separable convolution layer

In the channel-by-channel convolution, the feature graphs of input M channels are convolved with M convolution kernels in one-to-one correspondence, and M feature graphs are output. In channel-by-channel convolution, the extraction of spatial plane features is mainly concerned. Convolution calculation only occurs inside each channel, and the information between channels is not fused. In point-by-point convolution, M feature graphs output by channel convolution and N 1×1×M convolution kernels finally output feature graphs of N channels through convolution operation. In this process, the feature fusion of the input feature graph at channel level is realized by point-by-point convolution. Deeply separable convolution mainly changes the internal implementation of traditional convolution, while externally maintaining the previous characteristic dimensions of input and output. Therefore, the corresponding traditional convolution can be replaced by deeply separable convolution without modifying the network hierarchy.

When $D_k \times D_k$ is used to represent the size of the convolution kernel, $D_F \times D_F$ is used to represent the size of the input feature graph, M and N respectively represent the number of channels of the input and output. When the step size is 1 and there is a filling layer, the total amount of standard convolution calculation is:

$$F_1 = D_K \times D_K \times M \times N \times D_F \times D_F \quad (1)$$

The total calculation amount of depth separable convolution is:

$$F_2 = D_K \times D_K \times M \times D_F \times D_F + M \times N \times D_F \times D_F \quad (2)$$

The computation is reduced to the original:

$$\frac{D_K \times D_K \times M \times D_F \times D_F + M \times N \times D_F \times D_F}{D_K \times D_K \times M \times N \times D_F \times D_F} = \frac{1}{N} + \frac{1}{D_K^2} \quad (3)$$

When 3×3 convolution is used as an example, the depth-separable convolution reduces the number of parameters to 1/9 of the traditional convolution. Using depth-separable convolution can reduce the computational cost of the network model and improve the efficiency of the network model.

3. LIGHTWEIGHT OF YOLOv4 NETWORK

3.1. Lightweight Trunk Network Structure

The number of parameters contained in CSPDarkent-53, the backbone network of YOLOv4, accounts for a large proportion in the whole model. In order to make the backbone network structure lightweight, MobileNetV2 is selected in this paper to replace CSPDarkent-53 in YOLOv4. MoblienetV2 is mainly based on MobileNet, and it is mainly stacked by residual structures composed of deeply detacheable convolution. The main unit of MoblienetV2 is Bottleneck. The residual structure is composed of a convolution kernel with a size of 1×1 convolution layer and a depth separable convolution layer. The structure reduces dimension first and then increases dimension. When the network layer deeps, the number of model parameters can be effectively reduced and the gradient propagation can be enhanced. The structural parameters of MoblienetV2 are shown in Table 1, t is the width multiplier, c is the number of output channels, s is the step length of convolution, and n is the number of repetitions of this operation.

Table 1 MoblienetV2 structural parameters

Input	Convolution	t	c	n	s
416×416	Conv2d	—	32	1	2
208×208	Bottleneck	1	16	1	1
208×208	Bottleneck	1	24	2	2
104×104	Bottleneck	6	32	3	2
52×52	Bottleneck	6	64	4	2
26×26	Bottleneck	6	96	3	1
26×26	Bottleneck	6	160	3	2
13×13	Bottleneck	6	320	1	1
13×13	Bottleneck	6	1280	1	1

Replacing the trunk network with MobileNetV2 can greatly reduce the number of parameters, but it will reduce the accuracy of the model due to the reduction of the number of parameters. In order to extract more features of traffic signs from the feature map and reduce the loss of effective feature information, Add attention mechanism (SENet) to each of the MoblienetV2 networks, as shown in figure 3. The input feature map is first upgraded through a 1×1 volume, followed by SENet, then convolved through a 3×3 space, and then reduced by a 1×1 convolution to restore the channel, which is added to the input.

SENet focuses on the relationship between feature image channels and can automatically learn the weight of each channel. The useful features are enhanced according to the weights, and the features with lower weights are suppressed. Firstly, the input feature graph of W×H×C is compressed into a feature graph of 1×1×C. Secondly, weight is generated for each feature channel through parameter W. Finally, each feature channel is multiplied by different weight coefficients to pay attention to the feature information of each channel.

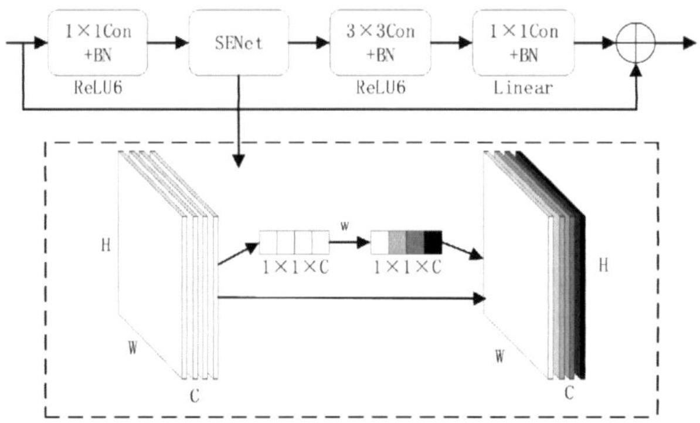

Figure 3 Adding the Bottleneck structure of SENet

The feature pyramid network mainly transfers the high-level features extracted from the deep network from the multi-scale feature graph top-down to the low-level for feature fusion, enhancing the semantic information but ignoring the location information. The enhanced feature extraction network Path Aggregation network (PANet) of YOLOv4 adopts the bottom-up strategy to solve this shortcoming. Firstly, the output of SPPNet after multi-scale maximum pooling is up-sampled to enhance feature fusion, and then down-sampled to transfer the information extracted from the shallow network to the deep network to better retain the object location information. A series of up-sampling and down-sampling make the training parameters of YOLOv4 mainly concentrated in PANet structure blocks. In order to reduce the model complexity and ensure the detection accuracy of small targets, this paper does not consider simplifying the model by reducing the network layer, but refers to the deeply detractable convolution in MobileNetV2. The 3×3 ordinary convolution in the original PANet network was changed to the depth separable convolution to obtain faster detection speed. The comparison before and after the change of convolution is shown in Figure 6. The purpose of changing the activation function to ReLU6 is to enhance the decimal expression ability of floating point number, reduce precision loss and prevent numerical explosion.

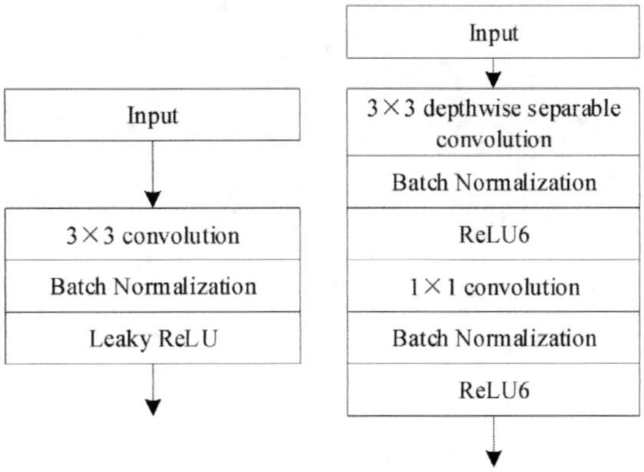

Figure 4 Convolution structure comparison diagram

The loss function of the improved YOLOv4 consists of the loss of object positioning offset, loss of object confidence and loss of object classification, as shown in Formula (4), where λ is the balance coefficient, L is the loss of image positioning offset, L_{cla} is the loss of object confidence, and L_{loc} is the loss of object classification.

$$Loss = \lambda_1 L_{conf} + \lambda_2 L_{cla} + \lambda_3 L_{loc} \tag{4}$$

The calculation formula of object positioning offset loss is as follows:

$$L_{conf} = -\sum (obj_i \ln(p_i) + (1 - Obi_i)\ln(1 - p_i)) \tag{5}$$

In Formula (5) represents whether there is an object in the prediction target bounding box I, and the result value is 0 or 1. Represents the probability that a real object is present in the prediction box. The probability value is calculated by the Sigmoid function.

The calculation formula of object confidence loss is as follows:

$$L_{cla} = -\sum_{i \in Box} \sum_{j \in class} \left(O_{ij} \ln(p_{ij})\right) + (1 - O_{ij})\ln(1 - p_{ij}) \tag{6}$$

In Formula (6), and respectively represent whether there are J-class objects and probabilities in the prediction boundary box I.

The calculation formula of object classification loss is as follows:

$$L_{loc} = 1 - IOU(A, B) + \frac{d_{AB}^2(A_{ctr}, B_{ctr})}{l^2} + \alpha v \tag{7}$$

The improved YOLOv4 algorithm uses CIOU algorithm to calculate the target positioning offset loss, and calculates the aspect ratio and euclidian distance of the center point (A_{ctr} , B_{ctr}) of the predicted boundary box A and the true boundary box B.

4. EXPERIMENT AND ANALYSIS

4.1. Experimental Environment And Parameter Setting

In this experiment, TT100K data set was used for the performance test. The dataset has 221 annotation categories, including 6099 image training sets and 3071 image test sets, covering traffic sign images under different weather conditions and different lighting conditions. The resolution of the image is 2048×2048. In this experiment, the original image is trimmed, and the size of the trimmed image is 416×416. Finally, 45 traffic signs with more than 100 cases were selected for detection in this experiment. In order to enlarge the data set, the training images were processed by rotation Angle, saturation adjustment, exposure adjustment and tone adjustment, so as to enhance and expand the images in the training data set. The selected 45 traffic sign categories are as follows: i2, i4, i5, il100, il60, il80, io, ip, p10, p11, p12, p19, p23, p26, p27, p3, p5, p6, pg, ph4, ph4.5, ph5, pl100, pl120, pl20, pl30, pl40, pl5, pl50, pl60, pl70, pl80, pm20, pm30, pm55, pn, pne, po, pr40, w13, w32, w55, w57, w59, wo.

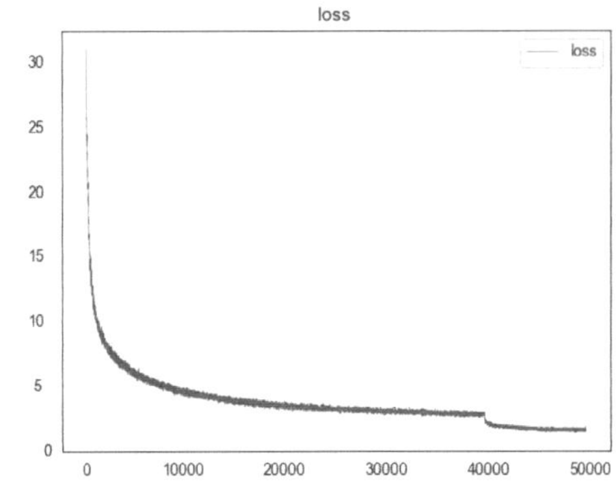

Figure 5 Improved model training Loss

This experiment is based on Pytorch framework, using i5-10600KF processor, 16G memory, GeForceGTX3070 video card and 8G video memory for training. In this experiment, the pre-training weight file on COCO dataset was used for training. A total of 50000 iterations of sample training were carried out, in which the batch size was set as 16, the initial learning rate as 0.001, the momentum as 0.949. When the number of training iterations is 40000 and 45000, the learning rate is reduced to 0.001 and 0.0001 respectively, which further converges the loss function. The improved model loss function figure is shown in Fig. 5.

The performance comparison between the improved YOLOv4 algorithm and YOLOv4 algorithm is shown in Tab 2. Comparison results show that the improved model size decreases by 79.5% and mAP decreases by 2.5%.

Table 2 Overall performance comparison of algorithms

Arithmetic	Parameter	Model size/MB	mAP/%
YOLOv4	64040001	244.3	84.8
Improved YOLOv4	13133414	50.1	82.3

The overall detection performance comparison results of the four algorithms are shown in Tab 3. The experimental results show that the improved YOLOv4 algorithm mAP can reach 82.3% in terms of detection accuracy, which is 17.4% higher than SSD300, but 2.5% and 4.1% lower than YOLOv4 and Faster R-CNN algorithms. In terms of detection speed, the improved YOLOv4 algorithm is better than the other three algorithms, which is 74.8, 105.1 and 49.7 higher than SSD300, YOLOv4 and Faster R-CNN, respectively, but there is still a certain gap with SSD300 algorithm. In general, the improved YOLOv4 algorithm improves the detection speed greatly while ensuring certain detection accuracy.

Table 3 Overall performance comparison of algorithms

Arithmetic	mAP/%	FPS	Model size/MB
SSD300	64.9	59.4	98
YOLOv4	84.8	84.5	244.3
Faster R-CNN	86.4	29.1	187

Improved YOLOv4	82.3	134.2	50.1

The improved YOLOv4 algorithm was compared with YOLOv4, SSD300, and Faster R-CNN to conduct a comparison experiment on the selected 45 categories of traffic signs, and the AP values of each type of traffic signs were compared one by one, and the best performance of each category was bolded. The results are shown in Tab 4.

Table 4 AP value comparison of the four algorithms

traffic sign	i2	i4	i5	il00	il60	il80	io	ip	p10	p11	p12	p19
SSD300	65.7	70.2	75.8	72.4	74.3	74.7	75.8	74.3	66.2	55.7	52.9	69.6
YOLOv4	80.5	90.7	92.6	89.1	87.9	91.8	86.7	87.8	84.3	88.2	86.8	83.6
Faster R-CNN	81.7	91.5	92.6	89.9	88.6	92.4	88.3	92.8	84.3	94.2	86.8	89.6
Improved YOLOv4	79.3	87.8	88.6	87.4	85.5	88.7	83.7	84.5	79.4	86.8	84.9	81.0

traffic sign	p23	p26	p27	p3	p5	p6	pg	ph4	ph4.5	ph5	pl100	pl120
SSD300	78.5	60.3	77.6	64.3	59.8	50.4	80.8	55.9	76.1	42.7	91.2	78.7
YOLOv4	84.3	88.7	90.5	85.8	89.1	82.7	90.4	73.7	79.4	91.2	93.8	90.0
Faster R-CNN	84.3	89.9	91.5	87.2	90.1	83.8	92.3	80.7	80.2	91.2	93.8	90.1
Improved YOLOv4	81.7	87.8	88.6	83.5	87.1	79.3	78.4	74.7	77.1	88.4	91.8	88.9

traffic sign	pl20	pl30	pl40	pl5	pl50	pl60	pl70	pl80	pm20	pm30	pm55	pn
SSD300	44.6	60.1	53.4	67.8	62.6	59.5	52.3	55.2	60.6	75.3	76.4	77.8
YOLOv4	79.2	84.5	78.9	82.5	91.9	84.8	81.7	90.2	79.5	85.7	84.6	89.9
Faster R-CNN	79.2	87.3	80.6	84.2	92.3	89.9	83.5	92.1	80.8	86.4	85.6	90.2
Improved YOLOv4	77.3	81.6	75.7	79.8	89.5	81.4	79.1	92.3	77.6	84	81.7	86.9

traffic sign	pne	po	pro40	w13	w32	w55	w57	w59	wo
SSD300	64.9	45.8	85.9	42.5	62.3	66.7	70.1	58.5	39.5
YOLOv4	84.8	77.5	95.8	83.4	87.6	85.9	66.7	68.7	58.6
Faster R-CNN	86.4	79.6	96.3	84.1	89.9	88.2	68.5	69.7	58.6
Improved YOLOv4	82.3	80.4	93.1	82.7	89.2	83.4	59.7	62.4	53.5

The comparison results of the check results of the four algorithms are shown in Fig. 6, in which error and missed detection occurred in SSD300 and YOLOv4 networks respectively. In the detection results of SSD300, the traffic sign type was w32, but the detection results were wrong, and the missed detection of traffic sign pl40 occurred in YOLOv4 network. It can be seen that the algorithm proposed in this paper has a good effect on the detection of traffic signs in the actual environment.

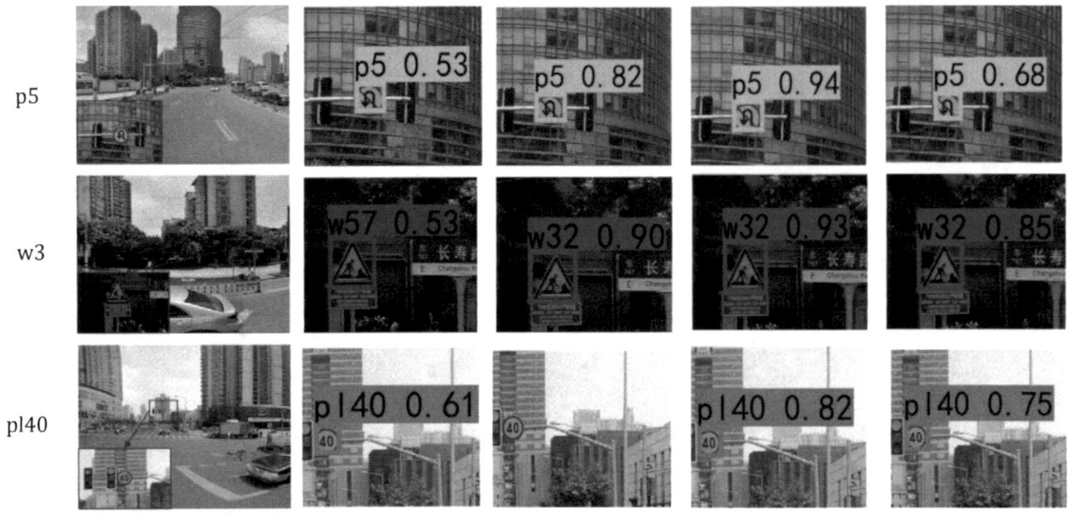

(a)Image (b) SSD300 (c)YOLOv4 (d) Faster R-CNN (e)Improved YOLOv4

Figue. 6 Comparison of traffic sign detection effects of four algorithm

5. CONCLUSION

According to the characteristics of large number of parameters and slow detection speed of current traffic sign detection algorithms, this paper proposes an improved lightweight YOLOv4 traffic sign detection algorithm based on YOLOv4. In order to reduce the number of parameters of the model, In this paper, the feature extraction network is changed into a lightweight network, MobileNetV2, and attention mechanism is added to reduce model parameters and improve the ability of network feature extraction. To simplify and strengthen the feature extraction network, the 3×3 ordinary convolution used in the original PANet network was changed to the depth separable convolution by referring to the depth separable convolution in MobileNetV2 to achieve faster detection speed. Forty-five traffic signs with more than 100 types were screened from TT100K data set and expanded. Compared with other algorithms, the model size of the improved YOLOv4 algorithm was reduced by 79.5%, FPS was increased by 58.8% and mAP index was reduced by 2.5%, proving the effectiveness of the improved network. In the future, the traffic sign detection algorithm can be applied to more complex lighting and weather conditions, and the algorithm accuracy can be further improved to provide more technical support for unmanned driving.

References

[1] M. Manzoor et al., "RFCNN: Traffic Accident Severity Prediction Based on Decision Level Fusion of Machine and Deep Learning Model," in IEEE Access, vol. 9, pp. 128359-128371, 2021, doi: 10.1109/ACCESS.2021.3112546.

[2] G. Xie, A. Shangguan, R. Fei, X. Hei, W. Ji and F. Qian, "Unmanned System Safety Decision-Making Support: Analysis and Assessment of Road Traffic Accidents," in IEEE/ASME Transactions on Mechatronics, vol. 26, no. 2, pp. 633-644, April 2021, doi: 10.1109/TMECH.2020.3043471.

[3] S. Swetha and P. Sivakumar, "SSLA Based Traffic Sign and Lane Detection for Autonomous cars," 2021 International Conference on Artificial Intelligence and Smart Systems (ICAIS), 2021, pp. 766-771, doi: 10.1109/ICAIS50930.2021.9396046.

[4] J. Cao, J. Zhang and X. Jin, "A Traffic-Sign Detection Algorithm Based on Improved Sparse R-cnn," in IEEE Access, vol. 9, pp. 122774-122788, 2021, doi: 10.1109/ACCESS.2021.3109606.

[5] SOUANI C,FAIEDH H,BESBES K.Efficient algorithm for automatic road sign recognition and its hardware implementation[J]. Journal of Real-time Image Processing,2014,9(1):79-93.

[6] ESCALERA A D L,MORENO L E,SALICHS M A,et al.Road traffic sign detection and classification[J].IEEE Transactions on Industrial Electronics,1997,44(6):848-859.

[7] ZHU Z,LIANG D,ZHANG S H,et al.Traffic-sign detection and classification in the wild[C]//Proceedings of the IEEE Conference on Computer Vision and Pattern Recognition,Las Vegas,June 26-July 1,2016:2110-2118.

[8] ZHANG J M,XIE Z P,SUN J,et al.A cascaded R-CNN with multiscale attention and imbalanced samples for traffic sign detection[J]. IEEE Access,2020,8:29742-29754.

[9] TABERNIK D,SKOČAJ D.Deep learning for large-scale traffic-sign detection and recognition[J]. IEEE Transactions on Intelligent Transportation Systems,2020,21(4):1427-1440.

[10] L. Wang, K. Zhou, A. Chu, G. Wang and L. Wang, "An Improved Light-Weight Traffic Sign Recognition Algorithm Based on YOLOv4-Tiny," in IEEE Access, vol. 9, pp. 124963-124971, 2021, doi: 10.1109/ACCESS.2021.3109798.

[11] M. Tan, B. Wang, Z. Wu, J. Wang and G. Pan, "Weakly Supervised Metric Learning for Traffic Sign Recognition in a LIDAR-Equipped Vehicle," in IEEE Transactions on Intelligent Transportation Systems, vol. 17, no. 5, pp. 1415-1427, May 2016, doi: 10.1109/TITS.2015.2506182.

The single train's punctual and energy-saving operation scheme based on dynamic programming algorithm

Yong Ding[1], Tiesheng Wang[1], Kexin Zhang[2*], and Jie Chen[2,3]

[1] CRRC Changchun Railway Vehicles Co Ltd, Changchun, Jilin, 130011, China

[2] Department of Electrical Engineering, Beijing Jiaotong University, Beijing, 10044, China

[3] Rail Transit Electrical Engineering Technology Research Center, Beijing, 100044, China

*Corresponding author's e-mail: 20121526@bjtu.edu.cn

Abstract. A punctual and energy-saving operation model and some principles to realize energy-saving operation for train are analysed in this paper in order to reduce the energy consumption and obtain the energy-saving speed curve. According to the basic principles of dynamic programming algorithm, the train operation line is discretized. We reduce the space of state variables by using the maximum capacity operation curve and average speed operation curve. And the cost function for energy and time consumption is constructed according to requirements for punctual and energy-saving operation of trains. The simulation results using specific parameters show that the dynamic programming algorithm using reduced state space has high computational efficiency and effectively reduces the energy consumption of train operation.

1. Introduction

With the rapid development of high-speed railway, how to further reduce the energy consumption of train operation and realize the punctual and energy-saving operation of train has always been a focus of public concerns. At present, the research method of train energy-saving operation can be roughly divided into two directions: analytical method and numerical method[1].

Reducing the energy consumption of train operation is the main aim of analytical method. This method first uses train parameters and line parameters to set up the train model, and gets optimal operation conditions by maximum principle. Then the optimal train operation curve is obtained according to the value of adjoint variables[2-3]. The solution process of this method which can obtain the best solution theoretically has a large amount of calculation and involves a lot of differential operations.

Numerical methods include genetic algorithm, particle swarm optimization algorithm, dynamic programming algorithm and so on. The heuristic algorithms represented by genetic algorithm and particle swarm optimization algorithm have defects such as slow search speed and easy to fall into local optimal solution, so many scholars have put forward improvement measures[4-6]. In reference[7], ant colony algorithm, genetic algorithm and dynamic programming algorithm are used to solve the train speed curve and the results show that dynamic programming algorithm has the best effect in off-line solution. Gao Hao et al.[8] designed an efficient solving algorithm based on multi-objective optimization theory, but they didn't reduce the state variable space, so it is not suitable for long-distance operation such as high-speed railway. Kong Fanxin et al.[9] adopted the dynamic programming algorithm of state space reduction, and the results show that it can not only improve the search efficiency but also reduce energy consumption.

Based on the existing research, the train operation line is discretized according to the line parameters, and the problem of solving train energy-saving optimal operation curve is transformed into multistage decision problem. Combined with principles of energy-saving operation, the state space is reduced by using the maximum capacity operation curve and average speed operation curve. Finally, the single train's punctual and energy-saving operation curve is obtained by using the reverse order solution method of dynamic programming algorithm.

2. Train Model with Punctual and Energy-saving Operation

2.1. Train operation model

In consideration of high-speed railway which has long operating mileage and complex line conditions, the following assumptions are made in order to describe the train model with punctual and energy-saving operation in mathematical language: 1) Single mass point model is adopted to ignore the influence of train length; 2) Ignore the influence of catenary voltage on traction and braking characteristics; 3) The influence of electricity division on train operation is not considered; 4) During regenerative braking, all energy can be fed back to the power grid and consumed by the train.

Taking the train operation position x as the independent variable, the train operation model is expressed as

$$\min J = \int_0^S \mu_p f_p(v) - \alpha \mu_b f_b(v) dx$$
$$s.t.\begin{cases} M\dfrac{dv}{dx} = \dfrac{\mu_p f_p(v) - \mu_b f_b(v) - f_w(v,x)}{v} \\ \dfrac{dt}{dx} = \dfrac{1}{v} \\ 0 \le \mu_p \le 1, 0 \le \mu_b \le 1, \mu_p \cdot \mu_b = 0 \\ v(x) \le V_{\lim}(x) \\ v(0) = 0, v(S) = 0 \\ t(0) = 0, t(S) = T \end{cases} \tag{1}$$

In there: μ_q 、 μ_b are control coefficient of traction and braking conditions; $f_p(v)$ 、 $f_b(v)$ are maximum traction and braking force corresponding to current speed, which can be obtained by characteristic curves; $f_w(v,x)$=$f_0(v) + f_s(x)$ is train's total running resistance, which contains basic running resistance $f_0(v)$ and additional resistance $f_s(x)$ of train operation; M is train's mass; $V_{\lim}(x)$ is speed limit value corresponding to the current position of the train; S is line length; T is target runtime.

2.2. Principles of train energy-saving operation

When the train runs between two stations, there are multiple operation curves that meet the time requirements. The goal of punctual and energy-saving train operation is to obtain the operation curve with the minimum energy consumption meeting the time requirements under the condition of fixed distance between stations.

Four driving modes (maximum capacity traction, constant speed operation, coasting and maximum capacity braking) of train energy-saving operation had been pointed in literature[2]. Although using maximum capacity traction during traction acceleration will produce more energy consumption, the acceleration time is short and the maximum constant speed of train is reduced, resulting in lower overall operation energy consumption.

3. Dynamic Programming Algorithm

Dynamic programming algorithm was proposed by Behrman in the 1950s to solve multistage decision problem without aftereffect. The basic idea of the algorithm is transforming the problem into a multistage decision problem that will be solved in order of stages. There will be multiple local solutions when solving each subproblem, and the local optimal solution needs to be chosen through decision-making. The solution of the subsequent stage is no longer affected by the previous stage after the state of the current stage is determined. The optimality of the overall decision of the whole process is realized by ensuring the optimal decision is selected in each stage[11].

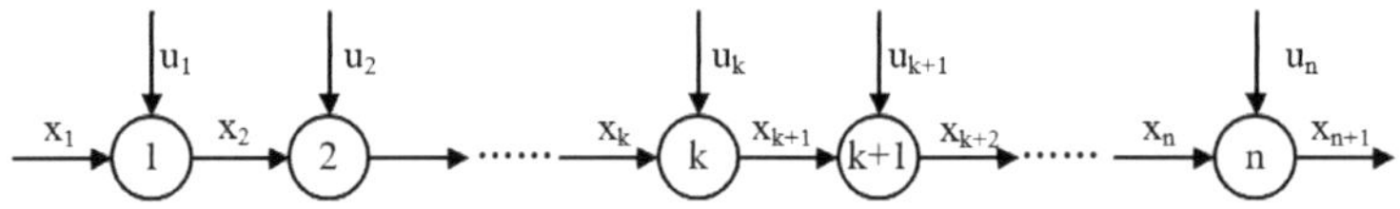

Figure 1. Decision Tree of Dynamic Programming Algorithm

Basic solution steps of dynamic programming algorithm are as follows:

1)Division stage: Division the problem into n stages according to time characteristics or space characteristics.

2)Define state variables: State is the initial condition for decision-making at each stage. The variables used to describe each state are called state variables. For example, the state variable in stage k is x_k ;

3)Determining decision variables and state transition equations: The process of determining the next stage through the current state is called decision making. The selected variable is called decision variable. The state variable of next stage can be obtained through the state transition equation as long as the current state variables and decision variables are determined. For example, u_k is selected to be the control variable of stage k and the state transition equation is $x_{k+1}=h(x_k,u_k)$;

4)Build index function: The index function value of a certain stage is determined by its initial state and decision variables. The optimal solution of the whole decision-making process can be obtained by controlling the index function value from each stage to the end point is the smallest. The index function for stage k is defined as $g_k(x_k,u_k)$.

5)Reserve order solution method: According to the basic idea of dynamic programming algorithm, the best solution from each point to the end point is obtained by backward recursion from the end point to the start point after the problem is divided into several subproblems. And the best solution is got when pushing back to the start point. The backward recursion method like figure 2 is called Reserve order solution method[12].

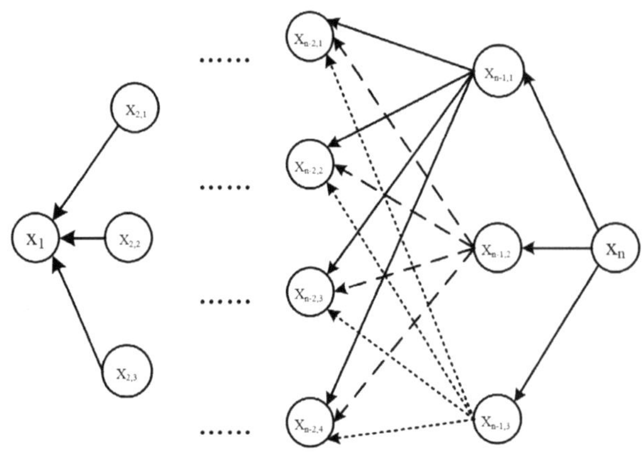

Figure 2. Reserve order solution method

4. The Single Train's Punctual and Energy-Saving Operation Based on Dynamic Programming Algorithm

4.1. Stage division

Firstly, divide the line evenly by using appropriate discretization distance. For the fixed line length, the smaller the discretization distance is, the more intervals will be divided, and the higher the solution accuracy will be, which will also increase the difficulty of solving the problem and reduce the calculation efficiency. Therefore, it is necessary to select the appropriate discretization distance according to the actual line conditions. Then, correct the above results according to the speed limit position of the line, the change position of the ramp and other information. If the speed limit change position or ramp parameter change position is in the middle of the above intervals, the intervals shall be re divided to ensure that the speed limit value and line additional resistance value of trains in each stage are unique. Finally, the line with length s is divided into n, and the length of each stage is Δs_k.

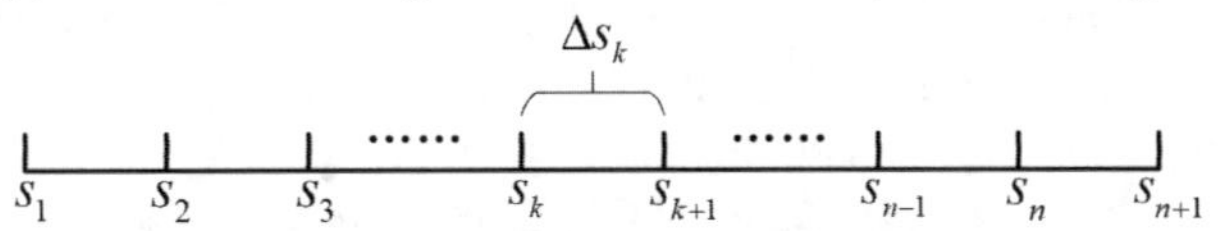

Figure 3. Line discretization results

4.2. Determining state variables

The speed value of each position needs to be planned to obtain the state variables after determining the positions of the line division stages. The traditional dynamic programming algorithm only considers the limit of maximum speed when discretizing the running speed at a fixed position, such as speed state set at s_k is $\{v_i \mid v_i = i \cdot \Delta v, 0 \le v_i \le V_{\lim}(s_k), i \in \mathbb{N}\}$. Considering the principles of train energy-saving operation, every speed value from 0 to the speed limit value will not be taken. Therefore, the above speed discretization method undoubtedly increases the search space of the algorithm and reduces the computational efficiency. In order to effectively reduce the state space, the maximum capacity operation curve and average speed operation curve are used to reorganize the speed state set.

The maximum capacity operation curve refers to the operation curve in which the train accelerates with the maximum traction force and decelerates with the maximum braking force under the restriction of speed limit, selects the speed limit value as the speed value for constant speed operation, and there is no coasting condition in the whole process. The average running speed can be obtained by using the line length and target running time. Taking this value as the maximum constant speed value and considering speed limit constraint and principles of train energy-saving operation, the average speed operation curve can be obtained. The speed values range between the maximum capacity operation curve and the average speed operation curve constitute the compressed speed state space set, and that are the filled circles in figure 4. Therefore, the speed state set at s_k is $\{v_i \mid v_i = i \cdot \Delta v, V_{\min}(s_k) \le v_i \le V_{\max}(s_k), i \in \mathbb{N}\}$, where $V_{\min}(s_k)$ and $V_{\max}(s_k)$ are the minimum possible speed and the maximum possible speed corresponding to the position respectively. And the ith state variable of the position is defined as $x_k^i = (s_k, v_i)$. Like stage division, the smaller the velocity interval selected, the higher the solution accuracy and the lower the calculation efficiency for velocity discretization. Therefore, it is also necessary to select the appropriate velocity interval to realize velocity discretization.

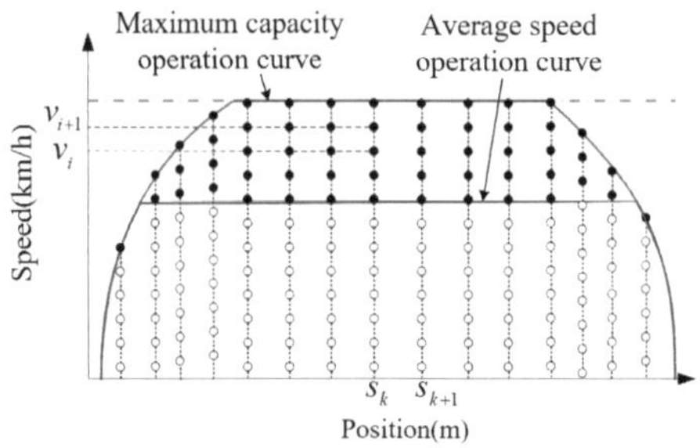

Figure 4. Speed status after compression

4.3. Determining Decision Variable and State Transition Equation

According to the principles of train energy-saving operation, the four working conditions of maximum capacity traction, constant speed, coasting and maximum capacity braking are taken as the decision variables. The state transition equation uses the position and velocity information of the current stage and the selected decision variables to calculate the state variable of the next stage. However, in the actual solution process, what is known is the state variable of the next stage, so there is no need to define the state transition equation. We just need to select the appropriate decision variable according to the state variables of the previous and subsequent stages, that is, determine the working condition of the corresponding speed connection of the previous and subsequent positions.

4.4. Building Speed Connection Array

According to the principle of reverse order solution, the reverse calculation is started from the last state variable, and the connection time, energy consumption and working condition information between adjacent state variables are recorded.

Take the k-stage speed connection as an example. Firstly, the possible speed range $\left\{v_{k\_j} \mid V_{\min}(x_{k+1}^{i}) \leq v_{k\_j} \leq V_{\max}(x_{k+1}^{i})\right\}$ of k position should be calculated according to the known state variables and the maximum traction and braking force of k+1 position. Then, the discrete speed value $\{v_{k\_j} \mid v_{k\_j} = j \cdot \Delta v, V_{\min}(s_k) \leq v_{k\_j} \leq V_{\max}(s_k), V_{\min}(x_{k+1}^{i}) \leq v_{k\_j} \leq V_{\max}(x_{k+1}^{i}), j \in \mathbb{N}\}$ within the above speed range is selected from the speed state set at position k, as show in figure 5. After that, the time, energy consumption and working conditions of k-stage speed connection are recorded in the three-dimensional speed connection array respectively that are $T(i,j,k)$, $E(i,j,k)$, $u(i,j,k)$.

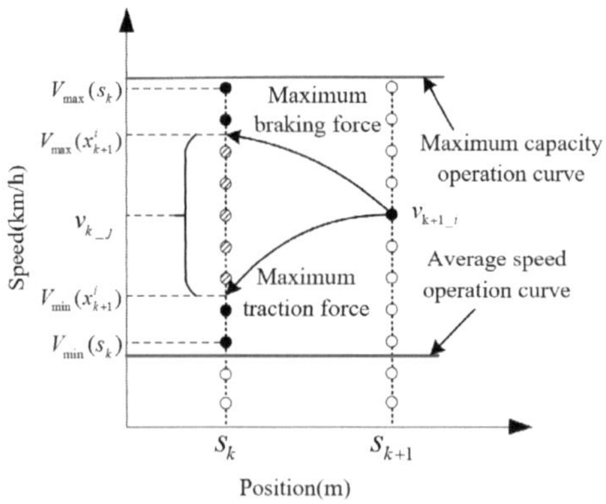

Figure 5. Speed connection range

4.5. Constructing Recurrence Formula

According to the train model with punctual and energy-saving operation in section 2, the optimization objective of the problem is to ensure the minimum operation energy consumption under time constraints. Therefore, the optimization objective can be defined as minimizing energy-time consumption cost:

$$G=\min\left\{\sum_{k=1}^{N}E_k+\lambda\cdot\left|\sum_{k=1}^{N}T_k-T\right|\right\} \tag{2}$$

In there: E_k and T_k are energy consumption and time of k-stage respectively; T is target running time of the whole process; λ is the calculation coefficient for energy-time consumption cost.

Recurrence formula is defined as:

$$G_k(x_k)=\begin{cases}\min(g_k(x_k)+G_{k+1}(x_{k+1})) & k=1,2,\ldots,n-1\\ g_k(x_k) & k=n\end{cases} \tag{3}$$

In there: g_k energy-time consumption cost of k-stage, and $g_k=E_k+\lambda\cdot T_k$; G_{k+1} is the smallest energy-time consumption cost from n stage to k+1 stage.

The whole line is divided into n stages and n+1 position points, and the last position point corresponds to the end position of the line. The train has only one state in n+1 stage, so the energy-time consumption costs corresponding to the connection between nth position and the terminal position are all the minimum energy-time consumption costs in n stage.

The reserve order solution method of dynamic programming algorithm is adopted to reverse search the minimum energy-time consumption cost from the end position of the line to the end position of each stage. Record the corresponding decision variables (operating conditions), time, energy consumption and speed, and substitute the results into the next stage for searching and calculation. The calculation coefficient for energy-time consumption cost is continuously adjusted according to real running time and target running time of each search and finally the energy-saving operation curve meeting the time constraints is found.

5. Simulation Verification

In order to verify the feasibility of the single train's punctual and energy-saving operation curve based on dynamic programming algorithm, the line parameters in document [13] and CRH2 train are used for verification.

Table 1. The main technical parameters

Parameter name	Parameter characteristics
Train mass(t)	408.5
Highest operating speed(km/h)	250
Regenerative braking utilization	0.9
Rotary mass coefficient	0.06

Firstly, the discrete value of simulation distance is defined as 50m, the discrete value of speed is defined as 2km / h, the target running time is defined as 22min=1320s, and the time error is set to 10 seconds According to the above information, the maximum capacity operation curve and average speed operation curve can be calculated. The time corresponding to the maximum capacity operation curve, that is, the shortest operation time in the whole process is 1296.30s and the maximum energy consumption is 631.32kWh.

The speed state set after compression can be obtained from the above two operation curves, and the optimal speed curve can be obtained according to the process of dynamic programming algorithm.

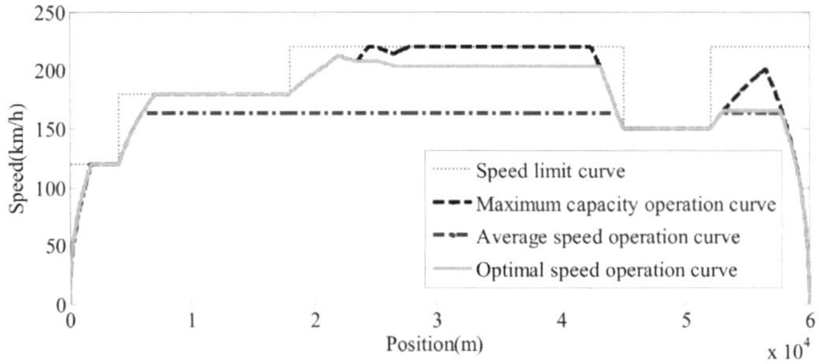

Figure 6. Maximum capacity operation curve and average speed operation curve

Table 2. Simulation result

Operation scheme	Running time/s	Energy consumption/kWh	Time margin /%	Energy saving rate /%
Maximum capacity operation	1296.30	631.32	-	-
Dynamic programming algorithm	1329.62	563.95	2.57	10.67

The simulation results show that under the time margin of 2.57%, the energy-saving rate of single train's punctual and energy-saving operation scheme based on dynamic programming algorithm can reach 10.67% compared with the maximum capacity operation scheme, and the program simulation time is 7.136s.

6. Conclusion

The single train's punctual and energy-saving operation scheme based on dynamic programming algorithm proposed in this paper not only adopts principles of train energy-saving operation, but also reduces the state space through the maximum capacity operation curve and average speed operation curve. The simulation results show that the scheme can reduce the energy consumption of train operation under the condition of setting a certain time margin, and also has high calculation efficiency. According to the optimal speed curve, it can be seen that the operating conditions of the train in the whole section are only traction, constant speed and braking, and there is no coasting condition. It is because that if the coasting condition is adopted, the speed change value will be less than the speed discrete value under the set distance discrete length. If the speed interval is reduced to meet the coasting requirements, huge storage space will be occupied in the generation stage of state space, and it will result in huge simulation time. Therefore, in this paper, the application of dynamic programming algorithm to reduce energy consumption is actually realized by reducing the running speed of some constant speed stages. In the follow-up research, we should consider combining the above algorithm with coasting condition, so as to find a more reasonable energy-saving operation scheme.

References

[1] Yang, J. (2017) Energy saving operation optimization and intelligent control method of freight train. Beijing Jiaotong University.

[2] Howlett P. (2000) The Optimal Control of a Train. Annals of Operations Research., 98(1/4): 65-87.

[3] Zhu, J.L., Li, H.C., Wang, Q.Y., Long, S. (2008) Optimization analysis of train energy saving control. China Railway Science., (02): 104-108.

[4] Ji, Y.X., Sun, P.F., Mao, C.H., Wang, Q.Y. (2018) Optimization of train operation curve based on Improved Genetic Algorithm. Computer and Modernization., (08): 1-7.

[5] Li, J.Y., Chen, G.W. (2021) Research on energy saving optimization of high speed train based on Improved Genetic Algorithm. Railway Computer Application., 30(03): 5-9.
[6] Ren, X.Y. (2015) Research on train energy saving strategy based on optimized particle swarm optimization. Beijing Jiaotong University.
[7] LU, S., HILLMANSEN, S., HO, T. K., et al. (2013) Single-Train Trajectory Optimization. IEEE Transactions on Intelligent Transportation Systems., 14(2): 743-750.
[8] Gao, H., Zhang, Y.D., Guo, J., Cha, Z. (2020) Optimization method of train energy saving operation based on dynamic programming. Journal of The China Railway Society., 42(08): 76-84.
[9] Kong, F.X., Chen, Y.Z., Zhan, J.Y., Shang, F. (2020) Optimization of train energy saving speed curve based on discrete dynamic programming. RAILWAY TRANSPORT AND ECONOMY., 42(08): 113-118.
[10] Albrecht, A., Howlett, P., Pudney, P., Vu, X., Zhou, P., Chen, D.W. (2014) Using maximum power to save energy. In: 17th International IEEE Conference on Intelligent Transportation Systems (ITSC). Qingdao, China. pp. 1205-1208.
[11] Xi, M. (2006) Solving discrete deterministic dynamic programming model by inverse order method. Science & Technology Ecnony Market., (12): 14-15.
[12] Qiu, H., Huang, J.Y., Huang, L.D. (2014) Research on two algorithms of shortest path problem in management operations research. Journal of Yuncheng University., 32 (02): 89-91.
[13] Liang, Z.C., Wang, Q.Y., He, K., Feng, X.Y. (2015) Energy saving operation of EMU based on maximum principle. Journal of The China Railway Society., 37(10): 16-25.

Research on hybrid collision detection algorithm based on separation distance

Anlin Hu[1a*], Yongyi He[1b*]

[1] School of Mechatronic Engineering and Automation, Shanghai University, Shanghai 200444, China

[a*]: huanlin@shu.edu.cn， [b*]: heyongyishu@163.com

Abstract. In order to improve the safety of digital subtraction angiography (DSA) equipment in the treatment of cardiovascular diseases, a hybrid collision detection algorithm based on separation distance is proposed by combining oriented bounding box (OBB) algorithm with GJK algorithm. Firstly, according to the structure of DSA equipment, the real-time position of DSA equipment is simulated in robot operating system (ROS); Then, the OBB is used to envelope the moving parts, and then the generated OBB model is applied to GJK algorithm to calculate the separation distance between the moving parts and quickly eliminate the detection pairs outside the threshold; Finally, for the detection pairs within the threshold, the real model is applied with GJK algorithm to calculate the separation distance between the real models and judge whether the models collide. The experimental results show that the average detection time of the hybrid collision detection algorithm is 15ms. The application of the algorithm in practical equipment has achieved good anti-collision effect.

1. Introduction

DSA is a new technology in medical imaging after X-ray and CT. It is also a breakthrough in medical imaging. Generally speaking, the contrast agent is injected into the blood vessels to be examined to expose the original shape of the blood vessels, and then the blood vessels are displayed more clearly through systematic processing, which is convenient for doctors to diagnose or perform surgery. DSA technology is the "gold standard" to solve cardiovascular diseases. In the process of diagnosis and treatment of cardiovascular diseases, in order to avoid collision between the mechanical arm and the patient, hospital bed and surrounding environment (DSA equipment is shown in Figure 1), it is very necessary to carry out real-time, fast and effective collision detection during the operation of the equipment by medical staff.

Collision detection algorithm is the basis of anti-collision function. In the field of robot application, collision detection is to detect whether there is a collision between the robot and the surrounding obstacles. At present, there are two types of collision detection algorithms, one is model-based geometric algorithm, the other is sensor-based collision detection method [1-2]. The collision detection method based on sensor can only detect the collision signal during the movement of the manipulator, and can't feed back the collision distance. Therefore, this method is only applicable to the field with low safety requirements, and can't be applied to the field of medical surgery.

Model based geometric algorithms can be divided into two categories in principle. The first is the collision detection algorithm based on model simplification, which simplifies the calculation of collision detection by approximating the complex model in space with simple geometry. Among them, various bounding box algorithms are more representative, such as spherical bounding box[3],

AABB[4], OBB[5]. The calculation of spherical bounding box is simple, but the envelope is poor. The generation of AABB bounding box is simple, but the generated model cannot change with the rotation of the model. The generation of OBB bounding box is relatively complex, but the envelope is good, and can rotate with the rotation of the model (see Figure 2). The above collision detection algorithms based on model simplification can quickly detect collisions, but these methods have the following problems: first, they can't feed back the real collision signal, and can only be used in the field where the accuracy of collision is not high; Second, the collision distance can't be fed back.

Another kind of algorithm is collision detection algorithm based on collision distance. It is mainly divided into two kinds. One is the algorithm based on Voronoi domain, such as Lin canny (LC) algorithm, v-clip algorithm, swift + + algorithm, etc. For example, LC [6] algorithm is a simple and effective algorithm to solve the nearest point between two convex polyhedrons. When performing a continuous curve movement, it can quickly detect the collision distance. This method is to solve the Voronoi domain between two convex objects to obtain the nearest feature (point, line and surface) between two convex bodies, and then obtain the collision distance. The other is a simplex based collision detection algorithm, such as Gilbert-Johnson-Keerthi (GJK) algorithm. GJK algorithm [7] is to find the nearest feature (point, edge or surface) between convex polyhedron A and B, and judge whether the set contains the origin by calculating the Minkowski difference. The algorithm designs a set of fast iterative process, and quickly converges to the nearest point through the support function. Dong Han et al. [8] proposed a method based on distance calculation and discrete detection based on GJK algorithm. The GJK algorithm is used to calculate the nearest distance between the manipulator obtained from the camera and the convex hull of any shape dynamic obstacle in real time, and detect whether the dynamic obstacle is located in the safe path of the manipulator, and good experimental results are obtained.

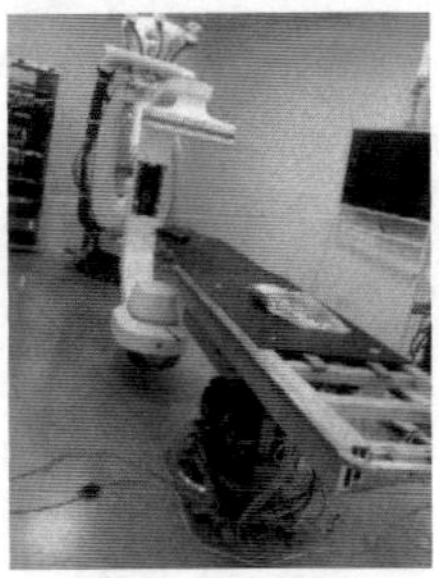

Figure 1. DSA equipment based on KUKA robot.

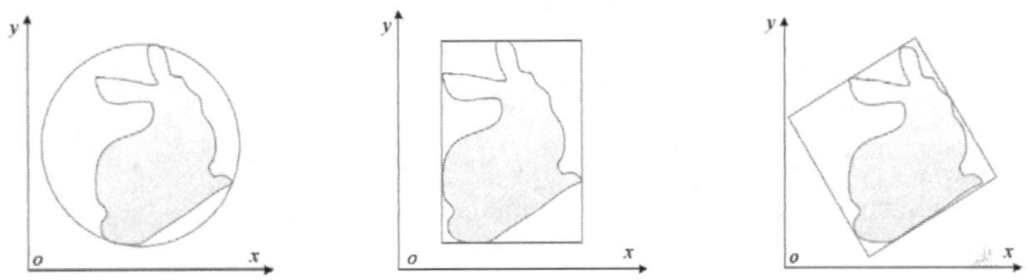

spherical bounding box AABB OBB

Figure 2. Common bounding box

2. Preliminary

The hybrid collision detection algorithm proposed in this paper is based on OBB algorithm and GJK algorithm. These two algorithms are briefly described below.

2.1. OBB algorithm

Bounding box is a simple geometric space, which contains objects with complex shapes. The purpose of adding bounding boxes to objects is to filter before fast collision detection or accurate collision detection. Compared with AABB, OBB always generates the smallest rectangular bounding box along the principal component direction of the object, which can rotate with the object and can be used for more accurate collision detection. At present, the application of OBB algorithm is to project OBB onto 15 potential separation axes. If any projection does not overlap, the two models will not collide. This method can only judge whether the two models collide, and can't calculate the separation distance.

At present, one of the common methods to describe three-dimensional models is the triangular patch network method. Triangular patch mesh refers to the surface model composed of triangular mesh, which uses points and edges to form many small triangular patches in space to approximate the surface of three-dimensional models. This approximation method is also the graphic representation used in most three-dimensional mesh files, almost all three-dimensional models can be approximated by triangles. Because the triangular patch mesh processing is simple and fast, the visualization effect is good and the performance is good.

OBB calculates the vertices of the object surface and obtains the feature vector through principal component analysis (PCA), that is, the principal axis of OBB. The calculation method of covariance is:

$$Cov(X_i, X_j) = E[(X_i - \mu_i)(X_j - \mu_j)] \tag{1}$$

The covariance matrix can be obtained through the calculation formula of covariance:

$$A = \begin{pmatrix} \operatorname{cov}(x,x) & \operatorname{cov}(x,y) & \operatorname{cov}(x,z) \\ \operatorname{cov}(x,y) & \operatorname{cov}(y,y) & \operatorname{cov}(y,z) \\ \operatorname{cov}(x,z) & \operatorname{cov}(y,z) & \operatorname{cov}(z,z) \end{pmatrix} \tag{2}$$

In the covariance matrix, the elements of the main diagonal represent the variance of the variable, and the larger elements of the main diagonal represent the strong signal. The elements of the non-principal diagonal represent the covariance between variables, and the elements of the covariance matrix are real and symmetrical. In the covariance matrix, the elements of the main diagonal represent the variance of the variable, and the larger elements of the main diagonal represent the strong signal. The elements of the non-principal diagonal represent the covariance between variables, and the elements of the covariance matrix are real and symmetrical. Calculate the eigenvalues v_x, v_y and V_Z of the covariance matrix(where $v_x \geq v_y \geq v_z \geq 0$).

The eigenvector of the covariance matrix represents the direction of the OBB. Obtain the eigenvectors corresponding to the eigenvalues v_x, v_y and v_z respectively:

$$AP_x = v_x A, AP_y = v_y A, AP_z = v_z A \tag{3}$$

P_x,P_y and P_z are the three coordinate axis vectors corresponding to the OBB bounding box. The points on the model are projected onto three coordinate axes respectively to obtain the maximum and minimum values of projection: $p_{x\max}$、$p_{x\min}$、$p_{y\max}$、$p_{y\min}$、$p_{z\max}$.

Thus, 8 vertices of OBB can be obtained: $(p_{x\max}, p_{y\max}, p_{z\max}), \cdots, (p_{x\min}, p_{y\min}, p_{z\min})$.

2.2. GJK algorithm

GJK algorithm is a method used to calculate the nearest distance between two convex set points in three-dimensional Euclidean space. It is often used in collision detection algorithm between convex bodies. Its principle is to calculate whether the Minkowski difference contains the origin through fast iteration, and then judge whether there is a collision between models.

$$Minkowski\_difference = \{a - b \mid a \in \mathrm{A}, b \in \mathrm{B}\} \tag{4}$$

Where, A and B represent two sets, and a and b represent the elements in the set. If the set and intersect, the origin is included. If the set and do not intersect, otherwise, the origin is not included

In addition, another important function of GJK algorithm is support function. The support function is defined as follows:

$$Support_{\mathrm{A}}(d) = \sup\{d \bullet a \mid a \in \mathrm{A}\} \tag{5}$$

Where $d \in R^n$. The key of GJK algorithm is to update the support function. Montanari [9] introduced a faster search strategy in this article. The GJK algorithm in this paper refers to this method.

3. Step of hybrid collision detection algorithm

The hybrid collision detection algorithm is based on the establishment of simulation model. Therefore, it is necessary to establish the kinematics model of DSA equipment on ROS, and then the s of hybrid collision detection algorithm

3.1. Kinematic modeling on ROS

The pose of the robot will change in real time during the movement. In order to update the pose of each moving part in real time, URDF description language is used in ROS to model the kinematics of DSA equipment. Its basic format is shown in Figure 3. The modelling and simulation results are shown in Figure 4.

ROS collects the joint angle of each moving part in real time, carries out the forward and inverse kinematics solution, and obtains the direct transformation matrix ${}_{world}^{i}T$ of each joint. All the transformation matrices are transformed into the world coordinate system. This information is released in the form of topic.

3.2. Hybrid collision detection algorithm based on separation distance

Firstly, read the static model and carry out OBB envelope according to OBB algorithm to convert the generated OBB into convex hull. Then, apply GJK algorithm to calculate the distance $d_{i,j}$ between two bounding boxes. The specific pseudo code is as follows:

Algorithm 1 Hybrid Collision Detection Algorithm

```
Input: joints : real time joint angle collection
Output: distance_min : minimum distance from collision
 set THRESHOLD = Value, distance_min = +∞ *Set collision threshold and initialize output*
 set Models = getModels() *Get static model*
 set InitOBBs = getOBB() *Calculate OBB bounding box of static model*
 T = getTransformMatrix(joints) *Calculate the transformation maxtix of each joint*
 for i = 1; i< N_numofjoints; i++ do *Calculate current model and OBBs*
   Model[i]=T[i] • Models[i] *Calculate current models*
   OBBs[i] = T[i] • InitOBBs *Calculate current OBBs*
 end for
 for i = 1; i< N_numofjoints; i++ do *Traverse all detection pairs*
   for j = i; j< N_numofjoints; j++ do
     distance_OBB = GJK(OBBs[i], OBBs[j]) *Calculate the distance of two OBBs with GJK*
     if distance_OBB < THRESHOLD then *OBBs detection is within the set threshold*
       distance = GJK(Models[i], Models[j]) *Calculate the distance from real model*
       if distance < distance_min then
         distance_min = distance *Record the minimum separation distance*
       end if
     end if
   end for
 end for
 return distance_min *Return the minimum separation distance*
```

ROS publishes a topic named "/ joint_states" in real time, and publishes a message named "joints". The collision detection algorithm subscribes to this topic and takes "joints" as a parameter to calculate the separation distance. Firstly, the coordinates of the OBB model and the actual model are updated according to the real-time joint angle; Then traverse all collision detection pairs, and use GJK algorithm to detect the separation distance between OBB model detection pairs. If the distance is less than the threshold, further use GJK algorithm to detect the separation distance between real models; Finally, the algorithm outputs the minimum separation distance after convenience.

```
<robot name="myrobot">
  <link name="link1">
    <visual>
      ...
    </visual>
    <collision>
      ...
    </collision>
  </link>
  <joint name="joint1">
    ...
  </joint>
</robot>
```

Figure 3. URDF basic syntax.

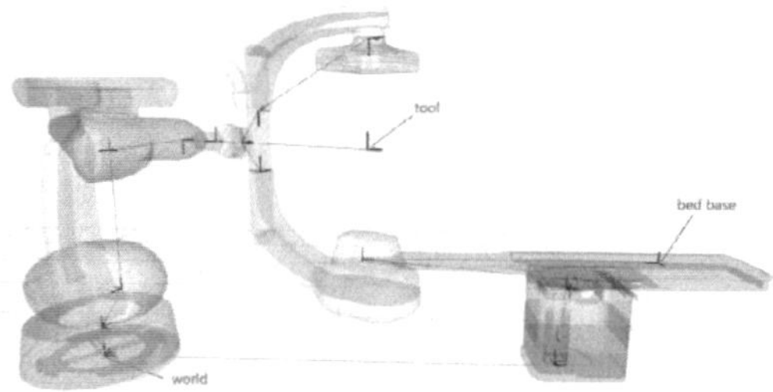

Figure 4. Kinematic model of DSA equipment.

4. Analysis of experimental results

In order to verify the effectiveness of the hybrid collision detection algorithm proposed in this paper, the actual simulation experiment is carried out on the embedded industrial computer. The specific operating environment is as follows: Ubuntu 18.04, ROS melodic, Beckhoff CX5140 industrial computer. TwinCAT is used as the motion control unit, and the separation distance fed back by the hybrid collision detection algorithm is applied to the motion anti-collision control of robot, C-arm and hospital bed to verify the reliability of the algorithm.

In this paper, the performance of hybrid collision detection algorithm and pure GJK algorithm are compared. Different locations are randomly selected. After hundreds of experiments, the specific results are shown in Table 1. The detection speed of the algorithm proposed in this paper is 8 times that of GJK algorithm.

Table 1. Comparison between GJK algorithm and the algorithm proposed in this paper.

	Average consumption	Detection speed
This paper	**15.381ms**	**fast**
GJK algorithm[9]	130.562ms	slow

We randomly tested five positions, each position repeated 10 experiments, and obtained the detection time of each location and the output distance fed back by the algorithm. The detection time of the hybrid collision detection algorithm proposed in this paper is shown in Table 2.

Table 2. Detection time at five different positions.

	Average consumption time	Separation distance of algorithm output	Collision(the threshold is set to 5cm)
Position1	**6.293ms**	+INF	No
Position2	20.205ms	3.52cm	Yes
Position3	**8.056ms**	6.08cm	No
Position4	**12.578ms**	7.13cm	No
Position5	18.433ms	4.52cm	Yes

It can be seen from the table that when there is no collision (the separation distance is greater than the threshold), the time spent by the algorithm is only half of that when there is collision, because the OBB algorithm eliminates a large number of unnecessary detection pairs and saves time.

Using hybrid collision detection algorithm, DSA equipment can realize anti-collision function. As shown in Figure 5 and Figure 6.

5. Conclusion and Prospect

5.1. Conclusion

（1） The hybrid collision detection algorithm based on separation distance designed in this paper realizes fast and accurate collision detection, and realizes the anti-collision function of DSA equipment combined with TwinCAT;

（2） Using OBB algorithm to quickly eliminate the detection pairs without collision can greatly improve the detection efficiency of the algorithm, especially when the objects do not collide, the detection time is faster.

（3） The collision detection algorithm proposed in this paper has high reference value for the application fields requiring high collision detection, such as medical treatment, human-computer interaction and so on.

5.2. Prospect

The algorithm in this paper runs in a single threaded environment. In the future, multi threads can be considered to split the detection tasks, which can double the detection speed of the algorithm.

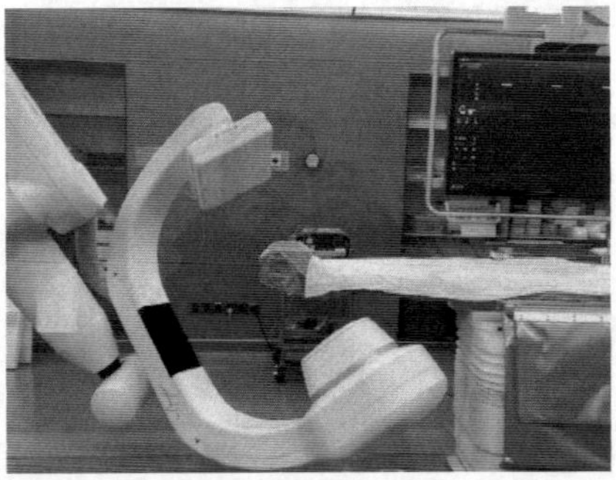

Figure 5. DSA equipment anti-collision effect.

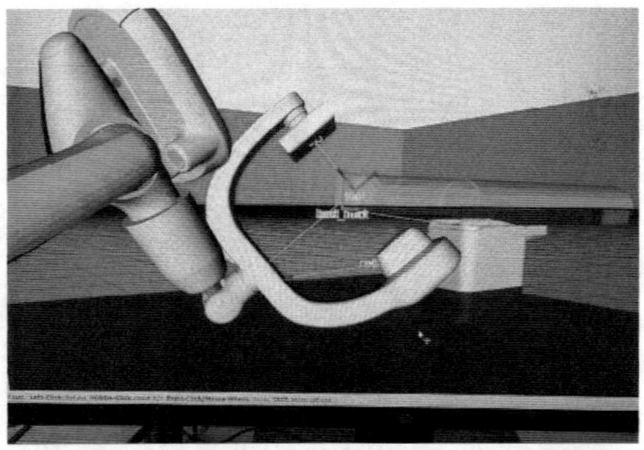

Figure 6. Collision detection based on simulation model.

References

[1] Zhang, Xiaojun, Jian Zhao, Minglu Zhang, and Xiaoyu Liu. "Disturbance Recognition and Collision Detection of Manipulator Based on Momentum Observer." Sensors (Basel, Switzerland) 20.15 (2020): 4187.

[2] Dalla Libera, Alberto, Elisa Tosello, Gianluigi Pillonetto, Stefano Ghidoni, and Ruggero Carli. "Proprioceptive Robot Collision Detection through Gaussian Process Regression." 2019 American Control Conference (ACC) (2019): 19-24.

[3] WANG, M., S. CHEN and Q. YANG, DESIGN AND APPLICATION OF BOUNDING VOLUME HIERARCHY COLLISION DETECTION ALGORITHM BASED ON VIRTUAL SPHERE. Journal of mechanics in medicine and biology, 2019. 19(7): p. 1940044.

[4] Xinlu Guo, Yakun Zhang, Rong Liu, and Yongxuan Wang. "Efficient Collision Detection with a Deformable Model of an Abdominal Aorta." 2016 IEEE International Conference on Information and Automation (ICIA) (2016): 927-32.

[5] Zhang, Rong, Xiaogang Liu, and Jianjun Wei. "Collision Detection Based on OBB Simplified Modeling." Journal of Physics. Conference Series 1213.4 (2019): 42079.

[6] Lin, M.C. and J.F. Canny, A fast algorithm for incremental distance calculation. 1991, IEEE Comput. Soc. Press. p. 1008-1014 vol.2.

[7] Gilbert, E.G., D.W. Johnson and S.S. Keerthi, A fast procedure for computing the distance between complex objects in three-dimensional space. IEEE journal of robotics and automation, 1988. 4(2): p. 193-203.

[8] Han, D., et al., Dynamic obstacle avoidance for manipulators using distance calculation and discrete detection. Robotics and computer-integrated manufacturing, 2018. 49: p. 98-104.
[9] Montanari, Mattia, Nik Petrinic, and Ettore Barbieri. "Improving the GJK Algorithm for Faster and More Reliable Distance Queries Between Convex Objects." ACM Transactions on Graphics 36.3 (2017): 1-17.

A Variable Step-size Normalized Least Mean Square Algorithm with Orthogonal Correction Factors

Ru LI [1a*], Shu-ying ZHAI [1b], Pan Wang[1c], Yong-feng ZHI [2]

[1]Xi'an Mingde Institute of Technology, xi 'an, shaanxi,710124,China

[2]Department of Automatic Control, Northwestern Polytechnical University, Xi'an 710072, China

[a*] liru0319@163.com, [b] 824701355@qq.com, [c] 2748876754@qq.com

Abstract—To improve the performance of the adaptive filters, this paper presents a variable step-size normalized least mean square algorithm with orthogonal correction factors, which uses an exponential function to obtain proper smoothing factors from the projected error norm. Simulation results show that the proposed algorithm is superior in terms of convergence rate and estimation error compared with traditional NLMS-OCF algorithm.

1. Introduction

Normalized Least Mean Square (NLMS) algorithm has been widely used due to its low computation and easy implementation. However, for highly autocorrelated input signals, the convergence speed of this algorithm shows a significant decline. Affine Projection Algorithm (APA) solves the above problems by repeatedly using past input signals [1-2]. However, the choice of step size will affect the relationship between algorithm convergence speed and misalignment. Traditional NLMS and APA algorithms can hardly find a balance point between fast convergence and low misalignment by using fixed step size, so variable step size is needed for adjustment. Literature [3] obtained the optimal step size of NLMS algorithm by minimizing the mean square deviation of each iteration. A Variable step-size Affine Projection Algorithm (VS-APA) Algorithm was proposed in literature [4], which achieved relatively high estimation accuracy but relatively slow convergence. Reference [5] proposed a Variable step-size Affine Projection Algorithm With Variable Smoothing Factor (VS-APA-VSF) Algorithm based on time-varying Smoothing Factor. The algorithm uses the norm of the projection error vector as the criterion to judge the optimal distance, but it needs to determine the "trigger point" according to the variation trend of the projection error norm, which increases the complexity of implementation. Reference [6] proposed VS-APA algorithm (Variable step-size Affine Projection Algorithm With Exponential Smoothing Factors), which directly obtains the smoothing factor from the projection error norm through the mapping of exponential function, and is easier to implement. Therefore, in order to improve the performance of NLMS-OCF algorithm[7-9], it is necessary to control the step size. In this paper, by studying the iterative step size of NLMS-OCF algorithm, a Variable Step-size Normalized Least Mean Square Algorithm With Orthogonal Correction Factors (VS-NLMS-OCF) algorithm is proposed ,.

In this article, we will adopt the following symbolic notation: || .||said Euclidean vector norm, $(.)^T$ is the transpose of a matrix or a vector. Scalars are lowercase or uppercase, all vectors are column vectors and lowercase bold, matrices are bold uppercase.

2. VS-NLMS-OCF Algorithm

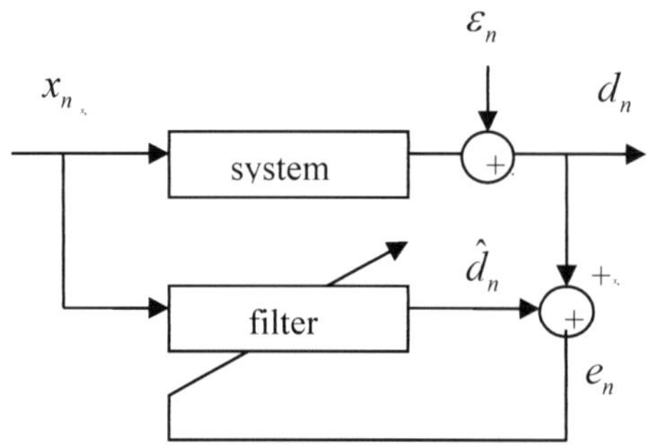

Fig.1 Adaptive system

Figure 1 shows an adaptive filter used in a system identification model. Filter output $d_n = \mathbf{x}_n^T\hat{\mathbf{w}}_n + \varepsilon_n$, where $\mathbf{x}_n =(x_n , x_{n-1} ,..., x_{n-N+1})^{\mathrm{T}}$ is the input signal (N is the dimension), d_n is the corresponding measurement output, $\hat{\mathbf{w}}_n$ is the unknown estimated weight vector, and the measurement noise ε_n is zero-mean white noise. Its objective is to estimate an n-dimensional weight vector $\hat{w}_n$ such that when $\mathbf{x}_n =(x_n , x_{n-1} ,..., x_{n-N+1})^{\mathrm{T}}$ is the input vector at the n moment, the estimated output $\hat{d}_n = \hat{\mathbf{w}}_n^T \mathbf{x}_n$ is as close to the measured output d_n as possible in the sense of mean square error.

NLMS-OCF algorithm is summarized in literature [7]. It can be seen from steps (3) and (8) that NLMS-OCF algorithm adopts fixed step size, which is difficult to balance convergence speed and misalignment. In order to solve this problem, this paper studies NLMS-OCF algorithm. A NLMS-OCF algorithm with variable step size is proposed, which uses the projection weight error vector norm to get the appropriate smoothing factor by mapping the exponential function. The algorithm is simple to implement, and the relationship between smoothing factor α and $\|\mathbf{p}_k\|^2$ is as follows:

$$\alpha_k = \exp(-\|\mathbf{p}_k\|^2) \tag{1}$$

Where, α_k represents the α value of the kth iteration, and $\exp(\cdot)$ represents the exponential function. α_k between 0 and 1, the bigger $\|\mathbf{p}_k\|^2$ is, the closer α_k is to 0, and vice versa, the closer α_k is to 1. Therefore, the exponential update strategy meets the smoothing factor requirement of the algorithm.

Therefore, if the idea of variable step size based on exponential smoothing factor is applied to NLMS-OCF algorithm, step (3) [7] can be changed to:

$$\alpha = \exp(-\|\mathbf{p}_n\|^2) \tag{2}$$

$$\mathbf{p}_n^1 = \alpha\mathbf{p}_n + \frac{(1-\alpha)\mathbf{x}_n e_n}{\|\mathbf{x}_n\|^2} \tag{3}$$

$$\bar{\mu} = \mu_{\max}\frac{\|\mathbf{p}_n^1\|^2}{\|\mathbf{p}_n^1\|^2 + C} \tag{4}$$

$$\mu_0 = \frac{\bar{\mu} e_n}{\|\mathbf{x}_n\|^2} \tag{5}$$

Step (8) [7] can be changed to:

$$\alpha = \exp(-\|\mathbf{p}_n^k\|^2) \tag{6}$$

$$\mathbf{p}_n^{k+1} = \alpha \mathbf{p}_n^k + \frac{(1-\alpha)\mathbf{x}_n^k e_n^k}{\|\mathbf{x}_n^k\|^2} \tag{7}$$

$$\bar{\mu} = \mu_{\max} \frac{\|\mathbf{p}_n^{k+1}\|^2}{\|\mathbf{p}_n^{k+1}\|^2 + C} \tag{8}$$

$$\mu_k = \begin{cases} \dfrac{\bar{\mu} e_n^k}{\|\mathbf{x}_n^k\|^2} & if \ \|\mathbf{x}_n^k\| \neq 0 \\ 0 & otherwise \end{cases} \tag{9}$$

Accordingly, step (10) [7] can be changed to:

$$\mathbf{p}_n = \mathbf{p}_n^{M+1} \tag{10}$$

$$\hat{\mathbf{w}}_{n+1} = \hat{\mathbf{w}}_{n+1}^{M+1} \tag{11}$$

Therefore, the VS-NLMS-OCF algorithm was obtained by optimizing NLMS-OCF algorithm. Where, $\alpha \in (0,1)$ is the smoothing factor; the larger the value of $\|\mathbf{p}_k\|^2$, $\mathbf{w}_{k+1}$ the farther away $\mathbf{w}_0$, and the closer μ_k is to $\mu_{\max}$; otherwise, μ_k is closer to 0.

3. Simulation Results

In this part, the NLMS-OCF algorithm and VS-NLMS-OCF algorithm are simulated and compared respectively in MATLAB, and the simulation results are analyzed. The mean square error (MSE) learning curves of the simulation results of the two algorithms are obtained by averaging 100 independent learning curves. Where, dimension N=32, $\mu_{\max} = 1$, $\mathbf{p}_0 = I_{N\times 1}$.

Example 1: Consider the highly correlated input signal model expressed as $x_n = -0.95x_{n-1} - 0.6x_{n-2} + z_n$, where z_n is gaussian white noise signal with mean equal to 0 and orthogonal correction factor M=2. As can be seen from Figure 2, compared with NLMS-OCF algorithm (step size is set to 0.1, 0.5 and 1 respectively), VS-NLMS-OCF algorithm shows better convergence speed and steady-state error.

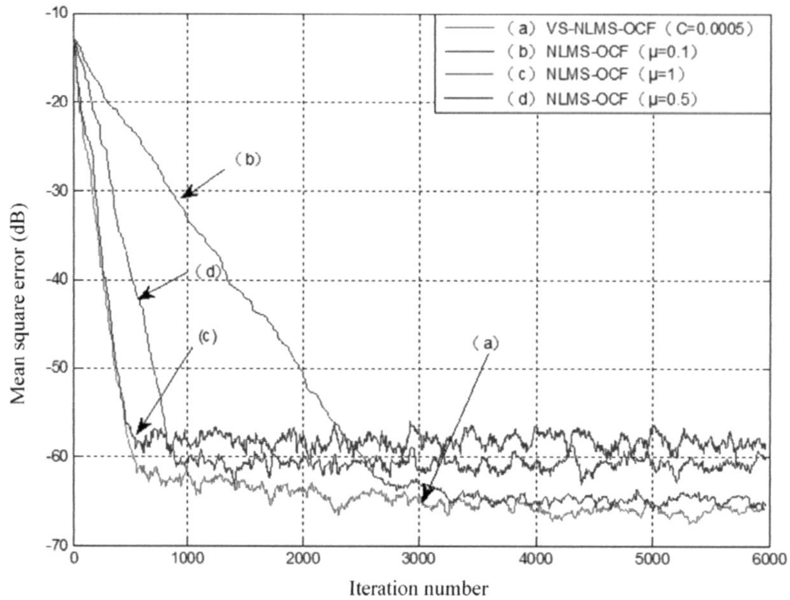

Fig.2 Mean square error learning curve

Example 2: Consider the highly correlated input signal model expressed as $x_n = -0.95x_{n-1} - 0.6x_{n-2} - 0.3x_{n-3} + z_n$, where z_n is gaussian white noise signal with mean equal to 0 and orthogonal correction factor M=3. As can be seen from Figure 3, compared with NLMS-OCF algorithm, the established algorithm achieves better convergence speed and steady-state performance.

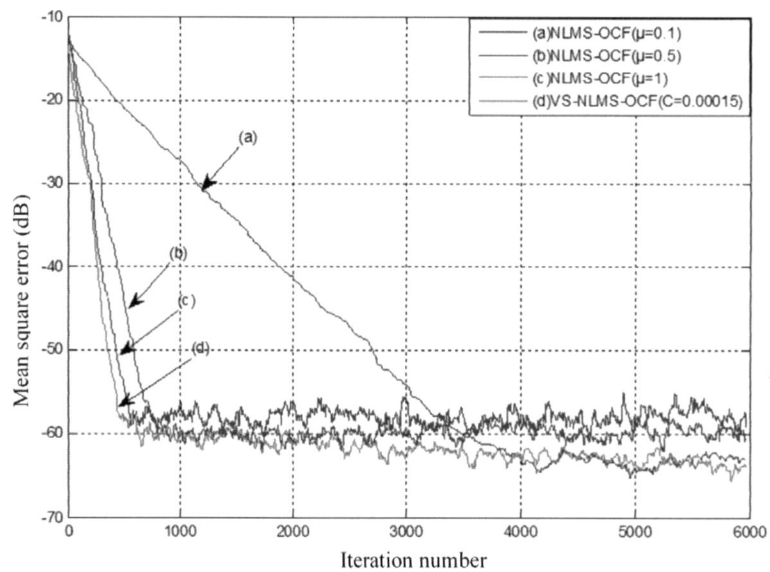

Fig.3 Mean square error learning curve

4. Conclusion

Aiming at the problem of fixed step size in NLMS-OCF algorithm, this paper proposes a variable step size NLMS-OCF algorithm, which uses the norm of the projection weight error vector to get the smoothing factor through the mapping of the exponential function. Compared with the fixed step

NLMS-OCF algorithm, the improved algorithm has faster convergence speed and smaller steady-state error. The performance of adaptive filter is improved. Algorithms that are easy to implement on hardware are the main trend of future research. The theoretical research of adaptive equalization algorithm should be further combined with hardware implementation.

Acknowledgments
This work was financially supported by the Fund Project:
1.Special Scientific Research Program of Shaanxi Provincial Education Department （21JK0813）
2.Scientific Research Foundation of Xi 'an Mingde Institute of Technology （2020XY01L03）
3. Shaanxi Innovation and Entrepreneurship Training Program （S202113894006）

References
[1] B. Widrow and S. D. Stears, Adaptive Signal Processing. Englewood Cliffs,NJ:Prentice-Hall,1985.
[2] S.Haykin, Adaptive Filter Theory,4th ed. Upper Saddle River,NJ:Prentice-Hall,2002.
[3] A. Mader, H. Puder, and G.U.Schmidt," Step-size control for acoustic echo cancellation filters—An Overview," Signal Process.,vol.80,pp.1697-1719,Sept.2000.
[4] Shin, H.C., Sayed, A.H., and Song, W.J., "Variable step-size NLMS and affine projection algorithms," IEEE Signal Process. Lett,2004,11,(2),pp.132-135.
[5] Dai, T., and Shahrrava, B. ,"Variable step-size NLMS and affine projection algorithms with variable smoothing factor," IEEE 48th Midwest Symp. Circuits and Systems, 2005,Vol.2, pp.1530-1532.
[6] Y. Fan and J. Zhang, "Variable step-size affine projection algorithm with exponential smoothing factors," The Institution of Engineering and Technology 2009, doi:10.1049/el.2009.1403.
[7] S. G. Sankaran and A. A. (Louis) Beex," Fast Generalized Affine Projection Algorithm," International Journal of Adaptive Control and Signal Processing, February 2000.
[8] Xiu Zhan, Ge Li. Research on IMU Signal Denoising Based on Variable Step Size LMS Algorithm [J]. Telemetry and Remote Control,2021,42(02):35-41.
[9] Zhang Jirong, Zhang Tian. An Improved Variable Step Size LMS Adaptive Filtering Algorithm [J]. Journal of Xi 'an University of Posts and Telecommunications,2021,26(01):7-12.

Mask wearing detection algorithm based on improved YOLOv4

Qing Ye [1*], Yuqi Zhao[2]

[1,2]School of Information Science and Technology, North China University of Technology, Beijing 100144, China

[1*]yeqing@ncut.edu.cn, [2]yoki185@163.com

Abstract—Since the outbreak of COVID-19, the domestic situation has gradually improved, but small-scale outbreaks are still frequent, the current situation of COVID-19 cannot be eased up. Using target detection algorithms to identify mask wearing to remind people to pay attention to the epidemic, can reduce the probability of cross infection in public places. YOLOv4 is a real-time target detection algorithm with fast detection speed and small size, but it has problems such as insufficient positioning of the bounding box and easy to miss the occluded targets. Based on this, we improve the algorithm and integrates the CBAM attention module into the backbone network. At the same time, the depthwise over-parameterized convolution (DO-Conv) is used to replace the ordinary convolution, which improves the network fitting ability without adding additional parameters. The experimental results show that the improved algorithm has improved the positioning accuracy of the bounding box and detection accuracy.

1. Introduction

In recent years, artificial intelligence puts glorious greatly in various industries, and target detection is one of the important areas. Due to the advantages of deep learning algorithms such as high scalability, strong learning ability and end-to-end learning, a series of target detection algorithms based on deep learning have been proposed and are widely used in various fields such as traffic safety, medical health and safety protection. With the emergence of COVID-19, social and economic development and people's life affected by different degree, according to the requirement to wear a mask is essential for epidemic prevention and control. Wearing masks in public places has become something that everyone should observe, it is very necessary to conduct mask wearing tests in places to strengthen the public's awareness of actively wearing masks.

At present, there are not many specialized researches on mask wearing detection algorithms. Baidu has open sourced the PyramidBox algorithm[1], which is a context-assisted single-step detection algorithm used for face detection. After improvement, it has obtained good results in mask wearing detection. DiDi proposed a mask wearing recognition algorithm combined with attention technology[2]. The algorithm is divided into two steps. First, the accurate positioning of the face region is realized based on feature fusion and segmentation supervision, and then the mask recognition technology is used to further determine whether the face is wearing a mask in the face region, which effectively solves the impact of environmental factors on mask detection. Huang et al. achieved good results by improving RetinaNet network and using the idea of transfer learning to detect mask wearing[3]. Yu et al. based on Efficient Det network, used soft-NMS and combined with multi-attention mechanism to eliminate inappropriate target detection box, and achieved good detection accuracy in mask wearing detection[4].

According to the various mask detection algorithms mentioned above, mask detection is different from other target detection tasks. It needs to detect the face and then recognize whether the mask is worn. Therefore, in order to pay more attention to faces, we construct a mask detection algorithm by improving the YOLOv4 network. The CBAM attention mechanism is integrated in the feature extraction network to make the detection box more accurately fall on the face, which can extract facial features more effectively and improve the detection accuracy. At the same time, we use depthwise over-parameterized convolution (DO-Conv) instead of the traditional convolution block. As DO-Conv can be folded into a general convolution in the inference stage, it has exactly the same number of parameters as a general convolution block, and the learnable parameters can be increased. While improving the performance of the algorithm, the number of parameters does not increase.

2. Our model

2.1 Improved YOLOv4 network

"You Only Look Once" (YOLO) is an object detection algorithm proposed by Redmon et al. in a research paper in 2016. The YOLOv4 algorithm is improved based on YOLO, optimizing data processing, activation function, backbone network, network training, loss function and other aspects. It combines the best optimization strategies in the CNN field in recent years, and also obtains the effect of SOTA. This paper mainly improves its CSPDarknet53 and PANet parts. First, the CBAM attention module is integrated in the residual module of the main part of the original network, and then the general convolution in PANet is replaced with DO-Conv. The improved YOLOv4 network structure diagram is shown in Fig.1.

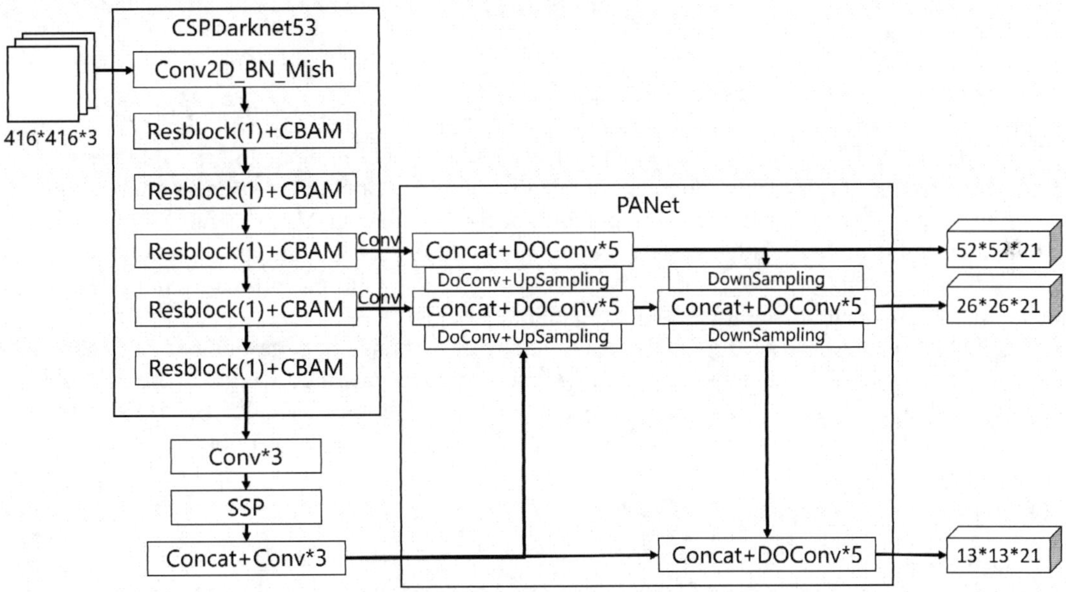

Fig.1 Improved YOLOv4 network structure diagram

2.2 CBAM module

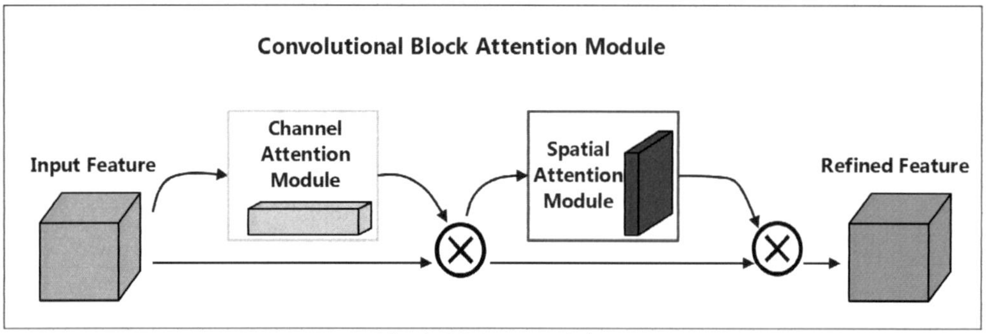

Fig.2 CBAM module

The CBAM module structure is shown in Fig.2. It consists of Channel Attention Module (CAM) and Spatial Attention Module (SAM) two modules. This structure does not require many additional parameters and will not cause calculation burden on the network. The calculation formulas are respectively formula (1) and (2) shows:

$$\begin{aligned} \mathrm{M}_c(F) &= \sigma\left(MLP\left(AvgPool(F)\right) + MLP\left(MaxPool(F)\right)\right) \\ &= \sigma\left(W_1\left(W_0\left(F^c_{avg}\right)\right) + W_1\left(W_0\left(F^c_{\max}\right)\right)\right) \end{aligned} \quad (1)$$

$$\begin{aligned} \mathrm{M}_s(F) &= \sigma\left(f^{7\times7}\left(\left[AvgPool(F); MaxPool(F)\right]\right)\right) \\ &= \sigma\left(f^{7\times7}\left(\left[F^s_{avg}; F^s_{\max}\right]\right)\right) \end{aligned} \quad (2)$$

2.3 Depthwise over-parameterized convolution module

Depthwise over-parameterized convolution (DO-Conv) adds a deep convolution on the basis of conventional convolution for over-parameterized, and its purpose is to obtain more parameters to speed up network training. With the development of deep learning, researchers are accustomed to constantly deepen the network to improve the learning ability of the network, making the current network model more and more bloated. The advantage of DO-Conv is that it uses a multi-layer structure to speed up network training during training, and folds it into a single layer during inference, and the amount of calculation is equivalent to conventional convolution.

The PANet module of the original YOLOv4 network contains a large number of convolutional modules. The continuous superposition of convolution modules can increase the expression ability of the network, but at the same time it also increases the amount of network calculations. Therefore, this paper changes the general convolution in PANet into DO-Conv. Although cannot reduce the amount of network parameters, the feature extraction ability and convergence speed of the network are further improved.

3. Experiments

At present, there is no large public data set in the field of mask wearing detection. This article uses the mask dataset collected and annotated by the AIZOO team, which were taken from the MAFA dataset and the WIDER FACE dataset respectively. We screened the pictures and removed some of the inappropriate data. Finally, we select about 4000 pictures for experiment, the pictures examples are shown in Fig.3. We organize it into a VOC data set format, with a training and testing ratio of 9:1. The experimental environment uses the Windows 10 operating system, Pytorch architecture, and the GeForce GTX 1080Ti graphics card and 12G memory for calculations.

Fig.3 Dataset pictures examples

In the network model training stage, the batch size is set to 4, the total number of iterations is 50, and the initial learning rate is set to 0.0001. The obtained loss function convergence diagram is shown in Fig.4. It can be seen that the loss function of the improved YOLOv4 network has smaller fluctuation and faster convergence speed compared with the original network. After calculation, the number of parameters in YOLOv4 is 255.8M, and the number of parameters in the improved network is only 256.7M, which is almost the same.

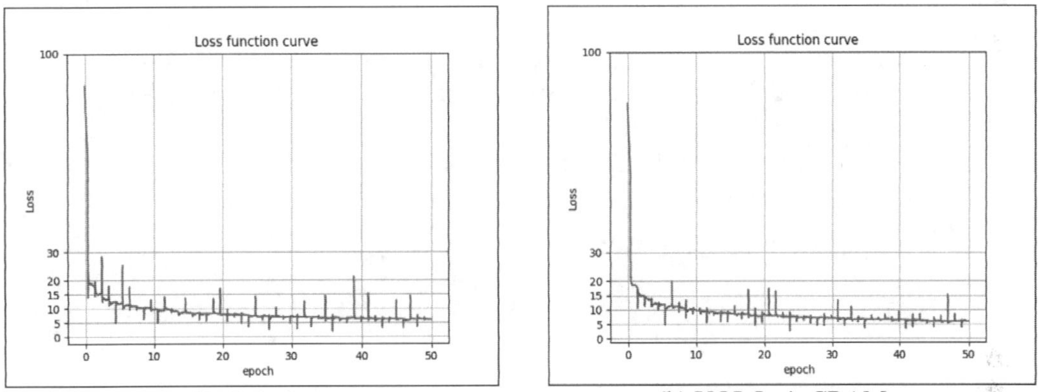

(a) YOLOv4 (b) YOLOv4+CBAM

Fig.4 Convergence graph of loss function

In order to better see the effect of the improvement, we use the original network and the improved network for training and testing. The recognition accuracy comparison chart is shown in Fig.5. "mask" means wearing a mask correctly, and "face" means not correct wear a mask.

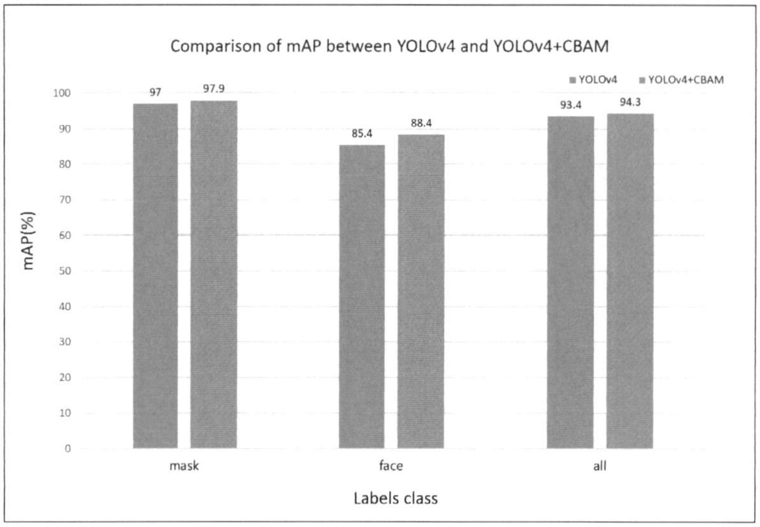

Fig.5 Comparison of mAP between YOLOv4 and YOLOv4+CBAM

It can be seen from the comparison chart that the improvement method proposed in this paper has been significantly improved compared with the original network. In addition, in order to better analyse the effectiveness of the mask wearing detection proposed in this paper, we have found some recent mask wearing recognition algorithms for comparison. The accuracy comparison results are shown in Table 1:

Table 1 Comparison of recognition accuracy with other algorithms

Algorithm	Accuracy (%)
Cheng [9]	88.4
Xiao [10]	89.0
Tan [11]	92.4
Cao [12]	93.3
ours	94.3

4. Conclusion

This paper proposes a mask wearing detection algorithm based on improved YOLOv4 network. The CBAM attention module is integrated in the backbone network to deal with the impact of complex detection environment on detection, so that the network can better focus on the face area and improve the detection accuracy. Then we use DO-Conv to replace the general convolution in PANet, which can increase the learnable parameters of the network, and further improve the fitting ability of the network.

The final experimental results show that the mask wearing detection algorithm based on the improved YOLOv4 proposed in this paper is significantly improved compared to the original algorithm, and it has achieved higher recognition accuracy than many current mainstream algorithms.

Acknowledgments

This paper is supported by Technology Project of Beijing Municipal Education Commission (No. SQKM201810009002), National Natural Science Foundation of China (No.61371143), National Natural Science Foundation of China (No.61806008), Beijing Natural Science Foundation (No.4192022), Ministry of Education Science and Technology Development Center Project (No. 2018A03029), Beijing Innovation Team, Key scientific research direction construction project of North China University of Technology. The authors also gratefully acknowledge the helpful comments and suggestions of the reviewers, which have improved the presentation.

References

[1] Tang, X., DU, Daniel, K., He, Z., et al. (2018) PyramidBox: A Context-assisted Single Shot Face Detector. In: Proceedings of the European Conference on Computer Vision (ECCV). Munich. pp.797-813.

[2] Zhang, X. B., Lin, Z. Y., Tian, W. X., et al. (2020) Mask-wearing recognition in the wild (in Chinese). Sci Sin Inform, 50: 1110–1120.

[3] Deng, H. X. (2020) Method of mask wearing detection based on transfer learning and RetinaNet. Electronic Technology & Software Engineering, 05:209-211.

[4] Yu, A. X., Li, C. R., Yu, S. Y., Li, H. J. (2021) Multi-Attention Mechanism of Mask Wearing Detection Network. Journal of Nanjing Normal University (Engineering and Technology Edition), 21:23-29.

[5] Redmon, J., Divvala, S., Girshick, R., Farhadi, A. (2016) You Only Look Once: Unified, Real-Time Object Detection. In: IEEE Conference on Computer Vision and Pattern Recognition (CVPR). Las Vegas. pp. 779-788.

[6] Bochkovskiy, A., Wang, C. Y., Liao, H. (2020) Yolov4: Optimal Speed and Accuracy of Object Detection. arXiv.org/abs/2004.10934

[7] Woo, S., Park, J., Lee, J. Y., Kweon, I. S. (2018) CBAM: Convolutional Block Attention Module. In: Proceedings of the European Conference on Computer Vision (ECCV). Munich. pp. 3-19.

[8] Cao, J., Li, Y., Sun, M., et al. (2020) Do-Conv: Depthwise Over-parameterized Convolutional Layer. arXiv:2006.12030

[9] Cheng, K. X., Wang, Y. D. (2021) Algorithm of Mask Wearing Detection in Natural Scenes Based on Improved YOLOv3. Computer Systems & Applications, 30:231-236.

[10] Xiao, J. J., (2020) Masked Face Detection and Standard Wearing Mask Recognition Based on YOLOv3 and YCrCb. Computer Engineering & Software, 41:164-169.

[11] Tan, S. L., Bie, X. B., Lu, G. L., Tan, X. H. (2021) Real-time Detection for Mask-wearing of Personnel Based on YOLOv5 Network Model [J]. LASER JOURNAL, 42:147-150.

[12] Cao, C. S., Yuan, J. (2021) Mask Wearing Detection Method Based on YOLO-Mask. Laser & Optoelectronics Progress, 58:211-218.

Research on Optimizing Multimodal Transport Path under the Schedule Limitation Based on Genetic Algorithm

Cheng Jiang [1a*]

[1] College of Transport and Communications, Shanghai Maritime University, Shanghai, 201306, China

[a*]202030610128@stu.shmtu.edu.cn

Abstract—Multimodal transportation is getting more and more attention due to its economical and efficient characteristics. Choosing a reasonable transportation plan can effectively control transportation costs. Considering that there are fixed schedules for railway and waterway transportation in real life, this article will establish a multimodal transportation route optimization model with the goal of minimizing the total transportation cost consisting of transportation cost, transit cost, waiting cost and carbon emission cost. The genetic algorithm based on preservation strategy and immigration strategy is used to solve the model. The effectiveness of the model is proved by the case study, and it can provide decision support for the choice of multimodal transportation scheme.

1. Introduction

Compared with the traditional single transportation method, multimodal transportation can give play to the advantages of various transportation methods, making cargo transportation more economical and flexible. It is currently being vigorously promoted all over the world, and the problem of multimodal transport route optimization has also received more and more attention. Soysal[1] established a mathematical model for the beef logistics network problem, and solved the model by using the ε-constraint method. Cheng[2] studied the impact of different carbon emission policies on the choice of multimodal transport routes under the condition of road congestion. Zhang[3] studied the optimization of multimodal transportation routes when the demand for goods is different. Dong[4] proposes a new mixed integer programming model designed to balance economic and environmental benefits. Pei[5] designed a dual pheromone ant colony algorithm to solve the multimodal transport path optimization problem of comprehensive satisfaction. Lei[6] established a multimodal transport route reuse planning model and solved the problem of maximizing the ratio of income to investment for large cargo. Zhang[7] established a multimodal transport model including environmental cost considering the customer service time window and solved it. Xie[8] used genetic algorithm to solve the multimodal transportation problem of vehicle logistics with time window. Although Lu[9] and Peng[10] considered the impact of the schedule on the choice of multimodal transport routes, the schedules of all transit nodes are the same.

In real life railway transportation and waterway transportation usually have fixed schedules, and once they are missed, they will have to wait for the next departure time. And the schedule may be different when using different modes of transportation between different cities. Therefore, this paper will establish a multimodal transport route optimization model under the schedule limitation. Considering the superiority of intelligent algorithms in such problems, a genetic algorithm with preservation strategy and immigration strategy is designed to solve the model.

2. Problem description and model building

2.1. Problem description

There is a batch of goods in the starting city *O* that needs to be transported to the destination city *D* within the specified time window. There are multiple node cities in the middle, and there is at least one transportation mode among roads, waterways and railways between each city. When you choose road transportation, you can leave directly. But when you choose railway transportation and water transportation, the departure time is restricted by fixed schedules. Now we will determine the multimodal transportation plan with the goal of minimizing the total transportation cost.

The assumptions are as follows: The goods cannot be separated during transportation; The conversion of the transportation mode only occurs in the node city and the transshipment operation starts immediately after reaching the city; The speed and cost of various modes of transportation are the same among different cities.

2.2. Optimization model

The optimization model of the multimodal transport route with schedule restrictions is shown below:

2.2.1. Objective function

$$MinZ=\sum_{i,j\in N}\sum_{K\in M}C_{ij}^{k}d_{ij}^{k}x_{ij}^{k}+\sum_{i\in N}\sum_{k,l\in M}C_{i}^{kl}y_{i}^{kl}+\sum_{i\in N}\sum_{k,l\in M}C_{i}^{l}t_{i1}^{l}y_{i}^{kl}+\sum_{i,j\in N}\sum_{k\in M}C_{e}E_{ij}^{k}x_{ij}^{k} \quad (1)$$

The objective function (1) minimizes the total cost including transportation cost, transit cost, waiting cost and carbon emission cost.

2.2.2. Constraints

$$\sum_{k\in M}x_{ij}^{k}\leq 1 \quad \forall i,j\in N \quad (2)$$

$$\sum_{k\in M}\sum_{l\in M}y_{i}^{kl}\leq 1 \quad \forall i\in N \quad (3)$$

$$\sum_{j\in N}\sum_{k\in M}x_{ij}^{k}-\sum_{j\in N}\sum_{k\in M}x_{ji}^{k}=\begin{cases}1 & i=O\\ -1 & i=D\\ 0 & i\in N\setminus\{O,D\}\end{cases} \quad (4)$$

$$\sum_{h\in N}x_{hi}^{k}+\sum_{j\in N}x_{ij}^{l}\geq 2y_{i}^{kl} \quad \forall i\in N;k,l\in M \quad (5)$$

$$t_{ij}^{k}=\frac{d_{ij}^{k}}{v^{k}} \quad (6)$$

$$t_{i1}^{l}=\begin{cases}a_{i1}^{l}-t_{i} & t_{i}<a_{i1}^{l}\\ 0 & a_{i1}^{l}\leq t_{i}\leq t_{i}+t_{i}^{kl}\leq b_{i1}^{l}\\ a_{i2}^{l}-t_{i} & \mathrm{b}_{i1}^{l}<t_{i}+t_{i}^{kl}\end{cases} \quad (7)$$

$$t_{i2}^{l}=\begin{cases}b_{i1}^{l}-a_{i1}^{l}-t_{i}^{kl} & \mathrm{t}_{i}<a_{i1}^{l}\\ b_{i1}^{l}-\mathrm{t}_{i}-t_{i}^{kl} & a_{i1}^{l}\leq t_{i}\leq t_{i}+t_{i}^{kl}\leq b_{i1}^{l}\\ b_{i2}^{l}-a_{i2}^{l}-t_{i}^{kl} & \mathrm{b}_{i1}^{l}<t_{i}+t_{i}^{kl}\end{cases} \quad (8)$$

$$T=T_{O}+\sum_{i,j\in N}\sum_{k,l\in M}\left(t_{ij}^{k}x_{ij}^{k}+t_{i}^{kl}y_{i}^{kl}+t_{i1}^{l}y_{i}^{kl}+t_{i2}^{l}y_{i}^{kl}\right) \quad (9)$$

$$T_E \leq T \leq T_L \tag{10}$$

$$x_{ij}^k \in \{0,1\}; y_i^{kl} \in \{0,1\} \tag{11}$$

Constraint (2) means that any two nodes can only be transported by at most one mode of transportation. Constraint (3) means that at any one node, only one transportation mode conversion can be carried out. Constraint (4) represents the node cargo flow balance, the origin city is *O*, and the destination city is *D*. Constraint (5) means that the goods at node *i* enter in transport mode *k*, and when they leave in transport mode *l*, the transportation mode conversion occurs at node *i*, which ensures the continuity of the transportation mode. Constraint (6) indicates the transportation time between different cities using transportation mode *K*. Constraint (7) represents the time for the goods to wait for the transit operation at node *i*. It means that if the goods reach node i earlier than the transit operation time, the waiting fee will be paid. Constraint (8) represents the time to wait for transport mode *l* to leave when the cargo completes the transit operation at node *i*. Constraint (9) represents the total transportation time of the goods. The time for goods to leave one node and move to the next includes transportation time, waiting time for transit, transit time, and time waiting to leave after reloading. Constraint (10) to ensure that the goods are delivered to the destination city *D* within the specified time window. Constraint (11) means that the decision variables x_{ij}^k and y_i^{kl} are 0-1 variables.

3. Design algorithm

The multimodal transport route optimization problem is an NP-hard problem, which can be solved by an intelligent algorithm. The genetic algorithm has a good global search ability and is more convenient when solving this problem. Therefore, the genetic algorithm will be used to solve the above model in this paper. Compared with the traditional genetic algorithm, the preservation strategy and immigration strategy are adopted when designing the algorithm. The preservation strategy can ensure that the dominant individuals in the population can continue to exist in the next generation, so that the entire optimization process always develops in a better direction. The immigration strategy can ensure the diversity of the population and avoid prematurely falling into the local optimal solution. The specific design ideas are as follows.

3.1. Encoding and decoding

Considering that the problem of multimodal transportation path optimization is a combined optimization problem, the choice of transportation path and transportation mode will affect the result, so the double-layer coding structure is adopted, and both are real number coding. The first layer is the encoding of the transportation path, the length is N. The second layer is the encoding of the mode of transportation, the length is N-1. The two parts of the code form a complete chromosome code, as shown below.

$$X = \{x_1, x_2, \cdots, x_N\} \mid \{y_1, y_2, \cdots, y_{N-1}\} \tag{12}$$

When decoding, the first part of the code is expressed as the number of the nodes that the goods pass through in turn, and the second part of the code is expressed as the mode of transportation selected between adjacent nodes. It should be noted that the nodes after the destination city D in the encoding are all invalid nodes, so only the previous nodes are decoded.

3.2. Generate initial population

Regarding the transportation route, the first number is fixed at 1, which represents the starting city, and then the remaining nodes are sorted randomly. Regarding the transportation method, a number is randomly generated based on the transportation methods that can be selected between neighboring cities, and the numbers after the destination city are all 0. According to the above principles, the initial individuals are generated until the population quantity requirements are met.

3.3. Calculate fitness

Since what this paper seeks is the minimization problem, and the principle of taking the larger is used in the subsequent selection operator, the reciprocal of the total cost calculated by each individual is taken as the fitness. In addition, in order to ensure that the goods are delivered on time, a sufficiently large penalty value is added to the target value of the plan that arrives outside the time window. This means that goods cannot arrive earlier or later than the specified time period.

3.4. Genetic operation

The new population will come from three parts. Firstly, the number of individuals in each new population is determined according to the number of individuals in the initial population, the preserving ratio P_F, the immigration ratio P_Y, and the genetic ratio P_G. Then select some individuals with higher fitness in the parent population to obtain the first population G_1. According to the previous principle, a certain number of new individuals are generated to form the second population G_2. The parent population is selected, crossed, and mutated to form the third population G_3. The selection operator uses the roulette method of selection. Crossover operator uses single point crossover method. The mutation operator uses two-point exchange mutation. Some infeasible individuals will be generated in the new population, so finally the correction operator is used to modify the transportation mode coding of the infeasible individuals to make it feasible. When the specified number of iterations is reached, the target value of the current optimal individual is output, and the algorithm ends.

4. Case study

There are 20 tons of goods that will be transported from the starting city *O* to the destination city *D* using a 20-foot container. There are 9 node cities in the middle. The distance between the cities and the schedule of rail and water transportation will be randomly generated. The carbon emission cost C_e is 0.1 yuan/kg, the receiving time window is [45, 55], and the values of other relevant parameters are shown in Table 1 and Table 2.

Table 1. Values of relevant parameters of various modes of transportation.

Parameters(units)	Highway	Railway	Waterway
v^k (km/h)	80	60	28
C_{ij}^k (yuan/TEU*km)	6	3.2	0.9
C_i^l (yuan/TEU*h)	0	30	20
E_{ij}^k (kg/t*km)	0.071	0.042	0.012

Table 2. Values of relevant parameters of transshipment.

Parameters(units)	Highway-Railway	Highway-Waterway	Railway- Waterway
C_i^{kl} (yuan/ TEU)	100	150	190
t_i^{kl} (h/TEU)	0.1	0.15	0.2

According to the established multimodal transport route optimization model, the genetic algorithm of preservation strategy and immigration strategy is adopted, and the solution is solved by Matlab2016b software programming. The relevant parameters in the genetic algorithm are as follows: the population size is 100, the number of iterations is 500, and the preservation ratio P_F is 0.1, the immigrant ratio P_Y is 0.1, the genetic ratio P_G is 0.8, the crossover probability is 0.8, and the mutation probability is

0.5. The calculated results are as follows: the transportation route is 1(O)-9-10-11(D), the transportation mode is railway-roadway-railway, the total transportation cost is 6722.328 yuan, and the total transportation time is 54.2167 h. The program iteration process is shown in Figure 1.

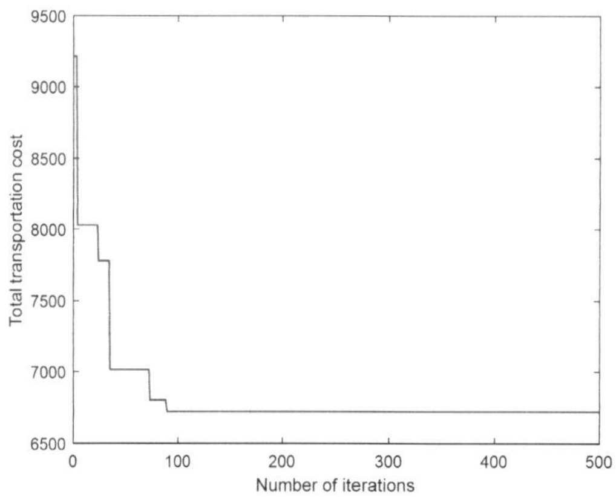

Figure 1. The change trend chart of the global optimal solution.

In order to verify whether the departure time will affect the choice of transportation plan, different departure times will be selected for calculation next. The calculation results are shown in Table 3. We can see that when the departure time changes, the choice of transportation plan will also change. Properly early departure time can effectively reduce transportation costs.

Table 3. Transport schemes selected at different departure times.

Departure time	Transportation plan	Total cost(yuan)
-4	$1 \xrightarrow{waterway} 3 \xrightarrow{roadway} 4 \xrightarrow{railway} 11$	5908.8136
-2	$1 \xrightarrow{waterway} 3 \xrightarrow{roadway} 4 \xrightarrow{railway} 11$	5868.8136
0	$1 \xrightarrow{railway} 9 \xrightarrow{roadway} 10 \xrightarrow{railway} 11$	6722.3280
2	$1 \xrightarrow{railway} 9 \xrightarrow{roadway} 10 \xrightarrow{railway} 11$	6662.3280
4	$1 \xrightarrow{railway} 9 \xrightarrow{roadway} 10 \xrightarrow{railway} 11$	6602.3280

5. Conclusion

Determining a reasonable multimodal transport plan is the key to giving play to the advantages of multimodal transport. The multimodal transport route optimization model established in the article takes into account the impact of fixed schedules on transportation time and transportation costs, so it is more practical. After many calculations and tests, the genetic algorithm designed in the article can effectively avoid falling into the local optimal solution and obtain the result quickly, which proves the feasibility of the algorithm and can provide a reference for the multimodal transport operator to determine the transportation plan. However, there are still some shortcomings in the research, such as failure to take into account the delay of the schedules and other uncertain factors. These will be the subject of further research in the future.

References

[1] Soysal M, Bloemhof-Ruwaard J M, J.G.A.J. van der Vorst. (2014) Modelling food logistics networks with emission considerations: The case of an international beef supply chain. International Journal of Production Economics.

[2] Cheng XQ, Jin C. (2019) Path selection of multimodal transport considering road congestion under low-carbon policy. Operations Research and Management, 28(04): 67-77.
[3] Zhang X, Jiang YD, Yuan XM. (2020) Low-carbon multimodal transport path optimization under uncertain demand. Fuzzy Systems and Mathematics, 34(05): 163-174.
[4] Dong B, Christiansen M, Fagerholt K, et al. (2020) Design of a sustainable maritime multi-modal distribution network – Case study from automotive logistics. Transportation Research Part E: Logistics and Transportation Review, 143.
[5] Pei X, Lu YP, Zhang CZ. (2020) Multimodal transport path model and algorithm based on comprehensive satisfaction. Traffic Information and Safety, 38(01): 136-144.
[6] Lei DY, Zeng BX, Wang Z. (2019) Planning Model and Algorithm of Multimodal Transport Path Reuse for Long-Range Cargo. Chinese Journal of Railway Science and Engineering, 16(07): 1810-1819.
[7] Zhang DZ, He RZ, et al. (2017) A multimodal logistics service network design with time windows and environmental concerns. PLOS ONE, 12(9).
[8] Xie XM, Yang JQ. (2017) Research on vehicle multimodal transport route optimization model with time window. Journal of Wuhan University of Technology (Transportation Science and Engineering Edition),41(06):1061-1065.
[9] Lu XW, Yang B, Huang ZD. (2018) Research on Optimal Route Selection of Multimodal Transport under Mixed Time Window Constraints. Railway Transport and Economy, 40(08): 6-11.
[10] Peng Y, Xiao YP, Zhou X, Liu S. (2021) Multimodal transportation path optimization based on time window and schedule constraints. Chinese Science and Technology Paper, 16(02): 211-216.

A hierarchy superpixel algorithm dealing with stripe

ZhenDong Zhou[1a*], Yu Chen[1]

[1]School of Computer Science and Engineering,Sichuan University,Chengdu,China

[a*]zhouzhendong@stu.scu.edu.cn

Abstract—The hierarchy superpixel algorithm,in computer vision, aims to generate multi-scale superpixels with boundary perception.The commonly used strategy to obtain hierarchical structure is utilize the similarity between adjacent finer superpixels,then iteratively generate coarser superpixels based on finer ones. However,these methods may not achieve desired result when dealing with stripe targets.The alternating parts of the stripes are too different in image features,thus computer believes these parts are unrelated.In human visual perception,however, these parts actually belong to one striped object .We propose a new ISF-based method to solve this problem.Firstly,we present a improved ISF algorithm in finest scale segmentation with the help of sliding window strategy,which runs faster without reducing accuracy.Then,in order to generate hierarchy superpixels of stripe images,we design a new cross-neighbor path-cost function and add a stripe-pattern detector.Experimental results demonstrate that the proposed algorithm is accurate and efficient.

1. Introduction

Superpixel algorithms,which have being of prime importance in computer vision,are studied for the purpose of grouping similar pixels.The concept of superpixel was first proposed by Ren et al.[1] in 2002.After about two decades of development, research achievements in this filed scramble to appear.Combining pixels into superpixels can reduce computational costs and reduce the effect of noise,so the superpixel technique are widely applied in lots of vision tasks,such as depth estimation[2],saliency detection[3],medical image segmentation[4],image matching[5],etc.A good superpixel algorithm must possess three properties[12].Firstly,superpixles should capture image boundaries.Secondly,the algorithm is acceptable with low computational costs.Thirdly,superpixels should also have regular shape.Recently,researchers have paid more attention on hierarchical superpixel segmentation methods which generate hierarchical structure with multi-scale superpixels. Hierarchy superpixel algorithms have better performance in visual tasks duo to their new property of the hierarchical representation which is close to the human vision cognition.Numerous works can benefit from this property [6][15][16].

The Iterative Spanning Forest(ISF) [7] superpixel framework is a graph-based algorithm,which treats the image as a undirected graph and spans regions based on edge-weights of adjacent pixels.Researchers illustrate that ISF achieves good results than SLIC[8] and LSC[9] in image structure representation and no need for extra post-processing[7].However,the price is more running times.Essentially,the reason is that the priority queue data structure used in ISF is not a linear complexity structure.The larger the image size,the more times the algorithm consumes.We improved the ISF algorithm by the local sliding window strategy that expounded in part 2.

Despite the existence if many works with good results,almost all hierarchy superpixel algorithm are based on the similarity(color or texture) of adjacent superpixels,such as SH[10],dense-sparse hierarchy[11] and RISF[17].Thus,it is inevitable that the stripe problem(see figure 1) will happen,if the background superpixel is more similar to *Sa* than *Sb*,the next hierarchy segmentation iteration will merge the target object and the background, causing wrong results to affect subsequent processes.By the way,some hierarchy superpixel algorithms also take advantage of a contour detector to cover up the stripe problem[7][13][14],but there is no algorithm specifically for this question.

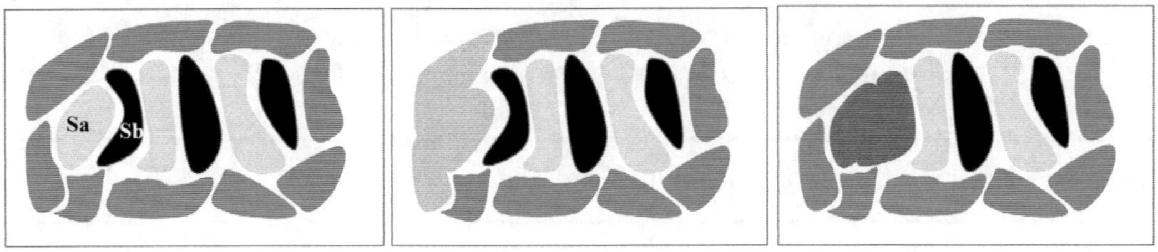

(a)Initial superpixel graph with stripe (b) The error merging (c) The ideal region merging
Fig. 1 Explanation of Stripe Problem

In this paper, we present a ISF-based hierarchy superpixel algorithm to solve the stripe problem.We use the improved ISF algorithm to generate the finest level,and use a region merging strategy with new path-cost function and a stripe-pattern detector to get the whole hierarchy segmentation of stripe image.Besides,extensive experiments shows that our algorithm's performance is also in line with expectations in terms of accuracy and efficiency for common pictures.

2. Method

2.1. The improved ISF

The one iteration of ISF framework has four main steps:(a)define the image as an undirected graph $\boldsymbol{G=(I,A)}$,whose I is image domain and $\boldsymbol{A \subset I \times I}$ is the adjacency relation(e.g.,pixel pairs $\boldsymbol{(s,t) \in A}$ when $\boldsymbol{s}$ is adjacent to $\boldsymbol{t}$).The path $\boldsymbol{\pi_t = <t_1,t_2,...t_n=t>}$ is a pixel path which end with pixel $\boldsymbol{t}$.(b) Sample the pixel seeds with some strategy.(c) Choose a path-coat function to compute superpixels by the IFT[18] algorithm. The mean path-cost function is

$$f(\langle s \rangle)=\begin{cases} 0 & if\ s \in \mathrm{S} \\ +\infty & else \end{cases} \tag{1}$$

$$f(\pi_s \cdot \langle s,t \rangle) = f(\pi_s) + \left(\alpha \left\| I(t), M(R(\pi_s)) \right\| \right)^{\beta} + \left\| s,t \right\|, \tag{2}$$

where $\boldsymbol{\alpha}$ and $\boldsymbol{\beta}$ are parameters that control the regularity if the superpixels,|| , || is the Euclidean distance,$\boldsymbol{I(t) = [l_t\ a_t\ b_t]^T}$ is the color vector at pixel $\boldsymbol{t}$,and $\boldsymbol{M(R(\pi_s))}$ is the mean color vector of the superpixel in the previous iteration,whose root is the seed $\boldsymbol{R(\pi_s)}$.(d)Recompute the new seeds according to the superpixels just got.

The core ideal to accelerate ISF algorithm is reducing the size of the priority queue.Noticing that the superpixels in fine scale do not have a large span on the image,we use the local sliding window strategy showed in figure 2.By limiting IFT to a local square scope and sliding the local window to all the seeds,we can get a faster ISF operation in finest level.Algorithm 1 presents the proposed procedure.

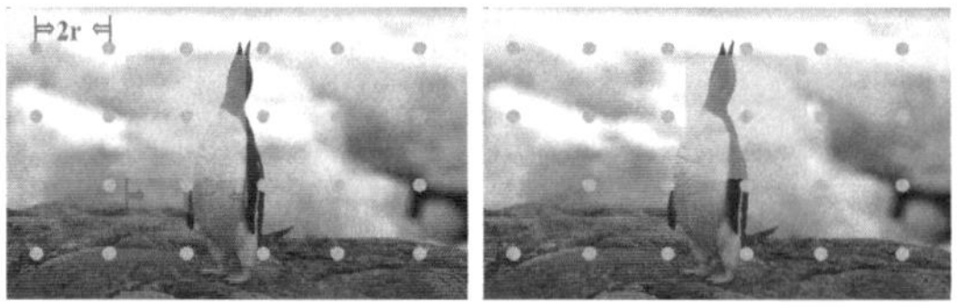

(a) (b)

Fig. 2 (a) The yellow square is the local window of a seed (blue spot) whose side length is $\boldsymbol{L}$, the grid radius is $\boldsymbol{r}$.(b)The local window slides to another seed.

Algorithm 1 Improved Iterative Spanning Forest

INPUT: Graph $\boldsymbol{G}$=(**I**,A), number of regions $\boldsymbol{n}$, the parameters $\boldsymbol{\alpha} \geq 0$ and $\boldsymbol{\beta} \geq 1$,and the ***MaxIters***≥1.

OUTPUT: Superpixel label map ***Ls***.

AUXILIARY:Seeds set ***Seed***,state map ***S***,priority queue ***Q***,cost map ***C***,root map ***R*** and superpixel mean array ***M***.

iter ← 0, compute ***Seed*** with the sampling strategy
While *iter* < ***MaxIters*** , do
    For each $t \in$ **I** , **do**
        R(t) ← t, ***Ls***(t) ← t , ***C***(t) ← $+\infty$
    label ← 1
    For each $t \in$ ***Seed*** , **do**
        C(t) ← 0 , ***Ls***(t) ← *label* , *label* ← *label* + 1 , ***R***(t) ← t
        For each t_2 in the window region round t, **do**
            S(t_2) ← *White*
        Insert t to ***Q*** , ***S***(t) = *Gray*
        While ***Q*** is not empty , **do**
            Remove s from ***Q*** such that ***C***(s) is minimum
            S(s) ← *Black*
            For each $t_3 \in$ A(s), such that ***S***(t_3) ≠ *Black* and t_3 in the window area , **do**
                Compute c with Equation (2)
                If c < ***C*** (t_3),**then**
                    R(t_3) ← ***R***(s), ***Ls***(t_3) ← ***Ls***(s), ***C***(t_3) ← c
                    If ***S***(t_3) = *Gray* , **then**
                        Update position of t_3 in ***Q***
                    Else
                        Insert t_3 in ***Q*** , ***S***(t_3) ← *Gray*
    Seed , ***M*** ← *RecomputeSeeds* (***Seed*** , ***G***, ***Ls***)
    iter ← *iter* + 1
Return ***Ls***

2.2. Hierarchy superpixel segmentation with striped targets

Striped objects are very common in both human society and nature.These objects are characterized by two or more kinds of similar components,then these parts are regularly arranged on the picture.The previous content has explained why just using adjacent similarity causes stripe problems.And it is worth noting that each component in stripe can always find its similar from its neighbor's neighbor,or in a nearby location.This observation allows us to find seeds that locate at stripe position ,which means these seeds are classified as stripe-pattern.We add a detector to test the similarity of seed nodes and its neighbors.As shown in figure 3,***Nei1*** is the first level neighbour,***Nei2*** and ***Nei3*** are second level neighbour and third level neighbour respectively.If ***Nei1*** is similar with ***Nei3*** and the seed ***t*** is also similar with ***Nei2***,we can simply detect the stripe-pattern.In order to avoid the case where all the neighbor are similar ,we add another condition that ***Nei1*** and ***Nei2*** must be significantly different.

If a seed ***t*** is detected to have stripe-pattern,we add the features of ***Nei1*** and ***Nei2*** to a vector ***SP(t)***.By replace the similarity of adjacent superpixels with the similarity between the superpixel being processed and each stripe components in ***SP(t)***,the stripe problem is solved.Although the alternating parts of the stripes are quite different,they can find similar parts in ***SP***.

Fig. 3 The stripe-pattern detector

Then we propose a new path-cost function used in ISF algorithm

$$f(\pi_s \cdot \langle s,t \rangle) = f(\pi_s) + \left(\alpha \times CrossNei(t,\pi_s) \right)^{\beta} + \| s,t \| \tag{3}$$

$$CrossNei(t,\pi_s) = \begin{cases} \| I(t), M(R(\pi_s)) \| & if \ \ Dist(t,\pi_s) > TH \\ Dist(t,\pi_s) & else \end{cases} \tag{4}$$

$$Dist(t,\pi_s) = \underset{Sp_k \in SP(R(\pi_s))}{Min} \left(\| I(t), I(Sp_k) \| \right) , \tag{5}$$

where ***Min(.)*** is the minimization function, ***SP(s)*** is the superpixel set of ***s***,and ***TH*** is the threshold value.If seeds are detected to have stripe-pattern,this new path-cost function could solve the stripe question by helping algorithm to merge parts of stripe targets and not to affect the merging of other parts of image.Then a multi-scale segmentation can be made by adopting the RISF algorithm based on the finest superpixels in section 2.1. All the processes are shown in Algorithm 3.

Algorithm 2 ISF dealing with stripe

INPUT: Graph ***G***=(**I**,A), number of regions ***n***, the parameters $\alpha \geq 0$ and $\beta \geq 1$,and the ***MaxIters***≥1.
OUTPUT: Superpixel label map ***Ls***.
AUXILIARY: Seeds set ***Seed***,state map ***S***,priority queue ***Q***,cost map ***C***,root map ***R*** , superpixel mean array ***M***, flag map ***F*** and vector ***SP***.

iter ← 0, compute ***Seed*** with the sampling strategy
While *iter* < ***MaxIters*** , do
    For each $t \in$ **I** , **do**
        R(*t*) ← *t*, ***Ls***(*t*) ← *t* , ***C***(*t*) ← + ∞,***S***(*t*) ← *White*,***F***(*t*) ← *false*
    label ← 1
    For each $t \in$ ***Seed*** , **do**

$C(t) \leftarrow 0$, $Ls(t) \leftarrow label$, $label \leftarrow label + 1$, $SP(t) \leftarrow \emptyset$
Insert t to Q , $S(t) = Gray$
For each $t_{nei1} \in A(t)$,**do**
If find t_{nei2}, t_{nei3} in the direction of ray form t to t_{nei1} , **then**
If $\| t, t_{nei1} \| > TH$ and $\| t, t_{nei2} \| \leq TH$ and $\| t_{nei1}, t_{nei3} \| \leq TH$ **then**
Add t and t_{nei1} to $SP(t)$, $F(t) \leftarrow true$
Break For
While Q is not empty , **do**
Remove s from Q such that $C(s)$ is minimum
$S(s) \leftarrow Black$
For each $t \in A(s)$, such that $S(t) \neq Black$,**do**
If $F(R(s)) = true$ **then**
Compute c with Equation (3)
Else
Compute c with Equation (2)
If $c < C(t)$,**then**
$R(t) \leftarrow R(s)$, $Ls(t) \leftarrow Ls(s)$, $C(t) \leftarrow c$
If $S(t) = Gray$, **then**
Update position of t in Q
Else
Insert t in Q , $S(t) \leftarrow Gray$

$Seed$, $M \leftarrow RecomputeSeeds$ ($Seed$, G, Ls)
$iter \leftarrow iter + 1$
Return Ls

Algorithm 3 Hierarchy superpixel segmentation dealing with stripe

INPUT: Graph$G$=(**I**,A), sequence $\{(ISF_1, n_1), (ISF_2, n_2), \ldots, (ISF_k, n_k)\}$, ISF_k is an ISF method((ISF_1 is presented in **Algorithm 1**,the else ISF methods are presented in **Algorithm 2**) ,$n_k > n_{k+1}$ is the seeds number ,and n_k is the number of regions in the coarsest level.

OUTPUT: Hierarchy superpixel segmentation H.

1. $H \leftarrow \emptyset$
2. **For** k =1 to K **do**
3. **If** k =1, **then**
4. $L \leftarrow ISF_k(G, n_k)$
5. **Else**
6. Build the new region adjacency graph A
7. $L \leftarrow ISF_k(G, n_k)$
8. $H \leftarrow H \cup L$
9. Return H

3. Experimental Result

All experiments are running on an i5 cpu,we present the segmentation results of both striped images and common images comparing with state-of-art algorithms.

At finest segmentation level,we generate 1500 superpixels with algorithm 1 and use a constant reduction factor(=1.5) to control the superpixel number in the next RISF iteration.The seed sampling is simple grid sampling and the seed recomputation strategy is choosing the centroid .The parameters are setting by α=0.05,*MaxIters* =10 , β =2 ,*TH*=30 and the window region is a square with side length $3\times r$.Then we get hierarchy segmentation by treating the superpixels generated in previous iteration as new single nodes and re-establish the adjacency relationship in algorithm 3. As shown in figure 4,it is obvious that our algorithm works dealing with strip,the striped targets are fully represented in a coarse superpixel and the background region is properly segmented with accuracy and regularity.The efficient of improved ISF is shown in table 1,it shows that the smaller the window area,the faster the algorithm speed up.If the size of local window is too large,the acceleration will disappear,because it will cause more repeated operations of overlapping area nodes.L must be greater than $2\times r$,otherwise the segmentation result will be nothing but a original grid map.

Table 1.Average processing time for superpixel segmentation in BSDS500

Times (sec)	Improved ISF (L=2×r)	Improved ISF (L=3×r)	Improved ISF (L=4×r)	Improved ISF (L=5×r)	Classical ISF
n=1000	0.625	0.951	1.411	1.589	1.432
n=2000	0.677	1.043	1.379	1.667	1.467
n=3000	0.698	1.055	1.421	1.669	1.519
n=4000	0.711	1.068	1.451	1.672	1.531

We use the Berkeley Segmentation Dataset(BSDS500) to evaluate our algorithm with benchmark of under-segmentation error(UE) and the boundary recall(BR) which are mentioned in[10].Then we use 300 training data to tune the parameters and the rest 200 images to test result .The accuracy of our methods is as well as other state-of-art algorithms(in Figure 5),such as RISF,SH,SLIC and LSC.For each method we quote,the parameters is as described in the original article.The LSC and SH get better performance in experiments results while LSC can not generate hierarchy segmentation and SH is hard to control the regularity of superpixels.

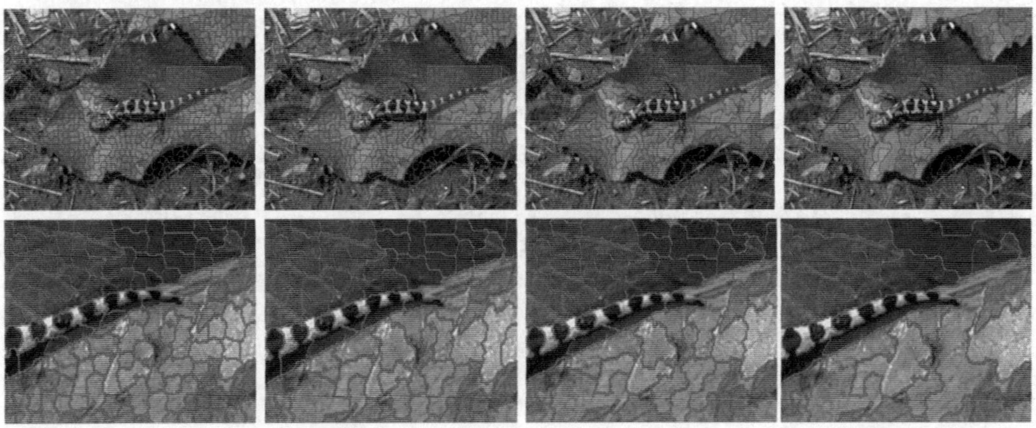

(a) n=900 (b) n=500 (c) n=300 (d) n=100

Fig. 4 The hierarchy segmentation results of our method

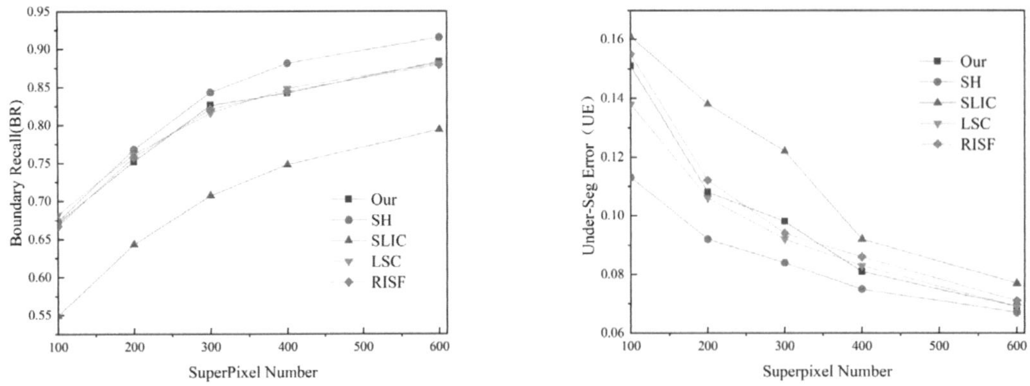

Fig. 5 The Segmentation accuracy evaluation on the BSDS500

4. Conclusion

In this work,we come up with a stripe problem and propose a ISF-based solution.Experiments show that our algorithm is a desirable solution to stripe targets.In order to reduce computation cost,the improved ISF algorithm is more efficient with accurate by adopting the local window strategy.Then, our method preforms as well as other advanced algorithms dealing with common pictures.

Since we use a simple method to detect stripe pattern of seeds, there are more smarter and faster solution worth investigation,like cooperation with deep learning network.The stripe pictures we take in to account are simple(only contain two components and regularly shaped),so dealing with complicated striped images also faces challenges.

References

[1] Ren X, Malik J. Learning a classification model for segmentation. In: Proceedings of the IEEE international conference on computer vision. IEEE; 2003.p. 10–17.

[2] C. L. Zitnick and S. B. Kang, “Stereo for image-based rendering using image over-segmentation,” Int. J. Comput. Vis., vol. 75, no. 1, pp. 49–65,Oct. 2007.

[3] Z. Liu, X. Zhang, S. Luo, and O. Le Meur, “Superpixel-based spatiotemporal saliency detection,IEEE Trans. Circuits Syst. Video Technol.vol. 24, no. 9, pp. 1522–1540, Sep. 2014.

[4] W. Wu, A. Y. Chen, L. Zhao, and J. J. Corso, “Brain tumor detection and segmentation in a CRF (conditional random fields) framework with pixel-pairwise affinity and superpixel-level features,” Int. J. Comput.Assist. Radiol. Surg., vol. 9, no. 2, pp. 241–253, Mar. 2014

[5] R. Giraud, M. Boyer, M. Clément Multi-scale superpatch matching using dual superpixel descriptors.Pattern Recognition Letters, 133 (2020), pp. 129-136

[6] Q. Yan, L. Xu, J. Shi, and J. Jia, “Hierarchical saliency detection,” in Proc. CVPR, Jun. 2013, pp. 1155–1162

[7] J.E. Vargas-Muñoz, A.S. Chowdhury, E.B. Alexandre, F.L. Galvão, P.A.V. Miranda,A.X. Falcão, An iterative spanning forest framework for superpixel segmentation, IEEE Trans. Image Process. (2019)

[8] R. Achanta, A. Shaji, K. Smith, A. Lucchi, P. Fua, and S. Süsstrunk,“SLIC superpixels compared to state-of-the-art superpixel methods,”IEEE Trans. Pattern Anal. Mach. Intell., vol. 34, no. 11, pp. 2274–2282,Nov. 2012.

[9] J. Chen, Z. Li, and B. Huang, “Linear spectral clustering superpixel,”IEEE Trans. Image Process., vol. 26, no. 7, pp. 3317–3330, Jul. 2017

[10] X. Wei, Q. Yang, Y. Gong, N. Ahuja, M.-H. Yang, Superpixel hierarchy, IEEE Trans. Image Process. 27 (10) (2018) 4838–4849

[11] Felipe Lemes Galvão, Silvio Jamil Ferzoli Guimarães, Alexandre Xavier Falcão,Image segmentation using dense and sparse hierarchies of superpixels,Pattern Recognition,Volume 108,2020

[12] Y. -J. Gong and Y. Zhou, "Differential Evolutionary Superpixel Segmentation," in IEEE Transactions on Image Processing, vol. 27, no. 3, pp. 1390-1404, March 2018
[13] Arbelaez, M. Maire, C. Fowlkes, and J. Malik, "Contour detection and hierarchical image segmentation," IEEE Trans. Pattern Anal. Mach.Intell., vol. 33, no. 5, pp. 898–916, May 2011.
[14] P. Arbeláez, J. Pont-Tuset, J. Barron, F. Marques, and J. Malik, "Multiscale combinatorial grouping," in Proc. CVPR, Jun. 2014, pp. 328–335.
[15] C. Gu, J. J. Lim, P. Arbelaez, and J. Malik, "Recognition using regions,"in Proc. CVPR, Jun. 2009, pp. 1030–1037.
[16] R. Girshick, J. Donahue, T. Darrell, and J. Malik, "Rich feature hierarchies for accurate object detection and semantic segmentation,"in Proc. CVPR, Jun. 2014, pp. 580–587.
[17] F.L. Galvão, A.X. Falcão, A.S. Chowdhury, RISF: Recursive iterative spanning forest for superpixel segmentation, in: Graphics, Patterns and Images (SIBGRAPI), 2018 31st SIBGRAPI Conference on,IEEE, 2018,pp. 408–415
[18] A. X. Falcão, J. Stolfi, and R. D. A. Lotufo, "The image foresting transform: Theory, algorithms, and applications," IEEE Trans. Pattern Anal. Mach. Intell., vol. 26, no. 1, pp. 19–29, Jan. 2004

GA-based fuzzy neural network control for ROV

Ziling Lu[1], Shaojie Xin[1*]

[1]Department of Mechanical Engineering,Shanghai Dianji University, Shanghai 201306, China

*Corresponding author's e-mail: luziling640@163.com

Abstract. Aiming at the problem of uncertainty and static instability of underwater robot system, a motion control algorithm of underwater robot based on fuzzy neural network optimized by genetic algorithm is proposed in this paper. The algorithm makes use of the characteristics of neural network which can predict and approximate the target value well and the complementarity of fuzzy function to neural network, and uses genetic algorithm to optimize the parameters of fuzzy neural network membership function, so as to reduce the computation of neural network. The experimental results show that the controller has good robustness and enhances the stability of the control system.

1. Introduction

The control of Remotely operated vehicles (ROV) is one of the most urgent problems to be solved. Therefore, many control methods are used in ROV control systems. Literature [1] designs a small ROV controller based on fuzzy PID. Literature [2] proposed the direct adaptive control method of underwater robot based on generalized dynamic fuzzy neural network, and the direct adaptive controller of underwater robot was designed based on this method. Literature [3] modeled the industrial ROV as a prototype, and then linearized it and used it for linear quadratic regulator (LQR) development. In reference [4], cascade PID controller is designed. Literature [5] proposed a new control algorithm (PID-ESDOBC) based on proportional integral differential and extended state state observer, and designed a PID-ESDOBC controller. In literature [6], a BP neural network S-plane controller was designed. The model simulation of micro intelligent underwater robot is used to verify that the control method has the characteristics of small overshoot and steady-state error, and can quickly recover stability under interference conditions. However, the control effect is not particularly ideal because of the lag of BP neural network learning.

In view of the lag of BP neural network learning in literature [6], this paper improves the method of S-plane controller of BP neural network. Considering that the selection of adaptive weight of BP neural network has a great influence on its fitting, the parameters of membership function of fuzzy neural network (FNN) are optimized by using the complementarity of fuzzy function and neural network, and genetic algorithm is used to accelerate the convergence of FNN and reduce the computation of neural network. The feasibility of the control method is verified by simulation analysis of ROV.

2. Dynamic modeling of ROV

In this paper, the 6-DOF spatial dynamic equation of ROV is given according to reference [7]. The 6-DOF dynamic equation of ROV in the carrier coordinate system is:

$$M(\dot{v})+C(v)v+D(v)v+g(\eta)+\tau_d=\tau \tag{1}$$

$$\dot{\eta}=J(\eta)v \tag{2}$$

where M is the mass inertia matrix, $C(v)$ is the matrix of coriolis force and centripetal force terms, $D(v)$ is the hydrodynamic damping matrix, $g(\eta)$ is the gravity and buoyancy vector, τ_d is the vector of external disturbance force and torque, τ represents the control vector acting in the vector coordinate system of the underwater vehicle.Transform equation (1) into a mathematical model in the inertial coordinate system:

$$M_\eta(\eta)\ddot{\eta}+C_\eta(v,\eta)\dot{\eta}+D_\eta(v,\eta)\dot{\eta}+g_\eta(\eta)+\tau_{d\eta}=J^{-T}(\eta)\tau \tag{3}$$

3. Design of control algorithm

3.1. Fuzzy neural network control algorithm

The FNN proposed in this paper is actually a fuzzy system with double inputs and double outputs, whose inputs are position error e and velocity error $\dot{e}$ respectively, and whose outputs are the forces or torques required to act on each degree of freedom of ROV.

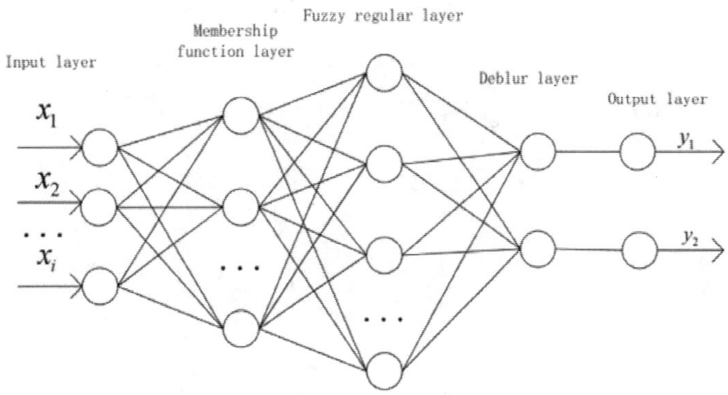

Figure 1. Structure of fuzzy neural network

There are five layers. The first layer is the input layer. This paper actually has two inputs, namely $x_1=e$ and $x_2=\dot{e}$. The output of the first layer is denoted as $f^{(1)}(x)=x_i$, and the superscript indicates the number of layers. The second layer is the membership function layer. The output of the second layer is:

$$f^{(2)}(i,j)=\exp[-\frac{(x_i-a_{ij})^2}{b_{ij}{}^2}] \tag{4}$$

where a_{ij}, b_{ij} are the center and width parameters of the membership function respectively.The third layer is the fuzzy regular layer, which uses AND operation. The inference operation is as follows:

$$f^{(3)}(i,j)=f^{(2)}(1,j)\bullet f^{(2)}(2,j) \tag{5}$$

The fourth layer is the defuzzification layer. This paper uses the weighted average method to solve the fuzzy:

$$f^{(4)} = u_k = \frac{\sum_{i,j=1}^{n} f^{(3)}(i,j) \bullet \omega_{ij}^{(3)}}{\sum_{i,j=1}^{n} f^{(3)}(i,j)} \tag{6}$$

where $i = 1,2$, $j = 1,2\ldots,n$, $\omega_{ij}^{(3)}$ indicates the connection weight coefficient between layer 3 and layer 4 in the subnet.In this paper, n is 13.

The fifth layer is the output layer of FNN, which outputs the forces or torques required by each degree of freedom acting on the ROV

$$y_k = f^{(5)} = \sum_{i=1}^{2} \omega^{(4)} \bullet u_k \tag{7}$$

where $\omega^{(4)}$ represents the weight coefficient between the fourth and fifth layers, u_k is the network output of the fourth layer.

3.2. Genetic optimization algorithm

The parameters to be optimized by GA are center position a_{ij} and width b_{ij} of gaussian function of the second layer in fuzzy neural network.The steps of genetic algorithm to optimize FNN are shown in Figure 2:

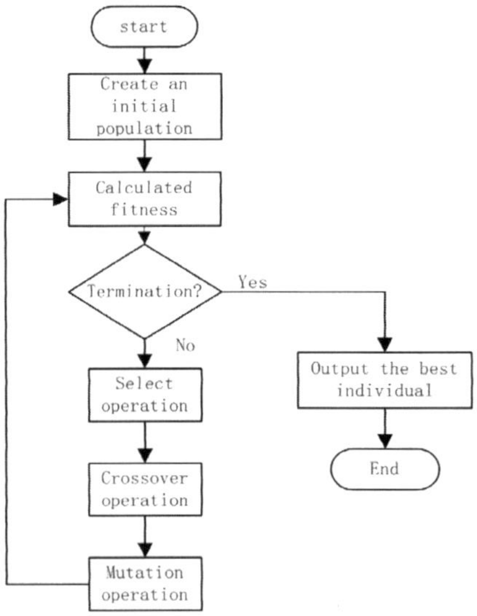

Figure 2. Genetic algorithm optimization steps diagram

Step 1: Decide on the encoding strategy and determine the size of the initial population.A total of 53 parameters are optimized.Due to the symmetry of membership function, the optimization parameters are reduced to 13.Floating-point encoding is adopted.The initial population number is 100.

Step 2: Calculate fitness.A discrete objective function is used to evaluate the performance of the control system, and the smaller the index, the better the performance of the control system

$$ITAE = \sum_{i=1}^{k} T^2 |e_i| \tag{8}$$

where e_i is the deviation value at the ith moment, k is the number of computer simulations.

Step 3: Select operation.The method of tournament selection is adopted, which selects some individuals from the parent population each time, compares the fitness values among individuals, and selects the individuals with the highest fitness values to enter the offspring group.

Step 4: Crossover operation.Global arithmetic crossing based on floating-point encoding is adopted.

Step 5: Mutation operation.The method used is single-point mutation. P_m is generally between 0.001 and 0.1,This paper takes $P_m = 0.05$.

Step 6: Terminates the algebra T to determine whether the operation terminates.

4. System controller design

Assume that the desired trajectory of the ROV is bounded

$$\left\| [\eta_d(t) \quad \dot{\eta}_d(t) \quad \ddot{\eta}_d(t)]^T \right\| \le \eta_B \tag{9}$$

where $\eta_d(t)$ is the expected position vector in the static coordinate system, $\dot{\eta}_d(t)$ is the expected velocity vector in the static coordinate system, $\ddot{\eta}_d(t)$ is the expected acceleration vector in the static coordinate system, η_B is a normal number.

Define ROV tracking error e and filtered tracking error $r(t)$:

$$e = \eta_d - \eta \tag{10}$$

$$r = \dot{e} + \Lambda e \tag{11}$$

where $\Lambda > 0$ is the positive definite parameter matrix.The error dynamics equation of ROV system is summarized as follows:

$$M\dot{r} = -Cr + f(x) + \tau_d - \tau \tag{12}$$

The unknown nonlinear function in ROV dynamics is:

$$f(x) = M(\eta)(\ddot{\eta}_d + \Lambda e) + C(\dot{\eta})(\dot{\eta}_d + \Lambda e) + D(\dot{\eta})\dot{\eta} + g(\eta) \tag{13}$$

Define $x = [e^T, \dot{e}^T, \eta_d, \dot{\eta}_d, \ddot{\eta}_d]$.On this basis, a general approximate controller is proposed:

$$\tau = \hat{f} + k_v r - v(t) \tag{14}$$

where $\hat{f}$ is an estimate of $f(x)$, $v(t)$ is the auxiliary signal to maintain robustness, The multi-loop control structure is shown in Figure 3.ROV system closed-loop error dynamic control equation is:

$$M\dot{r} = -Cr - k_v r + \tilde{f} + \tau_d + v(t) \tag{15}$$

where $\tilde{f}$ is the approximation error of function $f(x)$ .Using GA-FNN for online learning approximation of $f(x)$, then

$$f(x) = W^T \phi(x) + \varepsilon \tag{16}$$

where ε is the function approximation error, $\|\varepsilon\| < \varepsilon_N$, ε_N is a normal number. W is the ideal FNN weight number.Design input control quantity is:

$$\tau = \hat{W}^T \phi(x) + k_v r - v \tag{17}$$

Define the weight deviation as $\widetilde{W} = W - \hat{W}$.The closed loop error dynamics equation is:

$$M\dot{r} = -(C + k_v)r + \widetilde{W}^T \phi(x) + (\varepsilon + \tau_d) + v(t) \tag{18}$$

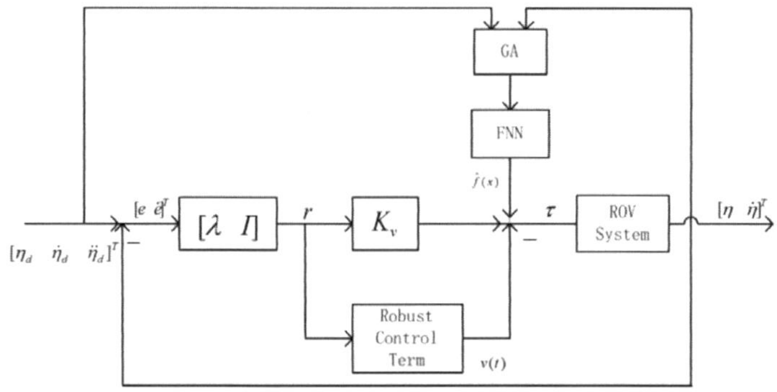

Figure 3. ROV control schematic diagram based on GA-FNN

5. Simulation experiment results and analysis

This paper analyzes the simulation experiment of small ROV in laboratory.As shown in Figure 4, its propulsion system adopts four thrusters, two of which are arranged in parallel on both sides of the ROV.The ROV is designed with four degrees of freedom, namely forward and backward/diving and floating/yaw/roll.The reference input is set to 5m longitudinally, and the initial state of the system is 0.The vertical direction is 5m, and the initial state of the system is 1m.Yaw angle is 45°, and the initial state of the system is 0.The initial roll angle is 0.2rad, and the expected roll angle is 0rad.The adaptive weight of NN is 0.7 and the simulation step is 0.5s.

Figure 4. Small ROV

The following is the control simulation curve of the controller optimized by GA on the four degrees of freedom of ROV:

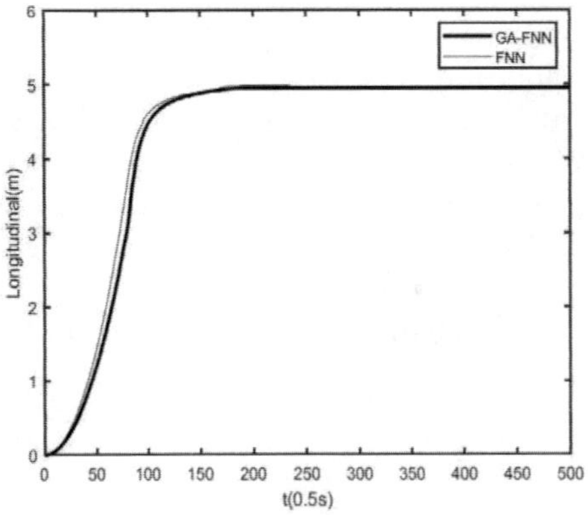

Figure 5. Longitudinal tracking trajectory curve

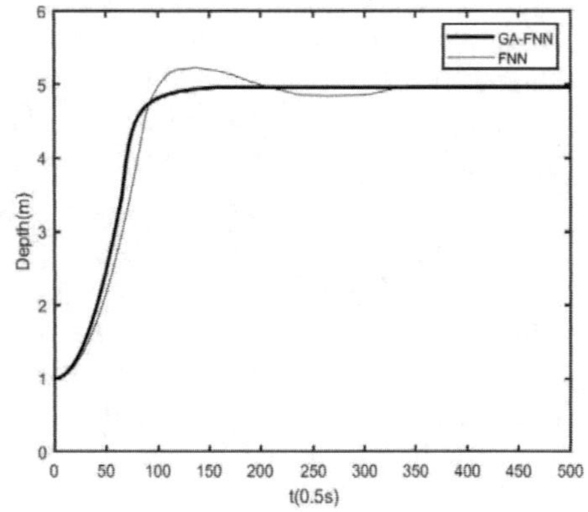

Figure 6. Depth tracking trajectory curve

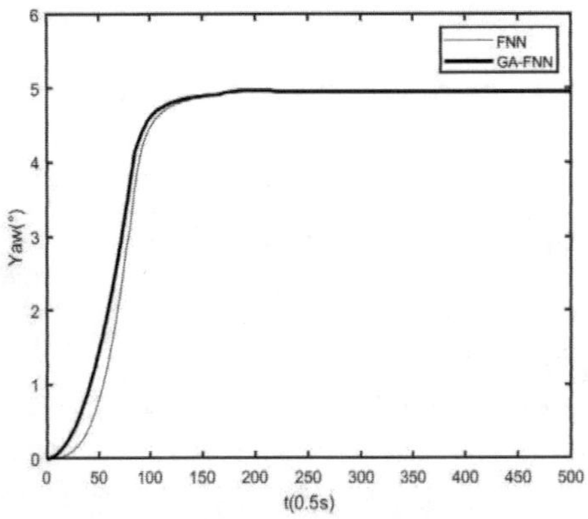

Figure 7. Yaw tracking trajectory curve

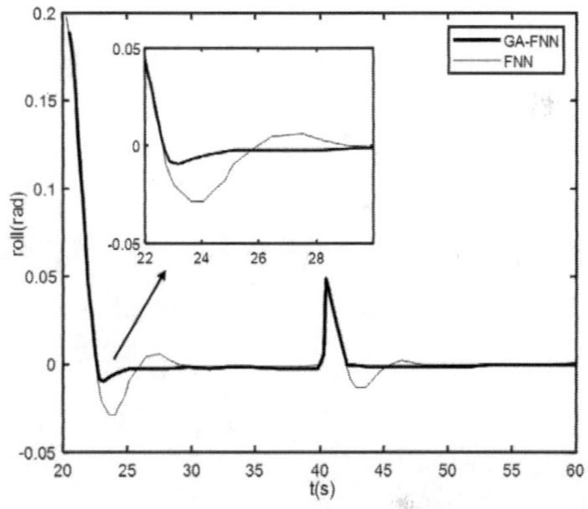

Figure 8. Roll tracking trajectory curve

From the tracking trajectory of the four degrees of freedom, it can be seen that the simple FNN has a certain overshoot in the longitudinal and depth control. After the GA optimization, the convergence speed of the FNN becomes faster, and the control effect is improved to a certain extent. And in the rolling motion, adding external disturbance, Figure 8 shows that the ROV can always be stable quickly in the presence of water flow disturbance.

6. Conclusion

This paper proposes a control algorithm based on FNN for ROV trajectory control. The algorithm uses GA to optimize the parameters of the Gaussian function of FNN to accelerate the convergence speed of FNN. A small ROV is taken as an example to verify the proposed algorithm. Simulation results show that under GA-FNN control, ROV has a good tracking effect of four degrees of freedom, which can accelerate the response speed of the system and improve the overall robustness and stability of the system.

References

[1] Qi, S.B., Yin, B.A., Su, Z.K. (2020) Simulation of depth determination motion control for small ROV based on fuzzy PID. J. Modern Electronic Technology., 43(02): 20-23+28
[2] Yu, J.C., Zhang, A.Q., Wang, X.H.,Su, L.J. (2007) Direct adaptive control of underwater robot based on fuzzy neural network. J. Journal of Automatica Sinica., (08): 840-846

[3] Mai, C., Pedersen, S., Hansen, L. (2017) Modeling and Control of Industrial ROV's for Semi-Autonomous Subsea Maintenance Services. J. IFAC-PapersOnLine., 50(1): 13686-13691.
[4] Wang, J.H.,Song, Y., Wei, G.L., Yuan, B. (2017) Application of cascade PID control in pitch control system of underwater robot. J. Journal of University of Shanghai for Science and Technology., 39(03): 229-235
[5] Gong, Y.H.,Li, L., Qi, S.B. (2021) Enhanced disturbance observer-based robust yaw servo control for ROVs with multi-vector propulsion. J. Industrial Robot: An International Journal., 48(3): 366-377.
[6] Wan, L., Tang, W.Z., Li, Y.M. (2019) S-plane control of intelligent underwater robot by BP neural network. J. Industrial Instrumentation & Automation., (02): 13-17
[7] Yuh, J. (1990) Modeling and control of underwater robotic vehicles. J. Systems Man &Cybernetics IEEE Transactions on., 20(06): 1475-1483

Overview of Crow Search Algorithm

Tingting Wang, Chengming Zhang, Ankun He and Wei Dong*

College of Information Science and Engineering, Shandong Agricultural University, Taian, Shandong, 271000, China

Wei Dong*: dongwei@sdau.edu.cn

Abstract-Crow Search Algorithm (CSA) is a swarm intelligence optimization resolving schemes. Its optimization method simulates the behavior of crows tracking each other. The paper elaborates the basic principle of the algorithm, discusses the main parameters affecting the algorithm performance, and analyzes the search mechanism of the algorithm. The improvement strategies of the algorithm are summarized. Examples are given to illustrate the application of the algorithm in 0-1 knapsack problem, Image processing, Scheduling issues, feature selection and parameter optimization. Finally, according to the research and application status of crow search algorithm, the future research and development direction are prospected.

1. Introduction

The optimization problem reflects people's pursuit of the optimal solution. The traditional optimization methods are complex and only suitable for small-scale problems, such as linear programming, dynamic programming, integer programming and branch and bound in operational research. With the optimization problems proposed in the fields of science, engineering and economy becomes more complex, it is more difficult to establish accurate mathematical models. Some problems have large dimension of variables, high order, many objective functions, and complex constraints. Even if the math models are established, they are difficult to solve. The traditional optimization algorithm faces great challenges.

In recent years, intelligent optimization algorithms have been rapidly developed and widely used. With the discovery of many optimization mechanisms in animals, plants and natural phenomena, hundreds of intelligent optimization algorithms have been developed. For example, algorithms simulating human intelligence include neural network algorithm, immune algorithm, artificial metabolism algorithm, endocrine algorithm, etc. The algorithms simulating the evolution process include genetic algorithm, differential evolution algorithm and DNA calculation, etc. Natural simulation optimization algorithms include simulated annealing algorithm, chaotic optimization algorithm, intelligent water droplet optimization algorithm, cosmic explosion algorithm, etc. The algorithms of simulating plant growth include artificial algae algorithm, tree root optimization algorithm, flower pollination optimization algorithm, etc. Swarm intelligence optimization algorithm, such as ant colony algorithm, particle swarm optimization algorithm, artificial bee colony algorithm, artificial fish swarm algorithm, wolf swarm algorithm, grey wolf optimization algorithm, cuckoo optimization algorithm, firefly algorithm. The crow search algorithm (CSA) studied in this paper is a swarm intelligence optimization algorithm proposed by Alireza Askarzadeh in 2016. It simulates the social behavior of crow groups to track each other and steal food, so as to simulate the optimization process.

This paper introduces the crow search algorithm, analyzes the search mechanism of the algorithm and the main parameters affecting the performance, and summarizes the improvement and application of the algorithm. Finally, the development prospect of the algorithm is prospected.

2. Crow Search Algorithm (CSA)

2.1 CSA Basic Description

Crows are considered to be the smartest birds. They hide food and take it back when needed. They follow each other, observe and track other crows' food storage sites, and steal food once the owner leaves. When crows find themselves being tracked, they will move the hiding place to guard against theft. Based on the behavioral habits of crows, the algorithm has four basic principles:

- Crows live in groups.
- Crows can remember the location of hiding food.
- Crows follow each other to steal food.
- Crows have perceptual ability. When they perceive themselves being tracked, they will change their food hiding place to prevent theft.

Based on the above principle, suppose there is a group of crows in the d-dimensional search space, the number is *N*, each crow represents a feasible solution to the problem, *d* represents the number of decision variables. Let $x^{i,iter} = (x_1^{i,iter}, x_2^{i,iter}, \ldots, x_d^{i,iter})$denote the position of crow *i* obtained at the iteration numbered *iter*. The update formula of crow position can be expressed as follow:

$$x^{i,iter+1} = \begin{cases} x^{i,iter} + r_i \times fl^{i,iter} \times (m^{j,iter} - x^{i,iter}) & ,r_j \geq AP^{j,iter} \\ random\, position, & r_j < AP^{j,iter} \end{cases} \tag{1}$$

Where $fl^{i,iter}$ represents the flight length of crow *i* in the iteration numbered *iter*, and $AP^{j,iter}$represents the awareness probability of the crow *j* in the iteration numbered *iter*, $m^{j,iter}$ is the best location for the crow *j* to hide food after the iteration numbered *iter*, r_i is a random number subject to uniform distribution in [0,1]. The lower the $AP^{j,iter}$ is, the probability that crow *j* perceives to be tracked is smaller. Therefore, when $r_j \geq AP^{j,iter}$, crow *i* tracks crow *j* to a new hiding-food location according to Formula 1. When $r_j < AP^{j,iter}$, crow *j* feels to be tracked by followers and will fly to a non-hiding food location. The position of *j* loses the tracking value, so a new position is randomly generated for crow *i*. At the end of each iteration, the matrix for storing the hiding food locations is updated as follow:

$$m^{i,iter+1} = \begin{cases} x^{i,iter+1}, & f(x^{i,iter+1}) \text{ is better than} f(m^{i,iter}) \\ m^{i,iter}, & O.W. \end{cases} \tag{2}$$

Where $f(\cdot)$ is the objective function, $m^{i,iter} = (m_1^{i,iter}, m_2^{i,iter}, \ldots, m_d^{i,iter})$ stores the crow *i*'s optimal storage position after the iterations numbered *iter*.

2.2 CSA Pseudo-code

In the d-dimensional search space, *Crows* is the initial location matrix, whose values are generated randomly. *Memory* is the matrix for hiding-food location, $iter_{max}$ represents the maximum number of iterations.

$$Crows = \begin{bmatrix} x_1^1 & x_2^1 & \ldots & x_d^1 \\ x_2^2 & x_2^2 & \ldots & x_d^2 \\ \vdots & \vdots & \vdots & \vdots \\ x_1^N & x_2^N & \ldots & x_d^N \end{bmatrix}, \quad Memory = \begin{bmatrix} m_1^1 & m_2^1 & \ldots & m_d^1 \\ m_2^2 & m_2^2 & \ldots & m_d^2 \\ \vdots & \vdots & \vdots & \vdots \\ m_1^N & m_2^N & \ldots & m_d^N \end{bmatrix}$$

while $iter<iter_{max}$

for i=1: N

randomly choose a crow *j*

define Awareness Probability $AP^{j,iter}$

if $r_j \geq AP^{j,iter}$
$x^{i,iter+1} = x^{i,iter} + r_i \times fl^{i,iter} \times (m^{j,iter} - x^{i,iter})$
else
$x^{i,iter+1} = random\ position$
end if
end for
calculate fitness function value,
evaluate crow ' s new position,
update *Memory* according to formula 2.
end while

2.3 CSA Optimization Analysis
The crow search algorithm has three key parameters, the r_i is a random number evenly distributed between 0 and 1, which is used to debug the randomness of each iteration. $fl^{i,iter}$ is the flight length, it' s value is different in each iteration and determines the relative position between *i* and its follower *j*. The randomness of $fl^{i,iter}$ diversifies the randomness of relative positions between crows, and its length affects both global and local strategies.

$AP^{j,iter}$ is the perceptual probability of crow *j* at the time numbered *iter*, reflects its sensitivity to the follower. The condition $r_j \geq AP^{j,iter}$ indicates that crow *j* does not found follower, and crow *i* can follow *j* to arrive the new location. This process reflects the convergence. On the contrary, $r_j < AP^{j,iter}$ indicates that crow *j* perceives to be followed, and crow *i* abandons following and selects a new location randomly. This process focuses on diversity. The algorithm balances the local and global search strategies of the algorithm by comparing the sizes of the two parameters.

2.4 CSA test results
Four test functions are given to test CSA. The algorithm is implemented by python. Let N = 100, $iter_{max}$ = 1000, $fl = 2\exp(-(iter/iter_{max})^2\pi)$, *AP* and *r* are all random numbers between (0,1). Table 1 shows the test functions. Fig.1 to 4 are the image and convergence rate of the corresponding functions.

Tabel 1.Benchmark Function

Name	Function	Range	Avg	Std	Fmin
Sphere	$f_1(x) = \sum_{i=1}^{n} x_i^2$	[-100,100]	2.623e-145	2.064e-289	0
Rosenbrock	$f_2(x) = \sum_{i=1}^{n} (100(x_{i+1} - x_i^2)^2 + (x_i - 1)^2))$	[-20,20]	1.327e-16	8.801e-32	0
Ackley	$f_3(x) = -20\,exp\left(-0.2\left(\frac{1}{n}\sum_{i=1}^{n} x_i^2\right)^{\frac{1}{2}}\right) - exp\left(\frac{1}{n}\sum_{i=1}^{n} cos(2\pi x_i)\right) + 20 + e$	[-20,20]	-2.069e-17	1.067e-07	0
Six-Hump_ Camel-Back	$f_4(x) = 4x_1^2 - 2.1x_1^4 + \frac{1}{3}x_1^6 + x_1x_2 - 4x_2^2 + 4x_2^4$	[-5,5]	-1.0316284	0	-1.03163

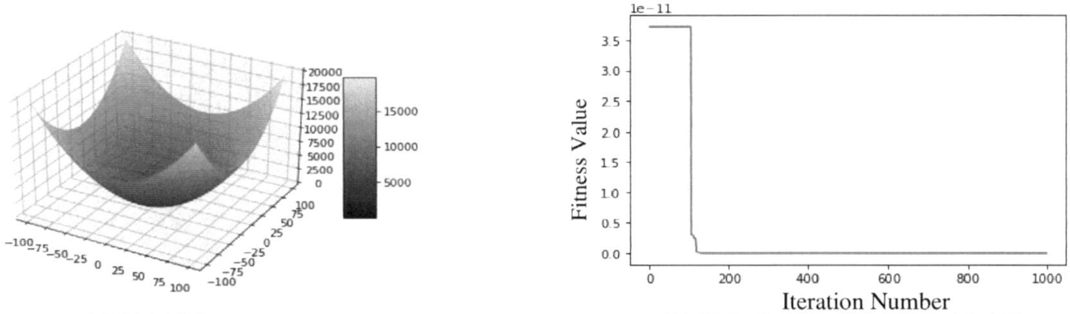

(a)$f_1(d=2)$ Image (b)CSA Convergence Rate(d=10)

Fig.1 Function 1 Image and Convergence Rate

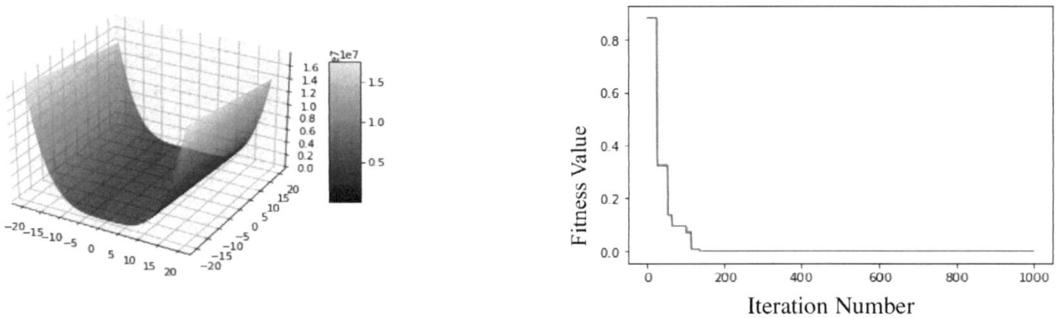

(a)$f_2(d=2)$ Image (b)CSA Convergence Rate(d=10)

Fig.2 Function 2 Image and Convergence Rate

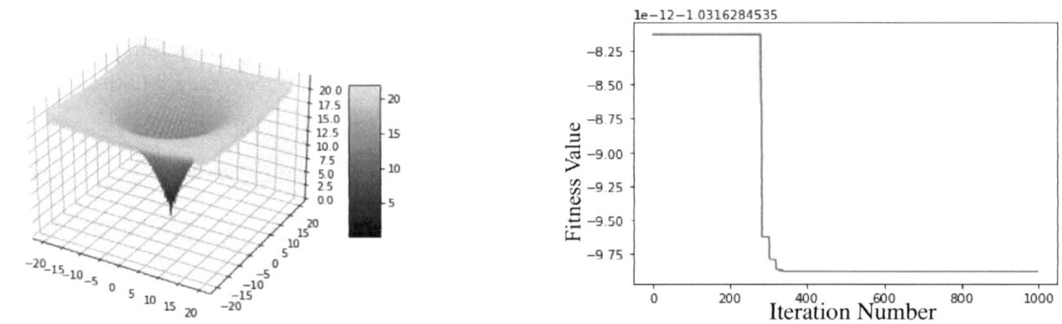

(a)$f_3(d=2)$ Image (b)CSA Convergence Rate(d=10)

Fig.3 Function 3 Image and Convergence Rate

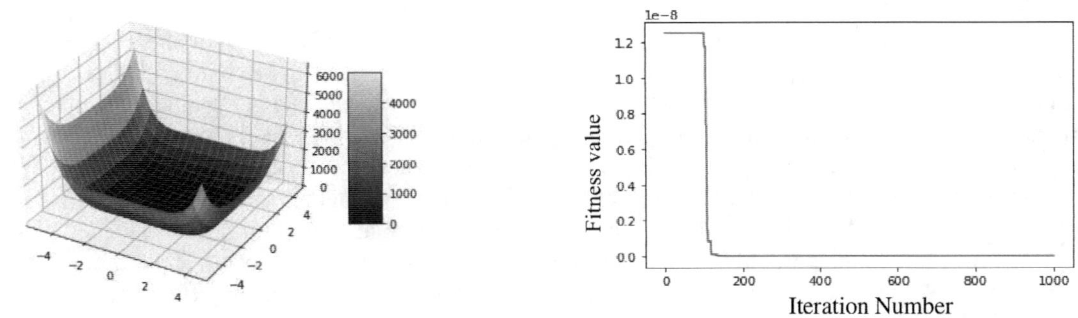

(a)f_4(d=2) Image (b)CSA Convergence Rate(d=10)

Fig.4 Function 4 Image and Convergence Rate

3. CSA Improvement

3.1 Initial group strategy

The distribution of initial solution affects the convergence rate and the solution quality. The original algorithm generates the initial crow position randomly. In order to make the location distribution of the initial positions more uniform and reasonable, the researchers use the following two strategies to improve the initial solution. Reference [4] and [5] adopt reverse search strategy. In the solution space, the solutions which are the opposite values of initial solutions are considered, so the coverage range of the solutions is expanded and the better initial group is obtained. Chaos map can traverse all states without repeating in a certain range according to its own law. Chaos trajectory can avoid the search process falling into local optimum and realize the search space more effectively. Reference [6],[7],[8] all use chaotic mapping to generate the initial solutions. In [6], logistics mapping function is used, circle mapping is used in [7] and chebyshev mapping is used in [8].

3.2 Improvement of flight strategy

Like other intelligent optimization algorithms, we hope that the algorithm tends to global search in the early iteration, which can prevent premature convergence and reduce the probability of local optimization. In the later stage, it tends to local search to reduce the probability of algorithm speed decline caused by slow convergence. Therefore, the focus of algorithm improvement is to better balance global and local search and improve convergence speed.

3.2.1 Optimal Position and Perturbation Strategy

In order to improve the convergence speed of the CSA, many references adopt the optimal crow position guidance strategy.In [9], when the crow's position is updated, the influence of the global optimal crow's position is added while considering the crow's position being tracked. In this way, the leadership of the optimal hiding-food position is increased. At the same time, adding the optimal particle factor in the early stage increases the risk of prematurely falling into local optimum, Therefore, this paper introduces a perturbation mechanism to disturb the global optimal position according to the normal random distribution with adjustable variance, The new position obtained after disturbing is added to the position update as a factor. In [10], cauchy mutation operator is introduced as perturbation strategy. In [11], The inheritance idea is introduced, and the optimal position of the previous generation of crows is considered when updating the position. By introducing the weighted learning factor, the algorithm expands the search domain of the potential food source and improves the population diversity. The weight of inheritance position is controlled by a parameter decreasing with iteration number. When the parameter value is large, the offspring is guaranteed to follow the inheritance position information more. With the decrease of parameter value, the offspring is guaranteed to inherit more the optimal crow's position information brought by the previous generation.

3.2.2 Introducing other flight strategies

Lévy flight is a random search path that follows the Lévy distribution. It is a short-range search and an occasional long-distance walk alternatively. When the solution is sparse and randomly distributed in the solution space, Lévy flight is a very ideal search strategy, which can increase population diversity, expand the search range and help to jump out of local optimum. Reference [12] introduces the variable scale coefficient to adjust the randomness of Lévy flight when the position is updated, and also introduces the optimal position guidance information. While Lévy flight provides global search diversity, the blindness of flight is reduced. Reference [13] Use differential evolution strategy to update the tracking position, and the levy flight is implemented according to a certain probability (literature 0.5 for example). In [6], the chaotic map value generated by using logistics as the mapping function is added to the position update as the weight of flight. In [10], the awareness probability of tracking crow i is introduced. When crow j knows that he is tracked, it is divided into two situations to deal with. That is, crow i knows that j already knows that it is tracked, and crow i does not know that j already knows that it is tracked. If crow i does not identify that it has been found by guide j, the location is randomly updated; If crow i perceives that the guide j has been found to be tracked, it will search for better positions near the current hiding position by Gaussian mutation. This reference assumes that the probability of occurrence of the two cases is 50 %.

3.2.3 Position adjustment after flighting

In [5], Using the idea of genetic algorithm, the positions obtained after flight are crossed and mutated, and the positions before and after calculation are evaluated and compared to obtain the optimal position of this iteration. In [14], the multi-probability walking strategy is used to obtain the new position after the crows flighting, which prevents the solution from falling into local optimum too early and improves the probability of jumping out of local optimum. In [15], sine chord algorithm is applied to the crow position update as a local optimization operator, and the crow position obtained after flight is adjusted again.

3.3 Dynamic progressive processing for algorithm parameters

In [11] and [16], the nonlinear decreasing functions of AP and fl are processed. With the increase of iteration times, the function value shows a curve decrease. The decrease of fl reflects the characteristics of early global search and late local convergence. The decrease of AP reflects that in the early iteration, it tends to randomly generate positions in the solution space to ensure diversity, and in the late iteration, it tends to follow other solutions to imitate the convergence process. The figure 5 and 6 are the progressive curve of the two parameters in [16].

$$fl = \frac{1}{2}(fl_{max} + fl_{min}) \exp\left(-\frac{iter^2}{iter_{max}^2}\pi + fl_{min}\right) \tag{3}$$

$$AP = AP_{max} - \frac{1}{2}(AP_{max} + AP_{min})\left(\frac{2iter}{iter_{max}} - \frac{iter^2}{{iter_{max}}^2}\right) \tag{4}$$

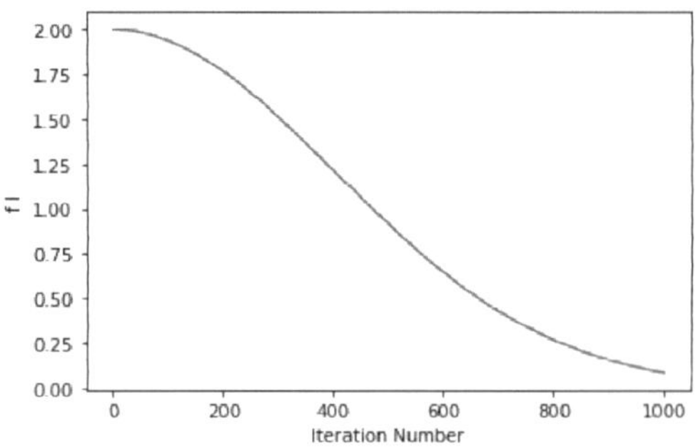

Fig.5 function image of *fl*

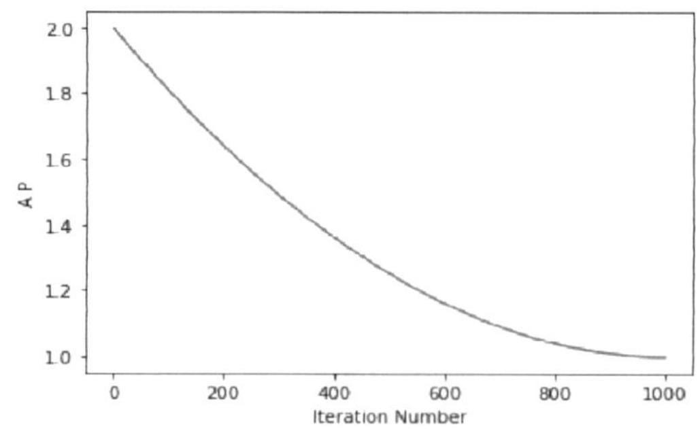

Fig.6 function image of *AP*

4. CSA application

Combined with the research status at home and abroad, the main applications of crow search algorithm are classified as follows.

4.1 0-1 knapsack problem

Reference [7],[8] and [17] fully studied how to effectively solve 0−1 knapsack problem using the CSA. The traditional CSA is used to solve the optimization problem of continuous space. After using hybrid coding strategy to encode discrete data, CSA can also solve the discrete space problem.

4.2 Image processing

In [12] and [18], The improved CSA is effectively applied to multi-threshold image segmentation by introducing levy flight mechanism and Kapur entropy as fitness function. In [19], a new method of automatic neural structure search image classification without manual intervention is proposed by combining CSA with binary network representation. In [20], an effective deep belief network (CSA-DBN) based on CSA is proposed to embed information into medical images.

4.3 Scheduling issues

In the task scheduling of cloud computing, References [5] proposes an improved CSA task scheduling method with the completion time and cost of the task as the optimization goal. In the optimization of power system dispatching, Reference [21] uses the minimum operation cost and the minimum diesel generator treatment as the objective function, and comprehensively considers the constraints such as power balance and pumped storage. Compared with other intelligent algorithms, CSA is more effective to deal with the multi-objective optimization scheduling model of islanded microgrid constructed in this paper. Job-shop scheduling problem (JSSP) is a typical NP problem. Reference [22] adds uncertainty conditions to JSSP problem, adopts process-based coding method and designs repair operator, and introduces chaos mechanism into CSA, which effectively improved the solving speed of the problem.

4.4 Feature selection and Parameter optimization

Reference [5] uses CSA algorithm to select the optimal feature subset, and simultaneously optimizes the parameters of kernel extreme learning machine classifier. The linear weighted objective function designed in this paper not only considers the maximum classification accuracy, but also reduces the false alarm rate and the number of feature subsets as much as possible. Reference [19] introduces reverse strategy learning, the lévy flight strategy and greedy strategy into traditional CSA to solve feature selection problem, which greatly reduces the scale of feature subset and improves the classification accuracy.

In [20], the improved CSA is used to select the feature subset of vulnerability classification and optimize the KNN classification parameters, which reduces the dimension and computational complexity of processing data, and improves the accuracy of classification model. Reference [23] proposes a short-term load interval prediction model based on kernel extreme learning machine (KELM) to support the decision-making of power system. It applies the improved CSA to optimize the weights of KELM output, and prediction quality of the model is improved. The selection of parameters has a crucial impact on the classification accuracy and generalization ability of SVM. Reference [24] uses the classification accuracy as the objective function, and adopts CSA to obtain the optimal parameter combination of SVM. Compared with genetic algorithm and particle swarm algorithm, the CSA achieves higher optimization ability and convergence speed in this paper.

5. Conclusion

The crow search algorithm (CSA) needs few parameters, and its structure is simple and easy to implement. It has been successfully applied to many fields mentioned in this paper, and the practical application effect is good. The CSA has the disadvantages of slow convergence speed and prematurely falling into local optimum like other intelligent optimization algorithms. Although the existing references have improved the performance of CSA from various aspects, there is still room for further improvement. In the current known research, the CSA draws lessons from the particle swarm algorithm, sinusoidal sine chord algorithm and genetic algorithm, and the research on reference and combination to other algorithms is not sufficient. Therefore, further research on algorithm combination can be carried out to obtain further improvement in population diversity, balance global and local search ability and convergence speed.

In the application of the algorithm, the application range of the algorithm is not wide enough, most of which are used to solve deterministic optimization problems. The application of uncertain optimization, such as robust optimization, chance constrained optimization, stochastic optimization, dynamic multi-objective optimization and so on, still needs further research and development.

References

[1]Li,S.,Li,Y.,Lin,Y.M. (2019) Intelligent optimization algorithm and emergence calculation. Tsinghua University Press, Bei Jing.

[2]Liu,Y., Ma,L., Zhang,H.Z., Wei,X.(2019) Intelligent Optimization Algorithm. Shanghai People 's Publishing House, Shang Hai.

[3]Alireza,A.(2016) A novel metaheuristic method for solving constrained engineering optimization problems: Crow search algorithm.Computers and Structures, 169:1-12.

[4]Wang,Y.,Cao,J.,Qiu,Z.(2019)A New Feature Selection Algorithm Based on Crow Search Algorithm.Journal of Jilin University (Science Edition), 57:869-874.

[5]Lin,T.,Hao,Z.X., Feng,J.K.(2020)Research on cloud computing task scheduling based on improved crow search algorithm. Microelectronics and computer,37:20-24.

[6]Ma,C.(2019) Parallel intrusion detection method based on ReliefF and improved crow search optimization.Computer application research, 36:3063-3068.

[7]Liu,X.J.,He,Y.(2018)Chaos crow algorithm based on differential evolution strategy for discounted{0-1} knapsack problem.Computer application, 38:137-145.

[8]Liu,X.J.,He,Y.C.(2018)Chaos binary crow algorithm for solving 0-1 knapsack problem.Computer engineering and application.China, 54:173-179.

[9]Xin,Z.Y., Zhang,D.m.,Chen,Z.Y., Zhang,Z.J., Yan,W.(2020) Multi - stage Perturbation Shared Raven Algorithm.Computer Engineering and Applications, 56:55-61.

[10]Huo,L.,Guo,Y.r.,Tan,Z.j.(2020)Cauchy mutation crow algorithm with adaptive step size.Computer science, 47:218-225.

[11]Zhao,S.J, Gao,L.F., Yu,D.M, Tu,J.(2019)Improved CSA algorithm based on variable factor weighted learning and neighborhood dimension crossover strategy. Acta Electronica Sinica, 47:40-48.

[12]Chang,J.J., Li,D.X., Zhong,X., Du,W.H., Wang,Q.N.(2019)Improved crow algorithm to optimize multi-threshold image segmentation.Packing engineering,42:238-246.

[13]Liu,X.J., He,Y.C.(2018)Differential crow algorithm based on Lévy flight for discounted {0-1} knapsack problem.Computer application,38: 433-442.

[14]Wang,S.X., Wen,H.(2020)Modeling of three-dimensional cargo packing problem and optimization of crow search algorithm.Journal of Hunan University (Natural Science Edition),47:21-30.

[15]Xiao,Z.Y., Liu,S., Han,F.F,, Yu,J.F.(2019)Research on crow search algorithm guided by sine strin. Computer engineering and application,55:52-59.

[16]Fan,Y., Zhang,D.M., Chen,Z.Y., Wang,Y.R, Xu,H, Wang,Q.Q.(2020)Spectrum allocation scheme for vehicular networks based on improved crow algorithm.Computer science, 47:273-278.

[17]Liu,X.J, He Y.C.,Wu,C.C.(2018)Quadratic greedy mutation crow algorithm for knapsack problem of ensemble coalition.Microelectronics and compute,35:13-19.

[18]Kang,L.F., Wu,F.(2021)Multi-level threshold image segmentation method based on crow search optimization algorithm.Journal of Southwest Normal University (Natural Science Edition),46:38-43.
[19]Wang,Y., Cao,J. Qiu,Z.Y.(2019)A new feature selection algorithm based on crow search algorithm.Journal of Jilin University (Science Edition),57: 867-874.
[20]Zhou,J.(2020)Improved crow search algorithm to optimize KNN parameters and feature selection and its application in vulnerability classification.Guangxi University,32:65-79.
[21]Huang,J.G., Chen,B., Lin,X.N.,Wu,W.,Yu,N.,Ye,Y.(2020)Multi-objective optimal scheduling of islanded microgrid based on crow search algorithm.High voltage apparatus, 56:162-168.
[22]Liu,K., Huang,H.X, Zhao,J.(2021)Chaotic crow search algorithm for fuzzy job shop scheduling problem.Sensors and micro systems, 40:110-117.
[23]Yue,Y.J, Liu,Y.H.(2020)Short-term load interval prediction based on pole symmetric mode decomposition-dispersion entropy and improved crow search algorithm-kernel extreme learning machine.Science, Technology and Engineering,22:9036-9042.
[24]Wang,L.T., Zhang,J.X., Zhang,J.H.(2019)The application of crow search algorithm in SVM parameter optimization.Computer engineering and application,55:214-219.

Journal of Physics: Conference Series 2258 (2022) 012018 doi:10.1088/1742-6596/2258/1/012018

Application of UAV path planning based on parameter optimization GA-PSO fusion algorithm

Chaoxiong Lin[1], Xuecong Zhang[2]

([1,2]Department of Mathematics, School of Information Science and Technology, Jinan University, Guangzhou, GuangDong, 511443, China)

[1]First author's e-mail: a1256267454@stu2019.jnu.edu.cn

[2]Corresponding author's e-mail: cong200124@stu2018.jnu.edu.cn

Abstract. Aiming at the complexity of the unmanned aerial vehicle (UAV) path planning problem and the great influence of genetic algorithm parameters on the stability of the results, a fusion algorithm based on parameter optimization is proposed in this paper. In the iterative process, the GA-PSO fusion algorithm uses particle swarm optimization algorithm to search the optimal value of crossover rate and mutation rate in genetic algorithm, which makes the algorithm convergence speed is fast and search ability is strong. In addition, the core part of the algorithm fusion framework is realized by introducing the optimal adaptive value of the population after crossover and mutation as the adaptive value of the parameter particle. Finally, we design two groups of experiments and compare the proposed fusion algorithm with the classical particle swarm optimization algorithm (PSO), ant colony algorithm (ACA), genetic algorithm (GA), artificial fish swarm algorithm (AFSA), Wolf pack algorithm (WPA), artificial bee colony algorithm (ABC) and the improved algorithm through experimental simulation. The experimental results show that: In general, the path planned by the GA-PSO fusion algorithm in this paper is 10% shorter than that planned by other algorithms on average. Simulation results also show that the convergence speed of the fusion algorithm in this paper is faster, and the final search path is smoother.

1. Introduction

UAV path planning is mainly optimized according to the objectives of shortest distance and highest efficiency [1], which has been applied in disaster relief and agricultural pesticide spraying, etc. [2]. At present, the path planning algorithms mainly include A*, artificial potential field method and breadth first search algorithm [3-4]. In recent years, many scholars have applied intelligent optimization algorithms, such as GA, ACA, and PSO [5-7]. However, the traditional intelligent algorithm has obvious disadvantages, such as slow running speed and many parameter Settings. ACA is easy to fall into local optimization and stagnation. Most of the algorithms in previous studies are only applied to the raster geographic model, which is not conducive to the simulation of UAV flying in the air. Reference [8] introduced the elite potential field strategy into the ACA to enhance the leading effect of excellent individuals on the population, accelerate the convergence speed of the algorithm and improve the search efficiency of the algorithm. Reference [9] introduces exponentially decreasing inertia weight factor to accelerate convergence speed and calculation accuracy of FASA algorithm to solve robot path planning of MAKLINK graph. Reference [10] improves selection operator, crossover operator and mutation operator to plan smooth and flyable path.

In intelligent algorithms, parameter setting is an important and difficult problem. Appropriate parameters not only make the solution better, but also make the algorithm convergence faster. There has not been a good theoretical study on how to select appropriate parameters. At present, the commonly used methods to solve the problem of parameter selection are comprehensive test method and orthogonal test method [11-12]. These two methods need to select appropriate parameters in a large parameter space, resulting in low accuracy. There are also many scholars in the academic circle who are engaged in the research on the parameter setting of various intelligent algorithms. Among them, Hu Wenbin [13] systematically studied the influence of the initial population size, crossover rate and mutation rate in genetic algorithm parameters on the convergence and search ability of the algorithm, but only gave the function of parameters from the level of qualitative analysis and made selection. Yan Lijun et al. [14] proposed a hybrid optimization algorithm of nest partition based on sequential optimization, which sampled all feasible Spaces and reduced the best region by repeated partitioning in the "most promising region", but this method was only effective for a small range of parameter regions. Reference [15], powered solution the problem of automatic selection of parameters in genetic optimization methods. Adaptive variants of the classical genetic algorithm and parallel genetic algorithm are developed. In this paper, particle swarm optimization algorithm is used to optimize the parameters in genetic algorithm, and the fitness of the population after a certain iteration in genetic algorithm is used as the adaptive value of the parameter particles to guide the updating direction of the parameter particles, so as to timely select the optimal parameters in the running process of genetic algorithm. In addition, the algorithm framework in this paper can analyze the changes and effects of parameters of other intelligent algorithms in the iterative process, providing scholars with the direction of parameter updating strategy and reducing the uncertainty of parameter qualitative analysis. Finally, this article applies the algorithm to the model of UAV path planning. The comparison of experimental results shows that the algorithm in this paper improves by 10% on average compared with other algorithms.

2. Model building

2.1. Eestablishment of geographic model

The UAV path planning process needs to build a geographical model, a good geographical model can effectively improve the efficiency of path planning. In this paper, the original terrain, obstacle area and other factors are considered to model the mountain environment, and the exponential function is used to describe it. The mathematical model is as follows:

$$Z(x,y) = \sum_{i=1}^{n} h_i \exp\left[-(\frac{x-x_i}{x_{si}})^2 - (\frac{y-y_i}{y_{si}})^2\right] \tag{1}$$

(x_i, y_i) is the central coordinate of the i-th peak; h_i is topographic parameter, Control the height; x_{si} and y_{si} are respectively the attenuation amount and control slope of the i-th peak along the X-axis and Y-axis directions; n is the total number of peaks.

2.2. UAV path planning model

UAV flight path representation: In this paper, a finite and bounded real number sequence is used to represent UAV flight path, such as $\{S, W_1, W_2, \ldots, W_{n-1}, G\}$. $S(x_0, y_0, z_0)$ represents the starting point of UAV. $G(x_n, y_n, z_n)$ represents the destination of UAV. $W_i(x_i, y_i, z_i)$ Represents the point in flight. The sequence is connected into broken lines, and the method of cubic spline interpolation is used for smoothing in order to get close to the actual flight path in real life.

Objective function: This paper takes the planned path length as objective function.

$$min\ fitnesslong = \sum_{i=1}^{n} L_i \tag{2}$$

L_i is the length of the curve $(x_{i-1}, y_{i-1}, z_{i-1})$ through (x_i, y_i, z_i).

Constraint condition 1: Ensure that the planned path can avoid all obstacles, that is, the uav flies above the obstacles.

$$Z_i(x_i, y_i) > Z(x_i, y_i); i = 1,2, \dots, n \quad (3)$$

$Z(x_i, y_i)$ is the altitude corresponding to (x_i, y_i) in the geographic model, and $Z_i(x_i, y_i)$ is the altitude corresponding to (x_i, y_i) in the flight path.

Constraint condition 2: Restrict UAV movement to one area.

$$\begin{cases} 0 < x_i < posBound(x) & posBound(x) = 100 \\ 0 < y_i < posBound(x) & posBound(y) = 100 \\ 0 < z_i < posBound(x) & posBound(z) = 100 \end{cases} \quad (4)$$

$posBound$ is the boundary of the flight. For example, $posBound(x)$ is the maximum selected boundary on the x-coordinate.

3. Algorithm design

3.1. Principles of genetic algorithm and particle swarm optimization

The main idea of GA is based on Darwin's biological evolution and Mendelian genetics. GA is a computational model of natural evolution system and a general adaptive search method to solve optimization problems. The algorithm constructs a fitness function according to the objective function of the problem, evaluates, conducts genetic operation and selects a population composed of multiple solutions, and obtains the individual with the best fitness as the optimal solution of the problem after multiple generations of reproduction.

PSO is a simplified social model, it simulated the behavior of birds and fish in the nature characteristics, according to the flock of group formation, the fish can benefit from a group of other individuals in the process of looking for food detection and past experience, namely the same biological information between social sharing can bring benefits to the individual. The basic principle of the algorithm is: a group of M particles flying in D - dimensional space at a certain speed search. When each particle searches, it considers the historical best position of its own search and the historical best position of other particles in the group, and finally changes its position on this basis.

3.2. Implementation of fusion algorithm based on parameter optimization

Genetic algorithm has the characteristics of potential search parallelism and robustness, easy to combine with other algorithms, algorithm parameters, different parameters have a great influence on the results. Particle swarm optimization (PSO) is easy to implement without gradient information and has the advantages of few parameters and little influence on the results. This paper presents a framework of parameter optimization GA-PSO fusion algorithm based on genetic algorithm and particle swarm optimization algorithm.

(1) Encoding method: This paper needs to encode the path of solution as well as the parameters.

Path coding method: This paper uses real coding method for path population. This paper adopts 5 real points in three-dimensional space to express the path:

$$Hi = \{S(x_0, y_0, z_0) \sim W_1(x_1, y_1, z_1) \sim W_2(x_2, y_2, z_2) \sim W_3(x_3, y_3, z_3) \sim G_2(x_n, y_n, z_n)\} \quad (5)$$

Encoding method of parameters: for parameters such as crossover rate, mutation rate and reversal rate in genetic algorithm, real coding method is used to encode: such as$m_i = \{m_i^1, m_i^2, \dots, m_i^l\}$, $m_i^j, j = 1,2, \dots l$ represents each parameter in the genetic algorithm. In this paper, crossover rate and variation rate are optimized, so the parameter particle can be expressed as $m_i = \{m_i^1, m_i^2\}$, where m_i^1 represents crossover rate and m_i^2 represents variation rate.

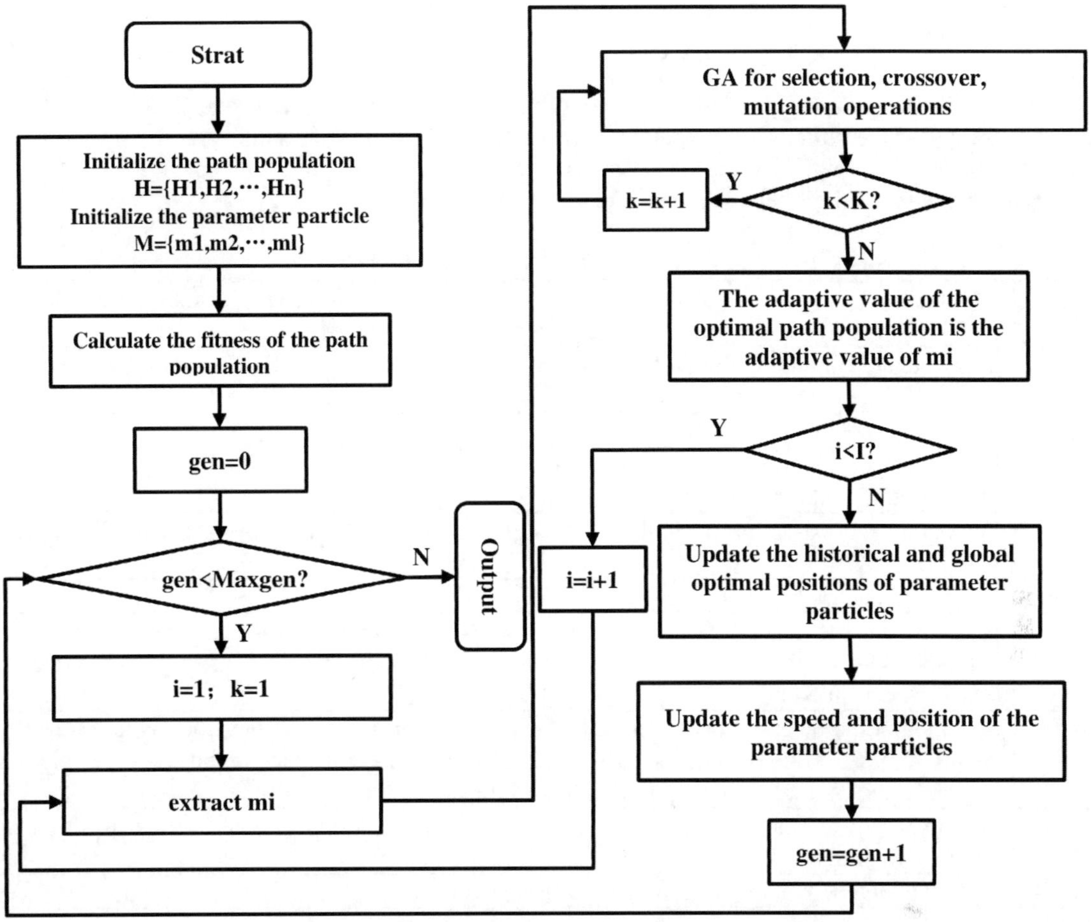

Figure 1. Flow chart of GA-PSO fusion algorithm.

(2) Fitness function: in this paper, path length is used as fitness function of path population, and the curve is divided into small segments of straight lines to calculate the length. An implicit function $F(x,H,K)$ is designed as the adaptive value function of the parameter, which is related to the parameter itself, the path population and the number of iterations K of the path population in the inner cycle, that is, the optimal adaptive value of the population after K iterations under the parameter is the adaptive value of the parameter.

(3) Selection operator: select individuals from the old population to the new population in the way of roulette wheel method. The probability of individual selection is related to the adaptive value. The greater the individual adaptive value, the greater the probability of being selected.

(4) Crossover operator [16] : referring to the single point crossover and uniform crossover operation of binary coding, the corresponding parameter x_i, x_j in two individuals $W_i(x_i, y_i, z_i)$ and $W_j(x_j, y_j, z_j)$ is finally operated into x_i', x_j', y_i, y_j into y_i', y_j', and z_i, z_j into z_i', z_j', which meets the following formula.

$$\begin{cases} x_i + x_j = x_i' + x_j' \\ y_i + y_j = y_i' + y_j' \\ z_i + z_j = z_i' + z_j' \end{cases} \tag{6}$$

Based on the above two points, the crossover method of this paper is as follows:

$$\begin{cases} x_i' = (1-\alpha)\times x_i + \beta x_j & if\ x_i'(x_j') < L\ then\ x_i'(x_j') = L \\ x_j' = (1-\alpha)\times x_j + \beta x_i & if\ x_i'(x_j') > L\ then\ x_i'(x_j') = L \end{cases} \tag{7}$$

Where α, β is the uniformly distributed random number on $(0, r), r \leq 1$, L and R are the left and right boundaries of the optimization parameters respectively, L=0, R=100. Adjusting the size of r can adjust the range of crossover operation. Using such a cross method, fully embodies the process of "information exchange" between two individuals.

Table 1. Steps of GA-PSO fusion algorithm.

Step1	initialization parameter, path population size is n=100, parameter particle size is L=20, K=5~10, Maxgen=100, speed of initialization parameter particle.
Step2	initialize the path population and parameter particles according to the constraints, and use the objective function to calculate the fitness f1 of each path individual.
Step3	set the loop count variable gen=0.
Step4	judge whether the maximum number of iterations is reached, if so, output the results and end the program; Otherwise, skip to Step5.
Step5	Set parameter extraction variable I =1, and parameter particle fitness calculation cyclic variable K =1.
Step6	Extract mi parameter particles.
Step7	perform genetic operation (selection, crossover, mutation) on the path population corresponding to this parameter particle until the end of K cycles, then jump to Step8, otherwise repeat Step7.
Step8	take the optimal adaptive value of path population after K times of genetic operation as the adaptive value of this parameter particle f2, return to Step6 to calculate the adaptive value of all parameter particles, and skip to Step9 until the end.
Step9	Update the historical individual optimal position and historical global optimal position of parameter particles.
Step10	Update the velocity and position of parameter particles.
Step11	gen=gen+1, return to Step4.

(5) Mutation operation [16]: refer to binary coded mutation operation, that is, an individual operation becomes another individual in its neighborhood, so this paper adopts such mutation operation:

$$\begin{cases} x_i' = x_i' + k\times(R - x_i)\times\gamma, if\ rand > 0.5 \\ x_i' = x_i' - k\times(x_i - L)\times\gamma, if\ rand \leq 0.5 \end{cases} \tag{8}$$

γ, k are random numbers between 0 and 1.

(6) Parameter particle speed and position updates

$$v_{iD}^{k+1} = v_{iD}^{k} + c_1\xi(pbest_{iD}^{k} - m_{iD}^{k}) + c_2\eta(gbest_{D}^{k} - m_{iD}^{k}) \tag{9}$$

$$m_{iD}^{k+1} = m_{iD}^{k} + v_{iD}^{k+1} \tag{10}$$

Where $D = 2$ represents the dimension, v_{iD}^{k} represents the speed of the i-th parameter particle in the k-th iteration, c_1, c_2 are learning factors, $\xi, \eta \in [0,1]$ is random number, $pbest_{iD}^{k}$ represents the historical optimal position of the i-th parameter particle in the k-th iteration, and $gbest_{D}^{k}$ represents the historical optimal position of the population of parameter particles in the k-th iteration (adaptive value).

4. The experimental simulation

In order to test the superiority of the proposed algorithm, two groups of experiments were designed to compare the results of the GA-PSO fusion algorithm with those of six traditional algorithms and the algorithm in reference [10] for five times respectively

Simulation experiment 1: the starting point is (1,1,1), the end point is (100,100,80), and the straight-line distance between the two points is 160.7576.

Simulation experiment 2: the starting point is (1,1,1) and the end point is (90,65,5), which is on the surface. The straight-line distance between two points is: 109.695.

Table 2. Comparison the path length of the six traditional algorithms, the improved algorithm [10] and the GA-PSO fusion algorithm in two simulation experimentals.

algorithm	**Simulation experiment 1**			**Simulation experiment 2**		
	Mean path length	Maximum path length	Shortest path length	Mean path length	Maximum path length	Shortest path length
PSO	166.9435	167.2597	165.5763	158.6354	164.3658	153.4933
ACA	190.934	199.5236	179.4763	160.8796	161.7895	150.7426
GA	173.0136	173.4628	171.3425	169.2469	173.4623	166.5397
AFSA	179.4236	182.1079	165.5819	169.4685	170.3954	167.4528
WPA	166.4636	166.3927	165.6789	172.3694	173.6987	164.5896
ABC	165.9965	166.1787	165.5311	187.9463	196.4763	182.5027
Reference [10]	166.8694	168.6517	165.0378	160.9875	164.4986	159.6237
GA-PSO	**165.7742**	**167.0036**	**164.5871**	**147.1937**	**156.4632**	**144.7288**

The path graph and convergence curve planned by the genetic algorithm, the algorithm in reference [10] and the GA-PSO fusion algorithm in this paper are given below.

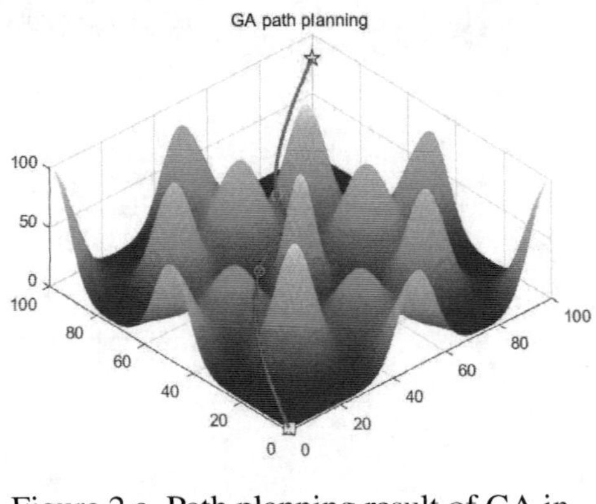

Figure 2.a. Path planning result of GA in simulation experiment 1.

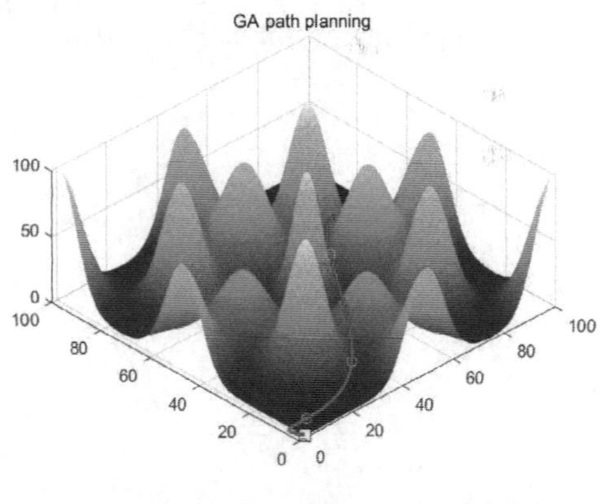

Figure 3.a. Path planning result of GA in simulation experiment 2.

By comparing Figure 2.a, Figure 2.b and Figure 2.c; Figure 3.a, Figure 3.b and Figure 3.c, the path curve planned by the GA-PSO fusion algorithm in this paper is smoother compared with the other two. Compared with the traditional GA, the nodes of sharp turns are reduced, which is more conducive to the flight of UAV. In addition, the path planned by the GA-PSO fusion algorithm is shorter, which saves the flight time and flight cost of UAV. In a word, the GA-PSO fusion algorithm is superior in the performance of results.

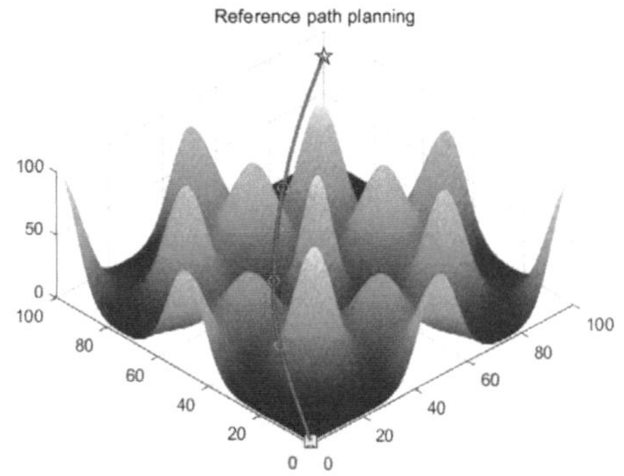

Figure 2.b. Path planning result of reference [10] algorithm in simulation experiment 1.

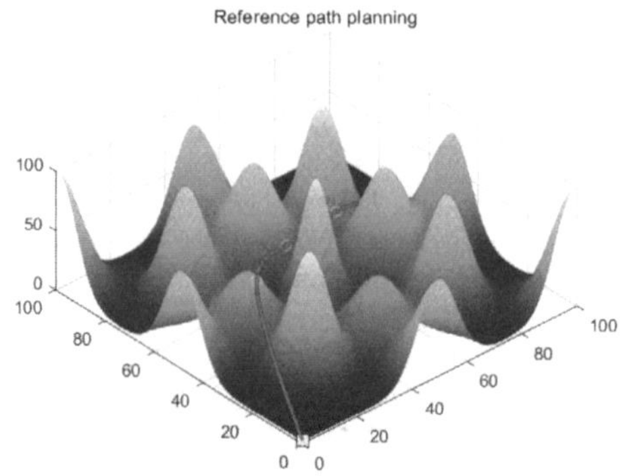

Figure 3.b. Path planning result of reference [10] algorithm in simulation experiment 2.

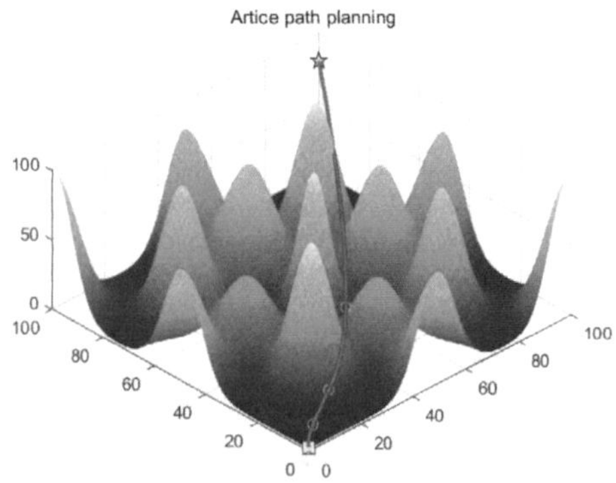

Figure 2.c. Path planning result of GA-PSO fusion algorithm in simulation experiment 1.

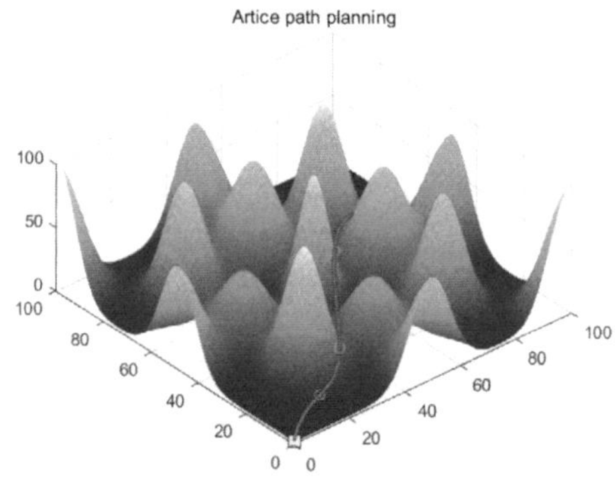

Figure 3.c. Path planning result of GA-PSO fusion algorithm in simulation experiment 2.

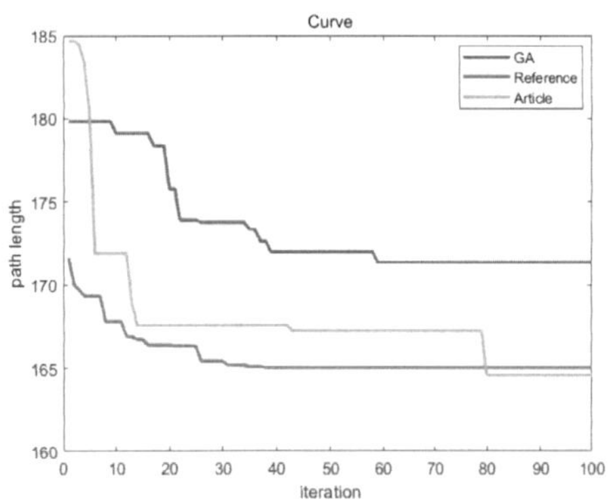

Figure 2.d. Convergence comparison of algorithms

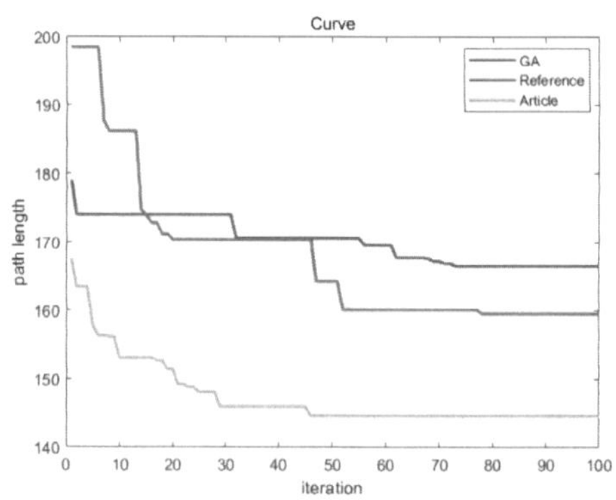

Figure 3.d. Convergence comparison of algorithms

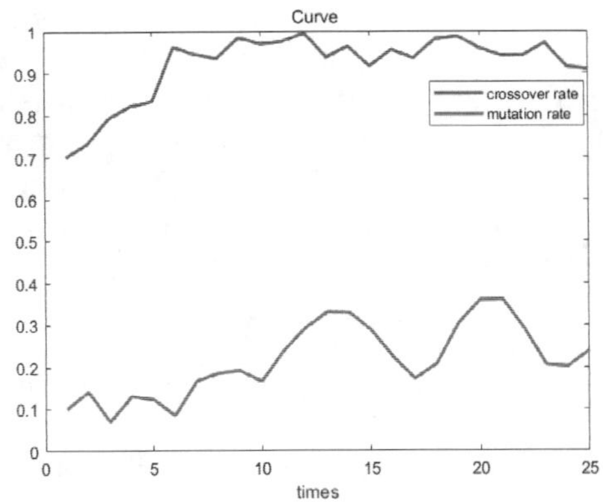

Figure 2.e. Change curve of crossover rate and variation rate in simulation experiment 1.

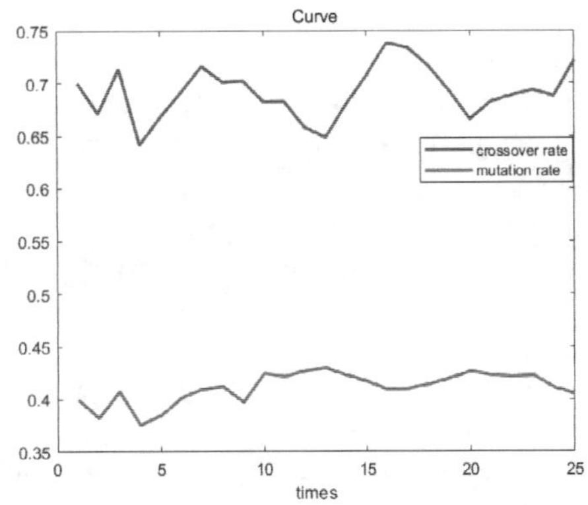

Figure 3.e. Change curve of crossover rate and variation rate in simulation experiment 2.

According to the convergence curve figure2.d, it can be seen that the convergence speed of the traditional GA and the improved algorithm in reference [10] is slow, and it falls into the local optimal solution after 40-60 iterations. By analyzing the cause of trapped in local optimal solution, and we found that after 40 times of iteration, path population has stable state, the population of individual genes have roughly the same. Constant crossover and mutation rates do not significantly change the genes of the population, do not bring good genes to the population, and thus do not improve the final result. the GA-PSO fusion algorithm in this paper the convergence curves of convergence speed is fast, after 20 iterations has reached a relatively optimal results, the iteration is 40 and 80 iterations, the GA-PSO fusion algorithm, breakthrough the limitation of the local optimal solution, to the direction of the global optimal convergence, get an improved algorithm compared with the traditional GA and reference [10] better results. In connection with Figure 2.e, the path population is analyzed. It can be seen that when iteration=40, that is, before times=10; Iteration =80, i.e. before times=20), the crossover rate and mutation rate of the GA-PSO fusion algorithm in this paper are significantly changed. The changed crossover rate and mutation rate make the path population more categories and richer, so that the path population can obtain more excellent genes, break through the local optimal solution, and search for shorter paths. Similarly, the same conclusion can be obtained by analyzing Figure3.d and Figure 3.e.

According to Figure 2.e and Figure 3.e, crossover rate and mutation rate increase with the increase of iteration times in general trend, which exactly verifies the role of crossover rate and mutation rate in the iteration process in reference [13]. As the iteration goes on, the path population falls into local optimum, and the richness of the population decreases. Increasing the crossover rate and mutation rate can enlarge the search solution space and get out of the local optimum solution. These two images also verify that the framework of the GA-PSO fusion algorithm in this paper can analyze the changes and functions of the parameters of the intelligent algorithm in the iterative process.

5. Conclusion

In this paper, a GA-PSO fusion algorithm based on genetic algorithm is proposed, that is, particle swarm optimization is used to optimize the crossover rate and mutation rate of genetic algorithm. The core of the algorithm is: after each iteration, we can choose the parameters that have the greatest effect on the convergence and searching ability of genetic algorithm. In addition, the GA-PSO fusion algorithm in this paper has the significance of universality: for each proposed intelligent algorithm, the parameter optimization framework in this paper is used to observe the changes of algorithm parameters in the iterative process, obtain the best parameters for the algorithm, and clarify the role of the parameters of the intelligent algorithm in the optimization process, so as to provide a strategic idea for the scholars to update the parameters.

UAV route planning to verify the superiority of the algorithm in this paper, first of all, the original terrain and obstacles simulated reality of geographical environment, and then GA-PSO fusion algorithm in this paper, and the remaining six kinds of traditional algorithm and reference [10]. This paper presents two groups of simulation experiment schemes and simulation results. The simulation results show that the path obtained by the fusion algorithm is shorter and the optimal value of parameters can be obtained during algorithm iteration. At the same time, the fusion algorithm in this paper also has the advantages of stronger convergence, better search effect and stronger robustness.

Reference

[1] Zhu, D.Q., Yan, M.Z. (2010) Survey on technology of mobile robot path planning. J. Control and Decision., 25(7): 961-967.

[2] Yang, L.Q., Guo, L., Zhu, J.F., et al. (2017) The development situation and prospect of agricultural UAV in China. J. Journal of Agricultural Mechanization Research., 39(8):6-11.

[3] Zhang, C.F., Wang, C.L. (2018) Service robot path planning based on improved Dijkstra algorithm. In: i-CREATe2018 Proceedings of the 12th International Convention on Rehabilitation Engineering and Assistive Technology. Shanghai: Assistive and Rehabilitative Technologies (START) Centre. 77-80.

[4] Chen, T.B., Zhang, Q.S. (2013) Robot motion planning based on improved artificial potential field. In: Proceedings of the3rd International Conference on Computer Science and Network Technology. Dalian: IEEE. 1208-1211.

[5] Ni, J., Wang, K., Huang, H., et al. (2016) Robot path planning based on an improved genetic algorithm with variable length chromosome. In: The 12th International Conference on Natural Computation, Fuzzy Systems and Knowledge Discovery (ICNC-FSKD). Changsha: IEEE. 145-149.

[6] Liu, J.H., Yang, J.G., Liu, H.P., et al. (2017) An improved ant colony algorithm for robot path planning. J. Soft Computing., 21(19): 5829-5839.

[7] Tharwat, A., Elhoseny, M., Hassanien, A., et al. (2018) Intelligent bézier curve-based path planning model using chaotic particle swarm optimization algorithm. J. Cluster Computing., 1-22.

[8] Li, X.X., Hu, F.C., Liu, C., Jin, M., Zhu, Y.H. (2021) Robot 3D Path Planning Algorithm Based on Improved Elitist Potential Field Ant Colony Algorithm .J. Computer Science and Application., 11(4): 849-858.

[9] Guo, W., Qin, G.X., et al. (2020) Robot path planning based on improved artificial fish swarm algorithm mand MAKLINK graph. J. Control and Decision., 35(9): 2145-2152.

[10] Huan, S.Z.，Tian, J.W.，Qiao, L.，Wang, Q.，Su, Y. (2021) Unmanned aerial vehicle path planning based on improved genetic algorithm. J. Journal of Computer Applications., 42(2):390-397.

[11] JACQUEZ, J.A. (1998) Design of experiments. J. Journal of the Franklin Institute., 335(2):259-279.

[12] Liang, R.J. (2008) Orthogonal test design for optimization of the extraction of polysaccharides from Phascolosoma esulenta and evaluation of its immunity activity. J. Carbohydr Polym., 1009,3(5):433-437.

[13] Hu, W.B., Han, P., Sun, M. (2015) Selection Method of Control Parameters in Binary - coded Genetic Algorithm. J. Computer Simulation., 32(03):447-450.

[14] Yan, L.J., Li, Z.B., Wei, J.H. (2008) Study on Parameter Setting Method for Simulated Annealing Algorithm. J. Journal of System Simulation., (01):245-247.

[15] Chekanin, V.A., Kulikova, M.Y. (2017) Adaptive adjustment of parameters of the genetic algorithm. J. Vestnik MGTU Stankin., (03):85-89

[16] Zhang T., Zhang H., Wang Z.C. (2000) Float encoding genetic algorithm and its application. J. JOURNAL OF HARBIN INSTITUTE OF TECHNOLOGY., (04): 59-61.

Optimization of flow shop scheduling based on genetic algorithm with reinforcement learning

Xiaoyu Gao[1], Shipin Yang[1], Lijuan Li[1*]

[1] College of Electrical Engineering and Control Science, Nanjing Tech University, Nanjing, 211800, China

*E-mail: ljli@njtech.edu.cn

Abstract. Genetic algorithm, as a kind of evolutionary algorithm, has the characteristics of easy operation and global search, but its stochasticity is relatively strong and highly susceptible to parameters. When facing a large-scale scheduling problem, a strategy is needed to improve the parameter adaptability to make its solution more effective. Reinforcement learning, as an optimization method, has a strong autonomous learning capability. Therefore, this paper proposes a genetic algorithm based on reinforcement learning, which uses Q-learning to self-learning the crossover probability and improve the generalization ability of genetic algorithm, so as to achieve the solution of large-scale replacement flow shop scheduling problem.

1. Introduction

The Permutation Flow-shop Scheduling Problem (PFSP), a classical shop scheduling problem, is widely used in realistic production models. The methods for solving the shop scheduling problem are mainly divided into optimal and approximate methods. Optimal solution methods include branch-and-bound [1-2], dynamic programming [3], integer programming [4-6], and Lagrangian relaxation [7]. Since the 21st century, approximate solution methods have been widely studied, mainly because the time cost of optimal solution methods to solve large-scale complex shop floor scheduling problems is high, while approximate solution methods can quickly obtain the required satisfactory solutions. Approximate solution methods include constructive heuristic, enhancement heuristic, and metaheuristic.

Genetic algorithm, as a metaheuristic algorithm, has achieved remarkable results in both discrete and continuous problems, and is one of the main methods to study shop scheduling problems. In addition, many other evolutionary algorithms have evolved from genetic algorithms, including genetic programming [8], evolutionary strategy [9], and evolutionary programming [10], which have made great breakthroughs in the shop floor scheduling problem. Su [11] discovered efficient allocation rules for solving job shop scheduling problems using genetic programming, and Karunakaran [12] obtained allocation rules for dynamic job shop scheduling problems in uncertain environments based on hyper-heuristic genetic programming. Siqueira [13] showed an algorithm based evolutionary strategy to solve a hybrid flexible job shop scheduling problem. Wang [14] used improved evolutionary programming to solve the flow shop scheduling problem.

The variation law of the parameters of genetic algorithm is usually predetermined, or the process of optimizing parameters is added to achieve parameter control, but it cannot well combine the individual characteristics in the population, and it is difficult to obtain the parameters more suitable for the evolution of the population. Due to this limitation, the algorithm solving ability and generalization ability still need to be improved. Reinforcement learning has the feature of strong self-learning ability, which can realize self-optimization according to historical data, reward and punishment mechanism.

Based on this, a parametric self-learning strategy using Q-learning algorithm is designed in this paper. Individuals of the population in the genetic algorithm are regarded as agents, and the state and action space are defined reasonably with parameter learning as the goal. A reward mechanism is set according to the goal for trial-and-error learning, which drives the agent to learn in the expected direction to obtain better parameters.

2. Genetic algorithm with reinforcement learning

The reinforcement learning mechanism is shown in figure 1. The feedback from the environment is used as input and learning by trial and error in the interaction with the environment, using feedback signals with evaluative to achieve optimization of the decision. Q-learning is a heterogeneous strategy reinforcement learning algorithm that updates by the action value function Q as shown in equation (1).

$$Q^{\pi}(s,a) = E_{\pi}\{R_t \mid s_t = s, a_t = a\} = E_{\pi}\{\sum_{k=0}^{\infty} \gamma^k r_{t+k+1} \mid s_t = s, a_t = a\} \tag{1}$$

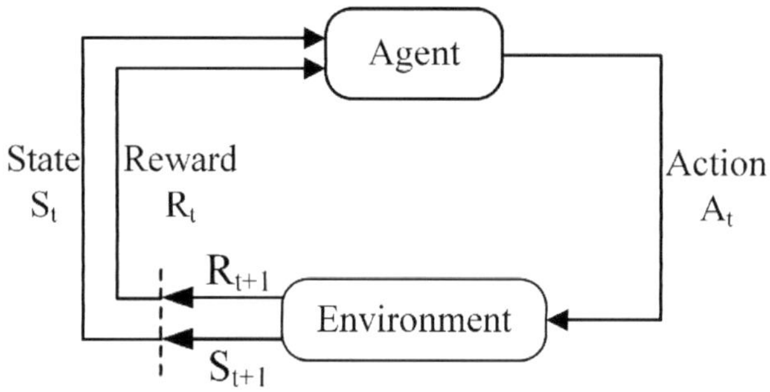

Figure 1 Reinforcement learning

Different from traditional search algorithms, genetic algorithm starts searching from a set of randomly generated initial solutions, which are mainly implemented by crossover, variation, and selection operations. There are three major operators in genetic algorithms: selection operator, variation operator and crossover operator. In the crossover operator, the population crossover probability is a key parameter that affects the efficiency and performance of the algorithm. Reinforcement learning is applied to the crossover probability self-learning, and the dynamic updating of the parameters is also realized. The process is as follows.

Step 1: Population initialization. Generate N individuals randomly.

Step 2: Fitness assessment. Individuals were assessed for fitness, and the FER (Fitness Euclidean-distance Ratio) index in equation (2), which defines the fitness function of population individuals integrated with the Euclidean distance between individuals, is used to select individuals with high quality in the neighborhood.

$$FER_{(j,i)} = \frac{f(a_j)-f(a_w)}{\|a_j-a_i\|} \tag{2}$$

where $f(\cdot)$ denotes the fitness function, a_i is the optimal individual, and a_w is the worst individual in the current population.

Step 3: Selection. All individuals in each generation still undergo the selection operation, which is designed as in equation (3).

$$p_i = \frac{f_i}{\sum f_i} \tag{3}$$

The probability of each individual being selected is p_i. Before each individual is selected, a number between 0 and 1 is randomly generated. If the number is less than p_i, the individual is selected, otherwise it is not selected.

Step 4: Crossover. The crossover parent is selected by "shuffling strategy".

Step 5: Crossover probability self-learning based on Q-learning algorithm. Considering the chromosomal characteristics of the population, the reward and punishment mechanism is set based on: the target value corresponding to the optimal individual of the population, the average target value, the genotypic variety of the population, and the standard deviation of the target value of the individuals of the population.

The crossover probability in the genetic algorithm constitutes the state space, and the crossover probability consists of two-digit decimal numbers in the interval [0.7,0.99]. {-1,0,1} constitutes the action space, which indicates decrease, no change and increase. The action change basis is 0.01, i.e., the next state and the state at this time satisfy equation (4).

$$s_{t+1} = s_t + action \cdot 0.01 \quad (4)$$

A four-digit number represents four evaluation indicators, one and two digits indicate the optimal target value and the average target value, respectively, and three and four digits indicate the population genotype type and population standard deviation. The number 1 indicates that the evaluation indicator has become better, 0 indicates that it has not become better, and 2 indicates that there is no requirement. The Q-learning reward and punishment mechanism is set up as follows:

(1) When the evolutionary algebra<60: (i) If 1122, $reward = +15$; (ii) If 1022, $reward = +5$; (iii) If 0122, $reward = +3$; (iv) Otherwise, $reward = -10$. (2) When the evolutionary algebra>60: (i) If 1111, $reward = +30$; (ii) If 1122, $reward = +12$; (iii) If 1022 or 0122, $reward = +5$; (iv) If 2210 or 2201 or 2211, $reward = -5$; (v) Other, $reward = -25$.

Step 6: Mutation. A single point mutation was used, where a position in the chromosome was first randomly selected for mutation.

3. Experiment and result analysis

Permutation Flow-shop Scheduling Problem (PFSP) is a typical problem in shop scheduling. A production line processes n workpieces, each with m processes, and each process is processed on m different machines. Each workpiece i is processed on m machines in turn. The order of workpiece processing can be arbitrary and can only be determined by the processing time t_{ij} of workpiece i on machine j, where all workpieces arrive at zero moment. The scheduling problem in the flow shop is to study the optimal processing order of all workpieces so that the maximum completion time is minimized.

For PFSP, the optimization metric is based on the total time to completion of all workpieces, denoted as $C_{i,j}$, denotes the completion time of workpiece i on machine j.

$$C_{\pi(i),j} = max\{C_{\pi(i),j-1}, C_{\pi(i-1),j}\} + t_{i,j} \quad (5)$$

where $C_{\pi(i),0}$=0, $C_{\pi(0),j}$=0, $\forall i \in \{1,2,\dots,n\}$,$j \in \{1,2,\dots,m\}$,$t_{i,j}$ denotes the machining time of workpiece i on machine j. The optimization objective is to minimize the maximum completion time, i.e., $minC_{max}$. Where, $C_{max} = C_{\pi(n),m}$.

Based on this, the mathematical model of the replacement flow shop scheduling problem is established as follows.

(1) Decision variables

$$X_{i,k} = \begin{cases} 1 , & Workpiece\ i\ at\ position\ k \\ 0 , & \text{otherwise} \end{cases} \quad i,k = \{1,2,\dots,n\} \quad (6)$$

$$C_{k,j} = Completion\ time\ of\ the\ workpiece\ at\ position\ k\ on\ machine\ j \quad j = \{1,2,\dots,m\} \quad (7)$$

(2) Optimization objective

$$minC_{max} = C_{\pi(i),j} \quad (8)$$

(3) Constraint conditions

$$\textstyle\sum_{k=1}^{n} X_{i,k} = 1, i = \{1,\dots,n\} \quad (9)$$

$$\textstyle\sum_{i=1}^{n} X_{i,k} = 1, k = \{1,\dots,n\} \quad (10)$$

$$C_{k,1} \geq \textstyle\sum_{i=1}^{n} X_{i,k} \cdot t_{i,1}, k = \{1,\dots,n\} \quad (11)$$

$$C_{k,j} \geq C_{k,j-1} + \textstyle\sum_{i=1}^{n} X_{i,k} \cdot t_{i,j}, k = \{1,\dots,n\}, j = \{2,\dots,m\} \quad (12)$$

$$C_{k,j} = C_{l,j} + \textstyle\sum_{i=1}^{n} X_{i,k} \cdot C_{i,j}, k = \{2,\dots,n\}, l = \{1,\dots k-1\}, j = \{1,\dots,m\} \quad (13)$$

$$C_{k,j} \geq 0, k = \{1, \dots, n\}, j = \{1, \dots, m\} \quad (14)$$

$$X_{i,k} \in \{0,1\}, j, k = \{1, \dots, n\} \quad (15)$$

The relevant parameters are determined as: Genetic algorithm (GA) population size N=100, initial variance=0.05, initial crossover probability=0.85, number of population iteration steps=500. Reinforcement learning (RL) learning rate μ=0.05, discount factor γ=0.9, the ε-greedy strategy of ε=0.85.

In order to verify the performance of the algorithm for solving the flow shop scheduling problem, the standard test instance of PFSP, Car and Rec, is applied and compared with the general GA and the adaptive genetic algorithm (AGA). As shown in the table 1, Best, BRE and ARE are used as evaluation metrics. C* represents the optimal solution of the test instance.

The figure 2 shows the change of the optimal value of 3 algorithms during the actual evolution of the test instance rec21, and it can be seen from the figure that the optimal value obtained by the proposed RL GA algorithm is better than the other two algorithms. According to the figure 3-5 comparing the BRE and ARE of each algorithm on all standard cases and the average value, the average BRE and ARE of the proposed algorithm RL GA are lower than the other two algorithms, which proves that the proposed algorithm has strong convergence and stability and can solve PFSP more effectively.

Table 1. Best[a], BRE[b] and ARE[c] of three algorithms.

Test Instance	n*m	C*	RL GA			GA			AGA		
			Best	**BRE**	**ARE**	**Best**	**BRE**	**ARE**	**Best**	**BRE**	**ARE**
Car1	11*5	7038	7038	**0**	**0.011**	7038	**0**	0.233	7038	**0**	0.021
Car2	13*4	7166	7166	**0**	**0**	7166	**0**	0.21	7166	**0**	0.009
Car3	12*5	7312	7399	**0.012**	**0.016**	7543	0.032	0.298	7400	**0.012**	0.022
Car4	14*4	8003	8003	**0**	**0.048**	8003	**0**	0.231	8003	**0**	0.055
Car5	10*6	7720	7720	**0**	**0.004**	7749	0.004	0.166	7821	0.013	0.017
Car6	8*9	8505	8505	**0**	**0.006**	8505	**0**	0.235	8505	**0**	0.006
Car7	7*7	6590	6590	**0**	0.015	6590	**0**	0.196	6590	**0**	**0.01**
Car8	8*8	8366	8366	**0**	**0**	8366	**0**	0.154	8366	**0**	**0**
Rec01	20*5	1247	1251	**0.003**	**0.02**	1263	0.013	0.2	1291	0.035	0.055
Rec03	20*5	1109	1125	0.014	0.048	1116	**0.006**	0.184	1122	0.012	**0.015**
Rec05	20*5	1242	1269	**0.022**	**0.022**	1281	0.031	0.2	1269	**0.022**	0.024
Rec07	20*10	1566	1619	**0.034**	**0.037**	1619	**0.034**	0.185	1619	**0.034**	**0.037**
Rec09	20*10	1537	1577	**0.026**	**0.031**	1632	0.062	0.293	1610	0.047	0.068
					......						
Rec33	50*10	3114	3267	**0.049**	**0.052**	3351	0.076	0.255	3290	0.057	0.057
Rec35	50*10	3277	3344	**0.02**	**0.034**	3438	0.049	0.167	3359	0.025	0.037
Rec37	75*20	4951	5464	0.104	0.111	5703	0.152	0.263	5455	**0.102**	**0.11**
Rec39	75*20	5087	5466	**0.075**	**0.079**	5672	0.115	0.221	5510	0.083	0.086
Rec41	75*20	4960	5508	0.11	0.114	5508	0.11	0.114	5448	**0.098**	**0.105**
AVG				0.033	0.04		0.055	0.227		0.04	0.048

[a] **Best** represents the optimal solution obtained by the algorithm.

[b] **BRE**=(Best-C*)/C* represents the relative error between Best and C*.

[c] **ARE**=(Average-C*)/C* represents the relative error between the Average and C*.

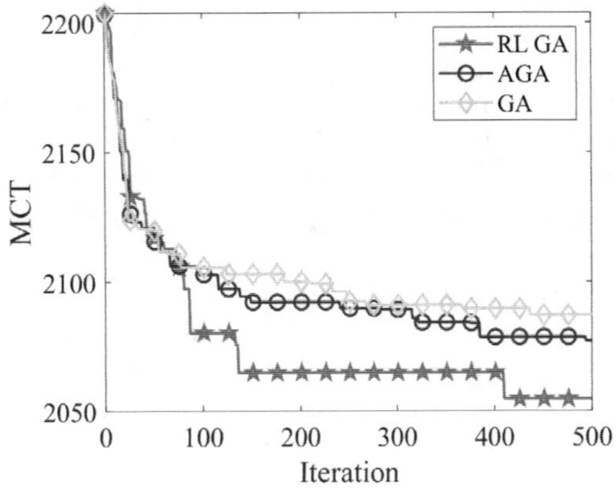

Figure 2 The change of the optimal value

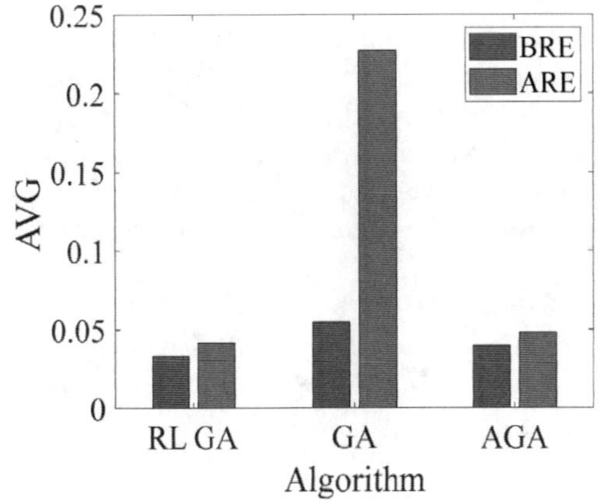

Figure 3 The AVG of BRE and ARE

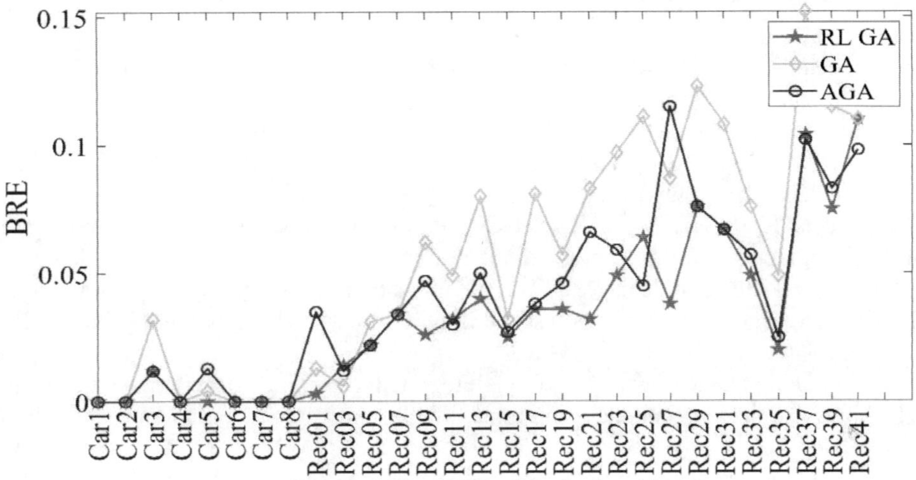

Figure 4 The BRE of each algorithm

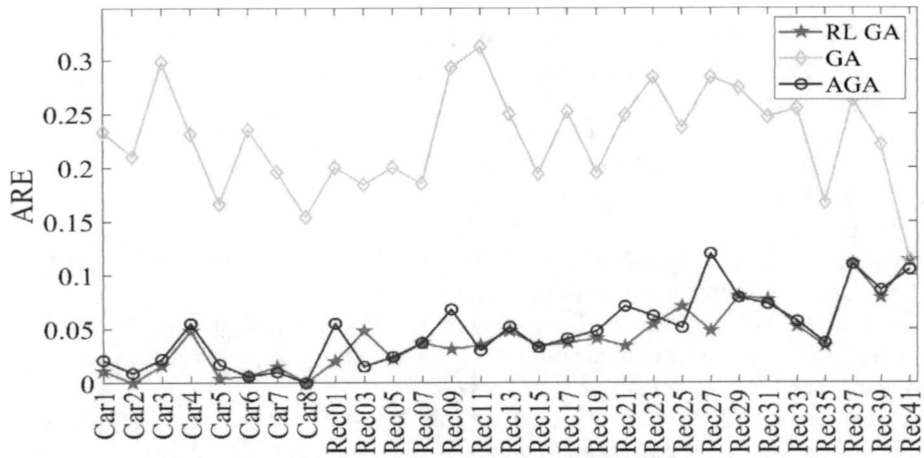

Figure 5 The ARE of each algorithm

4. Conclusion

In this paper, a genetic algorithm with reinforcement learning is proposed to overcome the stochasticity and instability of genetic algorithms. The algorithm implements a self-learning strategy for cross-rate of genetic algorithm based on reinforcement learning, and uses Q-learning method to learn and explore the cross-rate as the state space by setting the reward and punishment mechanism that facilitates population evolution and merit search. According to the characteristics of the replacement flow shop scheduling problem, the reward and punishment values adapted to the problem are set and the subpopulation is dynamically updated. Through the test instance of PFSP, the maximum completion time and average completion time are used as evaluation indexes, and the comparison experiments with traditional GA and AGA are conducted, and the results prove that the algorithm in this paper is more superior.

Acknowledgments

This work was financially supported by the National Natural Science Foundation of China (Grant Nos. 61873121 and 61973334) and the Natural Science Foundation of Jiangsu Province, China (General Program) (Grant No. BK20181376).

References

[1] Khoudi, A, Berrichi, A. (2018) Minimize total tardiness and machine unavailability on single machine scheduling problem: bi-objective branch and bound algorithm. Operational Research, 20(3): 1763-1789.
[2] Toumi, S., Jarboui, B., Eddaly, M., et al. (2017) Branch-and-bound algorithm for solving blocking flowshop scheduling problems with makespan criterion. International Journal of Mathematics in Operational Research, 10(1): 34-48.
[3] Ozolins, A. (2018) Bounded dynamic programming algorithm for the job shop problem with sequence dependent setup times. Operational Research, 20(3): 1701-1728.
[4] Zeng, C.K., Tang, J.F., Fan, Z. P., et al. (2019) Scheduling of a job-shop problem with limited output buffers. Engineering Optimization, 52(1): 53-73.
[5] Pei, Z., Zhang, X.F., Zheng, L., et al. (2020) A column generation-based approach for proportionate flexible two-stage no-wait job shop scheduling. International Journal of Production Research, 58(2): 487-508.
[6] Giglio, D., Paolucci, M., Roshani, A. (2017) Integrated lot sizing and energy-efficient job shop scheduling problem in manufacturing/remanufacturing systems. Journal of Cleaner Production, 148: 624-641.
[7] Fonseca, G.B., Nogueira, T.H., Ravetti, M.G. (2019) A hybrid Lagrangian metaheuristic for the cross-docking flow shop scheduling problem. European Journal of Operational Research, 275(1): 139-154.
[8] Koza, J.R., Poli, R. (2005) Genetic Programming. Springer, Boston.
[9] Beyer, H.G., Schwefel, H.P. (2002) Evolution strategies - a comprehensive introduction. Natural Computing, 1(1): 3-52.
[10] Bèack, T. (1998) Evolutionary Algorithms in Theory and Practice: Evolution Strategies, Evolutionary Programming, Genetic Algorithms. Springer, Berlin.
[11] Su, N., Zhang, M.J., Johnston, M., et al. (2019) Genetic Programming for Job Shop Scheduling. Springer, Cham.
[12] Karunakaran, D., Yi, M., Gang, C., et al. (2017) Dynamic Job Shop Scheduling Under Uncertainty Using Genetic Programming. Springer, Cham.
[13] Siqueira, E.C.D., Souza, M.J.F., Souza, S.R.D., et al. (2013) An algorithm based on Evolution Strategies for makespan minimization in hybrid flexible flowshop scheduling problems. In: Evolutionary Computation. Cancun. pp. 989-996.
[14] Wang, L., Zheng, D.Z. (2003) A modified evolutionary programming for flow shop scheduling. The International Journal of Advanced Manufacturing Technology, 22: 522-527.

Layered Convolutive Nonnegative Matrix Factorization for Speech Separation

Wang Yao[1], Danjv Lv[2*], Xin Huang[3], Jiali Zi[4], Mingyuan Gao[5], Rui Xi[6] and Yan Zhang[7]

[1,2*,3,4,5,6] College of Big Data and Intelligent Engineering, Southwest Forestry University, Kunming, Yunnan, 650224, China

[7] Institute of Mathematical, Southwest Forestry University, Kunming, Yunnan, 650224, China

[*]Corresponding author's e-mail: lvdanjv@swfu.edu.cn

Abstract. Nonnegative matrix factorization (NMF) has attracted significant attention for its good performance in single-channel speech separation. The improved algorithms of NMF have become research hotspots. Layered NMF (LNMF), an improved algorithm, can express the source signal more accurately for its multilayer structure. However, LNMF sometimes performs poorly because it ignores the short-term correlation of speech signals. Based on LNMF and the advantages of Convolutive NMF (CNMF), we proposed a Layered Convolutive NMF(LCNMF) algorithm for single-channel speech separation. The LCNMF corporates the multilayer structure into the NMF and expands the convolution of the top-level NMF model. During the training, NMF is used to learn the non-top-level basis matrices, and CNMF is used to learn the top-level basis matrix, then combined with each single-layer of basis matrix. During the prediction, CNMF is used to separate mixed signals. The results on the dataset MIK-1K showed that LCNMF outperformed NMF and LNMF for separating the mixture of single-channel speech signals. LCNMF improved by 0.019, 1.049dB, 1.305dB, and 0.851dB on average compared with NMF, and improved by 0.007, 0.172dB, 0.090dB, and 0.366dB on average compared with LNMF in sort-term objective intelligibility (STOI), Source to Distortion Ratio (SDR), Source to Interference Ratio (SIR) and Source to Artifacts Ratio (SAR) measures.

1. Introduction

The research of speech separation technology plays a vital role in signal processing. Its purpose is to recover the original signal from the mixed signal generated by multiple sound sources [1]. Speech separation can effectively improve the perceptual quality of speech signals [2] or working as a front-end in automatic speech recognition systems to help improve the recognition accuracy [3]. Over the years, miscellaneous methods for speech enhancement have been proposed, such as spectral subtraction (SS) [4], Wiener filtering [5], and minimum mean square error (MMSE) estimators [6]. These methods do not require any prior knowledge about the speech signals, nor any training stage beforehand, so they are highly flexible and allow implementation in various situations. However, those algorithms usually assume that the signals are stationary which are not good at dealing with non-stationary signal types, especially under the low signal-to-noise (SNR) conditions [7].

Compared with the above unsupervised method, the supervised method learns the relationship between mixed speech, target speech and interference signal to separate [8], which overcomes the shortcomings of the unsupervised speech separation method and has significantly better separation performance [9]. NMF is still a popular and effective supervised speech separation method. Since based on a family of statistical models [10], the NMF is easy to extend in many ways, such as adding sparsity [11], dynamics across time [12-13], and phase awareness [14]. However, these extensions are still limited to monolayer structures, which are difficult to extract non-linear features in speech [15]. According to the big success of deep neural networks in speech recognition, the hierarchical features can represent speech signals well [16]. Roux et al. proposed a Deep NMF model, which unfolds the NMF iterations and unties the NMF parameters for optimal separation performance [17]. Layered NMF (LNMF) stacks multiple layers of standard NMF blocks, and then combine sparse parts-based representation learned by each single-layer to interpret the speech data differently [18]. The experimental results showed that the hierarchical NMF model could significantly enhance speech separation performance [17-18].

Based on the multi-layer NMF model, the researchers carried out a further extension. Naiyang Guan proposed a deep transductive NMF model (DTNMF), which incorporated a multilayer structure into NMF and learned a shared dictionary on the source signal of each speaker and the mixture signal to be separated [15]. Chung-Chien Hsu developed the DLNMF by extending the generative bases in LNMF to the discriminative bases which were estimated according to a discriminative criterion [19]. Although LNMF outperformed NMF in speech separation, both the NMF and its LNMF variants were unable to capture the continuity of speech signals yet, that is, the learned dictionary can describe the variation in frequency of the spectrum well, but not well in time of the spectrum. Smaragdis proposed the convolution NMF (CNMF) algorithm, which used the result of the convolution summation between a set of basis matrices and its weight matrix to represent the matrix to be decomposed. Each basis matrix learned by CNMF described the spectrum of the signal at a certain time, and all the basis matrices jointly described the spectrum of the signal in a short time range. CNMF successfully modelled the continuity of speech and improved speech separation performance [20].

The LNMF could be improved by considering the continuity of the changes of the speech signal. Therefore, based on LNMF and the advantages of CNMF, LCNMF algorithm was proposed by extending the top-level NMF model to the CNMF model in this paper. LCNMF algorithm used hierarchical structure to extract speech information and considered the short-term correlation of the speech signal. we carried out a single-channel speech separation experiment on the open dataset MIK-1K. Four evaluation indexes including STOI, SDR, SAR and SIR were used to evaluate the experimental results.

2. NMF and LNMF

2.1. NMF for Single-channel Speech Separation

Nonnegative matrix factorization (NMF) preserves the non-negativity property of the magnitude spectrogram of the speech signal [15]. Given a sequence of time-domain signal v(t), its spectrogram is expressed by formula (1):

$$V = |Y| \tag{1}$$

where Y denotes the frequency-domain signal obtained by STFT on v(t), and the operator $|\cdot|$ outputs the modulus. NMF decomposes the spectrogram V of v(t) by formula (2):

$$V \approx WH \tag{2}$$

Figure 1 shows the flow of the NMF method for speech separation. Given $K \geq 2$ speakers and their training speech signals. NMF is utilized to learn several basis matrices for each speaker. Let $V_k \in R_+^{m \times n_k}$ denote the spectrogram of the training speech signals of the k-th speaker. NMF is performed on each V_k by formula (3):

$$V_k \approx W_k H_k \tag{3}$$

$W_k \in R_+^{m\times r}$ denotes the basis matrix learned by the k-th speaker. Spectral bases of all speakers are concatenated to construct W^m as expressed by formula (4):

$$W^m = [W_1, ..., W_K] \tag{4}$$

Let $v^m(t)$ denote the time domain mixed signals, and $V^m \in R_+^{m\times n}$ denotes the spectrum of the mixed signal, the NMF-based method decomposes it by: $V^m \approx W^m H^m$, where H^m denotes the activation matrix. H^m is decomposed by formula (5):

$$H^m = [H_1^{mT}, ..., H_K^{mT}]^T \tag{5}$$

the activation matrix of each speaker is obtained. The spectrum of the k-th speaker's speech signal is reconstructed by formula (6):

$$\hat{V}_k = W_k H_k^m \tag{6}$$

finally, the speech signal of the k-th speaker is obtained by using the inverse short-time Fourier inversion (ISTFT) transforms the spectrogram $\hat{V}_k$ to the time domain $\hat{v}_k(t)$.

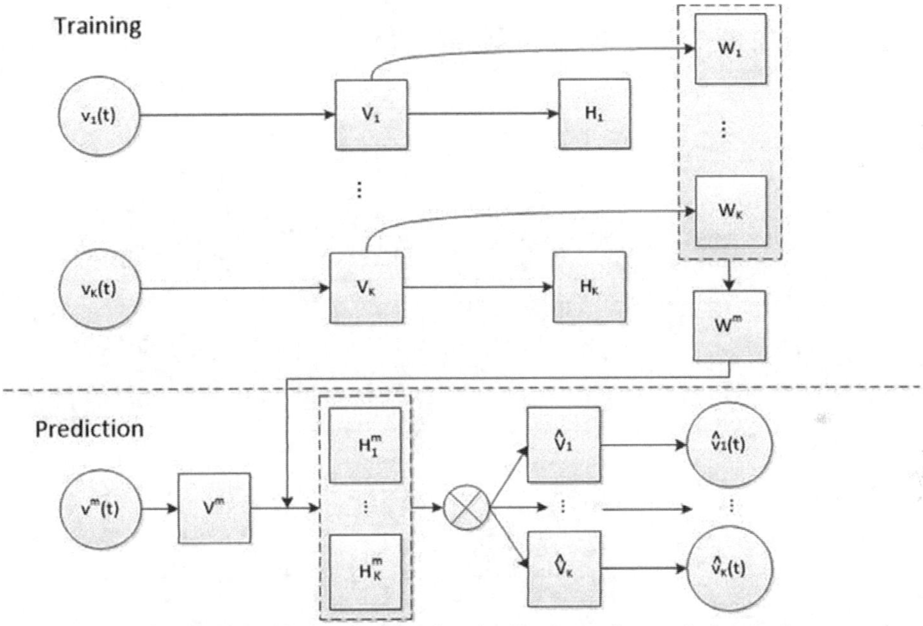

Figure 1. Flow chart of NMF for single-channel speech separation.

2.2. LNMF for Single-channel Speech Separation

According to the big success of deep neural networks in speech recognition, the hierarchical features can represent speech signals well. Chung-Chien Hsu incorporated the hierarchical architecture into the standard NMF and proposed a layered NMF(LNMF) [19]. The spectrum V_k of each speaker is decomposed into L layers by hierarchical structure. LNMF with L layers is performed by formula (7):

$$V_k \approx \hat{V}_k = (\prod_{l=1}^{L} W_k^{(l)}) \cdot H_k^{(L)} \tag{7}$$

Figure 2 shows the flow of LNMF for speech separation. During the training, the factors $\{W_k^{(l)}\}$ and $H_k^{(L)}$ are initialized layer by layer. In the first step, the standard NMF is performed to factorize: $V_k = W_k^{(1)} H_k^{(1)}$, where $W_k^{(1)} \in R_+^{m\times r_1}$ and $H_k^{(1)} \in R_+^{r_1\times n}$,Then the same factorization is performed on the result obtained from the first step as $H_k^{(1)} = W_k^{(2)} H_k^{(2)}$, where $W_k^{(2)} \in R_+^{r_1\times r_2}$ and $H_k^{(2)} \in R_+^{r_2\times n}$, The factorization continues to pre-train all layers until $H_k^{(L-1)} \approx W_k^{(L)} H_k^{(L)}$,where $W_k^{(L)} \in R_+^{r_{L-1} \times r_L}$ and $H_k^{(L)} \in R_+^{r_L \times n}$. After the initialization, the parameters of all layers $\{W_k^{(l)}\}$ and $H_k^{(L)}$, to reduce the total reconstruction error via formula (8):

$$\min_{\{W_k^{(l)}\}, H_k^{(L)} \geq 0} D\left(V_k \,||\, \left(\prod_{l=1}^{L} W_k^{(l)} \right) \cdot H_k^{(L)} \right), \forall l = 1, \ldots, L. \quad (8)$$

The dictionary of speech signal is formed by combining the basis matrix of each layer, and the prediction stage is the same as that of traditional NMF.

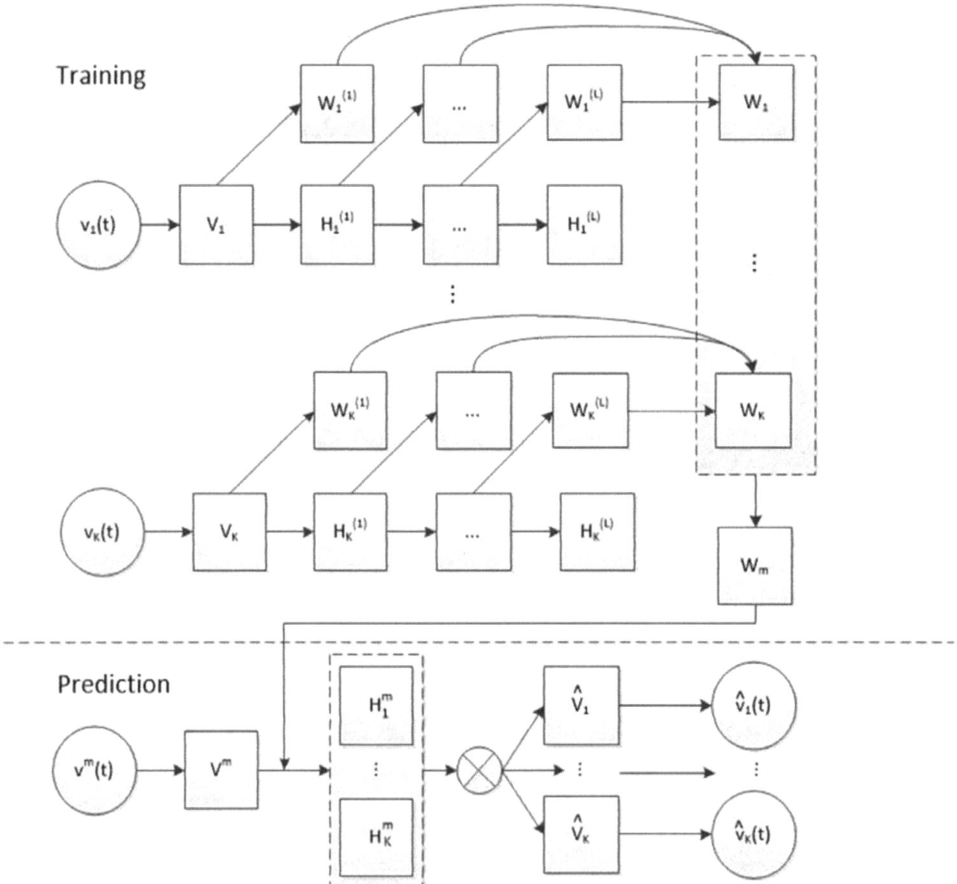

Figure 2. Flow chart of LNMF for single-channel speech separation.

3. LCNMF for single-channel speech separation

Based on LNMF and the advantages of CNMF, the convolution expansion of the NMF model on the top of LNMF is performed to obtain LCNMF. Since LNMF algorithm learns dictionary by combining basis matrix layer by layer during the prediction, it can only expand convolution expansion for the top-level NMF model. In this paper, LCNMF is used in supervised speech separation tasks between multiple speakers.

Assuming the mixture speech v^m is composed of speeches from K speakers, V_k represents the spectrum of the training speech of the k-th speaker. During the training, LCNMF decomposes V_k into L layers through hierarchical structure, and the mathematical model is expressed as formula (9):

$$V_k \approx \hat{V}_k = \left(\prod_{l=1}^{L-1} W_k^{(l)} \right) \cdot \sum_{t=0}^{T-1} W_k(t)^{(L)} \cdot \overset{t\rightarrow}{H_k^{(L)}} \tag{9}$$

LCNMF for speech separation can be illustrated in Figure 3. Firstly, the matrices are initialized layer by layer, and the initialization method from layer 1 to Layer L-2 is consistent with the LNMF algorithm. The matrix performs the standard NMF to obtain the basis matrix and activation matrix of each layer. CNMF is executed when initialization to layer L-1: $H^{(L-1)} = \sum_{t=0}^{T-1} W_k(t)^{(L)} \cdot \overset{t\rightarrow}{H_k^{(L)}}$,where $W_k(t)^{(L)} \in R_+^{r_{L-1} \times r_L}$ is the base matrix set, contains T matrices, $t \in [0, T-1]$, H is the weight matrix. The $\overset{t\rightarrow}{(\cdot)} \left(\overset{\leftarrow t}{(\cdot)} \right)$ denotes a column shift operator that moves its argument k places to the right(left); as each column is shifted off to the right(left) the leftmost columns are zero filled. After the initialization, the parameters of all layers to reduce the total reconstruction error via formula (10):

$$\min_{\{W_k^{(l)}\}, W_k(t), H_k^{(L)}} D\left(V \,\|\, \prod_{l=1}^{L-1} W_k^{(l)} \right) \cdot \sum_{t=0}^{T-1} W_k(t)^{(L)} \cdot \overset{t\rightarrow}{H_k^{(L)}}, \forall l = 1, \dots, L. \tag{10}$$

We use k-L divergence as the objective function, the multiplier updating formula of LCNMF model parameters was obtained by optimizing the objective function. Update $W_k^{(l)}$ from Layer 1 to Layer L-1 as formula (11):

$$W_k^{(l)} \leftarrow W_k^{(l)} \odot \frac{X_k^{(l)} \cdot \frac{V_k}{\hat{V}_k} \cdot Y_k^{(l)}}{X_k^{(l)} \cdot 1 \cdot Y_k^{(l)}} \tag{11}$$

Where $X_k^{(l)} = W_k^{(l-1)T} \cdot W_k^{(l-2)T} \cdots W_k^{(l)T}$, $Y_k^{(l)} = \left(\sum_{t=0}^{T-1} H_k^{(L)T} \cdot W_k^{(L)T} \right) \cdot W_k^{(L-1)T} \cdots W_k^{(l+1)T}$, $\odot$ denotes element-wise multiplication, and 1 is an $M \times N$ matrix with all its elements equal to unity. Update $W_k^{(L)}$ and $H_k^{(L)}$ of layer L by formula (12) and (13):

$$W_k^{(L)} \leftarrow W_k^{(L)} \odot \frac{X_k^{(l)} \cdot \frac{V_k}{\hat{V}_k} \cdot \overset{t\rightarrow}{H_k^{(L)}}}{X_k^{(l)} \cdot 1 \cdot \overset{t\rightarrow}{H_k^{(L)}}} \tag{12}$$

$$H_k^{(L)} \leftarrow H_k^{(L)} \odot \frac{W_k^{(L)T} \cdot X_k^{(l)} \overset{\leftarrow t}{\left[\frac{V_k}{\hat{V}_k} \right]}}{W_k^{(L)T} \cdot X_k^{(l)} \cdot 1} \tag{13}$$

Combined with the base matrix of each layer: $W_k(t) = \prod_{l=1}^{L-1} W_k^{(l)} \cdot W_k(t)^{(L)}, t \in [0, T-1]$,forming the dictionary for the k-th speaker voice signal.

In the prediction, the dictionaries of all speakers are combined into a whole matrix $W^m(t) = [W_1(t), \dots, W_K(t)]$, Keep $W^m(t)$ fixed perform the CNMF on spectrum M of mixed signals, update formula of H^m :

$$H^m \leftarrow H^m \odot \frac{W^{m(t)T} \cdot \left[\frac{\overset{\leftarrow t}{V^m}}{\hat{V}^m}\right]}{W^m(t)^T \cdot 1} \tag{14}$$

Update H with T $W(t)$ respectively and assign the average of all the update results to H. CNMF learned the activation matrix H^m of the mixed signal. By decomposing into $H^m = [H_1^{mT}, \ldots, H_K^{mT}]^T$,the activation matrix of each speaker is obtained. The spectrum of each speaker's speech signal is reconstructed: $\hat{V}_k = \sum_{t=0}^{T-1} W_k(t) \cdot \overset{t\rightarrow}{H_k^m}$. The speech signal of each speaker is obtained by using the inverse short-time Fourier inversion (ISTFT) transform the spectrogram $\hat{V}_k$ to the time domain $\hat{v}_k(t)$.

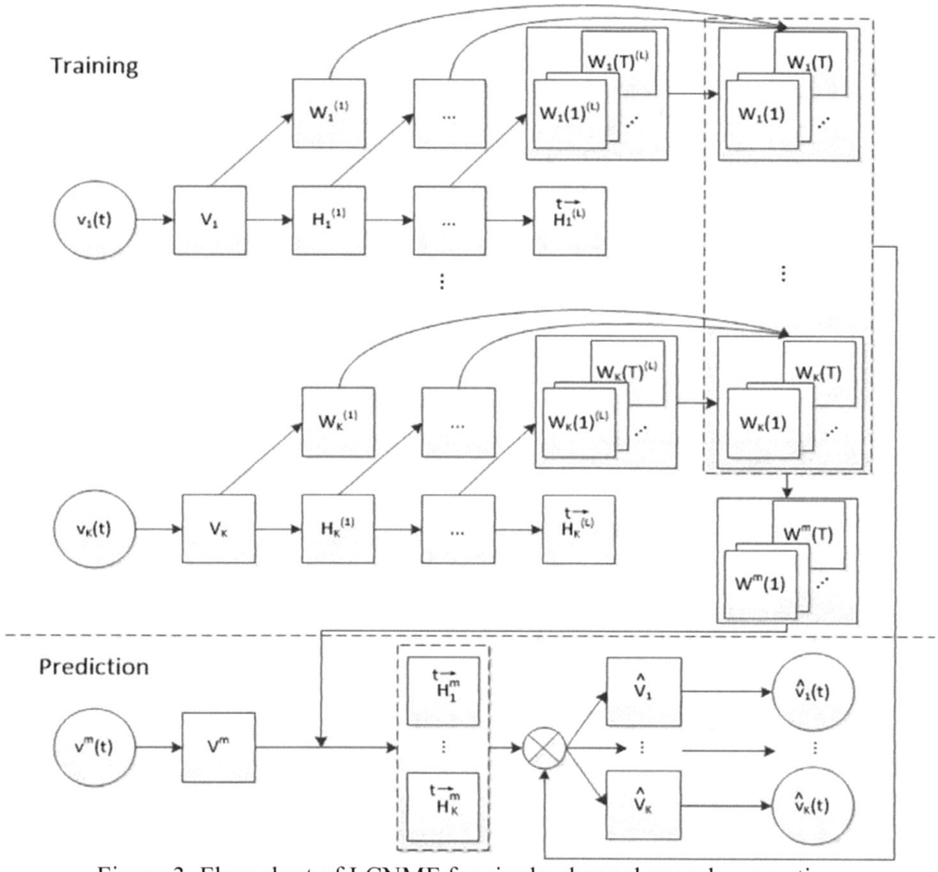

Figure 3. Flow chart of LCNMF for single-channel speech separation.

4. Evaluation of Speech Separation

In this paper, four evaluation indexes are used to evaluate the speech separation: short-term objective intelligibility (STOI), Source to Distortion Ratio (SDR), Source to Interference Ratio (SIR), and Source to Artifacts Ratio (SAR).

STOI is an objective index used to evaluate the intelligibility of speech. Its value ranges from 0 to 1. The larger the value is, the higher the intelligibility of speech is, indicating that the separation effect is better.

SDR, SIR and SAR can be used to evaluate the separation performance of signal separation algorithm. SDR is the ratio of the power of the input signal to the power of the difference between input and reconstructed signals. SIR reports errors caused by failures to remove the interfering signal fully, and SAR reports errors caused by extraneous artifacts introduced during the source separation procedure. Higher values of SDR, SAR, SIR, and STOI represent higher separation quality.

5. Experiment

5.1. Experimental Data and Parameters

In the experiment, the speech data are selected from MIK-1K database, which contains the clean speech and songs with accompaniment, and all sounds were sampled at 16000Hz. The speech separation experiment showed that the NMF can achieve better separation performance when the prediction and training time are longer [21], so each speaker's training and prediction speech duration was set to 12 seconds. We randomly mixed the clean sounds and songs of a male and a female, denoted by M and F respectively, generated mixed signals. Similar to the signal-to-noise ratio (SNR), we defined FMR as quotient of the strength of the sentence of female speaker divided by the strength of the sentence of male speaker to simulate the condition of signal mixtures. The FMRs were from -6dB to 6dB and the step interval was 3dB. A total of 180 speech separation tasks were constructed. Hamming window was used to calculate the spectrogram, FFT=1024. We executed our proposed LCNMF algorithm by comparing both TNMF and NMF. For LNMF and LCNMF, we both chose a two-layer structure, LNMF with more layers did not significantly improve the separation results [18]. The number of speech basis vectors K1 and K2 of the first and second layers was set as 120 and 30 respectively. Reference [22] shows that the separation effect of CNMF algorithm is better when the time span T=5, so the time span in LCNMF was set to 5.

5.2. Experimental Results

Table 1 showed the average training time and predicted time for NMF, LNMF and LCNMF.

Table 1. Average training time and prediction time for NMF, LNMF and LCNMF.

	NMF	LNMF	LCNMF
Average training time(s)	3.007	7.041	15.405
Average prediction time(s)	1.063	1.063	6.527

Compared with NMF, the training time of LNMF was longer, because the two-layer basis matrix was learned during the training, and the prediction time was the same, LNMF used the traditional NMF method to separate during the prediction. Compared with NMF and LNMF, the training time of LCNMF was longest, for not only trained the two-layer basis matrix, but also combined the operation method of convolutive summation of CNM. During the prediction, CNMF was used for separation and the calculation of convolution summation was still carried out, so the prediction time was much higher than NMF and LNMF. In general, the time complexity of LCNMF was much greater than that of NMF and LNMF.

This paper used STOI, SDR, SIR and SAR to measure the clean speech and song separation performance of NMF, LNMF and LCNMF algorithms.

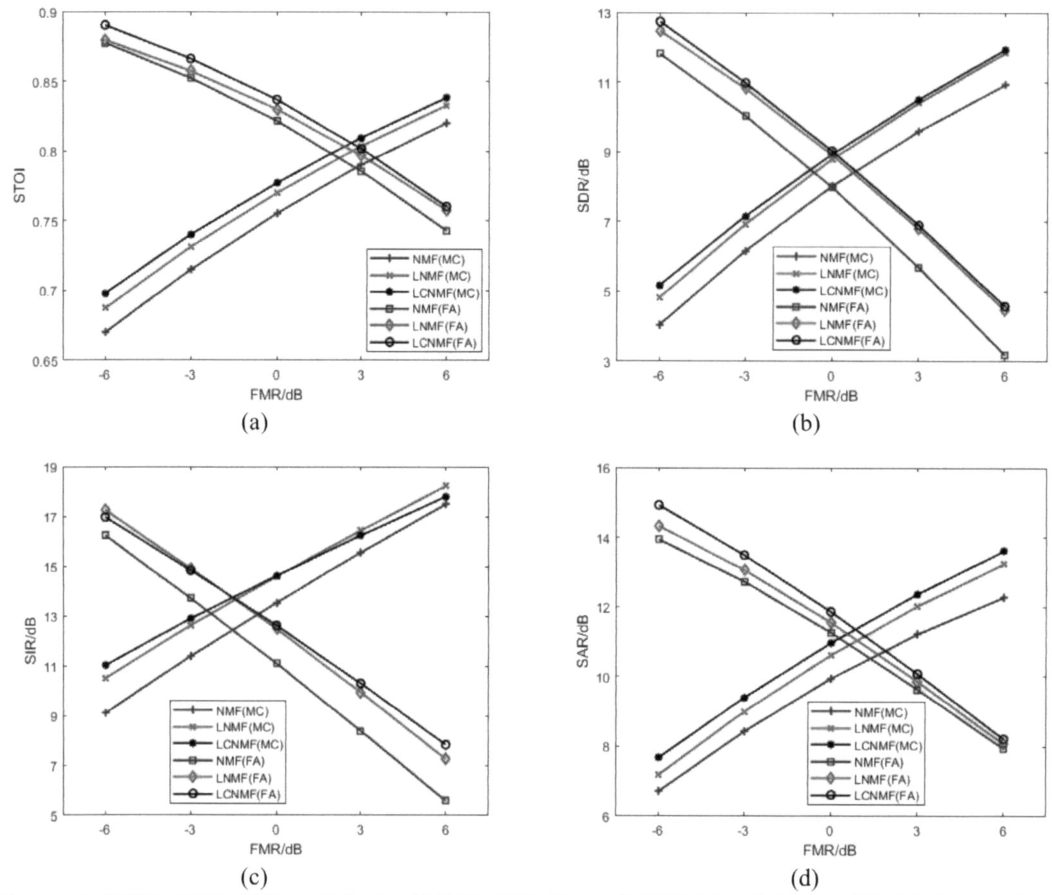

Figure 4. STOI, SDR, SIR, and SAR of NMF, LNMF and LCNMF at 5 kinds of FMRs in the clean speech separation tasks.

Table 2. The average STOI, SDR, SIR and SAR of NMF, LNMF and LCNMF at 5 kinds of FMRs in the clean speech separation tasks.

Method	Male				Female			
	STOI	SDR	SIR	SAR	STOI	SDR	SIR	SAR
NMF	0.750	7.754	13.411	9.714	0.816	7.745	10.998	11.104
LNMF	0.765	8.560	14.472	10.410	0.824	8.693	12.367	11.377
LCNMF	0.773	8.749	14.510	10.803	0.831	8.848	12.507	11.717

In the clean speech separation task, LCNMF significantly improved all speakers' average STOI, SAR, SIR, and SAR by 0.019, 1.049dB, 1.305dB and 0.851dB, respectively. It indicated that the separation performance of LCNMF was significantly improved compared with NMF. This was due to the LNMF algorithm has the ability to extract the multi-level features of the signal and the ability to capture the continuity of the speech signal combined with the CNMF method. Compared with LNMF, the STOI and SAR of LCNMF improved significantly, while the SDR improved slightly, the SIR only improved at low SNR, but when the speaker was at high SNR, the SIR decreased. For all speakers, the average STOI, SAR, SIR, and SAR increased by about 0.007, 0.172dB, 0.090dB, and 0.366dB, respectively. On the basis of LNMF, the CNMF method was combined, which could retain the advantages of the original hierarchical structure, and model the continuity of speech and further improve the performance of speech separation.

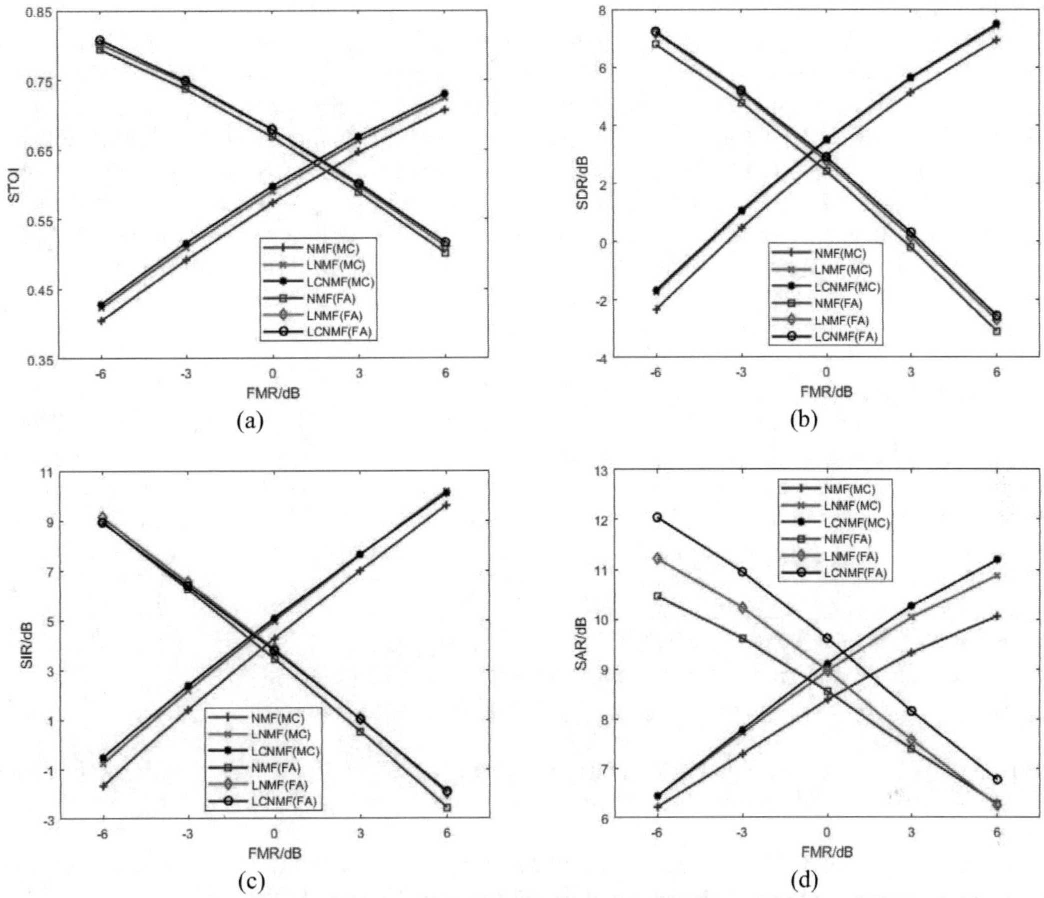

Figure 5. STOI, SDR, SIR, and SAR of NMF, LNMF and LCNMF at 5 kinds of FMRs in the song separation tasks.

Table 3. The average STOI, SDR, SIR and SAR of NMF, LNMF and LCNMF at 5 kinds of FMRs in the song separation tasks.

Method	Male				Female			
	STOI	SDR	SIR	SAR	STOI	SDR	SIR	SAR
NMF	0.564	2.626	4.112	8.246	0.658	2.132	3.322	8.453
LNMF	0.582	3.157	4.823	8.798	0.667	3.508	3.702	8.843
LCNMF	0.588	3.201	4.925	8.950	0.671	3.615	3.638	9.501

In the song separation task, the average STOI, SAR, SIR and SAR measurements of all speakers increased by 0.018, 0.529dB, 0.565dB and 0.876dB, respectively. Compared with LNMF, the STOI, SDR and SIR measurements of LCNMF showed a small increase, but the SDR measurements showed a large increase. The average STOI, SAR, SIR and SAR measurements of all speakers increased by about 0.005, 0.076dB, 0.019dB and 0.405dB, respectively. LCNMF still achieved high separation performance, indicating that the LCNMF algorithm was less affected by musical accompaniment and had good robustness. In general, compared with NMF and LNMF, LCNMF algorithm has improved in the clean speech and song separation tasks with various FMR.

6. Conclusion

This paper proposed a LCNMF algorithm for single-channel speech separation. LCNMF adopted hierarchical structure to express the speech signal more accurately at the cost of time. Meanwhile, the convolution extension of the top-level NMF model expressed the time-frequency domain correlation of the speech signal, which effectively represented the structural characteristics of the original speech signal. Experimental results on the MIK-1K data set showed that LCNMF had better separation performance than NMF and LNMF in the speech separation task, and had different degrees of improvement in the score of four evaluation indexes. However, this algorithm can only expand the convolution of the NMF model at the top of the LNMF, resulting in insufficient combination of the advantages of LNMF and CNMF. Future work will solve this problem and study the optimum structure of CNMF to fulfill faster training and prediction.

Acknowledgments

This research was funded by National Natural Science Foundation of China (Grant No. 31860332, No.61462078), Major Special Projects in Yunnan Province (key technology research and application demonstration of blockchain serving key industries, 202002AD080002), Major scientific and technological projects in Yunnan Province (Grant No. 202002AA10007) and Kunming Forestry Information Engineering Technology Research Center (Grant No.2015FIA02).

References

[1] Wang, X., Bao, C. (2019) Mask estimation incorporating phase-sensitive information for speech enhancement[J]. Applied Acoustics, 156: 101-112.
[2] Benesty, J., Sondhi, M. M., Huang, Y. (2008) Springer Handbook of Speech Processing. Springer, New York.
[3] Huang, X., Acero, A., Hon, H.W. (2011) Spoken Language Processing: A Guide to Theory, Algorithm, and System Development, 1st ed. Upper Saddle River, NJ, USA: Prentice-Hall.
[4] Kamath, S., Loizou, P. (2002) A multi-band spectral subtraction method for enhancing speech corrupted by colored noise. In: Proc. ICASSP. pp. 4164–4167.
[5] Le Roux, J., Vincent, E. (2012) Consistent Wiener filtering for audio source separation[J]. IEEE signal processing letters, 20(3): 217-220.
[6] Ephraim, Y., Malah, D. (1985) Speech enhancement using a minimum mean-square error log-spectral amplitude estimator. IEEE Trans Acoustic Speech, Signal Process, vol. 33, no. 2, pp. 443–445.
[7] Sun, M., Li, Y., Gemmeke, J. F. (2015) Speech enhancement under low SNR conditions via noise estimation using sparse and low-rank NMF with Kullback–Leibler divergence[J]. IEEE/ACM Transactions on Audio, Speech, and Language Processing, 23(7): 1233-1242.
[8] Zengxi, L. (2019) Research on Single-channel Speech Separation Method based on Autoregressive Deep Neural Network [D]. University of Science and Technology of China.
[9] Changchun, B., Zhigang, B. (2020) Speech Enhancement Based on Non-negative Matrix Factorization [J]. Signal Processing, 36(6): 791-803.
[10] Weninger, F., Hershey, J. R., Roux, J. L., Schuller, B. (2014) Discriminatively trained recurrent neural networks for single-channel speech separation. In: Proc. Global SIP. Atlanta, GA, USA. pp. 577–581.

Journal of Physics: Conference Series 2258 (2022) 012020 doi:10.1088/1742-6596/2258/1/012020

[11] Eggert, J., Korner, E. (2004) Sparse coding and NMF. In: Proc. IJCNN, vol. 4. Budapest, Hungary. pp. 2529–2533.
[12] Févotte, C., Roux, J. Le., Hershey, J. R. (2013) Non-negative dynamical system with application to speech and audio. In: Proc. ICASSP. Vancouver, Canada. pp. 3158–3162.
[13] Smaragdis, P., Févotte, C., Mysore, G., Mohammadiha, N., Hoffman, M. (2014) Static and Dynamic Source Separation Using Nonnegative Factorizations: A unified view. IEEE Signal Processing Magazine, vol. 31, no. 3, pp. 66–75.
[14] Kameoka, H., Ono, N., Kashino, K., Sagayama, S. (2009) Complex NMF: A new sparse representation for acoustic signals. In: Proc. ICASSP. Taipei, Taiwan. pp.3437–3440.
[15] Liu, Y., Guan, N., Liu, J. (2017) Deep transductive nonnegative matrix factorization for speech separation[C]. In: 2017 16th IEEE International Conference on Machine Learning and Applications (ICMLA). IEEE: 249-254.
[16] Hinton, G., Deng, L., Yu, D., Dahl, G., Mohamed, A., Jaitly, N., Senior, A., Vanhoucke, V., Nguyen, P., Sainath, T., Kings-bury, B. (2012) Deep neural networks for acoustic modeling in speech recognition: The shared views of four research groups. IEEE Signal Processing Magazine, vol. 29, no. 6, pp. 82–97.
[17] Le Roux, J., Hershey, J. R., Weninger, F. (2015) Deep NMF for speech separation. In: Acoustics, Speech and Signal Processing (ICASSP). pp. 66-70.
[18] Hsu, C.C., Chien, J.T., Chi, T.S. (2015) Layered nonnegative matrix factorization for speech separation. In: Proc. of Annual Conference of International Speech Communication Association. pp. 628–632.
[19] Hsu, C. C., Chi, T. S., Chien, J. T. (2016) Discriminative Layered Nonnegative Matrix Factorization for Speech Separation. In: INTERSPEECH. pp. 560-564.
[20] Smaragdis, P. (2007) Convolutive speech bases and their application to supervised speech separation. IEEE Transactions on Audio, Speech, and Language Processing, 15(1):1-12.
[21] Mak, Y. D. (2016) Monophonic speech separation based on non-negative matrix factorization [D]. National University of Defense Technology.
[22] Xinglei, D., Ying, H., Hao, H. (2020) High-noise Monophonic speech Separation based on Convolution Non-negative Matrix Partial Joint Decomposition [J]. Acta Automatica Sinica, 46(6): 120-1209.

The empirical analysis of the two types of change points estimation algorithms based on MOSUM

Dan Zhao[1], **Xuemin Zi**[1*]

[1]School of Science, Tianjin University of Technology and Education, Tianjin 300222, China

[*]Corresponding author's e-mail: zi_xuemin@aliyun.com

Abstract. In order to make big data play a greater value in financial markets, exploring a new data processing method to improve the ability to detect structural mutations is necessary for the diagnosis and monitoring of financial data and investment direction. This article applies financial time series data, based on the moving sum (MOSUM) statistics, empirically analyzes changes of stock data in Kweichow Moutai in the past ten years. Detecting the feasibility of two multiscale MOSUM algorithms of bottom-up merging and local pruning and visualizing the results by numerical simulation, combined with the fact of the time point, we proved the consistent advantages of the two algorithms in detecting the accuracy and precision of variable points.

1. Introduction

The problem of change points has been considered as an attractive topic in the fields of statistics for years, and the statistical test and estimation of suddenly changes in time series structure are of great significance in all application[1]. In the development of the change point problem, researchers have proposed detection methods such as maximum likelihood method, least squares, minimum absolute length and minimum descriptive length[2][3][4]. However, the above method is limited to specific conditions. To further improve the efficiency of variable point detection and reduce the computational complexity, Frick[5] put forward the synchronous multiscale variable point estimate SMUCE. Ciuperca[6] proposed a model selection criterion based on lasso type or adaptive lasso method. Kirch [7] used MMT method to improve the efficiency of detecting variable points. Fryzlewicz[8] proposed wild binary segmentation (WBS), etc. Despite scholars has been invested a lot of research in the detection of multiple change points, it is still difficult to consider an estimation method that reduces computational complexity and improves detection efficiency simultaneously.

Therefore, taking into account the diversity of time series change points, this paper uses moving sum statistics (MOSUM) to make a consistent estimate of the number and location of variations in the earnings data of the stock market in Kweichow Moutai in the past decade. Numerical results show that the proposed method is superior to the existing method, not only effectively detect the mutation of data structure and reduce the computational complexity, but also be well applied to the processing of like financial time series data.

2. Two algorithms based on MOSUM process

The process of detecting multiple variable points by MOSUM is actually the process of finding the maximum of the MOSUM statistic. Considering the significant environment (l,r) of change point

k, $l<k<r$, selecting the largest point in each important environment as the change point, i.e.

$$k^{\circ}_{(l,r)} = \arg\max_{l\le k\le r} \frac{|T_G(k)|}{\hat{\sigma}_k} \tag{1}$$

In practice, however, the MOSUM signal is affected by noise, which causes the signal to exceed the threshold again within one or two points after the threshold is lowered to make false estimates. Therefore, with the assurance that the critical neighborhood is large enough, using two criteria ε and η to reduce overestimation of the change point[9].

The multiscale MOSUM process algorithms of bottom-up merging (Algorithms 1) and local trimming (Algorithms 2) are given in the form of flowcharts.

Algorithm 1: multiscale MOSUM process with bottom-up merging.

function multiscale.bottomUp $(X, G, \cdots)$
  $P \leftarrow K \leftarrow \varnothing$
  while $G_i \in G$, $\hat{k} \in P_{G_i}$ **do**
    $(\hat{k}, G_i) \in P$ // Produce a set candidate P with change points P_G
  end while
  while $(\hat{k}_0, G_i) \in P$ **do**
    If $\min_{\hat{k}\in K} |\hat{k}_0 - \hat{k}| < \eta G$, **then**   // Merging in the increasing order with the size of G_i
      $\hat{k}_0 \in K$
    end if
  end while
  return K
end function

Algorithm 2: multiscale MOSUM process with localized pruning

function multiscale.localPrune $(X, G, \cdots)$
  $P \leftarrow C \leftarrow K \leftarrow \varnothing$
  while $G_i \in G$, $\hat{k} \in P_{G_i}$ **do**
    $(\hat{k}, G_i) \in P$ // Produce a set candidate P with change points P_G
  end while
  while $P \neq \varnothing$ **do**
    $(\hat{k}_{\circ}, G_{\circ}) \leftarrow \arg\max_{(\hat{k},G)\in P} c(\hat{k}, G)$
    $\hat{k}_L \leftarrow \max\{\hat{k} \in C \cup \{0\}, \hat{k} < \hat{k}_{\circ} : \hat{k} \in K \text{ or } \mathrm{I}(\hat{k}) \cap \mathrm{I}(\hat{k}_{\circ}) = \varnothing\}$
    $\hat{k}_R \leftarrow \min\{\hat{k} \in C \cup \{n\}, \hat{k} > \hat{k}_{\circ} : \hat{k} \in K \text{ or } \mathrm{I}(\hat{k}) \cap \mathrm{I}(\hat{k}_{\circ}) = \varnothing\}$
    $D \leftarrow \{(\hat{k}, G) \in P : \hat{k}_L < \hat{k} < \hat{k}_R\}$ // Perform exhaustive search in elements of D
    $R \leftarrow \{(\hat{k}_{\circ}, G_{\circ})\} \cup \{(\hat{k}, G) \in D : \hat{k} \in [\min \hat{A}, \max \hat{A}]\}$
    if $\hat{k}_L \in K \cup \{0\}, \hat{k}_R \in K \cup \{n\}$ **then**
      redefine R, update $P \leftarrow P / R$, $C \leftarrow P \cup A$

```
        end if
    end while
    return K
end function
```

Note that Algorithms 1 is only for multiple symmetrical bandwidths, and its process is easy to implement and computationally efficient. In the MOSUM package, the MOSUM program for implementing the method by the multiscale. bottom-up function. Algorithms 2 detects change points within the local interval, which allows for asymmetric bandwidth compared to Algorithm 1[9].

3. Simulation

The following set of data with a time scale of 550 is used for change points detection. Because this set of data comes from an unknown distribution, we consider using MOSUM statistics and the multi-change points position model[7][9]:

$$T_G := \max_{1\le k\le n} \frac{|T_G(k)|}{\hat{\sigma}_k} \tag{2}$$

$$X_t = f_t + e_t = \sum_{j=1}^{N+1} \mu_j \tag{3}$$

Based on the above two multiscale change points detection methods, simulated in the change point position of 50, 100, 150, and jump magnitudes is one in situation of mean change points, allowing to detect the estimated number of change points and the location of the change point with the default bandwidth G , see Figure 1.

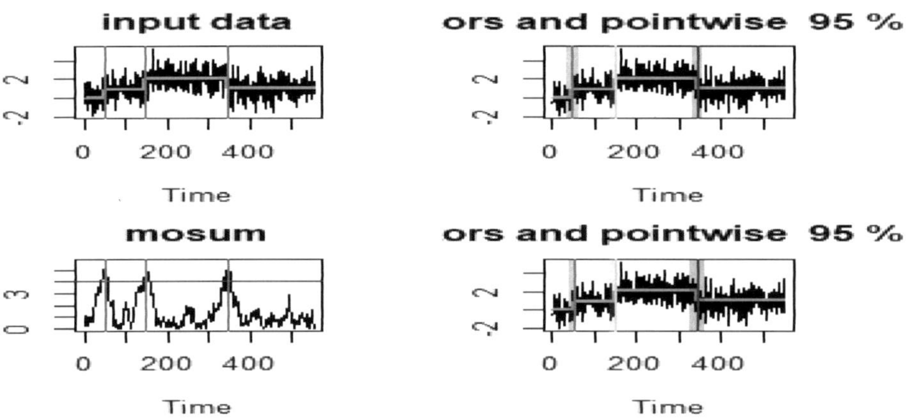

Figure 1. Change point detection by algorithms 1 and algorithms 2.

Figure 1: The time series data (top left) and the absolute MOSUM detector detected the change points situation (bottom left), where the rejection field threshold for α=0.05 (horizontal line). The number and position of change points consistently estimated by Algorithm 1 (top right) and Algorithm 2 (bottom right).

The change point positions estimated by Algorithm 1 are 48, 147 and 345, the change point positions estimated by Algorithm 2 are 50, 147 and 345. The position of change points is estimated by two Algorithms is relatively accurate, and it can be seen from the figure that bottom-up merging and the local trimming MOSUM algorithm can correctly detect the number and position of change points. From the calculation time point of view, algorithm 1 is faster than the calculation rate of algorithm 2.

4. Real data example

In order to verify the validity of the two multiscale MOSUM algorithms in detecting the consistency of the number and position of change point, and to judgment which algorithm can provide more computational efficiency, we selected the stock data of Kweichow Moutai for nearly a decade (2010.9-2021.8), which contains the opening price, the highest price, the closing price, the lowest price, the volume and the re-right closing price. The 503 weeks of data from volume can be detected by using two algorithms based on MOSUM detector, we analyzed the volatility of the Kweichow Moutai stock market.

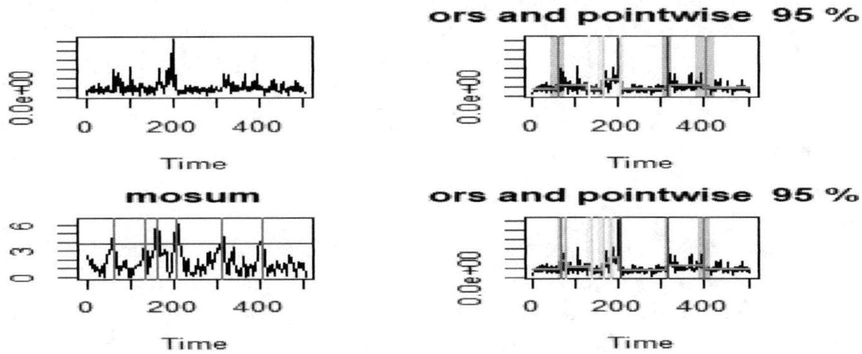

Figure 2. Kweichow Moutai turnover change point detection.

Figure 2: The change point situation detected by the absolute MOSUM detector (bottom left) at the bandwidth G=26, the detection of 6 and 8 change points separately by using the bottom-up merging multiscale MOSUM method (top right) and the local pruning multiscale MOSUM method (bottom right) under the default bandwidth, given a confidence interval of 95%. See Figure 3 and Figure 4 for details.

```
> summary(mbu)
change points detected at alpha = 0.1 according to eta-criterion
 with eta = 0.4

  cpts  G.left  G.right   p.value   jump
1   61      26       26  0.003660  1.436
2  133      26       26  0.050900  1.133
3  162      26       26  0.000277  1.730
4  206      26       26  0.000129  1.818
5  314      26       26  0.006990  1.362
6  403      26       26  0.042300  1.155
```

```
> summary(mlp)
change points detected at alpha = 0.1 according to eta-criterion
 with eta = 0.4

  cpts  G.left  G.right  p.value   jump
1   63      10       10  0.05530  1.903
2   78      10       20  0.03850  1.665
3  135      10       10  0.00206  2.433
4  162      10       10  0.06210  1.884
5  182      50       20  0.04730  1.072
6  205      20       20  0.03010  1.376
7  314      10       10  0.02480  2.034
8  400      20       20  0.04130  1.336
```

Figure 3. Bottom-up merging detects change points. Figure 4. Local pruning detects change points.

Figure 3: The time corresponding to the six change points is respectively is 2012.11.16, 2014.4.11, 2014.10.31, 2015.9.2, 2017.10.20, 2019.7.26. Figure 4: The time corresponding to the eight change points is respectively is 2012.11.30, 2013.3.22, 2014.4.25, 2014.10.31, 2015.3.20, 2015.8.27, 20217.10.20, 2019.7.26.

From the number and position of change point estimation, the multiscale MOSUM method of local pruning detects change point more accurate, but there may be exit some false estimates or overestimation of data. From the running time comparison, algorithm 1 runs 0.01 to 0.02 seconds less than algorithm 2. So, in improving the accuracy of time the bottom-up merging is more priority the local pruning multiscale MOSUM method. In the default bandwidth situation, the local pruning multiscale MOSUM method is well. Furthermore, this method is easier to calculate than the WBS method (Fryzlewicz, 2014), has better progressive properties, and further extends the combined bandwidth method (Eichinger and Kirch, 2018).

Through the test we can get the change of the stock turnover of Kweichow Moutai in the past ten years. During to the central government's "eight regulations" to rectify the market in 2012, Kweichow Moutai stock price fell from 266 yuan to 118 yuan, resulting in Moutai wine appeared a large backlog of inventory. Because the state to punish corruption to lead to Moutai suffered a crisis especially between 2013 and 2015, the stock price continued to fall and shareholders bought a lot at low prices. Moutai stock turnover hit a new high On July 23, 2015. Many unstable factors also indirectly led to the instability of the stock market during this period, and the phenomenon of volume and price divergence occurred. As the continuous easing of the monetary policy of the central bank of China after 2015, the domestic economy heated up so that the investment of market participants rose, the trading volume of Kweichow Moutai suddenly increased and Moutai stock price rose well, the trading volume of stocks tended to be stable with economic uncertainty. As people's living standards improved in 2017, the demand for Moutai wine increased. Because the market is in short supply, Moutai stock price continues to rise and the stock trading volume is also increasing. The trading volume of stocks has been in a relatively stable state from 2019 to now.

5. Conclusion

Based on the principle of MOSUM statistics, this paper applies bottom-up merging method and local pruning method to detect the number and position of change points in the time series data. The above method has played a promoting role in the detection of multiple variable point theory, and can be well applied to the analysis of data. Such as stock data and quality control in the industrial production process, we can use multiscale MOSUM method to detect mutation process, analysis of the relevant reasons and provides a reliable basis a reasonable for operating mechanism. However, there are many inconveniences in the use of the MOSUM method to improve the consistency estimation of the number and location of variable point detection in the case of further hypothetical relaxation, and how to realize the multidimensional variables detection by using MOSUM is the problem that we need to study next.

Acknowledgments

The authors would like to thank all the anonymous reviewers for their many constructive comments and suggestions on the paper, which have helped us to greatly improve the paper. This paper was supported by the NNSF of China, grants No 11771332 and 11771220.

References

[1] Messer, M., Albert, S., Schneider, G. (2018) The multiple filter test for change point detection in time series. Metrika., 81.
[2] Yao, Y.C., Au, S.T. (1989) Least-squares estimation of a step function. Sankhyā Ser. A., 51: 370–381.
[3] Hawkins, D.L., Gallant, A.R., Fuller, W. (1986) A simple least squares method for estimating a change in mean. Communications in Statistics-Simulation., 15: 523-530.
[4] Davis, R.A., Lee, T.C.M., Rodriguez-Yam, G.A. (2006) Structural break estimation for nonstationary time series models. J. Amer. Statist. Assoc., 101: 223–239.
[5] Frick, K., Munk, A., Sieling, H. (2014) Multiscale change point inference. J. Roy. Statist. Soc. Ser. B. Stat. Methodol., 76: 495–580.
[6] Ciuperca, G. (2014) Model selection by LASSO methods in a change-point model. Statist. Papers., 55: 349–374.
[7] Eichinger, B., Kirch, C. (2018) A mosum procedure for the estimation of multiple random change points. Bernoulli., 24: 526-564.
[8] Fryzlewicz, P. (2014) Wild binary segmentation for multiple change-point detection. Ann. Statist., 42: 2243–2281.
[9] Alexander, M., Claudia, K., Haeran, Cho. (2021) mosum: A package for moving sums in change-point analysis. J. Statist. Soft., 97: 1-42.

Journal of Physics: Conference Series 2258 (2022) 012022 doi:10.1088/1742-6596/2258/1/012022

Data Analysis Methods of Natural Fragment Spatial Distribution

Juan Lv[1*],Zhenyu Kang[1],Yuwen Li[2],Shuqiang Yuan[1],Xiangyu Cheng[1],Juan Du[3]

[1]Inner Mongolia Metal Material Research Institute,Ningbo 315103,China

[2]Liaoshen Industries Group Co.,Ltd Shenyang 110045,China

[3]Changchun Equipment and Process Institute,Changchun 130012,China

[*]Corresponding author's e-mail:52suo@norincogroup.com.cn (lvjuan1016@126.com)

Abstract. This paper aims to explore the simulation method of random spatial distribution of natural fragments. The transformation between three-dimensional spatial coordinates and two-dimensional planar coordinates of fragments is realized by coordinate mapping and inverse mapping, and the random uniform distribution of fragments in the two-dimensional plane is realized by the Monte Carlo stochastic points method, so as to obtain the spatial position distribution of fragments.

1. Introduction

Projectile, as suppression weapon, has an important place in conventional war, while the fragments, as the main damage element of a projectile, play a decisive role in its damage power[1]. For a long time, the killing power of ammunition can only be verified by the way of live ammunition target, and the Mott distribution is used to describe the rule of fragmentation formation, which cannot form the theory of fragmentation distribution in space, so it is impossible to verify and calculate the killing power of ammunition theoretically. At present, a statistical model based on experimental experience is widely used in fragment research. The findings on fragment distribution come down toa theoretical model after approximate treatment of physical phenomena under ideal conditions, through which the spatial position distribution of fragments cannot be accurately obtained [2-9].

The present study is to simulate the spatial distribution of natural fragments by the Monte Carlo stochastic points method. On the basis of the existing prediction of fragment quality and quantity distributions[10], the coordinate mapping method is used to realize the mutual transformation of three-dimensional (3D) coordinates and two-dimensional (2D) coordinates of fragments. Then through the method of random variable generation, the random uniform distribution of fragments in two-dimensional plane is realized, and converted into three-dimensional spatial coordinates of fragments. Thus, the spatial distribution of fragments can be obtained to provide support for the prediction of ammunition power.

2. Mapping and inverse mapping between three-dimensional spatial coordinates and two-dimensional coordinates of fragments

In this paper, the random number generator is used to simulate the random position coordinates of the fragments formed by the projectile. In order to reduce the computation and solution, the coordinate mapping is used to convert the three-dimensional spatial coordinates of the fragments into two-dimensional planar coordinates. After the random number generator is used to simulate the 2D planar random coordinates, the 3D spatial coordinates of the fragments are obtained through inverse mapping method.

Through comparative analysis of several common coordinate mapping methods, such as mapping from three-dimensional Cartesian coordinate to two-dimensional planar coordinate, mapping from cylindrical coordinate to two-dimensional planar coordinate, mapping from spherical coordinate to two-dimensional planar coordinate[11-12], and combining with the follow-up work needs of the mutilate prediction of the projectile, the spatial distribution of fragments is studied by using the mutual transformation of three-dimensional (3D) Cartesian coordinate and two-dimensional (2D) planar coordinate in this paper.

2.1 Relevant assumptions and parameter setting

2.1.1 Relevant assumptions

The simulation method of spatial position distribution of natural fragments studied in this paper is designed based on the generally accepted assumptions in natural fragment damage assessment:

1. The projectile shell is fully broken at the moment of explosion, forming natural fragments.
2. The part of the shell where each natural fragment forms is randomly located.
3. All fragments in each section are evenly distributed at the circumferential surface of the projectile shell.

2.1.2 Parameter setting

Based on the characteristics of the projectile's geometric structure, the shell is divided into *n* segments along the projectile axis, and each segment can be approximately a circular truncated cone or cylinder, as shown in Figure 1. The right-handed Cartesian coordinate system (Figure 2) is employed in the three-dimensional space of projectile fragments. The detonation point of the projectile is located at the origin of the coordinate system,and the projectile axis is the positive Z axis.The distance from the start point section and the endpoint section of each segment to the coordinate origin is designated as z_{i-1} and z_i respectively, and the section radius (taking the average value of the internal and external diameters of the projectile) is designated as r_{i-1} and r_i ($1 \leq i \leq n$). When $r_{i-1} = r_i$, the segment *i* is a cylinder, and otherwise, it is a truncated cone.

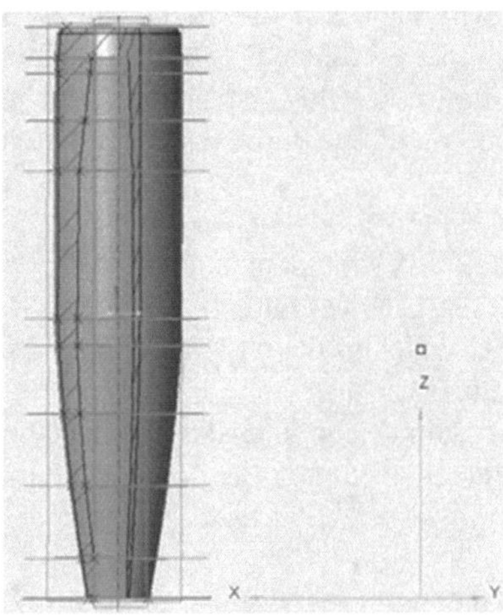

Figure 1. Diagram of projectile structure division

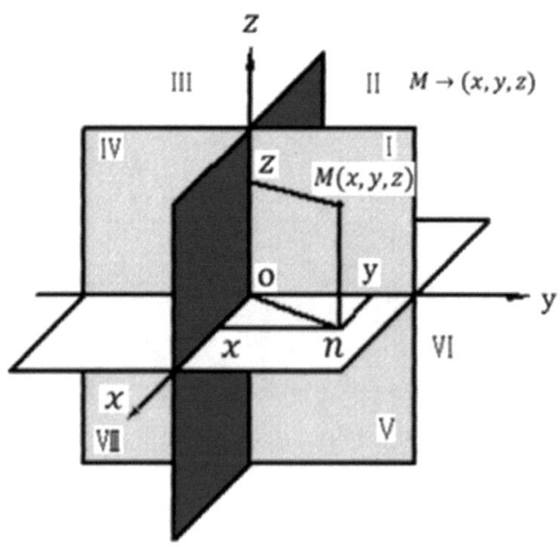

Figure 2. Right-handed Cartesian coordinate system

2.2 Mapping and inverse mapping of truncated cone projectile

2.2.1 Mapping from 3D spatial coordinates to 2D planar coordinates
For the convenience of calculation, let's take $r_{i-1} < r_i$ in the case of the circular truncated cone *i* as an example.The 2D planar coordinate system adopts polar coordinates. The truncated cone *i* is cut along the first octant section generatrix of the XZ plane, and the first, second, third and fourth octant is expanded in turn around the Z axis along the direction of the right index finger to form a sector, whose radian is β_i (Figure 3). If the sector center is the polar coordinate origin, then the generatrix length of the truncated cone *i* is:

$$l_i = \sqrt{(r_i - r_{i-1})^2 + (z_i - z_{i-1})^2} \tag{1}$$

the corresponding small cone's generatrix length:

$$l_{\min,\ i} = \frac{r_{i-1} * l_i}{(r_i - r_{i-1})} \tag{2}$$

the corresponding small cone's height:

$$h_{\min,\ i} = \frac{r_{i-1} * (z_i - z_{i-1})}{(r_i - r_{i-1})} \tag{3}$$

the angle between the corresponding small cone's generatrix and projectile axis:

$$\alpha_i = \sin^{-1} \frac{r_{i-1}}{l_{\min,\ i}} \tag{4}$$

and the sector radian:

$$\beta_i = \left(\frac{r_{i-1}}{l_{\min,\ i}}\right) * 2\pi \tag{5}$$

The vertex of the truncated cone *i* is translated along the Z axis to the origin (Figure 4), then the ordinates z_{i-1} and z_i become z'_{i-1} and z_i' after translation, and the translation distance:

$$d_i = z_{i-1} - h_{\min,\ i} \tag{6}$$

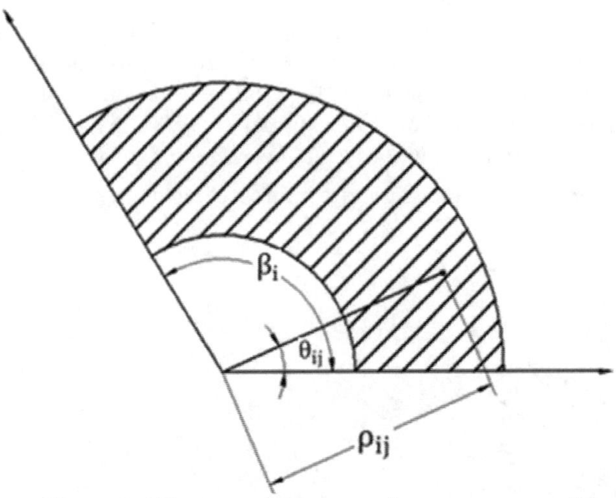

Figure 3. Diagram of the expande truncated cone *i*

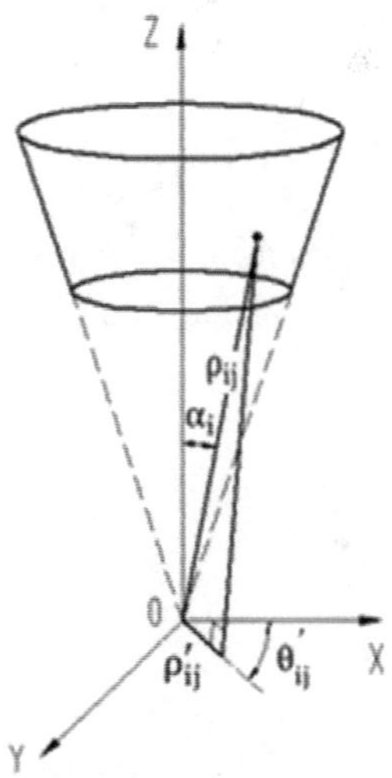

Figure 4. Diagram of coordinate conversion

2.2.2 Inverse mapping from 2D planar coordinates to 3D spatial coordinates
The total number of fragments formed after the detonation of the truncated cone *i* is m_i, and the polar coordinates of any fragment *j* ($1 \le j \le m_i$) mass-point in the expanded sector are (ρ_{ij}, θ_{ij}), where

$(l_{\min,\ i} \le \rho_{ij} \le l_{\min,\ i} + l_i,\ 0 \le \theta_{ij} \le \beta_i)$, then method for conversion from polar coordinates to rectangular coordinates $(x_{ij},\ y_{ij},\ z_{ij})$ (see Figure 4) is as follows:

Z axis coordinates after translation:

$$z'_{ij} = \rho_{ij} * \cos \alpha_i \tag{7}$$

the segment length of the projection of ρ_{ij} in the XY plane:

$$\rho'_{ij} = \rho_{ij} * \sin \alpha_i \tag{8}$$

The central angle of the conic projection of θ_{ij} on the XY plane is:

$$\theta'_{ij} = \frac{2\pi * \theta_{ij}}{\beta_i} \tag{9}$$

X-axis coordinates of the fragment *j*:

$$x_{ij} = \rho'_{ij} * \cos \theta'_{ij} = \rho_{ij} * \sin \alpha_i * \cos \frac{2\pi * \theta_{ij}}{\beta_i} \tag{10}$$

Y-axis coordinates of the fragment *j*:

$$y_{ij} = \rho'_{ij} * \sin \theta'_{ij} = \rho_{ij} * \sin \alpha_i * \sin \frac{2\pi * \theta_{ij}}{\beta_i} \tag{11}$$

Z-axis coordinates before translation of the fragment *j*: $z_{ij} = z'_{ij} + d_i = \rho_{ij} * \cos \alpha_i + d_i$.

$$z_{ij} = z'_{ij} + d_i = \rho_{ij} * \cos \alpha_i + d_i \tag{12}$$

2.3 Mapping and inverse mapping methods for cylindrical projectile

2.3.1 Mapping from 3D spatial coordinates to 2D planar coordinates

The plane rectangular coordinate LZ'axis is employed as the 2D coordinate system of the cylindrical projectile *i*. The cylinder *i* is cut along the section generatrix of the first octant in the XZ plane, and the first, second, third and fourth octant is expanded in turn around the Z axis along the direction of the right index finger to form a 2D plane-rectangle. The X/Z axis corresponds to the L/Z'axis respectively. In the plane rectangular coordinate system LZ', the side of the rectangle corresponding to the generatrix of cylinder *i* coincides with the Z'axis in the positive direction, and the side of the rectangle corresponding to the expanded circumference of the cylinder *i* is parallel to the X axis, i.e., it coincides with the extension line of the axial section of the cylinder *i* in the rectangular coordinate system LZ'(Figure 5). The coordinates of any point in the rectangle are expressed as $(l_{ij},\ z'_{ij})$, where $0 \le l_{ij} \le 2\pi r_{i-1},\ z_{i-1} \le z'_{ij} \le z_i$.

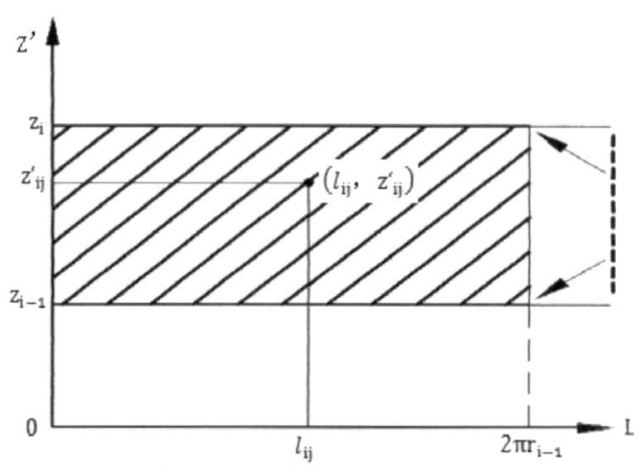

Figure 5 Diagram of the expanded cylinder *i*

2.3.2 Inverse mapping from 2D planar coordinates to 3D spatial coordinates

The total number of fragments formed after detonation of cylinder i ism_i, and the plane rectangular coordinates of any fragment j ($1 \leq j \leq m_i$) mass point in the expanded rectangular region are (l_{ij}, z'_{ij}). Then the method for conversion from 2D planar coordinates to 3D spatial rectangular coordinates (x_{ij}, y_{ij}, z_{ij}) is as follows:

The included angle between the radius projection of the fragment j in the XY plane and the X axis:

$$\theta'_{ij} = \frac{l_{ij}}{r_{i-1}} \tag{13}$$

X-axis coordinates of the fragment j:

$$x_{ij} = r_{i-1} * \cos\theta'_{ij} = r_{i-1} * \cos\frac{l_{ij}}{r_{i-1}} \tag{14}$$

Y-axis coordinates of the fragment j:

$$y_{ij} = r_{i-1} * \sin\theta'_{ij} = r_{i-1} * \sin\frac{l_{ij}}{r_{i-1}} \tag{15}$$

Z-axis coordinates of fragment j before translation:

$$z_{ij} = z'_{ij}. \tag{16}$$

3. Method of fragment random uniform distribution

Under the assumption that natural fragments of a projectile obey uniform distribution, we adopt the classical method of random variable generation in statistical theory - inverse transformation to solve the problem of random uniform distribution of fragments in 2D plane region[13].

3.1 Random uniform distribution method for two-dimensional sector region of truncated cone projectile

3.1.1 Probability density function of random uniform distribution of fragments in sector region[14]

The distribution of any natural fragment j (ρ_{ij},θ_{ij}) in the truncated cone i follows the uniform distribution in the sector region $D = \left\{ (\rho_{ij}, \theta_{ij}) \,\middle|\, l_{min,\ i} \leq \rho_{ij} \leq l_{min,\ i} + l_i,\ 0 \leq \theta_{ij} \leq \beta_i \right\}$ in the corresponding 2D planar coordinate system, where the density function of ρ_{ij} and θ_{ij} are:

$$f_\rho(\rho_{ij}) = \begin{cases} \beta_i \rho_{ij} \frac{1}{s}, & l_{min,\ i} \leq \rho_{ij} \leq l_{min,\ i} + l_i \\ & 0, \text{other} \end{cases} \tag{17}$$

$$f_\theta(\theta_{ij}) = \begin{cases} \frac{1}{\beta_i}, & 0 \leq \theta_{ij} \leq \beta_i \\ & 0, other \end{cases} \tag{18}$$

and ρ_{ij} and θ_{ij} are independent of each other, where $s = \frac{\beta_i}{2}\left\{\left(l_{min,\ i} + l_i\right)^2 - l^2_{min,\ i}\right\}$, β_i is the sector radian, l_i is the generatrix length of the truncated cone *I*, and $l_{min,\ i}$ is the corresponding small cone's generatrix length.

3.1.2 Fragment random uniform distribution algorithm[15]

According to the above random uniform distribution function in the 2D planar coordinate system, the calculation method to obtain the random distribution points formed by any fragment j of the truncated cone i is:

1) U (0,1) uniform random numbers $\varphi_{ij}^{(1)}$ and $\varphi_{ij}^{(2)}$ are generated independently, $j =$

1, 2, 3⋯, m;

2) The random points of $\{\theta_{ij}\}$ uniformly distributed on $[0, \beta_i]$ are obtained from

$$\theta_{ij} = \varphi_{ij}^{(1)}\beta_i \tag{19}$$

3) The random points of $\{\rho_{ij}\}$ uniformly distributed on $[l_{min,\ i}, l_{min,\ i} + l_i]$ are obtained from

$$\rho_{ij} = \sqrt{\varphi_{ij}^{(2)}\{(l_{min,\ i} + l_i)^2 - l_{min,\ i}^2\} + l_{min,\ i}^2} \tag{20}$$

According to the above algorithm, the coordinates of the fragment in two-dimensional planar coordinate system are obtained, and then the spatial distribution coordinates of the fragment are obtained through inverse mapping.

3.2 Random uniform distribution method for two-dimensional rectangular region of cylindrical projectile

3.2.1 Probability density function of random uniform distribution of fragments in rectangular region[14]

The distribution of any natural fragment *j* (l_{ij}, z'_{ij}) in the cylinder *i* in the corresponding 2D planar coordinate system obeys the uniform distribution in the rectangular region $D = [0, 2\pi r_{i-1}] \times [z_{i-1}, z_i]$, where $L \sim U[0, 2\pi r_{i-1}]$, $Z' \sim U[z_{i-1}, z_i]$, and L and Z' are independent of each other.

3.2.2 Fragment random uniform distribution algorithm[15]

According to the above random uniform distribution function in the 2D planar coordinate system, the calculation method to obtain the random distribution points formed by any fragment *j* of the cylinder *i* is:

1) U (0,1) uniform random numbers $\varphi_{ij}^{(1)}$ and $\varphi_{ij}^{(2)}$ are generated independently, $j =$ 1, 2, 3⋯, m;

2) The random points of $\{l_{ij}\}$ uniformly distributed on $[0, 2\pi r_{i-1}]$ are obtained from

$$l_{ij} = \varphi_{ij}^{(1)} 2\pi r_{i-1} \tag{21}$$

3) The random points of $\{z'_{ij}\}$ uniformly distributed on $[z_{i-1}, z_i]$ are obtained from

$$z'_{ij} = \varphi_{ij}^{(2)}(z_i - z_{i-1}) + z_{i-1} \tag{22}$$

According to the above algorithm, the coordinates of the fragment in two-dimensional planar coordinate system are obtained, and then the spatial distribution coordinates of the fragment are obtained through inverse mapping.

4. Computing verification

In this paper, a certain type of simulating projectile taking the fragments of a certain mass segment as an example was selected to verify the fragment spatial location distribution algorithm. The fragment parameters of this mass segment are shown in Table 1.

Table 1. The parameters of fragment

Number of projectile sections	Quality range(g)	Fragments quantity(piece)
10	1.0g~4.0g	2601

The algorithm of random uniform distribution of fragments studied in this paper was converted into computer program file by professional programming language (Figure 6). Thus, the spatial coordinate data output file of the fragment in this quality segment can be obtained, as shown in Figure 7.

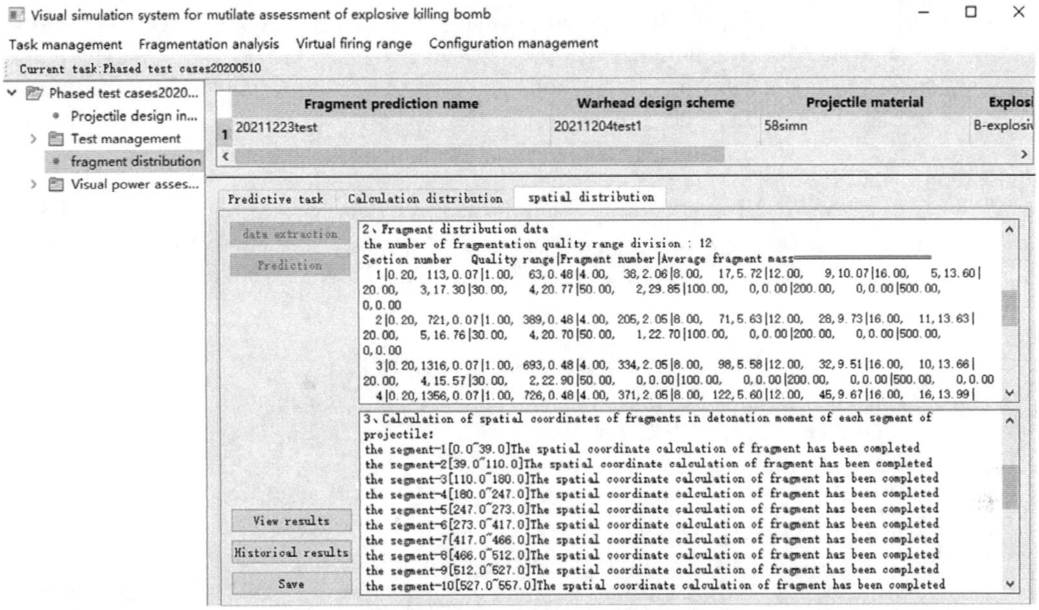

Figure 6 Computer program

	A	B	C	D	E	F	G	H	I	J	K	L
1	The spatial distribution table of the third mass rang fragment at the moment of detonation											
2	section number	mass rang number	fragment centroid coordinates			radius	quality	initial velocity	fragment shape			
3			X	Y	Z				parameter 1	parameter 2	parameter 3	parameter 4
4	1	3	0.69	-26.77	-11.32	29.06	2.06	1586.6	1	12	6	3
41	1	3	38.59	31.42	-8.68	32.6	2.06	1586.6	1	14	5	2
42	2	3	39.63	9.42	31.3	32.69	2.05	1586.6	1	10	7	3
246	2	3	109.43	18.81	38.1	42.49	2.05	1586.6	1	14	5	2
247	3	3	110.2	11	41.14	42.59	2.05	1660.45	1	14	5	2
580	3	3	179.53	49.41	3.59	49.54	2.05	1660.45	1	8	8	4
581	4	3	180.28	-23.82	43.51	49.6	2.05	1660.45	1	10	7	3
951	4	3	246.89	-38.15	37.85	53.74	2.05	1660.45	1	8	8	4
952	5	3	247.05	46.88	26.29	53.75	2.04	1586.6	1	16	6	2
1105	5	3	272.9	-27.76	-46.43	54.1	2.04	1586.6	1	13	8	2
1106	6	3	273.15	-51.51	16.54	54.1	2.05	1507.67	1	12	6	3
1958	6	3	416.78	18.26	-50.02	53.25	2.05	1507.67	1	16	6	2
1959	7	3	417.04	-13.45	51.52	53.25	2.05	1281.55	1	18	5	2
2196	7	3	465.87	48.29	-14.3	50.36	2.05	1281.55	1	13	8	2
2197	8	3	466.33	-4.66	50.11	50.33	2.06	1056.39	1	8	8	4
2402	8	3	511.98	-36.18	-30.99	47.64	2.06	1056.39	1	13	8	2
2403	9	3	512.04	-14.43	45.4	47.64	2.05	535.98	1	12	6	3
2568	9	3	526.8	-46.29	6.58	46.76	2.05	535.98	1	8	8	4
2569	10	3	527.21	-42.6	19.23	46.74	2.04	535.98	1	13	8	2
2604	10	3	555.99	40.26	-20.59	45.22	2.04	535.98	1	12	6	3
2605												

Figure 7 Fragment spatial coordinates data output file

It can be seen from figure 5 and figure 6 that the spatial location coordinates of fragments can be accurately obtained through the spatial location distribution method studied in this paper, thus providing technical support for the prediction of projectile power.

5. Conclusions

In this study, the Monte Carlo stochastic points method was used to solve the uniform distribution of fragment spatial position through the mutual transformation of 3D spatial position coordinates and 2D planar position coordinates.An efficient digital simulation algorithm for random uniform distribution of natural fragment spatial position was developed to realize the visual simulation of natural fragment spatial distribution, which is more consistent with the spatial distribution of real projectile fragments. The following conclusions are drawn:

1. For projectile segments with different shapes, the 3D spatial position and 2D plane position of fragments can be exchanged by coordinate mapping and inverse mapping.
2. The Monte Carlo stochastic points method was used to solve the problem of random uniform distribution of fragments in 2D plane region.

Acknowledgments

This work was supported by the stability support special open innovation fund from Bureau of Science,Technology and Industry for National Defense in 2020(project number:WDZC2020JJ002) and Venture Fund of Ningbo Research Institute in2021(project number: NBFJ2021-21).

References

[1] Wang, Sh.Sh. (2019) Terminal Effects. Science Press, Bei Jing.
[2] Gold, V.M., Baker, E.L., Poulos, W.J., (2007) Modeling fragmentation performance of natural and controlled fragmentation munitions. In: 23RD international symposium on ballistics. Tarragona. pp. 193-200.
[3] Gold, V.M. (2017) Fragmentation model for large L/D (Length over Diameter) explosive fragmentation warheads. J. Defence Technology., 13: 300–309.
[4] Huang, G.Y., Li, W., Feng, Sh.Sh. (2015) Axial distribution of Fragment Velocities from cylindrical casing under explosive loading. J. International Journal of Impact Engineering., 76: 20–27.
[5] Deng, Y.F., Yuan, J.J., Lu, M.J. (2017) Numerical investigation on failure mode and 1aw of different shape fragments impacting aluminum alloy target. J. 0rdnance Material Science and Engineering., 40: 81–85.
[6] Miao, Ch.Zh., Liang, Z.Y., Deng, D.Zh., Zhao, W.J., Wang, Y.Q. (2018) Research on Formation and Damage Effects of Pre-formed Fragments in Different Shapes. J. Journal of Gun Launch & Control., 39: 11–15.
[7] Prytz, A.K., Odegardstuen, G.(2011) 26THINTERNATIONAL SYMPOSIUM ON BALLISTICS. Miami. pp. 12-16.
[8] Huang, J.W., Li, W.B., Zheng, Y., Chen, W. (2013) Formation of natural fragments from large caliber shells. J. O. I. Automation., 32: 20–23.
[9] Gold, V.M., Baker, E.L., Poulos, W.J., Fuchs, B.E. (2006) PAFRAG modeling of explosive fragmentation munitions performance. J. International journal of impact engineering., 33: 294–304.
[10] Kang, Zh.Y., Yuan, Sh.Q. (2019) Reconstruction and redevelopment of prediction software for projectile fragment quality and quantity distribution. J. Journal of projectiles,rockets,missiles and guidance., 39: 83–87.
[11] Baike baidu, (2008). Three dimensional coordinate system. https://baike.baidu.com/item/ three dimensional coordinate system /10252312.
[12] Baike baidu, (2021). Cartesian coordinate system. https://baike.baidu.com/item/ cartesian coordinate system /4522878.
[13] Zhou, Y.D., He, P., Ning, J.H. (2021) Methods and applications of random simulation. Higher Education Press, Bei Jing.
[14] Li, J., Chen, J.B., Zhang, L.L. (2010) Probability density function estimation of stochastic processes. J. CHINESE JOURNAL OF APPLIED MECHANICS., 27: 481–485.
[15] Zhang, Y., Cheng, Sh.J. (2017) Probability theory and mathematical statistics. Tsinghua University Press, Bei Jing.

Distribution analysis method of key targets based on big data analysis engine for maritime space-time trajectory

Yue Xu[1a*], Zhaozhe Yang[1b*], Xinyu Zhang[1c]

[1]Systems Engineering Research Institute of CSSC, Beijing 100094, China

[a*]xuyue_331@163.com, [b*]2431195401@qq.com, [c]469873028@qq.com

Abstract. Now a large amount of maritime space-time trajectory data has been collected, but its deep value is unable to be tapped, and the analysis of the distribution of key targets has also encountered bottlenecks. The emergence of big data technology provides us with a solution. This paper proposes a big data analysis engine for maritime space-time trajectory, which makes full use of the data characteristics and improves the overall computing performance of the engine. Based on this engine, a dispersal analysis algorithm for key military targets at sea is proposed to achieve rapid retrieval and dispersal analysis of key marine targets. It has been tested and verified that this engine has excellent performance and can become an efficient query and analysis engine for maritime space-time trajectory data. The key target distribution analysis algorithm has certain research significance and practical value.

1. Introduction

Space-time trajectory is the data that continuously samples the moving process of moving objects, mainly including sampling time, object position information, speed, direction, etc. Maritime space-time trajectory mainly refers to the navigation trajectory of surface, air and underwater military targets or civil aircraft and ships, which can be obtained by military shipborne detection equipment or AIS and ADS-B data. Among them, the space-time trajectory of military maritime targets has purposeful, regular, regional and obvious motion characteristics. In-depth mining can obtain greater intelligence information.

The amount of space-time trajectory data of military maritime targets is growing rapidly. In the past, the analysis of data by intelligence departments stayed in small-scale and nearby time, with low analysis level and poor analysis effect. However, with the emergence and in-depth application of big data technology, it is possible to deeply mine the motion patterns of maritime targets, especially to obtain the motion information of enemy targets.

In order to improve the efficiency of big data query and analysis, major companies have launched technical research. In the second wave of Google's technology, based on hive and Dremel, the emerging big data company cloudera opened the big data query and analysis engine impala, hortonworks opened stinger, and fackbook opened presto. Similar to pregel, UC Berkeley amplab laboratory has developed spark graph computing framework and open source big data query analysis engine shark with spark as the core.

Unfortunately, even with the technical support of many big data query and analysis engines mentioned above, the mining and analysis of massive space-time trajectory data is still limited by performance. The essence is that the major general big data query engines are aimed at one query or analysis task. However, when mining space-time trajectory data on the sea, they are faced with multiple operations and multi-layer relationship mining, Moreover, data sources may involve many types, such as local files, Oracle, HDFS, etc. In the face of multi-source heterogeneous massive spatio-temporal trajectory data, it is an urgent problem to quickly obtain the target that meets the specific retrieval conditions.

Therefore, this paper proposes a big data analysis engine for space-time trajectory on the sea (BDAEMST), which provides a concise query analysis language like SQL and a high-performance parallel analysis model. Based on this model, a dispersion analysis algorithm for key military targets on the sea is proposed, Realize the rapid retrieval and dispersion analysis of key offshore targets.

2. Big data analysis engine for maritime space-time trajectory

The basic idea of BDAEMST engine is:

(1) Minimize the IO times and strive to perform more tasks in one data traversal;

(2) It can be flexibly used on different data storage systems (including local files, mysql, Oracle, HBase, HDFS, etc.);

(3) Based on spark, a new query language for marine space-time trajectory data is generated, named trackQL;

(4) Not limited to SQL statements, expand the application scope of trackQL, so that data mining algorithms can be expressed.

2.1. trackQL

TrackQL is a unified query language for querying and analyzing marine space-time trajectory data. It is encapsulated on spark. The expression is defined as follows.

Definition 1 (trackQL expression)

EXECUTE BLANKSPACE functions BLANKSPACE FROM BLANKSPACE tables BLANKSPACE WHERE BLANKSPACE filters BLANKSPACE TO tables。

The reserved words are execute, from, where, and, or and to.

Table 1. TrackQL lexical definition

<table>
<tr><th>TrackQL lexical definition</th></tr>
<tr><td>functions : functions COMMA function | function
function : function_name LPAREN parameters RPAREN
function_name : ID
parameters : parameters COMMA parameter | parameter
parameter : ID
tables : tables COMMA table | table
table : ID DOT ID
filters : filters BLANKSPACE AND BLANKSPACE filters | filters BLANKSPACE OR BLANKSPACE filters |
LPAREN filters RPAREN | filter_exp
filter_exp : field_name COMPOP value
field_name : ID
value : NUMBER | STRING | DATETIME
ID : [a-zA-Z][a-zA-Z_0-9]*
NUMBER : (\-|\+)?\d+(\.\d+)?
STRING : \"(\\.|[^"\\])*\"
DATETIME : \d{4}-\d{1,2}-\d{1,2}
COMMA : >|>=|<|<=|=|<>
COMPOP : ,
DOT : .
BLANKSPACE : \s+
LPAREN : \(
RPAREN : \)
EXECUTE : 'EXECUTE' | 'execute'
FROM : 'FROM ' | 'from'
WHERE : 'WHERE' | 'where'
AND : 'AND' | 'and'
OR : 'OR' | 'or'
TO : 'TO' | 'to'</td></tr>
</table>

TrackQL example: execute sum(attr) from HBase.tracks where TN='010101' and TimeStamp > '20180101' to MySQL.result。

The reserved words are execute, from, where, and, to;The task functions to be executed are sum, the parameter parameters are attr, and the query library table tables are HBase.tracks. The filter conditions are TN ='010101' and timestamp > '20180101'. Save the results to the result table of MySQL.

TrackQL is a kind of SQL fast data analysis language provided by BDAEMST engine for upper layer big data analysis application services, which makes analysts pay more attention to data analysis rather than code writing.

2.2. Engine architecture
BDAEMST engine is built on spark and mainly consists of trackQL syntax parsing component, task execution component, data access component, preprocessing component and data output component.

2.2.1. trackQL Syntax parsing component
The trackQL statement of the application layer is parsed into query analysis task functions and its parameters, which are mainly realized by Lex and yacc. Each syntax rule is defined as a python method, and the statement of the method implements the semantic behavior of the corresponding rule.

After the syntax parsing component, trackQL outputs the task functions to be executed, parameter parameters, Query Library tables, filter criteria filters and save result location tables.

Through the reflection mechanism, you can call the "data access component" according to the query library table tables to load the data of various storage methods, call the "task execution component" according to the functions, parameters and filters for parallel calculation, and call the "data output component" according to the saved result location tables to save as multiple storage methods.

2.2.2. Data access component
The data access component connects spark with marine space-time trajectory data with different storage methods, so that spark can access it directly.Different storage methods include local files (. TXT,. XSL,. XML, etc.), relational databases (mysql, Oracle, etc.), distributed storage (HDFS, HBase, etc.), network transmission format (Jason, etc.).

2.2.3. Pretreatment component
In order to unify the marine space-time trajectory data from different sources and formats, after loading the data with different storage methods, preprocessing operation is required to convert the space-time trajectory data with different formats and coding specifications into the same format, so as to ensure the standardization and consistency of the data. In this paper, data from different sources are preprocessed as follows:

(1) Unify the unit and accuracy of data.

(2) The data in different coding formats are decoded and uniformly encoded.

(3) For data with inconsistent message frequency, it is converted into one message per second by linear interpolation.

(4) Re batch the target batch number for reuse;

(5) Unify the attributes of targets with inconsistent attributes.

2.2.4. Task execution component
The task execution component encapsulates a series of mining model interfaces for marine space-time trajectory data on the basis of spark. This paper introduces the task component with the key target walking analysis model as a typical application. See Chapter 3 for details.

Each row of space-time trajectory data has a recording unit, which means that when parallel computing is performed on space-time trajectory data, multiple logical operations can be performed on each row in RDD without multiple retrieval and calculation like spark SQL, so as to save computing resources and improve the efficiency of parallel computing.

2.2.5. Data output component

The data output component connects the calculated output results of the data preprocessing component and the task execution component with different storage methods, so that spark can output directly. Corresponding to the data access component, different storage methods include local files (. TXT,. XSL,. XML, etc.), relational databases (mysql, Oracle, etc.), distributed storage (HDFS, HBase, etc.), network transmission format (Jason, etc.).

2.3. Execution process

The execution process of BDAEMST engine is as follows:

(1) The trackQL is parsed by the syntax parsing component to obtain the parse result parse_result;

(2) Create shared variables through sparkcontext. Parse can be used on different nodes after one resolution_result;

(3) From parse_Obtain the task function list, data source, filter conditions, etc. in result;

(4) Create an action sequence according to the analysis result, LoadAction - > TaskAction - > StoreAction;

(5) Execute LoadAction and return RDD (K, V);

(6) Filter RDD from RDD according to the filter conditions of trackQL to obtain the results;

(7) Execute different task functions in the task execution component according to functions;

(8) Write the execution result to the specified structure through the data output component.

3. Key target dispersion analysis algorithm

Assuming that a target tar has appeared n times, that is, there are N tracks, its historical track set is expressed as $R = (R_1, R_2,..., R_i,..., R_n)$, where $R_i = r_1r_2...r_m$, R_i represents the track point of this track when the timestamp is I, and its information includes time, longitude, latitude, altitude, speed, azimuth, distance, etc. Through the preprocessing component of the BDAEMST engine, the timestamp of the historical track has been synchronized, so the timestamp misalignment problem will not occur.

For the target tar, the calculation formula of its importance imp is as follows:

$$Imp = f(type) + f(max\_speed) + f(min\_distance) + \sum_{k=1}^{n} f(R_k) \tag{1}$$

$$f(type) = \begin{cases} 1, & dangerous\ type \\ 0, & common\ type \end{cases} \tag{2}$$

$$f(max\_speed) = \begin{cases} 1, & max\_speed >= S \\ 0, & max\_speed < S \end{cases} \tag{3}$$

$$max\_speed = max(max(speed_{R_1}), max(speed_{R_2}),..., max(speed_{R_n})) \tag{4}$$

$$f(min\_distance) = \begin{cases} 1, & min\_diatance <= D \\ 0, & min\_diatance > D \end{cases} \tag{5}$$

$$min\_distance = min(min(distance_{R_1}), min(distance_{R_2}),..., min(diatance_{R_n})) \tag{6}$$

$$f(R_k) = \begin{cases} 1, & Number\ of\ dangerous\ actions > 0 \\ 0, & no\ dangerous\ action \end{cases} \tag{7}$$

Among them, S and D are common parameters, and dangerous actions refer to behaviors unfavorable to us, such as accompanying us, quickly crossing our supervision, etc.

According to the importance Imp and the needs of the commander, the importance of historical objectives is divided into different levels. Here, we define the importance level as four levels, of which level 1 is the highest level, level 2 is the second highest level, level 3 is the general level and level 4 is the lowest level. The calculation formula of importance level is as follows:

$$\text{rank} = \begin{cases} 1, Imp >= a \\ 2, b <= Imp < a \\ 3, c <= Imp < b \\ 0, Imp < c \end{cases} \tag{8}$$

Among them, a, b and c are common parameters, and a > b > c.

In order to facilitate the subsequent dispersion analysis of different targets, we convert the importance level into the importance value for statistics, as shown below:

$$f(rank) = \begin{cases} 100 & rank = 1 \\ 50 & rank = 2 \\ 1 & rank = 3 \\ 0 & rank = 0 \end{cases} \tag{9}$$

The distribution analysis of key targets is to count the historical trajectories of all targets with important levels 1 and 2 in terms of azimuth, distance and height, so as to find the distribution law of key targets in azimuth, distance and height. The statistics are classified according to surface targets, underwater targets and air targets. Azimuth and distance statistics can be carried out for key targets on the water surface; Azimuth and distance statistics can be carried out for underwater key targets; Statistics of azimuth, distance and altitude can be made for key targets in the air.

For convenience, azimuth statistics specifies the size of azimuth interval and divides 360 ° evenly. For example, the pre specified partition angle is a °, will be generated 360/a intervals. Distance statistics specifies the distance interval and the maximum distance. Under the maximum distance, the data in each distance interval will be counted. Similarly, height statistics specify the height interval size and maximum height.

$$F\_direction(i) = \sum_{x=1}^{Q} f(rank) \tag{10}$$

Where I is the ith azimuth interval, and Q is the number of targets with an important level of 1 or 2 in the azimuth interval data source. The statistical values of all azimuth intervals are calculated and displayed in the statistical chart, that is, the azimuth dispersion analysis of key targets is completed.

Similarly, the calculation formula of distance and height dispersion analysis of key targets is as follows.

$$F\_distance(i) = \sum_{x=1}^{U} f(rank) \tag{11}$$

$$F\_height(i) = \sum_{x=1}^{V} f(rank) \tag{12}$$

The task execution component of BDAEMST engine provides target importance level calculation interface and key target azimuth, distance and height statistics interface, which can quickly complete target dispersion analysis.

4. Experiment

In order to test the efficiency of the calculation model of key objectives, the simulation data is generated. The simulation data include three types of models: surface target, underwater target and air target. The surface target adopts the four degree of freedom dynamic model, and the air target and underwater target adopt the six degree of freedom dynamic model. In order to simulate the detection of various types of targets in real data, water surface, underwater and air targets are generated according to 5:1:4 probability. Among them, the goal of reaching important level 1 accounts for about 10%, and the goal of reaching important level 2 accounts for about 30%.The azimuth of each type of target is distributed at 0 ~ 360 degrees, the height range is 0 ~ 10000 m (meter), and the distance range is 0 ~ 200 N miles. It is set that all types of targets are distributed more in the azimuth range of 240 ~ 300 °, the distance range of 0 ~ 10 N mile and the height range of 0 ~ 1000 m, and are evenly distributed in other ranges. The original data contains 24 fields such as target nationality, friend or foe attribute, type, model, longitude and latitude. A total of 1 TB (terabyte) and about 6 000 pieces of data were generated.

In this paper, the big data test cluster is composed of one master node and three slave nodes; Each computer has the same configuration, 32 GB (Gigabit) memory and Intel Core i7-7700 processor. The running test time for key target calculation under different data volumes is given below.

Table 2. Algorithm run time test.

Data volume / GB	Number of tests	Average running time / MS
10	100	51 277
100	100	332 043
500	100	1 553 310
1000	100	3 142 606

This paper uses the open source visualization library echarts to visualize the results. The results of dispersion analysis of simulated data are given below. In statistical analysis, the lowest importance level of the target is set as level 2, that is, the targets with importance levels of level 1 and 2 are regarded as key targets. The results of azimuth statistical analysis of water surface targets are displayed in the form of radar chart, as shown in Figure 1.

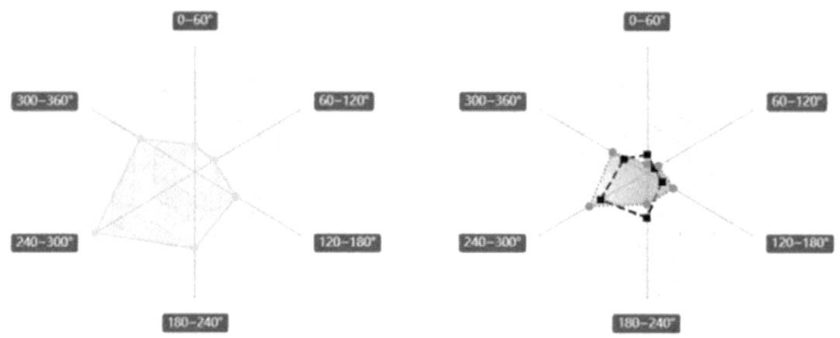

Fig. 1 Radar map of bearing statistical

In Figure 1, the right figure draws level 1 key objectives and level 2 key objectives in the figure at the same time. Among them, the black rectangle as the vertex is the secondary key target, and the blue circle as the vertex is the primary target. It can be seen that the distribution of key targets is the most at about 240 ~ 300 °. The figure on the left shows the total important value of the target. It can be seen that the total important value of the target is more affected by the primary target and is closer to the distribution of the important value of the primary target. The above analysis results are consistent with the azimuth setting of simulated data.

The results of statistical analysis of water surface target distance are displayed in the form of stacked histogram, as shown in Figure 2.

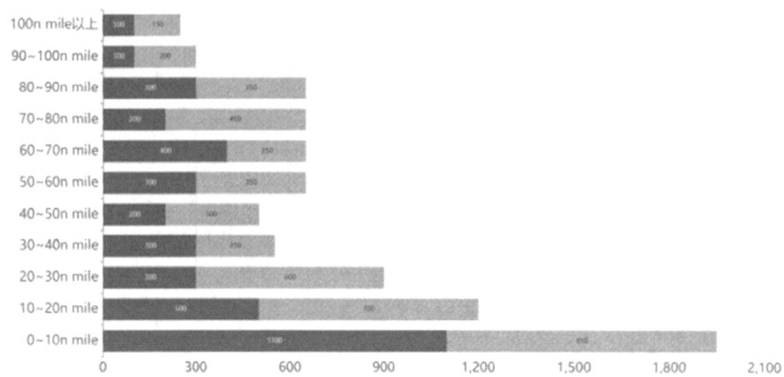

Fig. 2 distance statistics stack histogram

In Figure 2, the blue column at the bottom is the primary key target, and the green column at the top is the secondary key target. It can be seen that the total importance value of the target is large in the range of 0 ~ 10 N miles; On the whole, the importance value decreases with the increase of distance. The above analysis results are consistent with the simulation data distance setting

The results of statistical analysis of air target height are displayed in the form of stacked histogram, as shown in Figure 3.

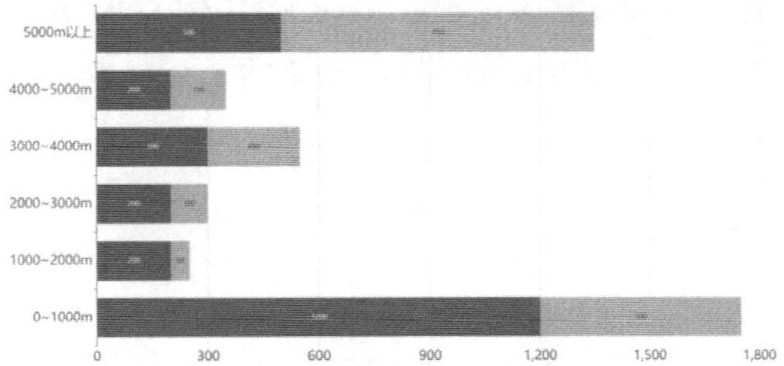

Fig. 3 height statistics stack histogram

In Figure 3, the blue column at the bottom is the primary key target, and the green column at the top is the secondary key target. The target importance values above 0 ~ 1000 m and 5000 m are higher.

5. Conclusion

This paper presents a big data analysis engine for marine space-time trajectory data, which is expected to be applied to massive space-time trajectory data intelligence mining of marine military targets. This engine provides a concise query analysis language similar to SQL and a high-performance parallel analysis model. Based on this engine, a dispersion analysis algorithm for maritime key military targets is proposed to realize rapid retrieval and dispersion analysis of maritime key targets. From the test results, the big data analysis engine and key target dispersion analysis algorithm proposed in this paper have certain research significance and practical value.

In the future, we will increase research on the optimization and application of the engine, further improve its performance and stability, and propose more valuable maritime space-time trajectory mining algorithms.

References

[1] Qiang G., Fengli Z., Ruijin W., Fan Z.. (2017). Trajectory Big Data: A Review of Key Technologies in Data Processing. Journal of Software, 28(4), 34.

[2] Xindong W., Shengya J.. (2018). Comparison of Mapreduce and Spark for Big Data Analysis. Journal of Software, 29(6), 22.

[3] Li, Z. , Fei, W. , & Crofoot, M. C. . (2013). Mining Following Relationships in Movement Data. IEEE International Conference on Data Mining. IEEE.

[4] Zaharia, M. , Chowdhury, M. , Franklin, M. J. , Shenker, S. , & Stoica, I. . (2010). Spark: Cluster computing with working sets.

[5] Jiuxuan L.. (2015). A Massive Spatial Remote Sensing Data Retrieval System Based on HBase. (Doctoral dissertation, Zhejiang University).

[6] Chang, F. . (2006). Bigtable : A Distributed Storage System for Structured Data. 7th USENIX Symposium on Operating Systems Design and Implementation (OSDI), 2006.

[7] LarsGeorge, George, Jie J., Jia L, Zhiyuan D. (2013). HBase Authoritative Guide. People's Posts and Telecommunications Publishing House.

[8] Shun L, Shenghua Z., Meizheng Z., Long G.. (2020). Research and Implementation of Space-Time Trajectory Data Acquisition and Compression Technology. Information Technology, 44(1), 5.

Journal of Physics: Conference Series 2258 (2022) 012024 doi:10.1088/1742-6596/2258/1/012024

Research on power load forecasting algorithm of power system based on artificial firefly

Haomin Chen[1*] and Jinzhao Liang[1]

[1] Digital Grid Research Institute,CSG, Shenzhen, China

[*]Corresponding author's e-mail: Chenhm@csg.cn

Abstract: The power system is constantly upgraded with the development of information technology. The digital management of power system effectively improves the operation and maintenance efficiency of power system. Although the current power system algorithms have made great progress, there has been no effective prediction algorithm in the field of load forecasting. Load forecasting is a key operation and planning approach of power system. It affects many decisions of power system, such as economic dispatching, automatic generation control, safety evaluation, maintenance dispatching and energy commercialization [1]. In view of the above problems, the power load forecasting algorithm of power system based on artificial firefly will be studied. In the existing resource management framework of power system digital grid platform, the power load business of power system is studied. The artificial firefly algorithm is used to complete the research of power load forecasting algorithm. Simulation results show that the established algorithm has good time complexity and space complexity, and the accuracy of power load forecasting results is improved by at least 22.8%.

1. Introduction

Load forecasting has always been a key operation and planning approach of power system. It affects many decisions of power system, such as economic dispatching, automatic generation control, safety evaluation, maintenance dispatching and energy commercialization. Accurate power load forecasting can economically and reasonably arrange the start and stop of power system generator units, which plays an important role in maintaining the safety and stability of power grid operation, maintaining the normal production and life of society, and effectively reducing power generation costs [1-2]. Driven by information science and technology, especially the development of China Southern Power Grid Digital Grid technology, the power system data acquisition and monitoring technology has developed unprecedentedly, and more and more load data are collected. How to mine the effective information contained in these data to improve the accuracy of power load forecasting, It has become one of the hot issues in the field of power system research [3].

The accuracy of power load forecasting directly affects the economic benefits between power grid and power plant. Uncertainty and randomness are one of the remarkable characteristics of load. At the same time, the influence of natural environment and weather factors should be considered. Therefore, the power load forecasting is determined by multi factor influencing factors. How to improve the accuracy of power load forecasting has become an urgent problem to be solved. Long Jinlian, Zhang Yufen and others proposed a short-term power load forecasting method based on Improved Particle Swarm Optimization LSSVM algorithm, Using particle swarm optimization (PSO) to optimize the non optimal particle problem caused by the gradient direction of velocity change in least square support vector machine (LSSVM), a short-term power load forecasting method based on adaptive particle swarm optimization LSSVM parameters is proposed [4] 。 Wang Baoyi and Zhao Shuo, in their paper on distributed power load forecasting algorithm based on cloud computing and limit learning machine [5], proposed online sequence optimization of limit learning machine forecasting algorithm according to the characteristics of power load data; The idea of distributed and multi-agent is introduced to improve the prediction accuracy of load forecasting algorithm; The proposed algorithm model is parallelized and improved by using the MapReduce programming framework of cloud computing to improve its ability to process massive high-dimensional data [6]. The above methods are optimized through the group optimization algorithm and a certain operation framework. In this paper, a power system power load forecasting algorithm based on artificial firefly is proposed.

2. Artificial firefly algorithm

Artificial firefly swarm optimization (GSO) is a new swarm intelligence optimization algorithm proposed by Indian scholars K.N. krishnand and D. ghose at the IEEE Swarm Intelligence Conference in 2005. In 2009, the Cambridge scholar Xin She Yang proposed the firefly algorithm according to the luminous behavior of fireflies in nature (Firefly algorithm, FA). Since the two firefly algorithms were proposed, scholars from various countries have studied, improved and applied the two algorithms. After several years of development, the firefly algorithm has good application prospects in the optimization process of continuous space and some production scheduling. In this paper, the artificial firefly swarm algorithm is used to improve the power load forecasting algorithm [7] 。 In the existing resource management framework of power system digital grid platform, the power load business of power system is studied. The artificial firefly algorithm is used to complete the research of power load forecasting algorithm.

2.1. Basic flow of artificial firefly algorithm

The artificial firefly swarm optimization (GSO) algorithm walks in the solution space for each artificial firefly. These fireflies carry fluorescence and have their own line of sight, It is called decision domain （local-decision range）. Their brightness is related to the target value of their position. The brighter the firefly, the better its position, that is, it has a better objective function value. The firefly will look for a set of neighbors in the decision domain. In the set, the brighter the neighbors have a higher attraction to attract the firefly to move in this direction and fly in each direction The orientation will change with the selected neighbors [8]. In addition, the size of the decision domain will be affected by the number of neighbors. When the neighbor density is lower, the firefly's decision radius will increase to find more neighbors; When the neighbor density is higher, its decision radius will be reduced. Finally, most fireflies gather in multiple locations, that is, reaching extreme points. In short, GSO algorithm mainly includes four stages: Firefly deployment (initialization), fluorescein update, location update and decision radius update.

The basic flow of the algorithm is as follows:

1. Initialize each parameter;
2. Randomly initialize the position of the i(i=1,2,...,n)i(i=1,2,...,n) firefly within the search range of the objective function;
3. The fluorescein value li(t)li(t) of firefly ii at tt time was calculated;
4. Each firefly selects individuals with higher fluorescein value than itself to form its domain set Ni(t)Ni(t), in which 0<rid(t)≤rs0<rdi(t)≤rs. rsrs is the perception radius of firefly.
5. Calculate the probability pij(t)pij(t) of firefly ii moving to individual jj in the neighborhood set;
6. Use roulette to calculate the individual jj, then move and update the position;
7. Update the value of firefly dynamic decision domain radius;
8. Whether the maximum number of iterations or the required accuracy is reached. If this is reached, turn to the next step, otherwise turn to step 4;
9. Output global extremum and optimal individual value.

2.2. Power load forecasting algorithm based on artificial firefly

When applying artificial firefly algorithm to power load forecasting, it is necessary to use limit learning function in power load forecasting to improve the accuracy and efficiency of power load and reduce the problems caused by difficult prediction parameters [9]. Suppose that the power load sample of the power system is $[x_i, y_i]$, the number of hidden layer units of the limit learning machine is k, and the excitation function is $g(x)$. through artificial intelligence, the output of the limit learning machine is:

$$O_i = \sum_{j=1}^{k} \beta_j g\left(a_j x_i + d_j\right) \tag{1}$$

In the above formula, β_j is the weight connecting the j th hidden layer node and the output node, a_j is the weight matrix connecting the j th hidden layer node and the input node. d_j is the offset value of the j th hidden layer node. $g(x)$ is set as RBF function in this paper. In the training process of limit learning, the above parameters are solved. The equation of relation between parameters is transformed into matrix form, and the parameters are solved by minimal 2-norm least squares [10].

The improved parameter optimization of limit learning machine is carried out by using artificial firefly. In the parameter training stage of limit learning machine, NP fireflies are randomly generated. The fluorescein Volatilization Coefficient of fireflies is ρ, the extraction proportion of parameter fitness is γ, the field change rate is β and the decision radius is r_d. Then the update rules of fireflies are as follows [11]:

$$l_i(g) = (1-\rho) * l_i(g-1) + \gamma * f\left(x_i(g)\right) \tag{2}$$

Calculate the movement probability of the firefly within the neighbor radius of the firefly: the i firefly moves by comparing the fluorescence brightness value of the firefly in its neighbor set. If it is less than the fluorescence brightness value of the neighbor, otherwise it does not move. The probability of firefly movement is as follows [11]:

$$P_{i,j} = \left(l_j(g) - l_i(g)\right) \Big/ \sum_{k \in N_i(t)} l_k(t) - l_i(t) \tag{3}$$

After determining to move a unit step to the neighbor firefly, recalculate the perception radius of the moved firefly. Calculate the parameter fitness value under the new position to judge whether it is global optimal. If it is globally optimal, the corresponding value of firefly at this time is the optimal parameter set of limit learning machine; Otherwise, continue the firefly algorithm. After training, the output result of the improved limit learning machine is the short-term load forecasting value of the power system.

3. Simulation experiment

This section will verify the effectiveness of the power system load forecasting algorithm based on artificial firefly through a series of comparative simulation experiments. According to the actual operation of the power system, the overall scheme of the comparative simulation experiment is designed, and the relevant algorithm input data and comparative experiment parameters are set. All the algorithms involved in the algorithm performance test in the simulation experiment are implemented in Java language and run in the same running environment. In this simulation experiment, the algorithm is applied to the computer simulation platform configured with Intel I core I i7-8543qp CPU with main core frequency of 3.75ghz, running memory of 16GB and operating system of 64 bit windows operating system, so as to ensure the consistency of the overall experimental environment of the simulation experiment and avoid the interference of uncontrollable factors on the experimental results [12].

3.1. Experimental scheme design

In this simulation experiment, it is compared with the power system load forecasting algorithm based on artificial firefly established in this paper. Compared algorithms include short-term power load forecasting algorithm based on Improved Particle Swarm Optimization LSSVM algorithm [4] (recorded as algorithm 1); distributed power load forecasting algorithm based on cloud computing and limit learning machine [5] (recorded as algorithm 2). The two algorithms and the power system power load forecasting algorithm based on artificial firefly proposed in this paper are applied to the same series of power system digital grid platform resource management framework to compare the forecasting accuracy and running time of the load forecasting algorithm, that is, the time complexity and space complexity of the algorithm [13].

On the computer simulation platform, a simulation processing environment is established with reference to the actual digital power grid platform resource management framework. All parameters in the simulation environment are set according to the actual operation parameters of the digital power grid platform resource management framework, and the parameters can be adjusted according to the experimental requirements [14]. In this experiment, it is set that the resource management framework of the digital power grid platform in the experimental environment has four functional partitions, corresponding to the main power system operation and maintenance functional modules of power system, such as load forecasting, fault monitoring and alarm processing, power resource scheduling and so on. Each partition corresponds to the next level power system with different scale and different power generation level. There are differences between the total power use and power dispatching change data tables in the partition. According to the relevant regulations of the power industry, the power system of the same or similar scale is divided into the same order of magnitude. In this experiment, the power system is divided into five different orders of magnitude. The specific algorithm simulation process is as follows:

Connect another computer with the same configuration to the simulation processing environment built by the experiment, and the computer will simulate the load change data sent by the power grid and power generation change to the platform. The computer is used to send different amounts of load data to the in the simulation experimental environment. Algorithm 1, algorithm 2 and the algorithm in this paper are used to process the corresponding load data and forecast, and the accuracy of load forecasting is counted. At the same time, the overall time-consuming of the algorithm in this process is counted to evaluate the time complexity and performance of the algorithm. Change the experimental parameters in the simulation environment, set the same experimental time, and use algorithm 1, algorithm 2 and the algorithm in this paper to process the requests received by the resource management framework of the analog digital power grid platform in the environment. Using the data monitoring software on the computer simulation platform, the space proportion temporarily occupied by the three algorithms during operation is obtained to analyze the space performance of the algorithms. Analyze the content of two performance indexes of the algorithm and evaluate the effectiveness and feasibility of the algorithm [15-16].

3.2. experimental results and analysis

Figure 1 below shows the time complexity statistical results of the three algorithms. The three-dimensional coordinate axes in the figure respectively represent the magnitude of the power system, the number of instantaneous service requests received by the digital grid platform in the simulation environment, and the time-consuming of the three algorithms in processing related services.

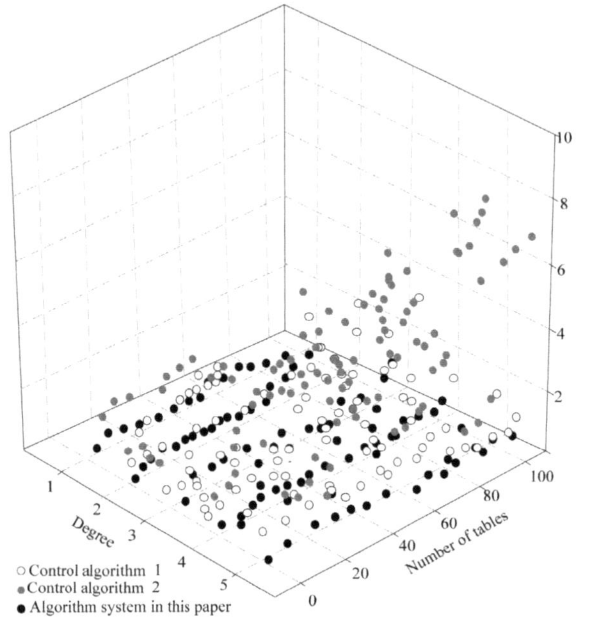

Fig. 1 Comparison of algorithm time complexity

According to the analysis of Figure 1 above, when the power system order is the same and the instantaneous service requests are different, the time-consuming of the algorithm in this paper is less than that of algorithm 1 and algorithm 2; When the number of service requests is the same in the digital grid platform, the time-consuming of algorithm 1 and algorithm 2 is higher than that of this algorithm to varying degrees. The overall analysis of the distribution of data points in the figure shows that the processing time of this algorithm is significantly shorter than the other two. The time-consuming data points in this algorithm not only correspond to small values, but also are evenly distributed. The above shows that the time complexity of the algorithm in this paper is better than algorithm 1 and algorithm 2 [17].

Table 1 below shows the space complexity comparison of the three algorithms under the condition that the analog digital power grid platform receives different service requests in the simulation experiment environment, that is, the comparison of the temporary occupation proportion of the algorithm to the space.

Table 1 Comparison of spatial complexity of algorithms

Instantaneous service requests	Algorithm 1	Algorithm 2	Paper algorithm
50	10.83	11.74	8.67
80	11.64	12.88	9.85
110	13.72	16.97	10.06
150	15.31	22.25	11.14
200	19.55	26.74	12.35
250	21.78	29.81	13.66
300	25.56	31.89	15.71
400	28.33	34.02	18.28

According to the data analysis in Table 1 above, when the instantaneous service requests are less than 110, the proportion of temporary space occupied by the three algorithms is relatively small. With the increase of instantaneous service requests, the proportion of temporary space occupied by algorithm 2 first increases, and always increases with the increase of instantaneous requests; When the instantaneous service requests of algorithm 1 are greater than 200, there is a rapid increase trend, but the increase range is less than that of algorithm 2. In contrast, the algorithm in this paper occupies less space when processing related requests. Although the temporary space occupied by the algorithm also increases with the increase of instantaneous service requests, the overall space occupation ratio does not exceed the maximum value of algorithm 1. It shows that the spatial complexity of the algorithm established in this paper is better than the other two algorithms [18].

Figure 2 below shows the comparison of the accuracy results of algorithm power load forecasting under the condition that the power generation and power load in the simulation digital grid platform are sent out by the power system of the same order of magnitude [19-20].

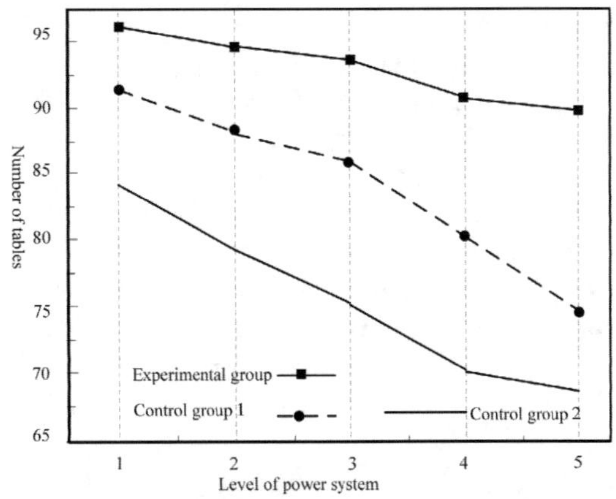

Fig. 2 Comparison of accuracy of power load forecasting algorithm

According to the analysis of the curve in Figure 2 above, when the power system order of power generation and power load is low, the prediction accuracy of the three algorithms is close to more than 85%; When the power system level is raised to level 3, the accuracy of load forecasting results of algorithm 1 and this algorithm is higher than 55%, and the accuracy of load forecasting results of algorithm 2 is slightly less than 50%. When the magnitude of power system is increased to the maximum, the accuracy of load forecasting results of this algorithm is about 22.8% larger than that of algorithm 1. The results show that the accuracy of load forecasting results of this algorithm is higher.

Summing up the above numerical experimental test data analysis of all algorithms, it can be seen that the performance of all algorithms in the power system power load forecasting algorithm based on artificial firefly established in this paper has been improved by different degrees compared with the performance of its comparison algorithm, which shows that the algorithm established in this paper is accurate when applied to power system power load forecasting. When dealing with the power load forecasting in the digital grid platform, the algorithm runs time-consuming, that is, the balance of time complexity and space complexity of the algorithm is better, and the algorithm takes less time.

4. Conclusion

This paper studies the proposed power system load forecasting algorithm based on artificial firefly, improves the accuracy and efficiency of power load forecasting by using limit learning function, and solves the problems caused by the difficulty of forecasting parameters. At the same time, the algorithm and comparison algorithm are used in the same series of power system digital grid platform resource management framework to compare the prediction accuracy of load forecasting algorithm and operation time, that is, the time complexity and space complexity of the algorithm. Simulation results show that the accuracy of power load forecasting results is improved by about 22.8%. Through the numerical experiment of the algorithm, it is verified that the constructed power system load forecasting algorithm is effective, and the performance of the algorithm is significantly improved compared with the traditional algorithm, that is, the research on power system load forecasting algorithm based on artificial firefly has higher practical application value.

Reference

[1] Mao Lifan, Yao Jiangang, Jin Yongshun, Chen Hualin, Li Wenjie, Guan Shilei. (2010) Theoretical research on medium and long term power combination forecasting model [J] Chinese Journal of electrical engineering, 30 (16): 53-59 DOI:10.13334/j.0258-8013. pcsee. 2010.16. eight.

[2] Li Yuancheng, Liu Kewen. (2010) Fast power load forecasting method based on core vector regression for large-scale samples [J] Chinese Journal of electrical engineering, 30 (28): 33-38 DOI:10.13334/j.0258-8013. pcsee. 2010.28. sixteen.

[3] Chai Yuanbin . (2015) Power load forecasting based on improved particle swarm optimization algorithm and least squares support vector machine [J] Electrical application, 34 (12): 46-49.

[4] Long Jinlian, Zhang Yufen, Lu Jiaxuan. (2016) Short term power load forecasting based on Improved Particle Swarm Optimization LSSVM [J] Electromechanical information, 2016 (36): 25 + 27 DOI:10.19514/j.cnki. cn32-1628/tm. 36. Fourteen.

[5] Wang Baoyi, Zhao Shuo, Zhang Shaomin Distributed power load forecasting algorithm based on cloud computing and limit learning machine [J] Power grid technology, 2014,38 (02): 526-531 DOI:10.13335/j.1000-3673. pst. 2014.02. thirty-nine.

[6]Mkrttchian V . (2021) Avatars-Based Decision Support System Using Blockchain and Knowledge Sharing for Processes Simulation: A Natural Intelligence Implementation of the Multi-Chain Open Source Platform[J]. International Journal of Knowledge Management, 17(1):72-92.

[7] Zhao Guangwei, Zhou Yongquan. (2016) The improved artificial firefly swarm optimization algorithm is used to solve the nonlinear equations [J] Practice and understanding of mathematics, 46 (01): 176-186.

[8] Liu Jiakun, Zhou Yongquan. (2011) An artificial firefly algorithm with maximum and minimum luciferin value [J] Computer application research, 28 (10): 3662-3664.

[9] Zhang Wentao, Ma Yongguang, Dong Zijian Short term power load forecasting based on optimized kernel limit learning machine [J] Computer simulation, 2019,36 (10): 125-128 + 135.

[10] Yu Xiaojuan. (2018) Research on heat load forecasting and optimal regulation strategy based on support vector machine [D] Hebei University of technology, DOI:10.27105/d.cnki. ghbgu. 2018.001448.

[11] Li Zhenwei, Su Tao, Zhang Lili. (2020) Application analysis and Prospect of artificial intelligence technology in smart grid [J] Communication power technology, 37 (05): 152-153 DOI:10.19399/j.cnki. tpt. 2020.05. sixty-three.

[12] Tian Jun. (2020) Smart grid big data platform and its key technology analysis [J] Science and technology innovation guide, 17 (04): 138 + 140 DOI:10.16660/j.cnki. 1674-098X. 2020.04. one hundred and thirty-eight.

[13] Weekend, Jin min. (2017) Short term power load forecasting method based on multi algorithm, multi model and online second learning [J] Computer applications, 37 (11): 3317-3322.

[14] Li tsunamis. (2021) Research on wireless sensor network location technology for intelligent factory [D] University of Chinese Academy of Sciences (Shenyang Institute of computing technology, Chinese Academy of Sciences), DOI:10.27587/d.cnki. gksjs. 2021.000001.

[15] Tang Dengping, Zhu Yanqin, Yang Zhe, Cao Guoping, Xiao Guangdi. (2016) Experimental design of computer network management simulation platform accessing the Internet [J] Journal of Changshu Institute of technology, 30 (02): 73-78 DOI:10.16101/j.cnki. cn32-1749/z.2016. 02.017.

[16] Li Xiang. (2016) Design of computer control technology simulation experiment platform [J] Journal of Xiangnan University, 37 (05): 31-36.

[17] Mei Rui, Yu Kun, Chen Xingying Operation topology identification of distribution network based on node injection power [J] Power construction, 2017,38 (11): 41-47.

[18] Li Zhenxiang, Li Quanfu. (2013) Experimental teaching reform and practice of Microcomputer Principle and interface technology based on CDIO mode [J] Laboratory science, 16 (01): 65-68.

[19] Zhang Chenyu. (2019) Application of machine learning and network embedding algorithm in power system transient stability and voltage stability evaluation [D] Zhejiang University.
[20] Wang Ling. (2009) Application of computer simulation technology in relay protection development [D] Shandong University.

Research on flow distribution optimization method of fluid loop system based on generalized Lagrange multiplier method

Lu Wang[1,2a], Changshuai Du[1], Yufeng He[1b*]

[1]Technology and Engineering Center for Space Utilization,Chinese Academy of Sciences, Beijing 100094, China

[2]University of Chinese Academy of Sciences, Beijing 100049, China

[a]wanglu194@mails.ucas.ac.cn, [b*]heyufeng@csu.ac.cn

Abstract—With the development of space science in China, the scale of space application payloads continues to increase, and the demand for heat dissipation is also gradually increasing. The fluid loop thermal control is a more efficient way of heat dissipation.In order to optimize the heat dissipation effect of the fluid loop system, this paper establishes a thermal model of the fluid loop system and converts it into a constrained optimization problem.The generalized Lagrange multiplier method is used to calculate the optimal flow distribution strategy, and the influence of the weight is analyzed, which provides a reference for the setting of the weight in practical applications.

1. Introduction

With the development of space science in China, the scale and number of space application payloads continues to increase, and the power is getting higher and higher. Passive thermal control can no longer meet the current heat dissipation needs. Therefore, active thermal control is needed to help the payloads dissipate heat. Fluid loop thermal control [1] is a more efficient way of active thermal control, which provides a unified fluid heat dissipation channel for the load equipment in the cabin. The entire application fluid loop[2,3] system forms a closed circuit by itself, and the main engine circulating pump group of the drive component provides the power source for the entire fluid loop circulation. The cooling medium is used to realize the collection, transportation, and aggregation of the heat of each load. Then, the heat is delivered to the platform's medium-temperature liquid-liquid heat exchanger, and the exchanger takes the heat away, providing favorable thermal environmental conditions for payload thermal control and fine temperature control.The application fluid loop system contains dozens of branches, each with a different load, and the total heat dissipation power consumption of the fluid loop is limited, which cannot meet the heat dissipation requirements of all branches at the same time. Therefore, it is necessary to dynamically adjust the flow of each branch according to the working requirements of each load to meet the heat dissipation requirements of the operating load. If the flow is too small, the heat dissipation requirements of the branch load will not be met, affecting the working state of the load, and even cause failure; if the flow is too large, although the heat dissipation requirements of the branch load can be met, it will cause waste of flow. Therefore, it is necessary to carry out proper flow distribution for each branch according to the working state of each load.

At present, there have been some studies on the optimization of parallel thermal networks. Zhang et al. [4] established a physical and mathematical model for the flow and heat transfer of the thermal network. The model takes minimizing the total heat transfer area of the system as the optimization objective function.They used the extreme value rational method to solve the transcendental equations, and obtained the relationship between the optimal flow distribution rate and the performance of each thermal component. This method requires that the objective function is continuous and derivable, and the calculation is complicated. Cheng et al. [5] also took minimizing the total heat transfer area as the optimization objective function, proposed a method that can accurately solve the optimal distribution coefficient through the variable conversion of the distribution coefficient and the steepest descent method. He used the Lagrange multiplier method to solve the problem [6].In 2011, Cheng et al. [7] introduced the minimum entropy generation principle and calculated the minimum entropy generation by Newton's method, optimized the fluid and heat transfer area distribution. Zhang et al.[8]took minimizing the annual cost as the optimization goal, modeled the thermal network problem as a constrained multivariable mathematical problem, optimized it with the Lagrange multiplier method, and realized the global optimization of the thermal network.But the Lagrange multiplier method [9] is suitable for solving constrained optimization problems that only contain equality constraints. Penalty function method and generalized Lagrange multiplier method can solve optimization problems with mixed constraints[10,11,12]. But the penalty function method is more difficult to select the initial point and when it is close to the optimal solution, the nature of the penalty function deteriorates. In addition, these models all aim to reduce the launch cost, that is, the minimum cold plate area or the total mass of the system, to optimize the fluid loop system. They are suitable for the completion of the system design, and when the system is determined, it is also necessary to optimize the flow distribution according to the working status of the load on the branch. At present, the main methods of flow distribution are manual adjustment and table lookup method. Manual adjustment relies on experience to try each parameter. This method is inefficient and cannot achieve the optimal flow distribution effect. The table lookup method is to form a database through manual attempts. When encountering a control problem, the database is consulted to obtain the corresponding control strategy. This method requires a large number of experiments, it is difficult to cover all working conditions, and the efficiency is low too. Therefore, a new flow control algorithm is needed to automatically calculate the optimal branch flow distribution strategy according to the load status.This algorithm can not only reduce the number of flow adjustments and improve the efficiency of staff control, but also optimize the control effect and ensure the normal operation of the load to the greatest extent. In this paper, the root mean square value of the difference between the actual operating temperature and the optimal operating temperature of each load is used as the objective function, and the fluid loop flow distribution problem is modeled as a mixed constrained optimization problem. The generalized Lagrange multiplier method is used to carry out the optimal flow distribution plan and the results and the influence of weights are analyzed.

2. Space fluid system modeling

The fluid loop system consists of multiple parallel branches. Each branch includes a flow regulating valve and a load. The flow regulating valve is used to regulate the heat dissipation working fluid passing through the branch. The load is used for scientific experiments. The circulating pump provides the power source for the entire fluid loop to circulate.The scientific load uses cold plate heat dissipation as the thermal control method. In order to simplify the loop system model, the series load thermal control of each branch is equivalent to the heat transfer relationship between a load and its cold plate.The simplified structure diagram is shown in Figure 1, where n is the number of branches. T_{in} and v are the inlet fluid temperature and total flow.v_i is the flow through each branch.Q_i and $T_{H,i}$ are the heat dissipation and working temperature when the load is working on the i-th branch. $T_{out,i}$ is the fluid temperature at the outlet of the i-th branch.

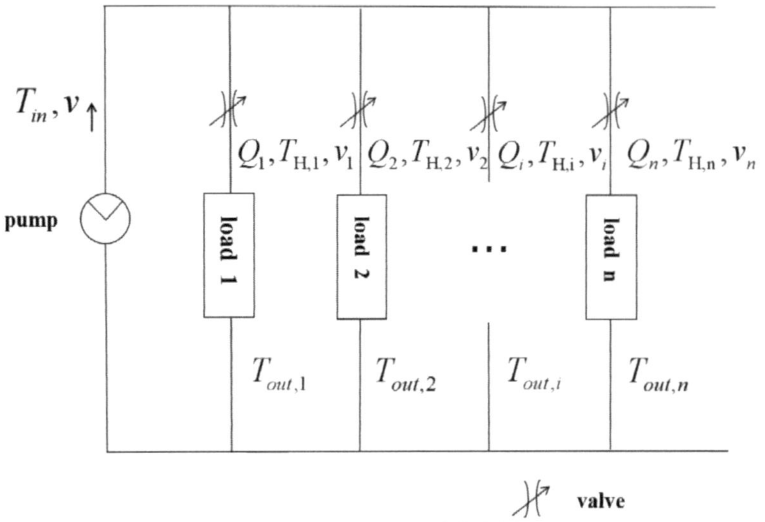

Figure 1 Schematic diagram of fluid loop system

Suppose the optimal working temperature of the load on the i-th branch is $T_{b,\,i}$. In order to make the scientific experiment of each load achieve the best experimental effect, that is, the actual working temperature of the load is the closest to the optimal working temperature. The root mean square value of the difference between the actual working temperature and the best working temperature is taken as the objective function, see formula (1). w_i is a weight coefficient set according to the sensitivity of each load to temperature.

$$\min \sqrt{\frac{\sum_i w_i (t_{H,i} - t_{b,i})^2}{n}} \tag{1}$$

The heat balance and heat transfer relationship of each load satisfies the relationship in formula (2) and (3).c is the specific heat capacity at constant pressure of the heat exchange working fluid, F_i is the area of the cold plate, k_i is the heat transfer coefficient, Δt_i is the logarithmic average temperature difference of the heat transfer of each load, expressed as formula (4).

$$Q_i = v_i c (T_{out,i} - T_{in}) \tag{2}$$

$$Q_i = k_i F_i \Delta t_i \tag{3}$$

$$\Delta t_i = \frac{T_{out,i} - T_{in}}{\ln \frac{T_{H,i} - T_{in}}{T_{H,i} - T_{out,i}}} \tag{4}$$

Ignoring the heat transfer resistance in the heat transfer process of the branch, it is assumed that the heat transfer coefficient satisfies the formula (5). k_0 is the proportional coefficient, α is the index.

$$k_i = k_0 v_i^{\alpha} \tag{5}$$

From formula (2) to (5), formula (6) can be obtained.

$$t_{H,i} = \frac{Q_i}{v_i c (1 - e^{-\frac{k_0 F}{c} v_i^{\alpha-1}})} + t_{in} \tag{6}$$

Incorporating formula (6) into formula (1), the final objective function formula (7) can be obtained.

$$f(x)=\sum_{i=1}^{n} w_i\left[\frac{Q_i}{x_i c\left(1-e^{-\frac{k_0 F}{c}x_i^{\alpha-1}}\right)}+t_{in}-t_{b,i}\right]^2 \tag{7}$$

Due to the characteristics of the fluid loop and each load itself, the flow distribution needs to meet certain constraints, including the temperature difference between the inlet and outlet fluid must not exceed the maximum temperature difference t_{max}.The flow of each branch cannot be lower than the minimum flow v_{min}. In addition, the sum of the flow of each branch should be equal to the total flow v, etc. Specifically as shown in formula (8).

$$\begin{cases} \frac{Q_i}{v_i c} \le t_{\max} \\ v_i \ge v_{\min} \\ \sum_{i} v_i = v \end{cases} \tag{8}$$

Then the flow distribution problem of the fluid loop is transformed into a mixed constrained optimization problem.The solution of this constrained optimization problem is the best flow distribution strategy.

3. Optimization algorithm for flow distribution of space fluid system

3.1. Generalized Lagrange multiplier method

The generalized Lagrange multiplier method combines the penalty function and the Lagrange function to construct a new objective function, so that when the penalty factor is appropriately large, the Lagrange multiplier can gradually reach the optimal solution of the original problem.For the constrained optimization problem of formula (9),construct the augmented Lagrange function as formula (10).

$$\begin{cases} \min f(x) \\ s.t.\ \ g_i(x)\le 0 \ \ i=1,2,\ldots,m \\ \qquad h_j(x)=0 \ \ j=1,2,\ldots,l \end{cases} \tag{9}$$

$$L(x,\sigma_k,w_n,v_n)=f(x)+\frac{1}{2\sigma_k}\sum_{i=1}^{m}\left\{\left[\max(0,\sigma_k g_i(x)-w_i^{(k)})\right]^2-(w_i^{(k)})^2\right\}$$
$$+\frac{\sigma_k}{2}\sum_{j=1}^{l}h_j^{\,2}(x)-\sum_{j=1}^{l}v_j^{(k)}h_j(x) \tag{10}$$

σ is the penalty factor, w and v are Lagrange multipliers, and the multiplier iteration formula is formula (11).See formula (12) for termination criteria.

$$\begin{aligned} w_i^{(k+1)} &= \max\left\{0,\sigma_k g_i(x_k)-w_i^{(k)}\right\} i=1,2,\ldots,m \\ v_j^{(k+1)} &= v_j^{(k)}-\sigma_k h_j(x_k)\ j=1,2,\ldots,l \end{aligned} \tag{11}$$

$$\sum_{i=1}^{m}\left\{\max\left[g_i(x_k),\frac{w_i^{(k)}}{\sigma_k}\right]\right\}^2+\left\|h(x_k)\right\|<\varepsilon \tag{12}$$

3.2. Algorithm optimization

In view of the flow optimization problem of the fluid loop, in some extreme cases, for example, the heat dissipation of a branch is too small, the problem of a negative number in x temporarily occurs during the iteration of the program. Obviously, the flow rate cannot be a negative number, and the negative value in x may increase the running time of the program, and even affect the normal operation of the program, resulting in the inability to output correct results. In order to solve the above problems for practical applications, a statement that judges the sign of x output in each round in the iterative process is added. When the flow of a branch is negative, the flow of the branch is reset to the minimum flow v_{min}. After testing, this method can well overcome the above problems, and can output the correct flow distribution strategy.

4. Application verification of flow distribution optimization algorithm

4.1. Numerical example

Take a fluid loop system with four branches as an example. The total flow is 660L/h, the minimum flow of the branch is 30L/h, the cold plate area is $0.12m^2$, the specific heat capacity is 1.05W/L℃, the inlet fluid temperature is 26℃, the maximum temperature rise is 10°C, and the heat transfer coefficient index is 0.5. The heat dissipation of the four branches are 100W, 500W, 1000W and 4000W respectively, with a weight of 0.1, and the best temperature is 30°C. The calculation results of the generalized Lagrange multiplier method are shown in Table 1. The calculation results conform to the trend that the greater the heat dissipation of the branch, the more the flow allocated to the branch.

Table 1 Flow distribution results of four branches

branch	heat dissipation(W)	weight	the best temperature(℃)	Flow(L/h)
1	100	0.1	30	29.93
2	500	0.1	30	74.04
3	1000	0.1	30	131.66
4	4000	0.1	30	424.38

The change of the augmented Lagrange function in the iterative process of the generalized Lagrange multiplier method is shown in Figure 2. As the number of iterations increases, the value of the function keeps decreasing. When the number of iterations is 0 to 4 times, the function value drops faster; from 4 to 30 times, the function value drops slowly. At about the 11th time, there is a sudden increase. After checking the flow value, it is found that when the point is iterated, the flow value of the first branch is less than the minimum flow value, and the "penalty" causes the objective function value to suddenly increase, but after adjustment, it quickly drops to the original level.After the number of iterations is greater than 30, the function value remains basically unchanged, and the iteration converges. At this time, the calculated flow value is the value that minimizes the objective function when the constraint conditions are met, which is the best flow distribution strategy.

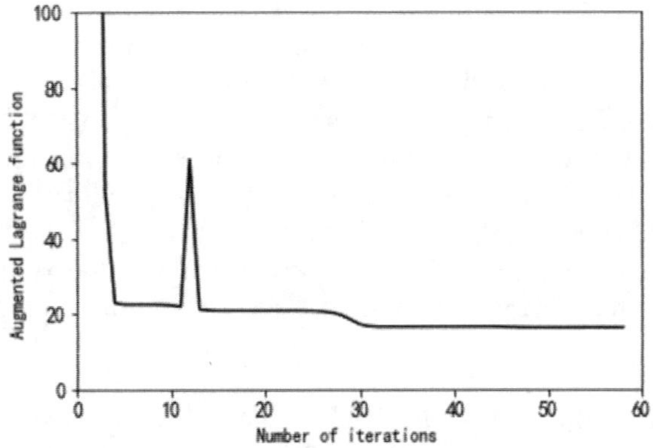

Figure 2 Augmented Lagrange function change graph

4.2. The effect of weight

The weight coefficient characterizes the relative sensitivity of the load on the branch to temperature.If the load has a high requirement on the accuracy of the temperature, a larger weight is given to it.Conversely,if the temperature has a small effect on the working state of the load, the weight of the load can be set smaller. For fluid loop systems with 2, 5, 8 and 10 branches, set the heat dissipation of each branch to 2000W, 800W, 500W and 400W, and the weight of the first branch is 0.01, 0.05, 0.1, 0.5, 1, 2, 5, and 10,respectively, the weights of other branches are all 1. The relationship between the flow proportion and weight of the first branch is shown in Figure 3.It can be seen that the overall trend of the curve is that the flow rate in the front section remains unchanged, the middle section rises faster, and the rising speed at the end section gradually slows down to a level, and finally basically unchanged. The front end of the curve remains unchanged because the flow reaches the critical value of temperature rise and cannot continue to decrease, so it is not affected by the weight.In the mid-term, the weight range is 0.05 to 5, as the weight increases, the traffic has changed significantly. When the weight is greater than 5, that is, it is greater than 5 times the weight of other branches, the influence of the weight is significantly reduced. After testing, even Increasing the weight to 100 or even 1000, the increase in the flow of this branch is also minimal, and even negligible in the later stage.Therefore, when setting the weight, the weight of each branch should be adjusted within 10 times, otherwise it may be meaningless and has little effect on the result.

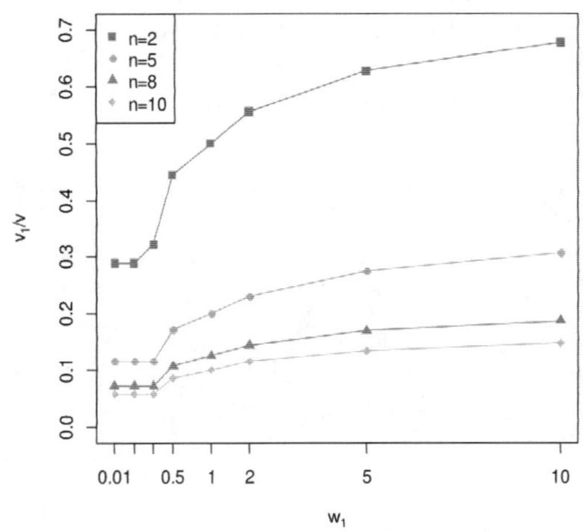

Figure 3 The effect of weight

5. Conclusion

In this paper, the mathematical and physical model of the fluid loop system is established, the optimal flow distribution strategy is calculated by the generalized Lagrange multiplier method, and the results are analyzed and discussed.The generalized Lagrange multiplier method can well solve the flow distribution problem of complex fluid loop systems, and can quickly obtain the best flow distribution plan when there are many branches in the system, which saves time for flow control in actual use and improves work efficiency.In addition, when setting the weight coefficient of each branch, it should be adjusted within the range of 10 times as much as possible. When the branch coefficient is adjusted beyond 10 times, it may be meaningless and will not have a great impact on the result.

References

[1] Fan Hanlin, Huang Jiarong, Liu Qing zhi. Scheme of fluid loop system on manned spacecraft for transport[J]. Chinese Space Science and Technology, 2007, 27(5):6.

[2] Zheng Tong, Zhao Liping, Liu Ronghui. Simulation and study on fluid loop of space utilization bypass control algorithm[J]. Control Engineering of China, 2020, 27(1):8.

[3] Shi Zhedong, Gu Yanping, Chen Gang. The simulation study of dynamic performance and closed loop control of spacecraft fluid loop[J]. Journal of Thermal Science and Technology., 2018, 17(6):8.

[4] Zhang Xinrong, Ren Jianxun, Liang Xingang, Guo Zengyuan. Optimization of the thermal network in parallel connection of manned spacecraft[J].J Tsinghua Univ (Sci & Tech),2002(04):462-465.DOI:10.16511/j.cnki.qhdxxb.2002.04.010.

[5] Cheng Xuetao, Xu Xianghua, Ren Jianxun, Liang Xingang. Optimization of the thermal network in parallel connection [J].Journal of Engineering Thermophysics,2008(12):2125-2127.

[6] Cheng Xuetao, Xu Xianghua, Ren Jianxun, Liang Xingang. Optimization of liquid cooling networks with couple parallel connections using Lagrange multipliers[J].J Tsinghua Univ (Sci & Tech),2008(08):1359-1361+1367.DOI:10.16511/j.cnki.qhdxxb.2008.08.019.

[7] Cheng Xuetao, Xu Xianghua Liang Xingang. Application of entransy to optimization design of parallel thermal network of thermal control system in spacecraft[J].Science China(Technological Sciences),2011,54(04):964-971.

[8] Zhang Chunwei, Cui Guomin, Chen Shang, Tao Jianan. Lagrange multiplier method combined with structure evolution strategy for heat exchanger network synthesis[J].Chemical Industry and Engineering Progress,2016,35(04):1047-1055.

[9] Bertsekas, DimitriP. Constrained optimization and Lagrange multiplier methods[M]. AP, 1982.

[10] Yuan Yaxiang, Sun Wenyu. Optimization theory and methods [M]. Science Press , 1997.

[11] Bertsekas D P , Nedic A , Ozdaglar A E . Convex Analysis and Optimization. Pitman Advanced Pub. Program, 2003.

[12] Chen Baolin. Optimization theory and algorithms[M]. Tsinghua University Press , 2005.

Journal of Physics: Conference Series 2258 (2022) 012026 doi:10.1088/1742-6596/2258/1/012026

A self-supervised feature selection algorithm for spectral embedding based on block HSIC lasso

Yajing Guo[1,2,3], Lijuan Duan[1,2,3*], Chenyao Xie[1,2,3] and Wenbo Zhang[1,2,3]

[1] Faculty of Information Technology, Beijing University of Technology, Beijing 100124, China

[2] Beijing Key Laboratory of Trusted Computing, Beijing 100124, China

[3] National Engineering Laboratory for Critical Technologies of Information Security Classified Protection, Beijing 100124, China

[*]Corresponding author's e-mail: ljduan@bjut.edu.cn

Abstract—With the rapid development of information technology, a large amount of unlabeled high-dimensional data has been generated. To be able to better handle these data, we propose a new self-supervised feature selection algorithm for spectral embedding based on block HSIC lasso (FSSBH). It innovatively applies the HSIC theoretical approach to unlabeled scenarios for feature importance assessment, and performs feature selection by self-supervised learning with the pseudo-label matrix formed by the spectral embedding. The algorithm is compared with five popular algorithms for classification experiments on seven publicly available datasets. The results show the superior performance of the FSSBH algorithm.

1. Introduction

With the development of technology, large amount of high-dimensional data has become a problem in many applications, such as computer vision [1, 2], data mining [3-5], and pattern recognition [6]. These high-dimensional data have a large amount of noise and redundant data, which not only increase the storage burden and processing time of computers, but also reduce the performance of algorithms [7-8]. Therefore, how to process and use these high-dimensional data more efficiently and conveniently has become a very important research direction. After the processing of high-dimensional data, the subsequent use of the data will become more efficient and convenient. Currently, feature extraction and feature selection are two important techniques for data dimensionality reduction. The feature extraction method does not consider whether the features are useful or not, but simply converts the original data into a specific form that can be easily recognized [9]. Feature selection selects the most representative and distinguishing features from the original data to generate a new subset of features, which does not change the original data representation. Feature selection is divided into supervised feature selection, semi-supervised feature selection, and unsupervised feature selection according to the presence or absence of label information in the data. The supervised feature selection method means that the label information of the data is known before the feature selection is performed [10]. The semi-supervised feature selection method means that only part of the data label information is known, so that the performance of the method can be improved by the known label information [11].Compared with the supervised and semi-supervised feature selection methods, the unsupervised feature selection method is more challenging due to the lack of label information of the data [12-14]. Because high-dimensional data usually do not contain label information, unsupervised feature selection is more practical and valuable for research in practical applications.

In order to solve the above problems and select rich, robust, and high-quality features in a more targeted manner in label-free scenarios, we introduce the Hilbert-Schmidt independence criterion(HSIC) theory into the algorithm to improve the ability to select high similarity and low redundancy features in data samples, and validate the algorithm effectiveness through classification experiments.

Our contributions are summarized as follows.

(1) We propose a new self-supervised feature selection algorithm FSSBH for high-dimensional data in unlabeled scenarios. The algorithm innovatively applies HSIC theory to the unlabeled feature selection method for constructing the evaluation function of feature importance.

(2) We propose to combine the pseudo-label matrix formed by the spectral embedding method with the feature importance evaluation module to select high-quality features using self-supervised learning.

(3) Experimenting on seven multi-source high-dimensional datasets, the FSSBH algorithm obtains excellent performance.

2. Related Work

According to the strategy of feature selection, the feature selection methods in unlabeled scenes are divided into three categories: filter, wrapper and hybrid.

The filter method[15-18] evaluate features based on the intrinsic properties of the data, which are usually intuitive and computationally efficient.He et al. proposed the Laplacian method, which calculates the score of each feature independently based on the ability to retain the intrinsic structure of the original data. The smaller the score, the stronger the local retention ability and more representative[18]. The advantage of such algorithms is that they have strong generalization ability and fast training speed, but there are some deviations in the performance and feature evaluation of learning algorithms.

The wrapper method[19] evaluates features based on the results of a specific clustering algorithm, and the results of feature selection perform well in the mining task. However, packaging methods usually have a large amount of calculation and weak generalization ability.

The hybrid method[20-28] makes full use of the advantages of the filter and wrapper method, paying special attention to the balance between the amount of calculation and the quality of the selected features. Because there is no need to evaluate the feature subset, it is more effective than the wrapper method . Therefore, the hybrid method has gradually become a hot spot. Li et al. Carried out joint iterative learning of clustering labels and feature selection matrix, and proposed nonnegative discriminant feature selection (NDFS)[25]. Yi et al. Proposed the unsupervised discriminative feature selection algorithm (UDFS), which combines discriminant analysis and l2,1-norm minimization into the joint framework of unsupervised feature selection to optimize the algorithm [26]. The above UDFS and NDFS methods only focus on the selection of discriminant features and do not consider the high correlation of the selected features, which has an adverse impact on the classification results. In order to select features with low redundancy, Deng et al. Proposed the multiclass cluster unsupervised selection algorithm (MCFS)[27] and Xie juanying et al. Proposed the unsupervised feature selection algorithm based on spectral clustering (FSSC-SD), which can find the most representative features in various clusters and add them to the feature subset[28].

3. Methods

Due to the advantages of hybrid methods in feature selection, this paper innovatively proposes a hybrid method for feature selection in unlabeled scenes - self-supervised feature selection algorithm for spectral embedding based on block HSIC lasso(FSSBH), and this section mainly introduces the principle of the algorithm.

The FSSBH algorithm innovatively applies the HSIC theory method to the unlabeled scenario and combines it with the self-supervised theory. The algorithm first uses the spectral embedding method to form a pseudo-label matrix by self-supervised learning of the data, and then uses the block HSIC lasso method to evaluate the importance of each feature, so as to select a subset of features with high similarity and low redundancy.

3.1.Determination of the pseudo-class label matrix

Consider a graph with n vertices, where each vertex corresponds to a data point. For each data point x_i , we can discover its p nearest neighbors and put an edge between x_i and its neighbors. There are many options for defining a weight matrix W on the graph.

Defining a diagonal matrix D whose entries are the sum of the columns or rows of W (since W is symmetric), we can compute the Laplacian $L = D - W$ of the graph. The "flat" embedding of the data points for the "unfolded" data stream shape can be found by solving the following generalized feature problem.

$$Ly = \lambda \mathrm{D}\mathrm{y} \tag{1}$$

Let $Y = [y_1, \cdots, y_c]$, where $y_c$is the eigenvector of the above generalized eigenvalue problem with respect to the minimum eigenvalue, and each row of Y is a "flat" embedding of each data point. c is the intrinsic dimensionality of the data, and each y_c reflects the distribution of the data along the corresponding dimension (topic, concept, etc.).When we try to cluster the data, each y_c reflects the distribution of the data on the corresponding cluster. Thus, if the number of clusters of the data is known, c is usually set equal to the number of clusters ,and in this experiment c is set to 10. By spectral embedding we can obtain the pseudo-class label matrix $Y = [y_1, \cdots, y_c]$ of the data X in a self-supervised manner.

3.2.Feature importance assessment

The goal of feature selection is to find the m features (m much smaller than d) that are most relevant to the output y of the predicted sample x.

$$\begin{aligned} HSIC(X,Y) &= E_{x,x',y,y'}[K(x,x')L(y,y')] \\ &+E_{x,x'}\left[K(x,x')E_{y,y'}[L(y,y')]\right] \\ &-2E_{x,y}\left[E_{x'}[K(x,x')]E_{y'}[L(y,y')]\right] \end{aligned} \tag{2}$$

Measuring the dependence between two random variables X and Y can be done by the Hilbert-Schmidt independence criterion HSIC(X, Y), as in Equation (2). K, L are positive definite kernels. E represents the expectation of independent pairs with the value of 0 if X and Y are independent, and non-negative otherwise.

For a given kernel matrix K_K (calculated from the k-th feature) and a given kernel matrix L of pseudolabels (calculated from the pseudolabel matrix), the normalized variables of the FSSBH are estimated using the v-statistic as follows, where f_k denotes the eigenvector of the k-th feature and y_c denotes the c-th resultant vector in the obtained pseudolabel matrix Y, as in Equation (3).

$$FSSBH_v(f_k, y_c) = \frac{1}{C}\sum_{c=1}^{C} tr(\overline{K_k L_c}) \tag{3}$$

$$\bar{K} = HKH/\|HKH\|_F \tag{4}$$

where H is a central matrix（$H_{ij} = \sigma_{ij} - 1/n$,where $\sigma_{ij} = 1$ when $i = j$,0 in other cases）, tr indicates tracking.Note that we use a normalized variant of the improved original empirical HSIC.Where, the larger the value of $FSSBH_v(f_k, y_c)$, the greater the relevance of the Kth feature to the result, i.e., the greater the feature importance of the Kth feature. Therefore, it is proposed that the features can be sorted in descending order of $FSSBH_v(f_k, y_c)$ for feature selection. Also, a vector α (feature weight value) is introduced for optimization in order to avoid the selection of multiple redundant features. Also the dataset is trained in chunks, which is convex and can be solved efficiently on a reasonably sized server, then the estimate can be written as Eq. (5).

$$FSSBH_b(f_k, y_c) = \frac{1}{C}\sum_{c=1}^{C} \frac{B}{n}\sum_{l=1}^{n/B} FSSBH_v\left(f_k^{(l)}, y_c^{(l)}\right) \tag{5}$$

where $f_k^{(l)}$ denotes the kth feature vector of the lth partition.

Since we need to calculate the importance of paired data with a fixed partition $\left\{\{(x_i^l, y_i^l)\}_{i=1}^{B}\right\}_{l=1}^{n/B}$,Therefore partitioning has a significant impact on performance.Therefore, we used the bagging method in the estimator. Given M random permutations of n samples, the new importance estimator is defined as Equation (6):

$$FSSBH_{bb}(f_k, y_c) = \frac{1}{C}\sum_{c=1}^{C}\frac{1}{M}\sum_{m=1}^{M}\frac{B}{n}\sum_{l=1}^{n/B} FSSBH_v\left(f_k^{(l,m)}, y_c^{(l,m)}\right) \quad (6)$$

where $f_k^{(l,m)}$ is the kth feature vector, restricted to the lth block of the mth alignment definition. This is the final feature importance evaluation criterion proposed in this chapter, because the larger the value of $FSSBH_{bb}(f_k, y_c)$, the greater the feature importance of the kth feature, so we sort the features in descending order of $FSSBH_{bb}$ for feature selection, and select the top d features to form a feature subset F.

3.3.Algorithm idea description

Algorithm 1: The FSSBH algorithm

Input:

Data matrix $X \in R^{m \times n}$; Parameters d,p,c

1:Construct the k-nearest neighbor graph and define the weight matrix W ;

2: $Y = [y_1, \cdots, y_c] \leftarrow Ly = \lambda Dy$

3:calculate X and Y 's kernel matrix K and L

4:$\bar{K} = HKH/\|HKH\|_F$, Centralized K and L

5:$FSSBH_v(f_k, y_c) = \frac{1}{C}\sum_{c=1}^{C} tr(\overline{K_k L_c})$

$FSSBH_{bb}(f_k, y_c) = \frac{1}{C}\sum_{c=1}^{C}\frac{1}{M}\sum_{m=1}^{M}\frac{B}{n}\sum_{l=1}^{n/B} FSSBH_v\left(f_k^{(l,m)}, y_c^{(l,m)}\right)$

6:Arrange the features in descending order of $FSSBH_{bb}$ value

7:Select the top d features

Output:

Feature set F composed of d features

4.Experiments and Analysis of Results

4.1.Data sets

Table 1.Description of the experimental data set.

Data Set	#Instances	#Features	#Classes
YaleB	2414	1024	38
warpPIE10P	210	2420	10
lung	203	3312	5
COIL20	1440	1024	20
Prostate-GE	102	5966	2
PCMAC	1943	3289	2
RELATHE	1427	4322	2

In this section, we conduct experimental studies on the algorithm in seven publicly available datasets, which can be obtained from FEATURE SELECTION DATASETS ,Cancer Gene Expression Data Sets database, and the dataset details are shown in Table 1, where YaleB and warpPIE10P are face datasets, COIL20 is a physical dataset, lung and Prostate-GE are biological datasets, and PCMAC and RELATHE are text datasets.

4.2.Experimental parameters

In order to verify the performance of the proposed FSSBH algorithm, experiments were conducted to compare the experimental results of FSSBH with five other unsupervised feature selection algorithms in selecting 2-100 features and in selecting all features. The compared unsupervised feature selection algorithms include FSSC-SD,Laplacian, MCFS, NDFS, and UDFS.

In this algorithm, Euclidean distance is used to calculate the inter-feature distance and measure the feature similarity, the intrinsic dimension value c is set to 10, the number of nearest neighbors K is set to 5, the bandwidth parameter t is set to 1, M is set to 3, the block size B is set according to the experiment, and the kernel of the block HSIC lasso kernel matrix is selected as RBF Gaussian kernel. Other parameters γ of NDFS algorithm are set to 10^8 and both α and β are set to 1. The regularization parameter of UDFS algorithm is set to 0.1 and the number of class clusters is set to 5.Same variables set to the same value.

For all the experimental data sets, the number of features selected is set as {2,4,6,... ,100}, the KNN algorithm is selected to classify the dataset, and the parameters in the KNN algorithm are set as K=5 and p=1. The 10-fold cross-validation method is used to divide the training set and the test set, and the maximization minimization method is used to normalize the data. 5 times of 10-fold cross-validation experiments are conducted, and the average of the 5 times results is taken as the experimental results to compare the performance of each algorithm. The evaluation criteria are ACC, AUC, and F-measure.

4.3.Analysis of experimental results

4.3.1 Accuracy comparison

Table 2 shows that the unsupervised feature selection algorithm not only reduces the feature dimension significantly, but also can improve the classification effect, especially the proposed FSSBH algorithm selects the number of features within 100, the maximum classification accuracy reaches the highest in the experimental dataset shown in Table 1.Compared with other algorithms, the average maximum classification accuracy is improved by 2.83% to 8.63%, which shows the the effectiveness and representativeness of the feature subset selected by the FSSBH algorithm.

Table 2.Maximum classification accuracy of unsupervised feature selection comparison algorithm.

Data Sets	All Feature	FSSC-SD	Laplacian	MCFS	NDFS	UDFS	FSSBH
YaleB	72.21	66.23	62.96	61.16	<u>66.99</u>	58.70	**80.23**
warpPIE10P	94.67	96.95	91.81	96.10	96.10	<u>97.33</u>	**99.33**
lung	95.78	<u>95.95</u>	90.54	94.99	95.18	90.15	**96.95**
COIL20	99.47	<u>99.32</u>	95.80	98.18	98.72	94.21	**99.57**
Prostate-GE	85.91	<u>87.69</u>	84.98	84.34	85.31	84.71	**87.78**
PCMAC	71.13	<u>66.25</u>	56.50	60.77	57.30	59.24	**67.98**
RELATHE	77.57	<u>73.26</u>	62.46	61.88	64.13	63.35	**73.62**
Average	85.25	<u>83.66</u>	77.86	79.63	80.53	78.24	**86.49**

4.3.2. Comparison of the number of selected features at maximum classification accuracy

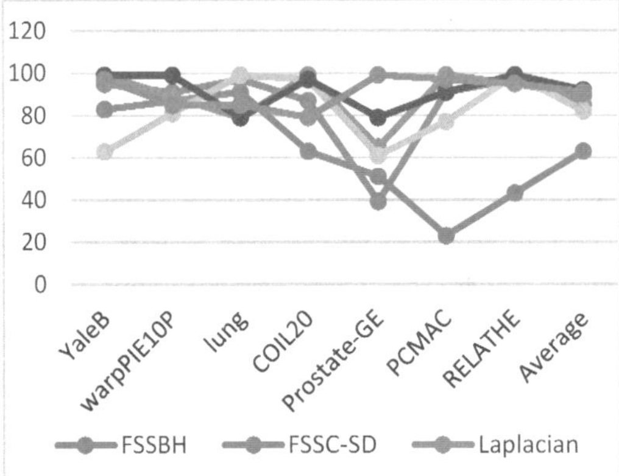

Fig.1 the number of features selected at the maximum classification accuracy.

Fig.1 shows that the high-dimensional dataset in Table 1. It can select features with much less than the original feature dimensions and with strong classification ability through feature selection. The average number of features selected is at least 63 on the basis of guaranteeing the maximum classification accuracy, which reflects the good performance of feature selection of this algorithm.

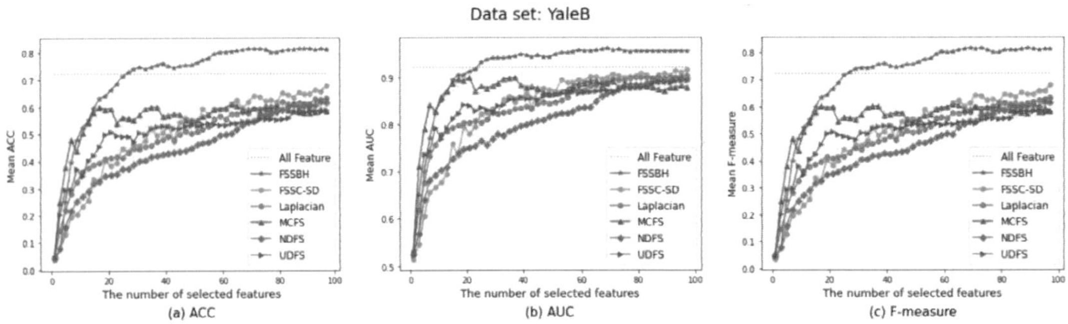

Fig.2 Experimental results on YaleB.

The results of Fig.2 show that the algorithm FSSBH has surpassed the results of selecting all the features after selecting more than 25 features for each index value and has a significant advantage over the other five algorithms.

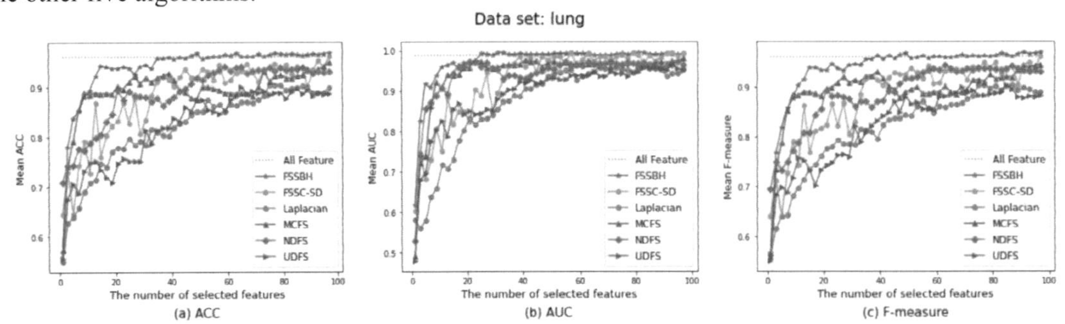

Fig.3 Experimental results on lung.

The results in Fig.3 show that the values of ACC and F-measure metrics of algorithm FSSBH exceed those of the other five algorithms after selecting more than 18 features, and the values of AUC are also comparable, which proves the superior performance of algorithm FSSBH.

4.3.3.Comparison of algorithm time consumption

Table 3.Time consumption of unsupervised feature selection comparison algorithm.

Data Sets	FSSC-SD	Laplacian	MCFS	NDFS	UDFS	FSSBH
YaleB	115.12	290.55	142.35	**114.25**	196.79	129.94
warpPIE10P	48.16	**14.31**	28.26	58.47	35.71	27.19
lung	60.95	**5.91**	20.05	58.06	220.91	20.66
COIL20	57.15	**46.84**	69.41	47.20	128.19	56.72
Prostate-GE	197.74	**8.26**	22.14	361.69	1371.33	20.87
PCMAC	118.95	**59.02**	172.65	2041.82	399.94	103.54
RELATHE	130.40	**35.15**	157.41	3172.13	754.27	99.28
Average	104.07	65.72	87.47	836.23	443.88	**65.46**

Table 3 shows the time consumption of 50 times of feature selection for each algorithm. In most cases, Laplacian 's time consumption is the least. However, it can be seen from table 2 that Laplacian 's classification accuracy is the lowest and the algorithm performance is not good. The time consumption of the proposed FSSBH algorithm in most datasets is second only to Laplacian algorithm, and the average feature selection time consumption is the lowest. Combined with its high classification accuracy in 4.4.1, it shows that FSSBH can not only ensure high classification accuracy, but also ensure less time consumption, which proves the efficiency of this algorithm.

5. Conclusion

In this paper, a new self-supervised feature selection method is proposed to construct a subset of features with strong classification ability by forming a pseudo-label matrix of the sample dataset through spectral embedding, and then evaluating the importance of each feature according to the defined importance evaluation function. After conducting various types of comparative experiments on seven publicly available datasets, this chapter confirms that the proposed algorithm can effectively improve the classification accuracy of the experiments and can ensure high classification accuracy while also ensuring less time consumption and reducing time cost. The performance on the feature selection task is excellent, and it is hoped that the proposed method will be useful in the field of data mining.

Acknowledgments

This work was supported in part by the National Natural Science Foundation of China under Grant 62176009, in part by the Key Project of Beijing Municipal Education Commission under Grant KZ201910005008, and in part by the Beijing Municipal Natural Science Foundation under Grant 4202025).

References

[1] Song, J. , Guo, Y. , Gao, L. , Li, X. , & Shen, H. T. . (2017). From deterministic to generative: multi-modal stochastic rnns for video captioning. IEEE Transactions on Neural Networks and Learning Systems.

[2] Li Xuelong，Chen Mulin，Nie Feiping，et al． A multiview-based parameter free framework for

group detection [C] / /Proceedings of the Thirty-First AAAI Conference on Artificial Intelligence. Palo Alto, USA: AAAI Press, 2017: 4147—4153.
[3] Li Xuelong, Chen Mulin, Nie Feiping, et al. Locality adaptive discriminant analysis [C] / /Proceedings of the 26th International Joint Conference on Artificial Intelligence. Palo Alto, USA: AAAI Press, 2017: 2201—2207.
[4] Zhang, R., & Tong, H. (2019). Robust principal component analysis with adaptive neighbors. Advances in Neural Information Processing Systems, 32.
[5] Wang Fei, Wang Quan, Nie Feiping, et al. Unsupervised linear discriminant analysis for jointly clustering and subspace learning [J] . IEEE Transactions on Knowledge and Data Engineering, 2021, 33(3) : 1276—1290.
[6] Song Jingkuan, Gao Lianli, Nie Feiping, et al. Optimized graph learning using partial tags and multiple features for image and video annotation [J] IEEE Transactions on Image Processing, 2016, 25(11) : 4999—5011.
[7] Lee P Y , Loh W P , Chin J F . Feature selection in multimedia: The state-of-the-art review[J]. Image and Vision Computing, 2017, 67(nov.):29-42.
[8] Rui Z , Fn A , Xl A , et al. Feature selection with multi-view data: A survey[J]. Information Fusion, 2019, 50:158-167.
[9] ZHAN Shanhua, WU Jigang, HAN Na, et al. Unsupervised feature extraction by low-rank and sparsity preserving embedding [J] . Neural Networks, 2019, 109: 56-66
[10] Jun Yin, Weiming Zeng, Lai Wei. Optimal feature extraction methods for classification methods and their applications to biometric recognition [J]. Knowledge Based System. 2019, 99(1):112–122.
[11] ZHANG Yan, ZHANG Zhao, QIN Jie, et al. Semi-supervised local multi-manifold isomap by linear embedding for feature extraction [J] . Pattern Recognition, 2018, 76: 662-678.
[12] ZHU Pengfei, ZHU Wencheng, HU Qinghua, et al. Subspace clustering guided unsupervised feature selection [J] . Pattern Recognition, 2017, 66: 364-374.
[13] XIANG Shiming, NIE Feiping, MENG Gaofeng, et al. Discriminative least squares regression for multiclass classification and feature selection[J]. IEEE Transactions on Neural Networks and Learning Systems, 2012, 23(11) : 1738-1754.
[14] ZHANG Huaqing, WANG Jian, SUN Zhangquan, et al. Feature selection for neural networks using group lasso regularization [J] . IEEE Transactions on Knowledge and Data Engineering, 2020, 32(4) : 659-673.
[15] Dai, J. , Chen, Y. , Yi, Y. , Lei, W. , & Gang, L. . (2018). Unsupervised feature selection with ordinal preserving self-representation. IEEE Access, PP(99), 1-1.
[16] Pengfei Zhu, Qian Xu, Qinghua Hu, Changqing Zhang. Co-regularized unsupervised feature selection [J]. Neurocomputing. 2018, 275:2855–2863.
[17] C. Lazar, "A survey on filter techniques for feature selection in gene expression microarray analysis," IEEE/ACM Transactions on Computational Biology and Bioinformatics, vol. 9, no. 4, pp. 1106–1119, 2012
[18] He X , Cai D , Niyogi P . Laplacian score for feature Selection[C]// Advances in Neural Information Processing Systems 18 [Neural Information Processing Systems, NIPS 2005, December 5-8, 2005, Vancouver, British Columbia, Canada]. 2005.
[19] M. M. Kabir, M. M. Islam, and K. Murase, "A new wrapper feature selection approach using neural network," Neurocomputing, vol. 73, no. 16-18, pp. 3273–3283, 2008.
[20] P. Chong, K. Zhao, Y. Ming, and C. Qiang, "Feature selection embedded subspace clustering," IEEE Signal Processing Letters, vol. 23, no. 7, pp. 1018–1022, 2016.
[21] Rui, Zhang, Feiping, Nie, Yunhai, & Wang, et al. (2019). Unsupervised feature selection via adaptive multimeasure fusion. IEEE transactions on neural networks and learning systems.

[22] Zhang, Y. , Zhang, Z. , Li, S. , Qin, J. , Liu, G. , & Wang, M. , et al. (2018). Unsupervised nonnegative adaptive feature extraction for data representation. IEEE Transactions on Knowledge & Data Engineering, 1-1.
[23] Du, X. , Nie, F. , Wang, W. , Yang, Y. , & Zhou, X. . (2018). Exploiting combination effect for unsupervised feature selection by 2,0 norm. IEEE Transactions on Neural Networks & Learning Systems, PP, 1-14.
[24] Li, X. , Zhang, H. , Zhang, R. , Liu, Y. , & Nie, F. . (2018). Generalized uncorrelated regression with adaptive graph for unsupervised feature selection. IEEE Transactions on Neural Networks and Learning Systems, PP, 1-9.
[25] Yi Y , Shen H T , Ma Z , et al. L21-norm regularized discriminative feature selection for unsupervised learning[C]// Proceedings of the Twenty-Second international joint conference on Artificial Intelligence - Volume Volume Two. AAAI Press, 2011.
[26] Z Li, Y Yang, J Liu,等. Unsupervised feature selection using nonnegative spectral analysis[J]. 2012.
[27] Deng C , Zhang C , He X .Unsupervised feature selection for multi-cluster data[C]// Proceedings of the 16th ACM SIGKDD International Conference on Knowledge Discovery and Data Mining, Washington, DC, USA, July 25-28, 2010. ACM, 2010.
[28] Juanying Xie,Lijuan Ding, Mingzhao Wang.Unsupervised feature selection algorithm based on spectral clustering [J].Journal of software, 2020, 031 (004): 1009-1024. (in chinese)

Particle swarm optimization algorithm with proportional factor based on Nash equilibrium

Wenxin Luo

Beijing University of Technology, Beijing, 100000, China

Email:l2512654684@163.com

Abstract: Based on the global gradient-less stochastic search method of traditional particle swarm optimization algorithm, an improved particle swarm optimization algorithm for finding mixed-strategy Nash equilibrium is proposed in view of the shortcomings of prone particle overflow and unconstrained relationship between individual dimensions. In the improved evolution equation, the position function is improved to conform to the concept and properties of the combination of strategies, and the proportional factor is introduced to ensure that the numerical value of particle iteration is mapped to the solution space of normal form game. Through the test of several groups of classical examples, the results show that the algorithm not only improves efficiency and accuracy, but also solves the problems of divergence of particle and low efficiency of iteration. The experiments show that the algorithm has good practical performance.

1. Introduction

Nash equilibrium is a strategy combination composed of each player's dominant strategy in non-cooperative games. Each player's strategy is the optimal response to the strategies of other players. In other words, any player can not improve his or her personal income by unilaterally changing their own strategy [1]. The study of non-cooperative equilibrium has great value in the fields of modern management, machine learning, economics and so on, the core of promoting R&D and technological progress in this field is the improvement of the method of solving Nash equilibrium. However, simple algorithms for solving the generalized Nash equilibrium problem (GNEP for short) in games can no longer meet the needs. For example, the once widely used graphical method [2], Lemke-Howson algorithm [3], homotopy method [4] has many disadvantages in solving multiplayer multi-dimensional non-cooperatives games [5].

In recent years, inspired by the idea of biological evolution, meta-heuristic algorithm using bionic phenomena has become another new way to solve GNEP, such as ant colony algorithm [6], genetic algorithm [7], etc. As a swarm intelligence algorithm that simulates the behavior of organisms, particle swarm optimization (PSO for short) algorithm [8] has strong processing skills in solving complex optimization problems with large amounts of nonlinearity, multi-peak, and nondifferentiation in data. PSO algorithm is applied to solve game theory problems, and great achievements and progress have been made [10].

However, the traditional PSO algorithm also has shortcomings in finding Nash equilibrium in game theory. For GNEP, the mixed strategy executed by the player represents the decision-making behavior, including the probability of each action taken by the player. It can be seen that the value in probabilistic scheduling of each behavior should be limited within the range of [0,1], and the sum of each probability is 1. The core idea of the traditional PSO algorithm is to adjust the individual position based on the evolution equation to fulfill design requirements through multiple iterations, that is, to find the optimal solution of the corresponding problem in the search space. However, there is a disconnection between the dimension values of particles in solving GNEP. Through the application of the evolution equation of traditional PSO algorithm in game theory, the strategy combination corresponding to particle position information can not meet the nature of the strategy as well as reflecting the probability of the player incorrectly. A large number of particles are invalid in the iterative process, which takes the consequence of poor performance and low time-efficiency of the algorithm. Aiming at the shortcomings mentioned above, this paper improves the PSO algorithm for solving GNEP by adding the proportional factor to the evolution equation and modifying the particle position function.

2. Particle swarm optimization algorithm based on solving Nash equilibrium problem

2.1. Traditional PSO algorithm

2.1.1. Evolution equation

PSO is a population-based stochastic optimization algorithm motivated by collective intelligence of some animals such as flocks of birds or schools of fish, which is successfully applied to a wide variety of problems (Neural Networks, Structural opt., Shape topology opts.) [11].

The population of PSO algorithm is composed of N particles, initialized with randomly generated population of particles and a random velocity, each particle has velocity and position attributes with dimension D, which represents the individual motion vector and a potential solution of the target problem respectively. In the PSO algorithm, the fitness function determines the individual fitness value to quantitative the pros and cons of particles , and guides the group to the optimization result over a number of iterations through the velocity and position function.

After the k-th iteration, each particle has a best "remembered" position attained by it in the past, which is called personal best position ($P_i^k = (P_{i_1}^k, P_{i_2}^k, \cdots, P_{i_D}^k)$). Each particle has its P_i and the particle with the best value of fitness is called global best particle ($G^k = (G_1^k, G_2^k, \cdots, G_D^k)$) [12]. In k+1-th iteration, the particle velocity $V_i = (v_{i_1}, v_{i_2}, \cdots, v_{i_D})^T$ of individual particles updated in the t($1 \le t \le D$) dimension is expressed as Eq. (1).

$$V_{i_t}^{k+1} = \omega V_{i_t}^k + c_1 r_1 (P_{i_t}^k - S_{i_t}^k) + c_2 r_2 (G_t{}^k - S_{i_t}^k) \tag{1}$$

Where ω is inertia weight that determines the degree to which the particle's previous velocity affects the current velocity [13], r_1 and r_2 are random numbers between 0 and 1, c_1 and c_2 are particle cognitive and social parameters, respectively.

The position $S_i = (s_{i_1}, s_{i_2}, \cdots, s_{i_D})^T$ of individual particles updated in the t ($1 \le t \le D$) dimension is expressed as Eq. (2).

$$S_{i_t}^{k+1} = S_{i_t}^k + V_{i_t}^{k+1} \tag{2}$$

2.1.2. Evolution equation analysis

In the traditional PSO algorithm, the position of particles represents the solution of the current problem, while the corresponding fitness determines its pros and cons in the population; the velocity vector affects the direction as well as the magnitude of change for individuals, which is not related to the measurement index of particles. Therefore, it is a fact that the concept of particle velocity is unnecessary [9].

Applying PSO algorithms to the normal form games, the position of the particle represents probability distribution over the player's actions, and the requirements are met$\sum_{j=1}^{D} s_j^i = 1$. But the traditional PSO algorithm does not have specific constraints on the scope of particle evolution in the search space, that is, it may appear the situation:$s_j^i < 0 \cup s_j^i > 1 \cup \sum_{j=1}^{D} s_j^i > 1$.Therefore, in the process of iteration, a large number of invalid particles that do not meet the probability range will be generated, resulting in problems such as slow convergence, particle divergence as well as poor evolution skills.

2.2. Design of improved PSO algorithm

2.2.1. Fitness function

There are many similarities between optimization problem and strategy equilibrium problem. The former is to solve the maximization and minimization problem of objective function; the latter is to seek the best beneficial effect of setting multiple objectives in the search space. Setting for solving Nash equilibrium problem associated with normal form game, the objective function of solving extreme value in optimization problem can be transformed into the function of achieving Nash equilibrium.

Suppose that the pure strategy set of player i ($1 \le i \le 2$) in a non-cooperative bimatrix game is $p^i = (p_1^i, p_2^i, \cdots, p_m^i)$, where m is the number of pure strategies of player i, and its probability distribution is $x^i = (x_1^i, x_2^i, \cdots, x_m^i)$, which must meet the conditions that $x_j^i \ge 0 \cap \sum_{j=1}^{m} x_j^i = 1$. If the player i selects the strategy a_j^i that indicates the probability x_j^i of choosing pure strategy p_j^i, the space of mixed strategy is as follows:

$$\begin{cases} A_1 = (a_1^1, a_2^1, \cdots, a_m^1), \sum_{j=1}^{m} a_j^1 = 1, a_j^1 \ge 0 \\ A_2 = (a_1^2, a_2^2, \cdots, a_m^2), \sum_{j=1}^{m} a_j^2 = 1, a_j^2 \ge 0 \end{cases}$$

Denote the normal form game as $\Gamma = (A_1, A_2; R_1, R_2)$, where $R_1 = (r^1)_{m \times m}$ and $R_2 = (r^2)_{m \times m}$ are the payment matrices of player 1 and player 2 respectively. And the fitness function for 2-person is expressed as Eq. (3). [13]:

$$f(A_1, A_2) = \max\left\{\max_{1 \le i \le m}\left(R_{1_i} A_2^T - A_1 R_1 A_2^T\right), 0\right\} + \max\left\{\max_{1 \le j \le m}\left(A_1 R_{2_j} - A_1 R_2 A_2^T\right), 0\right\} \quad (3)$$

The following mixed-strategy conditions must hold for all $a \in A$:

$$\begin{cases} \sum_{i=1}^{m} a_i^1 = 1 \\ \sum_{j=1}^{m} a_j^2 = 1 \\ a_i^1 \ge 0, i = 1,2, \cdots m \\ a_j^2 \ge 0, j = 1,2, \cdots m \end{cases}$$

To sum up, the Nash equilibrium for non-cooperative bimatrix game can be transformed into a single-objective optimization problem. Obviously, Nash equilibrium is obtained if the following condition of the fitness function is satisfied: $f(a_1, a_2) = 0$, $f(a_1^*, a_2^*) > 0, (a_1, a_2) \ne (a_1^*, a_2^*)$.

2.2.1.1. Proportional factor and improved position function

The improved PSO algorithm sets the proportional factor of particles, which is used to determine the changing value of particle position corresponding to the change of each dimension after iteration, so as to map the position values into the required solution of the normal form game. Based on the idea of simple particle swarm optimization [9], the improved PSO removes the velocity attribute of particles, so that the change of position is only determined by the personal best position and global best particle of the current dimension of particles, avoiding the tedious convergence of quantitative analysis of the algorithm. The equation also adds inertia factors ω_1, ω_2 to let the particle proportional factor always in the search space during the iteration. The proportional factor obtained from the t dimension of particle i after k+1-th iteration is expressed as $F_{i_t}^{k+1}$, the proportional factor of particles is as follows:

$$F_{i_t}^{k+1} = \omega_1 S_{i_t}^k + \omega_2\left[c_1 r_1\left(P_{i_t}^k - S_{i_t}^k\right) + c_2 r_2\left(G_t^k - S_{i_t}^k\right)\right] \tag{4}$$

For the property $\sum_{t=1}^{D} S_{i_t} = 1$ that the position of each dimension must satisfy, the improved PSO algorithm allocates the position values of each dimension according to the ratio of proportional factor in the total value $\sum_{j=1}^{D} F_{i_j}$, hence, the position is transformed into the strategy probability, which makes up for the limitations of the traditional algorithm in the normal form game, and let the position of each dimension mutually influencing be realized by implementing "information sharing" approaches. In order to ensure the feasibility of PSO algorithm for solving Nash equilibrium value, the position function of particles is changed as follows:

$$S_{i_t}^{k+1} = \frac{F_{i_t}^{k+1}}{\sum_{j=1}^{D} F_{i_j}^{k+1}} \tag{5}$$

To sum up, the proportional factor determines the state change of particles by weighing the self cognition and group learning of particles and constraining the change amplitude by relying on inertia factors ω_1, ω_2. The position function distributes the position values according to the proportion factor of each dimension, and successfully realizes the transformation of state quantities into fit combination strategies in the normal form game, so as to obtain the solution of Nash equilibrium problem. The improved PSO algorithm solves many problems caused by unshared information between individual dimensions through equations (4) and (5), and enhance the computational performance and efficiency of the algorithm in solving Nash equilibrium problems.

2.2.2. *Algorithm description*

The specific steps of the improved PSO algorithm for solving Nash equilibrium problems are as follows:

Step1: Initialize position: Randomly initialize the position of each dimension, while meeting the requirement $\sum_{j=1}^{D} s_j^i = 1$;

Step 2: initialize the personal best position as the initial position of the particle;

Step3: Evaluate all fitness values according to the fitness function and select the global best particle;

Step4: update the proportional factors and positions of each particle according to equations (4) and (5);

Step5: calculate the fitness value of each particle according to equation (3) and select the global best particle;

Step6: judge whether the number of search iteration reaches the preset number or not: If not, transfer to step 4; If so, output the result: the solution of the current problem evolved by the model is the corresponding position values of the global best particle.

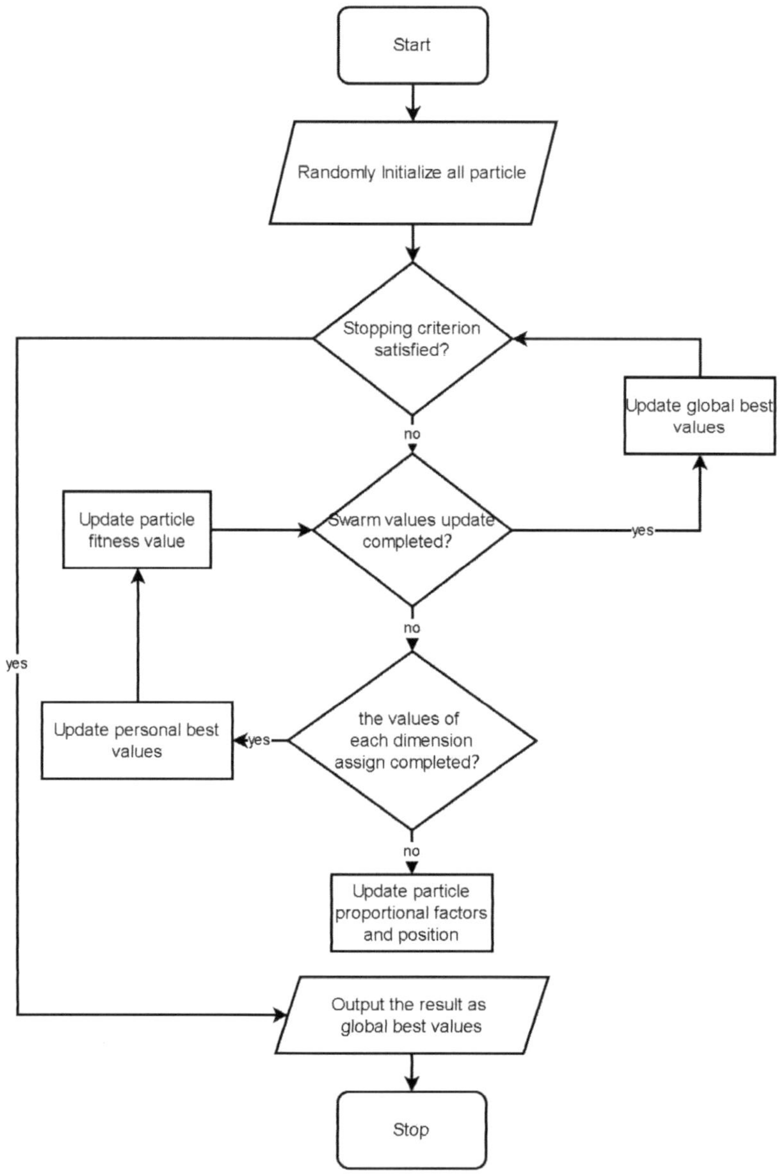

Figure 1 Flow chart of improved particle swarm optimization algorithm

3. Experimental results

In order to verify the feasibility of the improved PSO algorithm and evaluate its performance calculation, multi-dimensional bimatrix examples are used to carry out the experiment. The particle population number n = 1000 and the fixed number of iterations m = 50 are selected, inertia weights ω_1=0.8, ω_2= 0.45, the precision of the experimental result is 10^{-10}. The examples are evolved independently 5 times on eclipse (2020-12), and the average experimental result is used. Taking the fifth result, the individual fitness distribution box and whisker plot of PSO is presented through data visualization. And compared with the traditional PSO, off-line performance is used to test the convergence characteristics of the algorithms.

Example 1 a three-dimensional bimatrix game is given in reference [7] $\Gamma_1 = (x_1, y_1; A_1, B_1)$, the payoff matrix is as follows:

$$A_1 = \begin{bmatrix} 1 & 0 & 0 \\ 0 & 1 & 0 \\ 0 & 0 & 1 \end{bmatrix} \qquad B_1 = \begin{bmatrix} 0 & 1 & 0 \\ 0 & 0 & 1 \\ 1 & 0 & 0 \end{bmatrix}$$

Using the improved PSO algorithm, five experiments are carried out independently, and the unique Nash equilibrium $\left(\frac{1}{3},\frac{1}{3},\frac{1}{3};\frac{1}{3},\frac{1}{3},\frac{1}{3}\right)$ solution of this matrix is obtained after an average of 19 iterations, the best fitness value is 0.0. In the improved algorithm, as the number of iterations increases, the fitness distribution is changing, showing a trend of rapid aggregation in the early stage and slow fluctuation in the later stage. After 10 iterations, the swarm basically converges. The resulting fitness distribution box and whisker plot is shown in Figure 2.

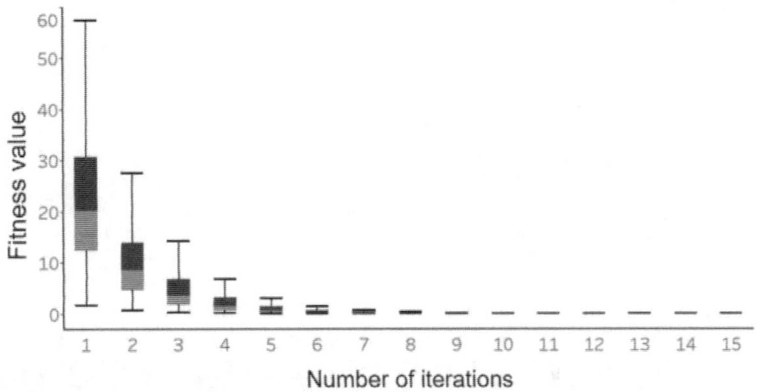

Figure 2 Box and whisker plot of particle swarm distribution for Example 1

In order to highlight the optimization performance of the improved algorithm, the improved PSO algorithm is compared with the traditional PSO algorithm, and the experiments are carried out with the same total number of particle and fixed iteration times. The off-line performance comparison chart of the algorithm is shown in Figure 3. After 15 iterations, the traditional PSO algorithm fell into the local optima, while the improved algorithm fluctuated slowly and reached the global optima after 17 iterations.

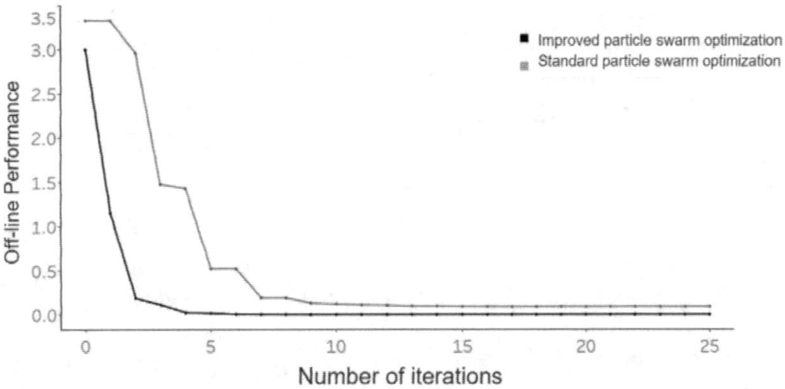

Figure 3 Comparison of off-line performance of algorithms for solving Example 1

Example 2 a three dimensional bimatrix game $\Gamma_2 = (x_2, y_2; A_2, B_2)$, the payoff matrix is as follows:

$$A_2 = \begin{bmatrix} 0 & -1 & 1 \\ 1 & 0 & -1 \\ -1 & 1 & 0 \end{bmatrix} \qquad B_2 = \begin{bmatrix} 0 & 1 & -1 \\ -1 & 0 & 1 \\ 1 & -1 & 0 \end{bmatrix}$$

Through the improved PSO algorithm, the unique Nash equilibrium solution $\left(\frac{1}{3},\frac{1}{3},\frac{1}{3};\frac{1}{3},\frac{1}{3},\frac{1}{3}\right)$ is obtained after an average of 22 iterations, and the best fitness value is 0.0. The box and whisker plot of the swarm average distribution of fitness obtained by the improved algorithm is shown in Figure 4. It can be seen that the particle swarm shows a discrete distribution at the beginning, after the iterative improvement, the maximum value is found and particles converge collectively.

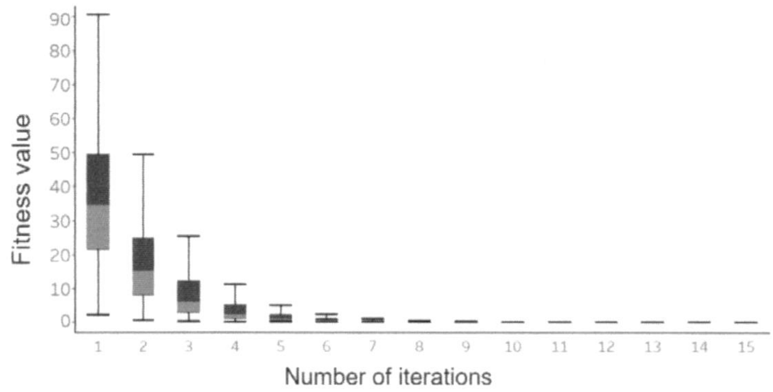

Figure 4. Box and whisker plot of particle swarm distribution for Example 2

The experimental result of this example is compared with the improved PSO algorithm and the traditional PSO algorithm, using the same total number and a fixed number of iterations to ensure fairness of the comparison. According to the off-line performance of the two algorithms, comparison chart is shown in Figure 5, which reflects the great computational performance of the improved algorithm.

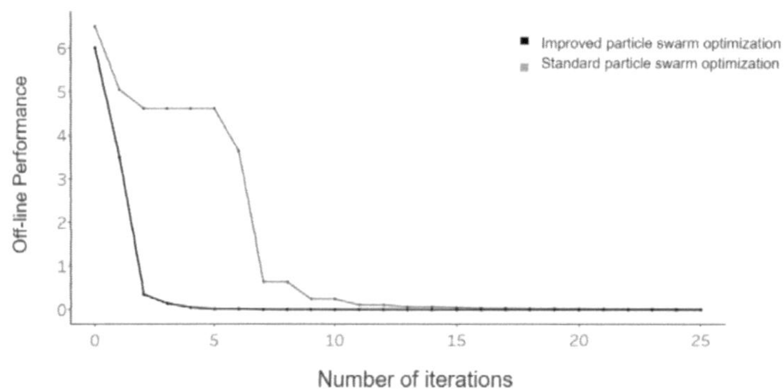

Figure 5 Comparison of off-line performance of algorithms for solving Example 2

Example 3 the Dekel-Scotchmer [15] four-dimensional bimatrix game $\Gamma_3 = (x_3, y_3; A_3, B_3)$, the payoff matrix is as follows:

$$A_3 = \begin{bmatrix} 1 & 235 & 0 & 0.1 \\ 0 & 1 & 235 & 0.1 \\ 235 & 0 & 1 & 0.1 \\ 1.1 & 1.1 & 1.1 & 0 \end{bmatrix} \qquad B_3 = \begin{bmatrix} 1 & 0 & 235 & 1.1 \\ 235 & 1 & 0 & 1.1 \\ 0 & 235 & 1 & 1.1 \\ 0.1 & 0.1 & 0.1 & 0 \end{bmatrix}$$

Through the improved PSO algorithm, the unique Nash equilibrium solution $\left(\frac{1}{3},\frac{1}{3},\frac{1}{3},0;\frac{1}{3},\frac{1}{3},\frac{1}{3},0\right)$ is obtained after an average of 20 iterations, the best fitness value is 0.0. The resulting of the fitness value of an individuals is shown in Figure 6. The result of particle swarm fitness value distribution is shown in box and whisker plot 3.5. It can be seen that the values corresponding to the individuals in the swarm are discretely distributed at the early stage, then the population is gradually gathered through iterative evolution, and the best fitness value is found within a small fluctuation range.

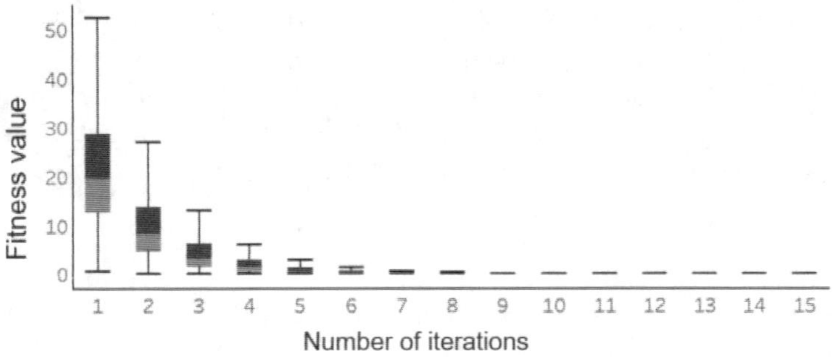

Figure 6 Box and whisker plot of particle swarm distribution for Example 3

The off-line performance of the improved PSO algorithm is shown in Figure 7. It can be seen that the improved algorithm has good computational performance in solving four-dimensional Nash equilibrium problems, which reflects the universality and practicability of the improved algorithm in dealing with multi-dimensional double matrix.

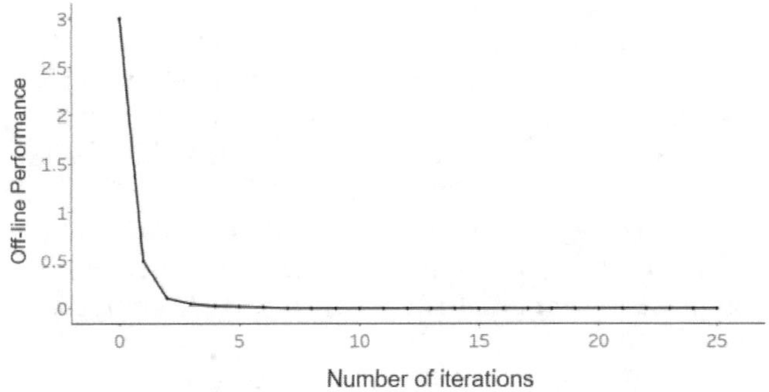

Figure 7 Off-line performance of improved PSO algorithms for solving Example 3

In summary, it can be known through the results of above examples that the improved PSO algorithm shows excellent convergence performance in solving the Nash equilibrium problem of normal form game. The algorithm basically reaches the target value after about 20 iterations in the three experiments, shows the good efficiency and high accuracy. The improvement not only simplifies the particle evolution equation, but also maps the numerical results of each dimension to the reasonable solution of the game strategy, which shows a good optimization effect to solve the Nash equilibrium problem.

4. Conclusion

Firstly, this paper introduces the equivalent transformation of objective function and Nash equilibrium from the perspective of theoretical analysis, let Nash equilibrium problem be transformed into an optimization problem. After pointing out the bad situations could arise when solving Nash equilibrium problem by using traditional PSO algorithm, such as particle position out-of-bounds, individual dimension without interactions, slow convergence of the overall algorithm, an improved PSO algorithm is proposed: the individual proportional factor is obtained by using the first-order evolution equation, and the position of each dimension is transformed into the probability distribution of the normal form game according to the position formula to determine the state quantity of the particle. Finally, three typical examples show the feasibility and efficiency of the improved PSO algorithm in solving the Nash equilibrium problem of multi-dimensional bimatrix game. Compared with the traditional PSO algorithm, the convergence speed and experimental accuracy of the algorithm are significantly improved, and the efficient computational performance is reflected.

The application of the improved PSO algorithm in the field of economics can improve the computing performance and accuracy in solving normal form game's problems, and the idea that the algorithm with proportional factor provides a new way forward for the iterative evolution of the swarm intelligence-based algorithm. Future work will focus on optimizing the setting of the fitness function, and do further research on the setting of inertia factors' values to improve the efficiency of population evolution.

References:

[1] Risk N A, Szafron D. Using counterfactual regret minimization to create competitive multiplayer poker agents[C]. International Conference on Autonomous Agents and Multiagent Systems. 2010:159-166.

[2] Nair K G, Tanjith G. Solution of 3×3 Games Using Graphical Method[J]. European Journal of Operational Research, 1999,112:472-478.

[3] Lemke C, Howson J. Equilibrium Points of Bimatrix Games[J]. Journal of Society of Industrial and Applied Mathematics, 1964,12:413-423.

[4] Herings P J J, Peeters R J A P. A Differentiable Homotopy to Compute Nash Equilibriua of n-Person Games[J]. Economic Theory, 2001,18:159-186.

[5] Yu Qian, Wang Xianjia. Evolutionary algorithm for solving Nash equilibrium based on particle swarm optimization [J]. Journal of Wuhan University (SCIENCE EDITION). 2006 (01).

[6] Wang Zhiyong, Han Xu, Xu Weisheng, et al. Nash equilibrium solution based on improved ant colony algorithm [J]. Computer engineering. 2010,36 (14).

[7] Chen Shijun, sun Yongguang, Wu Zongxin. A genetic algorithm for solving Nash equilibrium solution [J]. Systems engineering. 2001,19 (5): 67-70.

[8] Kennedy J,Eberhart R C. Particle Swarm Optimization [C]. Proceedings of IEEE International Conference on Neural Networks, Perth, Australia, 1995.1942-1948.

[9] Hu Wang, Li Zhishu. A more simplified and efficient particle swarm optimization algorithm [J]. Journal of software. 2007, (04).

[10] Pavlidis N G, Parsopoulos K E, Vrahatis M N. Computeing Nash Equilibria Through Computational Intelligence Methods[J]. Journal of Computational and Applied Mathematics, 2005,175:113-136.

[11] Wang, D., Tan, D. & Liu, L. Particle swarm optimization algorithm: an overview. Soft Comput 22, 387–408 (2018). https://doi.org/10.1007/s00500-016-24 74-6.

[12] Kumar A, Pant S, Ram M, Singh SB (2017b) On solving complex reliability optimization problem using multi-objective particle swarm optimization, mathematics applied to engineering. Elsevier, New York, pp 115–13.

[13] Dexiang Gao, Kou Li, Yuncheng Cai & Tao Wen. Predicting of landslide displacement based on time series and PSO-BP model in Three Georges Reservoir，China. Journal of Earth Science, doi:10.1007/s12583-021-1575-z.

[14] Liu Hao. Research on multi-agent game reinforcement learning algorithm and its equilibrium [D]. Xi'an University of science and technology 2020.

[15] Dekel E, Scotchmer S. On the Evolution of Optimizing Behavior[J]. Journalo f Economic Theory, 1992,57:392-406.

Video behavior recognition algorithm based on two-stream heterogeneous convolutional neural network

Qing ye[1*], Zhenghao Liang[1]

[1]School of Information Science and Technology, North China University of Technology, Beijing 100144, China

[*]Corresponding author's e-mail: yeqing@ncut.edu.cn

Abstract—In video behavior recognition, making full use of the spatio-temporal information contained in video frame is the critical point to further improve the recognition accuracy. In our study, we propose a video behavior recognition algorithm based on two-stream heterogeneous network, so that we can further extract the spatio-temporal feature information in the video frame, and finally take full advantage of the spatio-temporal features in the video sequence. In view of the characteristics of RGB and optical flow images, this paper uses DenseNet121 and Inception-V4 to structure a two-stream network, to fully extract the spatio-temporal feature information in the video. In this paper, UCF101 dataset is used to evaluate the algorithm, and the recognition accuracy is 91.7%, which verified the reliability of the proposed algorithm.

1. Introduction

Video behavior recognition is one of the key tasks in computer vision and video understanding. because of the vigorous development in the field of computer vision, video behavior recognition has been widely used in intelligent monitoring, video classification and other fields. With the updating of communication technology, the need to process a large amount of video information is also increasing. Due to the rapid rise of Internet technology and the rapid development of short video platforms, the number of videos has exploded. As the cost of acquiring information from various and miscellaneous videos becomes higher and higher, the demand for video information screening and classification is increasing, which makes the research on video behavior recognition more and more important[1].

Behavior recognition in video mainly has two directions: one is the traditional method based on artificial feature extraction, the other is the recognition method based on deep learning. Methods based on traditional feature extraction are easily limited by the extracted features. Although traditional methods have been developed for many years, they still have their limitations. In contrast, because deep learning method has achieved excellent results in the field of image processing, it has also been fully developed in the field of video. There are many convolutional neural networks for behavior recognition. Here, video behavior recognition in recent years is mainly divided into two types, one is typical two-stream network and its variants, the other is 3D convolutional neural network. Simonyan et al.[2] proposed a two-stream algorithm, which input RGB images one way to extract spatial information in video, input optical flow diagram one way to extract time information in video, and finally fused the extracted feature information to improve the recognition accuracy. Moreover, there are also a lot of work based on two-stream recognition network, such as Wang et al.[3], Girshick et al.[4], based on the two-stream network to further improve the recognition performance. 3D convolutional neural network uses 3D convolution kernel and pooling to extract feature information from video, such as the c3D-based recognition algorithm proposed by Tran et al.[5], or carira et al.[6] proposed to use 3D convolution kernel to construct two-stream recognition algorithm, etc. There are also network attempts to add a new time feature extraction module. For example, TSM[7], TANet[8] and TEA[9].

Although the current video behavior recognition technology has achieved some close combat, it still faces many challenges in practical application, such as insufficient feature information extraction.We propose a video behavior recognition technology based on two-stream heterogeneous convolutional neural network. By using different networks, the spatio-temporal features are further fully extracted and the performance of video behavior recognition is improved.

2. Proposed Method

Our recognition algorithm includes three parts: video image preprocessing, feature extraction, classifier classification. In the stage of image preprocessing, we will perform segmented random sampling of the input video, so as to reduce a lot of redundant information and retain valid information. We use UCF101 data set to verify the algorithm. Our algorithm framework is shown in Fig.1.

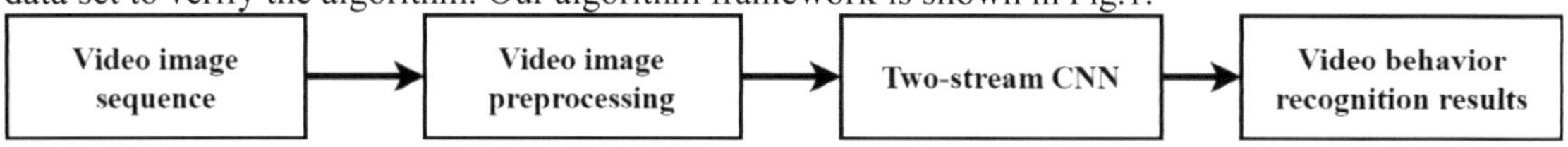

Fig.1 Video behavior recognition algorithm framework

2.1 Dateset

The UCF101 dataset contains five kinds of videos: Video of people interacting with things; Video of small movements of people; Video of action between people; Performance behavior video; Sports behavior video. UCF101 dataset collects data from YouTube. The dataset breaks the collected data into 101 behaviors and contains more than 13,000 video files. The 101 action videos in the dataset can be divided into 20 groups based on behavior type, with four to seven behavior categories in each group. UCF101 is more diverse than other video datasets, so it is one of the datasets frequently used in video recognition algorithm.

2.2 Image preprocessing

Before video behavior analysis, proper preprocessing of video data will have an important impact on the whole video behavior analysis. Since the data size in the UCF101 dataset is not uniform, the data needs to be standardized before being fed into the algorithm. This article uses the tools included with TorchVision to standardize the data. The standard deviation of data was set as [0.485,0.456,0.406], and the variance was set as [0.229,0.224,0.225]. The goal in the field of video behavior analysis technology, video tend to suffer when shooting direction, location, scaling, brightness and other factors, because these factors lead to the same goal, the feature information of the present is not the same, therefore, in order to improve the robustness of the algorithm model, half of the images were processed with random flipping and random Angle rotation to simulate complex scenes in practical application. In addition, segmented random sampling is performed to reduce computational redundancy and improve computational efficiency.

2.3 Two-stream heterogeneous network

Aiming at the complexity of video behavior feature information, this paper improved the traditional two-stream structure and proposed a two-stream heterogeneous convolutional neural network, which processed the temporal information and spatial information separately to fully extract feature information.

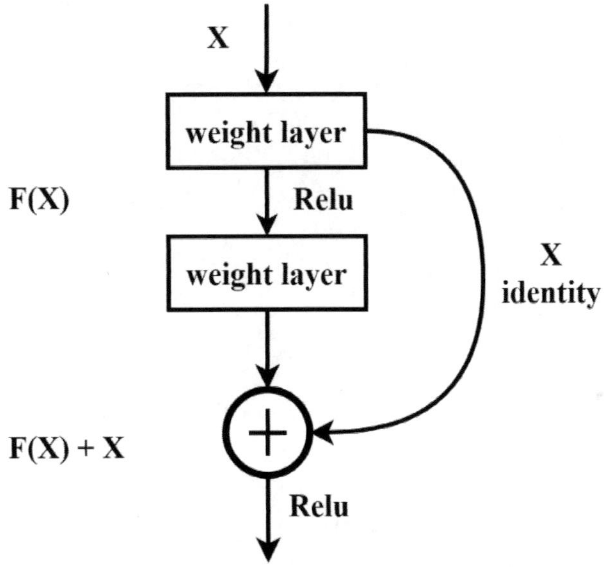

Fig.2 Residual learning unit

The proposed ResNet network and its improvement in recognition accuracy and training speed prove the importance of residual structure for the construction of deep network. Resnet network uses ReLU activation function to activate non-adjacent convolutional layers to build a shortcut structure, weaken the mutual influence of parameters in other layers, and finally reduce the overfitting chance[10]. The residual structure is shown in Fig.2. Based on ResNet network structure, DenseNet network connects the input of each network layer with the output of the network layer before this layer through shortcut connection, which greatly reduces the network parameters and reduces the complexity of calculation[11]. Also, the gradient disappearance is greatly reduced. The key structure of DenseNet can be known in Fig.3.

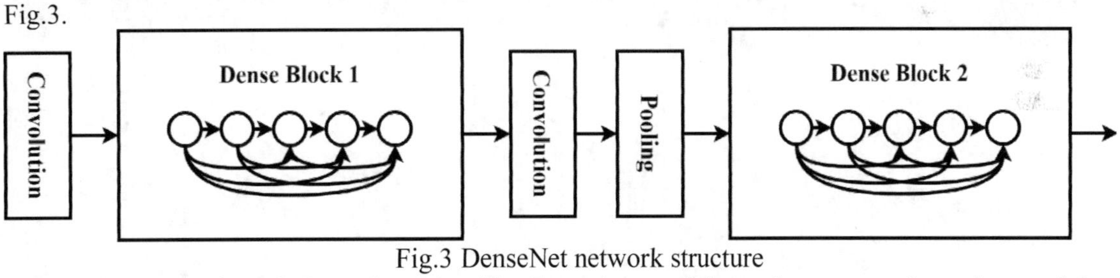

Fig.3 DenseNet network structure

Based on Inception-v3, Inception-v4 unifies Inception modules and uses more Inception modules, and further improves the layers and complexity of network modules on the basis of the original modules. In addition, parallel is also to reduce the computation and finally get better network performance.

Because the optical flow graph mainly contains the motion information in the image, and the feature information is relatively less, this paper chooses the Inception-v4 network when extracting the time information. The information in RGB image is much more than optical flow graph, so when extracting spatial information, this paper chooses DenseNet121 network and makes use of its deep network layer to extract the spatial features of RGB image to the maximum extent.

3. Experimental Results

In our study, the proposed two-stream heterogeneous convolutional neural network algorithm is validated on UCF101 dataset. To further observe the performance of densenet121 and Inception-v4 during training, we plot the changes in the validation probability and loss function of the two networks during training. It can be seen in Fig.4 and Fig.5.

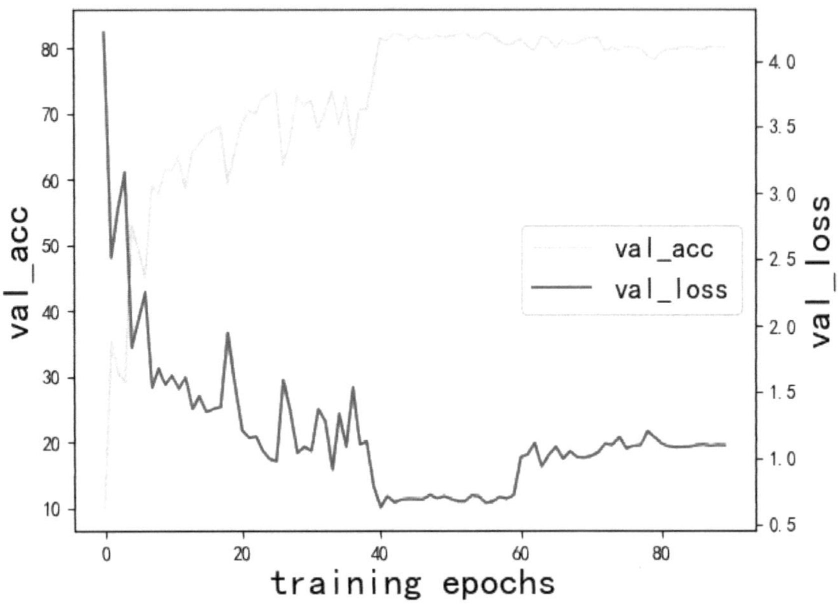

Fig.4 Overview of training process of Inception-v4

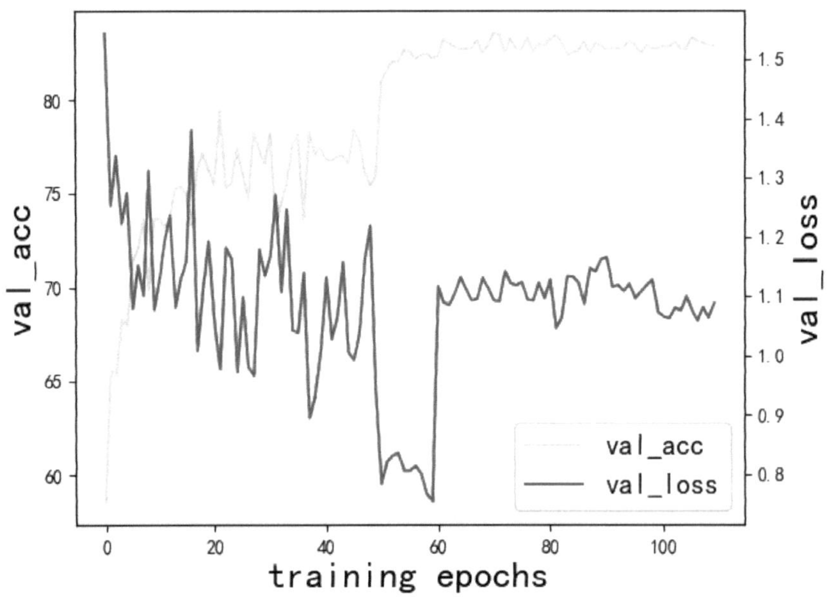

Fig.5 Overview of training process of DenseNet121

In our research, we compare the recognition accuracy of the algorithm with other methods. The comparison results can be seen in Table 5. Compared with other methods, the algorithm proposed in this paper achieves higher accuracy on UCF101 datasets, which proves the feasibility of algorithm proposed in this paper.

Table 1. Comparison of the accuracy of this algorithm with other algorithms

Methods	Accuracy rate (%)
mAP+SVM[12]	80.1
Two-Stream[2]	83.6
C3D with unequal time and space[13]	85.6
IP-LSTM[14]	88.5
T-VLAD[15]	89.0
Method of this paper	91.7

4. Conclusion

In our study, we propose a video behavior recognition algorithm based on two-stream heterogeneous convolutional neural network. According to the characteristics of RGB and optical flow, DenseNet121 network is selected as spatial network to extract spatial information, and Inception-v4 network is selected to extract time information, so as to fully extract the characteristic information contained in RGB and optical flow images and reduce the amount of computation. The accuracy of our method is 91.7% on UCF101 dataset, and the feasibility of the algorithm can be obtained. In practical application, the algorithm proposed in this paper can perform well in intelligent monitoring and other fields.

Acknowledgments

This paper is supported by Beijing Natural Science Foundation(No.4192022), Technology Project of Beijing Municipal Education Commission (No.SQKM201810009002), National Natural Science Foundation of China (No.61806008), National Natural Science Foundation of China (No.61371143), Ministry of Education Science and Technology Development Center Project (No.2018A03029). The author is also very grateful to the reviewers for their comments and help in improving the manuscript.

References

[1] G. Moon, J. Y. Chang and K. M. Lee, PoseFix: Model-Agnostic General Human Pose Refinement Network, IEEE Conference on Computer Vision and Pattern Recognition (CVPR), USA, 7765-7773, 2020.

[2] K. Simonyan, A. Zisserman. Two-stream convolutional networks for action recognition in videos[C]. Annual Conference on Neural Information Processing Systems, Montreal, Canada, 568-576, 2014.

[3] H. Wang and C. Schmid, Action Recognition with Improved Trajectories, 2013 IEEE International Conference on Computer Vision, Sydney, NSW, Australia, 3551-3558, 2015.

[4] R. Girshick, J. Donahue, T. Darrell and J. Malik, Rich Feature Hierarchies for Accurate Object Detection and Semantic Segmentation, 2014 IEEE Conference on Computer Vision and Pattern Recognition, Columbus, USA, 580-587, 2014.

[5] D. Tran, L. Bourdev, R. Fergus, L. Torresani and M. Paluri, Learning Spatiotemporal Features with 3D Convolutional Networks, 2015 IEEE International Conference on Computer Vision (ICCV), Santiago, Chile, 4489-4497, 2015.

[6] J. Carreira and A. Zisserman. Quo vadis, action recognition? a new model and the kinetics dataset. arXiv preprint arXiv:1705.07750, 2017.

[7] J. Lin, Chuang Gan, and Song Han. TSM: temporal shift module for efficient video understanding. In ICCV, pages 7082–7092, 2019.

[8] Z. Liu, L. Wang, W. Wu, C. Qian, and T. Lu. TAM: temporal adaptive module for video recognition. CoRR, abs/2005.06803, 2020.

[9] Y. Li, B. Ji, X. Shi, J. Zhang, B. Kang, and L. Wang. Tea: Temporal excitation and aggregation for

action recognition. In CVPR, pages 909–918, 2020.
[10] K. He, X. Zhang, S. Ren, et al. Deep Residual Learning for Image Recognition. IEEE Conference on Computer Vision & Pattern Recognition IEEE Computer Society, 2016.
[11] G. Huang, Z. Liu, L. Van Der Maaten and K. Weinberger, Densely Connected Convolutional Networks, in 2017 IEEE Conference on Computer Vision and Pattern Recognition (CVPR), Honolulu, HI, USA, 2261-2269, 2017.
[12] Soltanian, M. , and S. Ghaemmaghami . Hierarchical Concept Score Postprocessing and Concept-Wise Normalization in CNN-Based Video Event Recognition. IEEE Transactions on Multimedia , 21(1), 157-172,2018.
[13] C. Wu, J. Han and X. Li, Time-Asymmetric 3d Convolutional Neural Networks for Action Recognition, 2019 IEEE International Conference on Image Processing (ICIP), Taipei, Taiwan, 21-25, 2019.
[14] Yu S., Xie L , Liu L , et al. Learning Long-Term Temporal Features With Deep Neural Networks for Human Action Recognition[J]. IEEE Access, 8:1840-1850, 2020.
[15] Hbn A , Fmb C , Mhya C , et al. T-VLAD: Temporal Vector of Locally Aggregated Descriptor for Multiview Human Action Recognition[J]. Pattern Recognition Letters, 2021.

An effective metaheuristic for the robust distributed flowshop scheduling problem

Hengwei Guo[1], Hongyan Sang[1*], Xiaobin Sun[1], Yasheng Zhao[1], Mengxi Tian[1], Zhongkai Li[1]

[1] College of computer science, Liaocheng University, Liaocheng, Shandong, 252059, China

*Corresponding author's e-mail: 920563256@qq.com

Abstract. This paper discusses the distributed permutation flowshop scheduling problem under uncertain processing time and considers the sequence-dependent setup times (denoted as DPFSP_UPT_SDST). Uncertain processing times are described by interval scenarios. An expectation-standard deviation robust model is used to describe the DPFSP_UPT_SDST. Based on the characteristics of the DPFSP_UPT_SDST, four heuristics are proposed. In addition, a discrete Fruit Fly optimization algorithm with an age mechanism is proposed, which helps the algorithm to avoid local optimum and increases the diversity of solutions. The parameters of the proposed algorithm are calibrated by a detailed design experiment. Finally, the effectiveness of the proposed heuristics and metaheuristic is demonstrated by a comprehensive comparison with the closely related and state-of-the-art algorithms.

1. Introduction

In recent years, distributed factory production scheduling have received extensive attention. The distributed factory pattern can increase productivity, reduce delivery costs and manage risks by assigning tasks to different production units [1]. Among these distributed scheduling problems, the distributed permutation flowshop scheduling problem (DPFSP) has become a research hotspot, which assumes that all data (e.g. resources, processing times) are precisely known in advance and fixed [2,3], that is, the whole process is carried out in a deterministic environment. However, the real environment is full of uncertainties such as resource constraints, machine breakdowns, workers' proficiency, worker absences, order changes, and etc. The above uncertainty factors can lead to uncertain in processing times. If these uncertain factors are ignored, the results may deviate from expectations or even scheduling solutions may become infeasible. Therefore, the study of uncertain processing times has become a hot topic. In addition, the traditional DPFSP usually ignores setup times, which are derived from non-productive operations such as cleaning machines, replacing tools or changing molds [4]. Although these non-productive operations require no processing time, they must be performed before the coming job [5]. In a real-world environment, the setup times are influenced by the processing sequence of the job, and different scheduling sequences may result in different setup times, which are referred to as sequence-dependent setup times (SDST). Considering the SDST in scheduling decisions would generate significant economic benefits [4,6]. Therefore, we study the distributed permutation flowshop scheduling problem with the SDST under uncertain processing times (defined as DPFSP_UPT_SDST) and adopt intelligent algorithms to generate a reasonable scheduling sequence to minimize makespan.

If the probability distribution of uncertain factors can be obtained in advance from historical data and experience, uncertainty can often be resolved with some stochastic programming. However, in many cases, companies lack rich historical data and extensive experience, which means that methods for describing uncertainty through the probability distribution are not available [7]. As an alternative to stochastic programming, robust optimization (RO) techniques do not require information about probability distributions and have become a promising approach for risk-averse decision makers to deal with uncertainty [8].

For RO techniques, the scenario approach is one of the key tools for describing uncertain processing times, which uses variables in a given set of scenarios instead of random variables with known probability distributions [9,10]. In the robust scheduling, researchers usually use two methods to describe all possible scenarios: a set of discrete scenarios and a set of continuous interval scenarios [11]. The latter is actually used in this paper as the method of describing the scenario set. In an interval scenario, each parameter can take an arbitrary value between a given upper and lower bounds [10]. Thus an interval scenario set actually contains infinite number of scenarios [12]. In general, a solution can be considered as robust when it behaves "not too bad" in all scenarios [13]. Numerous methods have been proposed to deal with uncertainty, such as robust models, exact algorithms, heuristics, and so on. Kasperski et al. [14] modeled the two-machine permutation flowshop scheduling problem with uncertain processing time using discrete scenario with min-max and min-max regrets criterion. Allahverdi and Aydilek [15] solved the same problem as Kasperski, and designed five heuristics based on different weight coefficients, and used the Johnson's algorithm to generate the sequence. Feng et al. [7] presented two properties of the worst-case scenario and proposed an exact algorithm (iteration relaxation, IR) and two efficient heuristics. Wang et al. [10] developed a new robust optimization model for the job shop scheduling problem with uncertain processing time, and proposed a hybrid algorithm based SA and TS algorithm. Jing et al. [2] designed a robust model based on the expectation-risk rule and two metaheuristics to study uncertain processing times in a multi-factory environment. However, there are few multi-factory scheduling problems that consider uncertain processing times, the DPFSP_UPT_SDST contains multi-factory that needs to first consider the assignment of jobs to a certain factory, and then schedule the sequence of jobs assigned in each factory. Therefore, it is necessary to design new and effective methods according to the characteristics of the DPFSP_UPT_SDST.

Obviously, the decision-making process of the DPFSP_UPT_SDST is more complex than the traditional DPFSP. As we all know, the DPFSP is an NP-hard problem [3], so the DPFSP_UPT_SDST, as an extension of the DPFSP, is undoubtedly also an NP-hard problem. Therefore, designing effective algorithms including heuristics and metaheuristics for solving the DPFSP_UPT_SDST are the most promising methods.

Fruit fly optimization algorithm (FFO) was proposed by Pan [16] in 2012, it has been widely applied to various fields and obtained better results, such as function optimization [17], scheduling problems [18,6], parameter optimization [19], and so on. The FFO has shown excellent performance among existing methods for scheduling problems because of the simplicity of the computational process and the fewer control parameters [20]. Because of the excellent performance of the FFO, this paper uses the FFO to solve the DPFSP_UPT_SDST. Referring to the relevant literature, the uncertain processing times are described by the interval scenarios. In addition, the expectation-standard deviation method is used as a robust mathematical model for the DPFSP_UPT_SDST. Then we present four novel constructive heuristics and an improved discrete FFO (denoted as FFO_RO). An age mechanism is developed to FFO_RO for avoiding falling into local optimum.

The remainder of the paper is organized as follows. In Section 2, we describes the DPFSP_UPT_SDST and a robust model. In Section 3, we show four constructive heuristics and the FFO_RO in detail. The parameters are calibrated and test instances are designed in Section 4. In Section 5, several state-of-art metaheuristics are compared and evaluated. Finally, Section 6 concludes the work of this paper and presents an outlook for the future study.

2. Problem description and formulation

The DPFSP_UPT_SDST with makespan objective is described as follows: a set of jobs J need to be processed and each job contains M operations which corresponding to a set of machines $M = \{M_1, M_2, \dots M_m\}$. A set of factories $F = \{F_1, F_2, \dots F_f\}$ include f same factories, each of factory has the ability to process any one job independently. If job J_i is assigned to factory F_l, all operations for the job J_i will be done in factory F_l. When two jobs J_i, J_j are successively processed on machine M_k and job J_j is processed after job J_i, there exists a setup time $S_{i,j}$ that depends on the job sequence. In addition, this paper assumes that the SDST is only related to the job sequence. A job can only select one machine to process at one moment. At any time, the process of job cannot be interrupted and each machine can be occupied by no more than one job. In DPFSP_UPT_SDST, due to inevitable stochastic events or incomplete information, the processing times of all jobs are uncertain. The interval scenarios are used to describe uncertain processing times. Each scenario represents a unique set of job processing times. The robust model and parameters of DPFSP_UPT_SDST are as follows,

Notations

F	$F = \{F_1, F_2, \dots F_f\}$;
J	$J = \{J_1, J_2, \dots J_n\}$;
M	$M = \{M_1, M_2, \dots M_m\}$;
n_l	The number of job assigned to factory F_l.
$P_{i,k}$	The processing time of $J_i (J_i \in J)$ on $M_k \in M$.
$S_{i,j}$	The setup time between job J_j and job J_i. If J_j is the first job processed in a factory, then $i = j$.
π	A complete solution, $\pi = \{\pi_1, \pi_2, \dots \pi_f\}$.
π_l	The sequence assigned factory F_l, and contains n_l jobs.
$\pi_{l,k}$	The kth job in π_l, $k = 1,2, \dots n_l$.
Λ	The set of all scenarios.
$\lvert\Lambda\rvert$	The total number of scenarios.
μ	The total number of randomly selected scenarios from Λ, $\mu \le \lvert\Lambda\rvert$; the randomly selected μ scenarios are denoted as $\Lambda_1, \Lambda_2, \dots \Lambda_s$, respectively, $s = 1,2, \dots \mu$.
$P_{i,k}^s$	The processing time of $J_i (J_i \in J)$ on $M_k \in M$ for scenario Λ_s.
$S_{i,j}^s$	The setup time between job J_j and job J_i for scenario Λ_s.
$C_{l,s}$	The completion time of the last job on the machine M_m in factory F_l for scenario Λ_s.
$C(\Lambda_s)$	The maximum completion time of all jobs for scenario Λ_s.
$C^\mu(\Lambda)$	$C^\mu(\Lambda) = \{C(\Lambda_1), C(\Lambda_2), \dots C(\Lambda_\mu)\}$
$C_l^\mu(\Lambda)$	The completion times of factory F_l for μ scenarios.
$E[C^\mu(\Lambda)]$	The average makespan for μ scenarios.
$E[C_l^\mu(\Lambda)]$	The average completion time for factories F_l of μ scenarios.
$Std[C^\mu(\Lambda)]$	The standard deviation. $Std[C^\mu(\Lambda)] = \sqrt{\sum_{s=1}^{\mu} (C_l(\Lambda_s) - E[C^s(\Lambda)])^2 / \mu}$
$Std[C_l^\mu(\Lambda)]$	The standard deviation for factories F_l of μ scenarios. $Std[C_l^\mu(\Lambda)] = \sqrt{\sum_{s=1}^{\mu} (C(\Lambda_s) - E[C_l^s(\Lambda)])^2 / \mu}$
$Obj[C^\mu(\Lambda)]$	The objective value of all factories.
$Obj[C_l^\mu(\Lambda)]$	The objective value of factories F_l.

In fact, when $\lvert\Lambda\rvert = 1$, i.e. there is only one scenario. The DPFSP_UPT_SDST become a deterministic DPFSP/SDST. Therefore, we can use the model proposed by Naderi et al. [3] and Huang et al. [37] to optimize the makespan for each scenario. However, this does not mean that the DPFSP_UPT_SDST is a stack of many deterministic DPFSP/SDSTs. In the DPFSP_UPT_SDST, in order to obtain a robust solution, we consider all scenarios simultaneously and adopt the expectation-standard deviation model [2,23]. The objective is to minimize the $Obj[C^{\mu}(\Lambda)]$, calculated as in equation (1).

$$\min Obj[C^{\mu}(\Lambda)] = r * E[C^{\mu}(\Lambda)] + (1-r) * Std[C^{\mu}(\Lambda)] \quad (1)$$

where r is the preference coefficient and $r \in [0,1]$, the expectation-standard deviation model can balance performance and risk. In general, $E[C^{\mu}(\Lambda)]$ represents the average of the makespan for different scenarios, however, the makespan for different scenarios can vary significantly. Even if the mean makespan for all scenarios is good, the actual makespan for a certain scenario may be poor, resulting in unpredictable scheduling results and high risk for decision makers. The standard deviation $Std[C^{\mu}(\Lambda)]$ can reflect the fluctuation of makespan for all scenarios. The expectation-standard deviation model can balance performance and stability in production, where the value of r is determined by the decision maker. According to Zhang et al. [21] and Jing et al. [2], $r = 0.01$ is set in this paper.

3. Proposed heuristics and metaheuristic

3.1. Proposed constructive heuristics

Constructive heuristics can generate feasible solutions in a short time, because they rely on some problem-specific knowledge and specific rules, which is necessary for large-scale instances or real-time requirements [6]. For metaheuristics, adopting heuristics to generate a initial solution is promising [23]. The NEH heuristic is a simple and efficient method for solving PFSPs [24], which has been extended to solve the distributed flowshop scheduling problems [2,23]. In this section, we propose four heuristics, which are rNEH, sNEH_Imp, NEN_sum and NEH_ps.

In the basic NEH, all jobs are first ranked according to the LPT rule to generate an initial seed sequence. The priority of the job with the largest processing time is the highest. Then, the jobs in the seed sequence are inserted one by one into the best position. The best position can obtain the optimal objective function. Here, we extend the LPT rule into DPFSP_UPT_SDST, and named as sLPT. The total processing times for each job is calculated by equation (2). All jobs are ranked in non-increasing order according to the value of $TotalP_i, i = 1,2, \ldots n$.

$$TotalP_i = \sum_{s=1}^{\mu} \sum_{k=1}^{m} P_{i,k}^{s}, i = 1,2, \ldots n \quad (2)$$

Due to we need to determine the best position for the job in each factory, we introduce equation (3), which is a general extension of equation (1).

$$\min Obj[C_l^{\mu}(\Lambda)] = r * E[C_l^{\mu}(\Lambda)] + (1-r) * Std[C_l^{\mu}(\Lambda)], l = 1,2, \ldots f \quad (3)$$

(1) The first heuristic is a random heuristic based on the NEH, named rNEH. Firstly, a random seed sequence is generated. Then, the jobs in the seed sequence are inserted one by one into the best position which the smallest $Obj[C_l^{\mu}(\Lambda)]$ is obtained. Once all jobs in the seed sequence is assigned to the best position, the rNEH procedure terminated and the value of $Obj[C^{\mu}(\Lambda)]$ is calculated.

(2) The second heuristic is an improved version of the sNEH proposed by Jing [2] et al., named sNEH_Imp. Firstly, we adopt the sLPT to generate a seed sequence. Secondly, the jobs in the seed sequence are inserted one by one into the best position. After inserting a job into the best position, either the previous or following job (at random) is taken out, we reinsert the extracted job into the best position in the same factory.

(3) The third heuristic, known as NEH_sum, adopts a new approach to generate a seed sequence, it takes into account the uncertainty of the processing times. Firstly, the upper and lower bounds for each job in all scenarios are found, and then sum the upper and lower bounds for each job. The job with the largest summation has the highest priority, and the seed sequence is generated in non-increasing order. The all jobs of the seed sequence are inserted one by one into the best position.

(4) The last heuristic takes into account both the setup times and the processing time when generating the seed sequence, which is named NEH_ps. Firstly, the sum of the processing times for each job in all scenarios is calculated according to equation (2), and the mean setup times of the job is calculated by equation (4). Then, the seed sequence is generated by sorting the values of equation (5)

from largest to smallest. Finally, the process of inserting the job into the best position is the same as for NEH_Imp.

$$MeanST_i = \frac{1}{\mu * (2n-1)} \sum_{s=1}^{\mu} \sum_{j=1}^{n} (S_{i,j}^s + S_{j,i}^s), i \in \{1,2,\dots n\} \quad (4)$$

$$TotalP_i + MeanST_i, i \in \{1,2,\dots n\} \quad (5)$$

3.2. The improved FFO algorithm

As a metaheuristic, the FFO mimics the foraging behaviour of a fruit fly. The FFO optimization process consists of three main parts: initialisation stage, olfactory search stage, and visual flight stage [25]. This paper proposes a discretized FFO algorithm (named as FFO_RO) to solve the DPFSP_UPT_SDST.

3.2.1. Initialization. A good initial solution can provide a good start. As we can see in Section 5.1, the sNEH_Imp generates better solutions. Therefore, we adopt the sNEH_Imp to generate 2 solutions, the remaining *PSize*-2 solutions are generated randomly.

3.2.2. The olfactory search phase. During the olfactory search phase, the fruit fly actively explores new areas around it. How to generate new neighborhood solutions during the olfactory search phase is a key of FFO_RO. Therefore, two simple operators are proposed in this section to guide the fruit fly to explore.

The first operator is a random two-factory based insertion operation, named RTFac_Insert. Firstly, we randomly determine two factories. The factory with a larger $C_l^{\mu}(\Lambda)$ is taken as the origin factory F_{max} and the other one is taken as the target factory F_{min}. J_{rand} is selected randomly from factory F_{max} , then J_{rand} is inserted into the best position in F_{min}.

The second operator is a simple random single-factory insertion operator, named RSFac_Insert. Firstly, the factory F_{rand} is selected randomly. A job J_{rand} is taken out from F_{rand} and inserted into the position of the same factory where the $Obj[C_l^{\mu}(\Lambda)]$ is the smallest.

In order to enhance the search ability and maintain population diversity, the FFO_RO uses multiple different operators, which have a greater probability of producing different solutions than a single operator [25]. In the olfactory search phase, each fruit fly performs the two operators mentioned above in turn, i.e. the RTFac_Insert operator is performed first, and then the RSFac_Insert operator is performed.

3.2.3. Local search. In this paper, to further improve the exploitation capability of the individuals, we extend the LS3 method (referred as LS3_RO) proposed by Ruiz [26] et al. to suit the uncertain processing times. LS3_RO considers the dimensions of factory assignment and job sequencing simultaneously. Firstly, the factory with the largest $Obj[C_l^{\mu}(\Lambda)]$ is selected. Then, we extracted randomly a job from this factory. We inserted the extracted job into all possible positions in each factory. If the smallest $Obj[C^{\mu}(\Lambda)]$ in all these insertions is better than the starting $Obj[C^{\mu}(\Lambda)]$, the job is inserted into the best position with the smallest $Obj[C^{\mu}(\Lambda)]$ and the search starts again from the beginning, otherwise the job is reinserted back into its original position and the search continues. The above process is iterated until all jobs in the factory with largest $Obj[C_l^{\mu}(\Lambda)]$ have been tested.

3.2.4. The visual flight phase with age mechanism. In the visual flight phase, each individual will be guided towards a better position. In traditional FFO, all individuals fly towards the best individual in the visual flight phase. This approach tends to lead to poor performance because we do not know whether the current best individual has potential in the future. If there is still no improvement after several iterations, then the potential of the current best individual is limited. Therefore, we introduce an ageing mechanism in FFO_RO to escape from local optimum and premature convergence. In the FFO_RO, we adopt a simple flight approach that combines the age mechanism. In brief, if the age mechanism is not triggered, individuals move towards the newly explored region after each iteration; otherwise, the worst two individuals are replaced.

The process of the age mechanism is as follows: the *FFO_Age* is set as a variable to record the unimproved count of the global best individual. When the *FFO_Age* reaches the *gLimit* (*gLimit* is the upper bound of age). We use sNEH_Imp and rNEH to generate two new individuals and replace the worst two individuals in population. The value of *gLimit* will be calibrated in Section 4.2.

3.2.5. Pseudocode for the FFO_RO. Algorithm 1 is the pseudocode of the FFO_RO, π^* represents the best individual, which is a complete solution, and $func^*$ represents the best objective value according to equation (3).

```
Algorithm 1: FFO_RO(PSize,gLimit)
Pop = {π(1), π(2), ... π(PSize)}, FFO_Age=0
π(1), π(2) ← sNEH_Imp, π(3), π(4) ... π(PSize) ← rNEH //initialization (use heuristic)
π* ← Individual which obtains the best objective value
func* ← the best objective value
Perform LS3_RO to the best individual π* in Pop
while (time limit is not satisfied) do
  Pop': {π'(1), π'(2), ... π'(PSize)} ← Pop: {π(1), π(2), ... π(PSize)}
    for ind = 1 : PSize    //the olfactory search stage
   Perform the rdFac_Insert operator to π'(ind)
   Perform the rsFac_Insert operator to π'(ind)
    end for
  π** ← Individual which obtains the best objective value in Pop'
  func** ← the best objective value in Pop'
  if func** < func* then
    Update the global best individual and objective
    func* = func**, π* = π**, FFO_Age = 0
  Else
    FFO_Age++
  end if
  Perform LS3_RO to the best individual π** in Pop'
  if func** < func* then
    Update the global best individual and objective
    func* = func**, π* = π**, FFO_Age = 0
  else
    FFO_Age++
  end if
  if FFO_Age > gLimit then
    Perform sNEH_Imp and rNEH to the two worst individuals in Pop'
    FFO_Age = 0, Pop = Pop'
  else if
    Pop = Pop'
  end if
end while
Output π* and func*
```

4. Experiment settings and Parameters calibration

4.1. Instances design and experimental environment

There are no test instances can be used directly, due to the DPFSP_UPT_SDST is a new problem and has not been solved in previous study. Here, we design the DPFSP_UPT_SDST instances by referring to the relevant study [2,10,12]. The test instances consist of *f* factories, *n* jobs and *m* machines, where the number of factories $f \in \{2,3,4,5,6,7\}$, the number of jobs $n \in \{50,70,100,200,500\}$ and the number of machines $m \in \{2,4,6\}$. In addition, we set $Factor \in \{0.25,0.75,1.25\}$. Each possible processing time is generated identically from the uniform distribution on the interval [$p_{\min}$,$p_{\max}$], where $p_{\min}$ =10 and $p_{\max}$ =100. The setup times are generated by $(1 + \text{rand}()\%99) * Factor$. In test instances, the setup time and the processing time are integers and a scenario represents a set of setup times and processing times. We generate a total of $6 \times 5 \times 3 \times 3 = 270$ instances, and generate 30 scenarios for each instance to describe the uncertainty of processing times,. In addition, 20 calibration instances are randomly selected from the 270 instances for calibrating parameters, but the relevant data of 20 calibration instances are randomly generated again.

All algorithms will be coded in C++ language under the Visual studio 2015. The experiment is performed on the computer with Intel® Xeon® CPU @ 2.40GHz, 8GB RAM, and Windows 7 professional operating system.

4.2. Parameter calibration

In order to calibrate the proposed FFO_RO, we adopt the DOE technique and the ANOVA technique [1,5,19]. At the first, we carefully determine the range of parameters by referring to the relevant literature. For the FFO_RO, *PSize* at four levels: 3,5,7,9 and *gLimit* at four levels:6,8,10,12. Therefore, the tow parameters *PSize* and *gLimit* yield a total of $4 \times 4 = 16$ different configurations. Each configuration is used to solve the 20 calibration instances. Each calibration instance is repeated five times independent. In addition, the maximum elapsed CPU time $t = 200 \cdot m \cdot n$ (ms) is the criterion of the FFO_RO stop. The relative percentage deviation (*RPD*) is calculated using equation (6).

$$RPD = 100 \times \frac{Method_{sol} - Best_{sol}}{Best_{sol}} \tag{6}$$

where $Best_{sol}$ is the best objective value of the results for an instance, $Method_{sol}$ is the objective value of a certain algorithm for the same instance. There are a total of $16 \times 5 \times 20 = 1600$ *RPD*s.

All results are analysed by the ANOVA technique. The results of ANOVA are shown in Figure 1. Figure 1 shows the mean plots of relevant parameters and the 95% LSD confidence intervals is adopted. As can be seen in Figure 1(a), the *PSize* should be set to 5. As can be seen in Figure 1(b), the *gLimit*=8 would have better results. Therefore, the *PSize* and *gLimit* for FFO_RO will be set as 5 and 8 in Section 5.2.

5. Computational evaluation

5.1. Comparison of constructive heuristics

In order to evaluate the four heuristics proposed in this paper, we compare them with several heuristics: the LPT [27], the SPT [27] and the sNEH [2]. Note that in order to solve the robust scheduling problem, we have made a slight change to the LPT and SPT, named sLPT and sSPT. The sum of the processing times for each job is calculated by equation (2). The relevant statistical results are recorded in Table 1 and Table 2.

Table 1 records the average *RPD* values and the total CPU time to solve the 270 calibration instances and grouped by the number of factory. Obviously, although the sLPT and sSPT require less total CPU time, their results are very poor. For the remaining five heuristics, the sNEH_Imp obtains the best results with a slight increase in total CPU time compared to the sNEH, which is due to the fact that the sNEH_Imp performs an additional random reinsertion process after each insertion. The NEH_ps，NEH_sum、sNEH、rNEH are ranked second to fifth. There is no significant difference in total CPU time for the rNEH, NEH_ps, NEH_sum and sNEH. In order to determine whether the observed differences from Table 1 are indeed significantly, we carry out a multifactor ANOVA. The type of heuristics is considered as factor, but the sLPT and sSPT are not shown in Figure 2 due to the poor results. Figure 2 shows the mean plots of the type of heuristics. From Figure 2, the differences between the heuristics can be observed, indicating that the differences between the heuristics are significant.

we can find that the difference among the heuristics are significant, especially, the proposed the sNEH_Imp is the best, and the NEH_ps, NEH_sum, sNEH and rNEH are ranked sequentially. Therefore, the sNEH_Imp method will be adopted in the next section.

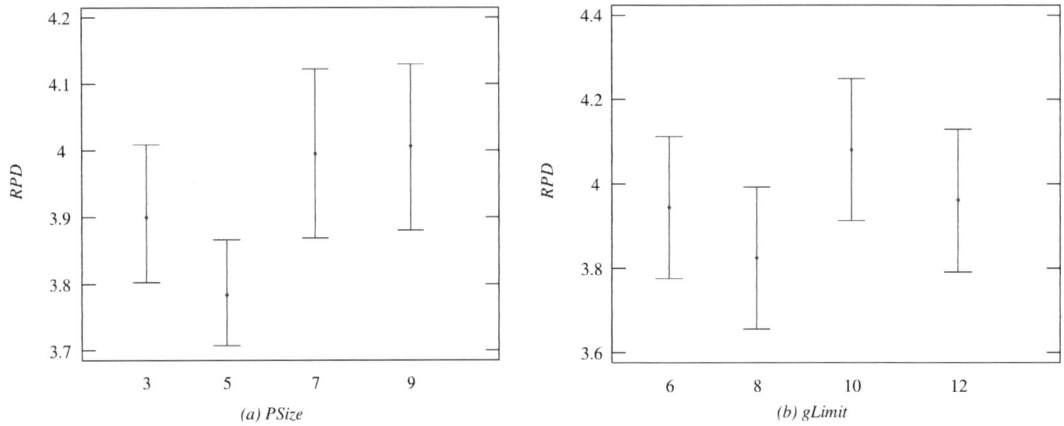

Figure 1. Means plot of the relevant parameters

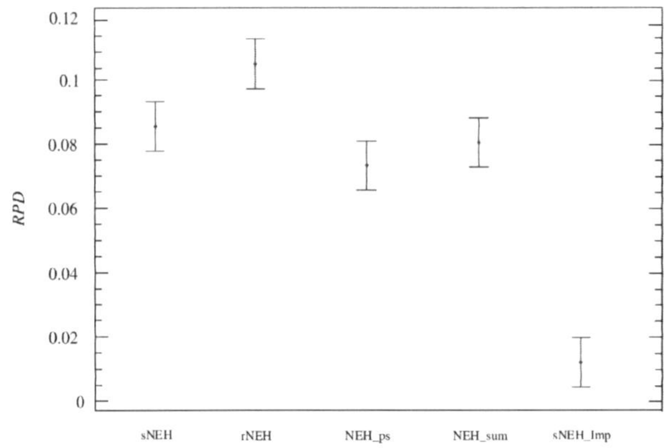

Figure 2. Means plots of the heuristics (95% least significant difference (LSD) confidence interval)

Table 1. Total elapsed CPU times and *RPD* for the heuristics

Heuristics	Average *RPD*							Total CPU Time (s)
	f=2	*f*=3	*f*=4	*f*=5	*f*=6	*f*=7	Mean	
sLPT	1.648	1.687	1.707	1.740	1.697	1.664	1.690	0.378
sSPT	1.696	1.733	1.735	1.691	1.735	1.672	1.711	0.338
sNEH	0.055	0.068	0.078	0.095	0.095	0.122	0.085	22.158
rNEH	0.070	0.082	0.093	0.131	0.143	0.133	0.108	23.309
NEH_ps	0.037	0.045	0.069	0.071	0.085	0.096	0.067	24.942
NEH_sum	0.052	0.068	0.074	0.073	0.076	0.114	0.076	23.966
sNEH_Imp	**0.012**	**0.017**	**0.010**	**0.012**	**0.013**	**0.005**	**0.012**	28.718

5.2. Comparison of metaheuristics

We evaluate the proposed FFO_RO through extensive experiments. We compare the FFO_RO against 3 state-of-the-art metaheuristics which are EA [28], DABC_h [22] and IGR [29]. The test instances adopt the 270 instances mentioned in Section 4.1. We perform five independent runs for each algorithm. In addition, we select the maximum elapsed CPU time $t = \omega \cdot m \cdot n$ (ms) as a termination criterion, where ω is set 200 and 300. For a fair comparison, all algorithms use the proposed sNEH_Imp to generate the initial solution. The relevant results are reported in Table 2 and Table 3 and grouped by the number of factory.

When $t = 200mn$ milliseconds, the average *RPD* values for the four algorithms are recorded in Table 2. We have marked the best result of each factory in bold. As we can see from Table 2, whatever the number of factory is, the FFO_RO always obtains the best result. The IGR is ranked second and the EA is ranked third with slightly worse results than IGR, the DABC_h obtains the worst average *RPD* values. Table 3 shows the average *RPD* values with $t = 300mn$ milliseconds. As can be seen from Table 3, the *RPD* values of all algorithms decrease to varying degrees as the elapsed CPU time increases. The FFO_RO still obtains the best results.

Further, in order to be able to observe the results in Table 2 and Table 3 more clearly, we further make an ANOVA analysis, and the type of comparison algorithms is selected as the factor. Figure 3 shows the mean plots with 95% LSD confidence interval for $t = 200mn$ and $t = 300mn$. From Figure 3(a), it is clear that the average RPD values generated by all algorithms are different. The best algorithm is FFO_RO, the IGR is ranked second, and the EA and DABC_h are the worst. Similarly, figure 3(b) shows that the FFO_RO is still the best, and then IGR, EA and DABC_h are ranked sequentially.

In a word, the proposed FFO_RO algorithm obtains the best result, the FFO_RO is better than the three existing metaheuristic in terms of average *RPD* values. What can be concluded is that the FFO_RO is an effective approach when it comes to resolving the DPFSP_UPT_SDST.

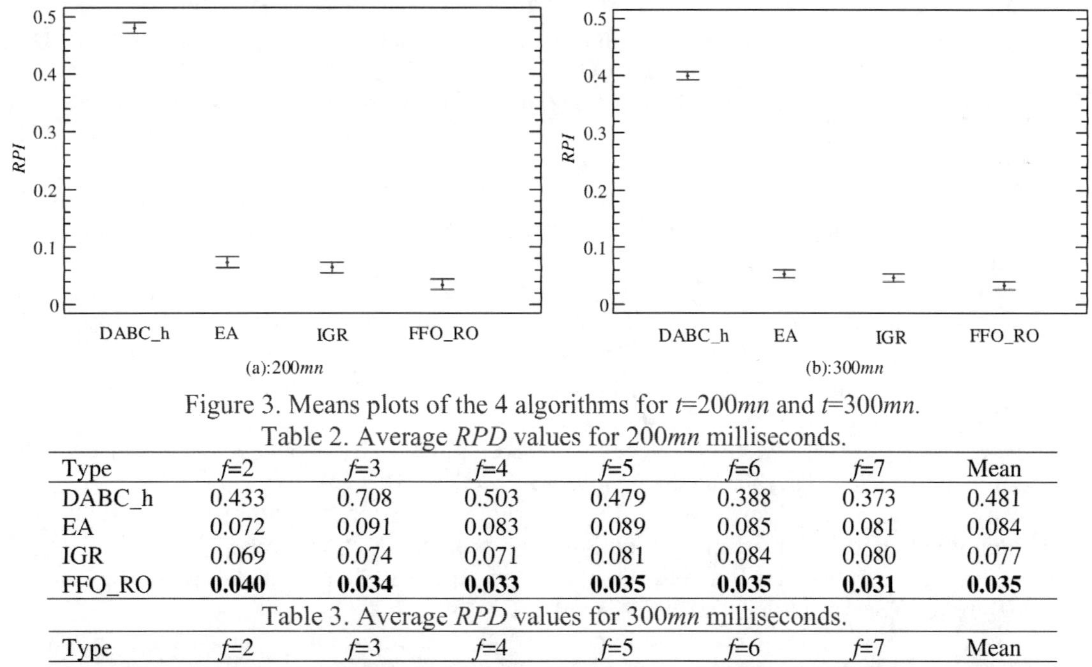

Figure 3. Means plots of the 4 algorithms for t=200*mn* and t=300*mn*.

Table 2. Average *RPD* values for 200*mn* milliseconds.

Type	*f*=2	*f*=3	*f*=4	*f*=5	*f*=6	*f*=7	Mean
DABC_h	0.433	0.708	0.503	0.479	0.388	0.373	0.481
EA	0.072	0.091	0.083	0.089	0.085	0.081	0.084
IGR	0.069	0.074	0.071	0.081	0.084	0.080	0.077
FFO_RO	**0.040**	**0.034**	**0.033**	**0.035**	**0.035**	**0.031**	**0.035**

Table 3. Average *RPD* values for 300*mn* milliseconds.

Type	*f*=2	*f*=3	*f*=4	*f*=5	*f*=6	*f*=7	Mean
DABC_h	0.441	0.497	0.339	0.347	0.385	0.388	0.400
EA	0.056	0.066	0.063	0.060	0.059	0.058	0.060
IGR	0.035	0.039	0.043	0.046	0.044	0.046	0.042
FFO_RO	**0.030**	**0.034**	**0.031**	**0.035**	**0.035**	**0.032**	**0.033**

6. Conclusions

This paper discusses the distributed permutation flowshop scheduling problem, and considers the sequence-dependent setup times and the uncertainty of processing times, where the uncertain processing times are described by the interval scenarios. It is very realistic to consider the uncertainty of processing times. We select the expectation-standard deviation model as the robust mathematical model. Firstly, 4 effective constructive heuristics are proposed, namely rNEH, sNEH_Imp, NEH_ps and NEH_sum. The sNEH_Imp obtains the best results. Secondly, we design a FFO with the age mechanism to avoid falling into a local optimum. Finally, the extensive experiments show that the proposed FFO_RO is an effective metaheuristic for solving DPFSP/SDST_UP. For the next research, it makes sense to develop simple heuristics based on the characteristics of uncertain environments if they can be explored. Of course, the application of uncertainty to other scheduling problems, such as blocking constraints, batch constraints, et al. is also worth investigating. In the future, we will consider studying more complex uncertainty problems, such as resource uncertainty. We will develop specific optimization algorithms based on the different characteristics of different problems. In addition, we will consider studying green scheduling in an uncertain environment. Finally, it would be an interesting direction to consider the use of reinforcement learning to solve scheduling problems in uncertain environments.

References

[1] Meng T, Pan Q K. A distributed heterogeneous permutation flowshop scheduling problem with lot-streaming and carryover sequence-dependent setup time[J]. Swarm and Evolutionary Computation, 2021, 60:100804.
[2] Jing X L, Pan Q K, Gao L. Local search-based metaheuristics for the robust distributed permutation flowshop problem[J]. Applied Soft Computing, 2021, 105(2):107247.
[3] Ruiz R, Naderi B. The distributed permutation flowshop scheduling problem[J]. 2010, 37(4): 754-768.
[4] Allahverdi A, Soroush H M. The significance of reducing setup times/setup costs[J]. European Journal of Operational Research, 2008, 187(3):978-984.
[5] Pan Q K, Tasgetiren M F, Suganthan P N, et al. A discrete artificial bee colony algorithm for the lot-streaming flow shop scheduling problem[J]. Information Sciences, 2011, 181(12):2455-2468.
[6] Shao Z S, Pi D C, Shao W S. Hybrid enhanced discrete fruit fly optimization algorithm for scheduling blocking flow-shop in distributed environment[J]. Expert Systems with Applications, 2019, 145:113147.
[7] Feng X, Zheng F F, Xu Y F. Robust scheduling of a two-stage hybrid flow shop with uncertain interval processing times[J]. International Journal of Production Research, 2016, 54(11-12):1-12.
[8] Mulvey J M, Vanderbei R J, Zenios S A. Robust Optimization of Large-Scale Systems[J]. Operations Research, 1995, 43(2):264-281.
[9] Schoemaker P J H, Multiple Scenario Development: Its Conceptual and Behavioral Foundation[J]. Strategic Management Journal, 1993, 14(3): 193-213.
[10] Wang B, Wang X Z, Lan F M, et al. A hybrid local-search algorithm for robust job-shop scheduling under scenarios[J]. Applied Soft Computing, 2018, 62: 259-271.
[11] Gerodimos A. Robust Discrete Optimization and its Applications[J]. Journal of the Operational Research Society, 1998, 49(12):1303-1304.
[12] Lu C C, Lin S W, Ying K C. Minimizing worst-case regret of makespan on a single machine with uncertain processing and setup times[J]. Applied Soft Computing, 2014, 23:144-151.
[13] Kala R, Lamboray C, Vanderpooten D, Lexicographic α -robustness: An alternative to min-max criteria[J]. European Journal of Operational Research, 2012, 220(3): 722-728.

[14] Kasperski A, Kurpisz A, Zieliński P. Approximating a two-machine flow shop scheduling under discrete scenario uncertainty[J]. European Journal of Operational Research, 2012, 217(1):36-43.
[15] Allahverdi A, Aydilek H. Heuristics for the two-machine flowshop scheduling problem to minimize maximum lateness with bounded processing times[J]. Computers & Mathematics with Applications, 2010, 60(5):1374-1384.
[16] Pan W-T. A new Fruit Fly Optimization Algorithm: Taking the financial distress model as an example[J]. Knowledge-Based Systems, 2012, 26(2):69-74.
[17] Sang H Y, Pan Q K, Duan P Y, Self-adaptive fruit fly optimizer for global optimization[J]. Natural Computing, 2019, 18(4),785-813.
[18] Han Y Y, Gong D W, Li J Q, et al. Solving the blocking flow shop scheduling problem with makespan using a modified fruit fly optimisation algorithm[J]. International Journal of Production Research, 2016:1-16.
[19] Zhou R H, Liu Q M, Wang J, et al. Modified semi-supervised affinity propagation clustering with fuzzy density fruit fly optimization[J]. Neural Computing and Applications, 2020(1):1-18.
[20] Li S B, Zhao H, Zhang CL, et al. Literature survey of fruit fly optimization algorithm[J]. Science Technology and Engineering, 2018, 18(1), 163-171.
[21] Zhang X C, Zhou H, Robust scheduling of multi-stage flflow shop under machine breakdowns following expect-risk rule[J]. Control and Decision, 2016, 31(6): 1087-1092.
[22] Huang J P, Pan Q K, Miao Z H, et al. Effective constructive heuristics and discrete bee colony optimization for distributed flowshop with setup times[J]. Engineering Applications of Artificial Intelligence, 2021, 97:104016.
[23] Pan Q K, Gao L, Wang L, et al. Effective heuristics and metaheuristics to minimize total flowtime for the distributed permutation flowshop problem[J]. Expert Systems with Applications, 2019, 124(JUN.):309-324.
[24] Rad S F, Ruiz R, Boroojerdian N. New high performing heuristics for minimizing makespan in permutation flowshops[J]. Omega, 2009, 37(2):331-345.
[25] Han Y Y, Gong D W, Li J Q, et al. Solving the blocking flow shop scheduling problem with makespan using a modified fruit fly optimisation algorithm[J]. International Journal of Production Research, 2016:1-16.
[26] Ruiz R, Pan Q K, Naderi B. Iterated Greedy methods for the distributed permutation flowshop scheduling problem[J]. Omega, 2019, 83(MAR.):213-222.
[27] Baker K R. Introduction to Sequencing and Scheduling[M]. Wiley, New York, 1974.
[28] Fernandez-Viagas V, Perez-Gonzalez P, Framinan J M. The distributed permutation flow shop to minimise the total flowtime[J]. Computers & Industrial Engineering, 2018, 118(APR.):464-477.
[29] Huang J P, Pan Q K, Gao L.An effective iterated greedy method for the distributed permutation flowshop scheduling problem with sequence-dependent setup times[J]. Swarm and Evolutionary Computation, 2020, 59:100742.

An application analysis method based on Clustering

Rui Duan[1*], Junde Chen[2]

[1]School of Informatics, Xiamen University, Xiamen 361005, China

[2]School of Informatics, Xiamen University, Xiamen 361005, China

[1*]23020191153234@stu.xmu.edu.cn, [2]23020180155666@stu.xmu.edu.cn

Abstract: Depending upon the load intensity and time variation information, we established the indicator system to analyze the of commercial halls, so that the further optimization actions can be taken in practical application scenarios. The representative clustering algorithms including the k-means and AC algorithm are chosen for the comparative analysis, and on this basis, an enhanced k-means algorithm is put forward in this study because the traditional method is sensitive to the initial clustering centroids and lacks the stability. A series of experiments are performed in this work, and the experimental findings reveal the validity of the proposed method. It provides a new idea for the efficiency analysis of commercial halls.

1. Introduction

A physical commercial hall is a place where companies directly implement specific commercial behavior such as commodity trading, business processing, service, and much more. With the acceleration of urbanization and rapid economic development, the problem of unreasonable resource allocation is becoming more and more prominent in many companies. For example, in some places, there are too many commercial halls resources, while in others, it is not enough. In reality, the deployment of new business halls (outlets), or the optimization of resource allocation on the existing outlets often needs to be decided by the company. Thus, how to measure the efficiency of commercial halls is becoming a crucial topic for the company.

To overcome the disadvantages of too strong subjectivity for human judgment, many scholars have done a lot of work for this kind of issue. German scholar Weber [1] first took the transportation cost, the range between the customers and warehouse into account, and applied the gradient-like method to study the issue of locations. Sanggyun Kang [2] considered the changes over time from central urban regions to city peripheries for the warehouse and probed the main effect factors of warehouse location choice. Anderson et al. [3] built a model to improve the efficiency of queuing service system for bank. Through a lot of data acquisition and demonstration, the optimal number of service windows was obtained. Geng Lin et al. [4] investigated the relationship between the street centrality and retail store. They pointed out that the street centrality also affects the types of retail stores in addition to the profound impact of the transportation network on the retailers' location. Based on the queueing model and the concept of column generation, Venkateshan et al. [5] developed a method for efficiently solving the problems that the enterprises needed to determine where to deploy their service centers, and determined the level of staffing in each center so that customers experienced a defined service level with minimum cost. Therefore, referring to the literature research, we introduced an evaluation method for commercial hall efficiency analysis. This method not only better performed the grade of business halls but also provided a preliminary optimal scheme for the geographic placements.

Depending upon the time change and load intensity information, we analyzed the daily operation features of the business hall like the ticket time, average waiting time, average load, and business types to establish the index system. On this basis, the relevant algorithms were implemented and the optimum business hall site planning was given as well.

The rest of this writing is structured with the following hierarchy. Section 2 mainly discusses the method to implement the task of commercial hall clustering as well as the relevant theory and the proposed method. Later in Section 3, experiments for investigating the performance of the proposed method are presented. Section 4 conducts the empirical study in practical scenarios and the efficiency analysis is performed for the entity business hall. Ultimately, the paper is concluded in Section 5.

2. Methodologies

2.1 Classical clustering

Clustering method plays a vital role in data analysis, data mining, engineering signal processing, and so on [6,7]. The applications of clustering algorithm range from image processing [8], pattern recognition [9,10], recommendation [11], and others. Generally speaking, clustering is to partition the data into diverseclasses or groups according to a similarity measurement. So far, researchers has proposed many clustering algorithms to improve the efficiency of data processing. Among them, k-means algorithm is one of the most popular methods owning to its effectiveness and simplicity. With the advantages of small-time and space complexity, this algorithm is often used in data mining and has achieved excellent results [12,13]. Whereas owning to different parameter settings and the arbitrary selection of initial clustering centroids, the traditional k-means is instable and may generate different clusters for a single dataset. In brief, there are two primary shortcomings for the classical clustering methods. One is the number of cluster that needs to be determined in advance for the classical clustering algorithms. The other is the choice of initial clustering centers, which is randomly determined by the classical partitional algorithms, such as the k-means method [14-16].

2.2 Enhanced k-means

The *k*-means [17] is a well-known algorithm used to cluster given dataset D ($\{x_i\}_{i=1,2,\ldots,N} \in \mathrm{R}^m$) into k categories, where k implies a predefined number. Then, the division is iterately implemented with the updating of the clustering centers. This process is ended until the global difference function is converged. Here, the squared error is calculated as the objective function. The formula is expressed below.

$$J_c = \sum_{i=1}^{k} \sum_{p \in C_i} \|p - M_i\|^2 \quad (1)$$

where M_i means the average value of category C_i, and p is the point in the category. The *k*-means algorithm demonstrates high efficiency and strong scalability. The major merits of this algorithm is simple, fast, and effective [35,36]. However, despite outstanding performances, the traditional *k*-means is sensitive for clustering centroids owning to their random selection, which makes the clustering results instable. For this reason, the paper proposes an enhanced *k*-means method that adopts the optimized initial custer centers to execute the clustering and grouping of business hall efficiency analysis. Initially, the pairs of the closest points are found out from the dataset, and then the mean value are calculated after the nearest points are categorized into a class, thereby obtaining the initial cluster centroid. Next, this procedure is iterated until all the initial clustering centroids are gained, and thus the *k*-means algorithm is implemented and obtaining the final clustering results. In summary, the enhanced k-means is implemented by applying the following processes.

1) For the sample set U, the similarity between the pairs is calculated using Euclidean distance, as defined in Eq. (2).

$$d(x,y)=\sqrt{\sum_{i=1}^{n}(x_i-y_i)^2} \quad (2)$$

Among them, $d(x, y)$ represents the similarity, x and y represent any two points.

2) The nearest point pair is formed a sample set $A_m\,(1 \le m \le k)$ and deleted from the whole sample set U.

3) Calculate the distances between the set A_m and sample points in set U. The closest points are moved into the set A_m from U, and the similarity between a set *c to a sample point i* is calculated in Eq. (3).

$$d(i,c)=\min(d(i,j),\, j \in c) \quad (3)$$

4) Repeat Step 3 until the number of samples in A_m reach the threshold ε, as expressed in Eq. (4).

$$\varepsilon = N / K \quad (4)$$

where N and k denote the number of samples and categories, respectively.

5) If m is less than k, set m = m+1. Return to the above-mentioned Step 2 and execute.

6) The arithmetic average value of the final k sets are calculated, and thus the clustering centers of all categories are obtained accordingly. The formula of the arithmetic average for the k-th sets is defined in Eq. (5).

$$AVR_k = \frac{sum(A_m)}{\varepsilon} \quad (5)$$

After that, using the initial clustering centroids, the k-means calculation is repeated until obtaining the final clustering results, as described in Algorithm 1.

```
Algorithm 1: Enhanced KM
Input:
Threshold ε
Data size: N
Begin
    The closest pairs are formed the set Am.
    The points in Am are deleted from U. Initialize t=0.
    Repeat {
        t ← t+1
        For each xj∈U do
            j* ← argmini{‖xj − Ci‖²} //xj is moved to the nearest cluster.
            Am = Am ∪ {xj},  |Am|≤ε
        For each i=1 to k do

            ci ← (1/|ci|) Σxj∈ci xj //Get the initial centroids

        Until Σk i=1|Ci| ≤ N   //Until all points are divided category.
        }
    Function KM(k, ci)        //execute the km algorithm, ci is the
        k-means ← ci do k-means   // computed centroids, k=N/ε.
    Output:
        The final results.
End.
```

2.3 AC algorithm

Analog Complexing (AC) algorithm is introduced by Laurence and is applied to the prediction initially [18]. It is relied on the assumption that the time process will be repeated in some form. In other terms, for a given process, the current period of state evolution may have one or more similar periods in history [19-21]. After continuous development and advancement, this algorithm has achieved great progress in theory and practice.

Similar patterns in different periods may have different mean and variance. To perform the similarity measure between patterns, it is necessary to find the transformation of a candidate pattern to the reference pattern to describe these differences, which is generally adopted the linear transformation. The major processes of the AC algorithm are described below.

1.Generating the alternate patterns

Given a m-dimensional dataset $x_t=\{x_{1t}, x_{2t}, \ldots x_{mt}\}(t=1,2,\ldots N)$, a pattern can be expressed in a table $P_k(i)$ with k rows (observation samples) starting at the i-th row. Here, k represents the length of patterns, $i=1,2, \ldots\ldots, N-k+1$.

$$P_k(i)=\begin{bmatrix} x_{1i} & \cdot & x_{li} & \cdot & x_{mi} \\ \cdot & \cdot & \cdot & \cdot & \cdot \\ x_{1,i+j} & \cdot & x_{l,i+j} & \cdot & x_{m,i+j} \\ \cdot & \cdot & \cdot & \cdot & \cdot \\ x_{1,i+k-1} & \cdot & x_{l,i+k-1} & \cdot & x_{m,i+k-1} \end{bmatrix}_{k\times m} \quad (6)$$

The set of all possible patterns $\{P_k(i), i=1,2, \ldots,l,\ldots, N-k+1\}$ is compared using the reference pattern P^R, and the similar patterns are to be found. The reference pattern can be any specific pattern. For the clustering, a variable or sample may be just a pattern.

2.Transformation of analogues

As mentioned previously, the patterns with similar forms may have different variance and mean value, which can be depicted by a transformation T_i. This implies that the similarity must be estimated between the transformed analog $T_i[P_k(i)]$ and the reference pattern. The transformed pattern $T_i[P_k(i)]$ is defined as a linear function, and expressed in Eq. (7).

$$T_i[P_k(i)]=\begin{bmatrix} x^*_{1i} & \cdot & x^*_{li} & \cdot & x^*_{mi} \\ \cdot & \cdot & \cdot & \cdot & \cdot \\ x^*_{1,i+j} & \cdot & x^*_{l,i+j} & \cdot & x^*_{m,i+j} \\ \cdot & \cdot & \cdot & \cdot & \cdot \\ x^*_{1,i+k-1} & \cdot & x^*_{l,i+k-1} & \cdot & x^*_{m,i+k-1} \end{bmatrix} \quad (7)$$

where $x^*_{l,i+j}=a^i_{0l}+a^i_{1l}x_{l,i+j}$, $i=1,2,\ldots,N-k+1$, $j=0,1,\ldots,k-1$, $l=1,2,\ldots,m$. a^i_{ol} is a hyper-parameter which can be depicted as the difference between the analogous pattern $P_k(i)$ and the states of reference pattern, and a^i_{1l} is an extension factor. a^i_{ol} a^i_{1l} can be evaluated for each pattern using a least-squares.

3. Determinationof most similar patterns

Generally, the distance between the reference pattern and the i-th pattern is written using Eq. (8).

$$d_i=\frac{1}{k+1}\sum_{j=0}^{k-1}\sqrt{\sum_{r=1}^{M}(x_{r,i+j}-x_{r,N-k+j})^2} \quad (8)$$

The similarity is estimated by the distance, and the similarity between the reference pattern and the i-th pattern can be written in $s_i=1/d_i$. Obviously, with a large distance comes a small similarity of the patterns.

The AC clustering estimates the similarity of the matrix. For a given matrix($n\times m$),every column j (j=1,2,…,m. sample space) and every row i (i=1,2,…,n. state space) is a reference pattern of $m(n)$ lenght, and all other columns (rows) are patterns. Their similarity measures, the S^1_{ik} (k=1， 2， …， n) or S^2_{jh} (h=1， 2， …， m), need to be estimated. Applying the threshold τ , a boolean structure matrix can be obtained from the similarity matrix S^1 (S^2), which is expressed as follows.

$$b_{ij}=\begin{Bmatrix} 1 & S_{ij}\geq\tau \\ 0 & S_{ij}<\tau \end{Bmatrix} \quad (9)$$

The AC clustering algorithm can be regarded as a sequence pattern recognition method for the complex objects, which can automatically generate the number of clusters through the computer and don't require too much human participation. Thus, the error caused by subjective reasons can be avoided and which makes the result more objective.

3. Experimental design

UCI Repository [22] is a publicly available dataset used for the machine learning(ML) algorithm test. To further probe the effect of the proposed method, a total of eight datasets, including balance scale, wine, breast-w, diabet, iris plant, etc., are downloaded from UCI datasets. They are commonly utilized as the criterion datasets in investigating the performances of methods. The computational experiments are primarily conducted in C language, and the software platform includes Codeblocks, Spss, Knowledgeminer, and so on.

The balance dataset has 525 samples, and for the traditional KM algorithm, the number of samples that the clustered category matching the real category is 271. The accuracy of the conventional k-means algorithm is 51.62%, while this value of the enhanced k-means achieves 54.86% with 288 correctly clustered samples. Likewise, the experimental results on other UCI datasets are also displayed in Table 1.

Table 1 The comparison on UCI datasets

Dataset	KM algorithm	Enhanced KM	Difference	Rank
Balance	51.62%	54.86%	3.24%	7
Breast	93.85%	93.99%	0.14%	2
Wine	57.30%	70.22%	12.92%	8
Diabet	66.28%	65.76%	-0.52%	4
Iris	89.33%	90.67%	1.34%	5
Hayes-roth	34.09%	34.09%	0.00%	1
Bupa	44.64%	44.93%	0.29%	3
Tic-tac-toe	54.02%	56.01%	1.99%	6

As seen in Table 1, the Wilcoxon signed-rank test may be employed to distinguish whether there is a obvious difference between algorithms. Let R^+ denote the sum of rank that the second algorithm is worse than the first algorithm, and R^- is in the reverse, which denote the sum of rank that the second algorithm is better than the first algorithm. The formula can be defined in the following Eq. (10).

$$\begin{aligned} R^+ &= \sum_{d_i>0} rank(d_i) + \frac{1}{2}\sum_{d_i=0} rank(d_i) \\ R^- &= \sum_{d_i<0} rank(d_i) + \frac{1}{2}\sum_{d_i=0} rank(d_i) \end{aligned} \quad (10)$$

where d_i means the the rank on the *i*-th dataset. Thus, computing the above datasets, we have:

$$\begin{aligned} R^+ &= 7+2+8+5+3+6= 31 \\ R^- &= 4+1=5 \end{aligned} \quad (11)$$

Suppose $T= min(R^+, R^-)$, and therefore the T value is 5. Subsequently, the Wilcoxon test is implemented. On condition that a = 0.05, the difference among the algorithms is significant if $T \leq 5$. Here, the T value is equal to 5, and it is just on the critical point. Other than that, from the comparison on the dataset, there are 6 datasets for the classical k-means is worse than the enhanced k-means. In terms of quantity, the enhanced k-means is more than that of the classical algorithm. Therefore, according to the comparative experiment, it can be assumed that the efficiency of enhanced k-means is extremely high. Additionally, Fig. 1 further depicts the clustering effect of each algorithm on wine dataset. For the conventional k-means and enhanced one, the clustering number is both assigned as k=3 and the largest number of iteration is 10. The AC algorithm is calculated as the similarity of 95%, and a total of 93 classes are clustered. Taking the number of real classes into account, there are a total of 3 classes for the dataset. The first and 81-th class with maximum number correspond to the real first and third category respectively, and the other classes are processed as the second one. The detailed description is presented as follows.

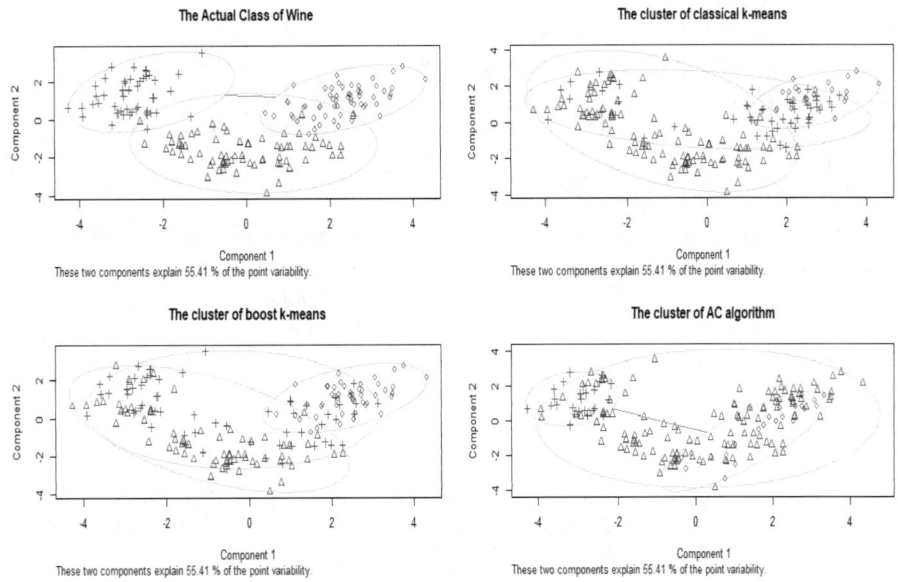

Fig. 1 The clustering chart of algorithms

It can be observed from Fig. 1 that the results for enhanced k-means clustering are most compatible with the real categories. The number of samples that the clustered class matches the actual class is 125, and thus the accuracy is 70.22%. The number of accurately clustered samples for AC algorithm is 103 and the accuracy reach 57%. For the classical k-means algorithm, the consistent number of samples is 102, and the accuracy attains 58%. Furthermore, apart from the comparison with the actual sample category, the compactness and separation effectiveness function are also employed to estimate the clustering performances [23, 24], as computed in Eq. (12).

$$s(U,k)=\frac{\frac{1}{n}\sum_{i=1}^{n}\sum_{j=1}^{k}(u_{ij})^2\left|x_i-c_j\right|^2}{\min_{p,q=1,2,\ldots,k}\left|c_p-c_q\right|^2} \tag{12}$$

where the c_q ,c_p and c_j refer to the cluster centroids, U denotes the sample set and k is the number of class, x_i is any sample in the dataset. Thus, the optimal cluster evaluation value (XB) can be defined in Eq. (13).

$$XB=\max_{k}\{\max_{\Omega} S(U,k)\}, k=2,3,\ldots,n-1 \tag{13}$$

where the value of XB is smaller, the quality of clustering is better. The XB value of different algorithms is calculated in Table 2.

Table 2 The XB evaluation value

Algorithms	Iterations /	The sample number of each			XB value
		C1	C2	C3	
k-means	8	27	102	49	0.70666
Enhanced k-means	4	47	69	62	0.70452
AC algorithm	95%	27	126	25	2.16944

In conclusion, there is a priori knowledge that the sample is of 3 categories, and the optimal clustering scheme is known as k=3 in advance. Whereas, in a practical application scenario, it is usually unknown or difficult to determine the optimal number of categories. Therefore, in this situation, the objective clustering methods can often play a role.

4. Business hall analysis

The phenomenon can always be found that some commercial halls are very busy, while others are very idle. The excessive or insufficient business hall is the embodiment of unreasonable resource allocation. Including the overlapped user coverage in unreasonable business hall location, different commercial halls, a lot of low-value enterprises, all of which are possible to cause the situation.

4.1 Establishment of the indicator system

Through the analysis of load, it can be found that the business halls with high or low loads are urgent to be optimized. Relied on the general waiting time, business types, etc., the average passenger carrying rate of each business hall can be analyzed, so that the relevant characteristic variables can be used to describe the passenger carrying situation of the commercial hall. Thus, by applying the cluster algorithms, the load category of each business hall can be obtained, which provides a basis for the location planning of business halls. Specifically, the following basic characteristics are extracted first.

1. Maximum business hall load(*M*)

$$M = I \times t / (P_1T_1 + P_2T_2 + \cdots + P_nT_n) \quad (14)$$

where *P* means the proportion of a specially designated business, *T* means the average time for employees to process business *i*, *n* refers to the number of business types, *t* is the working hours of employees, and *i* means the number of employees in the commercial hall.

2. The ratio of daily real load to maximum load (*K*)

$$K = A / M \quad (15)$$

where A is the actual load of the commercial hall during summit hours. This variable represents the ratio of the actual load of a business hall to its maximum load in a day. Thus, combined with the basic characteristics, historical data can be used to calculate indicators to prepare for model input. In this manner, the feature engineering of commercial hall performance analysis is built, and key performance indicators extracted are as follows.

(1). Ratio of average load to maximum load (*S*)

$$S = (\sum_{i=1}^{N} A_i) / (N \times M) \quad (16)$$

where *N* means the number of days (The following is the same.).

(2). Actual load trend (*Q*)

$$Load = b_0 + b_1T$$

$$Q = b_1 = \frac{\sum Load * T - n * \overline{Load} * \overline{T}}{\sum T_i^2 - n * \overline{T}^2} \quad (17)$$

where *Load* means the actual load, *T* is the time, and *n* is a number of statistical data. This indicator reveals that the load trend of the commercial hall will be declining, flat, or rising for a period of time.

(3). The ratio of high value business (*V*)

$$V = (\sum_{i=1}^{N} HBV_i) / (\sum_{i=1}^{N} TBV_i) \quad (18)$$

where HBV_i is the high value business volume and TBV_i is the total business volume.

(4). High load frequency (*FH*)

$$FH = count(K_i > h) / N \quad i = 1, 2, ..., N \quad (19)$$

This index denotes the frequency that *K* is greater than a certain threshold *h* for a period, and the high threshold *h* can be acquired from the overall analysis of historical data for commercial halls.

(5). Low load frequency (*FL*)

$$FL = count(K_i < l) / N \quad i = 1, 2, ..., N \quad (20)$$

This index denotes the frequency that *K* is less than a certain threshold *l* for a period, and the threshold *l* can be acquired from the overall historical data analysis for commercial halls.

(6). The latest time interval of high load (*RH*)

$$RH = time_{now} - time_{i=last(K_i>h)} \quad i = 1,2,...,N \tag{21}$$

where the $time_{now}$ is the current time, and $time_{i=last(Ki>h)}$ mean the time when the high threshold h is lower than the latest K. This index denotes the time interval from $time_{now}$ to $time_{i=last(Ki>h)}$.

(7). The latest time interval of low load (*RL*)

$$RL = time_{now} - time_{i=last(K_i<l)} \quad i = 1,2,...,N \tag{22}$$

Similar to the *RH*, this index denotes the time interval from the current time to the time when the last K value is lower than a certain threshold l.

4.2 Business hall clustering

Considering the time change load intensity information, we can divide a load of business halls into low, medium, high and the load trend is divided into decline, flatness, rise. In this manner, a nine-square grid map can be acquired as Fig. 2(a). Whilst, referring to the expertise of domain, the number of categories of subjective clustering method is finally determined as 9. Selecting the distance parameter as the Euclidean Distance, the maximum iteration times were set 500, and the seeds were set 10 for the enhanced k-means algorithm. Thereupon, the partial clustering outcomes were shown as the following Table 3 and Fig. 2(b).

Table 3 The clustering result table

Cluster #	Full Data	1	2	3	4	5	6	7	8	9
	(18)	(2)	(2)	(2)	(2)	(2)	(2)	(2)	(2)	(2)
Attribute					Cluster Centroids					
V	0.62	0.75	0.50	0.45	0.40	0.80	0.60	0.80	0.65	0.70
O	0.00	0.51	-	0.01	-0.50	0.50	0.53	-0.53	0.01	0.01
S	0.55	0.25	0.60	0.20	0.10	0.96	0.50	0.95	0.50	0.90
FL	0.35	0.85	0.10	0.80	0.95	0.02	0.25	0.10	0.06	0.02
FH	0.32	0.05	0.08	0.03	0.01	0.95	0.25	0.60	0.08	0.90
RL	0.51	0.30	0.10	0.04	0.98	0.90	0.50	0.08	0.80	0.90
RH	0.37	0.10	0.70	0.80	0.01	0.02	0.50	0.60	0.65	0.03

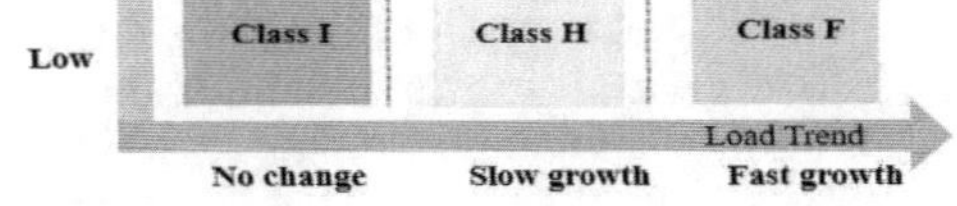

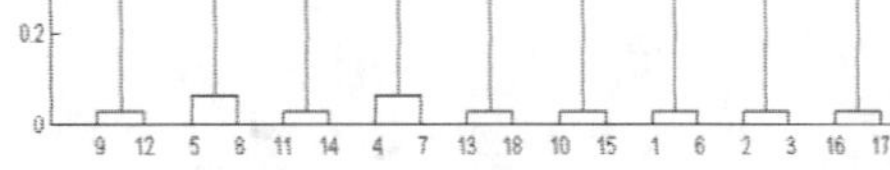

Fig. 2 (a)The Nine-square grid map; (b)The diagram of clustering results.

By contrast, unlike the subjective clustering method that determines the number of clusters in advance, the AC algorithm automatically confirms the number of clusters, which is computed based on the similarity between the samples. Here, the AC algorithm is conducted with a similarity of 95%, and the cluster results are exactly compatible with that of enhanced k-means algorithm. For instance, the 9-th and the 12-th samples, the 5-th and the 8-th samples are all clustered to the same categories, etc. Table 4 displays the partial results of AC algorithm.

Table 4 The clustering result of AC algorithm

Result.	V	Q	S	FL	FH	RL	RH
{C1}	0.8	0.5	0.95	0.01	0.95	0.9	0.02
{C2}	0.68	0.01	0.89	0.02	0.89	0.88	0.02
{C3}	0.79	-0.56	0.95	0.09	0.59	0.07	0.59
{C4}	0.6	0.5	0.5	0.25	0.26	0.5	0.5
…	…	…	…	…	…	…	…
{C9}	0.39	-0.49	0.09	0.93	0.01	0.97	0.01
{C9}	0.4	-0.5	0.1	0.95	0	0.98	0
{C8}	0.49	-0.51	0.58	0.09	0.07	0.09	0.69

The final clustering results acquired from the model give the load marks, which delivers a basis for the optimization of commercial hall location planning. Additionally, the results of the two algorithms are very close, which shows that the enhanced k-means algorithm is extremely effective and robust. On this basis, the business hall can be further optimized.

4.3 Business hall optimization

As shown in the bottom of Fig. 2(b), the 16-th and 17-th commercial halls are both in the class *I*. The present load of this category is low while the load trend remains unchanged, which reflects that the business hall resources are redundant in this region. By comparison, the category *A* is on the contrary. More business hall resources should be input to this region and the location optimation planning of commercial hall can be regarded as the following Eq. (23).

$$F_{\min} = \sum_{k=1}^{n} Q_k \Big/ n + \sum_{i=1} h_i d(v_i, x) \tag{23}$$

where n means the number of factors, Q is the quantitative value of factors that impact the rationality of business hall location Where Q represents the number of factors that affect the rationality of the location of the commercial hall, and n represents the number of factors, $d(v_i, x)$ is the distance function, h_i is the weight, and x is the target variable to be solved. Therefore, in this case, depending upon the relevant coordinate information of key units and objective function, we could calculate the optimum deployment location of increased business halls. Here, the optimum location was calculated as [100.0644,128.8199], and the optimal solution was 527.0368. Fig. 3 shows a detailed description of this phase.

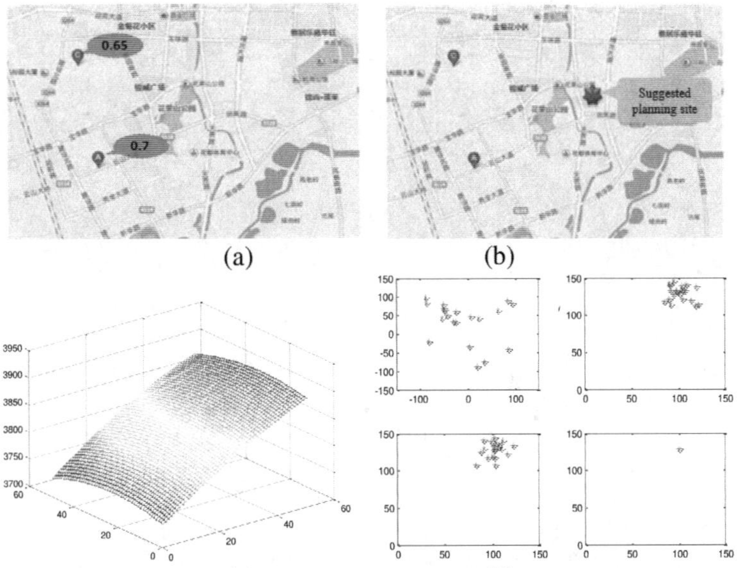

(c) (d)

Fig. 3 (a)The commercial hall location in category A; (b)The recommended location for supplementary business hall; (c)The objection function image; (d)The iteration process of computing.

The 9-th and 12-th commercial halls are in class *E*, which implies that the current load is normal and the trend is stable. Besides, the other categories are similar to the category *E*, and they are not the present focal point either. The major targets are categories *I* and *A*. In other terms, the insufficient or excessive commercial hall resources are primarily concened in these two categories, which are also the crucial points of optimization analysis.

5 Conclusion

The commercial hall efficiency analysis plays a crucial role in the evaluation and measurement of the performance of the physical business halls, where the clustering and optimization algorithms are applied in this study. Firstly, the paper discusses the clustering method in theory. The subjective clustering method represented by k-means and the objective clustering method represented by the AC algorithm is analyzed, respectively. Then, compared to the conventional k-means algorithm, an enhanced k-means algorithm is proposed and relevant experiments are performed as well. Experimental findings indicate the efficiency of the proposed method.

More than that, depending upon the load intensity and time change information of commercial halls, we established the indicator system and carried out further optimization actions. Referring to the business types, ticket time, and average waiting time, etc., the key feature indicators were extracted and the evaluation model was built for the performance analysis of business halls. In other words, using the outputs of business hall clustering, the optimization analysis of the commercial halls is implemented and relevant suggestion is given too. In this manner, the model acquired the load grading of each commercial hall. Finally, it provides a basis for the location optimization of the commercial hall.

Acknowledgments

This research is partly supported by the National Natural Science Foundation of China with the grant No. of 61672439. The authors wish to thank all the editors and anonymous reviewers for their constructive advice.

References

[1] Brandeau, Margaret L., and Samuel S. Chiu. "An overview of representative problems in location research." *Management science* 35.6 (1989): 645-674.

[2] Kang, Sanggyun. "Warehouse location choice: A case study in Los Angeles, CA." *Journal of Transport Geography* (2018).
[3] Anderson, David R., D. J. Sweeny, and Thomas Arthur Williams. "An Introduction to Management Science, Quantitative Approaches to Decision Making, 11th." *Ohio: South-Western* (2005).
[4] Lin, Geng, Xiaoxuan Chen, and Yutian Liang. "The location of retail stores and street centrality in Guangzhou, China." *Applied geography* 100 (2018): 12-20.
[5] Venkateshan, Prahalad, Kamlesh Mathur, and Ronald H. Ballou. "Locating and staffing service centers under service level constraints." *European journal of operational research* 201.1 (2010): 55-70.
[6] Monath N. , Zaheer M.,Silva D.,McCallum A. ,Ahmed A. "Gradient-based Hierarchical Clustering using Continuous Representations of Trees in Hyperbolic Space." *ACM SIGKDD International Conference on Knowledge Discovery & Data Mining (KDD)* (2019): 714-722.
[7] Portugal I. ,Alencar P.,Cowan D.. "A Framework for Spatial-Temporal Trajectory Cluster Analysis Based on Dynamic Relationships." *IEEE Access* (2020): 169775-169793.
[8] Zhang MX ,Jiao LC ,Shang RH, Zhang XR, Li LL. "Unsupervised EA-Based Fuzzy Clustering for Image Segmentation." *IEEE Access* (2020).
[9] Gao M , Shi GY. "Ship-handling behavior pattern recognition using AIS sub-trajectory clustering analysis based on the T-SNE and spectral clustering algorithms." *Ocean engineering* (2020).
[10] Lu, Jiwen, et al. "Reconstruction-based metric learning for unconstrained face verification." *IEEE transactions on information forensics and security* 10.1 (2014): 79-89.
[11] Chen Y . " Research on personalized recommendation algorithm based on user preference in mobile e-commerce." *Information system and e-business management* 18.4(2020): 837-850.
[12] Yu, Shyr-Shen, et al. "Two improved k-means algorithms." *Applied Soft Computing* 68 (2018): 747-755.
[13] Xu SL ,Feng L. ,Liu SL , Zhou J , Qiao H. " Multi-feature weighting neighborhood density clustering." *Neural computing & applications* 32.13 (2020): 9545-9565.
[14] Arthur, David, and Sergei Vassilvitskii. *k-means++: The advantages of careful seeding*. Stanford, 2006.
[15] Ismkhan, H. "I-k-means- plus : An iterative clustering algorithm based on an enhanced version of the k-means." *Pattern recognition* 79 (2018): 402-413.
[16] Erisoglu, Murat, Nazif Calis, and Sadullah Sakallioglu. "A new algorithm for initial cluster centers in k-means algorithm." *Pattern Recognition Letters* 32.14 (2011): 1701-1705.
[17] de Amorim RC., Mirkin B. "Minkowski metric, feature weighting and anomalous cluster initializing in K-Means clustering." *Pattern recognition*. 45.3.(2012):1061-1075.
[18] He, Yue, Zhenglin He, and Dan Zhang. "A study on prediction of customer churn in fixed communication network based on data mining." *2009 Sixth International Conference on Fuzzy Systems and Knowledge Discovery*. Vol. 1. IEEE, 2009.
[19] Karunanithi, Nachimuthu, and Yashwant K. Malaiya. "Neural networks for software reliability engineering." *Handbook of software reliability engineering* (1996): 699-728.
[20] Lemke, Frank, and J-A. Müller. "Self-organising data mining." *Systems analysis modelling simulation* 43.2 (2003): 231-240.
[21] He, Changzheng, et al."Customers' risk type prediction based on analog complexing." *Procedia Computer Science* 55 (2015): 939-943.
[22] Macia N., Bernado-Mansilla E. "Towards UCI plus: A mindful repository design." *Information sciences* (1998)：237-262.
[23] Liu YC,Li ZM , Xiong H ,Gao XD, Wu JJ,Wu S . " Understanding and Enhancement of Internal Clustering Validation Measures." *IEEE Transactions on cybernetics* 43.3 (2013): 982-994.
[24] Erlich, Zippy, Roy Gelbard, and Israel Spiegler. "Evaluating a positive attribute clustering model for data mining." *Journal of Computer Information Systems* 43.3 (2003): 100-108.

Landslide Hazard Identification Based on Deep Learning and Sentinel-2 Remote Sensing Imagery

Jun Xia[1a*], Hongjiang Liu[1b], Linfu Zhu[1c]

[1]School of Tourism, Leshan Normal University, Leshan 614000, China

[a*]8449146@qq.com, [b]hjliu040401@qq.com, [c]156340810@qq.com

Abstract—Landslide is one of the common geological disasters, which seriously threatens human life and property safety. It is particularly important to quickly identify landslide information. This paper takes the Wenchuan earthquake landslide area as the research area, and uses 7 deep learning methods(4-Layer-CNN, AlexNet, ResNet152V2, DenseNet201, InceptionV3, Xception and InceptionResNetV2) to discuss landslide detection methods based on Sentinel-2 remote sensing images. Using the marked landslide and non-landslide sample points, the Sentinel-2 remote sensing image was sliced into 80×80 pixel tiles, and then the deep learning method was used for model training, verification and testing. The results show that : (1) Among the 7 deep learning models, the F1-Score of the DenseNet201 model is the largest, reaching 0.8872, and the RMSE is the smallest 0.2503. It can be seen that the DenseNet model has a good recognition effect on landslide samples, with an accuracy of 0.9172; (2) Second It is InceptionResNetV2, the F1-Score is 0.8721, the RMSE is 0.2721, and the landslide sample recognition accuracy is 0.9012; (3) the worst effect is AlexNet, the minimum F1-Score is only 0.7263, the maximum RMSE is 0.4022, and the accuracy is 0.8295. It can be seen that the deep learning method is applied to Sentinel-2 remote sensing images for landslide image detection, and the accuracy can reach 91.72%, which can quickly and accurately identify landslide information, and improve the method reference and decision basis for disaster prevention and mitigation.

1. Introduction

Landslide is one of the most common geological disasters. Under the influence of heavy rainfall, river erosion and various artificial factors, the soil or rock mass originally in unfavorable geology, under the action of gravity, along the weak zone in the area. Or weak surface, there is a natural phenomenon of partial or whole sliding down. Worldwide, landslides cause huge casualties and economic losses every year. Landslide disasters have the characteristics of strong suddenness and wide influence, and are one of the most common natural disasters that cause the greatest harm to human beings. Therefore, it is necessary to quickly and accurately identify landslide hazards.

In recent years, with the development of remote sensing technology, statistical analysis and machine learning methods have been widely used in landslide image detection, such as fuzzy logic, analytic hierarchy process, multivariate statistical analysis, random forest, support vector machine and artificial neural network, etc.[1, 2]. With the rise of artificial intelligence, deep learning is applied to the extraction of landslide information, especially the convolutional neural network can fully exploit the hidden features of landslides[3-5]. Common convolutional neural network models used for image classification include: AlexNet, VGG, ResNet, DenseNet, UNet, and GoogLeNet, etc.[6-8], combined with multi-source remote sensing data, terrain and geological data, and applied to landslide image detection, which can Improve the accuracy of landslide identification and improve the efficiency of landslide detection[9-12].

Taking the Wenchuan earthquake and landslide as an example, this paper uses Sentinel-2 remote sensing data to compare 7 deep learning models, and discusses the method of quickly and accurately identifying landslide disaster information in areas with frequent landslide geological disasters, so as to provide method reference and decision-making for disaster prevention and mitigation. in accordance with.

2. Study Area

Wenchuan County is located on both sides of the Minjiang River in the northwest of Sichuan Province and the southeast of Aba Prefecture. The boundary is between 30°45′～31°43′ north latitude and 102°51′～103°44′ east longitude. The northeast is controlled by the Longmen Mountains, and the southwest is controlled by the Qionglai Mountains. The terrain slopes from northwest to southeast, with mountains above 3,000 meters above sea level in the west, and the exit of the Minjiang River in the Xuankou area in the southeast is only 780 meters above sea level. The Minjiang River enters from the north of the county, runs through the east, and is 88 kilometers long. Zagunao River, Caopo River and Shoujiang River are the main tributaries of the Minjiang River in the county.

Due to the "5.12" earthquake in Wenchuan in 2008, it affected the mountain structure and soil structure in the region, resulting in frequent landslide disasters, especially in the rainy season from July to August each year, which seriously threatened the lives and safety of local residents. Property security.

3. Data source and processing

3.1. Data source

Since the 2008 earthquake in Wenchuan, there have been many landslide geological disasters, especially the rainy season from July to August every year, when landslides occur frequently. Therefore, this paper uses the Sentinel-2 multispectral remote sensing image in August 2018 with a resolution of 10m and a cloud cover of less than 5%.

3.2. Landslide dataset

The Sentinel-2 satellite image was superimposed on the surface of 3D Google Earth, and with the help of Google Earth 3D terrain and high-resolution images, the geological disaster landslide points in Wenchuan County were marked, and a total of 889 landslide sample points were marked, as shown in the red circle in Figure 1.

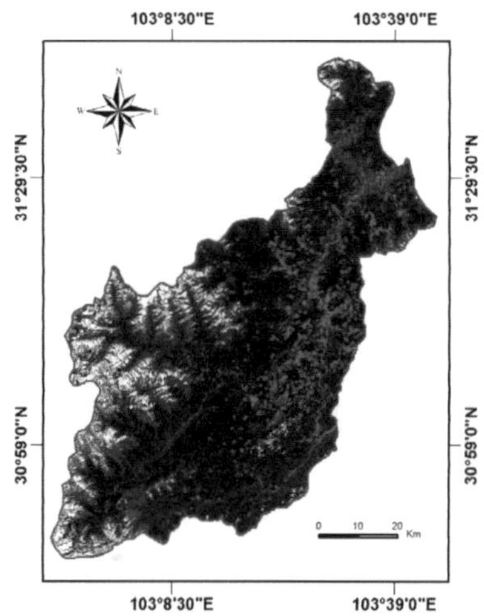

Fig.1 Distribution of landslide sample points in Wenchuan County

The landslide dataset includes landslide samples and non-landslide samples, including 889 landslide samples and 894 non-landslide samples. Each sample includes 3 bands of Sentinel-2 images. The input data of the deep learning model is a two-dimensional image, so a complete feature image needs to be cut into 80 × 80 Pixel tiles. Considering the insufficient number of landslide samples and the complexity of landslide features on remote sensing images, when using landslide sample points to generate feature images, offsets in four directions, up, down, left, and right, increase the number of samples, which is more realistic. The entire landslide dataset contains a total of 4445 landslide samples and 2738 non-landslide sample files, each of which contains 3 features (channels). The optical feature image of landslide mainly includes the landslide point and its surrounding landscape, and the non-landslide optical image includes vegetation, rivers, buildings, roads, etc.

4. Methods

There are many kinds of deep learning models. The most commonly used models and architectures are: Convolutional Neural Network (CNN), Deep Belief Network (DBN), Recurrent Neural Network (RNN), Recurrent Tensor Neural Network (RNTN), Autoencoder and Generative Adversarial Networks (GANs). Convolutional neural network is a common model in deep learning and is widely used in the field of image processing. CNN was proposed by Yann LeCun of New York University in 1998. It is essentially a multi-layer perceptron that uses local connections and shared weights. On the one hand, the reduced number of weights makes the network easy to optimize, and on the other hand, it reduces the risk of overfitting. CNN is a deep neural network with a convolutional structure, usually with at least two non-linear trainable convolutional layers, two pooling layers and a fully connected layer, with a total of at least 5 hidden layers. Commonly used classic models of CNN include: LeNet-5, AlexNet, VGGNet, GoogLeNet, SegNet, ResNet, etc.

In this paper, 7 CNN models including 4-Layer-CNN, AlexNet, ResNet152V2, DenseNet201, InceptionV3, Xception and InceptionResNetV2 are applied to the landslide image recognition of Sentinel-2 optical images, using Accuracy, Precision, Recall Recall, F1-Score and Root Mean Square Error (RMSE) evaluation indicators are used to evaluate the model results.

5. Results and Analysis

Upload the sample dataset to Google Drive, and use the tensorflow deep learning framework of the Google Colab platform to classify the samples. First, the entire data set is divided into training set, validation set and test set, that is, 20% of the entire data set is randomly selected as the test set, and then the training set and the validation set are randomly divided from the remaining 80% according to the ratio of 0.8:0.2 The number of samples in the training set is 5746, of which the number of samples in the test set is 1437.

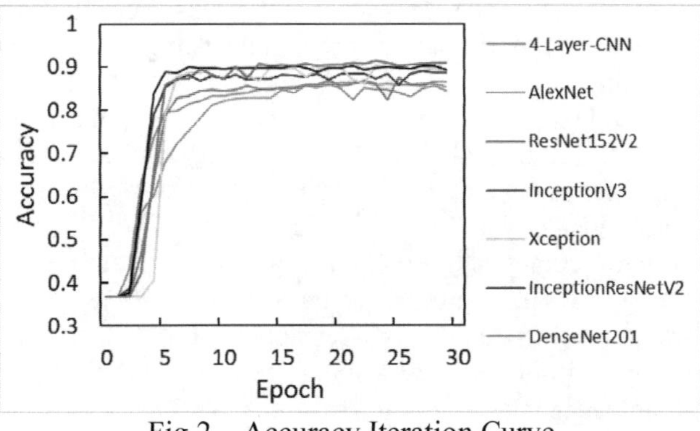

Fig.2 Accuracy Iteration Curve

Seven kinds of CNN network models are used to train the landslide dataset. The loss function is "SparseCategoricalCrossentropy", the optimizer is "Adagrad", the learning rate is set to 0.001, the number of batch samples is 32, and the number of iterations is 30. In order to expand the universality of the samples, the samples are randomly translated, rotated and scaled. The verification accuracy curve of the model is shown in Figure 2, as shown in the figure: the 7 models basically tend to be stable after 10 iterations, and the number of iterations is set to 15 here. Table 1 shows the verification and test results of 15 iterations of 7 models. Combining 7 evaluation indicators, the DenseNet201 model has the largest F1-Score (0.8872) and the smallest RMSE (0.2503), so the DenseNet model has the highest accuracy; followed by InceptionResNetV2, F1 -Score is 0.8721, RMSE is 0.2721; AlexNet is the worst, F1-Score is only 0.7263, and RMSE is 0.4022.

DenseNet201 is a CNN image classification algorithm based on the evolution of densely connected networks. This algorithm can effectively search the network structure space, and adaptively optimize the deep network structure and parameters based on limited computing resources. With the characteristics of strong anti-overfitting and generalization performance, it shows a strong potential in image classification. It has a good effect on landslide recognition on Sentinel-2 optical images, and the recognition accuracy of landslide samples is 91.72%.

Table 1. Model Validation and Test Results

Model	Validation		Test				
	Accuracy	RMSE	Accuracy	Precision	Recall	F1-Score	RMSE
4-Layer-CNN	0.8722	0.3437	0.8420	0.8371	0.7329	0.7815	0.3554
AlexNet	0.8670	0.3904	0.8295	0.9531	0.5866	0.7263	0.4022
ResNet152V2	0.8443	0.2867	0.8392	0.8124	0.7581	0.7843	0.3493
DenseNet201	0.9130	0.1845	0.9172	0.9341	0.8448	0.8872	0.2503
InceptionV3	0.8600	0.2288	0.8601	0.8624	0.7581	0.8069	0.3131
Xception	0.9104	0.1214	0.8873	0.9170	0.7780	0.8418	0.3009
InceptionResNetV2	0.9078	0.1196	0.9012	0.8705	0.8736	0.8721	0.2721

6. Conclusion

Taking the Wenchuan earthquake landslide as an example, this paper uses deep learning methods and remote sensing technology to explore the method of quickly and accurately identifying landslide disaster points. First, use the self-developed landslide disaster sample making software to slice Sentinel-2 remote sensing data as the input data of the deep learning model, and then use 7 deep learning models (4-Layer-CNN, AlexNet, ResNet152V2, DenseNet201, InceptionV3, Xception and InceptionResNetV2) for training and validation, and finally using an independent test dataset to test the model, the results show that:

(1) Among the seven deep learning models, the F1-Score of the DenseNet201 model is the largest, reaching 0.8872, and the RMSE is the smallest being 0.2503. It can be seen that the DenseNet model has a good recognition effect on landslide samples, with an accuracy of 91.72%.

(2) The second is InceptionResNetV2, the F1-Score is 0.8721, the RMSE is 0.2721, and the landslide sample recognition accuracy is 0.9012; the worst is AlexNet, the minimum F1-Score is only 0.7263, the maximum RMSE is 0.4022, and the accuracy is 0.8295.

(3) The Sentiel-2 remote sensing image has a resolution of 10m and a slice size of 80×80 pixels, which can contain complete landslide features. Using deep learning methods, the location of landslides can be quickly and accurately identified.

Since only 3 visible light bands of remote sensing data are considered, and disaster-causing factors such as vegetation index, topography, and geological structure are not combined, the further improvement of the recognition effect is limited.

Acknowledgments
This work was financially supported by the Sichuan Province Science and Technology Plan Project (2020YFS0354); the Project of Sichuan Tourism Development Research Center (LY21-19); the High-level Talents Introduction Project of Leshan Normal University (RC210121).

References

[1] Marjanović, M., Kovačević, M., Bajat, B., et al. (2011) Landslide assessment of the Strača basin (Croatia) using machine learning algorithms. Acta Geotechnica Slovenica, 8(2): 45-55.

[2] Goetz, J.N., Brenning, A., Petschko, H., et al. (2015) Evaluating machine learning and statistical prediction techniques for landslide susceptibility modeling. Computers & geosciences, 81: 1-11.

[3] Krizhevsky, A., Sutskever, I., Hinton, G.E. (2012) Imagenet classification with deep convolutional neural networks. Advances in neural information processing systems, 25: 1097-1105.

[4] Romero, A., Gatta, C., Camps-Valls, G. (2015) Unsupervised deep feature extraction for remote sensing image classification. IEEE Transactions on Geoscience and Remote Sensing, 54(3): 1349-1362.

[5] Ullo, S. L., Langenkamp, M. S., Oikarinen, T. P., DelRosso, M. P., Sebastianelli, A., & Sica, S. (2019) Landslide geohazard assessment with convolutional neural networks using sentinel-2 imagery data. In: IGARSS 2019 IEEE International Geoscience and Remote Sensing Symposium.Yokohama. pp. 9646-9649.

[6] Li, C., Yi, B., Gao, P., et al. (2021) Valuable clues for DCNN-based landslide detection from a comparative assessment in the Wenchuan earthquake area. Sensors, 21(15): 5191.

[7] Gao, X., Chen, T., Niu, R., et al. (2021) Recognition and mapping of landslide using a fully convolutional DenseNet and influencing factors. IEEE Journal of Selected Topics in Applied Earth Observations and Remote Sensing, 14: 7881-7894.

[8] Ullo, S.L., Mohan, A., Sebastianelli, A., et al. (2021) A New Mask R-CNN-Based Method for Improved Landslide Detection. IEEE Journal of Selected Topics in Applied Earth Observations and Remote Sensing, 14: 3799-3810.

[9] Ji, S., Yu, D., Shen, C., Li, W., & Xu, Q. (2020). Landslide detection from an open satellite imagery and digital elevation model dataset using attention boosted convolutional neural networks. Landslides, 17(6), 1337-1352.

[10] He, F., Liu, H., Liu, C., et al. (2021) Analysis of Radar Technology Identification Model for Potential Geologic Hazard based on Convolutional Neural Network and Big Data. Soft Computing. https://doi.org/10.21203/rs.3.rs-371548/v1.

[11] Catani, F. (2021) Landslide detection by deep learning of non-nadiral and crowdsourced optical images. Landslides, 18(3): 1025-1044.

[12] Del Rosso, M.P., Sebastianelli, A., Spiller, D., Mathieu, P.P., & Ullo, S. L. (2021) On-board volcanic eruption detection through cnns and satellite multispectral imagery. Remote Sensing, 13(17): 3479.

HRT-YOLO: A transformer-based high-resolution representation model for face flaw detection

Yage Sun[1*], Xueming Li[1], Xianlin Zhang[1]

[1] School of Digital Media and Design Arts, Beijing University of Posts and Telecommunications, Beijing; 100876, China

[*] Corresponding author's e-mail: 1269644380@qq.com

Abstract. Nowadays, it is imperative to accurately detect and repair face flaws to obtain high-resolution, detail-preserving face images in social life. However, there exists a significant gap between face flaws and natural scene features. In this paper, we propose a framework of HRT-YOLO to erase the problems of small scale, irregular shape, and regional overlap of face flaws. Based on YOLOv5, we introduce high-resolution representation, where the main network of high-resolution feature maps is gradually added in parallel to the sub-network of low-resolution feature maps, allowing for more accurate localization of flaws. We also implement transformer detection heads, which use a multi-head attention to mine contextual features and improve the defect-intensive regions detection. Finally, we present a high-resolution face flaw dataset to evaluate our approach. The evaluation results show that HRT-YOLO has excellent performance on face flaw detection, with 3.23% improvement in mAP and 5.33% improvement in accuracy compared to the baseline (YOLOv5).

1. Introduction

As mobile phones and cameras continue to improve their photographic capabilities, people can take high-definition portraits but flaws such as spots, acne, and pores on people's faces are clearly recorded. Having smooth, bright skin also enhances one's overall appearance in social life and improves the perception of others. As a result, more and more people want high-resolution images of faces that preserve detail and remove face flaws.

Traditional image processing usually used denoising algorithms to remove face flaws[1], but denoising algorithms do not distinguish well between face details (pores) and face flaws (spots, acne, scars, etc.) and could make the processed face image lose detail and give overly smooth distorted results. Overly smooth faces not only lose the texture of the skin itself, but also the character's own features. In addition, one can also use image processing software such as Photoshop to interactively select target areas for defect removal, which results in a natural and realistic image that retains detail in the face, but this method requires skill and experience and a lot of interactive work.

With the power of CNN networks, object detection based on deep neural networks have made significant progress in recent years. The YOLO series [2][3][4][5][6]play an important role in one-stage detectors, as they strike a great balance between accuracy and speed and are among the most outstanding object detection detectors. Consequently, we would like to replace manual defect detection with object detection, so that we can target the repair of flaws and obtain a face image that retains detail and removes flaws. However, major datasets in the field, such as MS COCO and PASCAL VOC, are designed based on natural scenarios. As a branch of object detection, face flaws detection has its particular characteristics: (1) face flaws have irregular shapes and obscure edges; (2) the scale of individual flaws is small, which is prone to the problem of missed detection; (3) face flaws often appear in patches, and there will be the problem of overlapping defect regions. This poses a great challenge to general object detectors based on ordinary convolutional networks.

To address the characteristics and difficulties of face flaws, this paper proposes a model HRT-YOLO based on YOLOv5 to solve the above three problems. We first perform feature extraction by means of a high-resolution encoder[7]. The high-resolution encoder uses a parallel approach to connect high-resolution and low-resolution subnets, maintaining high resolution throughout the process. It obtains more correct representations of the results by repeated multi-scale fusion, resulting in more authentic information on the spatial location of flaws. Secondly, we propose the Vision Transformer Detection Head (ViTDH) structure in Neck to collect dependencies between blocks of image features for dense flaw detection.

The main contributions of this paper are as follows:

- We perform feature extraction at the high-resolution encoder, and the main high-resolution feature map network is gradually added to the low-resolution feature map sub-network in parallel, with different networks achieving multi-scale fusion and feature extraction.
- We have integrated the ViTDH into YOLOv5 to improve the model's ability to detect areas of dense face flaws.
- We contribute a high-resolution face flaw detection dataset to train and evaluate our method.

2. Related work

Object detection models, from a component perspective, usually consist of two parts: a skeleton of the CNN that performs feature extraction, and a detection head that is used to predict the classes and enclose frames of the target. Object detection models often insert a network layer between backbone and head to collect different feature maps, which we refer to as the neck of the object detector.

2.1. Backbone

CNN skeletons commonly used are ResNet[8], EfficientNet[9], CSPDarkNet53[10], etc.

2.2. Neck

The neck is the top-down structure in object detection, which processes the features extracted from the backbone to derive different feature maps, and supplies the processed features to the head. Generally, the neck contains several bottom-up paths and several top-down paths. Its most common structures are FPN, PAN[11] and BiFPN etc. The most crucial operation of a neck is feature fusion, superimposing information from different size perceptual fields.

2.3. Head

The detection heads for object detection are mainly used to locate and classify targets by means of features extracted from the skeleton and are mainly of two types - one-stage and two-stage. Two-stage refers to the fact that the detection algorithm needs to be done in two steps, firstly to acquire the candidate regions and then to classify them. The most emblematic two-stage detection heads are the R-CNN family, which embraces fast R-CNN[12], faster R-CNN[13]. The YOLO series is one of the most promising one-stage detectors and the latest YOLOv5 proposes the structure of CSPDarknet53 with a focus layer as backbone, PANet as Neck, and YOLO detection head. To perform real-time defect detection and maintain good detection accuracy, the HRT-YOLO is developed using YOLOv5 as a baseline.

3. Proposed method

In this section we present a comprehensive discussion of the model architecture of HRT-YOLO and elaborate on our proposed high resolution hold feature extraction and ViTDH.

3.1. HRT-YOLO

This paper proposes a hybrid model HRT-YOLO based on YOLOv5 incorporating high-resolution representation and vision transformer for better detecting face flaws with inconspicuous edges, irregular shapes, and dense rows. The overall structure of HRT-YOLO has three parts: high-resolution encoder, PAN neck, and vision transformer detection head. The architecture of HRT-YOLO is shown in Figure 1. First, we perform feature extraction with a high-resolution encoder to obtain multi-scale, high-resolution fused features. Afterward, we feed the 1/8,1/16,1/32 resolution feature maps from the fourth stage of the high-resolution encoder into PAN Neck. Finally, the ViT-CSP module is added before each detection branch to enhance the boundary information, forming the ViTDH.

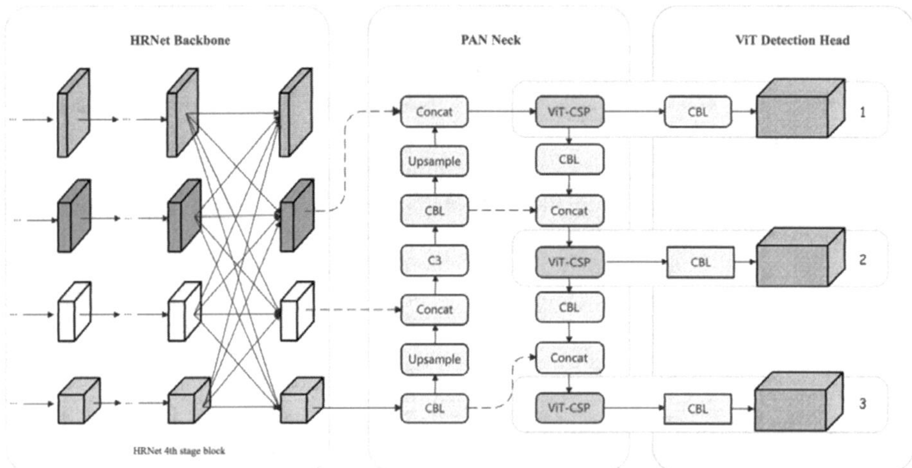

Figure 1. The architecture of the HRT-YOLO, which consists of high-resolution encoder, PAN neck, and vision transformer detection head.

3.2. High-resolution representations

We use the High-Resolution Network (HRNet) in the feature extraction phase of the model to better preserve the high-resolution information of the image. Most of the works encode images as low-resolution representations, form subnetworks by concatenating high-resolution to low-resolution convolutions and later recover high-resolution representations from low-resolution representations. HRNet, instead, uses a parallel approach to connect high-resolution to low-resolution subnetworks and performs iterative multi-scale fusion to obtain spatially more accurate results. Face flaws are location-sensitive and small in size, thus the introduction of High-Resolution representation is important to the accuracy of flaws detection.

The original YOLOv5 uses the PANet structure in the Neck for feature fusion via bottom-up and top-down paths. However, there is no feature fusion in the feature extraction phase, thus causing the network to miss the skin texture information. Therefore, we believe that repeated multiscale fusion in the feature extraction phase would better preserve the fine texture information of the skin for better defect localization. The multi-scale fusion repeated in HRNet[14] relies on the operations of stride convolution, up sampling, and summation. We illustrate feature fusion in HRNet with three resolution fusions as an example. As the input has three different resolution representations of $\{\boldsymbol{R}_r^i, r = 1,2,3\}$ the output has three corresponding resolution representations of $\{\boldsymbol{R}_r^o, r = 1,2,3\}$, where r denotes the resolution indicator. The formula is shown below.

$$\boldsymbol{R}_r^o = f_{1r}(\boldsymbol{R}_1^i) + f_{2r}(\boldsymbol{R}_2^i) + f_{3r}(\boldsymbol{R}_3^i) \quad (1)$$

When cross-stage fusion is performed, such as from stage 3 to stage 4, there is an additional output which is calculated as shown below.

$$\boldsymbol{R}_4^o = f_{14}(\boldsymbol{R}_1^i) + f_{24}(\boldsymbol{R}_2^i) + f_{34}(\boldsymbol{R}_3^i) \quad (2)$$

The x in the transformation function $f_{xr}(\cdot)$ represents the input resolution metric and the r represents the output resolution metric. If $x = r$, $f_{xr}(\boldsymbol{R}) = \boldsymbol{R}$. If $x > r$,$f_{xr}(\boldsymbol{R})$ is up-sampling the input R and adjusting the number of channels by a 1×1 convolution. If $x < r$,$f_{xr}(\boldsymbol{R})$ performs $(r - s)$ strided convolutions on the input R and thus downsamples it. By implementing High-Resolution representation in the feature extraction stage and feeding 1/8,1/16,1/32 resolution feature maps from the fourth stage of HRNet into the next stage of the network, we are able to improve the model's ability to locate face flaws by performing multi-scale feature fusion with high-resolution hold and repetition in the feature extraction stage.

3.3. ViTDH

For face flaws with irregular shapes and obscure edges, we need to collect dependencies between blocks of image features from large neighbourhoods and retain sufficient spatial information. The receptive field of common convolution depends on the size of the filter and the number of convolutional layers used, and there is always a limit and trade-off in the receptive field it can capture. In recent years, transformer[15] has achieved very satisfactory results in several tasks of natural language processing with its superior performance in handling complex dependencies between sequences by means of a multi-head attention. We therefore propose the Vision Transformer[16] Detection head (ViTDH) structure to better capture contextual information.

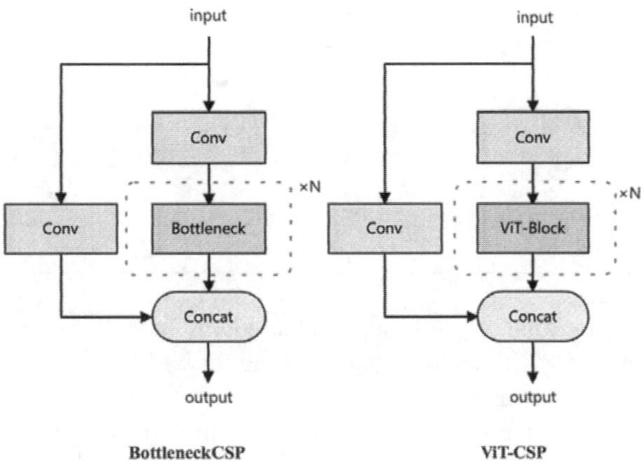

Figure 2. Left: A BottleneckCSP block Right: A ViT-CSP block.

From the first part of the network, we obtained multi-scale high-resolution fusion representations by means of high-resolution encoder. To process the extracted features at a better level, we have improved the BottleneckCSP, which consist of one 3x3 spatial convolution and one 1x1 spatial convolution, by proposing the ViT-CSP structure. Inspired by vision transformer[16], we embed vision transformer into the BottleneckCSP module to apply a multi-head attention on 2D feature maps. Transformer module consists of two main components: a multi-head attention and a feed-forward neural network (MLP), where the multi-head attention allows better mining of contextual features and enhances the ability to capture global features. The structure of the ViT-CSP is shown in Figure 2. The discrepancy between BottleneckCSP and ViT-CSP is that we replace Bottleneck with the vision transformer layer. We believe that adding ViT-CSP before each detection branch can effectively enhance the boundary information and that ViTDH can yield more meaningful contextual information about the spatial relationships between flaws, especially the regions of flaws that overlap each other.

4. Experiments

4.1. Datasets

As there is no publicly available dataset for the current task, we propose a face flaws detection dataset for training and evaluation of our method. We obtained high-resolution images of faces from 266 volunteers of all genders and ages using mobile phones. All images are annotated by hand, and the annotations mainly contain skin areas that are aesthetically displeasing, such as acne marks and acne, that differ significantly from the original skin of the face. We finally obtained 10986 face images of 640×640 and the corresponding annotations, and divided the data into a training set and a test set in the ratio of 8:2. To validate the performance of HRT-YOLO, mAP and accuracy are used as measures.

4.2. Experimental Results

We implement HRT-YOLO on Pytorch 1.7.0, and the models are trained and tested on an NVIDIA RTX1080Ti. Since the features of face flaws are quite different from those of natural scenes, none of the models used any pre-training weights during the training phase, and we trained the network from zero with a batch size of 2 to train 100 epochs.

Table 1. The comparison of the result on face flaws test set.

Methods	mAP (%)	Precision (%)
RetinaNet[17]	18.01	24.61
YOLOv3[4]	18.15	28.90
EfficientDet[9]	18.24	29.04
YOLOv5[6]	18.49	30.74
Ours	**21.72**	**36.07**

As shown in Table 1, multiple models are evaluated on face flaws test set for model results. Compared to these advanced algorithms, our model ended up with a mAP of 21.72% and an accuracy of 36.07%, which is 3.23% improvement in mAP and a 5.33% improvement in accuracy compared to YOLOv5. Figure 3 shows the subjective effect of YOLOv5 and HRT-YOLO.

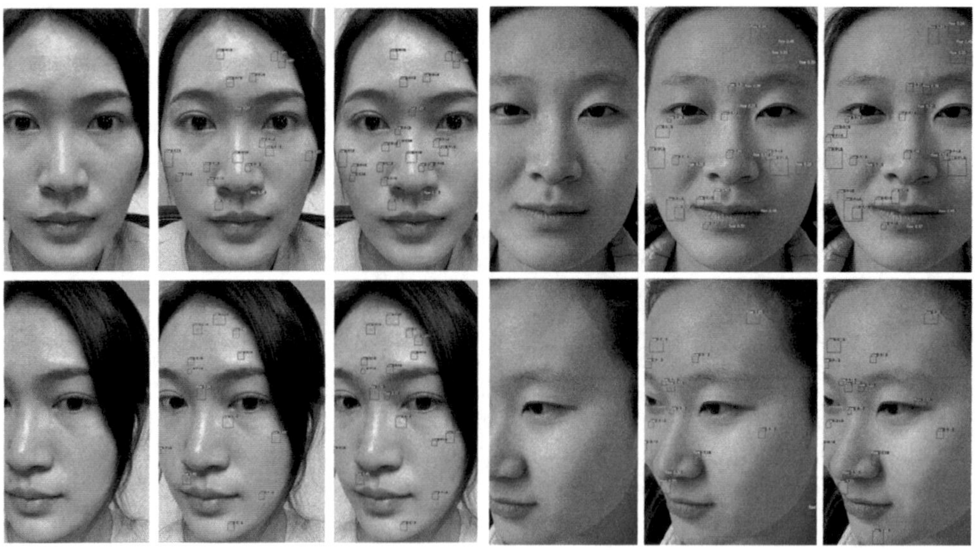

Figure 3. A comparison of the subjective results of YOLOv5 (green) and HRT-YOLO (purple).

4.3. Ablation study

We conduct ablation experiments to analyse the HRT-YOLO model on the test set of the face flaws dataset and the results are shown in Table 2. The impact of each component is listed in the table 2.

Table 2. Ablation Study on face flaws test set

Methods	mAP (%)	Precision (%)
YOLOV5	18.49	30.74
+ HRNet	20.76 (+2.27)	33.75 (+3.01)
+ ViTDH	21.72 (+0.96)	36.07 (+2.32)

5. Conclusion

In this paper, we present the high-resolution representation and ViTDH to form an advanced detector for face flaw detection - HRT-YOLO. On the face flaws dataset, our model achieves an excellent result compared to other state-of-the-art methods. The method in this paper allows for more accurate detection of face flaws. In the future we hope to carry out more realistic repair work on defects to obtain high-resolution, detail-preserving face images with flaws removed.

References

[1] Tomasi, C., & Manduchi, R. (1998, January). Bilateral filtering for gray and color images. In Sixth international conference on computer vision (IEEE Cat. No. 98CH36271) (pp. 839-846). IEEE.

[2] Redmon, J., Divvala, S., Girshick, R., & Farhadi, A. (2016). You only look once: Unified, real-time object detection. In Proceedings of the IEEE conference on computer vision and pattern recognition (pp. 779-788).

[3] Redmon J, Farhadi A. YOLO9000: better, faster, stronger[C]//Proceedings of the IEEE conference on computer vision and pattern recognition. 2017: 7263-7271.

[4] Redmon, J., & Farhadi, A. (2018). Yolov3: An incremental improvement. arXiv preprint arXiv:1804.02767.

[5] Bochkovskiy, A., Wang, C. Y., & Liao, H. Y. M. (2020). Yolov4: Optimal speed and accuracy of object detection. arXiv preprint arXiv:2004.10934.

[6] glenn-jocher, alexstoken, and Borda. (2021). YOLOv5. https://github.com/ultralytics/yolov5.

[7] Sun, K., Zhao, Y., Jiang, B., Cheng, T., Xiao, B., Liu, D., ... & Wang, J. (2019). High-resolution representations for labeling pixels and regions. arXiv preprint arXiv:1904.04514.

[8] He, K., Zhang, X., Ren, S., & Sun, J. (2016). Deep residual learning for image recognition. In Proceedings of the IEEE conference on computer vision and pattern recognition (pp. 770-778).

[9] Tan, M., Pang, R., & Le, Q. V. (2020). Efficientdet: Scalable and efficient object detection. In Proceedings of the IEEE/CVF conference on computer vision and pattern recognition (pp. 10781-10790).

[10] Wang, C. Y., Liao, H. Y. M., Wu, Y. H., Chen, P. Y., Hsieh, J. W., & Yeh, I. H. (2020). CSPNet: A new backbone that can enhance learning capability of CNN. In Proceedings of the IEEE/CVF conference on computer vision and pattern recognition workshops (pp. 390-391).

[11] Liu, S., Qi, L., Qin, H., Shi, J., & Jia, J. (2018). Path aggregation network for instance segmentation. In Proceedings of the IEEE conference on computer vision and pattern recognition (pp. 8759-8768).

[12] Girshick, R. (2015). Fast r-cnn. In Proceedings of the IEEE international conference on computer vision (pp. 1440-1448).

[13] Ren, S., He, K., Girshick, R., & Sun, J. (2015). Faster r-cnn: Towards real-time object detection with region proposal networks. Advances in neural information processing systems, 28, 91-99.
[14] Wang, J., Sun, K., Cheng, T., Jiang, B., Deng, C., Zhao, Y., ... & Xiao, B. (2020). Deep high-resolution representation learning for visual recognition. IEEE transactions on pattern analysis and machine intelligence.
[15] Vaswani, A., Shazeer, N., Parmar, N., Uszkoreit, J., Jones, L., Gomez, A. N., ... & Polosukhin, I. (2017). Attention is all you need. In Advances in neural information processing systems (pp. 5998-6008).
[16] Dosovitskiy, A., Beyer, L., Kolesnikov, A., Weissenborn, D., Zhai, X., Unterthiner, T., ... & Houlsby, N. (2020). An image is worth 16x16 words: Transformers for image recognition at scale. arXiv preprint arXiv:2010.11929.
[17] Lin, T. Y., Goyal, P., Girshick, R., He, K., & Dollár, P. (2017). Focal loss for dense object detection. In Proceedings of the IEEE international conference on computer vision (pp. 2980-2988).

Research on Vessel Behavior Modeling Based on MAXSim Simulation Platform

Weidong Guan[1a*], Yumei Wang[1b*]

[1]North China Institute of Computing Technology, Beijing 100083, China

[a*]hbhggwd@163.com, [b*]wym_email@263.net

Abstract—In a combat simulation system with the ocean as the main environment and vessels as the main platform, the accuracy of the behavior model of the computer-generated force will directly affect the confidence of the simulation system and the efficiency of simulation exercises. As a combat simulation application and development platform, the MAXSim simulation platform based on the Generic Black-Board structure has good secondary development. In order to research the behavior model of vessels and provide general modeling ideas and methods for the maritime combat simulation system, the architecture of the MAXSim simulation platform and the specific functions of the modules are firstly researched. Then, the common methods of behavior modeling by using MAXSim are researched from the level of weapons and equipment. Finally, starting from the system-level development of behavior models, the development methods of actions and conditions of periodic-driven and event-driven behavior models are researched, which improves the accuracy of Computer Generated Forces and supports a good basis for further improving the overall performance of the simulation system.

1. Introduction

With the development of modern war technology and the continuous improvement of performance requirements such as efficiency and accuracy in combat simulation training, Computer Generated Force(CGF) is widely developed in various countries to meet the needs of combat simulation and modeling[1][2]. In the combat simulation with the marine battlefield as the main simulation environment, it is more and more important to establish CGF model according to the characteristics of combat units. Structurally, CGF involves the design and implementation of models in behavior planning, tactical decision-making, motion control, etc., which are the key contents to be paid attention to when constructing CGF[3]. How to combine the real battlefield situation with the characteristics of the combat platform, and accurately establish the behavior models of weapons and equipment of various forces, determine whether it can provide stable support for the subsequent simulation battlefield environment operation[4]. In the marine environment, the most common combat platform is the vessels that moves on the surface or underwater. Although there are different types of vessels, they have certain commonalities in behavioral logic. Designing and implementing the model according to these common types of behavior can greatly reduce the redundant work of the combat simulation system in the modeling stage, as well as improve the efficiency of model designing.

MAXSim is a simulation application and development platform which is used in the combat field. It uses the Generic Black-Board(GBB) architecture to store the data generated during simulation, which makes the system have good secondary development. In MAXSim platform, there are database files dedicated to storing weapons and equipment and behavior models. Users can not only edit the corresponding configuration data in the editor of graphical interface according to actual needs, but also develop the program according to specific simulation requirements. Therefore, by using and developing MAXSim platform, it is feasible to meet the research needs of common properties in vessel behavior modeling.

2. Archetecture of Maxsim Simulation Platform

MAXSim simulation platform is a simulation engine based on distributed multi-Agent system framework with GBB architecture as its core. MAXSim can not only be used to build the actual distributed combat simulation system, but also complete the general development based on the simulation platform through secondary development, so as to prepare for the actual simulation application. The architecture of MAXSim simulation platform is shown in Figure 1.

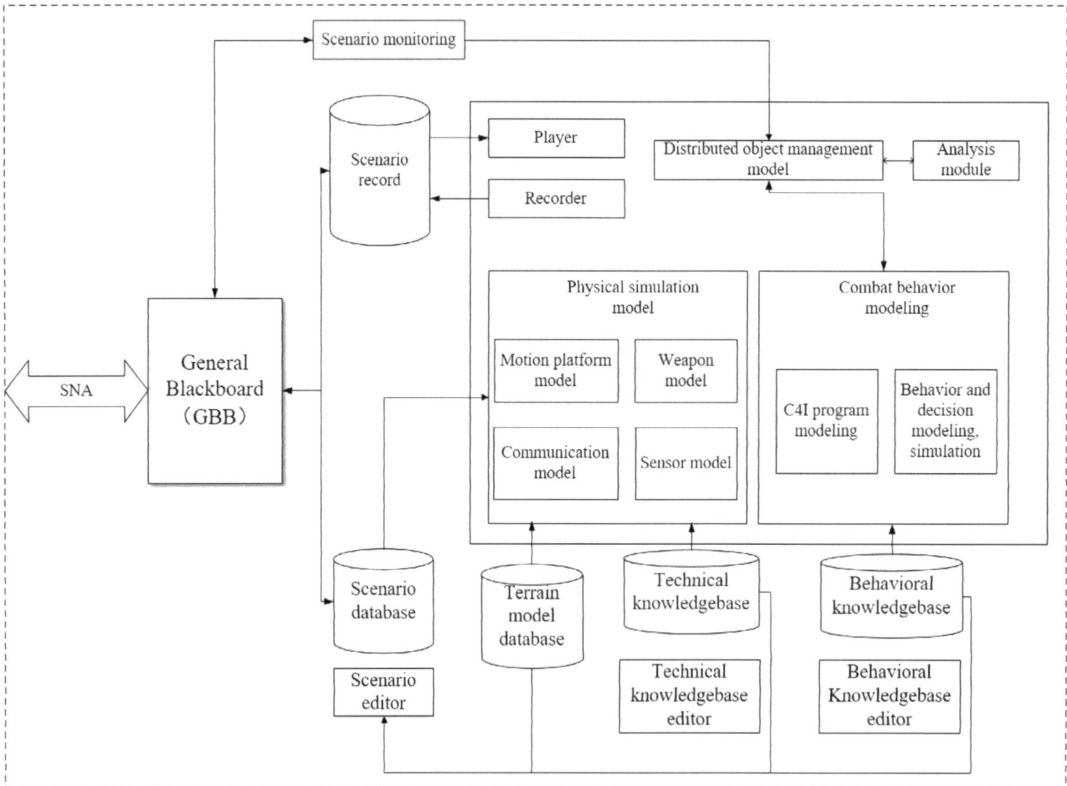

Fig.1 MAXSim platform architecture

The functions of several modules closely related to simulation application and model development are as follows:

1) Generic Black-Board:

In the simulation system, Agent is used to perform specific tasks or simulate specific entities[5][6]. GBB is used to store the application data generated by all Agents in the running process, and provide the data for Agents to get access to them and modify them[7][8]. The distribution and synchronization of these application data among nodes are completed by Serial Network Adapter(SNA).

2) Technical Knowledge Base:

Technical Knowledge Base(TKB) is used to store technical information related to weapons and equipment objects.

3) Behavioral Knowledge Base:

Behavioral Knowledge Base(BKB) is used to store the behavioral rules and constraints of CGF during simulation deduction. In MAXSim platform, each behavioral rule and constraint is called a Doctrine. The simplest doctrine consists of State, and the complex doctrine can nest several other doctrines.

3. Vessel Behavior Modeling Based on Equipment Level

The most common method to model the behavior of vessels is to directly use the existing data of technical knowledge base and behavior knowledge base, and combine the actual demand of vessel model in simulation system to create the weapon equipment model and set the behavior model[9].

By using the Technical Knowledgebase Editor, it is feasible to create a vessel equipment model that meets the actual needs. TKB file shows the equipment model system of MAXSim system. The file uses object-oriented method to classify all kinds of equipment layer by layer according to models and types[10]. The editor provides rich parameter fields, allowing users to modify the physical, dynamic, electronic jamming, carrying weapons and other parameters of vessels.

The classification of weapons and equipment in the Behavior Knowledge Base depends on the classification in the Technical Knowledge Base[11], so the user can set the behavior model only after the vessel equipment model is created and edited. In the behavior modeling based on equipment level, the main tool is Behavior Knowledgebase Editor. The behavior modeling in MAXSim platform adopts finite state machine, and each doctorine is connected by several states and state transition conditions. State and state transition conditions can be subdivided into Actions and Conditions respectively. Figure 2 shows the state and conversion conditions in the Behavior Knowledge Base of “launch vessel-to-vessel missiles” doctorine. In this doctorine, the “launch vessel-to-vessel missiles” and “mission finished” states are composed of maneuvers, target selection, weapons and sensors provided by the simulation platform.“?” surrounded by a circle represents the transition conditions between states, which is edited in the transition rule editor.

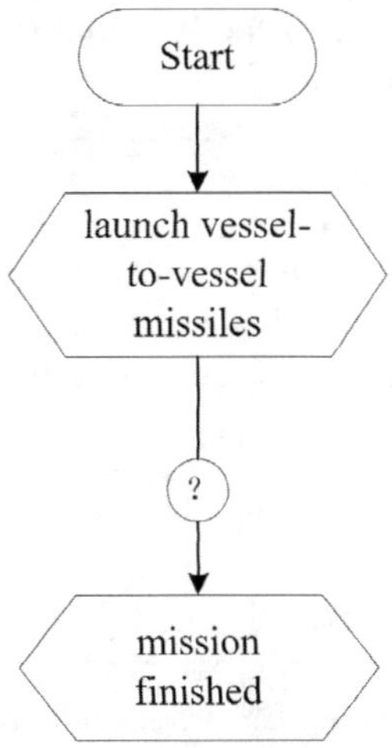

Fig.2 Doctorine of “launch vessel-to-vessel missiles”

4. Modeling of Vessel Behavior Based on Logic Control

In MAXSim, the main function of logic control behavior model is to convert logic instructions generated in the decision-making process of CGF brain model composed of doctorines into a series of practical instructions that can be executed[12][13]. According to the classification of the practical application of simulation, the logic control models include motion, weapons, sensors, target selection, etc. In order to realize this level of behavior modeling, it is necessary to expand and develop Actions and Conditions.

4.1. Extended development of Actions

In the Behavior Knowledge Base, the main function of the target selection model is to select the corresponding entities according to the content of the behavior model[14]. This section takes the target selection control model as an example to study its Action expansion development methods step by step, in order to realize the Action of "randomly selecting enemy air targets".

1) Update the extended configuration file of Action: The extended configuration file of Actions adopts XML document format, recording various attributes of Actions that need to be extended and developed, including type ID, name, various parameter ID and parameter name involved in Actions, etc. The tag "CGF" determines the applicable scope of this Action,whose parameter "Type" value is the ID value of the applicable platform; The ID of the ship tag can be looked up in the TKB configuration file, and then set the attribute of the Action in the Extended Actions configuration file, as is shown in Figure 3.

```
<ActionsData>
  <Actions>
    <Action ID="3510" Type="2" NameEN="Randomly select an enemy air target detected under #1meters">
    <Args>
      <Arg ID="1" Type="2" NameEN="#1"/>
    </Args>
    <CGFs>
      <CGF Type="101"/>
      <Arg ID="1" Default="0" Min="0" Max="20000"/>
    </CGFs>
  </Actions>
</ActionsData>
```

Fig.3 Action configuration

2) Implementing Actions: Actions need to be opened in projects in the corresponding development tools and implemented in the projects by coding. To create an Action belonging to the target selection model, it is necessary to create a derived class inherited from CbaseTargetSelectionAction, and implement a virtual function, which returns the ID and position parameters of the target, and its logic is shown in Figure 4.

3) Expand the deployment of Actions: After realizing the logic in the process of Actions, follow-up deployment work should be completed. The main process of this step is to find the target selection extension tag in the extension configuration file "ExternalConfig", and register the ID and class name of the Action in the corresponding file using the "RegisterActions" interface. Thus, when an entity needs to perform Actions, the corresponding instance can be found according to the ID, and the execution function of the instance object can be called to execute the action according to the set logic.

4.2. Extended development of Conditions

In the behavior model base of MAXSim platform, the role of Conditions is to form the transition conditions between different states in tactics according to the logical relations of intra-group logical AND and inter-group logical OR[15]. Take the Condition "distance to target (in nautical miles)" as an example, in the condition extension configuration file, the contents of the tag <Conditions> are: <Condition id = "1268" Type = "10" NameEN = "distance to target (nautical miles)" ValueType = "4" ValueUnit = "N Miles" Operators = "M; ME; L; LE; E">, where the value of "Type" determines the condition type, "ValueType" and "ValueUnit" determine the type and unit of the Condition input value, and "Operators" determine the operators that users can select when editing Conditions. M, L, E, LE, ME represent the size relationship. The implementation and deployment process of the extended Condition is similar to the Action process, except that the derived class corresponding to the extended Condition needs to inherit from the base class CbaseCondition, and the RegisterConditions interface needs to be used to complete the registration.

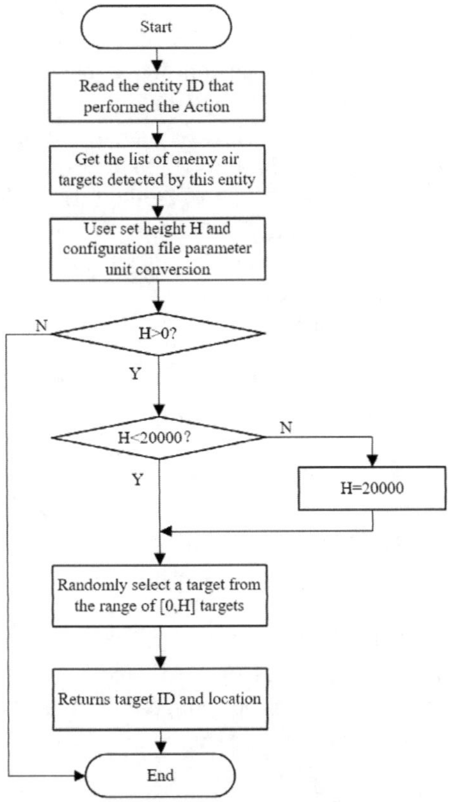

Fig.4 The logic of Action "target selection"

5. Operation and Test of Vessel Simulation Model

After the development and setting of Conditions and Actions, the vessel equipment parameters and behavior models have been established and can operate in the simulation system. In the scenario editor of MAXSim, you can set the goals and plans that the ship model needs to execute. After running the scenario, you can observe the ship's behavior, and try to read the latitude, longitude, speed and other information. During the running test, the vessel information data at a certain time is shown in Table 1.

Table 1. Parameter information of vessel operation test

Platform Type	ID	Latitude	Longitude	Speed	Course
Surface Vessel	101	N22°22'29''	E124°06'13''	30.0m/s	228.202°
Surface Vessel	103	N22°30'20''	E124°28'24''	30.0 m/s	149.269°
Submarine	2006	N23°23'27''	E124°35'03''	23.429 m/s	359.191°
Surface Vessel	111	N22°39'39''	E125°03'15''	24.0 m/s	88.803°

The data in Table 1 shows that the vessel behavior model using the extended Conditions and Actions, whether it is a surface vessel or an underwater submarine, can operate normally in the simulation system. The parameters such as latitude and longitude, speed, heading, etc. of these vessels indicate that their behavioral models function satisfactorily in the simulation system.

6. Conclusion

Taking the doctorine of vessels launching vessel-vessel missiles as an example, the doctorines are established by using MAXSim behavior modeling tool. It is introduced that the knowledge base and editor provided by MAXSim simulation platform can support users to edit the behavior model of vessel weapons and equipment at the state and doctorine levels. Through the research and development of logic controling behavior model at MAXSim platform model system level, it shows that the simulation platform supports the expansion of Actions and Conditions of composition states and conversion conditions. Developers can combine them to meet the requirements of doctorines, and further establish the behavior model of vessel CGF that meets the specific requirements of system simulation, so as to improve the fidelity of the simulation system.

The above results provide an improved idea and method for the accuracy of the behavioral model of the simulation model in the actual simulation exercise project, laying a solid foundation for the construction of a large-scale maritime combat simulation system. The next step is to study the reusability and scalability of the distributed simulation model in the system, and then realize a large-scale maritime combat simulation system.

Acknowledgments
This work was financially supported by the Combat Simulation System Project of North China Institute of Computing Technology.

References

[1] GAO A, DUAN L, ZHANG G H, et al. Development status of computer generated force behavior modeling[J]. Computer Engineering and Applications, 2019, 55(19): 43-51.
[2] XU K, ZENG J, YANG W L, et al. Research Review of Cognitive Behavioral Model Architecture for Comput-er-Generated Forces[J]. Journal of System Simulation, 2021, 33(02):239-248.
[3] LUOTSINEN L J，KAMRANI F，HAMMAR P，et al. Evolved creative intelligence for computer generated forces[C]// Proceedings of 2016 IEEE International Conference on Systems，Man，and Cybernetics，2016：3063-3070.
[4] TANG J B, ZHANG Q, JIAO P, ZHA Y B. A Preliminary Study on CGF Behavioral Modeling Check and Verifica-tion [J]. Journal of System Simulation, 2021, 33(02): 288-294
[5] DENG K, LONG F. Research on formation an-ti-submarine warfare simulation based on multi-agent[J]. Modern Defense Technology, 2016, 44(01):213-217+230.
[6] SUN Y X, LIU G F. Agent-based business modeling and simulation of submarine combat games[J].System Simu-lation Technology,2017,13(02):162-169.
[7] CORKILL D D. GBB: A generic blackboard development system[C]// AAAI. Vancouver, CAN:AAAI Press, 1986: 1008-1014.
[8] LIAO S Y, DAI J H. Agent-based modeling and simulation design mode and software framework for complex sys-tems[J]. Computer Simulation,2005(05):254-258.
[9] VSIHWAKARMA S, AGRAWAL A. A Survey on Activity Recognition and Behavior Understanding in Video Sur-veillance[J]. The Visual Computer (S0178-2789), 2013, 29(10): 983-1009.
[10] SHAO Z H, YOU D D, XIONG Z X. Design of surface warship anti-missile simulation model based on MAXSim[J]. Ship Electronics Engineering, 2017, 37(09): 59-64.
[11] LIU S, ZHANG H J. Architecture Design of Equipment System Combat Simulation System Based on MAXSim [J]. Computer Simulation, 2021,38(04):5-8+15.
[12] XU X，YANG M，Li G. Adaptive CGF commander behavior modeling through HTN guided Monte Carlo tree search[J]. Journal of Systems Science and Systems En-gineering，2018，27（2/3）：231-249.
[13] ZHANG Q，YIN Q J，HU Y. Modeling CGFs behavior by an extended option based learning behavior trees[C]// Proceedings of 2017 IEEE International Conference on Cybernetics and Intelligent Systems and IEEE Conference on Robotics，Automation and Mechatronics，2017：260-265.
[14] HAN X G, XU H G, WU X. Research on CGF force generation method of amphibious tank based on MAXSIM platform[J]. Computer and Digital Engineer-ing,2016,44(01):26-30+63.

[15] SHAO Z H, XIONG Z X, YOU D D. Simulation modeling of patrol tactics based on MAXSIM platform [J]. Com-mand Control and Simulation, 2016, 38(06): 90-92+104

Physiological Signal Emotion Recognition Based on Temporal Convolutional Networks

Yong Sheng[1*], Qinghong Hu[2], Jun Cao[3]

[1]Computer and information engineering college, Tianjin Normal University, Tianjin, 300387, China

[2]Computer and information engineering college, Tianjin Normal University, Tianjin, 300387, China

[3]Computer and information engineering college, Tianjin Normal University, Tianjin, 300387, China

[*]Corresponding author's e-mail: 18339906771@163.com

Abstract. Human emotion recognition has become an important research field. Because of its objectivity, physiological signals have become one of the most robust cues for emotion recognition. In recent years, deep learning methods have made great progress in the field of emotion recognition, especially the superiority of recurrent neural networks (RNN) in time series models, and more and more tasks are completed based on RNN. However, RNN has problems such as time-consuming and gradient disappearance and explosion, and the feature input and sentiment output of RNN are not aligned. To avoid these problems, this paper is based on the temporal convolutional networks (TCN) model and the connectionist temporal classification (CTC) algorithm to process the emotion recognition task. First, generate a spectrogram representation of the physiological signal in each channel; Second, use the TCN to learn long-term dynamic features, and use CTC to align the dynamic features and their sentiment labels, and then feed the learned deep features into the neural network to predict the sentiment of each channel; Finally, take the best result as the final emotion representation. Experimental results on the AMIGOS dataset show that the proposed method outperforms existing methods.

1. Introduction

Affective computing is mainly based on multiple modalities to identify human emotions. In recent years, affective computing has been widely used in many fields such as intelligent driving, smart city, and multimedia content recommendation.

Traditional emotion recognition mainly relies on facial expressions [1] and voice data [2] and other clues to analyze and infer the emotional state of humans. However, in practical applications, it is difficult to capture spontaneous emotional responses from these clues. In recent years, thanks to the objectivity of human physiological signals in emotional expression, electroencephalogram (EEG) [3], electromyography (EMG) [4], electrocardiogram (ECG) [5], are considered as emotion recognition the most powerful clue.

In previous work, methods for building emotion recognizers are usually divided into two steps: selecting classifiers and manually extracting features [6]. In terms of classifier selection, early work mainly utilizes shallow machine learning methods to train emotion recognition models based on handcrafted features. The design of handcrafted features requires a lot of domain expertise, and the extraction of certain features (such as linear features) is also time-consuming. With the success of deep learning techniques in classification tasks, more and more tasks begin to use deep learning models to extract features. Tripathi et al. [7] converted multi-channel EEG signals into two-dimensional features and used convolutional neural networks (CNN) to extract EEG signal features for classification; Kwon et al. [8] used the CNN model to improve the sentiment classification task and obtained good results. However, the EEG signal is a time-series signal, and CNN cannot extract time-series features in processing time-series signals, which limits the further improvement of emotion recognition performance.

Due to the advantages of recurrent neural networks (RNN) in sequence models, RNN is mainly used for processing time-series data. Li et al. [9] constructed an long short-term memory (LSTM) networks as a classifier to explore the temporal correlation of EEG signals. Zhang et al. [10] proposed a C-RNN model using both CNN and RNN and used multi-channel EEG signals to identify emotions. These works have achieved good results, and also reflect the superiority of RNN or LSTM in time series tasks. However, most of the tasks based on RNN are time-consuming, and there are problems such as gradient disappearance and explosion, which must have an impact on the final judgment of the model.

In addition, most of the final results of emotion recognition tasks have only one emotion label, while RNN has an output for each time point, which causes a situation that does not match the emotion label. Given the above problems, we use temporal convolutional networks (TCN) [11] to learn long-term dynamic features and use connectionist temporal classification (CTC) [12] to map the output of each time point to an emotion label to distinguish time points that are discriminative in emotion expression.

The main contributions of this paper are as follows:

1) The use of TCN to extract the time-series features in the physiological signal improves the accuracy of the emotion recognition model.

2) Use CTC to align emotional features and corresponding emotional labels, which is convenient for the model to judge the overall emotion.

2. Methodology

2.1. Problem overview

Considering there are N EEG channels, each channel can obtain $T \times f$ sampling values according to the sampling frequency f within the time T. Then, the physiological signals of these N channels can be expressed as $V = \{v_1, v_2, ..., v_N\} \in R^{N \times T \times f}$. Our problem is to judge the emotional state of users V. The corresponding emotion label $Y \in \{0,1\}^i$, where i is the dimension of the emotion model. The overall framework of the model is shown in Figure 1.

Perform short-time Fourier transform (STFT) [13] on the EEG signals of N channels to obtain the time-frequency data, and divide the obtained time-frequency data into units per second, and the data size obtained is $S = \{s_1, s_2, ..., s_N\} \in R^{N \times T \times F}$ (where F is the spectrogram feature at the corresponding time). That is the time and frequency characteristics of N channels.

2.2. Temporal convolutional networks (TCN)

We use a temporal convolutional network to obtain temporal features from the time-frequency data S. Compared with RNN, the advantage of TCN is that it can process sequences in parallel, and there is no problem of gradient disappearance and explosion. A TCN cell consists of three parts, namely causal convolution, dilated convolution, and residual convolution. Causal convolution is a one-way structure. It cannot see future data, and can only obtain the following results from the previous causes. Dilated convolution allows the input to be sampled at intervals during convolution, and its sampling rate is controlled by the dilation factor. And residual convolution can pass information across layers. We mainly use dilated convolutions to learn temporal features in time-frequency data.

For the data of each channel, it can be expressed as $\{x_1, x_2, ..., x_T\} \in R^F$, input it into TCN to get $\{\hat{y}_1, \hat{y}_2, ..., \hat{y}_T\} \in R^C$, C is the corresponding feature size, then for N channels, we will get $U = \{u_1, u_2, ..., u_N\} \in R^{N \times T \times C}$.

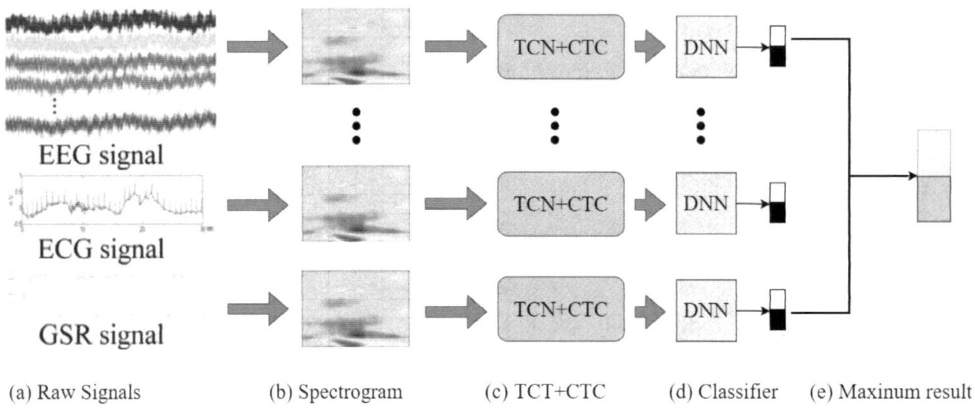

Figure 1. Emotion recognition framework based on temporal convolutional networks

2.3. Connectionist temporal classification (CTC)

CTC is mainly used for speech recognition tasks, which reduces the difficulty of recognition by aligning speech fragments with text. When we use physiological signals to identify emotions, there is also a situation where the input features and the final emotion are not aligned. For a physiological signal, we use an emotion label to represent the emotion of the signal. However, this emotion is not always maintained in the process. That is to say, some fragments of this physiological signal do not express this emotion, so using this data to make predictions will have an impact on the final results of the model.

The emergence of the CTC algorithm is to solve such problems. The CTC algorithm introduces a new placeholder for output alignment results. This new placeholder is called a blank placeholder, and this blank placeholder will be output in the alignment result. Using this blank placeholder, the input and output can be reasonably mapped. For the output of TCN, the CTC algorithm can be used to obtain a corresponding sentiment label for each time output. When we use the features extracted by TCN to make sentiment judgments, we can focus on those time points that are closely related to the final sentiment, thereby improving the final prediction results of the model.

3. Experiments and Analysis

3.1. Dataset

We use the data set AMIGOS [14] to evaluate the effectiveness of our model. The AMIGOS dataset is data generated by 40 participants watching long and short videos in two environments. Use relative equipment to record physiological signals to capture participants' brain activity, including 14-lead EEG (AF3, F7, F3, FC5, T7, P7, O1, O2, P8, T8, FC6, F4, F8, AF4), 2-lead ECG, and 1-lead skin signal, the sampling frequency is 128*HZ*. The experiment annotated the emotional status of participants through self-assessment and external annotations. Self-assessment includes: Arousal, Valence, Dominance, Liking, Familiarity, and seven basic emotions (neutral, disgust, happiness, surprise, anger, fear, and sadness), and external annotations are only Arousal and Valence. This article uses external annotations as data labels.

3.2. Experimental settings
We use leave-one-subject-out cross-validation (LOSOCV) to evaluate our method. At the same time, it is compared with some deep learning methods. We used Accuracy (ACC) and Macro-F1 score (F1-score) to evaluate the performance of the model. Further details on our proposed model are given below: a) learning rate: 1E-3; b) batch size: 100; c) activation function: LeakyReLu; d) optimizer: Adam.

3.3. Experimental result and discussion
Table 1 presents the experimental results of existing methods and our proposed method on the AMIGOS dataset. From the results we can draw the following conclusions: 1) CNN performed the worst, indicating that CNN is not suitable for processing sequence data; 2) The results of the method using TCN are better than other deep learning methods, indicating that TCN has great advantages in the feature extraction of sequence models; 3) The results of the model using the CTC algorithm are better, which shows that the CTC algorithm has greatly improved the accuracy of the model.

Table 1. Compared with other methods on AMIGOS dataset

Classifier	Arousal		Valence	
	ACC	F1-score	ACC	F1-score
CNN	0.787	0.711	0.779	0.723
LSTM	0.800	0.727	0.814	0.735
TCN	0.814	0.770	0.820	0.721
TCN+CTC(Ours)	**0.834**	**0.778**	**0.847**	**0.744**

In Table 2, we compare the results of other existing research methods for affective computing on the AMIGOS dataset. As can be seen from the table, our method has the best results compared to other methods. This is because our method focuses on features that have a large influence on mood, reducing the interference of irrelevant factors. This is very important for improving the performance of emotion recognition systems.

Table 2. Overview of the studies for emotion recognition on AMIGOS dataset

Reference	Evaluation Methods	Accuracy		F1-score	
		Arousal	Valence	Arousal	Valence
Correa [14]	LOSOCV	N/A	N/A	0.564	0.560
Chao [15]	LOSOCV	0.761	0.770	0.702	0.709
Li [16]	LOSOCV	0.833	0.764	0.716	0.698
Ours	LOSOCV	**0.834**	**0.847**	**0.778**	**0.744**

4. Conclusion
In this paper, we use the temporal convolutional network to extract the features of physiological signals and then use the CTC algorithm to align the features at different times with their emotional expressions, focusing on the distinguishing features, avoiding irrelevant information to the model. disturbance of the final result, thereby improving the performance of the emotion recognition system. The final experimental results show that our proposed model achieves good results on the AMIGOS dataset, which further proves the superiority of our model. In future work, the performance of the model will be further improved in order to enhance its generalization ability on different datasets.

Acknowledgments
Thanks to the teachers in the laboratory for their guidance and help, because of your guidance, I have learned a lot of professional knowledge and gradually become a qualified graduate student. At the

same time, I would like to thank Mr. Hu and Mr. Cao for their help in the experiment and writing of the paper. Without you, I would not be able to complete this paper. Thanks to my family for their silent support all the time, because of you I have been able to go all the way to this day, thank you, and wish you good health forever!

References

[1] Anderson, K., Mcowan, P.W. (2006) A real-time automated system for the recognition of human facial expressions.J. IEEE TRANSACTIONS ON CYBERNETICS., 36(1):96-105.

[2] Huang, X., Baker, J., Reddy, R. (2014) A historical perspective of speech recognition. Communications of the ACM., vol. 57, no. 1, pp. 94–103.

[3] Zheng, W. (2017) Multichannel eeg-based emotion recognition via group sparse canonical correlation analysis. IEEE Transactions on Cognitive and Developmental Systems., vol. 9, no. 3, pp. 281–290.

[4] Cheng, B., Liu, G. (2008) Emotion recognition from surface emg signal using wavelet transform and neural network. In: International Conference on Bioinformatics and Biomedical Engineering. Shanghai. pp. 1363–1366.

[5] Agrafioti, F., Hatzinakos, D., Anderson, A.K. (2012) Ecg pattern analysis for emotion detection. IEEE Transactions on Affective Computing., vol. 3, no. 1, pp. 102–115.

[6] Li, Y., Zheng, W., Wang, L. (2019) From Regional to Global Brain: A Novel Hierarchical Spatial-Temporal Neural Network Model for EEG Emotion Recognition.J. IEEE Transactions on Affective Computing., PP(99):1-1.

[7] Tripathi, S., Acharya, S., Sharma, R.D., Mittal, S., Bhattacharya, S. (2017) Using deep and convolutional neural networks for accurate emotion classification on deap dataset. In: AAAI Conference. San Francisco, California. pp. 4746-4752.

[8] Kwon, Y.H., Shin, S.B., Kim, S.D. (2018) Electroencephalography based fusion two-dimensional (2D)-convolution neural networks (CNN) model for emotion recognition system.J. Sensors., 18(5): 1383.

[9] Li, Z., Tian, X., Shu, L., Xu, X., & Hu, B. (2017) Emotion recognition from EEG using RASM and LSTM. In: International Conference on Internet Multimedia Computing and Service. Singapore. pp. 310-318.

[10] Li, X., Song, D., Zhang, P., Yu, D., Hou, Y., Hu, B. (2016) Emotion recognition from multi-channel eeg data through convolutional recurrent neural network. In: IEEE International Conference on Bioinformatics and Biomedicine (BIBM). Houston. pp. 352–359.

[11] Lea, C., Flynn, M.D., Vidal, R. (2017) Temporal convolutional networks for action segmentation and detection. proceedings of the IEEE Conference on Computer Vision and Pattern Recognition., pp: 156-165.

[12] Graves, A., Fernández, S., Gomez, F. (2006) Connectionist temporal classification: labelling unsegmented sequence data with recurrent neural networks. Proceedings of the 23rd international conference on Machine learning., pp: 369-376.

[13] Griffin, D.W., Lim, J.S. (1984) Signal estimation from modified short-time fourier transform. IEEE Transactions on Acoustics, Speech, and Signal Processing., vol. 32, no. 2, pp. 236–243.

[14] Correa, J.A.M., Abadi, M.K., Sebe, N. (2018) AMIGOS: A Dataset for Affect, Personality and Mood Research on Individuals and Groups.J. IEEE Transactions on Affective Computing., pp. 1-1.

[15] Chao, H., Liu, Y. (2020) Emotion Recognition from Multi-channel EEG Signals by Exploiting the Deep Belief-Conditional Random Field Framework.J. IEEE Access., PP(99):1-1.

[16] Li, C., Bao, Z., Li, L., Zhao, Z. (2020) Exploring temporal representations by leveraging attention-based bidirectional lstm-rnns for multi-modal emotion recognition. Information Processing & Management., vol. 57, no. 3, pp. 102185.

Comprehensive evaluation about the quality of steel strip based on combination weighting

DongXu Wang[1a*], GuiMei Cui[1b*], LiXiang Pi[1c], Wei Liu[1d]

[1]Inner Mongolia University of Science and Technology, Baotou 014017, China

[a*]1667267491@qq.com,[b*]WDX333666@126.com,[c]pilixiangzy@163.com

Abstract—In order to solve the problem that the quality of steel strip products is rough by manual evaluation, this paper will combine expert knowledge with quality data and establish four evaluation grades by taking surface quality, mechanical properties and dimensional accuracy as evaluation indicators. Analytic hierarchy process (AHP) and entropy weight method are used to confirm the weight of each evaluation indicator. The mechanical properties and dimensional accuracy indicators are confirmed by statistical method to find out the optimal range of indicator parameters, and the indicator deterioration degree is calculated. The deterioration degree of surface quality indicator is confirmed by fuzzy language quantization. The membership degree of each indicator parameter is calculated according to the membership degree function of ridge distribution, and then the fuzzy comprehensive evaluation result matrix of strip quality is obtained by synthesizing each indicator parameter step by step. Finally, the comprehensive score of strip steel quality is obtained by quantizing and summing the evaluation result matrix. The experimental results show that the evaluation result of the model is consistent with the actual data, so it has certain practicability.

1.Introduction

Hot-rolled steel strip product is widely used and it is one of the main products of steel enterprises, occupying an important position in the national economy and people's life. [1] With the rapid development of the economy, steel strip products are widely used in various manufacturing industries due to their smooth surface, precise dimensions and excellent mechanical properties. Therefore, high quality steel strip products are more competitive in the market. An excellent and rational evaluation of steel strip products is a necessary prerequisite for reducing product quality fluctuations and improving product quality. So, it improves the economic profit of the company in the aspect of production. At present, many steel enterprises use plate gauges, thickness gauges, iron loss gauges and other instruments to test quality data, and then the quality inspectors evaluate separately on a single indicator, which is somewhat subjective and rough. In addition, it is difficult to unify each indicator. What's more, it fails to consider the user requirements for product quality. Research about the evaluation of strip steel product quality is not complete yet. Wang Yonghui[2-3] et al. established a steel strip surface quality evaluation model by the defect information group decision method. They equated the evaluation process into a multi-attribute decision problem at first, and then ranked the steel strip products quality to obtain the evaluation results, which was consistent with the actual situation, but the method only evaluated the surface quality with a single evaluation indicator, thus limiting the application to a certain extent. Dong Dewei[4] et al. proposed an identification model with the properties of steel strip quality evaluation, established a quality level evaluation attribute set, calculated the attribute measures of quality indicators, and figured out the steel strip quality level according to the confidence criterion. The model evaluation results were comprehensive, but the implementation was more complicated.

This paper establishes a multi-level evaluation model by taking quality influencing factors as indicator set and quality level as evaluation set. At the same time, it calculates the information entropy of quality data, and combines expert experience to assign weights. Then it confirms the membership degree according to the degradation degree and the ridge-type function, and synthesizes the weights and the result matrix step by step to obtain the final evaluation results. Compared with other models, the advantages of this model are that it uses different assignment methods to combine the subjective experience of experts with quality data, reducing the roughness of manual evaluation, selecting multiple indicators affecting quality makes the evaluation results more comprehensive, and the continuous evaluation results calculated by the model can identify quality fluctuations in the production process.

2. Application of quality assessment

The main process of hot strip rolling is that continuous casting billets are heated by a heating furnace firstly and then phosphorus is removed coarsely by high pressure water. Next, it is coarsely rolled by a vertical roll coarse mill and removed phosphorus again by flying shears after trimming the head and tail. Then, it is rolled by a finishing rolling unit, and finally cooled by laminar flow and coiled to get a final strip product. It is clear from the process that hot strip rolling is a complex industrial process with multiple links. It is difficult for the production process to operate stably at a long term because of fluctuations in raw materials, unclear mechanisms and too many process parameters. [5] So it also leads to fluctuations in product quality. Scientific quality evaluation methods can control the quality timely and also provide guidance for the optimization of the production process. The current evaluation method is mainly based on a single indicator, which does not reflect the multi-faceted nature of quality and the impact among quality indicators, and the evaluation results are mostly in the discrete form of "qualified" and "unqualified", and most of the evaluation process is manual and subjective. To solve these problems, this paper builds an evaluation system that can reflect the comprehensive quality of strip steel products. The application of the model in the rolling process is shown in Figure 1. The evaluation results of the model are able to achieve continuous evaluation on the basis of discrete evaluation of levels and meet the diversity of user requirements. In addition, the evaluation results of the model can reflect the quality fluctuations of a batch of steel strip products in the production process. After improving the relevant process parameters and the production conditions where quality fluctuations occur, the direction of how to optimize operating each procedure and part to the utmost can be figured out, and the quality of the next batch of products can be optimally adjusted [6].

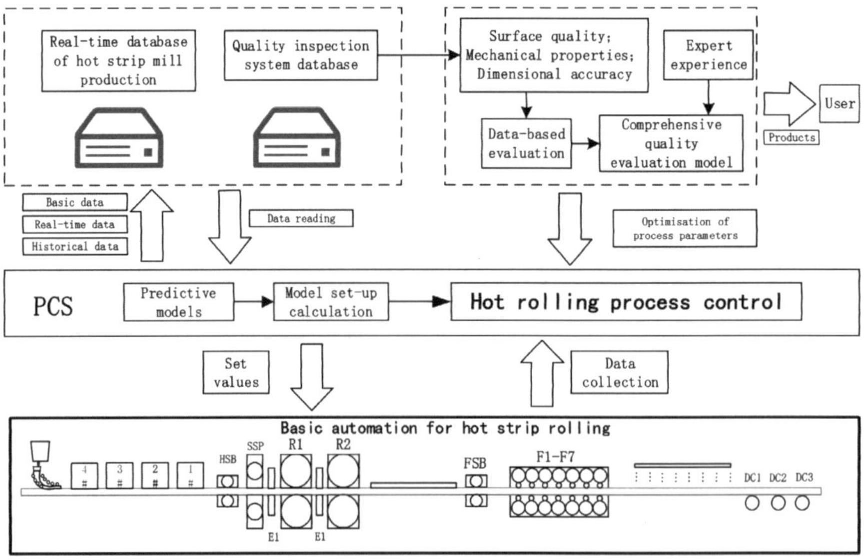

Figure.1 Application of quality evaluation in rolling process

3. Fuzzy integrated evaluation modelling approach

Fuzzy comprehensive evaluation is an evaluation method established by the theory of membership degree in fuzzy mathematics. For the event of many indicators and complexity, the fuzzy and unquantifiable problems will be shown systematically by the fuzzy comprehensive evaluation.[7-8] It has obtained good results in the field of photovoltaic array assessment and mining sustainability risk assessment because the advantage of the method is systematic and easy to calculate.[9-10] The fuzzy comprehensive evaluation mainly consists of evaluation set *L*, indicators set *V*, an evaluation matrix *R* and indicator weights *W*. The final evaluation results are obtained through layer-by-layer synthesis. The process of establishing the fuzzy comprehensive evaluation model is shown in Figure 2.

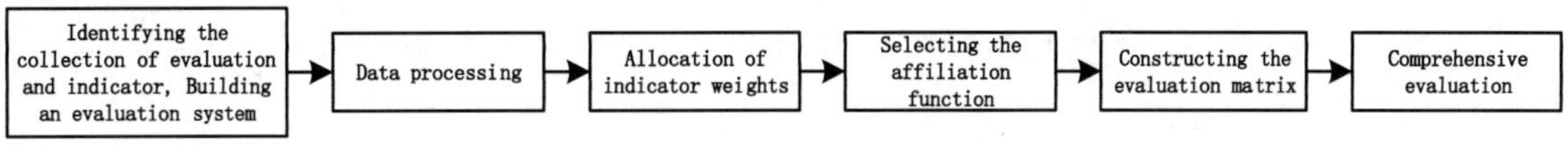

Fig.2 Flow chart of fuzzy comprehensive evaluation model

3.1 Defining a set of evaluation and indicator

The evaluation is divided into four levels, noted as evaluation set *L*= {level Ⅰ, level Ⅱ, level Ⅲ, level Ⅳ}. The level Ⅰ indicates the best and level Ⅳ indicates the worst. In order to transform the discrete level evaluation into continuous evaluation, the four levels of Ⅰ, Ⅱ, Ⅲ and Ⅳ are transformed into four score intervals of [100, 75, 50, 25], and the final evaluation results are expressed in the form of scores. The first indicators are identified as $V=\{V_1, V_2 \ldots V_m\}$, the secondary indicators $V_1=\{v_{11}, v_{12}, \ldots, v_{1n}\}$, $V_2=\{v_{21}, v_{22}, \ldots, v_{2n}\}$, ... $V_m=\{v_{m1}, v_{m2}, \ldots, v_{mn}\}$, and the N-level indicators $u=\{u_1, u_2, \ldots, u_n\}$, ...

3.2 Indicator data processing

To eliminate the influence of outliers on the modelling, the 3σ criterion is used to divide the data into reasonable intervals within (μ-3σ, μ+3σ), and the data are considered as outliers and will be removed if they are not within this interval. The x_k is the data sample, *N* is the number of samples, and μ and σ are the mean and standard deviation of the sample data.

$$\sigma = \sqrt{\frac{1}{N}\sum_{k}^{N}(x_k - \mu)^2} \tag{1}$$

The degree of deterioration can show the extent to which the data deviates from the good condition to the ultimate condition. [11] For the quality data in the evaluation of steel strip quality, it can eliminate the influence of dimension by means of the degree of deterioration analysis. And the degree of deterioration function is divided into the types of the larger the better and the smaller the better and so on.

The larger the better type function:

$$f(x) = \begin{cases} 1, & x < x_{\min} \\ \dfrac{x_{\max} - x}{x_{\max} - x_{\min}}, & x_{\min} \le x \le x_{\max} \\ 0, & x > x_{\max} \end{cases} \tag{2}$$

The smaller the better type function:

$$f(x) = \begin{cases} 0, & x < x_{\min} \\ \dfrac{x - x_{\min}}{x_{\max} - x_{\min}}, & x_{\min} \le x \le x_{\max} \\ 1, & x > x_{\max} \end{cases} \tag{3}$$

where $x_{\min}$ is the minimum value of the data and $x_{\max}$ is the maximum value of the data.

3.3 Determining the weight vector

The first weights in the evaluation system are assigned according to user requirements and expert experience, and the weights can be adjusted according to the actual situation. Different assignment methods are used for different forms of indicators. Surface quality uses AHP to confirm weight values. Mechanical properties and dimensional accuracy use entropy weight method to confirm weight values. The combination of subjective and objective assignment can synthesize weight information to ensure the rationality and objectivity of the evaluation results, also improving the accuracy of this model.[12-13]

AHP is based on expert experience to compare evaluation indicators with each other, and the importance of the *i*th indicator to the *j*th indicator is recorded a_{ij}, and the judgment matrix J is obtained.

Calculate the maximum characteristic root λ_{max} of the judgment matrix. The n represents the number of elements, ω represents the characteristic vector, and ω_i represents the element corresponding to the characteristic vector.

$$\lambda_{max} = \frac{1}{n}\sum_{i=1}^{n}\frac{J\omega}{\omega_i} \tag{4}$$

Calculate the consistency test indicator CI of the judgment matrix to confirm whether the matrix has consistency. The judgment matrix will be reconstructed if it doesn't have consistency.

$$CI = \frac{\lambda_{max} - n}{n-1} \tag{5}$$

The characteristic vector which is correspond to the maximum characteristic root λ_{max} will be normalized to obtain the weights $W = (W_1, W_2, \cdots, W_n)$.

The entropy weight method assigns weights according to the theory of different degrees of data change, different amounts of information and the different corresponding weights. It ensures the objectivity of weight. And meanwhile the entropy weight method can retain the original information of the data to the greatest extent and effectively avoid the influence of human factors. [14-15] Suppose there are n targets to be evaluated, m evaluation indicators. The z is the normalized matrix after analyzing by the degree of degradation and the p is the probability matrix after calculating. So the probability of each element is:

$$p_{ij} = \frac{z_{ij}}{\sum_{i=1}^{n} z_{ij}} \tag{6}$$

The information entropy e_j for the *j*th indicator is:

$$e_j = -\frac{1}{\ln n}\sum_{i=1}^{n} p_{ij}\ln\left(p_{ij}\right), \left(j = 1, 2, \cdots, m\right) \tag{7}$$

Calculate the utility value d_j of information:

$$d_j = 1 - e_j \tag{8}$$

An entropy weight for each indicator will be obtained by normalizing the information utility values:

$$\varphi_j = \frac{d_j}{\sum_{j=1}^{m} d_j}, \left(j = 1, 2, \cdots, m\right) \tag{9}$$

3.4 Membership function

The ridge-type membership function can transform the indicator degradation degree into the membership degree of the evaluation level. A descending ridge-type distribution is used for smaller and better data, an intermediate ridge-type distribution is used for intermediate data, and an ascending ridge-type distribution is used for larger and better data.[16] The membership functions of the degradation degrees at different levels are shown in Figure 3.

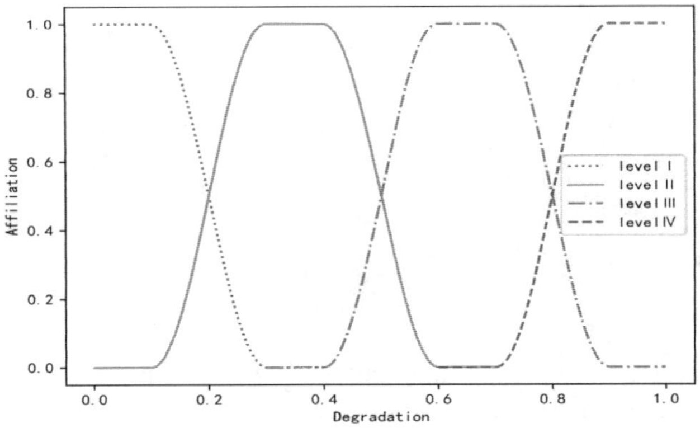

Figure.3 Ridge type membership function

3.5 Integrated evaluation

According to the synthesis operator, the weight and membership matrixes are synthesized step by step to obtain the final evaluation vector, and different synthesis operators are selected according to the different weight characteristics. This paper uses the main factor-determining operator $M(\wedge,\vee)$and the weighted average operator $M(\bullet,\oplus)$. The synthesis formula is as follows:

$$B = A \circ R \tag{10}$$

A represents the weight matrix of each grade; *R* represents the membership matrix of each indicator and *B* represents the final evaluation result matrix. The evaluation result matrix is normalized and then the score intervals for each grade are weighted and summed up to obtain the final score *S* for the steel strip. *Q*=[100 75 50 25] is the score interval.

$$S = B \cdot Q \tag{11}$$

4. Comprehensive evaluation of steel strip quality

Two steel coil samples X_1 and X_2 are selected from B Steel Mill, and the above method is used to obtain comprehensive evaluation according to the data and production status.

(1) The evaluation set *L*={level I, level II, Level III, level IV}.According to the actual situation of steel strip products, this paper establishes a two-level evaluation model. The first-level indicators are $V=\{V_1, V_2, V_3\}$, and V_1, V_2, V_3 represent surface quality, mechanical properties and dimensional accuracy respectively. The secondary indicators $V_1=\{v_{11}, v_{12}, v_{13}, v_{14}, v_{15}, v_{16}\}$, and v_{11}, v_{12}, v_{13}, v_{14}, v_{15}, v_{16} represent oxide skin, crack, roll mark, air bubble, pockmark, scratch respectively; $V_2=\{v_{21}, v_{22}, v_{23}\}$, and v_{21}, v_{22}, v_{23} represent tensile strength, yield strength, elongation respectively. $V_3=\{v_{31}, v_{32}\}$, and v_{31} and v_{32} represent thickness tolerance and width tolerance respectively. V_1 represents surface quality, V_2 represents mechanical properties and V_3 represents dimensional accuracy. The construction of steel strip quality evaluation system is shown in Figure 4.

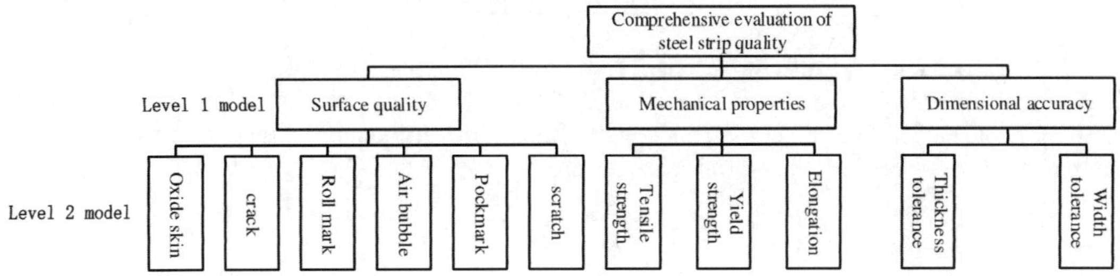

Figure.4 Fuzzy comprehensive evaluation system of strip quality

500 groups of quality data for a variety of strip steel in B Steel Mill are selected, and the data are

analyzed by means of probability statistics to eliminate outliers. The results are shown in Figure 5.

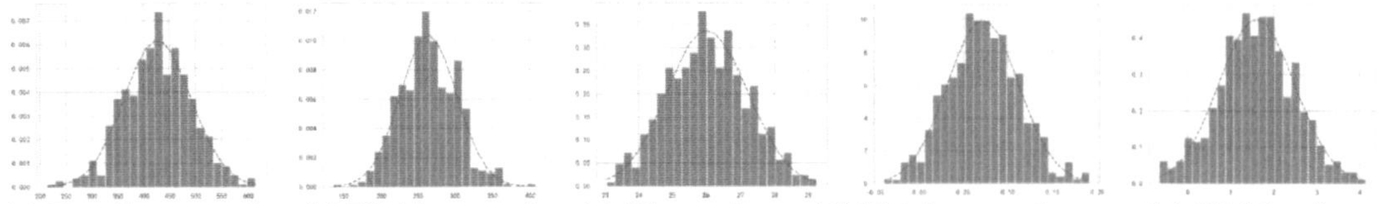

(a) Tensile strength (b) Yield strength (c) Elongation (d) Thickness tolerance (e) Width tolerance
Figure.5 Strip quality data distribution

According to the distribution of data from Figure 5, the optimal interval of indicators can be confirmed and the constraint range can be narrowed so that the values of indicators are more concentrated. The indicator intervals are confirmed by above analysis. The mechanical properties and dimensional accuracy of the degradation degree are calculated by the formula (2) and formula (3). The results are shown in Table 1. The mechanical properties are the larger the better type of data, and dimensional accuracy is the smaller the better type of data.

Table 1 Mechanical properties and dimensional accuracy indicator deterioration degree value

Indicator	Indicator interval	Degradation	
		X_1	X_2
Yield strength /(Mpa)	[195，325]	0.115	0.730
Tensile strength /(Mpa)	[315，540]	0.133	0.711
Elongation /(%)	[24，28]	0.275	0.850
Thickness tolerance /(mm)	[0，±0.15]	0.400	0.800
Width tolerance /(mm)	[0，±3]	0.267	0.600

The description for surface defects is {"none", "slight", "being", "obvious" "very obvious"}. It is quantified to {[0, 0.2], [0.2, 0.4], [0.4, 0.6], [0.6, 0.8], [0.8, 1]}. The degree of surface quality deterioration can be obtained by the degree of surface defects. The results are shown in Table 2:

Table 2 Deterioration degree of surface quality indicator

Indicator	Degradation	
	X_1	X_2
Oxide skin	0	0.68
Crack	0.13	0.43
Roll mark	0	0.60
Air bubble	0	0.70
Pockmark	0.28	0
Scratch	0.30	0.57

(3) Different users have different needs. The first-level weights are confirmed according to user needs. This paper confirms the weights F=[0.45,0.33,0.22] which correspond to the first level indicators, and the judgment matrix J which corresponds to the second level indicator , including air bubble, crack, scratch, oxide skin, pockmark and roll mark according to AHP and expert experience:

$$J = \begin{bmatrix} 1 & 1/2.5 & 2.8 & 2 & 3 & 2.2 \\ 2.5 & 1 & 3.5 & 1.8 & 4 & 3.2 \\ 1/2.8 & 1/3.5 & 1 & 1/3 & 1.7 & 1/1.5 \\ 1/2 & 1/2 & 3 & 1 & 3.5 & 2.9 \\ 1/3 & 1/4 & 1/1.7 & 1/3.5 & 1 & 1/2.3 \\ 1/2.2 & 1/3.2 & 1.5 & 1/2.9 & 2.3 & 1 \end{bmatrix}$$

According to formula (4), the surface quality indicator weight $G = [0.22, 0.32, 0.08, 0.23, 0.06, 0.1]$ will be obtained by calculating and normalizing. According to formula (5), the consistency test indicator is 0.065 which indicates that the judgment matrix has good consistency. According to formula (6), (7), (8), (9), 500 groups of data are analyzed to obtain entropy weights. The weights of yield strength, tensile strength, elongation are $U = [0.331, 0.300, 0.369]$, and the weights of thickness tolerance, width tolerance are $H = [0.489, 0.511]$.

(4) The membership matrix of each sample is calculated according to the degradation degree and the membership function. Suppose R_{ij} is the sample membership, i is the sample number, and j is the jth indicator. The membership matrixes of X_1 are R_{11}, R_{12}, R_{13}, and the membership matrixes of X_2 are R_{21}, R_{22}, R_{23}. And the results are shown below:

$$R_{11} = \begin{bmatrix} 1 & 0 & 0 & 0 \\ 0.946 & 0.054 & 0 & 0 \\ 1 & 0 & 0 & 0 \\ 1 & 0 & 0 & 0 \\ 0.024 & 0.976 & 0 & 0 \\ 0 & 1 & 0 & 0 \end{bmatrix} \quad R_{21} = \begin{bmatrix} 0 & 0 & 1 & 0 \\ 0 & 0.946 & 0.054 & 0 \\ 0 & 0 & 1 & 0 \\ 0 & 0 & 1 & 0 \\ 1 & 0 & 0 & 0 \\ 0 & 0.054 & 0.946 & 0 \end{bmatrix}$$

$$R_{12} = \begin{bmatrix} 0.986 & 0.014 & 0 & 0 \\ 0.934 & 0.066 & 0 & 0 \\ 0.038 & 0.962 & 0 & 0 \end{bmatrix} \quad R_{22} = \begin{bmatrix} 0 & 0 & 0.946 & 0.054 \\ 0 & 0 & 0.993 & 0.007 \\ 0 & 0 & 0.146 & 0.854 \end{bmatrix}$$

$$R_{13} = \begin{bmatrix} 0 & 1 & 0 & 0 \\ 0.066 & 0.934 & 0 & 0 \end{bmatrix} \quad R_{23} = \begin{bmatrix} 0 & 0 & 0.905 & 0.095 \\ 0 & 0 & 1 & 0 \end{bmatrix}$$

(5) The surface quality indicators use the operator determined by main factor because it has a large difference in the degree of influence on quality, while the other indicators use weighted average operator. The fuzzy matrix of secondary indicators is calculated and normalized according to formula (10). The results are shown as follows:

$$B_{11} = G \circ R_{11} = \begin{bmatrix} 0.773 & 0.227 & 0 & 0 \end{bmatrix} \quad B_{21} = G \circ R_{21} = \begin{bmatrix} 0.097 & 0.548 & 0.355 & 0 \end{bmatrix}$$

$$B_{12} = U \circ R_{12} = \begin{bmatrix} 0.599 & 0.401 & 0 & 0 \end{bmatrix} \quad B_{22} = U \circ R_{12} = \begin{bmatrix} 0 & 0 & 0.643 & 0.357 \end{bmatrix}$$

$$B_{13} = H \circ R_{13} = \begin{bmatrix} 0.062 & 0.938 & 0 & 0 \end{bmatrix} \quad B_{23} = H \circ R_{23} = \begin{bmatrix} 0 & 0 & 0.913 & 0.087 \end{bmatrix}$$

The fuzzy matrix of first level indicators is synthesized and normalized to obtain the matrix of evaluation results:

$$B_1 = \begin{bmatrix} 0.520 & 0.480 & 0 & 0 \end{bmatrix} \quad B_2 = \begin{bmatrix} 0.052 & 0.241 & 0.484 & 0.223 \end{bmatrix}$$

The final score of the sample is obtained according to formula (11):

$$S_1 = B_1 \cdot Q = 88 \quad S_2 = B_2 \cdot Q = 53.05$$

(6) Based on the evaluation results, it can be seen that sample X_1 scores 88 and sample X_2 scores 53.05. The comprehensive quality of X_1 is better than X_2. The actual data for the sample are shown in Table 3:

Table 3 Actual data of sample

	Os	Cr	Rm	Ab	Pm	Sc	Ys /(Mpa)	Ts /(Mpa)	El /(%)	Tt /(mm)	Wt /(mm)
X_1	none	slight	none	none	slight	slight	310	510	26.9	0.05	0.80
X_2	obvious	being	obvious	obvious	none	being	230	380	24.6	0.11	1.80

Note: Os(Oxide skin) Cr(Crack) Rm(Roll mark) Ab(Air bubble) Pm(Pockmark) Sc(Scratch) Ys(Yield strength) Ts(Tensile strength) El(Elongation) Tt(Thickness tolerance) Wt(Width tolerance).

Sample X_1 and sample X_2 are both qualified in the original system evaluation, but from the results of this model, it is clear that the comprehensive quality of X_1 is better than X_2, and it is more able to meet the needs of users. In the data, it can be seen that actual quality of the steel strip product is basically consistent with the model evaluation results. 70 groups of data are sampled from qualified products of the same steel strip species. Then they are analyzed by the evaluation model. The results are shown in Figure 6.

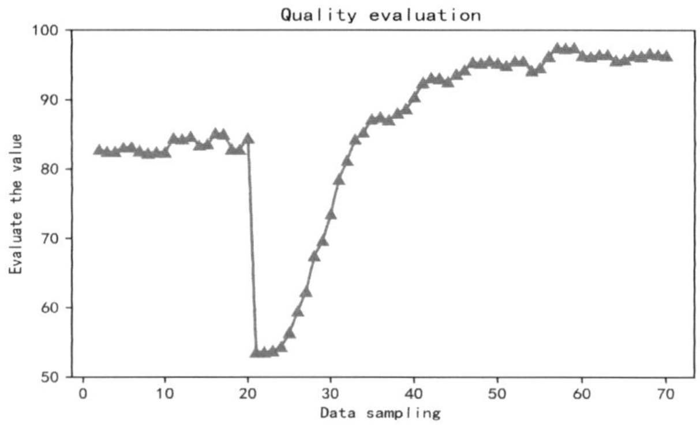

Fig.6 Analysis results of quality evaluation

In Figure.6, it is clear that during the production process, the evaluation value of samples No.1 to No.20 is stable between 80 and 85; the evaluation value of sample No.21 shows obvious quality fluctuations; the evaluation value of samples No.21 to No.31 gradually rises and becomes stable; the evaluation value of samples No.32 to No.70 is stable between 80 and 97. The above analysis indicates that it has a stable production state in the early stage, but the evaluation value of sample No. 21 descends due to the change of working conditions and the fluctuations of production process. It is eventually corrected at sample No.32. With reference to the results of this model, the same variety of steel strip in the next production should take the state of production of samples No.32 to No.70 as standard and avoid the same situation of sample No.21 to No.31 as far as possible.

5 Conclusion

(1) This paper proposes a fuzzy comprehensive evaluation model for the quality of steel strip, which solves the problems of single evaluation indicator, poor comprehensiveness and rough evaluation results. And it also proposes a combination of entropy weight method and AHP to assign weights. In order to obtain evaluation rules, the user requirements are quantified in the form of weight and the experience of field expert is qualified in the form of matrix. It not only improves the rationality and accuracy of the model, but also makes the quality evaluation of steel strip more hierarchical and structured. The experiments show that the evaluation results of the model are consistent with the actual situation and these results can optimize the production process.

(2) During the productive process, quality evaluation is one of the most important aspects of the intelligence of hot-rolled products. It is a necessary requirement to realize quality monitoring, diagnosis and optimization. But in this paper, the model uses offline data and the data sampling period is so long that the real-time capability of quality evaluation results tends to be worse. In the next step, we should

establish the information interaction with the quality database and improve the expert experience. What's more, the evaluation criteria for different steel grades should be enriched and the real-time capability and applicability of the evaluation model should be further improved. The cause of defective steel strip products will be found out according to the evaluation results and the relevant parameters which affect the quality will be corrected to achieve the goal of optimal control of the production process.

Acknowledgments

This work was financially supported by the National Natural Science Foundation of China (61763039).

References

[1] Wang, G, D.(2015) Technology innovation and development direction of iron and steel industry. Iron & Steel., 50(09):1-10.
[2] Wang, Y ,H., Yan, Y, H., Wu, Y, P., Liu, W, W.(2007) Surface quality evaluation method of strip steel based on defect information group decision making. Metallurgical Industry Automation., 05:30-33.
[3] Wu, Y, P., Yan, Y, H., Wang, Y, H., Zhang, Y. (2008) Application of fuzzy complementary matrix to surface quality evaluation of strip steel. Journal of Iron and Steel Research., 08:28-30.
[4] Dong, D, W., Yang, Y, H., Song, K, C., Peng, Y, S. (2013) Attribute recognition theoretical evaluation model of strip steel quality and its applications. China Mechanical Engineering., 23(15):1844-1847+1854.
[5] Ding, J, L., Yang, C, E., Chen, Y, D., Chai, T, Y. (2018) Research progress and prospects of intelligent optimization decision making in complex industrial process. Acta Automatica Sinica., 44(11):1931-1943.
[6] Zhang, D, H., Peng, W., Sun, J., Ding, J, G., Li, X. (2019) Key intelligent technologies of steel strip rolling process. Journal of Iron and Steel Research.,31:174-179.
[7] Xiao, Y, Q., Wang, K, P., He, G, J., Sun, Y, P., Yang, X, Y. (2014) Fuzzy comprehensive evaluation for operating condition of large-scale wind turbines based on trend predication. Proceedings of the CSEE., 34:2132-2139.
[8] Yang, G, J., Jiang, C, H., Gui, W, H., Yang, C, H., Xie, Y, F. (2015) Fuzzy synthesis evaluation method for position state of blast furnace cohesive zone based on entropy weight extension theory., Acta Automatica Sinica., 41:75-83.
[9] Zhu, X, R., Liu, Y, W. (2020) Fuzzy integrated judgment based photovoltaic array condition assessment. Acta Energiae Solaris Sinica., 41:103-111.
[10] Mithal, J.I., Cai, Q, Z., Zhou, W., Lu, X. (2020) Assessment of risks impeding sustainable mining in Pakistan using fuzzy synthetic evaluation. Resources Policy., 69:
[11] Cerrada, M., Chuan, L., Sánchez, R.V., Pacheco, F., Cabrera, D. (2018) A fuzzy transition based approach for fault severity prediction in helical gearboxes. Fuzzy Sets and Systems., 337:
[12] An, J., Xu, Y, X., Zeng, X., Li, Z, Q., Zhu, G, F. (2018) Equipment quality condition assessment under fusion information based on combination weighting. Control and Decision., 33:1693-1698.
[13] Dai, X, J., Tian, W. (2011) Choice of determining weights method in fuzzy synthetic evaluating model for water quality. Journal of the Graduate School of the Chinese Academy of Sciences., 28:169-176.
[14] Wang, Z., Xing, X, G., Yan, F. (2021) An abnormal phenomenon in entropy weight method in the dynamic evaluation of water quality indicator. Ecological Indicators., 131:
[15] Zhang, S., Zang, M., Chi, G, T. (2010) The Science and Technology Evaluation Model Based on Entropy Weight and Empirical Research. Chinese Journal of Management.,7:34-42.
[16] Li, L., Yuan, X, F., Wang, Y, L., Sun, B., Wu, D, Z. (2019) A two-layer fuzzy synthetic strategy for operational performance assessment of an industrial hydrocracking process. Control Engineering Practice., 93:

Hyperspectral image classification based on extrema morphological profiles and total variation

Tao Ma*, Weijian Huang, Chun Wang, Shuangquan Ge, Yunfei Liu

Institute of Computer Application, China Academy of Engineer Physics, Mianyang, Sichuan, 621900, China

*Corresponding author's e-mail: whumatao@163.com

Abstract. Hyperspectral image (HSI) contains spectral information and spatial information of a scene. Because of the unique spectral characteristics of materials, HSI has been proved to be particularly useful for distinguishing the sort of materials in the scene. In this paper, a HSI classification method is proposed. Firstly, we propose a new extrema morphological profiles (EMPs) to extract the spatial feature in HSI. Secondly, total variation (TV) is used to integrate the spatial feature into a fused feature map. Finally, support vector machine (SVM) is utilized to accurately classify the fused feature maps. In order to evaluate the proposed method, we take quantitative classification analyses on a HSI dataset, class accuracy (CA), overall accuracy (OA), average accuracy (AA), and Kappa coefficient are adopted as the metrics. The experimental results indicate that the proposed method can efficiently achieve high classification accuracy.

1. Introduction

Remote sensing image classification plays a fundamental role in many applications such as pattern recognition, agricultural monitoring, 3D reconstruction etc. HSI is suited to obtain accurate scene classification by combining spectral information and spatial information. Contiguous spectra can be produced by a hyperspectral sensor. Different from multispectral images, hyperspectral photographs provide a great deal of spectral information about a scene [1].

Due to the complementarity and redundancy between spectral bands of HSI, it is very important to make comprehensive use of useful information and eliminate redundancy. Because HSI is usually quite huge, analyzing it can be difficult [2]. To this end, many scholars focus their research on the feature extraction and feature fusion based on HSI. The purpose of feature extraction is to extract the features that can best express the scene structure from HSI, such as principal component analysis (PCA) [3], morphological profiles (MPs) [4], morphological attribute profiles (APs) [5] etc. The purpose of feature fusion is to combine the complementary information and reduce redundant information to one fused result [6-7].

In this work, a remote image classification method for HSI is proposed. This method consists of three main steps: feature extraction, feature fusion, and feature classification. In the first feature extraction step, a new structure feature called extrema morphological profiles (EMPs) is proposed to extract spatial and spectral information from HSI. In the second feature fusion step, a total variation (TV) model is used to fuse the extracted features from HSI. TV model is able to retain the spatial structures and keep piece-wise smoothness [8]. In the third feature classification step, support vector machine (SVM) is adopted to find accurate classification.

2. Method

In this section, we introduce the proposed HSI classification framework in detail. First, EMPs are constructed by applying several extrema morphological filters to each spectral channel. Second, a total variation strategy is used to fusion the EMPs of all spectrum channels. Finally, the SVM is used to process the fused EMPs to find the accurate classification.

2.1. Extrema morphological profiles (EMPs)

EMPs are modifications of MPs, which are widely used in remote sensing image classification tasks. EMPs produce a multilevel characterization of an image by applying morphological attribute filters in a sequential order that can be used to characterize various types of structural information. To extract context and geographic information from the input HSI, EMPs can be built via a series of thinning and thickening transformations with gradually increasing threshold values. Min-tree and max-tree provide thickening and thinning, respectively. Let f denotes the input grayscale image, the EMPs can be obtained as follows.

$$\begin{aligned} EMP(f) &= \left\{\tilde{\Pi}\phi(f), f, \Pi\gamma(f)\right\} \\ &= \left\{\phi^{b(n)}(f), \cdots, \phi^{b(1)}(f), f, \gamma^{b(1)}(f), \cdots, \gamma^{b(n)}(f)\right\} \end{aligned} \tag{1}$$

where $\Pi\phi(f)$ denotes the extremal morphology open profile, $\tilde{\Pi}\phi(f)$ is inverse order of the extremal morphological opening profile, and $\Pi\gamma(f)$ represents the extremal morphology close profile.

We perform morphological open and close reconstructions on image f to obtain a series of open and close reconstructed images. Multiple structural elements of the same shape and increasing size and extrema marker images are used in morphological reconstruction. Arranging these open reconstructed images and close reconstructed images in order to obtain extrema morphological open profiles and close profiles. The extrema morphological open profile can be defined as follows.

$$\Pi\phi(f) = \left\{\Pi\phi^{b}(f) : \phi^{b(i)}(f) = O^{R}_{b(i)}(g_{\max} \mid f), \forall i \in [1, \cdots n]\right\} \tag{2}$$

where $\{b(i) \mid i \in [1, \cdots n]\}$ represents the structure elements with the same shape and increasing size. $\phi^{b(i)}(f)$ is the morphological open reconstructed images and $O^{R}_{b(i)}(g_{\max} \mid f)$ is the morphological open reconstruction function. The maximum marker image $g_{\max}$ can be obtained by:

$$g_{\max}(k) = \begin{cases} f(k), & \text{if } O^{R}(f\Theta b \mid f)(k) = [f\Theta b](k), \\ & \text{and } [f\Theta b](k) \text{ is maximum.} \\ 0 & \text{others.} \end{cases} \tag{3}$$

where Θ represents the erosion operator. The extrema morphological close profile can be defined as follows.

$$\Pi\gamma(f) = \left\{\Pi\gamma^{b}(f) : \gamma^{b(i)}(f) = C^{R}_{b(i)}(g_{\min} \mid f), \forall i \in [1, \cdots n]\right\} \tag{4}$$

where $\gamma^{b(i)}(f)$ is the morphological close reconstructed images and $C^{R}_{b(i)}(g_{\min} \mid f)$ is the morphological close reconstruction function. The minimum marker image $g_{\min}$ can be obtained by:

$$g_{\min}(k)=\begin{cases} f(k), & \text{if } C^R(f\oplus b\mid f)(k)=[f\oplus b](k), \\ & \text{and } [f\oplus b](k) \text{ is minimum.} \\ \max(f), & \text{others.}\end{cases} \tag{5}$$

where $\oplus$ denotes dilation operator, and $\max(f)$ represents maximum of the input image.

2.2. Feature fusion using total variation

The EMPs of different channels of HSI retain the complementary information of the scene, and this complementary information can be synthesized through the fusion of feature images, while eliminating information redundancy. The feature map fusion can be modeled as follows.

$$A = FV + N \tag{6}$$

where $A=\left[\phi_{C1}^{b(i)}(f),\phi_{C2}^{b(i)}(f),\cdots\cdots,\phi_{Cn}^{b(i)}(f)\right]$ denotes the input multichannel feature maps, F is the output single channel feature map, V denotes the unknown coefficient matrix, and N indicates the deviation. In order to solve the output and constrain the fused image to retain the complementary information from the input image as much as possible, we express the fusion model as a total variation optimization model. as follows.

$$F^* = \arg\min_{F,V}\frac{1}{2}\|A-FV\|^2+\lambda\sum_{i=1}^{r}\left\|\sqrt{(\nabla_h F)^2+(\nabla_v F)^2}\right\|_1 \quad s.t.\ \mathrm{VV}^T=I_r \tag{7}$$

The final EMPs can be obtained by combing the grouped fused feature maps in sequence.

$$EMP_F(f)=\left\{\phi_F^{b(n)}(f),\cdots,\phi_F^{b(1)}(f),f_F,\gamma_F^{b(1)}(f),\cdots,\gamma_F^{b(n)}(f)\right\} \tag{8}$$

where $\phi_F^{b(i)}(f)$, $\gamma_F^{b(i)}(f)$, and f_F denote the fused results of open reconstruction feature maps, close reconstruction feature maps, and original images, respectively.

2.3. Feature maps classification

To find the accurate classification result, the fused EMPs are fed into a SVM framework. The classification step can be formulated as follows.

$$f_{class} = SVM(EMPs) \tag{9}$$

3. Experiments

3.1. Datasets and settings

To evaluate the proposed classification method, we conduct both qualitative and quantitative tests on a HSI dataset called Houston Dataset. This dataset is avaliable to access at http://www.grss-ieee.org/community/technical-committees/data-fusion/2013-ieee-grss-data-fusion-contest/. Some samples of the dataset are shown in Figure 1, from top to bottom: a color composite representation of the HSI, training samples, and test samples. This dataset has 144 spectral bands ranging from 0.38μm to 1.05μm. Each band of the HSI has dimensions 349×1905 with the spatial resolution of 2.5m. All methods are implemented with the MATLAB codes on a desktop with 16 GB memory and 3.6 GHz Intel Core i5 CPU.

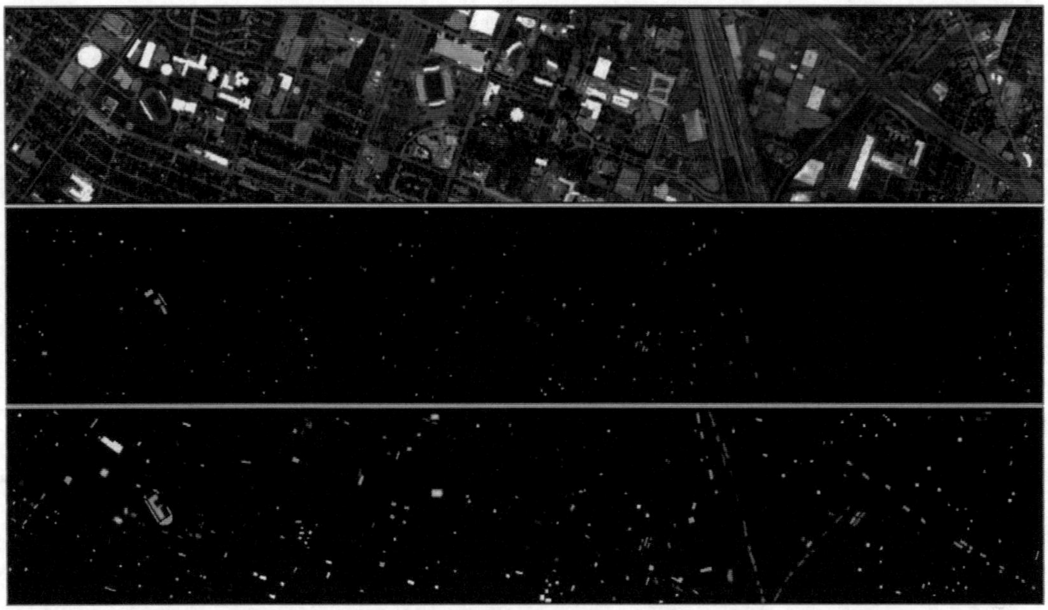

Figure 1. Houston dataset with HSI, training samples, and test samples.

In order to evaluate the classification performance of the proposed method more accurately, we use three metrics, i.e., OA, AA, and Kappa coefficient, to validate the classification results, they are defined as follows.

Overall accuracy (OA) is defined as the proportion of the sum of the correct classification numbers of all categories to the whole test set.

$$OA = \sum_{i=1}^{C} R_{ii} / N_{test} \tag{10}$$

where R_{ii}, denotes the number that class i can be correctly classified into class i. N_{test} represents the total number of samples in the test set.

Average accuracy (AA) is defined as the mean of the proportion that all categories are correctly classified.

$$AA = \left(\sum_{i=1}^{C} \left(R_{ii} / N_i \right) \right) / C \tag{11}$$

where N_i represents the number of samples of class i in the test set.

Kappa coefficient comprehensively considers the instability of classification results and reflects the error of classification results, which can be defined as follows.

$$Kappa = \left(N_{test} \sum_{i=1}^{C} R_{ii} - \sum_{i=1}^{C} \left(R_{i+} \bullet R_{+i} \right) \right) / \left(N_{test}^2 - \sum_{i=1}^{C} \left(R_{i+} \bullet R_{+i} \right) \right) \tag{12}$$

where R_{i+} denotes the sum of samples that misclassify the class i into other classes. R_{+i} denotes the sum of samples that misclassify other classes into class i.

3.2. Classification experiments

The classification results are displayed in Table 1 and Table 2. The detailed CA is shown in Table 1 to directly compared the performance of proposed EMPs . OA, AA, and Kappa coefficient are shown in

Table 2 to compared the statistical results. For all metrics, the large value indicate the good result and the best values are marked in bold.

From the qualitative comparison of CA shown in Table 1, it can be seen that the proposed method can achieve the best result by getting the largest values in 11 classes. The HSI PCA achieves the second best result by getting the largest values in 5 classes.

Table 1. Quantitative Comparison of Class Accuracy.

No. of class	Name of class	HSI	HSI PCA	Proposed EMPs
1	Healthy grass	0.762	0.673	**0.824**
2	Stressed grass	0.739	0.776	**0.939**
3	Synthetic grass	0.996	**1.000**	0.998
4	Trees	0.960	0.861	**0.989**
5	Soil	0.765	0.991	**0.999**
6	Water	0.881	**0.993**	0.951
7	Residential	0.844	0.811	**0.935**
8	Commercial	0.441	0.521	**0.955**
9	Road	0.479	0.731	**0.891**
10	Highway	0.447	0.645	**0.700**
11	Railway	0.600	0.895	**0.896**
12	Parking Lot 1	**0.885**	0.820	0.830
13	Parking Lot 2	0.459	**0.726**	0.621
14	Tennis Court	0.749	**1.000**	**1.000**
15	Running Track	0.861	**1.000**	**1.000**

Qualitative comparison of OA, AA, and Kappa coefficient shown in Table 2 demonstrates that our propose EMPs can obtain the best values on all metrics. These experimental results show that the proposed method is efficient and outperforms to the other state-of-the-art methods.

Table 2. Quantitative Comparison of OA, AA, and Kappa Coefficient.

	HSI	HSI PCA	Proposed EMPs
OA	65.182	79.673	**90.212**
AA	67.943	82.938	**90.198**
Kappa	0.626	0.774	**0.895**

4. Conclusion

In this paper, a HSI classification method based on EMPs and total variation is proposed. Firstly, the new proposed EMPs is utilized to extract the spatial feature from HSI. And then the total variation is used to formulate the feature map fusion problem, which is expressed as a non-convex optimization problem. Finally, the fused EMPs are fed into a SVM classifier to obtain the accurate classification. The quantitative comparisons on Houston HSI dataset is performed to evaluate the proposed method. The metrics are CA, OA, AA, and Kappa coefficient. The experimental results show that the proposed method is superior to the other method by achieving good classification accuracy. In the future work, the EMPs is extended to deal with the remote image classification for HSI and LiDAR data.

Acknowledgments
This work was supported by Foundation of Director of Institute of Computer Application under Grant SJ2021A05, and Fostering Program of Innovation Foundation of China Academy of Engineering Physics under Grant PY2021159.

References

[1] Rasti, B., Ghamisi, P., Gloaguen, R. (2017) Hyperspectral and LiDAR Fusion Using Extinction Profiles and Total Variation Component Analysis. IEEE Transactions on Geoscience and Remote Sensing, 55 (7): 3997-4007.

[2] Rasti, B., Sveinsson, J.R., Ulfarsson, M.O. (2014) Total variation based hyperspectral feature extraction, In: International Geoscience and Remote Sensing Symposium (IGARSS), pp.4644-4647.

[3] Green, A.A., Berman, M., Switzer, P., Craig, M.D. (1988) A transformation for ordering multispectral data in termsof image quality with implications for noise removal. IEEE Transactions on Geoscience and Remote Sensing, 26 (1): 65-74.

[4] Pesaresi, M., Benediktsson, J.A. (2001) A new approach for the morphological segmentation of high-resolution satellite imagery. IEEE Transactions on Geoscience and Remote Sensing, 39 (2): 309-320.

[5] Liao, W., Dalla Mura, M., Chanussot, J., Bellens, R., Philips, W. (2016) Morphological attribute profiles with partial reconstruction. IEEE Transactions on Geoscience and Remote Sensing, 54 (3): 1738-1756.

[6] Ma, T., Ma, J., Fang, B., Hu, F., Quan, S., Du, H. (2018) Multi-scale decomposition based fusion of infrared and visible image via total variation and saliency analysis. Infrared Physics and Technology, 92: 154-162.

[7] Ma, T., Ma, J., Yu, K., Zhang, J., Fu, W. (2021) Multispectral Remote Sensing Image Matching via Image Transfer by Regularized Conditional Generative Adversarial Networks and Local Feature. IEEE Geoscience and Remote Sensing Letters, 18 (2): 351-355.

[8] Fang, L., He, N., Li, S., Ghamisi, P., Benediktsson, J. A. (2018) Extinction Profiles Fusion for Hyperspectral Images Classification. IEEE Transactions on Geoscience and Remote Sensing, 56 (3): 1803-1815.

Dual-domain unsupervised network for removing motion artifact related to Gadoxetic acid-enhanced MRI

Chongxin Bai[1], Kewen Liu[1], Shi Chen[1], Zhao Li[2], Weida Xie[1], Qingjia Bao[3]* and Chaoyang Liu[3]

[1] Information and Communication Engineering, Wuhan University of Technology, Wuhan 430070, China

[2] The 38th Research Institute of China Electronics Technology Group Corporation, Hefei 230088, China

[3] State Key Laboratory of Magnetic Resonance and Atomic and Molecular Physics, Wuhan Centre for Magnetic Resonance, Innovation Academy for Precision Measurement Science and Technology, Wuhan 430071, China

*Corresponding author's e-mail: qingjia.bao@apm.ac.cn

Abstract—The transient severe motion may cause severe image degradation during gadoxetic acid-enhanced arterial phase imaging. This work proposes a new dual-domain unsupervised motion artifacts disentanglement network for motion correction related to gadoxetic acid-enhanced MRI. We assume that motion-free images and motion-corrupted images belong to the different domains, then the motion correction is converted to the image-to-image translation problem. The image-to-image translation within the same domain is designed to constrain autoencoders to learn the feature representation. And the cross-domain translation explores the cycle consistency in the absence of paired images. Experimental results demonstrate that our method can effectively reduce artifacts in the gadoxetic acid-enhanced images.

1. Introduction

Gadoxetic acid is a useful liver-specific magnetic resonance (MR) imaging contrast agent for liver metastasis[1], hepatocellular carcinoma[2], and other diseases[3]. However, despite the advantages of gadoxetic acid in MR imaging, previous reports[4][5] revealed a high incidence of severe image degradation during gadoxetic acid-enhanced arterial phase imaging. Therefore, a proper motion correction algorithm is essential to improve the quality of gadoxetic acid-enhanced arterial phase images.

Many prospective and retrospective strategies have been proposed to remove the motion artifacts. Prospective techniques perform a real-time update of the image acquisition by utilizing reacquired images from dedicated navigator scans[6]. However, they are subtle in liver MRI as non-rigid motion is very difficult to track. On the other hand, the retrospective motion correction method modifies the k-space or image data after the acquisition. These methods can be attractive, but they typically lead to a poorly conditioned and nonconvex optimization problem. Many researchers have also demonstrated the great potential of deep learning for MRI motion correction. Most of them are based on the supervised manner, which is not feasible for gadoxetic acid-enhanced images as the ground truth data is either extremely challenging to obtain or simply not available.

This work introduces a dual-domain unsupervised motion artifacts disentanglement network (DUMAD-Net) for the liver's gadoxetic acid-enhanced arterial phase images without any paired datasets. The motion correction is converted to the image-to-image translation problem, and experimental results demonstrate that our method can effectively reduce artifacts in the gadoxetic acid-enhanced images.

2. Related Work

Many deep learning methods can learn the mapping from motion-corrupted data to motion-free data. Armanious et al.[7] reported a feasibility study to correct motion artifacts by implementing a conditional generative adversarial network (MedGAN). Tamada et al.[8] introduced a multi-channel network (MARC) to reduce the motion artifacts caused by respiratory motion. Their method can work well with the artifact without changing the contrast ratios of the images. At the same time, several unsupervised methods have also been proposed to overcome the lack of paired data. Armanious et al. proposed Cycle-MedGAN[9] and Cycle-MedGAN V2.0[10], and they mainly focused on reducing rigid artifacts in brain MRI and achieved promising results. Ghodrati et al.[11] presented a framework based on the adversarial autoencoder for respiratory motion correction. Their method can effectively reduce blurring and artifacts associated with respiratory motion.

3. Proposed Method

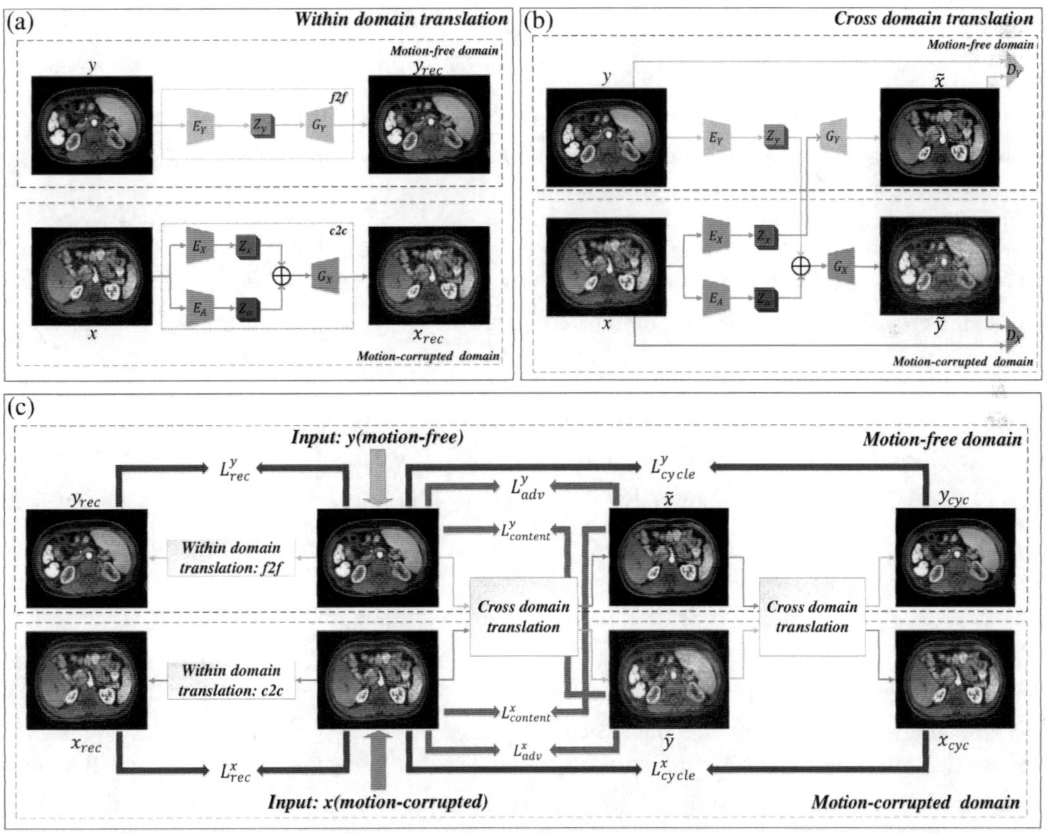

Fig 1.The proposed framework.(a)Within domain translation is used to learn the feature representation. (b)Cross-domain translation is designed to explore the cycle consistency in the absence of paired data. (c)The whole process. We can get the motion-corrected images by the cross-domain translation.

The overall architecture of the dual-domain unsupervised motion artifacts disentanglement network (DUMAD-Net) is shown in Fig 1. In the training stage, the inputs of DUMAD-Net are the unpaired images as shown in Fig 1(c). And in the test stage, only the motion-corrupted images are set as the input, and the translated motion-free images can be obtained with the cross-domain translation.

3.1 Within domain translation

The main function of within domain translation is to learn the content representation in the individual domain. For the motion-free images y, we use a content encoder E_Y to extract the most representative information z_y. Then the decoder G_Y is applied to reconstruct the motion-free images back with the z_y. This translation (f2f) is illustrated in the up part of Fig 1(a), and it can be described as follows:

$$Encoding : z_y = E_Y(y), Decoding : y_{rec} = G_Y(z_y) \quad (1)$$

where z_y is the extracted content information, and y_{rec} is the reconstructed motion-free image.

For the motion-corrupted images x, we assume that they are composed of a content component and a motion-artifact component. Hence, two kinds of encoders are employed to disentangle these two components. The first kind of encoder is the content encoder E_X which captures anatomical structures. And the second one is the artifact encoder E_A which captures artifacts. Then the decoder G_X ensembles the extracted anatomical content z_x and artifact features z_a from both encoders to reconstruct the motion-corrupted images back (c2c, as shown in the down part of Fig 1(a)). It can be described as follows:

$$Encoding : z_x = E_X(x), z_a = E_A(x), Decoding : x_{rec} = G_X(z_x \oplus z_a) \quad (2)$$

where z_x is the extracted content information, z_a is the artifact feature, x_{rec} is the reconstructed motion-corrupted image, and $\oplus$ represents the operation of channel concatenate.

3.2 Cross-domain translation

We propose cross-domain translation to explore the cycle consistency in the absence of paired images, as shown in Fig 1(b). For the motion-corrupted images x, the decoder G_Y rather than G_X is applied to generate the cross-domain translated motion-free image $\tilde{x}$ with z_x. It can be described as follows:

$$Encoding : z_x = E_X(x), Decoding : \tilde{x} = G_Y(z_x) \quad (3)$$

where $\tilde{x}$ is the cross-domain translated motion-free images of the original motion-corrupted images x.

For the cross-domain translation of motion-free images y, we use a content encoder E_Y to extract the content information z_y. Then we ensemble this anatomical content information z_y and artifact features z_a extracted from motion-corrupted images x to construct the translated motion-corrupted image $\tilde{y}$ via the decoder G_X. And it can be described as follows:

$$Encoding : z_y = E_Y(y), z_a = E_A(x), Decoding : \tilde{y} = G_X(z_y \oplus z_a) \quad (4)$$

where $\tilde{y}$ is the cross-domain translated motion-corrupted images of the original motion-free images y.

Meanwhile, to clear the domain boundaries between the motion-corrupted domain and motion-free domain, two discriminators: motion-corrupted domain discriminator D_X and motion-free domain discriminator D_Y are added to distinguish between real and fake images in the respective domain.

3.3 Loss function

The total loss function of DUMAD-Net contains four parts: the domain adversarial loss L_{adv}, the reconstruction loss L_{rec}, the cycle-consistency loss L_{cycle} and the proposed content consistency loss $L_{content}$:

$$L = \lambda_1 L_{adv} + \lambda_2 L_{rec} + \lambda_3 L_{cycle} + \lambda_4 L_{content} \quad (5)$$

where λ_1 , λ_2 , λ_3 and λ_4 are the loss weights used for controlling the contributions of the terms.

The first term L_{adv} in formula 5 is the domain adversarial loss to clear the domain boundaries:

$$L_{adv} = E\ \log D_X(x) + E\ \log(1 - D_X(\tilde{y})) + E\ \log D_Y(y) + E\ \log(1 - D_Y(\tilde{x})) \quad (6)$$

where E denotes the expectation operator.

The second term L_{rec} is the reconstruction loss in the within domain translation to ensure the consistency between an image and its reconstructed counterpart in the same domain:

$$L_{rec} = \mathrm{E}[\| x - x_{rec} \|_1] + \mathrm{E}[\| y - y_{rec} \|_1] \quad (7)$$

To deal with unpaired image data, we introduce the popular cycle consistency loss to calculate the difference between the cycle images and original images:

$$\begin{aligned} x_{cyc} &= G_X(E_Y(\tilde{x}) \oplus E_A(\tilde{y})), y_{cyc} = G_Y(E_X(\tilde{y})) \\ L_{cycle} &= \mathrm{E}[\| x - x_{cyc} \|_1] + \mathrm{E}[\| y - y_{cyc} \|_1] \end{aligned} \quad (8)$$

Due to the lack of effective supervised information for translated images, only relying on feature domain discriminators constraints would lead to domain shift problems inevitably in generated images. Furthermore, it is a fact that the motion artifact is always on the phase-encoding direction and the magnitude of artifacts is limited compared to the original anatomical content in the images. Specifically, although the paired motion-free image and motion-corrupted image are quite different due to the artifacts, the summations of the pixel value in every column along the phase-encoding direction are still very close. Therefore, we propose a content consistency module to penalize the discrepancy of summations in each column along the phase-encoding direction between the original images and their cross-domain translated images:

$$L_{content} = \sum_{i=1}^{n} (\| sum_i(x) - sum_i(\tilde{x}) \|_1 + \| sum_i(y) - sum_i(\tilde{y}) \|_1) \quad (9)$$

where n is the number of the column along the phase-encoding direction, and $sum_i(\bullet)$ is the summation of the pixel value in the ith column.

3.4 Experimental data acquisition

All the gadoxetic acid-enhanced arterial phase images are acquired from Wuhan Union Hospital. The institutional review board approved the study and each subject provided written informed consent. 150 subjects are collected in total, including 30 subjects without motion artifacts, 60 subjects with mild motion artifacts and 60 subjects with severe motion artifacts. To evaluate the motion correction results quantitatively, the simulation was performed to obtain the pairs of motion-free and motion-corrupted images. Similar to [8], the artifact simulation is implemented by adding the phase error in k-space lines.

4. Results

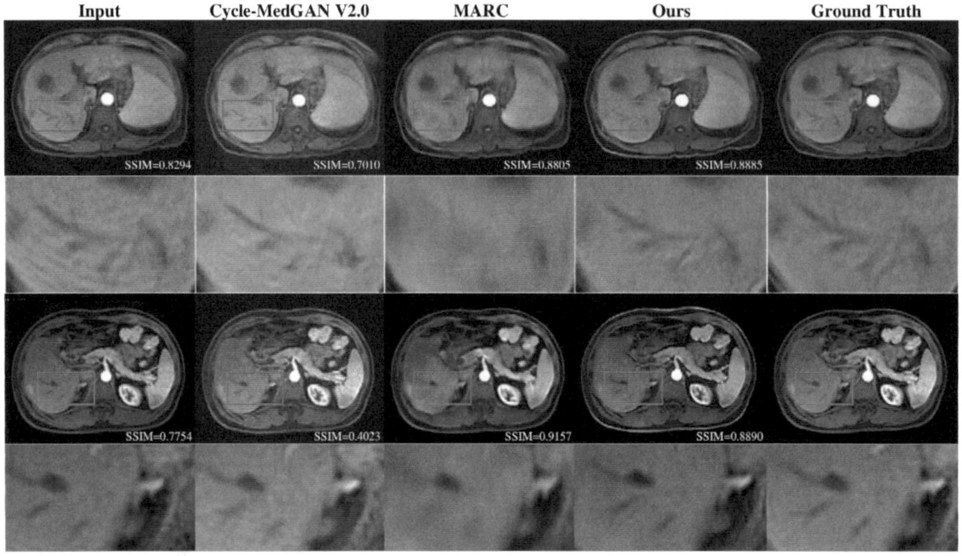

Fig 2. Motion artifacts correction results of simulated motion artifact. Row 1 and row 3 show the different slices and row 2 and row 4 are the corresponding local enlarged drawings.

Fig 2 shows the results with different correction methods for the simulation motion artifact of gadoxetic acid-enhanced arterial phase MRI. Our method successfully corrects motion artifacts and shows similar quantitative metric values as the supervised method MARC.

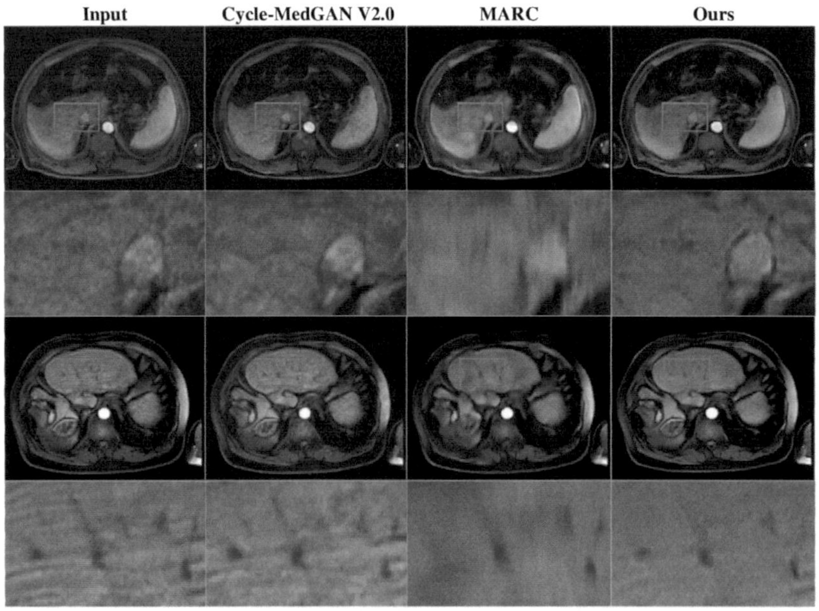

Fig 3. Motion artifacts correction results of in-vivo data. Row 1 and row 3 show the different slices. Row 2 and row 4 are the corresponding local enlarged drawings of the area of the red rectangle.

Fig 3 shows the motion correction results for in-vivo data. Especially, we tested real artifacts images on supervised method MARC that trained on paired simulated artifacts images, while the other two methods were trained on shuffled data. We can notice that DUMAD-Net can reduce motion artifacts effectively in liver MR images and keep the anatomical details. We also observed the 3D voxel of the same subject from different axial positions after motion correction, and Fig 4 shows the results. It is clear that our method achieved both motion artifact correction and restoration of high frequency details.

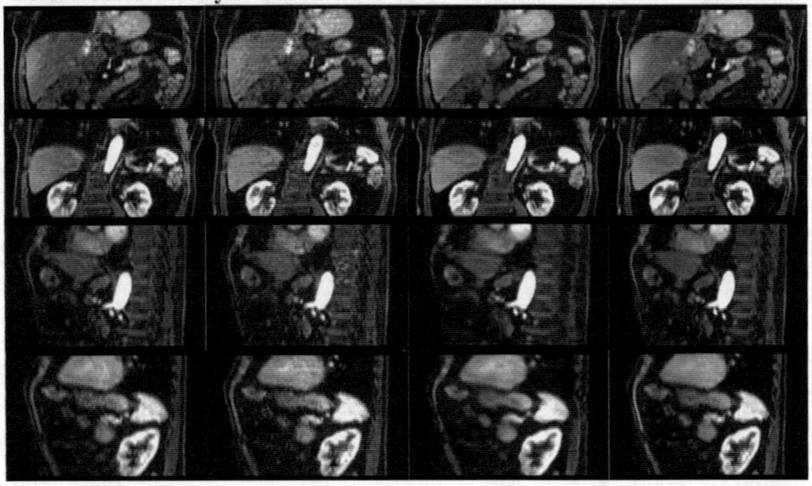

Fig. 4. The corrected results were observed from different axial positions. Row 1 and Row 2 are in the coronal view, and Row 3 and Row 4 are in the sagittal view.

5. Conclusion

In this paper, we proposed a dual-domain unsupervised motion artifacts disentanglement network (DUMAD-Net) for the liver's gadoxetic acid-enhanced arterial phase images without any paired datasets. The motion correction is converted to the image-to-image translation problem, and within domain and cross-domain translation are explored to constrain autoencoders to learn the feature representation and explore the cycle consistency in the absence of paired images, respectively. Moreover, the motion-corrupted images are disentangled into content and artifact space by corresponding encoders, and then the decoder is applied to obtain the motion-free image with the extracted content features. Both simulation and in vivo experiments demonstrate that our method can effectively reduce artifacts in the gadoxetic acid-enhanced arterial phase images, possessing a potential for facilitating the radiologists to perform a more accurate diagnosis.

Acknowledgments

No funding was received for conducting this study. The authors have no relevant financial or non-financial interests to disclose.

References

[1] Armbruster, M., Zech, C.J., Sourbron, S., Ceelen, F., Auernhammer, C.J., Rist, C., Haug, A., Singnurkar, A., Reiser, M.F., Sommer, W.H. (2014) Diagnostic accuracy of dynamic gadoxetic-acid–enhanced MRI and PET/CT compared in patients with liver metastases from neuroendocrine neoplasms. Journal of Magnetic Resonance Imaging, 40: 457–466.

[2] Kim, Y.Y., Park, M.S., Aljoqiman, K.S., Choi, J.Y., Kim, M.J. (2019) Gadoxetic acid-enhanced magnetic resonance imaging: Hepatocellular carcinoma and mimickers. Clinical and Molecular Hepatology, 25: 223–233.

[3] Koh, J., Chung, Y.E., Nahm, J.H., Kim, H.Y., Kim, K.S., Park, Y.N., Kim, M.J., Choi, J.Y. (2016) Intrahepatic mass-forming cholangiocarcinoma: prognostic value of preoperative gadoxetic acid-enhanced MRI. European Radiology, 26: 407–416.

[4] Davenport, M.S., Viglianti, B.L., Al-Hawary, M.M., Caoili, E.M., Kaza, R.K., Liu, P.S.C., Maturen, K.E., Chenevert, T.L., Hussain, H.K. (2013) Comparison of Acute Transient Dyspnea after Intravenous Administration of Gadoxetate Disodium and Gadobenate Dimeglumine: Effect on Arterial Phase Image Quality. Radiology, 266: 452–461.
[5] Motosugi, U., Bannas, P., Bookwalter, C.A., Sano, K., Reeder, S.B. (2016) An Investigation of Transient Severe Motion Related to Gadoxetic Acid-enhanced MR Imaging. Radiology, 279: 93–102.
[6] Ehman, R.L., Felmlee, J.P. (1989) Adaptive technique for high-definition MR imaging of moving structures. Radiology, 173: 255–263.
[7] Armanious, K., Jiang, C.M., Fischer, M., Kustner, T., Nikolaou, K., Gatidis, S., Yang, B. (2020) MedGAN: Medical image translation using GANs. Computerized Medical Imaging and Graphics, 79: 101684.
[8] Tamada, D., Kromrey, M.L., Ichikawa, S., Onishi, H., Motosugi, U. (2020) Motion Artifact Reduction Using a Convolutional Neural Network for Dynamic Contrast Enhanced MR Imaging of the Liver. Magnetic Resonance in Medical Sciences, 19 (1): 64-76.
[9] Armanious, K., Jiang, C.M., Abdulatif S., Kustner, T., Gatidis, S., Yang, B. (2019) Unsupervised Medical Image Translation Using Cycle-MedGAN. In: 27th European Signal Processing Conference (EUSIPCO). A Coruna.
[10] Armanious, K., Tanwar, A., Abdulatif S., Kustner, T., Gatidis, S., Yang, B. (2020) Unsupervised adversarial correction of rigid MR motion artifacts. In: IEEE 17th International Symposium on Biomedical Imaging (ISBI). Iowa. pp. 1494-1498.
[11] Ghodrati, V., Bydder, M., Ali, F., Gao, C., Prosper, A., Nguyen, K.L., Hu, P. (2021) Retrospective respiratory motion correction in cardiac cine MRI reconstruction using adversarial autoencoder and unsupervised learning. NMR in Biomedicine, 34(2): e4433.

An Effective Approach for Sonar Image Recognition with Improved Efficientdet and Ensemble Learning

Yunben Du[1a*], Wei Lin[2], Weibo Zhong[3b*], Yu Yuan[4]

[1]College of Electronics and Information, Jiangsu University of Science and Technology, Zhenjiang 212000,China

[2]College of Electronics and Information, Jiangsu University of Science and Technology, Zhenjiang 212000,China

[3]College of Electronics and Information, Jiangsu University of Science and Technology, Zhenjiang 212000,China

[4]College of Electronics and Information, Jiangsu University of Science and Technology, Zhenjiang 212000,China

[a*]202030025@stu.just.edu.cn, [b*]vebopost@sohu.com

Abstract: Precisely recognition of underwater targets plays a critical role in the field of underwater unmanned exploration. According to the characteristics of sonar image, in order to improve the accuracy of underwater automatic target recognition, this paper proposes an underwater target sonar image recognition method combined with improved Efficientdet and ensemble learning. According to the regional dominance principle and threshold segmentation method, the sonar image is preprocessed, and preliminary effective features extracted from preprocessed data by Efficientnet backbone, so as to balance the efficiency and power consumption of the algorithm. The feature fusion adopts the improved weighted bi-directional feature pyramid (BiFPN) structure to aggregate the global information and strengthen the feature representation of the shallow feature map, which optimizes the problem that the small feature extraction is not ideal due to the low resolution of the sonar image and less detail information. By building a two-level classifier to complete the learning task in an ensemble way, a better classification effect is obtained than a single classifier. This method is applied to the underwater intelligent perception group of the 10th National Marine Aircraft Design and Production Competition in 2021, and won the first prize in the country. The experimental results show that the designed ensemble network model is more accurate than the conventional convolution neural network in identifying underwater targets on the measured sonar image data set while ensuring a certain prediction speed.

1. Introduction

Sonar image recognition is one of the basic and challenging research issues in the field of underwater intelligent perception. Sonar image has a very wide range of applications in marine surveying and mapping and exploration, seabed rescue, marine warfare, and exploration of oil and gas. In recent years, deep learning has achieved great success in the field of optical image recognition, while there are few studies on the high-precision recognition of sonar images based on deep learning methods in China and abroad. Sonar images usually have poor quality, low resolution, small samples and other shortcomings[1]. In 2019, Liu Weber established a target recognition model of deep learning underwater sonar image based on YOLOv3 network structure, and analyzed the performance of the model. The average recognition rate of five kinds of targets is 0.79[2]. Zhu Zhaotong et al. proposed a sonar image recognition method based on fine-tuning convolutional neural network (CNN), and obtained a relatively high starting point for training CNN by fine-tuning the pre-trained CNN[3]. For the overall target detection and recognition model, the average recognition rate of the five types of targets is 0.79. In 2020, Jin et al. proposed the recognition method of target sonar image based on convolution neural network. According to the characteristics of sonar image, an underwater target recognition model integrating image salient region segmentation and pyramid pooling was designed[4]. The proposed method can achieve the correct recognition rate of 89.52 % on nine types of target recognition tasks. Based on the research status and development trend at home and abroad, the sonar image recognition of underwater targets based on deep learning method still needs to be further optimized to solve the problem of low recognition rate of underwater targets[5].

Based on the convolutional neural network model, this paper proposes an underwater automatic target recognition model based on improved EffcientDet and ensemble learning. While ensuring a certain prediction speed, the recognition accuracy is improved.

2. A Network Model for Sonar Image Recognition

Efficientdet is a target detection model proposed by Google Brain Team in 2020. It is a general term for a series of scalable and efficient target detectors[6]. Its network architecture is shown in Fig.1.

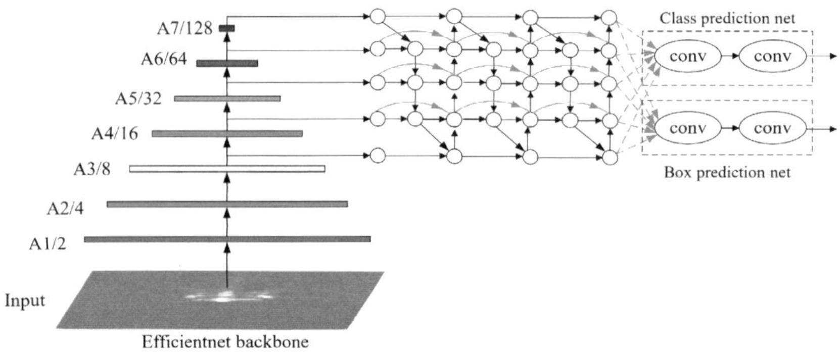

Fig.1 Efficientdet-d0 network structure

This paper presents an underwater target sonar image recognition method based on improved Efficientdet and ensemble learning. The network model is shown in Fig.2. Starting from the whole process of sonar image recognition, the data input module, feature extraction and fusion module and feature classification output module are improved. The improvements of these three parts are described below, and the design ideas and implementation steps of each part are given.

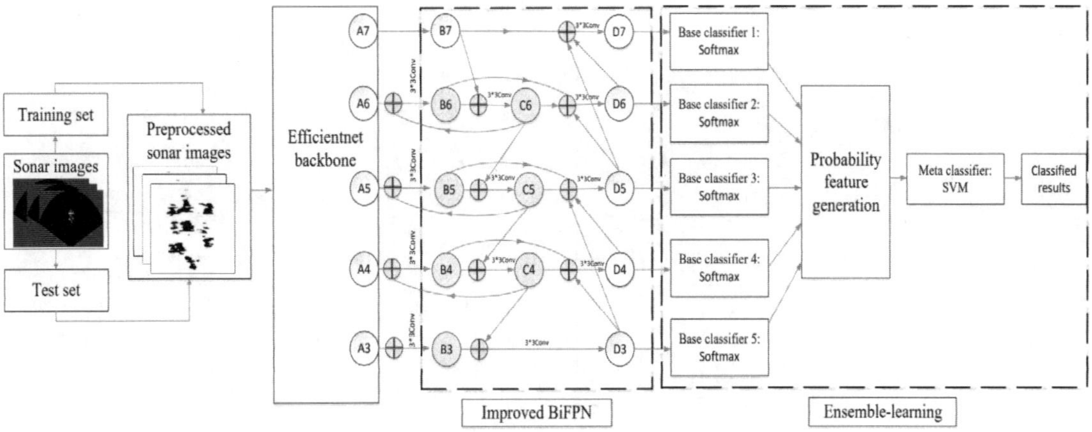

Fig.2 Illustration of the approach proposed in this paper

2.1. Data Input Module

In this paper, the saliency detection based on graph-based manifold ranking (GMR)[7], and the binary threshold segmentation method are used to detect the target area of sonar image, so as to accurately highlight the highlighted. The minimum enclosing rectangle is used to divide the binary image into target region and non-target region, and the target region in sonar image is further cut to reduce the interference of image background on target feature extraction process and improve the efficiency of data processing, as shown in Fig.3.

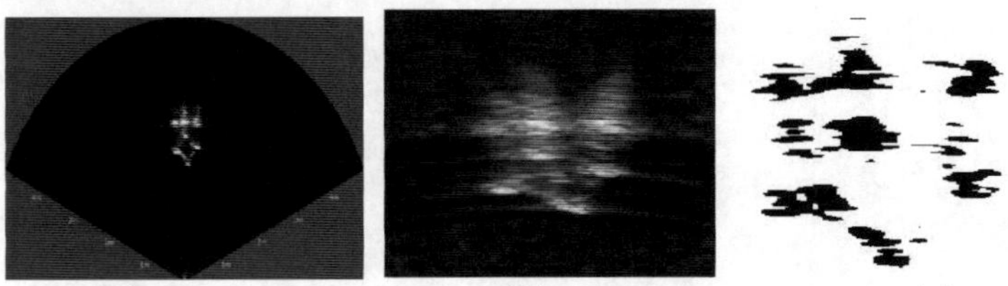

(a) (b) (c)

Fig.3 (a) is the original sonar graph, (b) is the explicit graph, (c) is the cut binary graph

2.2. Feature Extraction and Fusion Module

In this paper, Efficientnet backbone is used for feature extraction, and the improved BiFPN structure is used for feature fusion, as shown in Figure 4. The design of feature fusion module in this paper is based on BiFPN to perform feedback loop stacking and multi-layer fusion of features. The feature map information at different levels is fully utilized, and the extraction of multi-scale information in the feature map is strengthened. The global information is aggregated, and the feature expression of shallow feature map is strengthened. The problem that the small feature extraction is not ideal due to the low resolution of sonar image and less detail information is optimized.

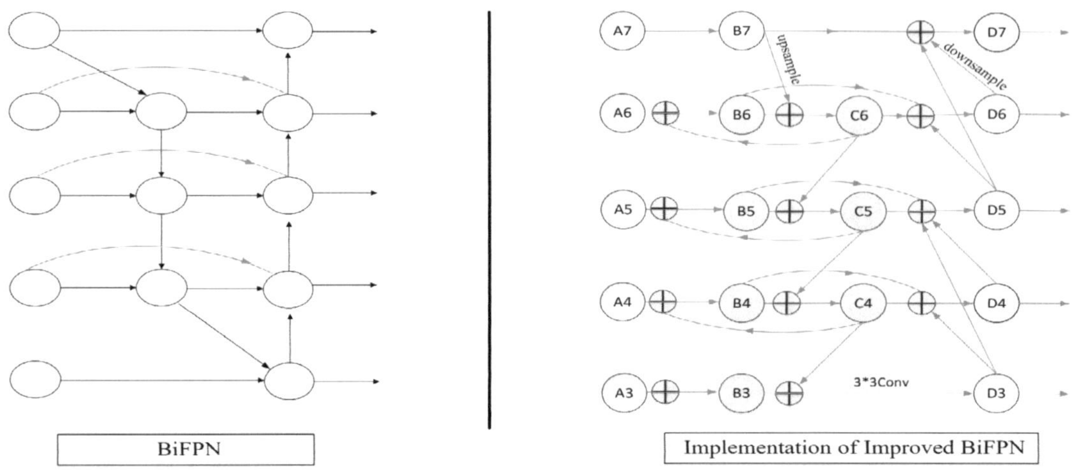

Fig.4 Feature fusion network diagram

The preprocessed image data are input into Efficientnet backbone feature extraction network to obtain the effective feature layer A_h of five stages, h = 3,4,5,6,7. The output feature layer A_h is used as the input of the feature fusion network, and the effective feature map D_n^{out} is finally output, n = 3, 4, 5, 6, 7. The reference calculation formula of D_n^{out} is as follows :

$$D_3^{out}=3*3\text{Conv}\ (\frac{\beta_1 B_3^{out}+\beta_2 C_{4\,upsample}^{out}}{\beta_1+\beta_2+\sigma}) \quad (1)$$

$$D_4^{out}=3*3\text{Conv}\ (\frac{\alpha_3 B_{4\,shortcut}^{out}+\alpha_1 C_4^{out}+\alpha_2 D_{3\,downsample}^{out}}{\alpha_1+\alpha_2+\alpha_3+\sigma}) \quad (2)$$

$$D_5^{out}=3*3\text{Conv}\ (\frac{\mu_3 B_{5\,shortcut}^{out}+\mu_1 C_5^{out}+\mu_2 D_{4\,downsample}^{out}+\mu_4 D_{3\,shortcut}^{out}}{\mu_1+\mu_2+\mu_3+\mu_4+\sigma}) \quad (3)$$

$$D_6^{out}=3*3\text{Conv}\ (\frac{\varphi_3 B_{6\,shortcut}^{out}+\varphi_1 C_6^{out}+\varphi_2 D_{5\,downsample}^{out}}{\varphi_1+\varphi_2+\varphi_3+\sigma}) \quad (4)$$

$$D_7^{out}=3*3\text{Conv}\ (\frac{\theta_1 B_7^{out}+\theta_2 D_{6downsample}^{out}+\theta_3 D_{5shortcut}^{out}}{\theta_1+\theta_2+\theta_3+\sigma}) \quad (5)$$

Among them, 3*3Conv represents a convolution kernel of 3*3 size; β_1、β_2、α_1、α_2、α_3、μ_1、μ_2、μ_3、μ_4、φ_1、φ_2、φ_3、θ_1、$\theta_2$and $\theta 3$ are the weights corresponding to the feature layer ; σ is to prevent denominator set to zero infinity, this value is 0.0001.

2.3. Feature Classification Module

Aiming at the problem that it is difficult to achieve high precision classification by using a single classification method for small sample sonar images, this paper chooses Stacking integration method[8]. Building a two-level classifier to complete the learning task in an ensemble manner can not only obtain better results than a single classifier, but also reduce the dependence of the training process on the required data, making the model more consistent with the sonar image of small samples[9]. In this paper, softmax regression is used as the base classifier, and SVM is used as the meta-classifier to construct the Stacking model. The algorithm is implemented as follows.

1.Construct the extracted valid features as dataset $D^{out}=\{D_3^{out}, D_4^{out}, D_5^{out}, D_6^{out}, D_7^{out}\}$. $D_n^{out}=\{(Y_{n,1}, y_{n,1}), (Y_{n,2}, y_{n,2}), \ldots (Y_{n,Q}, y_{n,Q})\}$; $(Y_{n,j}, y_{n,j})$ represents the feature and corresponding category information extracted by image j in D^{out}, n∈{3,4,5,6,7}, $y_{n,j}$ ∈{1,2,…,N};N is the number of image categories; Q is the total number of training samples.

2.Construct the base classifier B composed of five softmax regression models, and Bn={B3,B4,B5,B6,B7}. Five features in D^{out} are used to train a softmax regression model.

3.The probabilistic feature space is constructed and the meta-classifier M composed of a SVM is trained. Classifier Bn is used to predict any data $Y_{n,j} \in D_n^{out}$, and the probability distribution $P_{j,n} = \{d_{j,n}^1, d_{j,n}^2, \dots d_{j,n}^N\}$ is obtained, where $d_{j,n}^k$ is the probability of sample j passing through classifier Bn to predict that it belongs to category k. The data set $D_B = \{(F_1, y_1), (F_2, y_2), \dots, (F_Q, y_Q)\}$ is constructed according to probability distribution, where $F_j = \{P_{j,3}, P_{j,4}, P_{j,5}, P_{j,5}, P_{j,6}\}$;Thus, classifier M is trained by D_B.

4.The test uses the base classifier B to predict the five effective features extracted from the same image. After generating the probability features, the meta-classifier M is used to predict and finally the sonar image recognition results are obtained.

3. Experiment and analysis

3.1 Experimental Data and Environment

Due to the high cost of sonar image acquisition experiment and its military particularity, there is no open and unified sonar image data set in the field of underwater target recognition. The experimental data in this paper are as follows : five different categories of targets are put into water, and the image sonar is suspended at the same depth of the target, and the target is rotated by 360°; in the rotation process, the omni-directional sonar images of the target are recorded, and different types of underwater targets are classified and labeled. The labels are single, double, triple, quadruple, and T-shape, respectively, as shown in Fig. 5. A total of 1640 sonar images were obtained and divided into 1312 training sets without intersection and 328 test sets. The experimental platform of this paper is Windows10, Graphic Processing Unit (GPU) is NVIDIA RTX3070, mainly dependent on python3.7, numpy = = 1.19.5, matplotlib = 3.2.1, tqdm = 4.56.0, torch > = 1.7.1, torchvision > = 0.8.2, pycocotools, opencv4.4.1. Model periods set to 100, learning rate set to 0.01.

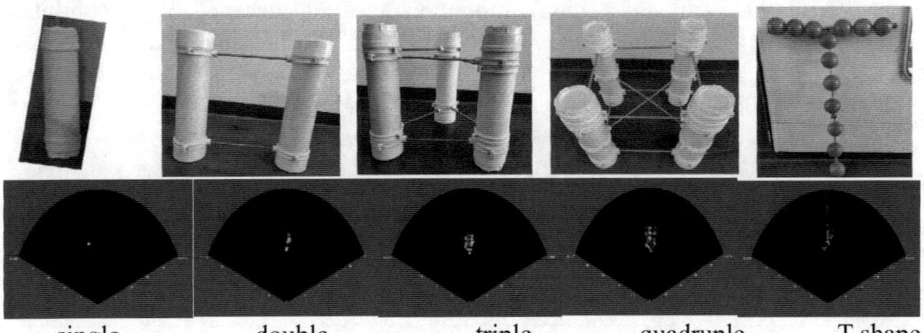

single double triple quadruple T-shape

Fig.5 Test target and sonar images

3.2 Identification Results

When training the underwater target sonar image recognition method based on improved Efficientdet and ensemble learning, the loss of the network on the training set and the test set and the recognition rate on the test set are counted, and the network training process curve is drawn after training. This article takes 51.25 iterations to complete a epoch. Figure 6 is the performance of the network on the training set. It can be seen from the graph that the loss of the training set decreases rapidly with the increase of the number of network training iterations, and tends to 0 after 40 times, indicating that the model has been convergent and fits the training set data well. Figure 7 shows the training curve of the network on the test set. It can be seen from Figure 7 that after 30 epochs, the loss function value tends to be stable as a whole. The recognition accuracy of the network on the test set increases rapidly in the first 20 epochs, and then tends to be stable, indicating that the generalization ability of the model is good, and there is no fitting phenomenon. By comparing all the prediction results of the model on the test set with the labels of the real image categories, the overall recognition accuracy of the image reaches 98.43 %. And on this experimental platform, the average time of network processing of an image is about 80 ms, indicating that the recognition model can ensure a certain prediction speed.

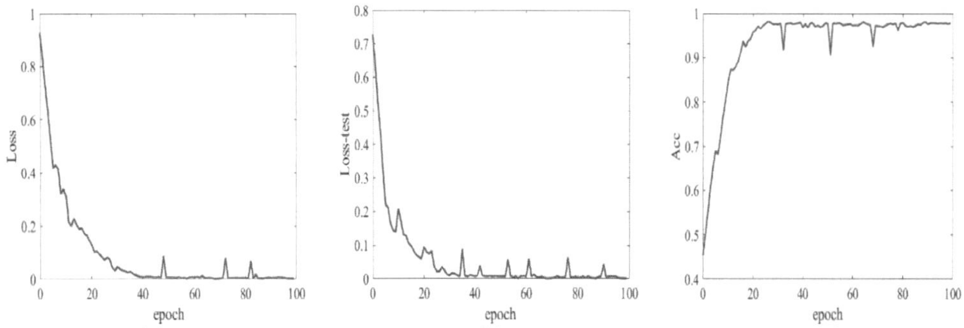

(a) (b) (c)

Fig.6 (a) is the loss of the model on the training set;(b) is the test set loss ; (c) Test set accuracy

3.3 Validation of Improved Network

The sonar image recognition network model designed in this paper is different from Efficientdet network model. The input data are processed by explicit segmentation. The preprocessed data are extracted by Effcientnet backbone, and then the improved BiFPN network is used for feature fusion. Finally, the learning task is completed by integrating two-level classifiers. In order to verify the effectiveness of this structural improvement, we compared it with the other three network structures : after data preprocessing, network A is an Efficientdet network, BiFPN does not improve, and uses a single classifier. Network B uses improved BiFPN for feature fusion, but uses a single classifier to verify the effect of improved BiFPN. Network C is the feature fusion part of BiFPN does not improve, but the use of two classifiers ensemble learning, mainly used to verify the effect of ensemble learning ; network D network model structure designed in this paper, the feature fusion module using improved BiFPN, and in the feature classification module using two classifiers ensemble learning. On the same sonar image data set, three experiments are repeated for each network structure, and the average recognition accuracy of the three test sets is taken as the final recognition result, which is listed in Table 1.

Table 1. Sonar image recognition results of four different networks

Network structure	Accuracy/%	Prediction time per image/ms
A (Efficientdet)	89.12	75.3
B（Improved BiFPN）	94.56	70.6
C（Ensemble-learning）	92.28	81.9
D（Improved BiFPN and Ensemble-learning）	98.43	79.3

The experimental results of table 1 show that the combined improved Efficientdet and ensemble learning network model designed in this paper has the highest recognition accuracy for sonar image dataset, which is 9.31 % higher than that of network A. Compared with network B and network A, the improved BiFPN increases the recognition accuracy by 5.44 %; compared with network C and network A, the integration method improves the recognition rate by 3.16 %.

4. Conclusion

According to the characteristics of sonar image, this paper proposes a sonar image recognition method combined with improved Efficientdet and ensemble learning, and designs an underwater target recognition model combined with improved BiFPN network and two-level classifier, which effectively improves the accuracy of underwater target recognition. Adam optimization algorithm is used to adaptively adjust the learning rate and accelerate the convergence of the network, and give full play to the ability of network self-learning image features. The experimental results on the measured sonar image data set show that the proposed method can achieve 98.43 % correct recognition rate on the five types of target recognition tasks, which is higher than that of the Efficientdet network without improvement, and the prediction time of a single image is equal to that of the network without improvement. The experimental results show that under the condition of ensuring the target image prediction speed, the network model designed in this paper improves the recognition rate of underwater targets.

References

[1]Peter, T., Kastner, R., Diamant, R. (2020) Target detection using features for sonar images.J. IET Radar, Sonar & Navigation., 14(12) : 1940-1949.

[2]Liu, W. (2019) Research on underwater target image recognition method based on deep learning.D.Universities of Electronic Science and Technology.

[3]Zhu, Z.T., Hu, Y.F. (2019) Sonar image recognition based on fine-tuned convolutional neural network. J. MATEC Web of Conferences., 283 : 04012.

[4]Jin, L.L., Liang, H., Yang, C.S. (2020) Underwater target sonar image recognition method based on convolutional neural network. J. Journal of Northwest University of Technology., 188(02) : 285-291.

[5]Tian, Y.Y., Lan, L.Y., Guo, H.T. (2020) A review on the wavelet methods for sonar image segmentation. J. International Journal of Advanced Robotic Systems., 17(4)

[6]Tan, M., Pang, R., Le, Q.V. (2020) EfficientDet: Scalable and efficient object detection. J. Proceedings of the IEEE Computer Society Conference on Computer Vision and Pattern Recognition., 10778-10787.

[7]Cheng, D. (2020) Saliency Detection via a Multiple Self-Weighted Graph-Based Manifold Ranking. J. IEEE Transactions on Multimedia., 22(4) : 885-896.

[8] Philip, M. (2021) COVID-19 Image Segmentation Based on Deep Learning and Ensemble Learning. J. Studies in health technology and informatics., 281 : 518-519.

[9] Nordhausen, K. (2013) Ensemble Methods: Foundations and Algorithms by Zhi - Hua Zhou. J. International Statistical Review., 81(3) : 470-470.

Journal of Physics: Conference Series 2258 (2022) 012039 doi:10.1088/1742-6596/2258/1/012039

A network security situation assessment model based on BP neural network optimized by D-S evidence theory

Kun Yin[1a*], Yu Yang [1b*], Jinwei Yang[1c], Chengpeng Yao [1d]

[1] Academy of Information Engineering, Engineering University of PAP University, Xi'an 710016, China

[a*]745703224@qq.com, [b*]miaoyude@163.com, [c]836993008@qq.com, [d]997792557@qq.com

Abstract—Traditional networks rely heavily on the distribution of expert experience when assessing complex network security situations, resulting in low assessment accuracy, which has been unable to adapt to the current network security needs of the big data era, and has unavoidable problems such as low efficiency and poor flexibility. In response to these problems, this paper proposes a network security situation assessment method based on D-S evidence theory to optimize neural networks. First establish the CS-BP neural network model, enhance the local search ability of the cuckoo algorithm through conjugate gradient calculation, and then introduce it into the BP neural network to improve the training convergence speed and overcome the local minimum problem; finally, in order to reduce the basic probability distribution (BPA) subjective impact, using DS evidence theory to optimize the CS-BP neural network, determine the degree of impact of each attack, and evaluate the value of the network security situation. The experimental results show that the network situation assessment model of CS-BP neural network optimized based on D-S evidence theory can effectively assess the network security situation in the environment of trusted equipment.

1. Introduction

With the increasingly severe form of network security, network security posture assessment has become an effective means of assessing the current network posture. It integrates the indicators affecting network security and provides network managers with decision-making advice to minimize the harm generated by the threat of network attacks [1]. Since the introduction of network security situational awareness by BASS [2], network security situational awareness has undergone a long period of development, and nowadays, many studies have been conducted in the area of situational assessment. The cybersecurity posture literature [3] proposed a security evaluation method of hierarchical analysis and gray correlation analysis by constructing a multidimensional system security evaluation model of environmental security, network security and vulnerability security. The literature [3] describes the network security posture evaluation model by means of attack diagrams. The literature [4] describes the vulnerability of the network and how to reduce the structural overhead of network security from the perspective of network topology. The literature [5] traps network attacks by means of honeynets to achieve network security posture assessment. The literature [6] combines support vector machines and improves the cuckoo algorithm to predict network security posture, and the method achieves high accuracy in performance on the KDD dataset.

In the context of big data, Xu and Ning [7], Zhai et al [8] and Allodi and Massacci [9] analyze cybersecurity based on attack scenarios, which are obtained by correlating alerts. There are techniques to integrate all available information to analyze cybersecurity.M2D2[10],task impact based correlation methods [11] and cvss based risk assessment methods [12] are of great help in our study. Based on the above background, the key techniques for the situational analysis based on trusted measures are investigated. By further analyzing the original security events of trusted measures, the burden of network administrators is reduced, the response processing capability is improved, and the early warning capability of threats is enhanced through automated information processing.

At present, the analysis technology and evaluation system of network security are becoming more and more mature, and the prevention of network attacks and network risks is also becoming more stable, but there are still the following shortcomings.

1) The singularity of data sources for network attack detection: At present, the detection of network security mainly focuses on the analysis of traffic data or network alarm log data, with relatively one-sided data sources and unbalanced data volume, making the lack of characteristic data of attacks and the accuracy of attack detection cannot be improved.

2) Network structure characteristics of network threat identification is ignored: in the identification of network threats, usually due to the propagation characteristics of its attacks make the network threat exhibit regional characteristics. The regional characteristics of the network threat are not only related to the characteristics of the network attack itself, but also have a great relationship with the network structure.

3) Lack of research on network topology prediction and security: Currently, the research on network security posture is focused on data mining, and lacks research on the correlation between network structure and security.

Traditional methods in evaluating complex network posture lead to low evaluation accuracy due to the problems of relying on expert experience assignment and having subjectivity. To address the shortcomings of the above methods, this paper proposes a D-S evidence theory network security posture evaluation method based on deep learning. In order to reduce the subjectivity of BP neural network processing results and improve the accuracy of evaluation, the conjugate gradient CS algorithm is selected for optimization. The effectiveness of the model is verified by building a test environment.

1) Construct a BP neural network, use sample data to train the network, and adjust the weight and threshold of the neural network.

2) Using the CS algorithm based on conjugate gradient to optimize the BP neural network。

3) Adjust the corresponding parameters of the neural network to the weights and thresholds optimized for training, and input the actual data into the model. At the same time, compare experiments with models of the same type.

4) Synthesize the basic probabilities using D-S evidence theory to obtain the final synthesis results.

Through experiments, it is proved that the method proposed in this paper can achieve real-time evaluation of network security, and the evaluation effect is more efficient and intuitive.

2. Based on D-S evidence theory to optimize BP neural network network security situation assessment model optimization

2.1The main Theoretical Basis

The credibility of the generated security event is generated by multiple sources of the same type and format. Therefore, D-S evidence theory has advantages in integrating multi-source data. In order to solve the problem that the basic probability distribution of the D-S evidence situation assessment process depends on expert experience and subjectivity, and to improve the network situation assessment ability, a DS evidence theory situation assessment model based on an optimized deep autoencoder is proposed. This article's research on the scenario assessment model is based on DS evidence theory. In order to reduce the subjectivity of basic probability distribution, some defects in the application of BP neural network are introduced and analyzed. Aiming at the limitation of BP neural network, a CS algorithm based on conjugate gradient is proposed to solve this problem, and the result of the combination of the two is analyzed.

D-S evidence theory is an efficient method for fusing uncertain information about supervisors, and has significant advantages in fusing multi-source heterogeneous data. D-S evidence theory can deal with the uncertainty caused by randomness, and it can distinguish between Unknown and Uncertain[13].

In the process of applying D-S evidence theory, we integrate information from two aspects of evidence establishment and evidence synthesis.

We define the three concepts of D-S evidence: suppose the set Ω is the sample space, Ω is composed of mutually exclusive sentences, and the expressions of various sentences are combined into a power set 2Ω.

Definition 1: Basic probability assignment, BPA

$$\mathrm{M}: 2^{\Omega} \rightarrow [0,1] \tag{1}$$

We assume that M satisfies the following two conditions: , Indicates the probability of an impossible event; A⊆Ω, The sum of all results is 1.

Definition 2: Trust function Bel

We use the trust metric Bel (A) to represent the sum of the basic probabilities of all subsets in A:

$$\mathrm{Bel}: 2^{\Omega} \rightarrow [0,1], \mathrm{Bel}(A) = \sum M(B), A \subseteq \Omega \tag{2}$$

According to the definition of probability distribution function:

$$\mathrm{Bel}(\phi) = M(\phi) = 0\mathrm{Bel}(\Omega) = \sum_{B \subseteq \Omega} M(B) \tag{3}$$

Definition 3: Likelihood function PI:

$$PI: 2^{\Omega} \rightarrow [0,1] PI(A) = 1 - \mathrm{Bel}(-A), A \subseteq \Omega \tag{4}$$

Suppose Bel1(A),Bel2(A),Bel3(A)…,BelN(A) is a trust function,m1,m2,m3...mn is the basic probability distribution.

If $\mathrm{Bel}_1(A) \oplus \mathrm{Bel}_2(A) \oplus \mathrm{Bel}_3(A) \ldots \oplus \mathrm{Bel}_N(A)$ exists, the combined trust function is

$$\mathrm{Bel}(A) = \Big(\big((\mathrm{Bel}_1(A) \oplus \mathrm{Bel}_2(A)) \oplus \mathrm{Bel}_3(A) \big) \ldots \Big) \oplus \mathrm{Bel}_N(A) \tag{5}$$

The synthesized BPA is

$$\begin{gathered} m(A) = K^{-1} \textstyle\sum_{\cap A_i = A} \bigcap_{1 \le i \le n} m(A_i) \\ K = \textstyle\sum_{\cap A_i \ne \emptyset} \bigcap_{1 \le i \le n} m(A_i) \end{gathered} \tag{6}$$

2.2 CS algorithm based on conjugate gradient

Cuckoo coding method[14]. Floating-point number coding is intuitive and has the characteristics of easy control of coding length, high coding accuracy and strong search ability in large spaces. It is conducive to processing multi-dimensional and high-precision continuous functions, and can reduce computational complexity and improve computational efficiency. Therefore, this article uses floating-point coding.

Design of fitness function. Fitness index is the standard to measure the quality of individuals in the population in the algorithm. Therefore, the higher the fitness of the population, the closer the individual is to the optimal solution. There are three design schemes of fitness function: the first is to directly use the objective function as the fitness function; the second is to add the objective function once; the third is to add the objective function once and then take the reciprocal.

As far as this state is concerned, another expression is: after the host finds the foreign eggs at a certain probability, the same number of new bird nests are used to replace the old bird nests of poor quality. In general, in order to simplify the problem, when applying the CS algorithm, one bird's egg is usually regarded as a solution. Each cuckoo lays one egg at a time, and there is only one egg in each nest. Based on the above situation, eggs and bird's nests It is no different from a cuckoo.

The steps of the cuckoo search algorithm are as follows:

Step 1: Set the population size of the host bird. Usually to avoid complexity, it is assumed that there is only one egg in each bird's nest.

Step 2: The cuckoo randomly searches the lair through Levy flight. In this step, the cuckoo uses Levi to fly to randomly selected lairs.

Step 3: After the cuckoos lay their eggs, calculate the fitness of the nest where they are located. The size of the fitness is expressed by the root mean square error, and it is compared with the fitness of the host bird's egg.

Step 4: If the fitness of cuckoo eggs is better than that of the host, replace the eggs of the host with cuckoo eggs.

Step 5: If the host bird detects foreign eggs, it abandons the nest and rebuilds the nest. Repeat steps 2 to 5 until the number of iterations meets the requirements.

It can be considered that a nest corresponds to an egg, which means a cuckoo. Cuckoo's location update algorithm is

$$x_{ij}^{m+1} = x_{ij}^{m} + \alpha \times \mathrm{levy}(\lambda) \tag{7}$$

The Levy (λ) distribution is Fourier transformed and simplified to obtain the probability density function in power exponential form:

$$\mathrm{Lev}\, y(\lambda) \sim \mathrm{u} = t^{-\lambda} 1 < \lambda < 3 \tag{8}$$

Yang et al. completed the programming implementation of cuckoo algorithm through the Levy flight jump path formula of Mantegna model. [15] The formula is:

$$s = \frac{\mu}{|v|^{1/\beta}} \tag{9}$$

We define Levy's flight jump path as Levy(λ); the relationship between the parameter β and λ is λ=β+1, the value interval of β is 0<β<2, and the parameter μv is a random number, which obeys the following normal distribution:

$$\begin{aligned} r \quad & \sigma_\mu < \frac{\Gamma(1+\beta)\sin\left(\frac{\pi\beta}{2}\right)}{\Gamma\left[\frac{1+\beta}{2}\right] 2^{\frac{\beta-1}{2}} \beta} > \\ & \sigma_v = 1 \\ r \quad & \mu \sim \mathrm{N}\left(0, \sigma_\mu^2\right) \\ & v \sim \mathrm{N}\left(0, \sigma_v^2\right) \end{aligned} \tag{10}$$

The implementation of Levy's flight relies on two random parameters. This algorithm can be combined with CS algorithm to have strong global optimization ability. At the same time, the weak local search ability of CS algorithm is improved by searching the conjugate direction near a certain point, and the local information is utilized to a greater extent. After testing, the method has good local search capabilities. Through the combination of the two experiments, after each calculation iteration is completed, the random walk changes the position, and finally the updated nest position is obtained.the gradient is slightly multiplied by the conjugate factor , and then added to the position of the negative gradient to construct a set of conjugate directions . Then by adjusting the search direction of the conjugate direction, the determination of the position of the parasitic nest is completed within a predetermined number of iterations, and the next iteration is performed after the new position of the parasitic nest is obtained [16].

2.3 Optimizing BP neural network based on CS algorithm

When initializing the BP neural network, the randomness of the weights and threshold assignments leads to a large amount of calculation and slower convergence speed. The cuckoo search algorithm based on conjugate gradient has strong global optimization ability, convergence ability and robustness, and it is suitable for the optimization of neural networks. Therefore, The optimized CS algorithm is used to adjust the initial value of BP neural network, which can improve the convergence speed and reduce the calculation cost.

The training time of the BP neural network under the random allocation of weights and thresholds is generally longer. The weights and thresholds obtained through training may not be optimal. In this paper, the CS algorithm is used to find the optimal weights and thresholds. CS algorithm is a typical search algorithm of swarm intelligence technology, which provides a new solution for BP neural network parameter optimization.

2.4 Situation Assessment Model

Data fusion is at the core of network security situation assessment. The various processes of situation assessment are inseparable from data fusion. It focuses on the association between multi-source data and is a process of mapping network security situation elements to the situation space.

Through the nonlinear mapping and self-learning capabilities of the CS-BP neural network, the trend elements of the input processing are mapped to the basic probability distribution of the output. Then, use the D-S evidence theory to fuse the basic probability distributions at different time points to obtain a more objective value.

1) Situation data input: Collect situation-related data of each node in the network in time intervals. The data collected at each time (situation data 1, situation data 2, ..., situation data n) are used as a set of data in the situation input part.

2) Situational data mapping: This part consists of three layers, an input layer that accepts situational data, a hidden layer for information processing, and an output layer that outputs situational values. The input layer obtains input data from the situational input part. These data are processed by the hidden layer to obtain the output of the hidden layer, and then the output of the output layer is obtained through the operation of the output layer, and the situational data mapping is completed.

3) Situational data output: Receive the value passed by the situational mapping part, which is the situational value of the network during a certain period of time.

The network data collected in this paper comes from trusted data sources, and the collected data has a uniform data format and the same type characteristics. Therefore, we use D-S evidence theory to calculate the network situation value. The implementation path is shown in Figure 1.

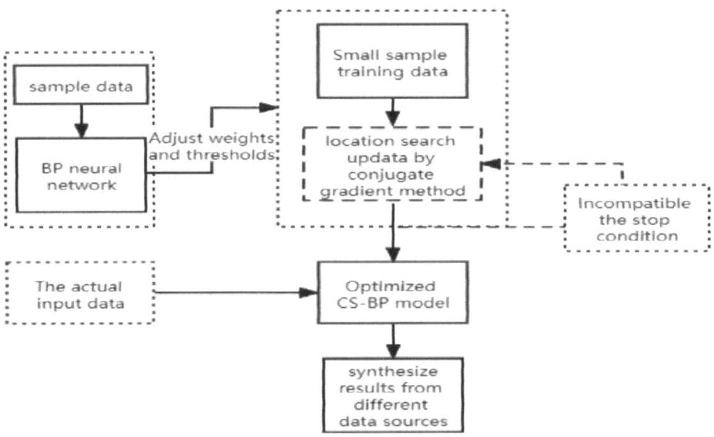

Fig.1 Flow chart of network security situation assessment based on D-S evidence theory to optimize CS-BP neural network

We count the number of trusted input security incidents. As shown in Table 1, the output data is the BPA corresponding to the unit time of the security event.

TABLE 1 INPUT PARAMETERS AND ATTACK TYPE

Input Parameters	**Situation Indicator**	**Discription(Unit Time)**
X_0	Dynamic measure	Number of collections
X_1	Static measure	Number of collections
X_2	Irregular peripherals	Number of collections
X_3	Internet connection	Number of collections
X_4	Forced access control	Number of collections

In order to eliminate the influence of dimensionality, we standardize it so that its value is in the range of [0,1]. The normalization formula of situation indicators:

$$X' = \frac{X-X_{min}}{X_{max}-X_{min}} \tag{11}$$

The result parameters of the situation level are shown in Table 2:

TABLE 2 OUTPUT PARAMETERS AND SITUATION LEVELS

Output Parameters	**Situation Level Name**
P	Normal situation
N	Abnormal situation
Unknown	Unknown situation

The number of neurons in the input layer of the BP neural network designed in this paper is 5, the number of neurons in the hidden layer is not fixed, and the number of neurons in the output layer is 3. The formula used is $M = \sqrt{m+n} + a$ (m, n is the number of neurons in the input layer and output layer, a is a constant in the interval [0, 10]).

Generally speaking, the number of neurons set in the hidden layer is positively correlated with the training effect, and negatively correlated with the training time. So we set up 9 neurons in the hidden layer.

Based on the research of computer trust enhancement technology system, we have established a network structure of trend index levels. Some training samples are shown in Table 3.

TABLE 3 TREND INDICATORS TO THE SITUATION LEVEL CORRESPONDING TABLE

Serial	**Situation Indicators**					**Situation Level**		
Number	***X0***	***X1***	***X2***	***X3***	***X4***	***P***	***N***	***Unknwon***
1	200	50	20	100	40	0.60	0.35	0.05
2	200	50	20	100	35	0.65	0.30	0.05
3	200	50	20	140	30	0.50	0.45	0.05
4	200	50	20	145	40	0.45	0.50	0.05
5	200	50	15	100	40	0.50	0.45	0.05
…	…	…	…	…	…	…	…	…
34	100	30	10	145	30	0.60	0.35	0.05
…	…	…	…	…	…	…	…	…

In order to map the situation indicators to the neural network, we follow the construction principle of the network-level trend indicators. In order to make the neural network have the ability to map the situation indicators, we then train the CS-BP neural network according to the corresponding relationship. When entering scenario indicators, substituting BPA into equation (14) to obtain the network security situation level at the corresponding point in time.

3. Experiment

In the experimental part of this article, we used the optimized CS-BP neural network, BP neural network and GA-BP neural network to train the sample data in Table 3. The training results are shown in Figure 2. We can see that with the number of samples With the increase of the situation, the situation value is gradually increasing.

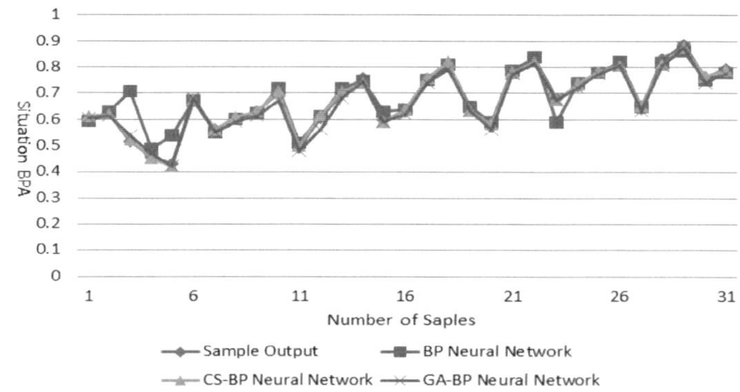

Fig.2 results contrast

We control the fault tolerance at 0.01 and the maximum number of iterations is 10,000. Using BP neural network, optimized CS-BP neural network and GA-BP neural network training sample data respectively, the corresponding training results of each neural network are obtained. As shown in Table 4.

TABLE 4 ALGORITHM EFFICIENCY COMPARISON

Algorithm	**Training Times**
back propagation	7056
Optimized	3510
GA-BP	3656

Through comparative experiments, the CS-BP neural network algorithm is optimized. Compared with BP neural network algorithm and GA-BP neural network algorithm, it has better performance in calculating BPA. It has the advantages of small error, fewer training times, and high computational efficiency. We set up an experimental environment as shown in Figure 3 to further verify the validity and reliability of the model. Refer to the fault diagnosis and prediction research method in the threat injection method, load the malicious program into the operating environment, and observe the influence of the network security situation value.

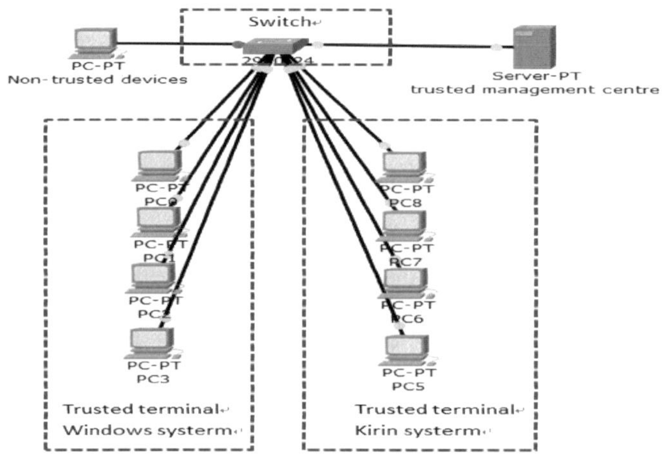

Fig.3 Experimental environment topology

We collect security events generated within 9 hours of each statistical period as experimental samples. In the second period, we load a malicious program-gray pigeon every 10 minutes, and visit untrusted devices in the fifth period. Untrusted devices and trusted terminal devices can Communication and other communication are carried out through FEIQ message, file transmission and other communication software 1~6, and at the same time, the experimental data obtained through the trusted management center is shown in Table 5.

TABLE 5 EXPERIMENTAL DATA

Serial Number	X0	X1	X2	X3	X4
1	97	20	16	65	30
2	245	12	11	75	26
3	123	25	8	71	20
4	156	21	10	52	23
5	116	31	12	106	22
6	114	18	16	62	16
7	123	29	6	57	19
8	115	12	8	63	23
9	106	29	8	56	20

First, normalize the statistical data of various security events, and then use the optimized CS-BP algorithm to process the normalized data to obtain the BPA value at each time point. As shown in Table 6.

TABLE 6 THE PROCESSED EXPERIMENTAL DATA.

Time Point	P	N	Unknown
1	0.753	0.192	0.055
2	0.621	0.312	0.067
3	0.775	0.172	0.053
4	0.806	0.142	0.052
5	0.673	0.316	0.051
6	0.915	0.032	0.053
7	0.832	0.124	0.044
8	0.903	0.052	0.045
9	0.672	0.261	0.067

As shown in Table 6, the network loads malware in the second cycle and accesses untrusted devices in the fifth and ninth cycles, which has an impact on the network security situation. BPA shows a significant downward trend under normal conditions, and a significant upward trend under abnormal conditions. As shown in Figure 4.

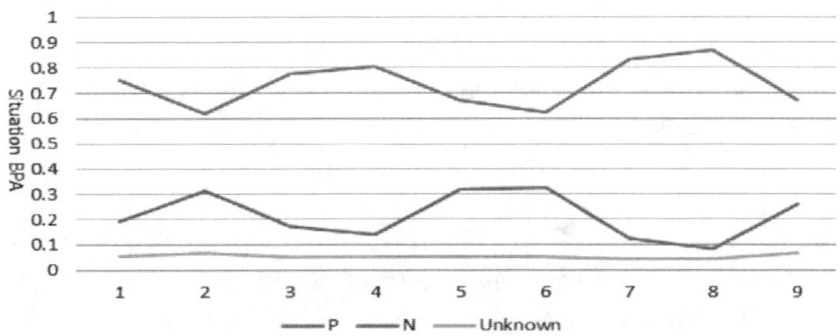

Fig.4 Experimental environment topology

We use the D-S evidence theory to fuse the situational values at each time point, and the result is shown in Figure 5.

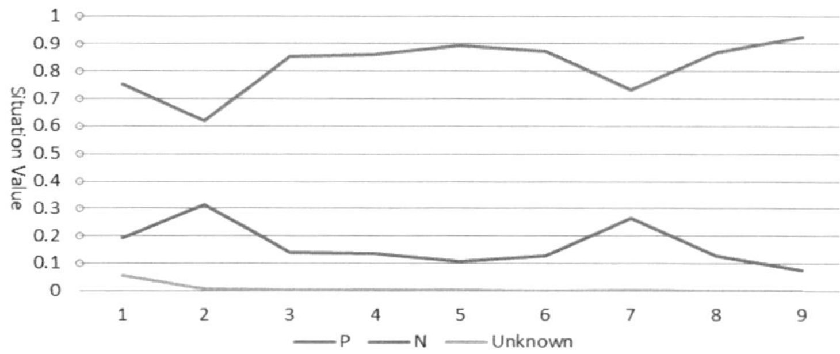

Fig.5 Experimental environment topology

Through the above experiments, the following conclusions can be drawn.

(1) When the experiment is conducted on a platform composed of trusted devices, the Unknown value is always higher, indicating that the network integrity is higher in a trusted environment.

(2) The decrease in the Unknown value indicates that the point-in-time based temporal BPA integration of D-S evidence reduces the uncertainty of the posture assessment.

(3) The situation when the malicious program is loaded in the second cycle and when the untrusted device is accessed for communication in the fifth cycle also has an impact. By observing the curves, it can be seen that there is a downward trend in the normal state values and an upward trend in the abnormal trend at these two time points. The training results of the model are consistent with the actual situation, verifying the feasibility and effectiveness of the network security situation assessment proposed in this paper.

4. CONCLUSION

According to the characteristics of situation assessment, when selecting indicators, follow the principles of easy quantification, clear hierarchy, and similarity, and choose situation assessment indicators suitable for the method in this paper. Combining the characteristics of network security situation assessment and existing network security situation assessment methods, an assessment model is designed. We proposed a CS-BPNN-based network security situation assessment method, introduced in detail the steps and processes of the method, and at the same time, in order to further solve the problem of long training time for BP neural network. In order to prevent the neural network from falling into a local minimum, that is, the problem of learning complexity. We introduced the Cuckoo algorithm (CS) based on conjugate gradient to optimize the initial parameters of the BP neural network and improve the training efficiency of the BP neural network. Finally, through experimental verification, the network security situation assessment model proposed in this paper can effectively assess the network security status in an environment composed of trusted devices. In the future, it is possible to further improve the convergence speed of the algorithm and the quality of the solution by adjusting the step size parameters of the CS algorithm, so that the CS-BP neural network situation assessment method can achieve better evaluation results.Acknowledgment

Acknowledgement

This work is supported by the research on key technologies of armed police optical cable network security situation awareness (WJY202130), the exploration and practice of the online and offline hybrid teaching mode of the "Computer Network" course (WJX2021120), and the establishment of a postgraduate funding project (XYYJS08).

References

[1] Yao J., Fan X., Cao N. (2019) Survey of Network Security Situational Awareness. In: Vaidya J., Zhang X., Li J. (eds) Cyberspace Safety and Security. CSS 2019. Lecture Notes in Computer Science, vol 11982. Springer, Cham. https://doi.org/10.1007/978-3-030-37337-5_4

[2] Bass T. Intrusion detection systems and multisensor data fusion[J]. Communications of the ACM, 2000, 43(4): 99-105.
[3] Zhao X., Xu H., Wang T., Jiang X., Zhao J. (2020) Research on Multidimensional System Security Assessment Based on AHP and Gray Correlation. In: Han W., Zhu L., Yan F. (eds) Trusted Computing and Information Security. CTCIS 2019. Communications in Computer and Information Science, vol 1149. Springer, Singapore. https://doi.org/10.1007/978-981-15-3418-8_13
[4] Lingyu Wang, Anoop Singhal, and Sushil Jajodia. 2007. Toward measuring network security using attack graphs. In Proceedings of the 2007 ACM workshop on Quality of protection (QoP '07). Association for Computing Machinery, New York, NY, USA, 49–54. DOI:https://doi.org/10.1145/1314257.1314273
[5] Jajodia S., Noel S., O'Berry B. (2005) Topological Analysis of Network Attack Vulnerability. In: Kumar V., Srivastava J., Lazarevic A. (eds) Managing Cyber Threats. Massive Computing, vol 5. Springer, Boston, MA. https://doi.org/10.1007/0-387-24230-9_9 [5] Barford P., Chen Y., Goyal A., Li Z., Paxson V., Yegneswaran V. (2010) Employing Honeynets For Network Situational Awareness. In: Jajodia S., Liu P., Swarup V., Wang C. (eds) Cyber Situational Awareness. Advances in Information Security, vol 46. Springer, Boston, MA. https://doi.org/10.1007/978-1-4419-0140-8_5
[6] J. Hu, D. Ma, C. Liu, Z. Shi, H. Yan and C. Hu, "Network Security Situation Prediction Based on MR-SVM," in IEEE Access, vol. 7, pp. 130937-130945, 2019, doi: 10.1109/ACCESS.2019.2939490.
[7] D. Xu and P. Ning, "Alert correlation through triggering events and common resources," 20th Annual Computer Security Applications Conference, 2004, pp. 360-369, doi: 10.1109/CSAC.2004.5.
[8] Y. Zhai, P. Ning, P. Iyer and D. S. Reeves, "Reasoning about complementary intrusion evidence," 20th Annual Computer Security Applications Conference, 2004, pp. 39-48, doi: 10.1109/CSAC.2004.29.
[9] Allodi L, Massacci F. Comparing vulnerability severity and exploits using case-control studies[J]. ACM Transactions on Information and System Security (TISSEC), 2014, 17(1): 1-20.
[10] Morin B., Mé L., Debar H., Ducassé M. (2002) M2D2: A Formal Data Model for IDS Alert Correlation. In: Wespi A., Vigna G., Deri L. (eds) Recent Advances in Intrusion Detection. RAID 2002. Lecture Notes in Computer Science, vol 2516. Springer, Berlin, Heidelberg. https://doi.org/10.1007/3-540-36084-0_7
[11] Porras, P .A., Fong, M.W., V aldes, A.: 'A mission-impact-based approach to INFOSEC alarm correlation'. Recent Advances in Intrusion Detection, 2002,pp. 95–114
[12] Doynikova, E., Kotenko, I.V .: 'CVSS-based probabilistic risk assessment for cyber situational awareness and countermeasure selection'. Int. Conf. Parallel, Distributed and Network-Based Processing, 2017, pp. 346–353
[13] Denoeux T. Analysis of evidence-theoretic decision rules for pattern classification[J]. Pattern recognition, 1997, 30(7): 1095-1107.
[14] Mantegna R N. Fast, accurate algorithm for numerical simulation of Levy stable stochastic processes[J]. Physical Review E, 1994, 49(5): 4677.Payne R B, Sorensen M D, Klitz K. The cuckoos[M]. Oxford University Press, 2005.
[15] Yang X S, Deb S. Cuckoo search via Lévy flights[C]//2009 World congress on nature & biologically inspired computing (NaBIC). Ieee, 2009: 210-214.
[16] Xiao, M., Liao, Y., Bartos, P. et al. Fault diagnosis of rolling bearing based on back propagation neural network optimized by cuckoo search algorithm. Multimed Tools Appl (2021).

Maritime search and rescue probability of containment model design and simulation

Jun Zhang[1*] , Yu-zhu Zhu[1]

[1] Navigation College,Dalian Maritime University, Dalian 116026, China

[*] Corresponding author's e-mail:zhangj@dlmu.edu.cn

Abstract—In maritime search and rescue operations, in order to improve the success rate of search and rescue, it is necessary to quickly determine the search area after target drift according to the existing accident and Marine environment information, and establish the probability of containment model, to provide information support for subsequent search and rescue planning. AP98 wind pressure drift model is used to model the drowning person at sea, and Monte Carlo simulation is used to simulate the floating situation of the drowning person. A large number of initial hypothetical points were arranged, the factors affecting the drift trajectory were quantitatively analyzed, and the probability density of the hypothetical points in the search area after a certain period of time was counted to form the inclusive probability model. Matlab software is used to test the model in different scenarios to prove the effectiveness of the proposed design model.

1. Introduction

Maritime search and rescue is the last line of defense to ensure the safety of life and environment at sea. The study of maritime search and rescue theory is to study how to use the search theory to determine the optimal search area, allocate search resources, develop search plans, so as to determine the location of people in distress at sea in the shortest time with less search resources.[1] In the search and rescue operation planning, the first thing to do is to quickly determine the search area after the target drift according to the existing accident and Marine environment information, and develop a reasonable and efficient search and rescue plan.[2] In the past, the conventional search and rescue probability model generally used analytical method to analyze the drift trajectory of the target in distress, which was too complicated and tedious, and could not feedback and update the effective information obtained in the search and rescue process.[3] Therefore, a search and rescue inclusion model based on Monte Carlo simulation is proposed to simulate it in different scenarios, and a real-time search and rescue probability of containment matrix is quickly established according to the target drift characteristics, providing theoretical basis for subsequent search and rescue operations.[4]

2. Marine target drift model modeling

Targets in distress at sea usually include drowning men, disabled ships, rafts, etc. These floating objects mainly move under the joint action of wind, wave and current during drifting. However, due to the complex action mechanism of wave on target, the current Marine target drift model usually only analyzes the influence of wind and current. The part below the sea surface is dominated by water flow, while the part above the sea surface is mainly affected by wind. Its drift track can be roughly regarded as the vector sum of flow pressure drift and wind pressure drift.[5] It is assumed that the target in distress falls vertically into the water, and the partial velocity on the two-dimensional surface of the sea is 0. According to the law of motion, when the object accelerates to a stable velocity drift, the two forces generated by the ocean current and the wind pressure difference and the resistance generated by the drift add up to zero. In addition, the duration of the search and rescue mission of the target in distress at sea is relatively long, usually up to several hours or even days. In contrast to the long drift, the target in distress will quickly reach its steady velocity with negligible acceleration time. The Lagrange integral of sea surface velocity can be used to calculate the flow pressure drift vector. However, it is difficult to simulate the wind pressure drift vector. Up to now, the third generation (the first generation GDOC model, the second generation CASP model) and AP98 wind pressure drift model (hereinafter referred to as AP98 model) have been developed.[6]

The AP98 model was developed by A.Allen of the United States Coast Guard. In "Review of Leeway" and "Leeway Divergence", he made a comprehensive analysis of the wind-pressure drift model and held that different types of sea flotsam would show different characteristics in the face of wind-induced drift. In the AP89 model, when wind acts on floating objects at sea, the actual wind pressure drift vector (L,Leeway) of floating objects can be decomposed into two directions: Downwind drift vector (DWL,Downwind Leeway) and vertical wind pressure vector (CWL, Crosswind Leeway), as shown in Fig. 1.

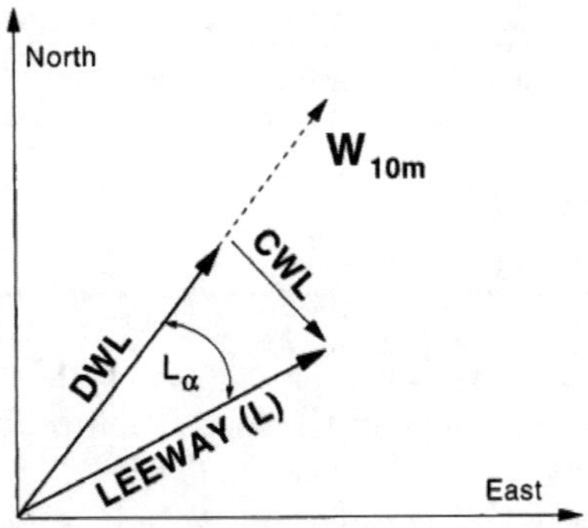

Fig.1 Wind-induced drift vector diagram

In the figure, W_{10m} is the wind speed vector at 10m height; L is the wind pressure drift vector; L_α is wind pressure difference Angle; DWL is the downwind drift vector; CWL is the drift vector of vertical wind direction. Due to the irregular shape of the floating object, the wind-induced drift track of the target usually has a certain Angle with downwind. This Angle is measured by CWL.It should be noted that in the actual use of the model, the probability of CWL appearing offset on both sides of DWL is uncertain and generally roughly equal.[7]

Allen first assumed that the slope of DWL and CWL regression curves (i.e. the wind pressure coefficients of the two components) was related to the slope of the overall wind pressure relative to the wind speed regression curve. The slope of regression between wind pressure difference and wind speed is the coefficient of wind pressure difference multiplied by wind speed in traditional wind pressure difference drift method. According to different wind pressure characteristics, different drifting targets at sea are classified, and the AP89 model is fitted with the regression coefficients of downwind and lateral wind of the wind pressure difference with 63 kinds of data obtained in the experiment.

When AP89 model is used to calculate the wind pressure drift vector of a specific drifting target, nine AP89 wind pressure drift parameters and the wind field information of sea state at the time of the incident need to be input. These nine parameters are the slope, intercept and standard deviation of regression equation in three directions obtained by decomposing wind-induced drift vector and linear regression in a large number of drift data of such floating objects. (Downwind direction is DWL, right-biased vertical wind direction is +CWL, left-biased vertical wind direction is -CWL) and the standard deviation follows normal distribution, so the following mathematical formula of AP89 model is obtained:

$$DWL: L_d = a_d \times W_{10} + b_d + \varepsilon_d$$

$$\pm CWL: L_{\pm c} = a_{\pm c} \times W_{10} + b_{\pm c} + \varepsilon_{\pm c} \quad (1)$$

In equations (1), L represents the fitting wind pressure drift vector, a represents the slope of each wind pressure drift equation; b represents the intercept of the wind pressure drift equation; ε is the regression deviation of each wind pressure drift equation, $\varepsilon = norm \times S_{y/x}$, $norm$ is the random number selected from the standard normal distribution function, and $S_{y/x}$ is the regression standard variance of each wind pressure drift equation.

Drowning is the most dangerous in maritime accidents. After falling into the water, the core temperature of a person cannot be maintained due to the influence of the sea surface environment, and the search and rescue response time is greatly shortened.So in this paper, a person in the water (PIW) wearing a life jacket lying flat on the sea surface is adopted as the tracking target of drifting. The data obtained according to a. Allen's experiment are shown in Table 1.[8]

Tab.1 AP98 parameters of PIW(horizontal position,survival suit)

Float category	Dependent variable	b (cm/s)	α	$S_{y/x}$ (cm/s)
PIW（horizontal position,survival suit）	DWL	1.12	1.71%	3.93
	+CWL	-3.30	1.36%	1.71
	-CWL	-2.65	-0.13%	1.62

Although AP98 model is the most accurate wind pressure drift model at present, it also has many limitations: CWL left-deviation probability and right-deviation probability must be equal; Once the CWL is determined to be left-biased or right-biased, the CWL will maintain its fixed attitude and will not reverse; At the same time, the model is only a strict function of the wind speed vector at 10m height.

3. Probability of Containment model based on Monte Carlo simulation

3.1. *Probability of Containment*

Probability of Containment (POC) refers to the Probability of the existence of the search target in the region, or in a partition of the region, and is a trade-off between the possibility that the search object is contained in a certain region.[9] In search and rescue operations, a good inclusion probability model can improve search efficiency, reasonably optimize the allocation of search resources, and provide a theoretical basis for making a reasonable search plan.

3.2. *Monte Carlo simulation*

Monte Carlo simulation in the drift model, its basic idea is to put the floater is defined as a single particle, the moving trajectory of the particles is a satisfying the probability distribution of random process, under the same rule, a large number of these particles, and the movement of these particles unrelated, so these particles in statistical area after a period of time is the probability of floater drift region.[10] In this paper, it is assumed that the position of the floating object on the sea meets the Markov process, that is, the probability density function of the future shape of the floating object depends only on the current external force state of the floating object, and has nothing to do with the special process of the object reaching the current state.[11] Compared with the conventional analytical method to determine the search area, Monte Carlo random particle method has many advantages. For example, when multiple types of distress targets need to be searched at the same time when a maritime accident occurs, or when the time and location of the accident are uncertain, the MONTE Carlo random example method can be used to quickly establish the POC model of the target. The reason is that the particle drift motion between Monte Carlo method is independent of each other, so you can according to different types, different time, different location of target set several simulated particle separately corresponding drift motion, and then to overlay all particles position, to get the overall target model, the analytical method is unable to do it.[12]

3.3. Establish POC method for Marine distress target

In the establishment of POC model of the shipwrecked target, the shipwrecked target is abstracted into N particles, and it is assumed that the motion of these particles is independent of each other, and they have all the drift characteristics of the shipwrecked target. If N particles are initialized at the last known position of the distress target, the distribution probability of particles after t time can represent the POC model after t time of drift of the distress target at sea.[13] The specific steps are as follows :

(1) Initialize N(N should be large enough, but should be even) particle positions at the last known location of the target, denoted as Loc_{0N}, $Loc_{0N} = [\ln g_{0N}, lat_{0N}]$, and $\ln g_{0N}, lat_{0N}$ represents the longitude and latitude of the initial position of N particles respectively.

(2) The particles drift in the sea area affected by the Marine environment and their drift characteristics are the same as those of the target in distress. The drift velocity of N particles conforming to drift model characteristics was randomly simulated by computer:For given wind speed W_{10m} and current velocity V_c at 10m height,N random numbers are selected from the standard normal distribution to generate N groups of wind speed in the downwind direction when calculating DWL. The corresponding N/2 group +CWL and N/2 group -CWL can be obtained by calculating CWL with the same method (probability of deviation from left to right is equal), which can be synthesized to simulate drift velocity V_N, $V_N = [v_1, v_2 ... v_N]$ Among them v_N for N drift vector velocity of particle.

(3) According to the initial position Loc_{0N} and drift velocity V_N of the particle, the position Loc_{tN} of the particle at time t can be obtained after the integral of the drift trajectory for a certain period of time. The minimum rectangular region G_t containing all particles is the region to be searched.

(4) The grid method is used to divide the search area G_t into several rectangular sub-areas g_{ij}. Where g_{ij} represents the sub-region of row i, column j.

(5) Retrieves the number of particles contained in g_{ij}, represented by $N_{g_{ij}}$. The following formula can be used to solve the ratio of the number of particles $N_{g_{ij}}$ containing the drifting target in sub-region $N_{g_{ij}}$ to the total number of particles N, that is, POC value, marked by P_{tij}. The probability matrix P_t of the target can be obtained by calculating the POC values of all sub-regions.

$$P_{tij} = N_{g_{ij}} / N \tag{2}$$

To sum up, area A to be searched and probability matrix B constitute the target POC model at time t. It should be noted that the value of the rectangular side length of the neutron region in step 4 is related to the accuracy of the established distress target POC model and the amount of data generated: If the side length value is too large, the division of the search area is not fine enough, and the accuracy of the target POC model will be affected. If the side length is too small, the precision of POC model can be improved to a certain extent, but redundant data will also be generated, which affects the calculation speed. Therefore, the side length must be set based on actual search and rescue mission scenarios.

4. POC simulation of Marine distress target in different scenarios

4.1. A single drowning person with known time and location of distress

Taking the search and rescue work of a single drowning person in the north of yellow Sea as the simulation scene. Suppose a fishing boat operating in the area breaks down, and the person in distress calls for help to the maritime rescue center and reports his current position, then abandons the ship wearing a life jacket and drifts. The total sea velocity at the time of the accident was 10 knots due northeast; Two knots at 10m, southwest wind.

Monte Carlo random particle method was used to establish the target POC model. The coordinate system was established with the origin of the last drowning position reported by the person in distress. The N=50000 particles were initialized at the origin, and the drift velocity was simulated according to the environmental information to obtain the drift trajectory of the particles. The POC model of the target in distress was established based on the position information of the particle at 1h.

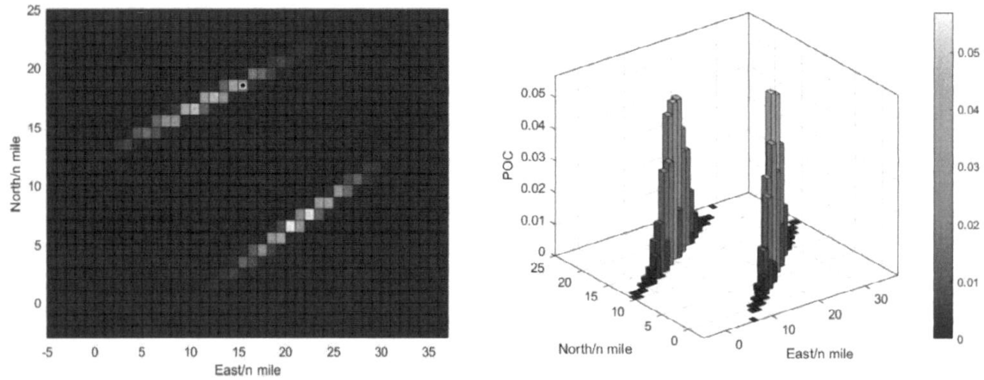

Fig.2 POC model 1 hour after drowning at sea

Fig. 2 shows the target POC model 1h after the sea personnel fell into the water, where each square represents a region to be searched, and different colors and levels in the square represent the size of target POC in the region.It can be seen that the possible areas mainly include two strip divergent areas due to the wind pressure deviation phenomenon left and right after the drift for 1h. The highest POC was 5.7% in the [(22,23),(7,8)] region, 22-23 nm east and 7-8 nm north of the initial alarm point.

4.2. Individual overboard with unknown time and location of distress

Assume that the maritime search and rescue center received a danger alarm at 10 o 'clock and passengers were missing from a cruise ship sailing in Bohai Bay. It is predicted that the person may fall into the water, but the specific time is unknown.

The missing passenger is known to have been seen eating at the restaurant by other passengers around 8am and is presumed to have fallen between 8pm and 10pm. Between 8 and 9 a.m., the ship was heading due northeast at 14 knots; Between 9 and 10, sail 30 degrees east and north at 10 knots. The total sea velocity at the time of the incident was 2 knots, due northwest direction; Wind 10 knots at 10 meters, southeast.

The initial position of the cruise ship at 8 o 'clock was set as the origin to establish the coordinate system. According to 8-10 time points, N=50000 particles are assigned with equal probability to represent the people falling into water at different time points (t_n), and then the initial positions of particles are calculated according to the navigation information of the cruise ship. According to the environment information to simulate the drift trajectory, the establishment of the distress target POC model. The results are shown in Fig. 3 and Fig.4.

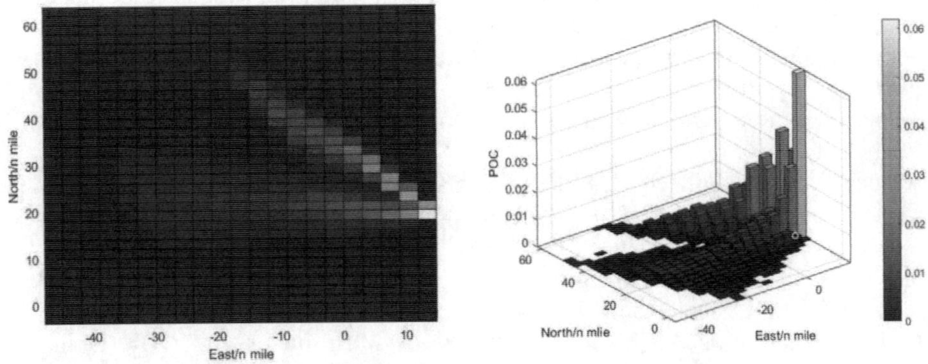

Fig.3 POC model of the person who fell overboard at 9 o 'clock

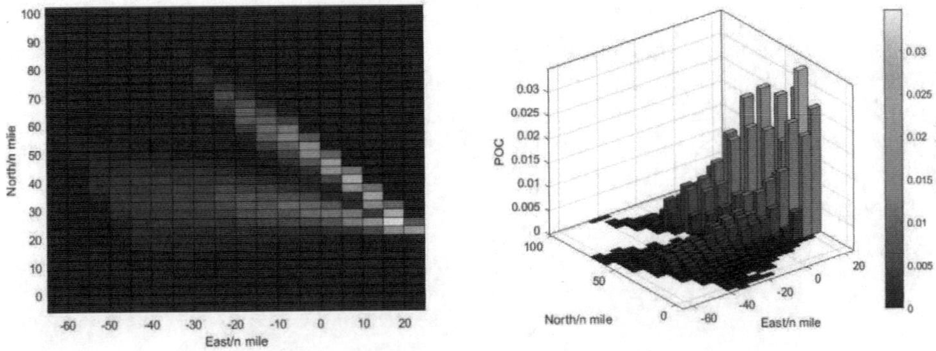

Fig.4 POC model of the person who fell overboard at 10 o 'clock

Since the time of the missing persons falling into the water is unknown, the model is divided into two time points: 10 o 'clock and 9 o 'clock.Common ground: POC is higher as you get closer to the route, and more scattered as you get farther away.Different ground: the more concentrated the model is at 9 o 'clock, the more dispersed the model is at 10 o 'clock, indicating that the earlier the search and rescue work, the higher the efficiency.

At 9 o 'clock, the highest POC region appeared 12 to 15 nautical miles east of the initial point and 18 to 20 nautical miles north, reaching 6.1%. At 10am, the highest POC region occurred 15-20 nm east of the initial point and reached 3.5% 24-27 nm north of the initial point.

5. Conclusion

In this paper, the inclusion probability model of the target is studied and established, and the POC model of the target is simulated and analyzed in different scenarios, and the results show that it has good applicability. The proposed model provides effective theoretical support for the initial decision-making process of maritime search and rescue.

References

[1] International Maritime Organization/International Civil Aviation Organization.International Aviation and Maritime Search and Rescue Manual [M]. Dalian: Dalian Maritime University Press, 2001.

[2] Fang-bing Xiao, Yong Yin, Yi-cheng Jin, Xin-yu Zhang. Journal of dalian maritime university,2011,37(02):58-63.

[3] Jun Pan. Determination of maritime search and rescue area model based on Monte Carlo method [J]. China water transport (second half),2012,12(09):43-44.
[4] Ya-dong Lu, Ya Zhou. Optimal Search and Rescue Model: Updating Probability Density Map of Debris Location by Bayesian Method. 2015, 1(1)
[5] Wei Pan. Selection and optimization of search area for Ships in distress [D]. Dalian Maritime University,2016.
[6] Lawrence D. Stone, Colleen M. Keller, Thomas M. Kratzke, et al. Search for the Wreckage of Air France Flight AF 447. 2014, 29(1):69-80.
[7] Allen A A . Leeway Divergence[J]. Leeway Divergence, 2005.
[8] Hong-qi Hu, Jian-hua Chen, Gang-qiang Li. Key words: Maritime Drowning Personnel search area Simulation; Monte-Carlo method [C]. Proceedings of the 17th Chinese.
[9] Hao Li. Research on Inclusion Probability of Maritime Search [D]. Dalian Maritime University,2011.
[10] Hua-lin Jiang, Zhao-chen Sun , Li Li, Bo LIU, Jun-ping Liu. Maritime search area determination model based on Monte Carlo method [J]. Waterway and port,2011,32(04):285-290.
[11] Jun Pan. Determination of maritime search and rescue area model based on Monte Carlo method [J]. China water transport (second half),2012,12(09):43-44.
[12] Ru-yi Wang, Hong-lie Li, Zhen Yang, Hang Yang. Sea water based on monte carlo method are drifting track prediction [J]. Journal of ship electrical engineering, 2020, 40 (12) : 48-50 + 60.
[13] Qian-wen Jia. Research on maritime distress target search algorithm based on agent [D]. Hainan University,2019.

Multi-view Attention Mechanism Learning for POI Recommendation

Fujian Zhu[1*], Shaojie Dai[2]

[1,2] Department of Computer Science and Technology, Ocean University of China, Qingdao, Shandong, 266100, China

[*]Corresponding author's e-mail: zfj@stu.ouc.edu.cn

Abstract. POI(point of interest) recommendation is a very necessary research field in both academic and commerce, however, predicting users' potential points of interest is always faced with the problems of data sparsity and context semantics. Some studies have shown that graph embedding technology alleviates the problem of data sparsity to a certain extent. However, neither graph embedding techniques nor unsupervised learning models can adaptively learn the different effects of multiple relations between users and POIs, respectively. In view of this, we leverage the contextual information of users and POIs to build the multi-view affinity graphs(e.g. User-User, POI-POI and User-POI), and learn the latent representations of users and POIs based on the **G**raph **E**mbedding technology and **A**ttention mechanism, namely the GEA model. In particular, we first construct multi-view affinity graphs by using user's social relationship, geographical distance and check-in behaviour, and embed them into a low dimensional shared space to learn the latent representation of users and POIs. Afterwards, in order to take advantage of the different effects of multiple relationships in the final recommendation task, we exploit the attention mechanism to obtain the fused latent representation and make recommendation according users' potential preferences. Finally, we design a multi-task objective function for joint optimization to obtain more accurate recommendation results. Extensive experiments on Gowalla have verified the effectiveness of our model.

1. Introduction

In recent years, the recommendation technology has played an increasingly important role in location-based social network services. To pursue a high-quality POI recommendation, it is crucial to analyse users' interests and make full use of users' social circle. Providing users with potential POI not only enriches users' life, but also brings more commercial value to real businesses, such as targeted advertising. Hence, it is essential to further explore users' POI.

With the rapid development of graph embedding technology, it brings a new perspective for further research on POI recommendation. For example, firstly, graph embedding technology has been verified and applied in many fields, such as emotion analysis, link prediction and text mining [1]. Then, in the field of POI recommendation, JLGE [2] model takes into account the relationship between user check-in periodicity and the location of interest points, and embeds the influencing factors into the potential space for joint learning; HREC framework improves the accuracy of recommendation by embedding learning and ranking module into heterogeneous graph to learn more potential information, and optimizing the objective function[3]. In addition, PRME-G model [4] embeds user preference and order transformation into two different spaces, which may lose some potential relationships between users and POIs. Finally, GE [5] verifies that using graph embedding technology and unified joint training of multiple factors can make more accurate location recommendation. PPR [6] model improve the accuracy of prediction by jointly embedding user and POI information and using spatiotemporal neural network framework. The research of GE and PPR shows that graph embedding technology has a good application effect in POI recommendation.

Despite their excellent performance, there are still some challenges for POI recommendation, which are summarized as follows: *1) Data sparsity problem.* The well-known short video recommendation system only needs online operation to complete data collection, because the types of videos that users like to browse can be regarded as their hobbies, and the same is true for music and e-book recommendation. However, a POI is a specific location in the real world. People must go to the location by himself and check-in, so as to generate data. The cost of data source is much higher than that of other recommendation systems, which also leads to the loss of advantages of recommendation algorithm based on collaborative filtering in POI recommendation. *2) Context information influence.* In location-based social network recommendation, in addition to personal interests and preferences, there are more potential relationships hidden in its context information. For example, in the social circle of users, users in the same circle may have more similar POI and common characteristics, and some potential points of interest of users may be excavated through the social circle. Afterwards, geographical factors(e.g. the distance between two places) will also affect users' preferences, so the distance between two POI should also be considered. *3) Personalized preferences.* Everyone is unique in this world and has different levels of favourite things. In the personalized recommendation results, the importance of each factor, such as social relationship and geographical location, is different. Additionally, it is necessary to find the optimal expression of users and preferences. We note that many studies ignore the different importance of each factor, which is obviously not conducive to users' personalized recommendation.

To overcome these challenges, we present a novel model GEA based on graph embedding technology which shows prefect results in the field of recommendation. Specifically, in the first, we build User-User graph, POI-POI graph and User-POI graph for users and POIs. The aforementioned three graphs are embedded into a low dimensional shared space for common learning to learn the latent representation of each vertex of users and POIs. Then, inspired by PPR, we use a rich graph connection method to alleviate the problem of data sparsity, that is, adding second-order neighbours to low in/out degree nodes. Finally, in order to improve the POI recommendation results, by adaptively learning the different effects of various relationships(i.e., User-User, User-POI, POI-POI) in the final recommendation task, a supervision objective function is designed to make the user more similar to the POI latent representation he/she has checked, so as to mine the potential preference of users. We conduct massive experiments on Gowalla dataset to verify the effectiveness of our model.

The main contributions are summarized as follows: 1) For LBSN based recommendation, we propose a GEA model. In this model, we carefully constructed three different relationship graphs, which combine the user's social relationships, geographical distance and the user's check-in behaviour to learn the latent representation between user and POI. 2) We use the attention mechanism that fuse different relations with users and POIs to adaptively capture the different importance of the three graphs(i.e., User-User, POI-POI and User-POI) in the final recommendation task. Meanwhile, we use the carefully designed supervised objective function to make the latent representation of users more similar to the POI they have visited, to capture the users' personalized preference. 3) We conduct extensive experiments on Gowalla dataset(*http://snap.stanford.edu/data/loc-Gowalla.html*) and compare with four variants(i.e., GEA-U, GEA-P, GEA-UP and GEA-A in Section5.3). Our model has better recommended results.

2. Related work

POI recommendation has been studied by many scholars because it is of great significance to business and academia, and is affected by many factors.

At the beginning, researchers use collaborative filtering(CF) and matrix factorization(MF) method for POI recommendation. Ye et al. [7] combines user-based collaborative filtering method for POI recommendation by analyzing geographic and social influence. Lian et al. [8] take into account geographic influences and use a weighted matrix factorization framework to address the problem of data sparseness; TSG-MF [9] integrates social regularization, geographic factor influence, and multi-label influence into a prefect matrix factorization framework, and simultaneously verifies the important roles of social relations and geographic factors in POI recommendation; Liu et al. [10] use a matrix factorization-based TL-PML model to analyze the matching degree between user interests and POI topics and word-of-mouth. Then, many scholars have studied the application of Markov chain model in POI recommendation, Cheng et al. [11] utilizes personalized Markov chain and users' movement constraints to complete the next POI recommendation. The FPMC model(including the public Markov chain model and matrix factorization model) proposed by Rendle et al. [12] is used for next-basket recommendation. However, these studies lack a comprehensive consideration of social influence, geographic location and contextual information of user check-in history. In addition, recommendation methods based on graph embedding techniques have improved the prediction accuracy in multiple domains, for example, Tang et al. [1] and Chen et al. [13] have validated graph embedding techniques in text classification and music prediction effectiveness. In terms of POI recommendation, Xie et al. [14] use a graph-based embedding model to embed the homogeneous graph for the next recommendation, and PRME-G model [4] use two spaces for training user preferences and temporal patterns. However, the above methods lack analysis and utilization of potential information about users and POIs. Recently, Wu et al. design a graph neural network based POI recommendation model that comprehensively utilizes collaborative filtering information and spatiotemporal information [15]. The PPR [6] model integrates graph embedding technology and spatiotemporal neural network to improve the accuracy of top-K POI recommendations. Inspired by previous research, we comprehensively consider the contextual information, influencing factors of user and POI and combine the attention mechanism to recommend a list of POIs for each user that he/she is interested but never visited.

3. Problem definition

In this section, we first describe the proper definition used in this paper. Then, the problem definition of POI recommendation is given.

Definition 1: (POI) A POI represents a real place in the real world, such as a ski resort, museum, etc. We use $v=(p, l)$ to represent a unique POI, where p is the name of the POI v, l represents the specific geographical location of v, expressed in longitude and latitude.

Definition 2: (Check-in behaviour) A check-in behaviour means that user u reaches the point of interest v at time t, which is represented by a four tuple (u, v, t, l_v).

Definition 3: (User activity) User activity represents a collection of user u check-in behaviours, denoted by C_u. It mainly includes the name, location and check-in time of a POI. In this paper, we use data set C to represent all C_u, that is, the set of check-in behaviour of all users.

Definition 4: (User-User graph) User-User graph is defined $G_{u,u} = (U, \varepsilon_{u,u})$, U and $\varepsilon_{u,u}$ respectively represent the set of users and the set of edges between users. When u_i and u_j have a social relationship(such as friends or colleagues, etc.), if they have visited a common POI v_k, that is, $v_k \in C_{u_i}$ and $v_k \in C_{u_j}$, there will be an edge e_{ij} from u_i to u_j, and the weight w_{ij} is used to indicate the strength of their relationship.

Definition 5: (POI-POI graph) POI-POI graph is represented by $G_{v,v} = (V, \varepsilon_{v,v})$, where V and $\varepsilon_{v,v}$ represent POI set and edge set between POI respectively. When a user access v_j after v_i, there is an edge between v_i and v_j.

Definition 6: (User-POI graph) User-POI graph is represented by $G_{v,v} = (U, V, \varepsilon_{u,v})$. When the user $u \in$ U has a check-in behaviour record at the POI $v \in V$, we connect an edge between u and v.

Problem 1:(POI recommendation) Given users' check-in records, POI recommendation aims at recommending a list of POIs for each user that he/she is interested but never visited.

4. Methods

In this section, we first describe how to build multi-view graph in GEA model, and then introduce the attention mechanism to adaptively learn the importance of multiple relationships. Lastly, we show that the optimization of the model can adaptively learn user preferences. The model framework is shown in figure 1.

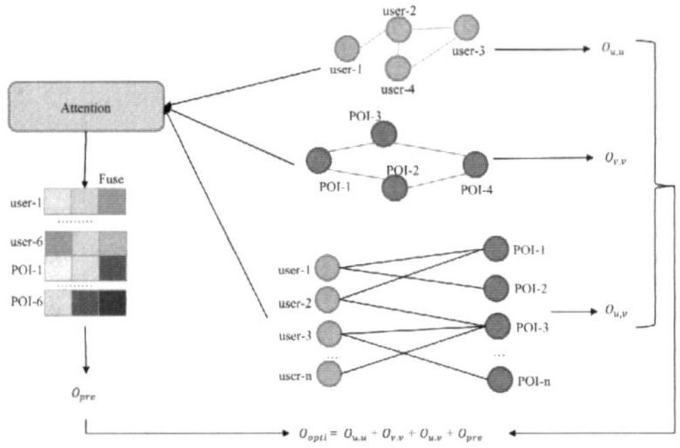

Figure 1. Framework diagram of GEA model.

4.1. Build multi-view graph

In order to make full use of the potential information between nodes, we carefully analyze a variety of factors and relationships between user and POI, use the user, POI and check-in information to build $G_{u,u}$, $G_{v,v}$ and $G_{u,v}$ graphs respectively, and embed them into low-dimensional shared space for learning.

Firstly, taking into account social relationships is conducive to better learning latent representation, which plays an important role in many user-oriented recommendation studies [9]. We carefully analyzed the user's social circle and check-in POI data, and build $G_{u,u} = (U, \varepsilon_{u,u})$, as described in Definition 4, when user u_i and u_j have a social relationship and have visited the common POI $v_k \in V$ there is an edge e_{ij} with a weight of w_{ij}, from u_i to u_j. We express the intersection number of POIs visited by u_i and u_j as $| C_{u_i}^r \cap C_{u_j}^r |$ and the union number as $| C_{u_i}^r \cup C_{u_j}^r |$, so the calculation formula of weight is:

$$w_{i,j} = \frac{|C_{u_i}^r \cap C_{u_j}^r|}{|C_{u_i}^r \cup C_{u_j}^r|} \tag{1}$$

where $C_{u_i}^r$ represents the set of C_{u_i} after removing the duplicate POIs.

Then, according to the user's check-in history, we first sort the check-in records of each user in ascending order by timestamp, and there is an edge between the POIs of two adjacent check-in record. As described in Definition 5, if there is an edge between v_i and v_j, it means that at least one user has accessed $v_j(v_i)$ after accessing $v_i(v_j)$. When selecting points of interest, users prefer to select POI closer to the current location. Therefore, we consider using the distance between POIs to represent the weight between two POIs. The weight w_{ij} is calculated as follows:

$$w_{i,j} = \frac{1}{d(mh) + k} \tag{2}$$

where, $d()$ denote the Manhattan distance between two POIs, and k is a small floating-point number. On the one hand, it can avoid calculation errors when the Manhattan distance is zero, and on the other hand, it can control the weight in a reasonable range.

Finally, we build the User-POI graph, that is $G_{u,v}$. As described in Definition 6, when the user $u_i \in U$ checks in at POI $v_i \in V$, there is an edge e_{ij} between u_i and v_j. We take the number of times the u_i accesses v_j as the weight of e_{ij}. Compared with other POI v_k $(v_k \in V, k \neq j)$, the more times a user u_i

accesses a POI v_j, it means that the u_i prefers v_j, and the greater the weight w_{ij} of e_{ij}. The w_{ij} is defined as follows:

$$w_{i,j} = freq(u_i, v_j) \tag{3}$$

where, $freq(,)$ indicates the number of u_i check-ins at v_j. Conversely, if the user u_i has not accessed v_j, there is no edge between u_i and v_j.

The above are three kinds of network graphs carefully designed from multi-view, all of which are undirected graphs. After building multi-view graph, the next part will introduce how to fuse three different relationships in different graphs for POI recommendation through adaptive learning with attention mechanism.

4.2. Fuse different relation with attention

Inspired by the LINE [16] model of learning homogeneous network relational representation, we extend it to the heterogeneous graph nodes obtained by building graphs from multi-view.

In graph $G_{u,u}$ and $G_{v,v}$, we treat each user as a node u_i^s and each POI as a node v_i^g respectively. In graph $G_{u,v}$, we treat each user as a node u_i^c and each POI as a node v_i^c. We embed the three graphs into the low dimensional space for joint learning, and use $\boldsymbol{u_i^s}$, $\boldsymbol{u_i^c}$, $\boldsymbol{v_i^g}$ and $\boldsymbol{v_i^c}$ to represent the embedding vectors of nodes u_i^s, u_i^c, v_i^g and v_i^c respectively, where node u_i^s and u_i^c represent the same user and node v_i^g and v_i^c represent the same POI. Table 1 also lists these variables.

Table 1. Main notations and definitions.

Notation	Definition
u_i^s, u_i^c	the node of user u_i in $G_{u,u}$ and $G_{u,v}$
v_i^g, v_i^c	the node of POI v_i in $G_{v,v}$ and $G_{u,v}$
$\boldsymbol{u_i^s}$, $\boldsymbol{u_i^c}$	embedding vectors of node u_i^s and u_i^c
$\boldsymbol{v_i^g}$, $\boldsymbol{v_i^c}$	embedding vectors of node v_i^g and v_i^c

For user node $u_i^s(u_i^c)$, we learn two latent representations of $\boldsymbol{u_i^s}$ and $\boldsymbol{u_i^c}$ from the $G_{u,u}$ and $G_{u,v}$ graphs. When using the two latent representations for calculation, we need to analyse their different effects on the recommendation results. To address this problem, we use the attention mechanism [17] to adaptively learn their different effects on the final recommendation result, and use $\boldsymbol{u_i^f}$ to represent the latent representation of node $u_i^s(u_i^c)$ after processing.

In the first, we use d to denote the embedding dimension to be learned, and use the following formulas to calculate e_1 and e_2 respectively.

$$e_1 = \boldsymbol{q}\tanh(\boldsymbol{W}\boldsymbol{u_i^s} + \boldsymbol{b})^T \tag{4}$$

$$e_2 = \boldsymbol{q}\tanh(\boldsymbol{W}\boldsymbol{u_i^c} + \boldsymbol{b})^T \tag{5}$$

where, $q \in R^{1\times d}$, **W** is the trainable matrix. Then, we can calculate the importance of the roles played by $\boldsymbol{u_i^s}$ and $\boldsymbol{u_i^c}$ as follows:

$$w_1 = \frac{\exp(e_1)}{\exp(e_1)+\exp(e_2)} \tag{6}$$

$$w_2 = \frac{\exp(e_2)}{\exp(e_1)+\exp(e_2)} \tag{7}$$

where, $exp()$ represents exponential function. Finally, using w_1 and w_2 we can calculate the final vector representation:

$$\boldsymbol{u_i^f} = w_1 \cdot \mathrm{u_i^s} + w_2 \cdot \mathrm{u_i^c} \tag{8}$$

where, $\boldsymbol{u_i^f}$ is the final latent representation of user node $u_i^s(u_i^c)$. For POI node $v_i^s(v_i^c)$, we use the same way to deal with it.

Now that we can effectively learn latent representations of users and POIs, in order to further satisfy personalized POI recommendations, we will introduce how to model users' personalized preferences.

4.3. Model users' personalized preferences

After 4.1 and 4.2, all latent representations of users and POIs are mapped into a shared space, however, their latent representations may ignore some hidden connections between users and POIs. Assuming that a person's preferences can be represented by the set of POIs he/she has visited, our purpose is to make a user more similar to the POIs he/she has visited, so as to improve the effect of user personalized preference recommendation. To this end, we construct the following objective function to train the model:

$$O_{pre} = -\sum_{u_i \in U} \sum_{v_j \in C_{u_i}} \langle \boldsymbol{u_i}, \boldsymbol{v_j} \rangle \tag{9}$$

4.4. Model optimization

There are three kinds of graphs embeddings in our GEA model, and the intuitive method is by minimizing the sum of the objective functions. Due to the heterogeneous graphs used in the model, we adopt the joint optimization [5], that is, alternatively sampling from the three sets of edges(i.e., $\varepsilon_{u,u}$, $\varepsilon_{v,v}$ and $\varepsilon_{u,v}$) respectively. In addition, we model the personalized preference in Section4.3 and obtain the objective function O_{pre}, so we also add it to the final objective function O as following:

$$O = O_{u,u} + O_{v,v} + O_{u,v} + O_{pre} \tag{10}$$

where,

$$O_{u,u} = -\sum_{e_{i,j} \in \varepsilon_{v,v}} \log_p \left(v_i^s \mid v_j^s \right) \tag{11}$$

$$O_{v,v} = -\sum_{e_{i,j} \in \varepsilon_{v,v}} \log_p \left(v_i^c \mid v_j^c \right) \tag{12}$$

$$O_{u,v} = -\sum_{e_{i,j} \in \varepsilon_{u,v}} \log_p \left(v_i^c \mid v_j^c \right) \tag{13}$$

where, $p(v_i \mid v_j)$ denotes that given a node v_j, the possibility of a real edge between node v_i and node v_j. The $O_{u,u}$, $O_{v,v}$ and $O_{u,v}$ aims to maximize the similarity between nodes if there is an edge exists. By minimizing the objective function O, we make the user more similar to the POIs he/she preferred, so that the more accurate recommendation results can be obtained.

5. Experiments

5.1. Dataset

We conduct a large number of experiments on the LBSN dataset Gowalla, which is a very popular location check-in website. The basic data statistics are shown in table 2. We select 251378 check-in record generated by 6846 users from Feb 2009 to Oct 2010 in Asia. Each check-in record stores the user-Id, check-in timestamp, POI-Id and POI location (i.e., longitude and latitude). During data pre-processing, we filter out POIs visited by less than 5 users and users with less than 10 check-ins.

Table 2. Basic statistics of Gowalla.

Dataset	Gowalla
users	6846
POIs	74856
check-ins	251378
time span	Feb 2009 – Oct 2010

5.2. Experiments result

In this paper, we use the two evaluation metrics mentioned in PPR [6] to evaluate the model, that is Precision(*Pre*) and Recall(*Rec*), and they are defined as follows:

$$Pre@k = \frac{1}{|D_T|}\sum\frac{|R_k^u \cap T_k^u|}{|R_k^u|} \tag{14}$$

$$Pre@k = \frac{1}{|D_T|}\sum\frac{|R_k^u \cap T_k^u|}{|T_k^u|} \tag{15}$$

where $|D_T|$ is the size of test data set, R_k^u is the top-K POIs of user u with the highest score after training, and T_k^u denotes the real POI recommendation of user u in the test set.

We use 80 percent of the data set as training data and the remaining as test data. In the experiments, we set learning rate to 0.025 in graph embedding, d of embedding dimension to 128, the number of negative samples to 5, and the batch size set to 2000. We run it repeatedly on the Gowalla dataset for 5 times, and calculate the average of *Pre* and *Rec* as the final result, as shown in table 3, the results show that our model achieves good results.

Table 3. Result of GEA.

	k=5	k=10
Pre	0.014	0.009
Rec	0.029	0.057

5.3. Ablation experiment

In order to further analyze the influence of user social relations, geographical, user check-in history and attention mechanism on the model, we carefully design four variants, i.e., GEA-U, GEA-P, GEA-UP and GEA-A:

1) GEA-U: the user's social relationship are not considered in the model, that is, we do not consider $O_{u,u}$, and other factors remain unchanged.

2) GEA-P: the distance relationship of POI is not considered in the model, that is, we do not consider $O_{v,v}$, and the rest of the factors remain unchanged.

3) GEA-UP: the user's check-in history is not considered in the model, that is, we do not consider $O_{u,v}$, and the rest of the factors remain the same.

4) GEA-A: the attention mechanism is not used in the model to obtain the vector representation of the fused user and POI, that is, we do not consider O_{pre}, and the rest of the factors remain the same.

Table 4. Result of GEA.

	GEA-U	GEA-P	GEA-UP	GEA-A
Pre@5	0.011	0.006	0.010	0.001
Rec@5	0.024	0.017	0.021	0.002

We show the experimental results for the four variants at k=5 in table 4, we can conclude that the GEA model get the best results in combination with the above factors. Both social relations and geographical relations have different effects on users' preference recommendation to different degrees. The attention mechanism part we designed plays a more important role in the model, which greatly improves the accuracy.

6. Conclusion

In this work, we propose a multi-view POI recommendation model combined with an attention mechanism. We build three graphs by analyzing the influence of User-POI relationships, geographic location, check-in sequence and social relationship. Then, we incorporated attention mechanism to fuse user and POI representations in different graphs. Furthermore, we jointly optimize the model through an unsupervised objective based on the topological structure according to the relationship between user and POI, and a supervised objective to capture users' personalized preferences. Finally, we verified the effectiveness of the model on the Gowalla dataset, and conducted ablation experiments to explore the influence of each component of GEA on the recommendation results. In the future work, we will further optimize the graph and objective function by taking into account the POI categories or the variation rules of user check-in time to.

References

[1] TANG, J., QU, M., MEI, X.Z. (2015) PTE: predictive text embedding through large-scale heterogeneous text networks. In: Proceeding of the Special Interest Group on Spatial Information. New York. pp.1165-1174.

[2] CHRISTOFORIDIS, G. , KEFALAS, P. , PAPADOP- OULOS, A., et al. (2018) Recommendation of Points-of- Interest using graph embeddings. In: DSAA. Italy. pp. 31-40.

[3] YUAN, F., GUO, G., JOSE, J., et al. (2016) Joint geo-spatial preference and pairwise ranking for point-of-interest recommendation. In: ICTA. San Jose. pp.46-53.

[4] FENG, S. S., LI, X. T., ZENG, Y. F., et al. (2015) Personalized ranking metric embedding for next new poi recommendation. In: AAAI. New York. pp.2069–2075.

[5] XIE, M., Yin, H., Wang, H., et al. (2016) Learning graph-based poi embedding for location-based recommendation. In: CIKM. Atlanta. pp. 15–24 .

[6] DAI, S. J., YU, Y. W., Fan, H,. (2021) Personalized POI Recommendation: Spatio-Temporal Representation Learning with Social Tie. In: DASFAA. Taipei. pp. 558-574.

[7] YE, M., YIN, P. F., LEE, W. C., et al. (2011) Exploiting geographical influence for collaborative point-of- interest recommendation. In: Proceedings of the 34th International ACM SIGIR Conference on Research and Development in Information Retrieval. Beijing. pp. 325-334.

[8] LIAN, D., ZHAO, C., XIE, X., (2014) Joint geographical modeling and matrix factorization for point-of-interest recommendation. In KDD. New York. pp. 831–840.

[9] ZHANG, Z., LIU, Y., ZHANG, Z., et al. (2019) Fused matrix factorization with multitag, social and geographical influences for POI recommendation. In: WWW. Shanghai. pp.1135–1150.

[10] LIU, B., XIONG, H. (2013) Point-of-interest recommendation in location based social networks with topic and location awareness. In SDM. Boston. pp.396–404.

[11] CHENG, C., YANG, H.G., MICHAEL, R,. et al. (2013) Where you like to go next: Successive point-of-interest recommendation. In: Proceeding of the International Joint Conference on Artificial Intelligence. Beijing. pp. 2605-2611.

[12] RENDLE, S., FREUDENTHALER, C., SCHMIDT- THIEME L. (2010) Factorizing personalized Markov chains for next-basket recommendation. In: Proceedings of the 19th International Conference on World Wide Web. ACM. Raleigh. pp.811-820.

[13] Chen, S., Moore, J. L., Turnbull, D., et al. (2012) Playlist prediction via metric embedding. In: KDD. Beijing. pp.714–722.

[14] XIE, M., Yin, H., XU, F., et al. (2016) Graph-based metric embedding for next poi recommendation. In: WISE. Shanghai. pp.207-222.

[15] WU, S. W., ZHANG, Y. X., GAO, C. L., et al. (2020) Anonymous recommendation of Point-of-Interest in mobile networks by graph convolution network. In: Proceedings of the Data Science and Engineering. Nanjing. pp.433-447.

[16] Tang, J., QU, M., Wang, M. Z., et al. (2015) Line: Large- scale information network embedding. In WWW. Florence. pp.1067–1077.

[17] Wang, X., Zhu, M., Bo, D., et al. (2020) Am-gcn: Adaptive multi-channel graph convolutional networks. In: KDD. Virtual Event CA. pp. 1243–1253.

Transfer learning with CNN-LSTM model for capacity prediction of lithium-ion batteries under small sample

Gang Zou[1,2], Zhen Yan[1], Chengying Zhang[1], Lei Song[1*]

[1]Key Laboratory of Space Utilization, Technology and Engineering Center for Space Utilization, Chinese Academy of Sciences, 100094, Beijing, China.

[2]University of Chinese Academy of Sciences, 100049, Beijing, China

*Corresponding author's e-mail: songlei@csu.ac.cn

Abstract—Nowadays, the application of lithium-ion batteries in various industries including our daily life is increasing, so the evaluation of the safety and reliability of lithium-ion batteries is essential. Assessing the performance of lithium-ion batteries accurately and in real-time has become a more critical research direction. However, in most application scenarios, the problem of small sample is often hidden in performance evaluation tasks of lithium-ion batteries. A model-based transfer learning framework to solve this problem is presented in this paper. Firstly, a CNN-LSTM fusion model is proposed by combining the solid spatial feature extraction capability of convolutional neural network (CNN) and the solid temporal dimensional feature extraction capability of long short-term memory network (LSTM); further, a model-based transfer learning approach is introduced to accomplish the performance evaluation task by pre-training and re-training. The experiments are conducted on NASA's public dataset of lithium-ion batteries, and multiple sets of comparison experiments are set up. The results show that the method can better solve the small sample problem encountered in the lithium-ion batteries performance evaluation task.

1. Introduction

With the continuous breakthrough of microelectronic technology in the 20th century, coupled with the increasing number of miniaturized equipment, people put forward high requirements for power supply. Power sources with high energy, small size, lightweight, safety and reliability are used as power sources for convenient portable electronic products and electric vehicles. Lithium-ion batteries use redox reactions to achieve the function of charging and discharging. With the advantages of high energy density, low self-discharge rate, high discharge voltage, and long life cycle (many times of charging and discharging), it has achieved rapid development [1] in recent years, and then entered the large-scale practical stage, is continuously improving and enhancing our quality of life. Especially for high energy density lithium-ion batteries, which can significantly reduce the weight and volume of energy storage systems in aerospace applications, so it has become the third generation of satellite power batteries [2].

The principle of lithium-ion battery is roughly based on the movement of lithium ions between the anode and cathode through the diaphragm and electrolyte to store and provide electricity. In the process of operation, some irreversible electrochemical reactions will occur in the battery, such as solid electrolyte growth and lithium-ion plating, which will lead to the reduction of battery capacity and shorten the service life of the battery. Therefore, as the number of charges and discharges of the battery increases, the lithium-ion battery will inevitably produce performance degradation or even failure due to these reactions [3,4]. To ensure the reliability of Li-ion batteries, an accurate state estimation of their performance degradation is required. Especially when lithium-ion batteries are used in significant fields such as electric vehicles, energy storage, and aerospace, it is also vital to guarantee the safety and reliability of lithium-ion batteries during operation. Therefore, it is necessary to perform a state assessment of the performance of Li-ion batteries [5]. Considering that many studies have taken the capacity of Li-ion battery as the measure of condition assessment [6-8], this paper will also take the battery capacity as the index parameter.

Through previous studies, it can be concluded that there are roughly three methods for state assessment of the capacity of lithium-ion batteries: direct measurement method, model-based method, and data-driven method [9,10]. The direct measurement method is mainly based on the terminal voltage, measured current, and impedance to estimate the battery capacity. The open-circuit voltage method [11]and the Coulomb counting method [12] have become the two most commonly used direct measurement methods due to their convenience and low computational complexity, which can obtain the maximum capacity of the current battery. The open-circuit voltage method requires the battery to stand for a long time to achieve a stable voltage. Still, it takes several hours or even a dozen hours for the battery to return to a steady state, which makes the measurement difficult, and the determination of the standing time is also a questionable issue. The Coulomb counting method is probably used to obtain the battery capacity by integrating the current in time during the charging and discharging process, which is also very time-consuming and does not easily overcome cumulative errors and sensor noise. The model-based method relies on the internal electrochemical failure mechanism of the battery. After the in-depth analysis of the influence of various degradation factors on the state variables of the battery, the performance degradation law during the battery operation is excavated, so as to establish the physical degradation model of the battery, and then carry out the state assessment of the lithium battery. Ashwin T R et al. [13] proposed a pseudo-two-dimensional electrochemical model to judge the degree of battery capacity decay and predict the remaining useful life of the battery. Xiong et al. [14] analyzed the electrochemical impedance spectra (EIS) under different conditions to extract the solid electrolyte interface resistance (SEI) to estimate the degradation of the battery. Su et al. [15] used a standard particle filter to estimate the parameters and used two empirical exponential models to track the degradation of battery capacity, which achieved good results. Although the model-based approach reveals the mechanism of battery performance degradation, it can accurately and effectively assess the current moment of capacity and serve as a real-time monitor. However, this method requires not only a complex measurement process and sophisticated instrumentation, but also requires in-depth understanding and learning of the battery mechanism studied and some professional knowledge applied. More importantly, the physical model built also does not have good generalization ability, and the model loses its accuracy for different models or batteries with varying mechanisms of production. These drawbacks are not suitable for practical applications, so more scholars have started to use data-driven method to evaluate the state of Li-ion battery capacity. The data-driven approach does not need to consider the complex electrochemical reactions and degradation mechanism inside the lithium-ion battery. It can perform the evaluation task well based on the historical degradation information only. The data-driven method mainly contains machine learning and deep learning methods. Li et al. [16] proposed a model based on average entropy and relevance vector machine (RVM) to evaluate the health status of batteries. Song et al. [17] proposed a new method for battery performance prediction based on recurrent neural network (RNN), which overcomes the drawback of dealing with long-term relationships in RNN. The RNN-GRU has a simple structure and high computational efficiency. Li et al. [18] used the convolutional neural networks to achieve the estimation of battery capacity by taking three features, namely, voltage, current, and temperature, during the charging process of lithium-ion batteries as the inputs to the model.

Although the data-driven method has achieved good results in battery performance evaluation tasks and has unique advantages that the previous two approaches do not have, its shortcomings are also evident. Data-driven methods that rely excessively on large data sets for training are "overshadowed" by the few training samples available. Therefore, the small-sample problem is a very challenging problem in data-driven methods. To solve this problem, a model-based transfer learning method for lithium battery performance evaluation is proposed. In the process of using source domain data to train the model, some parameters or super-parameters in the model can provide a positive effect for the training of target domain data, so as to achieve the purpose of improving the target task accuracy. Finally, through multiple groups of comparative experiments, it is confirmed that the model-based transfer learning method proposed in this paper can solve the problem of small samples encountered in the lithium battery status assessment task, and has achieved a good assessment effect.

The remainder of this paper is organized as follows. Section II describes the basic principles of LSTM, CNN, model-based transfer learning, and freezing mechanism. Section III presents the overall architecture of the proposed method and its detailed procedure. Section IV presents the experiments and results analysis, followed by the key conclusions summarized in Section V.

2. Methodologies

2.1 LSTM

The LSTM proposed by Hochreiter et al. [19] is essentially a variant of the RNN. The difference is that the LSTM layer can obtain both cell state and hidden state information from the previous moment. The gate mechanism is introduced to control the transmission and update of cell state information and hidden state information instead of the hidden layer in the RNN. According to different functions, the gate mechanism can be divided into "forget gate", "input gate", and "output gate". The LSTM structure is shown in Fig.1. C_{t-1} and C_t in the figure represent the cell state information at moment $t-1$ and moment t respectively, h_{t-1} and h_t represent the hidden layer state information at moment $t-1$ and moment t respectively.

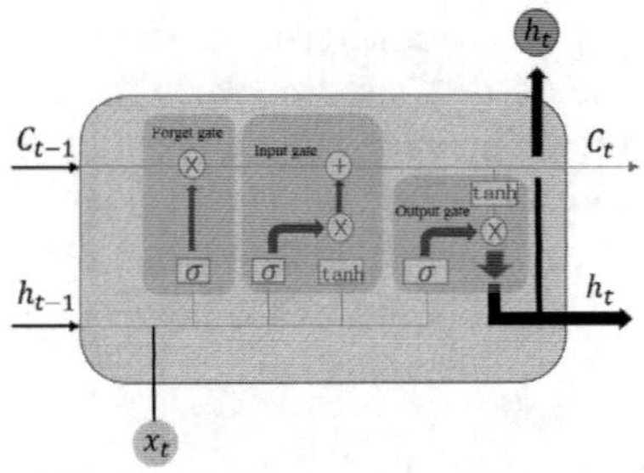

Fig.1 LSTM structure diagram

According to Fig.1, the core formula of the LSTM cell is shown below:

$$f_t = \sigma(W_f \cdot [h_{t-1}, x_t] + b_f) \quad (1)$$

$$i_t = \sigma(W_i \cdot [h_{t-1}, x_t] + b_i) \quad (2)$$

$$\tilde{C}_t = tanh(W_C \cdot [h_{t-1}, x_t] + b_C) \quad (3)$$

$$C_t = f_t * C_{t-1} + i_t * \tilde{C}_t \quad (4)$$

$$o_t = \sigma(W_o \cdot [h_{t-1}, x_t] + b_o) \quad (5)$$

$$h_t = o_t * \tanh(C_t) \quad (6)$$

Where f_t , i_t , and O_t are the forgetting gate, input gate, and output gate at moment t respectively; $\tilde{C}_t$ represents the candidate update information at moment t. σ is the Sigmoid activation function; x_t represents the cell input at moment t; W_f、W_i、W_C and W_o represent the weights; and b_f、b_i、b_C、b_o represent the bias.

The "forgetting gate" is responsible for deciding which information will be discarded from the cell state at the last moment and how much information can be retained, as shown in equation (1). The "input gate" determines which information from the current input is kept in the cell state, as represented by equations (2), (3), and (4). The final "output gate" determines which information will be output by the LSTM cell, which is represented by equations (5) and (6). With its unique "gate mechanism", LSTM effectively overcomes problems such as gradient disappearance or explosion, and solves the long-term dependency problem encountered by RNN. It is also widely used in natural language processing and performance evaluation tasks involving time-series data.

2.2 CNN

CNN has now become one of the most popular deep learning models in the field of computer vision. In 1990, LuCun et al. [20] first proposed convolutional neural networks. Then, to successfully solve the practical problem of recognizing handwritten digital characters, they improved it and presented the LeNet-5 model [21] in 1998, which is the original prototype of contemporary convolutional neural networks. Compared to the total connectivity in traditional neural networks, the neurons inside the CNN take a local connectivity approach. The locally connected approach means that the number of parameters for model training is smaller, and the required training time is shorter. Therefore, the use of convolutional kernels in CNNs has the characteristics of local connectivity and parameter sharing. In addition, its pooling layer is good at capturing translation invariance, which makes the abstract feature extraction more powerful. Considering the solid conceptual feature extraction ability of CNN, it is used in the performance evaluation task as the abstract feature extraction part of the pre-multidimensional information.

2.3 Model-based transfer learning

Transfer learning refers to learning a prediction function with better performance that can predict the knowledge of the target domain (label) by virtue of the knowledge of the source domain [22,23]. Combining the characteristics of the NASA lithium-ion battery dataset used in this paper and the available experimental conditions, it was decided to use a model-based transfer learning approach for subsequent experiments. The model-based transfer learning method uses the parameter information common between the source and target domains to achieve transfer. Therefore, the model-based transfer learning approach is also known as the parameter-based transfer learning approach. This approach assumes that specific parameters should be shared between models in the source and target domains. Model-based transfer learning focuses on constructing a pre-training set and a re-training set. The pre-training set is selected from the source domain based on the similarity between the data in the source domain and the target domain, and the criterion of similarity varies with the task. The re-training set selected the most similar data than the pre-training set.

2.4 Freezing mechanism

The freezing mechanism can be simply summarized as "freezing" part of the network structure during model training. It does not participate in the training of the model, and only training other networks that are not "frozen" to achieve fine-tuning. Therefore, this method is commonly used in transfer learning. During the pre-training process, the parameters of all model layers are changed with the training process. In contrast, during the re-training phase, some of the specified network layers are "frozen", and their parameters remain unchanged. Only the parameters of the unfrozen network layers are updated during training.

3. The Proposed Approach of Model-based Transfer Learning Framework

A model-based transfer learning framework is proposed. The methodological logic of the framework is divided into two main parts, as shown in Fig.2. The first part is the fusion model of CNN and LSTM. Firstly, CNN is responsible for extracting abstract features from the data and compressing the step length used, which reduces the computational pressure of the system; secondly, LSTM will use the internal unique gate mechanism to extract features from the data in time dimension; finally, there is the fully connected layer, which is usually used in combination with Dropout. The primary purpose is to prevent the model from overfitting during the training process. The fusion model has both the abstract feature extraction capability of CNN and the time-dimensional feature extraction capability of LSTM.

The second part introduces a model-based transfer learning method based on the fusion model, mainly divided into two steps: pre-training and re-training, with pre-training mainly feeding the pre-training set into the fusion model for training. When the training is finished, a pre-trained fusion model is obtained, which is also referred to as a pre-trained model. The model already has some performance evaluation capability, but the goodness of the model depends heavily on the dataset from the source domain used for the pre-training. Theoretically, the model improves as the proportion of data from the same domain in the pre-training set increases. However, considering the general situation, it is more challenging to know the ratio of same-condition data in advance. Therefore, the performance of the model can be volatile, and this situation is essentially a cross-case problem. The re-training step of the pre-trained model effectively solves this problem. The essence of re-training is to use some inheritable parameters or hyperparameters in the pre-trained model. Then, fine-tune the non-inheritable parts of the model through some more relevant data, so as to improve the performance of the model. In the re-training process, a "freezing mechanism" is introduced to "freeze" the LSTM layer of the pre-trained model, while the CNN and fully connected layers are kept updated. This is because for lithium-ion batteries, although the local decay of each battery varies greatly due to individual differences, the overall decay trend is more or less the same, and the LSTM can extract the long-term decay information of the data well. Because of this, the LSTM is less sensitive to the local decay caused by individual differences in batteries. Moreover, the data volume of the re-training set is much smaller than that of the pre-training set, which contains a lot of information about regional differences caused by themselves. If the LSTM layer is updated at the same time, it is likely that not only the local differences are not captured, but also the long-term decreasing trend learned previously will be affected. Therefore, the LSTM layer is "frozen" during the re-training process. The CNN layer and fully connected layer are fine-tuned with the training process to capture the characteristics of each cell and improve the accuracy of local decay evaluation.

In conclusion, the problem of small sample in lithium-ion battery performance evaluation can be better solved by training a large amount of source domain data in the pre-training stage of the framework and the fine-tuning the model in the re-training stage.

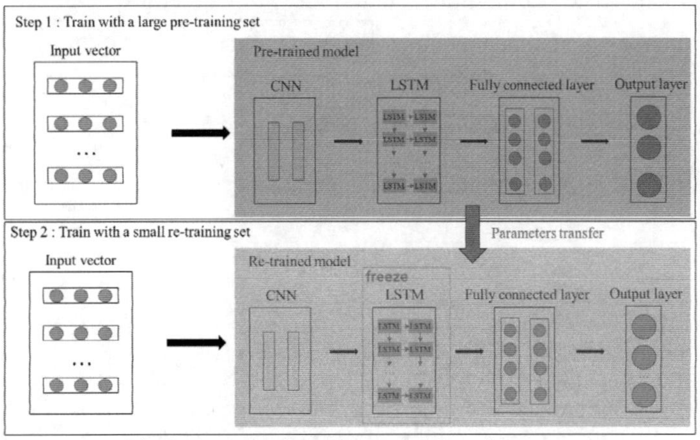

Fig.2 Transfer learning model architecture

4. Experimental Results and Discussion

In this section, the proposed model-based transfer learning framework is compared and experimented under different experimental conditions to verify the correctness and adaptability of the framework.

4.1.Data set description and evaluation criteria

4.1.1.Space lithium-ion battery cycling test data set

The data of 16 Li-ion batteries used in this experiment are from NASA public data set [24]. The original data of each battery consists of three parts, namely charge circle data, discharge circle data, and impedance data. According to the different operating conditions, the 16 batteries can be divided into four categories, and each of the four pieces belongs to the same working condition. Battery B0005, B0006, B0007, and B0008 belong to the first group with the experimental temperature of 24℃ and discharge current of 2A; Battery B0025, B0026, B002, and B0028 belong to the second group with experimental temperature of 24 ℃, discharge current amplitude of 4A, the duty cycle of 50% and frequency of 0.05Hz; Battery B0029, B0030, B0031, and B0032 belong to the third group with the experimental temperature of 43℃ and discharge current of 4A; Battery B0053, B0054, B0055, and B0056 belong to the fourth group with the experimental temperature of 4°C and discharge current of 2A. In addition, the discharge stopping voltage of each battery in the same group is also different to some extent.

According to the analysis, for the discharge period, the data under different working conditions have great differences in the discharge current size and experimental temperature. In addition, there are also differences in the discharge stopping voltage of different batteries under the same operating conditions, and such differences may cause some influence for the subsequent feature extraction. For the charging cycle, other experimental parameters remain consistent except for the experimental temperature. Therefore, combining various factors into consideration, this experiment intends to perform feature extraction on the voltage data of the charging cycle data.

4.1.2.Evaluation criteria

In this experiment, we use three evaluation criteria to measure and demonstrate the accuracy of the proposed method. Two of them are the most commonly used evaluation metrics for regression problems, root mean square error (RMSE), and mean absolute error (MAE) respectively. In addition, this paper also proposes a new evaluation index error sum (ES), which is suggested due to the situation that a model has a small MAE but a large RMSE in the experimental process. In this case, using either MAE or RMSE to evaluate the model performance is not comprehensive enough. Therefore, in this paper, the MAE and RMSE are summed to obtain the ES, which can be used to comprehensively evaluated the performance of the model. The definition formula of these three metrics is as follows:

$$RMSE = \sqrt{\frac{1}{m}\sum_{i=1}^{m}(y_i - \hat{y}_i)^2} \tag{7}$$

$$MAE = \frac{1}{m}\sum_{i=1}^{m}|y_i - \hat{y}_i| \tag{8}$$

$$ES = MAE + RMSE \tag{9}$$

4.2.Feature extraction

Taking battery B0005 as an example, the slope of its voltage change curve increases with the increase of charging cycle. To extract the characteristics of this local decay, we divided the voltage curve of the whole charging cycle (from the starting point to 4.2V) into four intervals according to the amount of voltage change. Since the slope of the curve shows a tendency to change from large to small and then from small to large, the division ratio of the four intervals was finally set as 2:3:1:3. As shown in Fig.3, after geometric analysis of the charging cycle curve, 12 features were finally selected as health factors in this experiment.

F1, F2, F3, F4: As the degradation of Li-ion battery performance increases, the starting voltage values of the four intervals assigned in the same proportion increase. F1 refers to the starting voltage of the first interval and also the initial voltage of the whole charging cycle. F2, F3, and F4 are the starting voltages of the second, third and fourth intervals, respectively.

F5, F6, F7, F8: The termination voltage of the charge remains constant while F1 increases. As a result, the duration of the four intervals will gradually decrease with the degradation of the Li-ion battery performance. F5, F6, F7, and F8 represent the time elapsed from F1 to F2, from F2 to F3, from F3 to F4, and from F4 to the termination voltage (4.2V), respectively.

F9, F10, F11, F12: According to Faraday's law of electromagnetic induction, the magnetic flux (MF) is equal to the integral of the voltage in the dimension of time, which can represent the power supply capacity of the supplier. For lithium-ion batteries, the magnitude of magnetic flux reflects the magnitude of the battery's electrical capacity. As the performance of a lithium-ion battery decays, its capacity decreases and the magnetic flux decreases. In this experiment, the flux is calculated using the equation (10) :

$$MF = \int_{T_j}^{T_{j+1}} V(t)dt \tag{10}$$

Among them, $j = 0, 1, 2, 3,$ the T_{j+1} and T_j refer to the time interval between the initial point of segment $j+1$ and the initial point of segment j. In addition, when $j = 3$, the T_{j+1} refers to the point in time when the voltage reaches the termination voltage (4.2V).

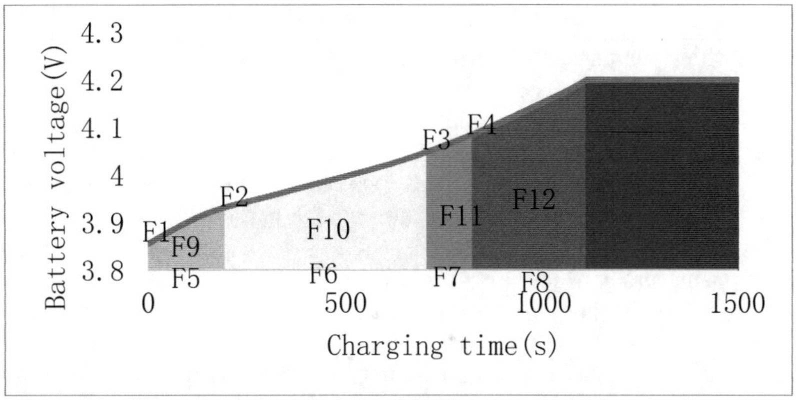

Fig.3 Image representation of the feature extraction process

4.3. Experiments and Results

In order to reflect the small sample problem that may occur in the lithium-ion battery performance assessment task and verify the effectiveness of the proposed model-based transfer learning framework, two groups of experiments with entirely different ideas are designed in this paper. The first set of experiments will exemplify the situation that may occur when the number of model training data is small, i.e., when the small sample problem is encountered. This set of experiments is called the traditional performance evaluation methods. The second group of experiments will try to solve the above problem using the transfer learning method, which is named as the transfer learning performance evaluation method. In the following, each group of experiments will be introduced in detail.

In the first group of experiments, battery B0005 was selected as the data used for this group of experiments. It contained a total of three different experiments. The specific experiments are arranged as follows:

A. After training the fusion model with 25% data of battery B0005, the model is used to evaluate the performance of the remaining 75% data.

B. After training the fusion model with 50% data of battery B0005, the model is used to evaluate the performance of the remaining 50% data.

C. After training the fusion model with 75% data of battery B0005, the model is used to evaluate the performance of the remaining 25% data.

To prove that the transfer learning method can indeed solve the small sample difficulty that appeared in the previous group of experiments, this paper divides the data to obtain the second group of experiments according to the method shown in table 1. Taking the experiment numbered A in the table as an example, the form corresponds to the case when the pre-training data are three sets of data with different working conditions from battery B0005, and the re-training data are the first 25% of B0005. At this point, if the performance of this model is better than that of the traditional performance evaluation method, it can be considered that the transfer learning performance evaluation method can solve the problem of small sample in the performance evaluation task of lithium-ion battery to a certain extent.

Table 1. Data division of Experiment 2

Experiment number	A	B	C	D	E	F	G	H	I	J	K	L
Pre-training data	B0025、B0029、B0053			B0006、B0025、B0029			B0006、B0007、B0025			B0006、B0007、B0018		
Re-training data	B0005 top 25%	B0005 top 50%	B0005 top 75%	B0005 top 25%	B0005 top 50%	B0005 top 75%	B0005 top 25%	B0005 top 50%	B0005 top 75%	B0005 top 25%	B0005 top 50%	B0005 top 75%

Fig.4 and table 2 show the results of three experiments A, B, and C in the conventional performance evaluation method. In Fig.4, the abscissa corresponds to the number of charging and discharging rounds of the battery, and the ordinate is the normalized capacity. The yellow line segment is the result of experiment A, the green line segment is the result of experiment B, the blue line segment is the result of experiment C, and the red line segment represents the real data. From table 2, it is easy to see that the MAE, RMSE, and ES of experiment A are the largest, the parameters of experiment B are smaller compared to the former, and the one of experiment C is the smallest. That is, these three evaluation metrics are in decreasing order, which also proves the hypothesis we proposed when designing the experiments: when the number of training data is small, the model trained with this data has poor performance in the lithium-ion battery performance evaluation task due to the insufficient total information contained in the data, and the phenomenon will gradually decay as the number of data increases. This also confirms that the small sample problem does exist in the lithium-ion battery performance evaluation task.

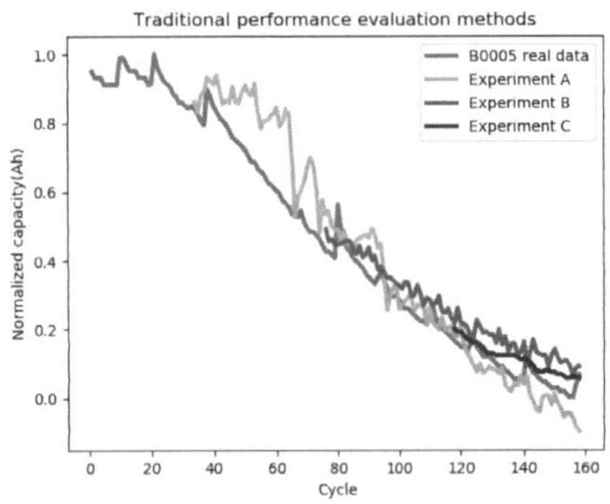

Fig.4. Result of traditional performance evaluation methods

Table 2. Result of Experiment 1

Experiment number	MAE	RMSE	ES
A	0.0775	0.1015	0.1790
B	0.0610	0.0667	0.1277
C	0.0338	0.0376	0.0715

Fig.5-8 and Table 3 show the results of the transfer learning performance evaluation method. The blue line in each figure corresponds to the performance evaluation results after re-training the model using the first 25% data of battery B0005; the green line has 50% of the re-trained data; the yellow line has 75% of the re-trained data; and the red line corresponds to the real data of battery B0005. To verify the effectiveness of transfer learning in solving small sample problems, Table 4 is drawn on the basis of Table 2 to Table3 to conclude more intuitively.

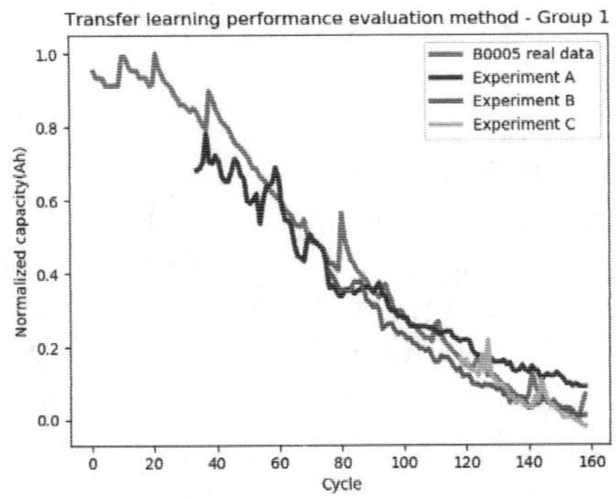

Fig.5 Result of transfer learning - A-C

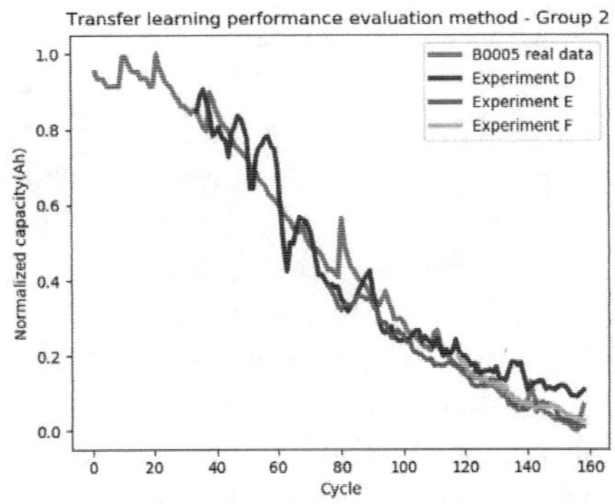

Fig.6 Result of transfer learning - D-F

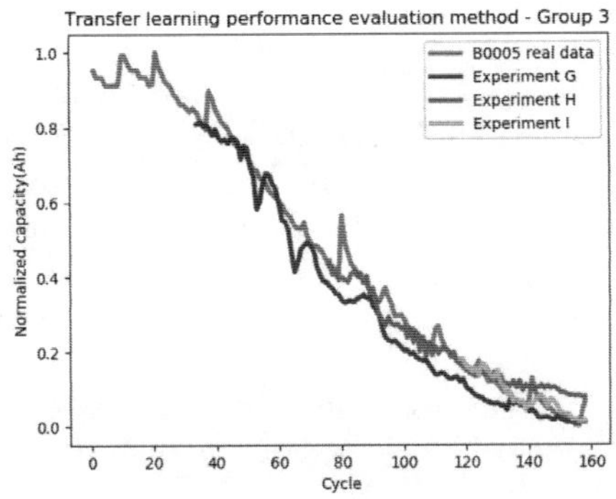

Fig.7 Result of transfer learning - G-I

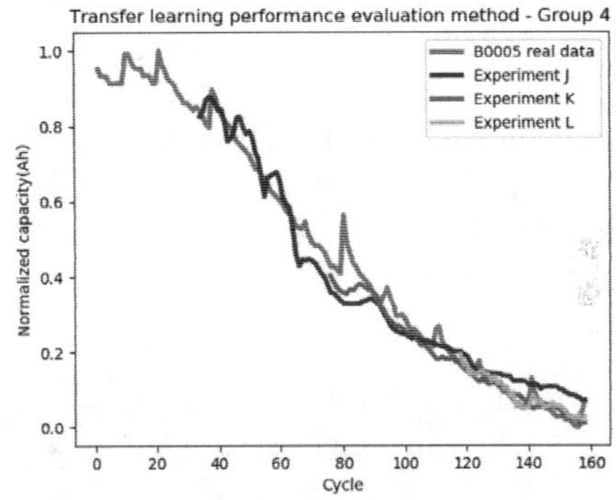

Fig.8 Result of transfer learning - J-L

Table 3. Result of Experiment 2

Experiment number	A	B	C	D	E	F	G	H	I	J	K	L
MAE	0.0576	0.0419	0.0207	0.0572	0.0367	0.0210	0.0510	0.0341	0.0177	0.0467	0.0283	0.0161
RMSE	0.0726	0.0540	0.0323	0.0689	0.0520	0.0241	0.0629	0.0445	0.0227	0.0575	0.0423	0.0195
ES	0.1303	0.0959	0.0531	0.1262	0.0887	0.0451	0.1139	0.0786	0.0404	0.1042	0.0706	0.0356

Table 4. Comparison Results of Experiment 1 and Experiment 2

Experiment number	ES	Experiment number	ES	ES improvement	ES average improvement
1-A	0.1790	2-A	0.1303	0.0488	0.0604
		2-D	0.1262	0.0529	
		2-G	0.1139	0.0651	
		2-J	0.1042	0.0749	
1-B	0.1277	2-B	0.0959	0.0319	0.0443
		2-E	0.0887	0.0391	
		2-H	0.0786	0.0491	
		2-K	0.0706	0.0571	
1-C	0.0715	2-C	0.0531	0.0184	0.0279
		2-F	0.0451	0.0264	
		2-I	0.0404	0.0311	
		2-L	0.0356	0.0359	

Several conclusions can be intuitively drawn by carefully analyzing Table 4 through comparative experiments: firstly, transfer learning can indeed solve the small sample problem that occurs in traditional performance prediction methods; secondly, the smaller the re-training data set, the more pronounced the effect of transfer learning is; finally, it can also be analyzed to know that under the same amount of data in the re-training data set, the higher the number of batteries in the same working condition in the batteries pack, the better the transfer effect. In summary, the framework has high accuracy for lithium-ion battery performance evaluation and can effectively solve the problem of small samples encountered in practice.

5. Conclusion and Future Work

In this paper, a model-based transfer learning framework is proposed to evaluate the performance of lithium-ion batteries. A multi-dimensional health factor is constructed as the input to the model. A CNN-LSTM fusion model is introduced to compensate for the weakness of LSTM in capturing the local degradation features of each battery when building a deep learning model. In the process of introducing transfer learning, this paper finds that the problem of small samples is prevalent in the lithium-ion battery performance evaluation task. Hence, it tries to use the model-based transfer learning method to solve the issues encountered problems. The effectiveness of the method on the NASA dataset is verified by setting up multiple sets of comparison experiments for comparison. The experimental results show that the model-based transfer learning framework can better solve the small sample problem arising in the lithium-ion battery performance evaluation task.

In this experiment, it is intuitive to see that the percentage of same-work data in the pre-training set greatly affects the evaluation results. In the future, the instance-based transfer learning method may become the focus. This method can determine the similarity of all data and the target data before training. In this case, data with higher similarity can be selected to form the pre-training set to improve the evaluation ability of the model. Meanwhile, the fusion of multiple models will also become the trend of lithium-ion battery capacity prediction.

References

[1] He, W., Williard, N., Osterman, M., & Pecht, M. (2011). Prognostics of lithium-ion batteries based on Dempster–Shafer theory and the Bayesian Monte Carlo method. Journal of Power Sources, 196(23), 10314-10321.

[2] Song, Y., Liu, D., Yang, C., & Peng, Y. (2017). Data-driven hybrid remaining useful life estimation approach for spacecraft lithium-ion battery. Microelectronics Reliability, 75, 142-153.

[3] Yoshino, A. (2012). The birth of the lithium - ion battery. Angewandte Chemie International Edition, 51(24), 5798-5800.

[4] Zubi, G., Dufo-López, R., Carvalho, M., & Pasaoglu, G. (2018). The lithium-ion battery: State of the art and future perspectives. Renewable and Sustainable Energy Reviews, 89, 292-308.

[5] Meng, H., & Li, Y. F. (2019). A review on prognostics and health management (PHM) methods of lithium-ion batteries. Renewable and Sustainable Energy Reviews, 116, 109405.

[6] Qu, J., Liu, F., Ma, Y., & Fan, J. (2019). A neural-network-based method for RUL prediction and SOH monitoring of lithium-ion battery. IEEE access, 7, 87178-87191.

[7] Oji, T., Zhou, Y., Ci, S., Kang, F., Chen, X., & Liu, X. (2021). Data-Driven Methods for Battery SOH Estimation: Survey and a Critical Analysis. IEEE Access, 9, 126903-126916.

[8] Yuchen, S. O. N. G., Datong, L. I. U., Yandong, H. O. U., Jinxiang, Y. U., & Yu, P. E. N. G. (2018). Satellite lithium-ion battery remaining useful life estimation with an iterative updated RVM fused with the KF algorithm. Chinese Journal of Aeronautics, 31(1), 31-40.

[9] Hannan, M. A., Lipu, M. H., Hussain, A., & Mohamed, A. (2017). A review of lithium-ion battery state of charge estimation and management system in electric vehicle applications: Challenges and recommendations. Renewable and Sustainable Energy Reviews, 78, 834-854.

[10] Zhang, R., Xia, B., Li, B., Cao, L., Lai, Y., Zheng, W., ... & Wang, W. (2018). State of the art of lithium-ion battery SOC estimation for electrical vehicles. Energies, 11(7), 1820.

[11] Weng, C., Sun, J., & Peng, H. (2014). A unified open-circuit-voltage model of lithium-ion batteries for state-of-charge estimation and state-of-health monitoring. Journal of power Sources, 258, 228-237.

[12] Ng, K. S., Moo, C. S., Chen, Y. P., & Hsieh, Y. C. (2009). Enhanced coulomb counting method for estimating state-of-charge and state-of-health of lithium-ion batteries. Applied energy, 86(9), 1506-1511.

[13] Ashwin, T. R., Chung, Y. M., & Wang, J. (2016). Capacity fade modelling of lithium-ion battery under cyclic loading conditions. Journal of Power Sources, 328, 586-598.
[14] Xiong, R., Tian, J., Mu, H., & Wang, C. (2017). A systematic model-based degradation behavior recognition and health monitoring method for lithium-ion batteries. Applied Energy, 207, 372-383.
[15] Su, X., Wang, S., Pecht, M., Ma, P., & Zhao, L. (2017). Prognostics of lithium-ion batteries based on different dimensional state equations in the particle filtering method. Transactions of the Institute of Measurement and Control, 39(10), 1537-1546.
[16] Li, H., Pan, D., & Chen, C. P. (2014). Intelligent prognostics for battery health monitoring using the mean entropy and relevance vector machine. IEEE Transactions on Systems, Man, and Cybernetics: Systems, 44(7), 851-862.
[17] Song, Y., Li, L., Peng, Y., & Liu, D. (2018, October). Lithium-ion battery remaining useful life prediction based on GRU-RNN. In 2018 12th international conference on reliability, maintainability, and safety (icrms) (pp. 317-322). IEEE.
[18] Chaoran, L., Fei, X., & Yaxiang, F. (2020). An approach to lithium-ion battery SOH estimation based on convolutional neural network. Transactions of China Electrotechnical Society, 35(19), 4106-4119.
[19] Hochreiter, S., & Schmidhuber, J. (1997). Long short-term memory. Neural computation, 9(8), 1735-1780.
[20] LeCun, Y., Boser, B., Denker, J., Henderson, D., Howard, R., Hubbard, W., & Jackel, L. (1989). Handwritten digit recognition with a back-propagation network. Advances in neural information processing systems, 2.
[21] LeCun, Y., Bottou, L., Bengio, Y., & Haffner, P. (1998). Gradient-based learning applied to document recognition. Proceedings of the IEEE, 86(11), 2278-2324.
[22] Pan, S. J., & Yang, Q. (2009). A survey on transfer learning. IEEE Transactions on knowledge and data engineering, 22(10), 1345-1359.
[23] Torrey, L., & Shavlik, J. (2010). Transfer learning. In Handbook of research on machine learning applications and trends: algorithms, methods, and techniques (pp. 242-264). IGI global.
[24] Liu, D., Xie, W., Liao, H., & Peng, Y. (2014). An integrated probabilistic approach to lithium-ion battery remaining useful life estimation. IEEE Transactions on Instrumentation and Measurement, 64(3), 660-670.

Optimal scheduling model based on unmanned collector truck arrival time variation field bridge under balanced strategy

JiaHong Chen[1a*], JiaJing Zhou[2b], YuTing Zheng[2c], RongZhe Ma[2d]

[1]School of Transportation and Logistics Wuhan University of Technology Wuhan, China

[2]School of Transportation and Logistics Wuhan University of Technology Wuhan, China

[a*]cjh2289207252@163.com,[b]whutzhoujiajing@whut.edu.cn, [c]whutzyting@whut.edu.cn, [d]whutmarongzhe@whut.edu.cn

Abstract—The scheduling of container yard bridges has an important impact on the operational efficiency of the terminal. For the future automated terminal, to reasonably schedule the field bridges, under a given batch of tasks, considering the change of the arrival time of unmanned trucks and the balance of the workload of each field bridge, a balanced strategy is established based on the change of the arrival time of unmanned trucks. The optimal scheduling model of field bridges is given, and an example of field bridge scheduling is given through experiments. Provides a scientific solution for the future automated dock, yard, and bridge scheduling

1. Introduction

The port side of the future automated container terminal consists mainly of yards, yard bridges, shore bridges, and unmanned trucks, and yard operations are one of the key factors governing terminal efficiency. In terminal operations, unmanned trucks are responsible for the horizontal transport of containers between the yard and the shore bridge. Due to a large number of unmanned trucks operating at the same time, battery capacity constraints, and navigation accuracy, there is uncertainty in the arrival time of some unmanned trucks at the yard, which leads to mutual waiting between the yard and the yard bridge and has a huge impact on the efficiency of the whole automated terminal operation. Therefore, how to reasonably carry out the scheduling of the yard and bridge is an important problem that needs to be solved in the future automated terminal, which is the key to ensuring the horizontal transportation of containers between the shore bridge and the yard, and is of great significance to improve the operational efficiency of the automated terminal.

The problem of scheduling field bridges was first investigated by Kim [1], who studied the optimization of static field bridge scheduling with a known card arrival time, to minimize card delays. Ng et al [2] studied the multi-field bridge field-bridge optimization scheduling problem in discrete-time based on an improvement of Kim's model. they upgraded the model to a multi-field bridge scheduling model while considering the effect of waiting time, designed a linear objective function in the form of discrete-time nodes and used an improved branch-and-bound algorithm to solve the problem. Ng and Mak et al [3] developed an integer programming model and Li et al [4] developed a discrete-time model to minimize the total delay waiting time in both loading and unloading processes, taking into account the interference problem between field bridges. was solved using a rolling time-domain approach. Since then Li et al [5] have upgraded the discrete model to a continuous-time model for more accurate simulation of the problem. Petering Matthew EH [6] developed a model of a real-time yard and bridge scheduling system that considered the efficiency of the entire system, including the efficiency of the shore bridge, yard bridge, and collector truck. Chang et al [7] first used a genetic algorithm to find the initial solution to the problem and then used a rolling time-domain approach to perform dynamic scheduling optimization, providing a fairly good solution to the field and bridge dynamic scheduling problem.

For the joint scheduling of field bridges and collector trucks, Kim [8] et al. studied the dynamic grouping of collector trucks to match the scheduling of field bridges to improve the efficiency of field bridges while reducing the occurrence of field bridge interference. lee et al [9] studied the integrated scheduling problem of field bridges and collector trucks and used a simulated annealing algorithm to solve the problem. Liang developed a system containing a shore bridge, field bridge, and collector truck The integrated scheduling system was divided into two subsystems, two models were generated and a coupled algorithm based on a genetic algorithm was selected to solve the problem. The consideration of the scheduling strategy, on the other hand, progresses from a static perspective at the beginning to dynamic planning for the scheduling problem. From the gradual development of these studies, it is the joint scheduling of comprehensive systems that is the trend of future research.

Based on existing research, this paper takes into account the dynamic scheduling of the yard and bridge due to the change in arrival time of unmanned trucks and the balance of the work tasks of the yard and bridge. With the application of Internet technology, terminals can obtain real-time predictions of the exact arrival time of unmanned trucks at the yard, which makes it possible to dynamically combine the scheduling of yard bridges and unmanned trucks. In this paper, the objective is to minimize the waiting time between the yard bridge and unmanned trucks and to establish an optimized scheduling model for the yard bridge based on the variation of the arrival time of unmanned trucks under the strategy of ensuring the balance of the workload of each yard bridge.

2. Description of the problem

The operation of the port section of the terminal is mainly divided into pick-up and loading and unloading and storage, and the whole operation system flow is shown in Figure 2.1. In the whole operation process, the operation time is mainly divided into three categories: time spent on the shore bridge, time spent on the unmanned container truck transportation, and time spent on the operation at the bridge end. In this paper, we mainly study the reasonable scheduling of the yard bridge and unmanned trucks and do not consider the scheduling of the shore bridge.

In this model, both processes, picking and loading, and unloading and storage are measured in terms of operating time and require both unmanned trucks and bridges to be ready for operation.

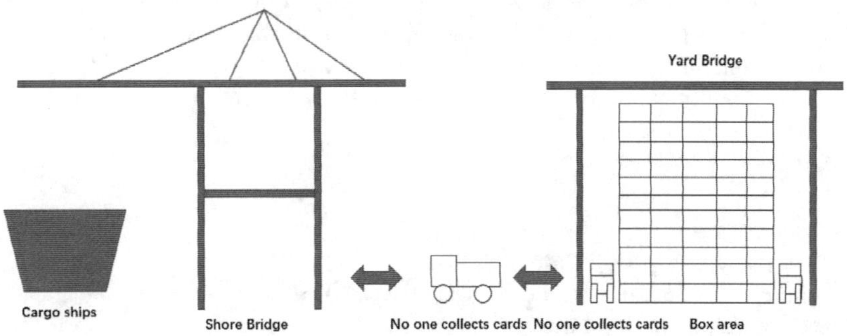

Fig.1 Field and bridge operating system

In Figure 2.2, you can see three zones and nine box areas, a box area can be operated by multiple yard bridges together, the red area in the figure indicates that there is a task at that box location after the unmanned collector truck arrives, the yard bridge corresponding to its task will start working to complete the pick-up or unloading. The yard will schedule the bridges according to the predicted arrival time of the trucks and the location of the task. If the bridges are not scheduled properly, there will be unmanned trucks waiting in the box area and the bridges will be empty. If the scheduling of multiple yards and bridges is not balanced, it can lead to empty yards and bridges or long running times and low efficiency, so it is very important to have reasonable scheduling of yards and bridges. In practice, unmanned trucks have unexpected conditions, and their actual arrival time at the yard changes in real-time, which the terminal can obtain in real-time using existing technologies such as GPS and the Internet, so that the scheduling plan for the yard bridge can be updated in real-time. In this paper, only orbital gantry cranes are considered for yard and bridge schedules. This model can solve the problem of the yard and bridge scheduling in the same area, and in theory, the number of box areas, the number of tasks, the number of yard and bridge, and the number of births can be adjusted at will.

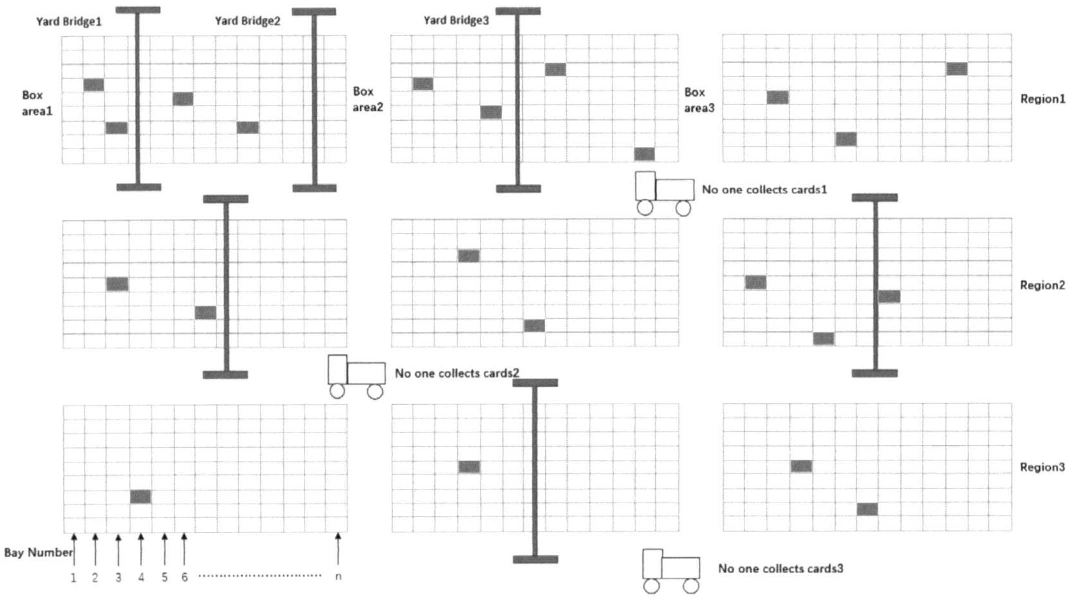

Fig.2 Diagram of yard bridge operations in the yard

Based on the arrival uncertainty of unmanned trucks, this paper constructs a dynamic scheduling model for yard bridges under a balancing strategy to dynamically optimize the task positions and corresponding processes of each yard bridge according to the changes in the arrival time of unmanned trucks, to minimize the mutual waiting time between the yard bridges and trucks, improve the efficiency of the yard bridges, and thus improve the efficiency of the whole terminal operation system.

3. Model building

3.1 Model Assumptions

1. The tasks take place in the same area (zone).
2. Each yard crane (YC) operates with the same efficiency and moves at the same speed.
3. The widths of the bays are the same.
4. The availability and workload of each bay is determined by the pick-up and loading
5. The field bridges cannot cross each other, and the safe distance between them is two bays.
6. The start time is counted as 0 hours.

3.2 Parameter settings

Table 1 Parameter settings

Parameters	Definition	Parameters	Definition
B	A collection of all the blocks in the task zone	N	Total number of assignments
L	A collection of all bays in the mission area	n	Work sequence, n=1,2,3,…,N
L_0	The set of initial bay positions where the field bridge is located	t	The time for the field bridge to drive over a bay,
T_a	Scheduled time of arrival for the set of cards at Bay a	W_a	Operating time (in minutes) for Bayit a
K	Total number of yard cranes	Q_k	Number of assignments for yard crane k
k	Numbering of field bridges, k=1,2,3,…, K	$\bar{Q}$	The average number of tasks on the yard crane

3.3 Variable settings

F_a^k: *The moment when field bridge k completes the work at bay position a*

RE_a^k : *The length of time that the field bridge k is earlier than the time when the set card reaches the bay position a*

RL_a^k: *The length of time that the field bridge k is later than the time when the set card reaches the bay position a*

A_a^k: *The moment when the field bridge k reaches the bay position a*

$$V_a = \begin{cases} 1, \text{ There is a task at bay } a \\ 0, \text{ There is no work task at bay position } a \end{cases}$$

$$y_{a,n,k} = \begin{cases} 1, \text{ If the field bridge } k \text{ processes the work task of the position } a \text{ in the nth process} \\ 0, \text{ else} \end{cases}$$

$$X_{n,k} = \begin{cases} 1, \text{If task } n \text{ is assigned to field bridge } k \\ 0, \text{ else} \end{cases}$$

3.4 The objective function and its constraints

$$Min\ f = \sum_{k=1}^{K}\sum_{a\in L} RE_a^k + \sum_{k=1}^{K}\sum_{a\in L} RL_a^k + \sum_{k=1}^{K}\left|Q_k - \bar{Q}\right| \quad (1)$$

$$\sum_{k=1}^{K}\sum_{n=1}^{N} y_{a,n,k} = V_a \quad a \in L; \quad (2)$$

$$\sum_{a\in L} y_{a,n,k} \le 1 \; n \in N, k \in K; \quad (3)$$

$$\sum_{n=1}^{N} y_{a,n,k} \le 1 \; a \in L; \; k = 1,2,3,...,K; \quad (4)$$

$$\sum_{k=1}^{K} X_{n,k} = 1 \text{ n=1,2,3,...,N}; \quad (5)$$

$$\sum_{n=1}^{N}\sum_{k=1}^{K} X_{n,k} = \sum_{a\in L}\sum_{n=1}^{N}\sum_{k=1}^{K} y_{a,n,k} = \text{N} \quad a \in L, k = 1,2,3,...,K, n = 1,2,3,...,N; \quad (6)$$

$$Q_k = \sum_{n=1}^{N} X_{n,k} \quad n = 1,2,3,...,N; \tag{7}$$

$$\bar{Q} = N / K \tag{8}$$

$$\sum_{a \in L} y_{a,n+1,k} \le \sum_{b \in L} y_{b,n,k} \quad n = 1,2,3,...,N; \ k = 1,2,3,...,K; \tag{9}$$

$$A_l^k >= y_{l,\mu,k} \bullet \left\{ \sum_{n=2}^{\mu} \left[t \bullet \left| \sum_{b \in L} (b \bullet y_{b.n.k}) - \sum_{a \in L} (a \bullet y_{a,n-1,k}) \right| + \sum_{a \in L} (W_a \bullet y_{a,n-1,k}) - W_l \right] \right\} l \in L \cap l \notin L_0;$$
$$k = 1,2,3,...,K; \mu\text{=2,3,4,...,N}; \tag{10}$$

$$\sum_{a \in L} (A_a^k \bullet y_{a,n+1,k}) - \sum_{b \in L} (A_b^k \bullet y_{b,n,k}) \ge t \bullet \left| \sum_{b \in L} (b \bullet y_{b,n,k}) - \sum_{a \in L} (a \bullet y_{a,n-1,k}) \right| + \sum_{a \in L} (W_a \bullet y_{a,n-1,k})$$
$$k = 1,2,3,...,K; n = 1,2,3,...,N; \tag{11}$$

$$F_a^k \ge A_a^k + W_a \quad a \in L; k = 1,2,3,...,K; n = 1,2,3,...,N; \tag{12}$$

$$F_a^k \ge T_a + W_a \quad a \in L; k = 1,2,3,...,K; \tag{13}$$

$$M \bullet W_a \ge RE_a^k \ge T_a - A_a^k \quad a \in L; k = 1,2,3,...,K; \tag{14}$$

$$M \bullet W_a \ge RL_a^k \ge A_a^k - T_a \quad a \in L; k = 1,2,3,...,K; \tag{15}$$

Eq. (1) is the objective function representing the minimization of the time spent waiting between unmanned collectors and field bridges for each other as well as the task balance between the field bridges. Eq. (2) represents the division of labor among the bridges to complete all tasks. Eqs. (3) and (4) indicate that the field bridges can only perform at most one task at a given process for a given bay. Eq. (5) indicates that a task is performed by only one field bridge. Eq. (6) shows the total number of tasks. Eqs. (7) and (8) define and respectively. Eq. (9) indicates that for each field bridge, only the nth process can exist before the n+1st process can exist. Eq. (10) indicates that the moment when the field bridge k is in position a is at least greater than the cumulative sum of its pure work and movement times in the previous processes. Eq. (11) means that the difference between the arrival times of the two adjacent processes of field bridge k must be at least greater than the sum of the movement times between the two processes and the working times of the earlier process. Eq. (12) indicates that for field bridge k, the difference between the moment of completion of the work task at bay a and the moment of arrival at bay a must be greater than the amount of work task at that location; Eq. (13) indicates that for field bridge k, the difference between the moment of completion of the work task at bay a and the moment of arrival at bay a by an unmanned set of cards must be greater than the amount of work at that location. Eqs. (14) and (15) indicate how to calculate the early arrival and late arrival times for a field bridge, where M is a sufficiently large positive number.

4. Case analysis

By collecting the actual operation data of a port, the validity of the model established in this paper is verified. Select 20 tasks of unmanned trucks that arrive within 60 minutes in two container areas of the port, and each container area has 30 bays. The data are shown in Table 2. In the model, it is assumed that the arrival time of unmanned trucks obeys a uniform distribution within [0-60], and the arrival time of changes is obtained in real-time by the scheduling system. When the card arrives, there are 4 changes; the number of workplace bridges is set to 3, assuming that the initial positions of the field bridges 1, 2, and 3 are at bay positions 15, 33, and 39, respectively. It takes 0.3 minutes for the field bridge to drive through one bezel bit; the basic unit of workload is 3 minutes; the program uses Lingo17.0 to develop the algorithm, the results are all in Intel(R) Core(TM) i5-9300H CPU @ 2.40GHz Measured on a PC with 2.40 GHz and 16 GB RAM.

Table 2 Example data

task	Bay	arrival time	workload	task	Bay	arrival time	workload
1	3	5	6	11	51	26	3
2	5	26	3	12	56	45	6
3	9	22	3	13	37	50	3
4	13	18	9	14	40	31	3
5	18	40	3	15	25	35	6
6	22	37	6	16	31	23	3
7	30	28	3	17	28	44	3
8	35	58	3	18	42	41	9
9	38	33	3	19	48	18	3
10	45	30	3	20	52	8	3

Due to the fast update speed of the arrival information of unmanned trucks, this paper assumes that the arrival time variation interval of unmanned trucks is 10 minutes, and makes predictions every 15 minutes to obtain the two changes of arrival time, that is, the corresponding field bridge operation sequence within 30 minutes. Variety. A summary of the forecast results of the arrival time of unmanned pickups is shown in Table 3.

Table 3 Prediction of the arrival time of unmanned pickups

Bay	0min	15min	30min	45min	60min
3	5				
5	26	26			
9	22	27			
13	18	20			
18	40	38	38		
22	37	37	40		
30	28	31	31		
35	58	56	54	49	
38	33	29			
45	30	31	31		
51	26	26			
56	45	42	51	51	
37	50	50	45	45	
40	31	29			
25	35	35	33		
31	23	23			
28	44	46	47	50	
42	41	45	47	47	
48	18	18			
52	8				

Take the position of the field bridge at each predicted time as the initial condition, input the corresponding arrival time of the truck, and get the task assignment and completion status of the field bridge at each time according to the value of the y in the model (represented by the number of the bay where the task is located) Summary As shown in Table 4 below.

Table 4 Field bridge operation sequence

Field bridge number	1	2	3
0min	15-3-13-9-5-30-18-22	33-31-38-25-28-37-35	39-52-48-51-45-40-42-56
15min	15-3-13-9-5-30-22-18	33-31-38-25-28-37-35	39-52-48-51-40-45-56-42
30min	15-3-13-9-5-30-18-22	33-31-38-25-37-28-35	39-52-48-51-45-42-56
45min	15-3-13-9-5-30-18-22	33-31-38-25-37-35-28-56	39-52-48-51-45-42
60min	15-3-13-9-5-30-18-22	33-31-38-25-37-35-28	39-52-48-51-45-45-56

From Table 4, we can see that the corresponding scheduling policy for each field bridge is updated in real-time as time changes, and the number of job bays for field bridges 1, 2, and 3 is 8, 7, and 7 respectively, which is a more even distribution of job tasks, in line with the balancing policy and improves the utilization of the field bridges

5. Conclusion

This paper aims to reduce the mutual waiting time between unmanned trucks and yard bridges by optimizing the work sequence of yard bridges in real-time based on the real-time updated arrival information of unmanned trucks for the future application of unmanned trucks in the pick-up or retrieval operations of horizontal transportation in automated terminals. Compared with previous studies, this paper contributes by considering the uncertainty of the arrival time of unmanned trucks in practice, which is more in line with the operational reality of terminals. A phased prediction of the arrival time of unmanned trucks and a corresponding change in the work order of the yard and bridge are given, making the scheduling strategy more in line with the actual requirements of automated terminals. The balance of tasks on the yard and bridge is taken into account to improve the utilization of the yard and bridge, making the yard scheduling scheme more scientific.

Acknowledgment

This research was partially supported by the National Col-lege Students Innovative Entrepreneurship Training Pro-gram(No.S202110497200).

References

[1] Kim Kap Hwan, Keung Mo Lee, Hark Hwang. Sequencing delivery and receiving operations for yard cranes in port container terminals[J]. International Journal of Production Economics,2003, 84(3): 283-292.

[2] Ng W C, Crane scheduling in container yards with inter-crane interference[J]. European Journal of Operational Research, 2005, 164(1): 64-78.

[3] Ng W C, K. L. Mak. Yard crane scheduling in port container terminals[J]. Applied mathematical modeling, 2005, 29(3): 263-276.

[4] Li Wenkai, et al. Discrete-time model and algorithms for container yard crane scheduling[J]. European Journal of Operational Research, 2009, 198(1): 165-172.

[5] Li Wenkai, et al. A continuous-time model for multiple yard crane scheduling with last-minute job arrivals[J]. International Journal of Production Economics, 2012, 136(2): 332-343.

[6] Petering Matthew EH, et al. Development and simulation analysis of real-time yard crane control systems for seaport container transshipment terminals[J]. OR spectrum, 2009, 31(4): 801-835.

[7] Chang Daofang, et al. Developing a dynamic rolling-horizon decision strategy for yard cranes cheduling[J]. Advanced Engineering Informatics, 2011, 25(3): 485-494.

[8] Kim Kap Hwan, Young-Man Park. A crane scheduling method for port container terminals[J]. European Journal of operational research, 2004, 156(3): 752-768.European Journal of operational research, 2004, 156(3): 752-768.

[9] Lee Dehong, Cao Zhi, Meng Qiang. Scheduling of two-transtainer systems for loading outbound containers in port container terminals with simulated annealing algorithm[J]. Int JProduct Econ, 2007,107: 115-124.

Person identification using a new CNN-based method and radar gait micro-Doppler signatures

Yu Huang[1a], Enshuo Jiang[1b*], Haodong Xu[1c], Guangbo Zhang[1d]

[1] School of Computer Science and Engineering, University of Electronic Science and Technology of China, Chengdu, Sichuan, 611731, China

*Corresponding author's e-mail: enshuojiang@std.uestc.edu.cn

Abstract-The use of gait micro-Doppler signatures to identify a person is a hot topic of research. In this paper, we present a new CNN-based method called *Multi-Scale CNN* (MS-CNN) to obtain features at multiple scales. It extracts shallow features at low-level multi-scale blocks by using multiple kernels at the same time, then extracts deep features and fuses multi-branch embedding features at high-level multi-branch blocks. Experimental results reveal that our method outperforms other commonly used CNN algorithms in terms of accuracy, allowing it to be used for personal identification.

1. Introduction

Personnel identification has very important applications in intrusion detection, surveillance, and intelligent transportation. Identification of humans based on human gait characteristics has been widely utilized because human gait characteristics are unique to each person, non-invasive, and difficult to disguise [1]. The majority of current gait identification algorithms rely on wearable devices [2] or visual sensors [3]. However, wearable technologies necessitate subject participation. Environmental elements like opacity and light can readily alter visual sensors, and there are issues such as privacy violations. In contrast, radar has been proven to be used for human gait analysis due to its environmental tolerance (for example, unaffected by light, weather, and clothing conditions) and privacy protection [4].

The characteristics linked with the target's behavior are known as micro-Doppler signatures (MDS) [5]. The motion of various regions of the body produces different MDS in the radar echo when a person is walking, and the radar system can extract the MDS from the received return signal. Because human walking is a complicated movement with a variety of walking patterns, the MDS of gait produced by different persons when walking is likewise diverse. As a result, the gait MDS can be used to identify the relevant person.

With the rise and development of deep learning, many scholars have begun to use deep learning to solve the above problems. Some researchers [6-8] have proposed some methods to preprocess the data, removed part of the noise, designed new CNN architectures, and obtained some improvements. Some researchers [9-12] use some existing well-known networks, such as ResNet50 [13], AlexNet [14] and VGG-16 [15] to apply to this task. They also obtained a great result. However, from our perspective, the dataset is small, and it is easy to underfit when using large networks directly. Although it can achieve better results than traditional methods, large networks consume more computing resources and have limited improvement. Although some researchers have proposed other methods, such as using convolutional neural network (CNN) to build the network, the constructed network is too simple, and it is difficult to achieve better results just by using CNN to stack in depth. Therefore, how to build a smaller and more efficient network to solve this task becomes our focus.

In this paper, we propose a novel approach named *Multi-Scale CNN* (MS-CNN) for person identification based on gait MDS. Since the difference between people is mainly due to different walking postures, such as different swing frequencies of arms and legs. MS-CNN considers the use of multi-scale kernels for modeling, in order to obtain a variety of features. On this basis, we use 3 CNNs to embed and extract high-level features. Furthermore, we do not just utilize the output of the last convolutional (Conv) layer. According to many papers on convolution layer visualization, we found that the low-level convolution layer extracts shallow features, while the high-level convolution layer extracts deep features and fusion features. Due to the small size of the dataset, we believe that multiple high-level features should be used, not just the output of the last layer, so we fuse the high-level Conv layer features and feed them to the classifier.

2. Related works

The radar echo signal caused by micro-motion is a time-varying frequency signal. From a time-varying frequency signal, time-frequency analysis can extract the relationship between the original signal frequency and time. The MD spectrogram created following time-frequency processing of human walking echo signals contains a wealth of human information in the study of human gait MDS.

Many current studies use deep learning to identify MDSs in MD spectrograms to address the problem of person identification based on human gait MDSs. A deep convolutional neural network (DCNN) with four Conv layers and two fully connected (FC) layers was used by Vandersmissen et al. to identify a group of people based on their gait MDS [6]. In [7], a CNN architecture that can utilize hierarchical features is proposed to identify MDSs related to personal identity. To learn and identify gait Doppler signatures, Xia et al. [8] created a lightweight multi-branch CNN. Some methods just use well-known networks for this task. In [9], Abdulatif et al. used MDS to train ResNet-50 to identify walking persons. In [10], Ni et al. used pre-trained, fine-tuned ResNet50 on an indoor gait dataset of 20 people, eventually achieving a high recognition rate. Qiao et al. [11] used AlexNet to enhance person identification accuracy on human micro-Doppler (MD) separation. A VGG-16 DCNN classifier was used to identify people in spectrograms of MDS recorded by an X-band monostatic CW radar [12].

3. Methods

Figure 1 illustrates the framework of the *Multi-Scale CNN*, which consists of three parts: low-level multi-scale block, high-level multi-branch block, and classifier block. The purpose of low-level multi-scale block is to embed the input with different kernels to form features of different fields of view. The high-level multi-branch block stacks 5 layers of Conv networks to further embed the multi-scale features, then fuses multi-branch embedding features into the high-level. Then we flatten the output of high-level block and use 3 full-connected layers as the classifier.

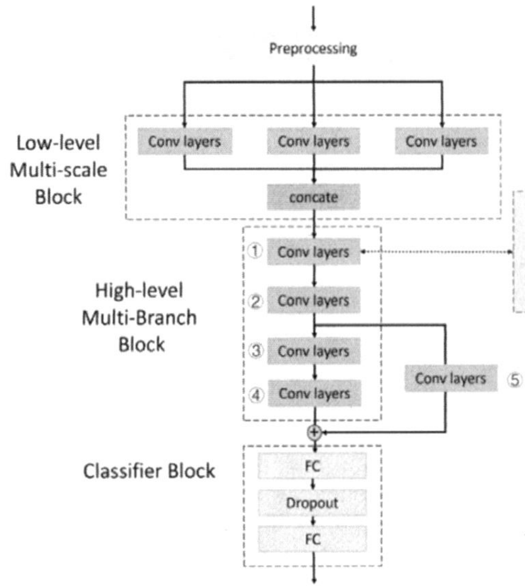

Figure 1. The framework of our proposed MS-CNN.

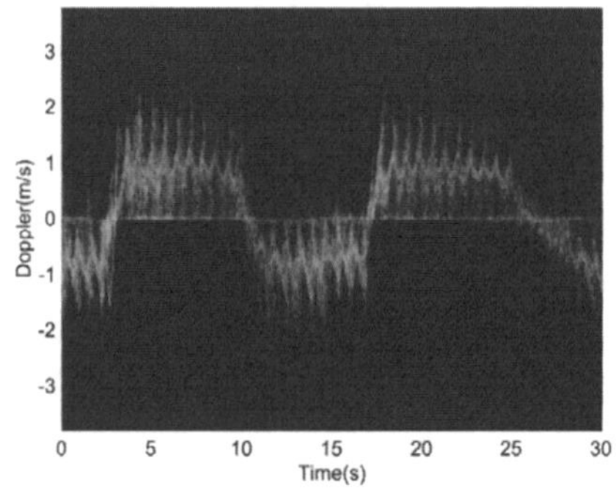

Figure 2. MDs of a walking person.

3.1. Preprocessing

In this paper, we use [6] to process the radar data frames in the IDRad data set (This data set can be found at https://www.imec-int.com/en/IDRad) to obtain the gait MDS. The following processes are used to compute the MDS: The range-Doppler map (RDM) is first created by applying a 2D fast fourier transform (FFT) to the FMCW radar data frame. Summing the absolute values of the RDM in the range axis generates an MDS vector. Then combine n MDS vector to reorganize the MDS. Finally, the noise is reduced from the MDS.

The frames are used to represent time steps in MDS, which are described by 256 Doppler channels. 1 second contains about 15 frames since an FMCW radar data frame contains 256 chirps and the duration of each chirp is $256\mu s$ (i.e., 1 second of MDS contains approximately 15 MDS vectors). According to [6], the center three Doppler channels denoting static objects, as well as the 24 outside Doppler channels on either side of the MDS, are removed, leaving each MDS vector with 205 Doppler channels. Figure 2 shows the MDs of a walking person for 30 seconds (i.e., 450 frames).

Each MDS sample in the training set, validation set, and test set is acquired using a 45-frame time window, resulting in a sample input size of 45×205. We use shift to indicate the number of frames that the window shifts backward when the next sample is obtained, and we adjust the value of shift to analyze the influence of the number of samples on identification accuracy. Finally, to obtain the next sample, we set the shift to 10, which overlaps 35 frames.

3.2. Low-level Multi-Scale Block

In previous works [6], they often used just a 3×3 kernel for convolution, however from our perspective, the larger the convolution kernel, the broader the receptive field, the more picture information you can perceive, and the greater global features you can obtain in a CNN. Thinking deeper, if we use various kernels at the same time, we could obtain multi-scale features, which will provide us multi-view information in low-level.

So, we propose the low-level multi-scale block. As Figure 1 shows, this block consists of several Conv layers. They use different kernel sizes to ensure capturing various views. Each Conv layer is composed of a Conv layer, an activation layer, and a pooling layer. Here we illustrate Conv layers with 3×3, 4×4 and 5×5 kernels. We use ELU as activation function and max pooling as pooling function. We compute the output of low-level multi-scale block with various kernels as follows:

$$f_{ms} = \left[Conv_3(X), Conv_4(X), Conv_5(X)\right] \quad (1)$$

f_{ms} is the output of a low-level multi-scale block. $[\cdots]$ denotes concatenate in the depth dimension, $Conv_i$ means doing convolution with kernel $i \times i$.

3.3. High-level Multi-Branch Block

We stack 5 Conv layers as high-level block, each Conv layers in Figure 1 is marked with number. Conv layers ①, ② and ③ all use 3×3 kernel for convolution, but they use different pooling size, the sizes are 2×3, 2×2, 1×2. Through this pooling, the final feature map size is 5×5 in each channel. Conv layer ④ use 1×1 filter kernel to decrease the channel of features, as well as fusion the information in 128 channels. We could also do feature fusion through FC layer, but as the dataset size is limited, we should design model which is small but smarter. 1×1 convolution could effectively decrease the number of parameters of our proposed model. The method described above can be summarized as follows:

$$f_{cl2} = Conv_3(Conv_3(f_{ms})) \tag{2}$$

$$f_{cl4} = Conv_1(Conv_3(f_{cl2})) \tag{3}$$

f_{cl2} and f_{cl4} are the result of Conv layer ② and Conv layer ④. There are many experiments to verify that different Conv layers extract different features, low-level Conv layers extract simple features, and deep Conv layers extract fusion features. Inspired by residual connection, we originally wanted to connect the outputs of the multi-branch Conv layers ② and ⑤, but their outputs have different sizes, so we added another Conv layer ④ to align their dimension, which has 1×1 filter kernel. Then we add the output of Conv layers ④ and ⑤ and take the average value as our final feature. This process is summarized as follows:

$$f_{cl5} = Conv_1(f_{cl2}) \tag{4}$$

$$f = \frac{1}{2}(f_{cl4} + f_{cl5}) \tag{5}$$

f is the output of the high-level multi-branch block. Through residual connection, we could obtain a variety of fusion features at different convolution level, as well as avoiding too much information loss.

3.4. Classifier Block

In Figure 1, we can see the classifier block has two FC layers and a dropout layer. Firstly, we flatten the feature from 3D to 1D, and then we use the first FC layer to fusion the feature into 128 dimensions. As we have 5 targets, the second FC layer maps the feature to 5 dimensions. We use cross-entropy loss as loss function.

4. Experiments

4.1. Experimental Setting and Dataset

The IDRad data set was collected using an INRAS FMCW radar in single-input single-output mode. This data set collects micro-Doppler data from 5 males with similar walking postures. They are between the ages of 23 and 32, weigh 60 to 99 kg, and stand 1.78 to 1.85 meters tall.

IDRad data set comprises 150 minutes of data, with the training set accounting for 100 minutes and the validation and test sets accounting for 25 minutes each. The data collection in the training set is not completed on the same day in order to simulate a more realistic setting. Data was collected first for each subject walking freely in the same room for 5 minutes, and then again two weeks later when they walked freely in the same room for 15 minutes. Furthermore, the validation and test sets are collected separately from the training set in a different room. In the validation and training sets, each person has 5 minutes of continuous walking data, and their creation times are two weeks apart.

We implement our model using PyTorch toolkit. We use Adam optimizer with a learning rate of equal to 1e-4 and a decay weight of 5e-4. We train the model for 300 epochs with a mini-batch size of 64, which takes roughly 19 minutes on Tesla V100. For a dataset, we average 3 times the testing result as the reported result.

4.2. Comparison with other methods
We compare our method to generally utilized methods [6], [9], [11] and evaluate its performance on the challenge of person identification based on gait MDS.

The performance comparison findings between our method and the methods in the aforementioned literature are shown in table 1. Our approach has a recognition rate of up to 88.57, which is 2% to 11% greater than other regularly CNN-based methods. In addition, our method has fewer parameters than most of the methods displayed in table 1.

Table 1. The MS-CNN and other methods were compared in terms of performance on the same test set.

Methods	Accuracy
Vandersmisse et al. [6]	79.06
Abdulatif et al. [9]	86.17
Qiao et al. [11]	77.81
MS-CNN (our method)	88.57

4.3. Ablation study

4.3.1. Effect of the multi-scale kernel
Figure 3 (a) illustrates the result of different settings on kernel size. It is obvious that a multi-scale kernel does better than a single-scale kernel. Among these settings, MS-CNN performs the best, which we believe is because the model can capture effective multi-scale features, such as person's height, walking posture, and stride length, etc. We can also see that it is not that the more scales, the better the model effect. After more than 3 scales, the accuracy of the model will not continue to increase, but there will be a downward trend, indicating that too many scales may cause feature redundancy, which is not conducive to model learning.

4.3.2. Effect of the shift size
In Figure 3 (b), we evaluate our model with different shift sizes. When the shift size gradually increases from 5 to 45, it can be seen that the training data becomes less and less due to the increasing offset. In the process of increasing from 20 to 45, the accuracy gradually decreased. We believe that part of the reason is the gradual reduction of training data. When shift=10 and shift=5, the training data has greatly increased, but the model accuracy has not increased. We believe that when the offset is too small, the training data is redundant, so the model cannot learn more information. Overall, when the training shift is set to 10-20, the model works best with the least training cost.

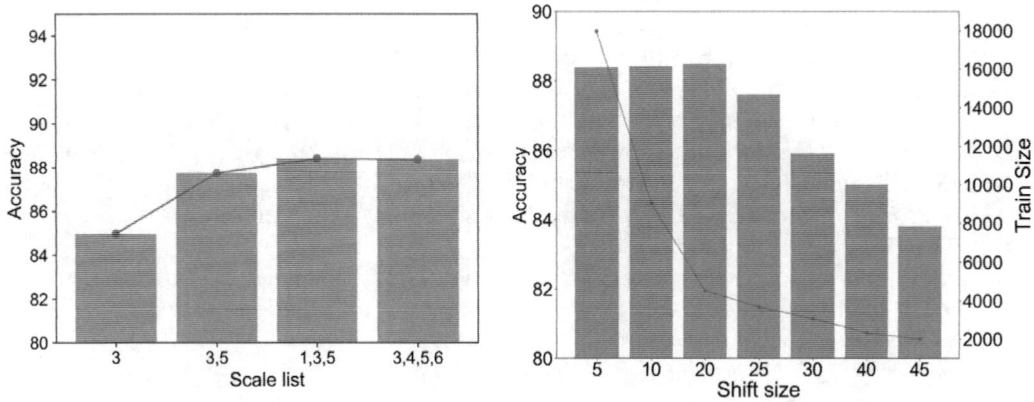

(a) Result of various scale (b) Result of various shift

Figure 3. Comparison of different settings on scale (a) and shift size (b).

5. Conclusion

In this paper, we propose a novel approach named Multi-Scale CNN (MS-CNN) for personal identification based on gait MDS. MS-CNN considers the use of multi-scale kernels for modeling in low-level, then stacking 3 layers of CNN to further embed the feature, then fusing multi-branch embedding features in high-level. We analyze the performance of our proposed approach using a more realistic and challenging dataset in which individuals can freely walk indoors, making it more adaptable to real-world circumstances. Extensive experiments show that our proposed MS-CNN approach achieves a substantial improvement.

References

[1] Pegoraro, J., & Rossi, M. (2021) Real-time People Tracking and Identification from Sparse mm-Wave Radar Point-clouds. IEEE Access, 9: 78504-78520.

[2] Marsico, M. D., & Mecca, A. (2019) A survey on gait recognition via wearable sensors. ACM Computing Surveys (CSUR), 52: 1-39.

[3] Kumar, M., Singh, N., Kumar, R., Goel, S., & Kumar, K. (2021) Gait recognition based on vision systems: A systematic survey. Journal of Visual Communication and Image Representation, 75: 103052.

[4] Wan, C., Wang, L., & Phoha, V. V. (Eds.). (2018) A survey on gait recognition. ACM Computing Surveys (CSUR), 51: 1-35.

[5] Chen, V. C., Li, F., Ho, S. S., & Wechsler, H. (2003) Analysis of micro-Doppler signatures. IEE Proceedings-Radar, Sonar and Navigation, 150: 271-276.

[6] Vandersmissen, B., Knudde, N., Jalalvand, A., Couckuyt, I., Bourdoux, A., De Neve, W., & Dhaene, T. (2018) Indoor person identification using a low-power FMCW radar. IEEE Transactions on Geoscience and Remote Sensing, 56: 3941-3952.

[7] Yang, Y., Hou, C., Lang, Y., Yue, G., He, Y., & Xiang, W. (2019) Person identification using micro-Doppler signatures of human motions and UWB radar. IEEE Microwave and Wireless Components Letters, 29: 366-368.

[8] Xia, Z., Ding, G., Wang, H., & Xu, F. (2021) Person Identification With Millimeter-Wave Radar in Realistic Smart Home Scenarios. IEEE Geoscience and Remote Sensing Letters, 19: 1-5.

[9] Abdulatif, S., Aziz, F., Armanious, K., Kleiner, B., Yang, B., & Schneider, U. (2019) Person identification and body mass index: A deep learning-based study on micro-Dopplers. In 2019 IEEE Radar Conference (RadarConf) pp. 1-6.

[10] Ni, Z., & Huang, B. (2020) Human identification based on natural gait micro-Doppler signatures using deep transfer learning. IET Radar, Sonar & Navigation, 14: 1640-1646.

[11] Qiao, X., Shan, T., & Tao, R. (2020) Human identification based on radar micro - Doppler signatures separation. Electronics Letters, 56: 195-196.
[12] Papanastasiou, V. S., Trommel, R. P., Harmanny, R. I. A., & Yarovoy, A. (2021) Deep Learning-based identification of human gait by radar micro-Doppler measurements. In 2020 17th European Radar Conference (EuRAD). pp. 49-52.
[13] He, K., Zhang, X., Ren, S., & Sun, J. (2016) Deep residual learning for image recognition. In Proceedings of the IEEE conference on computer vision and pattern recognition, pp. 770-778.
[14] Krizhevsky, A., Sutskever, I., & Hinton, G. E. (2012) Imagenet classification with deep convolutional neural networks. Advances in neural information processing systems, 25: 1097-1105.
[15] Simonyan, K., & Zisserman, A. (2014) Very deep convolutional networks for large-scale image recognition. arXiv preprint arXiv:1409.1556.

Journal of Physics: Conference Series 2258 (2022) 012045 doi:10.1088/1742-6596/2258/1/012045

Document-level Stance Detection Method for Specific Targets Based on XLNet

Yuwei Fang[1], Lei Zhang[1,2*], Fei Yang[3], Xiaoming Zhao[1], Xueqiang Song[1], Dong Li[1], Hong Li[4] and Haizhou Wang[3*]

[1]School of Computer Science & School of Software Engineering, Sichuan University, Chengdu, Sichuan, 610065, China

[2]Institude for Industrial Internet Research, Sichuan University, Chengdu, Sichuan, 610065, China

[3]School of Cyber Science and Engineering, Sichuan University, Chengdu, Sichuan, 610065, China

[4]College of Continuing Education, Sichuan University, Chengdu, Sichuan, 610065, China

[*]Corresponding author's e-mail: zhanglei@scu.edu.cn, whzh.nc@scu.edu.cn

Abstract. Social media and slef-media platforms are increasingly being used to discuss topics of varying polarities where different people take different stances. As far as we know, most of the current research focuses on short texts. However, there are relatively few studies on how to collect and extract people's opinions and stances from long texts such as review articles. This paper proposed a target-specific hierarchical attention model based on the XLNet pre-training model, named XLNet-THAN. The attention mechanism and topic embedding are utilized to obtain the semantic representation of the document. After contrast experiments with other popular text classification models, the result shows that the model proposed in this paper can process document-level text well, and achieves great performance in the task of document-level stance detection.

1. Introduction

With the rapid development of the Internet and the emergence of social media software, more and more netizens are using self-media platforms to express their views and stances by writing commentary articles. Compared with news articles of traditional official media, self-media authors' standpoints are clearer, and it often contains helpful advices on products, policies, current news and other information. These massive review articles are of great significance to public opinion mining, market analysis and other fields, and can provide decision makers with suggestions from the public.

The stance of the text refers to the understanding of a specific target or the attitude that the author of the text holds. The targets mainly include people, events, organizations, policies, products, etc. Stance detection is an important subtask in the field of text sentiment analysis. The difference between stance detection and sentiment analysis is that stance detection needs to discriminate not the positive or negative emotional polarity of the text content, but the more complicated "Favour", "Against" or "Neither" attitude of the discussion target. Emotion is usually used as a carrier for stance expression, and the same stance can be expressed through different emotional polarities.

Most of the current research in the field of stance detection focuses on the analysis of short text data sets and many algorithms have been proposed [1-6]. However, in the field of text detection at the document level, it is difficult for traditional stance detection algorithms to extract the semantic information in the text due to the length of the article.

Inspired by text classification tasks and sentiment analysis tasks, this paper proposed a target-specific hierarchical attention model based on the XLNet pre-training model, referred to as XLNet-THAN. We collected document-level text data under multiple topics from self-media platforms such as WeChat and Zhihu and finally forms a long text stance detection data set containing 4,086 articles after manual annotation. Then we compared the proposed XLNet-THAN model with a variety of popular traditional stance detection models, and achieves excellent performance. The F1-score of the model proposed in this paper reached 0.7637 on the data set, and the accuracy rate reached 76.38%.

2. Related Research

Early research on stance detection was mainly focused on debate texts. After that, stance detection experiments on public speeches, online debates, and tweets also emerged. Driven by the stance detection competition SemEval, NLPCC, etc., the development of stance detection technology for tweets and microblogs was very rapid. The stance detection of document-level text was originally a sub-task of FNC (Fake News Challenge), but how to detect the stance of an article (including title and content), which is similar to the task proposed in short text stance detection, the related research is relatively weak.

In the early stance detection work, rule-based classifier algorithms were mainly used, including supervised algorithms such as Naive Bayes (NB) [2], Support Vector Machines (SVM) [3], and Logistic Regression (LR) [4]; Later, deep learning methods were gradually applied to tweets stance detection task. Wei et al. [5] uses the convolutional neural network to obtain text features for classification. Yang et al. [7] uses the LSTM units and hierarchical attention mechanism to obtain the final tweets representation; Du et al. [6] propose a new attention mechanism to implant target information into the stance classification model to locate the most relevant part of the text. From 2018, pre-training models such as BERT [8], XLNet [9], ELMo have been proposed one after another, refreshed the optimal results time and time again in many text classification tasks.

3. Approach

3.1. XLNet pretrained model

XLNet is a new pre-training model proposed by Google Brain on the basis of BERT, which overcomes two shortcomings of BERT. XLNet adopts PLM (Permutation Language Model), arranges sentences randomly, uses autoregressive method for training, and learns the dependencies between tokens through a dual-stream attention mechanism.

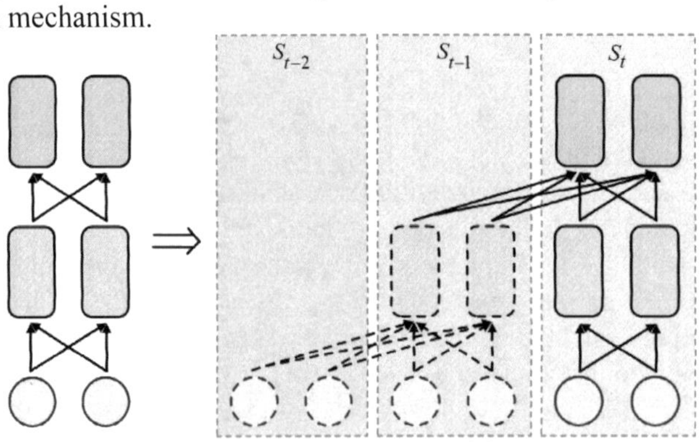

Figure 1. Transformer-XL model structure

The text length of the document-level stance detection task proposed in this paper is concentrated in 1500~3000 words. Limited by the size of system memory, the sequence length input by Transformer has an upper limit. In research, the overly long sequence is usually cut into segments with a fixed length N, and then input into the Transformer for calculation and characterization. As shown in the Figure 1, in order to allow Transformer to learn longer dependencies under memory constrains, Transformer-XL draws on the idea of TBPTT (Truncated Back-Propagation Through Time), caches the representation calculated from the previous segment, and adds it to the characterization calculation of the current segment. In addition, Transformer-XL introduces Relative Position Embedding, which solves the ambiguity of position coding between different segments, and solves the problem that as the number of layers deepens, the input position coding information is erased by excessive calculations. The application of Transformer-XL enables XLNet to model longer dependencies.

3.2. Bi-LSTM units

LSTM is a special recurrent neural network model. The network model is based on RNN, adds forgetting units and memory units to solve the problems of gradient disappearance and gradient explosion. LSTM is composed of three gate structures: forget gate, input gate and output gate. The LSTM architecture can be expressed as follows:

$$i_t = \sigma\left(W_i \cdot x_t + U_i \cdot h_{t-1}\right) \tag{1}$$

$$f_t = \sigma\left(W_f \cdot x_t + U_f \cdot h_{t-1}\right) \tag{2}$$

$$o_t = \sigma\left(W_o \cdot x_t + U_o \cdot h_{t-1}\right) \tag{3}$$

$$\tilde{c}_t = \tanh\left(W_c \cdot x_t + U_c \cdot h_{t-1}\right) \tag{4}$$

$$c_t = f_t \odot c_{t-1} + i_t \odot \tilde{c}_t \tag{5}$$

$$h_t = o_t \odot \tanh\left(c_t\right) \tag{6}$$

where $W_i, W_f, W_0, U_i, U_f, U_o$ are weight matrices; x_t, h_t are input state and hidden state at time step t respectively; σ is the sigmoid function; $\odot$ denotes element-wise product.

The bidirectional LSTM unit obtains the final feature representation by concatenating the vectors obtained in order and reverse order of the sequence, which can better capture and utilize contextual semantic information.

3.3. Attention mechanism

The Attention mechanism can better extract the semantic information of the input sequence by assigning different weights to the importance of each word or sentence in the input sequence. In attention layers, the vector representation is passed through a two-layer neural network to obtain the attention weights and construct the final feature vector representation. The calculation process can be expressed as follows:

$$\alpha_i = W_2 \cdot \tanh\left(W_1 \cdot v_{ji} + b_1\right) + b_2 \tag{7}$$

$$\alpha = softmax\left(\alpha\right) \tag{8}$$

$$R_j = \sum_{i=1}^{n} \alpha_i v_{ji} \tag{9}$$

where v_{ji} is the i-th word embedding vector of the input sequence V_j; W_1, W_2 are weight matrices; b_1, b_2 are bias; n is the sequence length and R_j is the final representation.

3.4. Model Design and Implementation

This paper proposed an XLNet-THAN stance detection model, which uses the segmented memory mechanism of XLNet to extract the feature vectors of words in the document-level long texts, and then inputs them to the bidirectional LSTM unit to further extract the sentence semantic information. After passing the word-level attention layer and segment-level attention layer, the stance of the article is determined in the end.

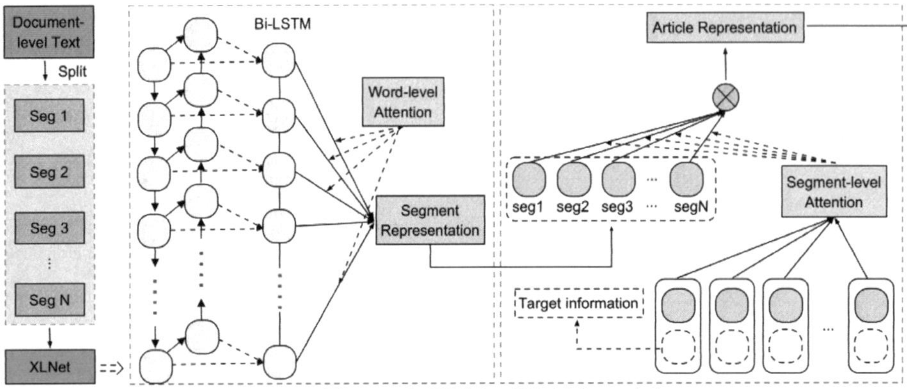

Figure 2. XLNet-THAN model structure

XLNet-THAN is composed of XLNet embedding layer, word-level attention layer, segment-level attention layer from the bottom up, as shown in the Figure 2. The XLNet embedding layer uses the pre-trained XLNet model to vectorize the long text. We first splice the article title and the article body, and divides each article into several segments and inputs them into the model. For each segment, word-level attention layer uses the bidirectional LSTM coding unit to extract the contextual semantic information contained in the feature vector, assigns weights through the word-level attention mechanism to each word, and obtains the final vector representation after calculation. Then segment-level attention layer combines the topic word information with the segment representation, and get the overall representation of the article after the segment-level attention mechanism. Finally, the stance classification result is output after the dropout and softmax layers. Our model embeds the target information discussed in the article into the calculation of the segment-level attention mechanism, which can better extract the important information related to the target, and predict the stance of the article more accurately.

4. Experimental process and result analysis

4.1. Experimental data
This paper uses keywords such as "Genetically Modified Food", "Feminism", and "Chinese Medicine" to collect relevant self-media commentary articles. The original data is pre-processed, including operations such as removing URL links, garbled characters, and extra spaces in the text, to obtain a cleaned data set. Since the understanding of the author's stance may be subjective, this paper defines four fine-grained stance categories: 0 (unrelated), 1 (favour), 2 (against), and 3 (neither). After counting the votes of multiple annotators, a document-level stance detection data set including 4,086 articles was finally established.

4.2. Model building and parameter setting
The model in this paper uses the Pytorch framework for experimentation, and is trained on NVIDIA Tesla V100 GPU with 16GB of RAM. In the training process, the Adam optimizer is used, the learning rate is set to 5×10^{-5}, and the batch size is 64. In addition, dropout is used in the BiLSTM layer to suppress the overfitting problem, and the value is 0.5.

4.3. Analysis of experimental results
This paper reproduces several short text stance detection models and popular models used in text classification tasks, and compares them with the proposed XLNet-THAN mode. The comparison models are as follows:

- **Bi-LSTM.** Bi-LSTM is a variant of recurrent neural network (RNN), which can effectively mine long-term dependencies in data. It is a popular baseline model in text classification and semantic understanding tasks [10].
- **Text CNN.** Text CNN extracts the n-gram features of text data by using convolution kernels of different sizes. It is also the most popular baseline model for text classification tasks [5].
- **DPCNN.** Text CNN is difficult to extract the long-distance dependence of text data. DPCNN [11] can effectively extract the long-distance relationship of text by increasing the depth of the network.
- **Target-Specific Attention Network.** TAN [6] embeds the relevant information of a specific target into the attention mechanism, obtains the guidance information of the article's subject words.
- **Hierarchical Attention Network.** HAN [7] uses the attention mechanism to perform hierarchical coding of sentences and articles in text data, which can extract document information well in text coding tasks.

Table 1. Results of different models on the dataset.

Model	Precision	Recall	F1-score	Accuracy
Bi-LSTM [10]	0.5267	0.5654	0.4977	0.5565
Text CNN [5]	0.7088	0.7123	0.7045	0.7099
DPCNN [11]	0.7078	0.7173	0.7080	0.7173
TAN [6]	0.7225	0.7156	0.7089	0.7173
HAN [7]	0.7320	0.7186	0.7220	0.7186
XLNet-THAN	**0.7674**	**0.7638**	**0.7637**	**0.7638**

As can be seen from Table 1, BiLSTM does not performs well on the data set, while Text CNN and DPCNN can achieve relatively high accuracy. This is because most of the articles in this data set are document-level texts of more than 1,000 words, it is difficult for the LSTM unit to extract such long-distance semantic dependence, but CNN can correctly predict the stance through the N-gram features obtained by convolution. The HAN segmented the article at first, and then put it into the BiLSTM unit to improve performance. The TAN only selected the first 800 words of the article for encoding, but proves that topic words can help the model predict the stance. Our model achieves the best F1-score and accuracy, which proves that extracting the semantic information of document-level text data is effective, and also shows that the target-specific attention mechanism can improve the performance in the task of detecting the stance of a document-level text.

5. Conclusion and Future Work

This paper constructs an annotated Chinese document-level stance detection data set to fill the gap in the field of Chinese long text stance detection. On the basis of this corpus, this paper proposed a target-specific hierarchical attention model based on the XLNet pre-training model. Experiments and evaluation results prove that the proposed XLNet-THAN model can accurately predict the stance of document-level text. In future work, we will conduct in-depth research on how to more accurately extract the author's attitude on the discussion topic, and consider adding the extended knowledge information of the article discussion target to the stance detection task.

Acknowledgments

This work is supported by the National Natural Science Foundation of China (NSFC) under grant nos. 61802270, 61802271, 81602935, and 81773548. In addition, this work is also partially supported by Joint Research Fund of China Ministry of Education and China Mobile Company (No. CM20200409), the Key Research and Development Program of Science and Technology Department of Sichuan

Province (No. 2020YFS0575), and the Sichuan University and Yibin Municipal People's Government University and City Strategic Cooperation Special Fund Project (Grant No. 2020CDYB-29).

References

[1] Wei, P., Mao, W., Zeng, D. (2018) A Target-Guided Neural Memory Model for Stance Detection in Twitter. In: Proceedings of the 15th International Joint Conference on Neural Networks. Rio de Janeiro. pp. 1–8.

[2] Walker, M., Anand, P., Abbott, R., Grant, R. (2012) Stance classification using dialogic properties of persuasion. In: Proceedings of the 8th Conference of the North American Chapter of the Association for Computational Linguistics: Human Language Technologies, 2012, Montreal, pp. 592–596.

[3] Hasan, K. S., Ng, V. (2013) Stance classification of ideological debates: data, models, features, and constraints. In: Proceedings of the 6th International Joint Conference on Natural Language Processing. Nagoya. pp. 1348–1356.

[4] HaCohen-Kerner, Y., Ido, Z., Ya'akobov, R. (2017) Stance classification of tweets using skip char ngrams. In: Proceedings of the 28th Joint European Conference on Machine Learning and Knowledge Discovery in Databases. Skopje. pp. 266-278.

[5] Wei, W., Zhang, X., Liu, X., Chen, W., Wang, T. (2016) A specific convolutional neural network system for effective stance detection. In: Proceedings of the 10th International Workshop on Semantic Evaluation. San Diego. pp. 384-388.

[6] Du, J., Xu, R., He, Y., Gui, L. (2017) Stance classification with target-specific neural attention networks. In: Proceedings of 26th International Joint Conference on Artificial Intelligence. Melbourne. pp. 3988–3994.

[7] Yang, Z., Yang, D., Dyer, C., He, X., Smola, A., Hovy, E. (2016) Hierarchical attention networks for document classification. In: Proceedings of the 11th Conference of the North American Chapter of the Association for Computational Linguistics: Human Language Technologies. San Diego. pp. 1480-1489.

[8] Devlin, J., Chang, M. W., Lee, K., Toutanova, K. (2018) Bert: Pre-training of deep bidirectional transformers for language understanding. https://arxiv.org/abs/1810.04805.

[9] Yang Z., Dai Z., Yang Y., et al. (2018) XLNet: Generalized Autoregressive Pretraining for Language Understanding. https://arxiv.org/abs/1906.08237.

[10] Augenstein, I., Rocktäschel, T., Vlachos, A., & Bontcheva, K. (2016) Stance detection with bidirectional conditional encoding. https://arxiv.org/abs/1606.05464.

[11] Johnson, R., Zhang, T. (2017) Deep pyramid convolutional neural networks for text categorization. In: Proceedings of the 55th Annual Meeting of the Association for Computational Linguistics. Vancouver. pp. 562–570.

Spatial-temporal graph attention network for traffic flow forecasting

Ling Wang[1,2a], **Gaofeng Jia**[1,2b*]

[1]School of Automation and Electrical Engineering, [2]Key Laboratory of Knowledge Automation for Industrial Processes, Ministry of Education, University of Science and Technology Beijing, Beijing 100083, China

[a]lingwang@ustb.edu.cn, [b*]gaofeng.jia@foxmail.com

Abstract—Traffic flow forecasting is one of the most challenging tasks in intelligent transportation system, which is very important and critical for many traffic services. Although there are existing works on forecasting the future traffic flow, the majority of them have certain limitations on modelling spatial-temporal correlations: they usually use different modules for spatial and temporal dependencies, and lack abilities of utilizing the historical periodic data. In order to overcome these problems, a novel Spatial-Temporal Graph Attention network (STGAT) is proposed. Considering the periodic characteristic of traffic data, the raw data is transformed into a three-dimensional tensor corresponded to the weekly period and recent period before inputting into the STGAT model. The random walk theory is used to construct a localized spatial-temporal graph to model the spatial-temporal dependencies, and an elaborately designed spatial-temporal graph attention mechanism is used to synchronously capture the complex spatial-temporal dynamic correlations. By calculating the similarity of spatial-temporal features of weekly and recent period input, the model integrates the periodic features into the target. Experimental results on two public real-world datasets demonstrate that STGAT achieves superior performance and consistently outperforms other baselines.

1. Introduction

Traffic flow forecasting is one of the most challenging tasks in intelligent transportation system (ITS), which is very important and critical for many traffic services, such as traffic flow control, road trip planning and navigation [1]. The prediction of urban traffic flow is of great significance in practical production. However, its performance is affected by the road network structure and complex and changeable actual traffic conditions, which is difficult to be accurately described by mathematical formulas.

In order to predict the traffic flow, the spatial and temporal dependencies should be jointly modeled to capture the complex spatial-temporal correlations of the traffic data. Traffic flow depends on its corresponding road network topological structure, i.e., spatial dependencies. It is intuitive that the traffic flow on one road is affected on its adjacent roads. Meanwhile, the traffic flow is also dynamically influenced by its historical condition, which consists of closeness and periodicity.

In recent years, many studies have used graph neural networks [2] to model spatial-temporal data and achieved impressive performance. Compared with traditional deep learning methods, graph neural network can more effectively model the spatial-temporal dependencies in data, so as to help improve the accuracy of prediction. However, these models still have some problems that limit their performance. In order to capture spatial-temporal correlations, Yu et al. [3] used spectral convolution, and Li et al. [4] used diffusion convolution. However, most of these graph convolution methods use simple dependency graph generated by distance or similarity measures. These kinds of graphs are normally incomplete, which are not suitable for direct use in prediction tasks. Several studies have been attempted to improve the dependency graph. For instance, Bai et al. [5] proposed a dynamic graph structure learning module to capture the complex relationships and uncertainties in spatial. Wu et al. [6] developed a novel self-adaptive dependence matrix to model the correlations variation between nodes and neighbors over time. However, they all use different modules to model temporal and spatial dependencies separately, which fragments the spatial-temporal correlations. Spatial-temporal correlation is limited by the spatial distances between nodes and the time range of time series, thus it is usually localized. Song et al. [7] proposed a novel spatial-temporal graph convolution module by using spatial-temporal subgraph of continuous time steps, which captures the spatial-temporal correlations synchronously. However, they solely consider static graphs, while the spatial dependencies in real-world are changing, and some random abnormal events may even lead to drastic changes in spatial, such as road maintenance, vehicle breakdowns, etc.

To capture the complex spatial-temporal correlations in the data, we propose a model named Spatial-Temporal Graph Attention network (STGAT). The contributions of this paper are concluded as follows:

(1) This paper designs a novel spatial-temporal graph attention mechanism to simultaneously capture the complex temporal and spatial dependencies in data.

(2) This paper proposes localized spatial-temporal graph to model the spatial-temporal information diffusion process in traffic data, and the spatial-temporal graph attention mechanism can effectively capture the spatial-temporal dependencies in the data.

(3) Extensive comparison experiments were conducted with several benchmark models on real-world traffic datasets, the result shows that STGAT achieves better prediction accuracy compared to existing benchmark models.

2. Problem definition

Given traffic data collected in urban road, its sensor network can be defined as a graph $G=(V,E,A)$, where $V=\{v_1,v_2,\cdots,v_N\}$ is the set of nodes, $|V|=N$ corresponding to the observation of N sensors/roads; E is the set of edges, which indicates the connectivity of nodes; $A\in\mathbf{R}^{N\times N}$ is the adjacency matrix of graph G. The traffic data is regarded as the graph signal sequence on the graph. For example, the traffic data x_t^i is regarded as the signal value of the node v_i at time t, and the data observed by the road network at time t is represented as the graph signal: $X_t=\left(x_t^1,x_t^2,\cdots,x_t^N\right)^{\mathrm{T}}\in\mathbf{R}^{N\times F}$, where F is the number of variables recorded for each sensor. Taking the traffic data in Figure 1 as an example, the traffic road network is represented as a graph, and the variable values of each node of the graph at each time step are represented as a signal sequence in Figure 1 (a). Each node in the graph records three variables' values: traffic flow, average speed and road occupancy.

Given F variables' observation of N sensors in past T_h time steps $\mathrm{X}=\left(X_{t-T_h+1},X_{t-T_h+2},\cdots,X_t\right)^{\mathrm{T}}\in\mathbf{R}^{T_h\times N\times F}$, the goal of spatial-temporal data forecasting is learning a function f, which is able to predict $Y=\left(y^1,y^2,\cdots,y^N\right)^{\mathrm{T}}\in\mathbf{R}^{N\times T_p}$ of the next T_p time steps, where $y^i=\left(y_{t+1}^i,y_{t+2}^i,\cdots,y_{t+T_p}^i\right)\in\mathbf{R}^{T_p}$ denotes the traffic flow of node i from $t+1$.

$$Y=f\left(\left[X_{t-T_h+1},X_{t-T_h+2},\cdots,X_t\right],G\right) \tag{1}$$

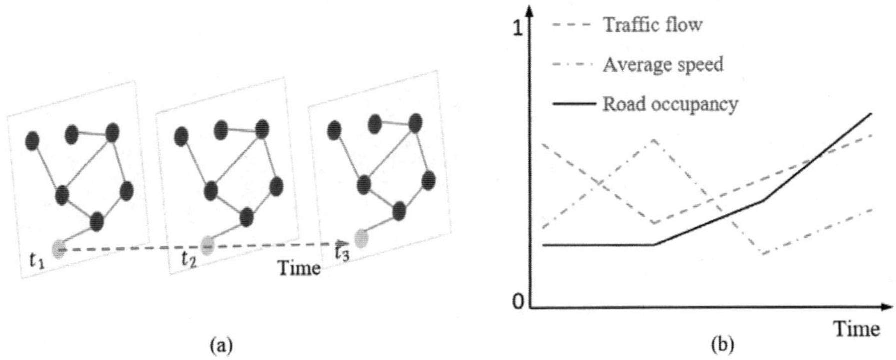

Figure 1. Graph signal representation of traffic data

3. Spatial-temporal graph attention network

Traffic data contains temporal and spatial dependency characteristics. Since the goal is to predict the traffic conditions of each road of the road network in the future, the key point is how to effectively capture the spatial-temporal dependencies and periodic characteristics in the data [8]. Therefore, this paper proposes STGAT model, the architecture is shown in Figure 2, where T_w、T_r and T_p denotes the time period of weekly-period data X_{week}, recent-period data X_{recent} and the predictions Y respectively. The weekly-period data can be considered as predictions with noise. The STGAT model using the self-attention mechanism to dynamically model the similarity of recent-period's spatial-temporal features to that of weekly-period to obtain the predictions, which effectively filter out the noise.

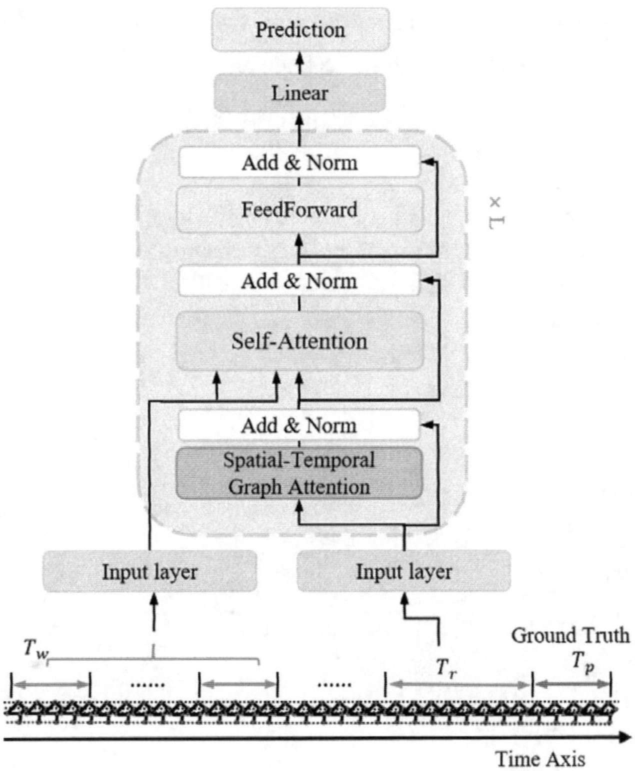

Figure 2. The architecture of STGAT model

The input of STGAT model contains data from two different periods as shown in Figure 2. Assume that the data is sampled at a frequency of q times per day, current time is t_0 and the size of forecasting window is T_p . Then, the recent-period data $\mathrm{X}_{recent} = \left(X_{t_0-T_r+1}, X_{t_0-T_r+2}, \cdots, X_{t_0}\right)^{\mathrm{T}} \in \mathbf{R}^{N\times F\times T_r}$ is the segment of historical time series directly adjacent to the forecasting period, the weekly-period data $\mathrm{X}_{week} = \left(X_{t_0-7*(T_w/T_p)*q+1}, \cdots, X_{t_0-7*(T_w/T_p)*q+T_p}, X_{t_0-7*(T_w/T_p-1)*q+1}, \cdots, X_{t_0-7*(T_w/T_p-1)*q+T_p}, \cdots, X_{t_0-7*q+1}, \cdots, X_{t_0-7*q+T_p}\right)^{\mathrm{T}} \in \mathbf{R}^{N\times F\times T_w}$ is the sampled data for the same forecasting time segment over past few weeks.

3.1. *Input layer*

In order to capture the spatial-temporal correlations efficiently, we first mapping the node representation of the input data $\mathrm{X}_{in} = \left(X_1, X_2, \cdots, X_{T_h}\right)^{\mathrm{T}} \in \mathbf{R}^{F\times N\times T_h}$ to a high-dimensional space as $\mathrm{X}_{emb} = \left(X_1, X_2, \cdots, X_{T_h}\right)^{\mathrm{T}} \in \mathbf{R}^{d_{emb}\times N\times T_h}$. The above operation is implemented the by using 1D convolution layer, where the input channel is F and the output channel is d_{emb} . Unlike long short-term momery (LSTM), which is inherently capable of retaining temporal causal information in the data. The temporal causal information will be ignored by directly feeding the above embedded data into the attention layer. Therefore, the position embedding is added to data to incorporate information about the order of the sequence. In detail, a learnable temporal embedding matrix $P \in \mathbf{R}^{T_h \times d_{emb}}$ is added to the input data X_{in} using the broadcast mechanism.

$$\mathrm{X}_{out} = Conv_{1\times1}\left(\mathrm{X}_{in}\right) + P \tag{2}$$

3.2. *Localized spatial-temporal graph*

The information propagation between nodes is a kind of diffusion process with Markov property. The study [4] utilize the diffusion convolution over a graph to capture the spatial dependencies, in which the diffusion process is characterized by a random walk with state transition matrix $D^{-1}A$, where $D = diag(A)$ is the degree matrix.

$$X_t \star f_\theta = \sum_{i=0}^{\tau-1}\left(\theta_{i,1}(D_O^{-1}A)^i + \theta_{i,2}\left(D_I^{-1}A^{\mathrm{T}}\right)^i\right)X_t \tag{3}$$

Where $\star$ is the diffusion convolution, τ is the diffusion step, $\theta \in \mathbf{R}^{\tau\times 2}$ is the learnable parameter of the convolution kernel f_θ; the outgoing diagonal matrix of the directed graph $D_O = diag(A)$, and the incoming diagonal matrix $D_O = diag(A^{\mathrm{T}})$. $D_O^{-1}A$ and $D_I^{-1}A^{\mathrm{T}}$ denote the state transfer matrix of the diffusion process and its inverse transfer matrix, respectively.

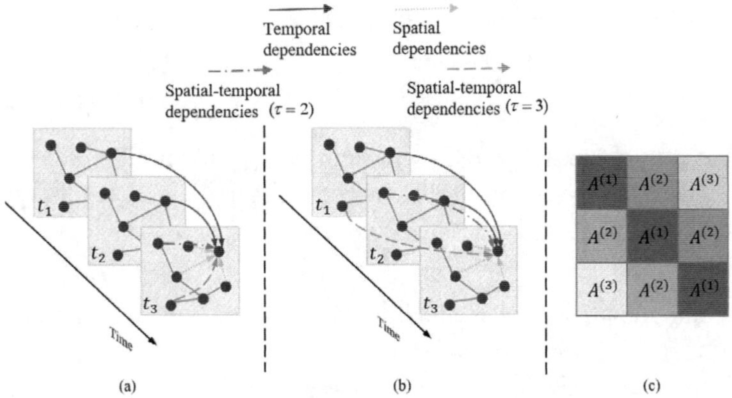

Figure 3. Localized spatial-temporal graph

However, the above diffusion convolution operation uses graph signal of single time step in different diffusion steps. In order to capture the spatial temporal correlations synchronously, it is necessary to aggregate graph information of corresponding time step for different diffusion steps. Therefore, inspired by the random walk theory, we design localized spatial-temporal graph $A_{ST} \in \mathbf{R}^{\tau N \times \tau N}$ by considering the spatial-temporal dependencies of consecutive time steps. Figure 3(a) shows the separation of temporal dependencies and spatial dependencies caused by the separate modelling of spatial-temporal dependencies of previous models such as [4]; Figure 3(b) shows the spatial-temporal dependencies of the localized spatial-temporal graph proposed in this paper, which captures graph signals at different times rather than at a single time through random walk; Figure 3(c) is the form of the corresponding localized spatial-temporal graph. Where $A^{(\tau)}$ represents the τ order adjacency matrix. For node i, its τ order adjacency represents all node sets that can be reached through at most τ node.

3.3. *Spatial-temporal graph attention mechanism*

Self-attention has been successfully used in a variety of NLP and CV tasks. The input of vanilla self-attention in [10] is a three-tuple (query, key, value), and the attention computation is performed in the form of scaled dot-product, which can be expressed as:

$$Attention(Q,K,V) = SoftMax\left(\frac{QK^{\mathrm{T}}}{\sqrt{d}}\right)V \tag{4}$$

where Q, K, $V \in \mathbf{R}^{T_h \times d}$ are query, key and value matrix respectively, and d is the input dimension. The attention score is computed by the part of softmax function, which is used as the weights to aggregate information from the corresponding value.

To model the spatial-temporal dependencies in the data, we propose spatial-temporal graph attention to capture the dynamic temporal association within nodes and the nonlinear spatial-temporal association features between nodes based on the localized spatial-temporal graph. Unlike the vanilla self-attention mechanism, we first use 1D convolution network (CNN) to capture the long sequence temporal correlation of each node. Because of the limitation of the spatial distances between nodes and the time range of the time series, complex spatial-temporal correlations are usually localized [7]. To model such sophisticated correlations, the multi-head attention mechanism is adopted. We split the Q, K and $V \in \mathbf{R}^{T \times N \times d}$ into multiple heads along the temporal dimension, and then perform the attention function in parallel to capture the localized spatial-temporal correlations.

$$MultiHead(Q,K,V) = Concat(head_1, \cdots, head_s) \tag{5}$$

In the spatial-temporal graph attention, each head corresponds to a set of adjacent spatial-temporal graph of τ consecutive time steps. The input of each head is $Q=\{q_1,\cdots,q_i,\cdots,q_{\tau N}\}\in\mathbf{R}^{\tau N\times d}$, $K=\{k_1,\cdots,k_i,\cdots,k_{\tau N}\}\in\mathbf{R}^{\tau N\times d}$ and $V=\{v_1,\cdots,v_i,\cdots,v_{\tau N}\}\in\mathbf{R}^{\tau N\times d}$. The attention coefficient matrix M is obtained by calculating the dot product (matrix multiplication) of Q and K. Each row of the matrix is calculated by the dot product of the corresponding i-th query for each key.

$$M=\frac{QK^{\mathrm{T}}}{\sqrt{d}} \tag{6}$$

In spatial-temporal graph attention module, the information of node i is updated by aggregating the information of its related nodes, $SoftMax$ function is used to calculate the weight. The attention weight can be calculated efficiently using sparse-dense matrix multiplication [4].

$$a_{ij}=SoftMax\left(M_{ij}\right)=\frac{\exp\left(M_{ij}\right)}{\sum_{p\in Mask_i}\exp\left(M_{ip}\right)} \tag{7}$$

$$Attention_i=\sum_{j=1}^{\tau N}a_{ij}v_j \tag{8}$$

3.4. Output layer

Finally, a linear layer is used as the output layer. Different from previous sequence to sequence models that output prediction step-by-step, STGAT outputs all the predictions of the target time period at one time, which alleviates the problem of error accumulation in the autoregressive model. And L2 loss is used as the loss function:

$$loss=\left\|Y-\hat{Y}\right\|^2 \tag{9}$$

Where $\hat{Y}$ is the model's predictions, Y is ground truth.

4. Experimental results and Discussions

In order to evaluate the performance of the model, comparative experimental results and detailed analysis on two real traffic datasets are presented in this section.

4.1. Dataset

The performance of STGAT was evaluated on two public real traffic datasets: PeMS08 and PeMS04 [9]. These two datasets were collected from California by the Caltrans Performance Measurement System (PeMS) at a frequency of sampling every 30 seconds. PeMS04 contains data from 307 sensors in the San Francisco Bay Area from January to February 2018, and PeMS08 contains traffic data from 170 sensors in San Bernardino from July to August 2016. Both of them contain three kinds of traffic variables: traffic flow, average speed and road occupancy. In this paper, the original data were aggregated into 5-minute intervals, which means that there were 288 records for the sampled time points in a day. In addition, the data from each sensor is normalized in the following way: $x'=\frac{x-mean(x)}{std(x)}$, where $mean(x)$ and $std(x)$ are the mean and standard deviation of x respectively.

Table 1. Dataset description information

Dataset	Nodes	Edges	Samples	Missing rate
PeMS08	170	295	17856	0.696%
PeMS04	307	340	16992	3.182%

4.2. Baseline

In the experiments, we use the following five baselines to validate the performance of our model. (1) LSTM: a sequence-to-sequence model, which uses a fully connected LSTM layer in both the encoder and decoder. (2) DCRNN: Diffusion convolution recurrent neural network [4], which uses random walk on the graph to simulate information diffusion process, and uses gated cyclic units to construct sequence-to-sequence model. (3) STGCN: Spatial-temporal graph convolution network [3], which uses 1D causal convolution to extract temporal dependencies instead of LSTM or other recurrent neural networks, and spectral graph convolution to extract spatial dependencies. (4) STSGCN: Spatial-temporal synchronization graph convolution network [7], which designs a spatial-temporal synchronization modelling mechanism to synchronously capture complex local spatial-temporal dependencies. (5) ASTGCN: attention based spatial-temporal graph convolution network [9], which uses three same components to extract spatial-temporal dependencies from input data in three different historical periods, and obtains prediction results through weighted fusion.

4.3. Parameter setting and evaluation criteria

Our model is implemented by Pytorch 1.6.0. For the lengths of the input and output data, we set them as $T_w = 12$, $T_r = 24$ and $T_p = 12$. In localized spatial-temporal graph, diffusion step $\tau = 3$. The input layer mapping dimensions d_{emb} is 36. In addition, we split the dataset with ratio 6:2:2 at PeMS04 and PeMS08 into training sets, validation sets and test sets. We train our model using Adam (adaptive moment estimation) optimizer with learning rate 0.001, the attenuation rate is 0.92, the batch size is set to 16, and training epoch is 40.

In the experiment, the average absolute error MAE and root mean square error RMSE are selected as the evaluation criteria. The smaller the two values, the better.

$$MAE = \frac{1}{N \times T_p} \sum_{i=1}^{N} \sum_{t=1}^{T_p} \left| y_t^i - \hat{y}_t^i \right| \tag{10}$$

$$RMSE = \sqrt{\frac{1}{N \times T_p} \sum_{i=1}^{N} \sum_{t=1}^{T_p} \left(y_t^i - \hat{y}_t^i \right)^2} \tag{11}$$

4.4. Experiment results

In the experiment, we compare the STGAT model with the above five models on the datasets PeMS04 and PeMS08. The MAE and RMSE of each model's predictions at one hour prediction level are shown in Table 2. It can be seen that STGAT model achieves the relatively best performance on the two datasets. Firstly, traditional deep learning model LSTM only captures the temporal dependencies in the data and ignores the spatial dependencies, so it achieves the worst performance in this forecasting task. In contrast, graph neural networks such as STGCN, DCRNN use spectral convolution or diffusion convolution to capture the spatial dependencies, and achieve better performance. However, most of them only use the original dependency graph with insufficient information, and their graph information aggregation mechanism cannot effectively extract the dependencies in the data. STSGCN, ASTGCN have better performance than STGCN and DCRNN. Among them, ASTGCN achieves relatively good predictions because it considers the periodicity of data, which verifies the importance of historical periodic data in spatial-temporal data forecasting task. The excellent prediction performance of STSGCN also proves the importance of synchronous capture time and space dependence. Considering the advantages of these two aspects, the STGAT model considers the periodicity and the synchronous extraction ability of spatial-temporal dependencies features in the data at the same time, so it further reduces the prediction error and achieves the relatively best prediction performance. Figure 4 show the predictions of STGAT and other models on the PeMS08 dataset, and STGAT obtains more reliable predictions. The X-axis in the Figure represents time (with a granularity of 5 minutes) and the Y-axis represents traffic flow.

Table 2. Performance comparison on PeMS04 and PeMS08

Method	PeMS08		PeMS04	
	MAE	RMSE	MAE	RMSE
LSTM	22.72	34.85	29.10	43.75
DCRNN	17.86	27.83	24.70	38.12
STGCN	21.49	32.40	22.70	35.55
STSGCN	17.13	26.80	21.19	33.65
ASTGCN	16.63	25.27	21.80	32.82
STGAT	**15.87**	**24.86**	**20.75**	**32.21**

To further validate the effectiveness of the proposed spatial-temporal graph attention mechanism, the following ablation comparison experiments are conducted: (1) VTE (Vanilla transformer's encoder): Using the encoder of the vanilla transformer model, stack three layers of self-attention and the feedforward, and finally a full connection layer is used as the output layer to obtain the predictions, and the input data is the recent time period data. (2) STG-attn: The self-attention layer of the above VTE model is replaced by the spatial-temporal graph attention module proposed in this paper to obtain the STG-attn model. (3) MSTGCN: Single component ASTGCN model obtained by using only recent time period data as input.

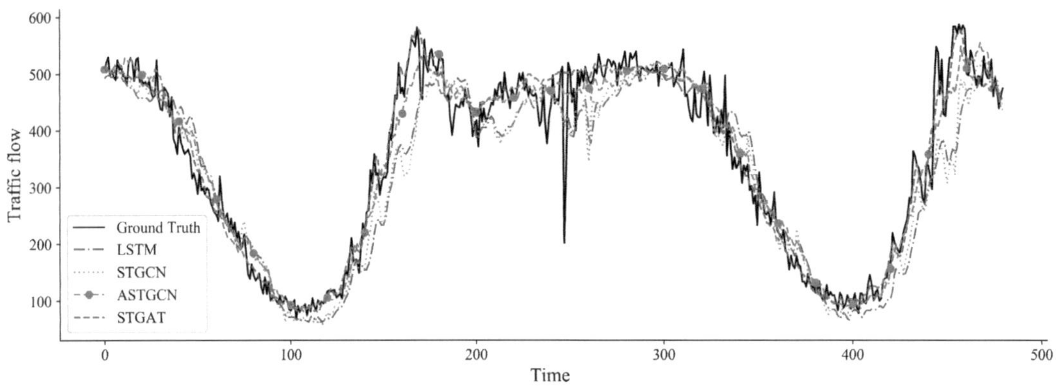

Figure 4. Visualization of traffic data predictions

The MAE and RMSE of the predictions for the above models as well as STGAT on the PeMS08 dataset are shown in Figures 5 and 6 respectively. From Figure 5, it can be seen that the MAE of VTE at the 15-minute (min) is very close to that of STG-attn. The MAE of STG-attn at 30 minutes is 16.65, which is only 0.54 different from its MAE at 15 minutes, while the MAE of VTE at 30 minutes is 18.11, which is 1.6 different from its MAE at 15 minutes of 16.51. This difference widens further at the 60-minute forecasting level. The RMSE comparison results in Figure 6 shows a similar phenomenon. The reason for above phenomenon is that VTE models each node in the graph separately, just taking intra-node temporal dependencies into account and ignoring information on spatial dependencies between nodes in the data. The state of node's neighbours at one moment will profoundly affect the node's state at the next moment, and this phenomenon begins to emerge as the forecasting level becomes longer. In addition, STG-attn obtains better predictions for both MAE and RMSE than MSTGCN, which also uses only recent time period data as input. The above comparison experiments validate the effectiveness of the spatial-temporal graph attention mechanism.

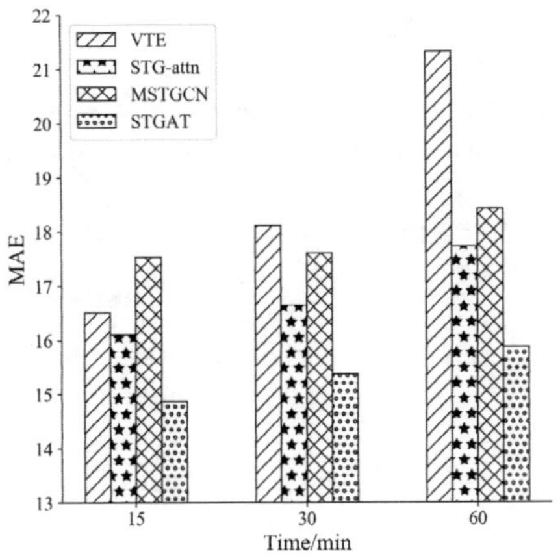

Figure 5. Comparison of each model's MAE

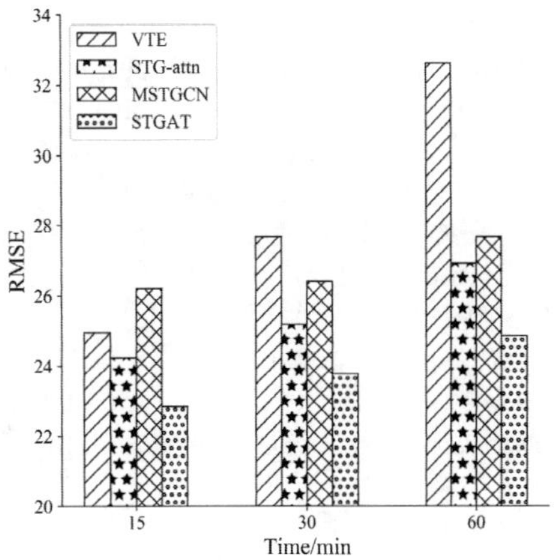

Figure 6. Comparison of each model's RMSE

5. Conclusion

To address the problem of predicting traffic data generated by urban road network, this paper proposes a new spatial-temporal graph attention network, STGAT, which is experimentally validated on two public transport datasets to obtain the following conclusions.

(1) With the help of localized spatial-temporal graph, spatial-temporal dependencies in the input data are captured using spatial-temporal graph attention for recent period, enabling adaptive learning of dependencies between nodes in both temporal and spatial dimensions.

(2) For the periodic features in the weekly period, the STGAT model using the self-attention mechanism to dynamically model the similarity of recent-period's spatial-temporal features to that of weekly-period to obtain the predictions, which extracts the periodic features and improves the prediction accuracy.

(3) The MAE and RMSE of the model's predictions on the two public transport datasets are 15.87, 24.86 and 20.75, 32.21 respectively, which obtains an average improvement of around 10% in prediction accuracy compared to other current state-of-the-art traffic flow forecasting methods in the field.

Acknowledgments

This research work was supported by the National Natural Science Foundation of China (Grant No.62076025 and No.61572073), the Fundamental Research Funds for the Central Universities (FRF-GF-19-014B).

References

[1] Zhang J.L., Chen F, Cui Z, et al. (2020) Deep learning architecture for short-term passenger flow forecasting in urban rail transit. IEEE Transactions on Intelligent Transportation Systems. 1-11.

[2] Zhou J, Cui G, Zhang Z, et al. (2020) Graph neural networks: a review of methods and applications. AI Open. 1: 57-81.

[3] Yu B, Yin H.T., Zhu Z.X. (2018) Spatio-temporal graph convolutional networks: a deep learning framework for traffic forecasting. In: Proceedings of the Twenty-Seventh International Joint Conference on Artificial Intelligence. Stockholm, Sweden. pp. 3634-3640.

[4] Li Y.G., Yu R, Shahabi C, Liu Y. (2018) Diffusion convolutional recurrent neural network: data-driven traffic forecasting. In: International Conference on Learning Representations,.

[5] Bai L, Yao L, Li C, et al. (2020) Adaptive graph convolutional recurrent network for traffic forecasting. In: Proceedings of the 34th Conference on Neural Information Processing Systems. MIT Press. pp. 17804-17815.

[6] Wu Z.H., Pan S.R., Long G.D., et al. (2019) Graph WaveNet for deep spatial-temporal graph modeling. In: Proceedings of the Twenty-Eighth International Joint Conference on Artificial Intelligence. Morgan Kaufmann. pp. 1907-1913.
[7] Song C, Lin Y.F., Guo S.N., et al. (2020) Spatial-temporal synchronous graph convolutional networks: a new framework for spatial-temporal network data forecasting. In: Proceedings of the AAAI Conference on Artificial Intelligence. Palo Alto. pp. 914-921.
[8] Chen C, Li K.L., Teo S.G., et al. (2019) Gated residual recurrent graph neural networks for traffic prediction. In: Proceedings of the AAAI Conference on Artificial Intelligence. Palo Alto. pp. 485-492.
[9] Guo S.N., Lin Y.F., Feng N, et al. (2019) Attention based spatial-temporal graph convolutional networks for traffic flow forecasting. In: Proceedings of the AAAI Conference on Artificial Intelligence. Palo Alto. pp. 922-929.
[10] Vaswani A, Shazeer N, Parmar N, et al. (2017) Attention is all you need. Part of Advances in Neural Information Processing Systems 31. Long Beach. pp. 6000-6010.

Hybrid Electric Vehicle Energy Online Allocation Strategy Based on Fuzzy Reinforcement Learning

Dong XU

(Automotive Technical College, Wuxi Vocational Institute of Commerce, Jiangsu Wuxi 214153, China)

xd1228@126.com

Abstract: To lessen fuel consumption of hybrid electric vehicle, energy online distribution strategy based on fuzzy reinforcement learning is proposed. Vehicle dynamic model and key component model are built, so that hybrid power system simulation model is given. Energy distribution problem of hybrid electric vehicle is transferred to constrained optimization problem of stochastic dynamic system through modelling, and solving problem based on fuzzy reinforcement learning is put forwarded. Fuzzy reinforcement learning is used to optimize fuzzy inference system real-time, which makes fuzzy inference system adjust with driving cycle adaptively, so that optimal control under any driving cycle come true. Clarified by simulation, fuel consumption under the control of rule-based is 3.89L/km, and it is 3.45L/km under the control of fuzzy reinforcement learning, which saves 11.31% fuel consumption compared with rule-based. The data above proves validity of fuzzy reinforcement learning on hybrid electric vehicle energy management.

1. Introduction

Safety, energy saving and environmental protection are the three eternal themes in the development of the automobile industry. Compared with traditional vehicles, hybrid electric vehicles have obvious advantages in these three aspects, but the energy saving potential of hybrid electric vehicles in China has not been fully tapped. Effective and accurate management and distribution of HEV energy is of great significance to improve energy efficiency, reduce environmental pollution and save energy [1].

Energy management strategies of hybrid electric vehicles can be divided into rule-based management strategies, optimization-based management strategies and learning-based management methods. The rule-based energy management method takes the steady-state characteristics of the engine or motor as the index, and refers to the driver's demand and the vehicle state for mode switching and energy distribution. In literature [2], mode-switching strategies and energy management strategies under different modes were formulated according to the cycle conditions, and the overall fuel consumption was 20% lower than that of the original vehicle. In literature [3], multi-chaos operator genetic algorithm is used to optimize fuzzy control membership function, which reduces fuel consumption and harmful gas emissions. However, the parameters of this method are fixed after optimization and cannot adapt to changes according to road conditions. Energy management based on optimization can be divided into global optimization method and instantaneous optimization method. Global energy optimization method includes dynamic programming, minimum principle, etc., and instantaneous energy optimization method includes equivalent fuel minimization method, etc. In view of the lack of theoretical basis for determining weights in dynamic planning, literature [4] proposed the coefficient determination method of secant iteration, which was verified in the comprehensive working conditions of Chinese cities and reduced fuel consumption. Literature [5] formulated a real-time optimization method of parameter matching based on the minimum principle, which reduced the energy consumption and exhaust emissions of vehicles. The above global optimization method has two problems: first, it needs to predict the future driving status of vehicles; second, it is difficult to apply it to the actual control of vehicles due to the large amount of calculation. Literature [6] using the improved grey prediction and kalman filter to achieve the short-term forecast of road conditions for the future, put forward the adaptive strategy, the minimum fuel consumption at an equivalent to improve fuel economy by about 6.5%, represented by the minimum fuel consumption of the instantaneous optimization method of small amount of calculation, high real-time performance, but can't achieve the global optimal. In order to adapt to the requirements of different working conditions, energy management strategies based on learning are proposed. In literature [7], the optimal objective function of energy consumption is established based on evolution-enhancement learning, and the optimal control strategy and optimal energy consumption value are solved by using evolutionary algorithm to reduce the oil cost. The energy management strategy based on learning can adjust the control strategy adaptively according to different driving conditions, so it has great development potential.

In this paper, the energy management of parallel hybrid electric vehicle is studied, and the fuzzy inference controller adaptively changes with driving conditions is obtained by using reinforcement learning real-time optimization fuzzy inference system, which achieves the purpose of reducing fuel consumption.

2. Hybrid electric vehicle model

2.1. Brief introduction of single shaft parallel hybrid power system

In this paper, the single-axle parallel hybrid electric vehicle is taken as the research object. Its power system structure is shown in Figure 1. Key components include power battery, motor/generator, engine, automatic clutch and gearbox.

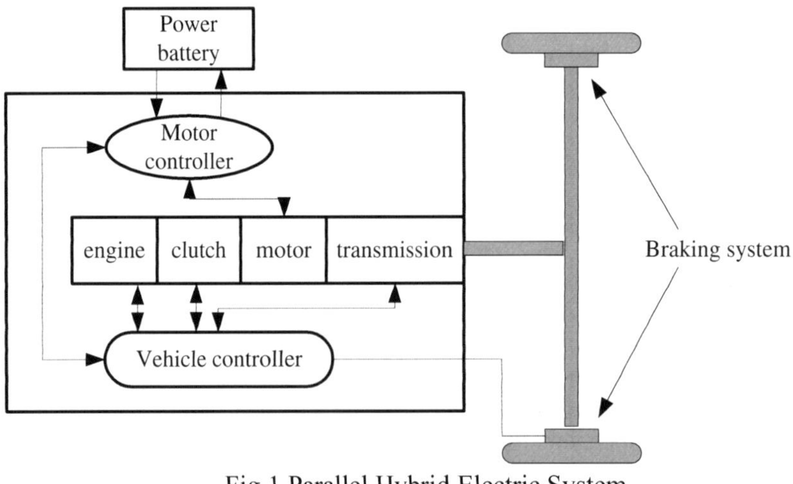

Fig.1 Parallel Hybrid Electric System

The working modes of parallel hybrid electric vehicles include pure electric mode, pure engine drive mode, joint drive mode, charging mode and regenerative braking mode. When the clutch disconnect for motor and motor model for pure electric mode, when the disconnect and clutch motor is generator for regenerative braking mode, when the clutch and the motor does not work for pure motor drive mode, when the clutch and motor for generator for charging mode, when the clutch and motor for motor to united drive mode. The motor controller in the figure is used to control the working mode of the motor. The vehicle controller selects the working mode of the vehicle according to the current driving condition of the vehicle, and distributes energy in the corresponding working mode to achieve the best fuel economy or emission performance.

2.2. Hybrid power system simulation model

This section establishes parallel hybrid power system simulation model, which is the basic work of energy management research, and establishes vehicle dynamics model and key component model.

(1) Vehicle dynamics model. Ignoring the lateral movement of the vehicle, given that the current speed is, and the road slope is, the torque demand on the wheel can be obtained according to the dynamics model:

$$T_{req} = r\left(\frac{C_d A v^2}{21.15} + fmg\cos\alpha + mg\sin\alpha + \delta\frac{dv}{dt}\right) \tag{1}$$

where:$T_{req}$is the wheel demand torque,$r$is the wheel radius,$A$is the windward area of the vehicle,$C_d$is the air resistance coefficient,$f$is the rolling damping coefficient of the road surface,$m$is the vehicle mass,$g$is the gravitational acceleration,$\delta$is the rotation mass conversion coefficient.

Remember the torque loss caused by the friction of the gearbox is$T_{loss}(\omega_w, \varsigma)$, then the torque demand at the input of the gearbox is:

$$T_{dem} = \begin{cases} \frac{T_{req}+T_{loss}(\omega_w,\varsigma)}{R(\varsigma)\eta(\varsigma)}, T_{req} + T_{loss} \geq 0 \\ \frac{T_{req}+T_{loss}(\omega_w,\varsigma)}{R(\varsigma)}\eta(\varsigma), T_{req} + T_{loss} < 0 \end{cases} \tag{2}$$

where:$T_{dem}$is the torque demand at the input end of the gearbox,$R$is the total speed ratio,$\eta$is the transmission efficiency from the gearbox to the driving wheel,$\omega_w$is the wheel angular speed,$\varsigma$is the corresponding speed ratio. The torque requirements of the gearbox are provided by the engine and motor, namely:

$$T_{dem} = T_e + T_m \tag{3}$$

where:$T_e$is the engine torque,$T_m$is the motor torque.

In Formula (2), the calculation method of wheel angular velocity is:

$$\omega_w = \frac{(1+s)v}{r} = \frac{\omega_{in}}{R(\varsigma)} \tag{4}$$

where:$s$is slip rate,$\omega_{in} = \omega_e = \omega_m$is gearbox input angular velocity,$\omega_e$is engine speed,$\omega_m$is motor rotation.

(2) Engine model. When the engine is preheated and the temperature effect of the engine is ignored, the fuel consumption rate of the engine is the sum function of engine torque and speed, which can be obtained by looking up the table, namely:

$$\dot{m}_f = \psi_1(T_e, \omega_e) \tag{5}$$

where:$\dot{m}_f$is fuel consumption rate,$\omega_e$is engine speed,$T_e$is engine torque,$\psi_1$is engine query function.

(3) Clutch model. The role of the clutch is to transfer the engine torque and speed to the gearbox, the clutch has three working states, respectively for separation, sliding friction, combined and so on. The engine torque in the separated state cannot be transmitted to the gearbox, part of the engine torque will be transmitted to the gearbox in the sliding wear state, and all the engine torque will be transmitted to the gearbox in the combined state. Due to the very short time of the clutch's sliding wear state, this paper sets the clutch to work in two states of separation and combination without affecting the overall analysis and control.

(4) Motor model. Ignoring the electromagnetic characteristics of the motor, the motor efficiency is expressed as a static function of the torque and speed of the motor, and the motor model is established by bench test as follows:

$$\eta_m = \psi_2(T_m, \omega_m) \tag{6}$$

where:$\eta_m$is the motor working efficiency,$\psi_2$is the motor query function,$T_m$is the motor torque,$\omega_m$is the motor speed.

(5) Gearbox model. The transmission gears are set as 1~5, and the transmission gears transfer is established as a discrete system model, namely:

$$g(t+1)=\begin{cases}5, g(t)+shift(t)>5\\ 1, g(t)+shift(t)<1\\ g(t)+shift(t), else\end{cases} \tag{7}$$

where:$g(t)$ is the gear position of the gearbox at the$t$moment,$shift(t)$ is the gear transfer at the moment,$shift(t)=0,-1,1$ which respectively means no transfer, down 1 gear and up 1 gear.

(6) Power battery model. The state of charge represents the electric quantity of the battery, which is one of the most important reference quantities in energy management. A simple internal resistance model is used to model the power battery as follows:

$$S(t+1)=S(t)-\frac{V_{oc}-\sqrt{V_{oc}^2-4R_{int}\ P_m(t)}}{2(R_{int}\ +R_t)Q_{max}} \tag{8}$$

where:$S(t)$ is the$t$moment charge state,$V_{oc}$is the open circuit voltage of the battery,R_{int} is the internal resistance of the battery,$P_m$is the power of the battery,$R_t$is the terminal impedance,Q_{max} is the maximum charging capacitance.

Based on the vehicle dynamics model and key component model established above, ADVISOR software can be used to establish the simulation model of single axle parallel hybrid electric vehicle.

3. Energy management strategies based on reinforcement learning

3.1. Problem modeling and transformation

In essence, the energy management and torque distribution of hybrid electric vehicles are optimal problems with constraints, and due to the randomness of driving conditions and driving behaviors, the energy management and torque distribution of hybrid electric vehicles are constrained optimization problems of stochastic dynamic systems. Therefore, the hybrid system can be discretized into a random dynamic system as follows:

$$x(t+1)=f(x(t),u(t),w(t)), t=0,1,\cdots,N-1 \tag{9}$$

where:$t$is the current sampling time,$N$is the total sampling time,$x(t)$ is the state variable,$u(t)$ is the control variable,$w(t)$ is the random variable. The number of states isN_1, the number of control isN_2, the number of random variables isN_3, then the state space is$X=\{x_1,x_2,\cdots,x_{N1}\}$, the control space is$U=\{u_1,u_2,\cdots,u_{N2}\}$, and the random variable space is$W=\{w_1,w_2,\cdots,w_{N3}\}$.

Select time to turn demand, speed and battery charge state as state vector, namely $x(t)=(T_{dem}(t),v(t),S(t))^T$; Select engine output torque and shift action as control variables, namely$u(t)=(T_e(t),shift(t))^T$. Assume that a control policy is used$\pi=\{u(0),u(1),\cdots,u(N-1)\}$ can be used to control the initial state to the final state, there must be an optimal control law to achieve optimal control. The evaluation function of automobile cycle conditions from the initial state to the end state is defined as:

$$J_\pi(s)=E\{L(x(N))+\textstyle\sum_{t=0}^{N-1}L(x(t),u(t),w(t))\,|x(0)=s\} \tag{10}$$

where: $J_\pi(s)$ represents the evaluation function value of the initial state under the action of the control law,$E$is the average function,$L(x(N))$ represents the evaluation function value of the car at the end of the state, $L(x(t),u(t),w(t))$ represents the evaluation function value of the car at the$t$time.

Formula (10) is the expression of the evaluation function in the finite time domain. In the infinite time domain, there is no need to calculate the terminal function value, and the infinite time domain is time-invariant. Therefore, the evaluation function is expressed in the infinite time domain, where:

$$J_\pi=\textstyle\sum_{t=0}^{\infty}\gamma^t L(t) \tag{11}$$

where:$J_\pi$is the evaluation function,$\gamma\in(0,1)$ is the discount factor to ensure the convergence of the infinite time domain cost function,$L(t)$ is the evaluation function of the car at the$t$time. In this paper, the weighted sum of fuel consumption and power consumption is taken as an evaluation function, namely:

$$L(t)=L_{fuel}(t)+a_1L_{SOC}(t)$$

where:$a_1$is the adjustment coefficient,$L_{fuel}(t)$ is the evaluation function of fuel consumption,$L_{SOC}(t)$

is the evaluation function of battery energy consumption. The two functions are respectively defined as:

$$L_{fuel}(t) = \begin{cases} -0.5, \dot{f}_{uel} > 0.88g/s \\ -1, 0.53g/s < \dot{f}_{uel} < 0.88g/s \\ 0.5, 0 < \dot{f}_{uel} < 0.53g/s \end{cases}, L_{SOC}(t) = \begin{cases} -0.3, \dot{S}_{OC} < 0 \\ 0, 0 < \dot{S}_{OC} < 0.001 \\ 0.3, \dot{S}_{OC} > 0.001 \end{cases}$$

where: $\dot{S}_{OC}$ is the battery discharge rate, $\dot{f}_{uel}$ is the instantaneous consumption rate of fuel.

It can be seen that the larger the evaluation function is, the better the control strategy is. In essence, the energy management problem of HEV is to obtain the control sequence corresponding to the optimization problem, namely:

$$J_{\pi^*}(x(0)) = J_{\pi}^*(x(0)) = maxJ_{\pi}(x(0))$$

In the process of energy optimization of parallel hybrid electric vehicle, the vehicle's dynamic performance constraints should be met, namely:

$$\begin{cases} \omega_{emin} \le \omega_e \le \omega_{emax} \\ S_{min} \le S(t) \le S_{max} \\ T_{emin}(\omega_e(t)) \le T_e(t) \le T_{emax}(\omega_e(t)) \\ T_{mmin}(\omega_m(t), S(t)) \le T_m(t) \le T_{mmax}(\omega_m(t), S(t)) \end{cases} \tag{12}$$

where: max , min is the maximum and minimum values of corresponding parameters.

3.2. Based on the framework of reinforcement learning energy management

In this paper, the Q-learning [8] method of reinforcement learning is used to optimize the fuzzy controller online, so that the controller can adapt to the change of working conditions and realize the optimal control adaptive to any working conditions. This optimization method is called fuzzy reinforcement learning algorithm in this paper.

The framework of fuzzy reinforcement learning algorithm is shown in Figure 2. This algorithm consists of two parts: one is the estimation network of reinforcement learning (Q-learning), and the other is the adaptive fuzzy reasoning system. The basic idea of this algorithm is to use BP neural network to estimate the functions of the system state-control pair in real time, and then use the functions to adjust the parameters of fuzzy controller online. The fuzzy controller used in this paper is Sugeno fuzzy controller.

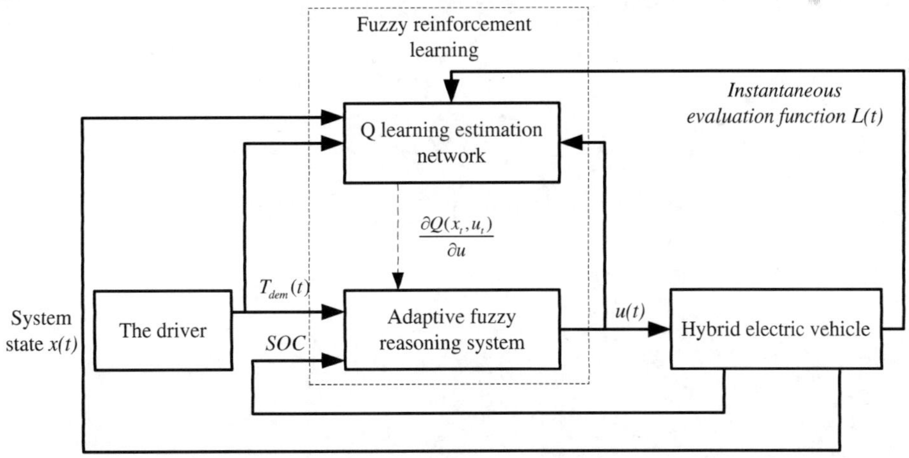

Fig.2 Fuzzy Reinforcement Learning

The fuzzy reinforcement learning algorithm combines the advantages of fuzzy reasoning system and reinforcement learning closely and forms the complementary advantages and disadvantages. The traditional Q-learning method selects the optimal action from the finite discrete action set, but the fuzzy reasoning method improves the generalization ability of state space, and the control action is continuous rather than discrete, which solves the dimension disaster of continuous system discretization. From the perspective of fuzzy reasoning, the introduction of Q-learning can adjust the parameters of fuzzy reasoning system online according to different driving conditions, make the fuzzy reasoning system adapt to the driving conditions, and realize the optimal control under any driving conditions, which solves the problem that the traditional fuzzy reasoning system only applies to the optimal driving conditions.

3.3. Q-learning estimation network

Since BP neural network has good approximation performance [9], this paper uses BP neural network to construct Q-learning estimation network and fit the optimal action function. Q-learning estimation network is shown in Figure 3, which consists of input layer, hidden layer and output layer. The input layer contains four nodes, the hidden layer contains ten nodes and the output layer has one node. The corresponding inputs of the four input layer nodes are speed $v(t)$, state of charge $S(t)$, torque demand$T_{dem}(t)$, control action respectively$u(t)$. The output corresponding to the output node is the state-control pair function$Q(x,u)$.

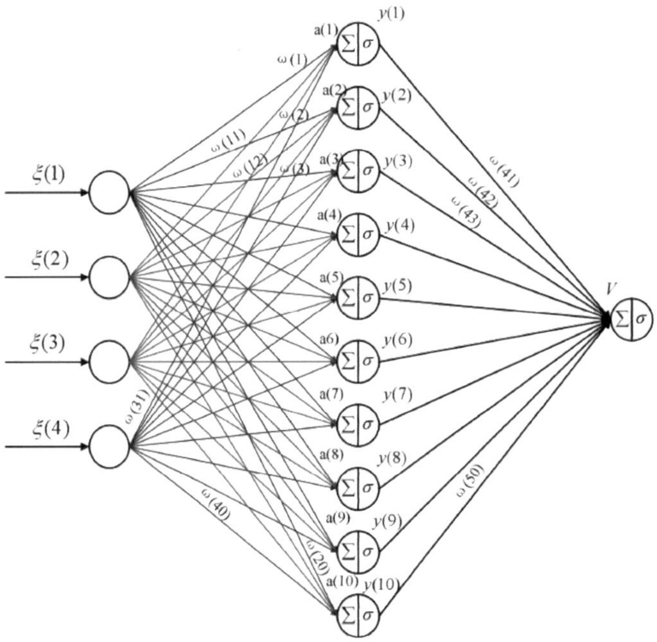

Fig.3 Q-Learning Estimation Network

Q-learning estimation network output is:

$$Q(x,u) = f(V)$$
$$V = \sum_{i=1}^{10} \omega(40+i)y(i)$$
$$y(i) = f(a(i))$$
$$a(i) = \textstyle\sum_{j=1}^{4} \xi(j)\omega(j-1,i) \quad (13)$$

where:$i = 1,2,\cdots,10$ is the node number of the hidden layer,$j = 1,2,3,4$is the node number of the input layer,$V$is the input quantity of the node of the output layer,$y(i)$ is the output quantity of the node of the hidden layer,$a(i)$ is the input quantity of the node of the hidden layer,$\xi(j)$ is the first j input,$w(i)$ is the transfer weight. The node activation function in the network uses the Sigmoid function, namely:

$$f(x) = \frac{1}{1+exp(-x)}$$

In order to reduce the time difference error (TD error) between the actual output function of the estimated network and the optimal action function, the generalized strategy iterative method is used to adjust the weights of the estimated network, so as to approach the optimal action function continuously. The weight adjustment target of the neural network is $min\ E = \frac{1}{2} \cdot \delta_t^2$, and the weight adjustment method obtained by using the gradient descent method is:

$$w(t+1) = w(t) + \eta \delta_t \frac{\partial Q(x_t, u_t)}{\partial w} \quad (14)$$

where: $\eta \in (0,1)$ is the learning rate, $\delta_t = \gamma Q^*(x_{t+1}, u_{t+1}) - Q(x_t, u_t)$ is TD error. $\frac{\partial Q(x,u)}{\partial w}$ can be obtained according to the chain rule, where:

$$\frac{\partial Q(x,u)}{\partial w(40+i)} = y(i)Q(x_t, u_t)(1 - Q(x_t, u_t)), (i = 1,2, \cdots, 10)$$

$$\frac{\partial Q(x,u)}{\partial w(j-1+i)} = w(40+i)\xi(j)Q(x_t, u_t)(1 - Q(x_t, u_t))y(i)(1 - y(i)),$$

$$(i = 1,2, \cdots, 10; j = 1,2,3,4)$$

The estimation results of Q-learning estimation network are used to adjust the quantity of fuzzy inference system online $\frac{\partial Q(x_t,u_t)}{\partial u}$ is:

$$\frac{\partial Q(x_t, u_t)}{\partial u} = Q(x_t, u_t)(1 - Q(x_t, u_t)) \times \sum_{i=1}^{10}[w(40+i)w(30+i)y(i)(1 - y(i))] \quad (15)$$

3.4. Adaptive adjustment method for parameters of fuzzy reasoning system

By adjusting the parameters of the fuzzy reasoning system to $Q(x, u)$ maximize, the fuzzy controller can be optimized. The parameter to be optimized for the fuzzy reasoning system is ϑ, and the gradient rule is used to optimize it as:

$$\vartheta(t+1) = \vartheta(t) + \beta \frac{\partial Q(x,u)}{\partial \vartheta}$$

$$\frac{\partial Q(x,u)}{\partial \vartheta} = \frac{\partial Q(x,u)}{\partial u} \cdot \frac{\partial u}{\partial \vartheta} \quad (16)$$

where, β is the gradient descent coefficient.

In the above equation, $\frac{\partial Q(x,u)}{\partial u}$ is given in Equation (15), and here we only need to calculate $\frac{\partial u}{\partial \vartheta}$. For the Sugeno type fuzzy controller used in this paper, the input quantity is $x = (x_1, x_2, \cdots, x_n)^T$, then the output is $u = K_0^l + K_1^l x_1 + \cdots + K_n^l x_n$, where $K_0^l \sim K_n^l$ is the constant coefficient of the fuzzy rule in article l. Single point fuzzification and center average method are used to defuzzify, and the output of the fuzzy controller is:

$$u(x) = \frac{\sum_{l=1}^{M}\left[\left(\prod_{i=1}^{n} \mu_l(x_i)\right)\left(\sum_{j=0}^{n} K_j^l x_j\right)\right]}{\sum_{l=1}^{M}\left(\prod_{i=1}^{n} \mu_l(x_i)\right)} \quad (17)$$

where, $i = 1,2, \cdots, n$ is the number of the input quantity, $l = 1,2, \cdots, M$ is the number of the fuzzy rule, $x_0 = 1$, $\mu_l(x_i)$ is the membership function of the input variable. Gaussian function is used in this paper:

$$\mu_l(x_i) = exp\left(-\frac{(x_i - c_i^l)^2}{\sigma_i^{l^2}}\right) \quad (18)$$

where, σ_i^l and c_i^l are the mean and variance of the Gaussian function respectively, and reinforcement learning is used in this paper for adaptive adjustment of these two parameters.

The product of different membership functions in the fuzzy rule l is $z_l = \prod_{i=1}^{n} \mu_l(x_i)$, and the output of the fuzzy rule l is $\bar{y}_l = K_0^l + K_1^l x_1 + \cdots + K_n^l x_n$. Let $a = \sum_{l=1}^{M} \bar{y}_l z_l$, $b = \sum_{l=1}^{M} z_l$, $u = \frac{a}{b}$, where a represents the sum of weighted outputs, b represents the sum of weights of all fuzzy rules, and u is the output. Then $\frac{\partial u}{\partial \vartheta}$ can be obtained as:

$$\frac{\partial u}{\partial c_i^l}=\frac{\bar{y}_l-u}{b}z_l\frac{2(x_i-c_i^l)}{{\sigma_i^l}^2}$$
$$\frac{\partial u}{\partial \sigma_i^l}=\frac{\bar{y}_l-u}{b}z_l\frac{2(x_i-c_i^l)}{{\sigma_i^l}^3} \quad (19)$$

Substituting Eq. (19) and (15) into Eq. (16), the optimization result of fuzzy inference system parameters adapting to the operating condition can be obtained.

3.5. Algorithm implementation

The essence of fuzzy reinforcement learning algorithm is to use reinforcement learning method to optimize the parameters of fuzzy reasoning system online, so that the fuzzy reasoning system can adapt to the driving conditions, and realize the optimal control of energy management under any driving conditions.

Fuzzy reasoning system mainly includes parameter fuzzification, fuzzy rules and defuzzification [10]. As mentioned above, gaussian membership function is used for parameter fuzzification. Firstly, an initial value is set for parameters σ_i^l and c_i^l for parameter fuzzification, and then reinforcement learning method is used to continuously optimize these two parameters. According to the practical experience of experts, fuzzy rules are formulated, and the torque demand T_{dem} and the state of charge $S(t)$ are blurred into 5 levels, respectively VS<S<M<B<VB, and the engine torque X is blurred into 8 levels, respectively VS<S<M<B<VB<SC<MC<BC. The fuzzy rules are shown in Table 1.

Tab.1 Fuzzy Rule

T_e		T_{dem}				
		VS	S	M	B	VB
SOC	VS	BC	BC	MC	SC	VB
	S	BC	MC	MC	SC	VB
	M	SC	SC	M	M	B
	B	S	S	M	B	B
	VB	VS	VS	S	S	M

The defuzzification of the fuzzy quantity is shown in Equation (17), thus the torque quantity T_e allocated to the engine is obtained, and the torque quantity undertaken by the motor is $T_m = T_{dem} - T_e$. At this time, the management and distribution of HEV energy are realized.

In order to more intuitively describe the torque distribution results and the working mode of the vehicle, the engine torque bearing rate $R_{Te} = \frac{T_e}{T_{dem}}$ is defined, so $R_{Te} = 0$ represents pure electric mode, $R_{Te} = 1$ represents pure engine mode, $0 < R_{Te} < 1$ represents hybrid drive mode, and $R_{Te} > 1$ represents charging mode.

4. Simulation analysis

In the Matlab simulation environment, the energy management method based on fuzzy reinforcement learning is verified by using UDDS (Urban Dynamometer Driving Scheme) cycle conditions. The parameters of the experimental vehicle are as follows: the peak power of the engine is 50kW, the peak power of the motor is 10kW, the battery capacity is 6.5Ah, the highest voltage is 154V, and the vehicle mass is 1000kg.

The parameters of the fuzzy reinforcement learning algorithm were set as follows: learning rate $\eta = 0.34$, gradient descent coefficient $\beta = 0.33$, discount factor $\varphi = 0.9$, weight $a_2 = 1$, initial SOC value was set as 0.7, and was limited to 0.5±0.05 at the end of the cycle. The initial value of the parameter σ_i^l,c_i^l to be optimized is set randomly, and then the reinforcement learning method is used for continuous iterative optimization. In order to form a contrast, both rule-based energy management method and fuzzy reinforcement learning algorithm are used for energy management and allocation.

In the UDDS cycle of 1369s, the speed change, battery charge change curve, demand torque, motor and torque distribution results obtained by fuzzy reinforcement learning are shown in FIG. 4.

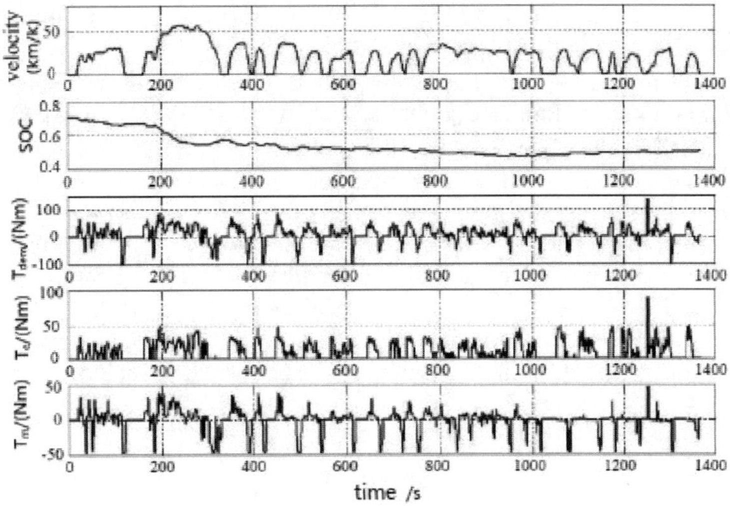

Fig.4 Control Result of Fuzzy Reinforcement Learning

According to statistics, when fuzzy reinforcement learning algorithm is used to control the vehicle under UDDS condition, the error between actual speed and set speed under UDDS condition is very small, and the maximum error of speed is 1.02km/h. As can be seen from Figure 4, the final value of SOC is stable at 0.505, meeting the set SOC limit condition. By comparing the torque demand, engine torque and motor torque in FIG. 4, there is no torque mutation, and the torque changes are relatively stable, maintaining the driving stability of the vehicle. By calculation, the fuel consumption of fuzzy reinforcement learning control is 11.31% lower than that of rule-based control method, which fully proves the effectiveness of fuzzy reinforcement learning in HEV energy management.

5. Conclusion

In this paper, the energy management and allocation of hybrid electric vehicle is studied, and a torque allocation method based on fuzzy reinforcement learning is proposed. The results show that :(1) the fuzzy reasoning system can be adjusted adaptively with driving conditions by optimizing the parameters of the fuzzy reasoning system online through reinforcement learning, and the optimal control of any driving conditions can be obtained. (2) Compared with rule-based control, fuzzy reinforcement learning algorithm has less fuel consumption.

References

[1] LI Jun, ZHU Ya-zhou, XU Yang-jiao, et al. Research on Control Strategy Optimization in Power Transmission System of Hybrid Electric Vehicle [J]. Machinery Design & Manufacture, 2014(3):138-141.

[2] ZHOU Qi-xun, CAO Shi-hong, JI Xin-jie. A Research on the Control Rules Optimization and Torque Distribution Strategy for ISG Hybrid Electric Vehicles [J]. Automotive Engineering, 2015，37(7)：751-756.

[3] LIANG Jun yi, ZHANG Jian long, MA Xue rui, et al. Control Strategy Optimization for Hybrid Electric Vehicle Based on Multi-Chaotic Operators Genetic Algorithm [J]. Shanghai Jiaotong Univ.(Sci.), 2015, 49(04):442-449.

[4] BA Te, GAO Yin-han, WANG Qing-nian, et al. Control Strategy Optimization for a Parallel Hybrid Electric Bus Based on Dynamic Programming Algorithm [J]. Automotive Engineering, 2015, 37(12):1359-1365.

[5] LING Bin, MA Mei. Study on Optimization Control of HEV Energy Conservation and Emission Reduction [J].Computer Simulation, 2015, 32(12):102-106.

[6] DENG Tao, LUO Yuan-ping, CHENG Dong, et al. Adaptive Equivalent Fuel Consumption Minimization Strategy for Hybrid Electric Vehicle Based on Road Condition Prediction [J]. Automobile Technology, 2019（6）:8-13.

[7] CHEN Zheng, LIU Ya-hui, YANG Fang. Energy Management Strategy for Plug-in Hybrid Electric Bus with Evolutionary-Reinforcement Learning Method [J]. JOURNAL OF MECHANICAL ENGINEERING, 2017， 53(16):100-107.

[8] Hung S M, Givigi S N. A Q-Learning Approach to Flocking With UAVs in a Stochastic Environment[J]. IEEE Transactions on Cybernetics, 2016, 47(1):1-12.

[9] Liu T, Yin S. An improved particle swarm optimization algorithm used for BP neural network and multimedia course-ware evaluation[J]. Multimedia Tools & Applications, 2017, 76(9):11961-11974.

[10] WANG Yu-gang, XIU Shi-chao. Evaluation Method of CNC Machine's Green Degree Based on Clustering and ANFIS [J]. China Mechanical Engineering, 2018, 29(23):77-81.

Time series prediction of cabin temperature based on mechanism model

Tanke Zhang [1,2], Kanjian Zhang [1], Xu Yao [3], Haikun Wei [1*], Shixiong Fang [1], Tao Wang [3]

[1]School of Automation, Southeast University, Nanjing City, Jiangsu Province, 211189, China

[2]School Of Cyber Science and Engineering, Southeast University, Nanjing City, Jiangsu Province, 211189, China

[3]Polar Research Institute of China, Shanghai City, 201209, China

[*]Corresponding author's e-mail: hkwei@seu.edu.cn

Abstract—Since the Antarctic scientific research device has been operating in Antarctica for a long time, the power generation cabin needs to continuously provide heat so that its equipment can work within a suitable temperature range. In view of the complexity of the environment and the high cost of equipment, it is difficult to test and evaluate, which can be well solved by numerical simulation test. The thermodynamic mechanism model of the Antarctic redundant cabin is established to simulate the temperature distribution of the Antarctic scientific research power cabin. The temperature data of the mechanism model is combined with the time series algorithm to predict the future high temperature accumulation in the engine room, thereby providing theoretical and data support for the decision-making of the temperature control system of the scientific research device.

Keywords-mechanism model; thermodynamics; time series

1. Introduction

To further optimize and improve the Antarctic scientific research support device and ensure the stable energy supply of the Antarctic cabin, the Antarctic scientific expedition team reproduced the layout of the Antarctic cabin, determined the appropriate control parameters through local debugging, and then used it in the polar regions to reduce the impact of new technologies. Use the possibility of causing an accident to the cabin. In terms of temperature control system, maintaining a suitable temperature in the cabin plays a key role in the stability of the energy supply system. In view of the remoteness of the polar region, the numerical simulation of the redundant cabin can be regarded as the field simulation of the Antarctic cabin, which is used for the decision-making of the temperature control system. Provide support, with strong practicality.

The mechanism model constructed by numerical simulation can be used to simulate the process of heat generation and heat transfer during the operation of the generator in the power generation cabin, and then calculate the temperature distribution data. The research of the school of energy, power and machinery of North China Electric Power University has realized the reconstruction of spatial temperature distribution by using the a priori data of a few nodes combined with the mechanism model

[1]. The mechanism model can also provide strong temperature data support for subsequent time series prediction, operation parameter adjustment and optimization and temperature control. Although the mechanism model can obtain data, it has high complexity, strong nonlinearity and long running time, and can't provide temperature data quickly. Therefore, the data-driven model is needed to support the subsequent rapid prediction [2]. Neural network meets this requirement very well.

As a typical deep learning method, neural network is also a typical nonlinear prediction method. It has very strong nonlinear fitting ability. At present, many scholars have improved it and applied many different forms of neural network. For predicting sea surface temperature, boiler temperature or greenhouse temperature. For example, in the prediction method of Zhang et al., the LSTM time series is used to model and predict the sea surface temperature series [3]. In view of the fact that the cabin temperature has not only the time dimension, but also the information in the spatial dimension, this paper draws on the method of precipitation prediction (Kim et al., 2017; Shi et al., 2015) model [4-5], using convolutional long-term and short-term The time series (ConvLSTM) obtains the spatial-temporal feature relationship to predict the temperature distribution in the cabin. In view of the instability of the data collected by the temperature sensor, it can be stabilized by modal decomposition to improve the prediction accuracy of the subsequent time series.

2. Methodology

To improve the prediction performance of cabin temperature distribution, this paper combines the numerical method with the neural network method. Therefore, the research method has two aspects: (1) Firstly, a CFD simulation model which can feed back the overall temperature distribution in the cabin is constructed to simulate the temperature value divided into regions during the heating period in the cabin without the fan on; (2) Secondly, the temperature value is smoothed, and the convolution long-term and short-term time series model is input to capture the space-time dependence of the cabin temperature. When the cabin temperature distribution is output end-to-end, the accuracy of the prediction results without smoothing and the prediction results after smoothing are compared.

2.1. Data acquisition

To obtain the value of cabin temperature distribution at any time, this paper relies on the past cabin data point set, sets the physical coefficient in fluent software according to the field conditions, and simulates the temperature distribution in the cabin when each generator generates electricity and heat in turn, and the fan turns on and off automatically according to the temperature threshold standard. The outer wall of the cabin is made of alloy, the filling material is foamed plastic, and the geometric dimension is 220mm × 560mm × 200mm, with the material physical properties as the standard, the small hole release method is used to simulate the external heat loss of the cabin. The minimum size of the grid is 0.1mm, and the specific number and size of the grid have been tested for grid independence. The physical parameters of the cabin are shown in Table 1. The established cabin grid model is shown in Figure 1. The temperature fitting results are shown in Figure 2.

Table.1 CFD physical parameters

Parameter	Value
Generator power	525W
Floor board	0.0003W/m2*K
External temperature	1W/m2*K
Air vent	0Pa
External ventilating fan	-0.25Pa
Foamed plastics	0.006W/m2*K

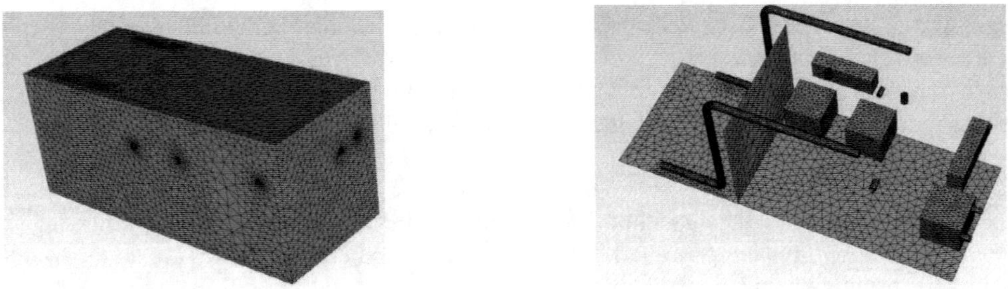

(a) External grid construction (b) Internal grid construction

Figure.1 Cabin grid model

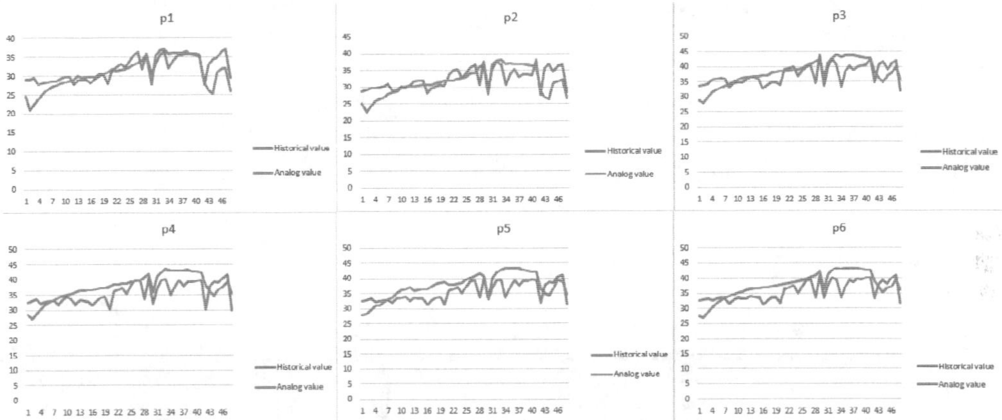

Figure.2 Temperature fitting effect

The temperature distribution calculated by CFD simulation and the temperature measured by 6 temperature sensors are within 8%. Therefore, it can be considered that the three-dimensional temperature distribution of the cabin obtained by CFD simulation can well represent the real temperature distribution. The temperature lattice output from the model can be used as the input of the time series model.

2.2. *Preprocessing*

Empirical mode decomposition (EMD) method is a signal decomposition method based on the time scale characteristics of the data itself. This method is different from Fourier decomposition and wavelet decomposition. It does not need to set the basis function. It is suitable for the decomposition of any type of signal, especially in dealing with non-stationary and nonlinear data. Due to the strong regional correlation of temperature series data, based on the sequence group with reduced stationarity comprehensively considers the spatial characteristics between different nodes. This smoothing treatment of temperature is of great significance to improve the accuracy of time series prediction of temperature distribution in the cabin [6].

The decomposition steps of EMD are as follows:

Step1. Input the time series, and calculate the maximum point and minimum point according to the input signal $\mathrm{i}(t)$.

Step2. Draw the upper and lower envelopes of the input sequence according to the maximum and minimum points respectively.

Step3. Draw the mean envelope $\mathrm{h}(t)$ according to the mean value of the upper and lower envelope.

Step4. Make a difference between the input signal $\mathrm{i}(t)$and the mean envelope $\mathrm{h}(t)$ to obtain the intermediate signal $\mathrm{m}(t)$, which is $\mathrm{m}(t) = \mathrm{i}(t) - h(t)$.

Step5. If $\mathrm{m}(t)$satisfies the conditions of IMF, then $\mathrm{m}(t)$ is an IMF. Otherwise, let $\mathrm{i}(t) = \mathrm{m}(t)$ and return step1 for execution.

Step6. When $\mathrm{m}(t)$ is satisfied as an IMF component, let $\mathrm{i}(t) = \mathrm{i}(t) - \mathrm{m}(t)$ and return step1 for execution. Until the decomposition is completed, several IMF and a residual res are obtained.

2.3. Prediction model

After being processed by EMD algorithm, five IMF and one res with low nonstationarity are obtained, and then these IMF and res are predicted. Finally, the predicted characteristics are added to obtain the temperature distribution results in the cabin.

Since the data form of eigenvalues obtained in the above experimental steps is matrix, in order to extract spatial features when extracting temporal features, ConvLSTM is used to train and predict the feature matrix sequence of subsequence. ConvLSTM actually replaces the matrix operation of each gate in LSTM neurons with convolution operation. The main difference is the input dimension. The input data of LSTM is one-dimensional, so the spatial characteristics of the data cannot be captured. The input of ConvLSTM is multidimensional data, which can not only express the spatial correlation of data, but also reflect the temporal correlation [7]. The operation process of ConvLSTM is shown in (1) - (6).In the formula σ is Sigmoid activation function.

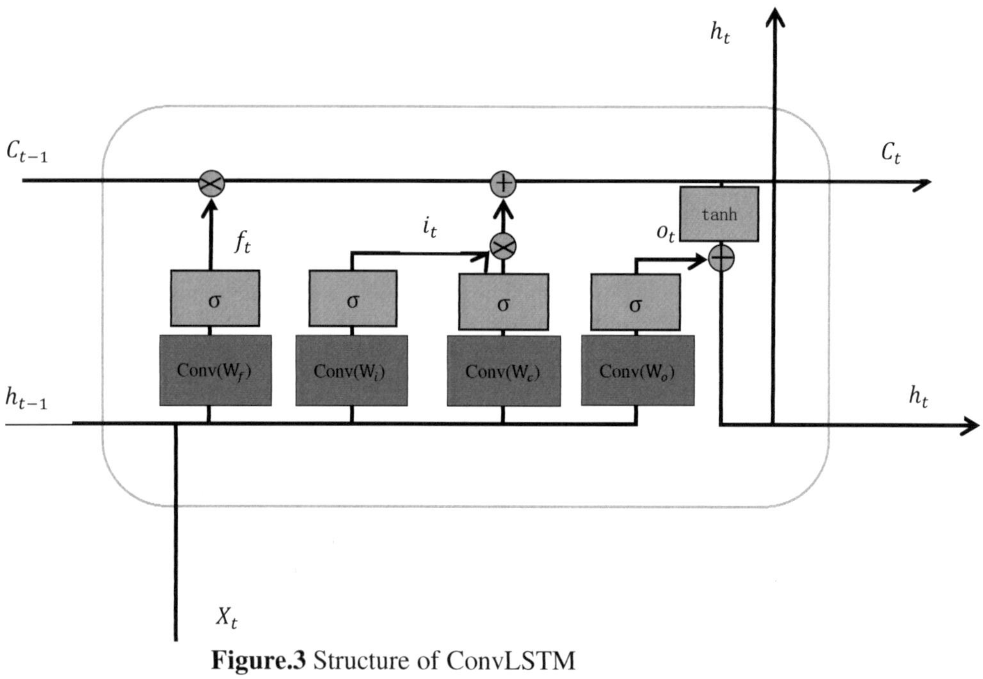

Figure.3 Structure of ConvLSTM

$$i_t = \sigma(W_{ii}x_t + b_{ii} + W_{hi}h_{t-1} + b_{hi}) \quad (1)$$
$$f_t = \sigma\left(W_{if}x_t + b_{if} + W_{hf}h_{t-1} + b_{hf}\right) \quad (2)$$
$$g_t = tanh\left(W_{ig}x_t + b_{ig} + W_{hg}h_{t-1} + b_{hg}\right) \quad (3)$$
$$o_t = \sigma(W_{io}x_t + b_{io} + W_{ho}h_{t-1} + b_{ho}) \quad (4)$$
$$c_t = f_t c_{t-1} + i_t g_t \quad (5)$$
$$h_t = o_t tanh(c_t) \quad (6)$$

The model has a strong ability to obtain spatiotemporal relations, and takes the previously obtained feature sequence as the input of ConvLSTM model [8]. The matrix obtained by directly adding the predicted value matrix obtained from the subsequence is the matrix of the predicted actual temperature. Convolution parameters are shown in Table 2.

Table.2 Convolutional layer related parameter information table

Name	Kernel	Step	Padding	Depth	Input	Output
Conv3D_1	3*3*3	1*1*1	same	1/150	5*14*10	5*14*10
Conv3D_2	2*2*2	1*1*1	valid	150/100	5*14*10	4*13*9
Conv3D_3	2*2*2	1*1*1	valid	100/100	4*13*9	3*12*8
Flatten	1*1*1	1*1*1	valid	100/1	3*12*8	28800
Dense_1	1*1*1	1*1*1	valid	1/1	28800	1000
LSTM_1	1*1*1	1*1*1	valid	1/1	1000	10*1000
LSTM_2	1*1*1	1*1*1	valid	1/1	1000	1000
Dense_2	1*1*1	1*1*1	valid	1/1	1000	7000
Reshape	1*1*1	1*1*1	valid	1/10	7000	10*5*14
UnConv3D_1	2*2*2	1*1*1	same	10/100	10*5*14	10*5*14
UnConv3D_2	2*2*2	1*1*1	same	100/100	10*5*14	10*5*14
UnConv3D_3	3*3*3	1*1*1	same	100/10	10*5*14	10*5*14

2.4. Performance evaluation

To comprehensively evaluate the effectiveness of the models built in this chapter, two indexes, root mean square error (RMSE) and prediction accuracy (PACC), are used to evaluate the performance of different prediction methods.

$$\mathrm{RMSE} = \sqrt{\frac{1}{N}\sum_{i=1}^{N}(y_i - \tilde{y}_i)^2} \quad (7)$$

$$\mathrm{PACC} = 1 - \frac{1}{N}\sum_{i=1}^{N}\left(\frac{|y_i-\tilde{y}_i|}{\tilde{y}_i}\right) \quad (8)$$

Where N represents the length of the predicted time series, y_i represents the predicted value, $\tilde{y}_i$ represents the true value. When forecasting the model, the smaller the RMSE index, the better the prediction effect and the better the model performance. PACC on the contrary, the larger the index value, the better the prediction effect and the better the model performance.

3. Test Results and Discussions

In this paper, the ConvLSTM model is used in the prediction part, in order to demonstrate that the modal decomposition of the original temperature matrix will improve the accuracy of the convolutional neural network prediction. This section compares the experimental conditions of predicting one time point and five time points when the input is k = 10 and K = 15 to fully illustrate the effectiveness of the experiment.

Table.3 Comparison between prediction results of EMD-ConvLSTM and ConvLSTM

Model	Evaluating indicator	One time point		Five time points	
		K=10	K=15	K=10	K=15
ConvLSTM	RMSE	0.28	0.37	0.54	0.62
	PACC	99.14	98.93	97.81	97.28
EMD-ConvLSTM	RMSE	0.22	0.33	0.45	0.48
	PACC	99.26	99.17	98.75	98.44

Table3 shows the comparison of the experimental results of the EMD-ConvLSTM and ConvLSTM models. The comparison shows that the prediction effect of the EMD-ConvLSTM model is better than that of the pure ConvLSTM model under different input and output conditions. The RMSE of the EMD-ConvLSTM model is the lowest at 0.22, 0.33, 0.45, and 0.48; its PACC is the highest at 99.26%, 99.17%, 98.75%, and 98.44%, respectively. The reason is that when the fan in the cabin is on, the airflow will disturb the temperature of different space nodes. At this time, the temperature value fluctuates. The modal decomposition method is used to reduce the interference effect of the airflow and extract the substantial law of temperature change to let the input of the ConvLSTM model more explanatory, so it can help the ConvLSTM model to extract more effective spatial features, which fully shows that the prediction effect has been improved and the effectiveness of the EMD-ConvLSTM model. In addition, the experimental results also show that when the prediction output length increases, the prediction accuracy decreases, and when the prediction input length increases, the prediction accuracy also decreases to varying degrees.

The experimental results confirm that the EMD-ConvLSTM model has better applicability for some prediction activities with higher non-stationarity and more obvious spatial correlation. It is confirmed that it has very broad application prospects in the prediction of temperature distribution and other fields such as meteorology.

4. Conclusion

This paper proposes a prediction model for temperature distribution of Antarctic cabins based on the mechanism model. The EMD algorithm is used to stabilize the temperature sequence output by the mechanism model, which improves the accuracy of the ConvLSTM prediction model for temperature distribution prediction. The temperature distribution state in the future can be used as data support for the decision-making of the temperature control system of the support device, and then the threshold standard for the opening of the cabin fan can be reasonably defined, and the method of how to ensure the appropriate temperature in the cabin. This also confirms that the combined algorithm of modal decomposition combined with convolutional neural network has very broad application prospects in the prediction of temperature distribution and other fields such as meteorology.

The future work will mainly focus on the following aspects: (1) Exploring the temperature accumulation, clarifying the high failure rate of generator and other equipment under which temperature accumulation, and avoid the failure caused by it through temperature control. (2) Aiming at the practical problem of the deviation between the simulation data and the actual data, the key parameters of the prediction algorithm for the mechanism model are adjusted to meet the prediction calculation requirements in the real problem scenario.

Acknowledgments

This work was supported in part by the National Natural Science Foundation of China under Grant 61973083, in part by Social Development Project of Science and Technology Innovation Action Plan of Shanghai under Grant 20dz1207102, in part by the Science and Technology Project of State Grid Corporation of China (Intelligent operation and maintenance technology of distributed photovoltaic system under Grant SGTJDK00DYJS2000148).

References

[1] Chen M., Liu S., Sun S., et al. (2020) Rapid reconstruction of simulated and experimental temperature distributions based on proper orthogonal decomposition[J]. Applied Sciences, 10(11):3729

[2] XU Guang, YU Huangjun, TAO Shaohui, et al. (2005) Modeling fermentation process based on hybrid support vector machines-kinetics mechanism[J]. Journal of Chemical Industry and Engineering(China), 56(4): 653-658(in Chinese).

[3] Zhang Q, Wang H, Dong J, et al. (2017) Prediction of Sea Surface Temperature using Long Short-Term Memory[J]. IEEE Geoscience and Remote Sensing Letters, 14(10):1745-1749.

[4] Kim, S., Hong, S., Joh, M., Song, S.-K., (2017) DeepRain: ConvLSTM Network for Precipitation Prediction Using Multichannel Radar Data. https://arxiv.org/abs/1711.02316

[5] Shi, X., Chen, Z., Wang, H., Yeung, D.-Y., Wong, W.-K., Woo, W.-C.(2015) Convolutional LSTM Network: A Machine Learning Approach for Precipitation Nowcasting. In: Cortes, C., Lawrence, N.D., Lee, D.D., Sugiyama, M., Garnett, R. (Eds.), Advances in Neural

Information Processing Systems 28, pp. 802–810.
[6] Yang H L, Lin H C. (2016) An Integrated Model Combined ARIMA, EMD with SVR for Stock Indices Forecasting[J]. International Journal of Artificial Intelligence Tools, 25(02):1650005.
[7] A.G.Salman, B.Kanigoro, and Y.Heryadi.(2016) Weather forecasting using deep learning techniques, ICACSIS 2015 - 2015 International Conference on Advanced Computer Science and Information Systems, Proceedings, vol. 00, pp. 281–285.
[8] M.A.Zaytar and C.El Amrani. (2016) Sequence to sequence weather forecasting with long short-term memory recurrent neural networks, International Journal of Computer Applications, vol. 143, no. 11, pp.7–11.
[9] J. Lee, C. G. Kim, J. E. Lee, N. W. Kim, and H. Kim.(2018) "Application of artificial neural networks to rainfall forecasting in the Geum River Basin, Korea," Water (Switzerland), vol.10, no.10.
[10] P.Hewage, M.Trovati, E. Pereira, and A. Behera. (2020) "Deep learning based effective fine-grained weather forecasting model," Pattern Analysis and Applications, no.0123456789. https://doi.org/10.1007/s10044-020-00898-1
[11] Sun S., Liu S, Chen M.et al. (2020) An optimized sensing arrangement in wind field reconstruction using CFD and POD[J]. IEEE Transactions on Sustainable Energy, 11(4): 2449-2456

A method of air target recognition based on combined-rules

ZhenZhang[1*], FengtangFang[2], YupengXing[3], and BinFu[4]

[1234]CSSC Systems Engineering Research Institute, Beijing, China

[*]Corresponding author's e-mail: 18801006023@139.com

Abstract. In order to improve the authenticity, reliability and accuracy of air target identification in complex situation, a processing method based on combined-rules is proposed to further enhance the automation degree of air target identification among various air target identification methods. Rule combination and conflict resolutionare used to reduce the workload of human intervention and manual judgment and make target recognition more rapid and accurate.

1. Introduction

With the intensification of maritime competition, the maritime environment faced by modern ships has become more and more complex[1]. In modern naval battles, with the application of high-tech means, the threat of offensive airborne precision weapons to the fleet is increasing[2]. One of the prerequisites for warships to complete airdefence and antimissile tasks is to have a comprehensive and accurate grasp of the battlefield situation[3]. The requirements for the ability to identify enemy or friend of aerial targets among ship commanders are getting higher and higher, and it is directly related to the commander's operational decision-making[4].

1.1. Technical defects

The traditional friend or foe recognizer usually uses the correspondence between a single task and a single model in which the recognizer recognizes one target at a time[5]. In the actual combat process, the commander is usually assigned several main recognition tasks, such as the task of identifying air targets. In the process of performing the main task, it is necessary to identify unknown targets, potential targets, disappearing targets, and so on. At this time, the recognition by the recognizer cannot completely solve all the recognition target problems[6]. At the same time, through the manual determination method, it takes a lot of time to think and judge to make a manual recognition result. Therefore, the timeliness and diversity of recognition must be taken into account[7].

1.2. Solution

This paper proposes an aerial target recognition method based on combined rules. Aiming at the real-time and adaptable requirements of aerial target recognition, by focusing on the action and rule theory of aerial targets, a set of reasonable recognition rules and rules for aerial targets are constructed. The description method can improve the automatic ability of the system in the recognition of air targets.

2. Target recognition scheme

2.1. combined-rulescomposition

The combined-rulesof airborne target recognition of ships are mainly composed of air recognition area rules, recognizer use parameter rules, recognition warning parameter rules, friend or foe evaluation parameter rules, friend or foe conflict and resolution parameter rules.

(1) Air recognition area rules: According to different combat missions, determine the relevant areas that need target recognition. Areas can be used for identification warnings, automatic friend or foe assessment, etc. The area parameters are defined in the following table 1:

Table 1. Air recognition area rules

Rule	Parameter name
Air recognition area rules	Area type
	Area shape
	Entry sign

(2) Parameter rules for the identifier of friend or foe: According to the ability of the identifier of friend or foe, determine the friend or foe inquiry mechanism and related parameters in different situations. The parameters are defined as table 2 follows:

Table 2. Parameter rules for the identifier of friend or foe

Rule	Parameter name
Parameter rules for the identifier of friend or foe	Location parameter
	Speed parameter
	Target attribute
	Response type

(3) Identify the alarm parameter rules: set the target characteristic parameters that need to be identified, such as CPA, altitude, speed, etc., and The parameters are defined as table 3 follows:

Table 3. Identify the alarm parameter rules

Rule	Parameter name
Identify alarm parameter rules	Recognition area
	Location parameter
	Speed parameter
	CPA parameter information
	Attribute value

(4) Parameter rule of friend or foe evaluation: set the automatic evaluation parameters of the target friend or foe attribute, such as CPA, altitude, speed, target type, etc. The parameters are defined as table 4 follows:

Table 4. Parameter rule of friend or foe evaluation

Rule	Parameter name
Parameter rule of friend or foe evaluation	Recognition area
	Location parameter
	Speed parameter
	CPA parameter information
	Radiation information

Rule	Parameter name
	Target type
	Target attribute

(5) Conflict between friend or foe and the rule of resolution table: According to the conflict between friend or foe and the conversion table, modify or alert related target attributes. The parameters are defined as table 5 follows:

Table 5. Conflict between friend or foe and the rule

Rule	Parameter name
Conflict between friend or foe and the rule of resolution table	Target type
	Target current attributes
	Target conflict resolution
	disposal method

2.2. combined-rulesuse

The requirements for the combined rules of air target identification friend or foe to the model application target identification friend or foe rule are as follows:

For air targets located in the recognition area, according to the parameters of the friend or foe identifier[8], one or more targets that meet the query conditions of the friend or foe identification will be automatically inquired, and the attributes of the target friend or foe will be modified according to the result of the inquiry[9].

When an air target is located in the recognition area (recognition warning area) and the friend or foe attribute is "pending", and the recognition alarm parameter or friend or foe evaluation parameter is met, the system prompts the commander to recognize the air targets[10].

When the air target's different information sources conflict with friend or foe, the result is obtained according to the conflict between friend or foe and the resolution table[11]. When the conflict can be automatically resolved, the target's friend or foe attributes are automatically set. If the conflict cannot be resolved automatically, the commander will be prompted to the target attribute conflict.

2.3. Combined-rules design

The requirements for the combined rules of air target identification friend or foe to the model application target identification friend or foe rule are as follows:

(1) Rule setting requirements

Identify and alert targets entering or located in the integrated combat area.

Identify and alert the targets that appear in the direction of high threats.

Identify and alert the targets that appear in the direction of the radiation source.

Identify and alert the targets near the civil waterway.

Identify and alert the target of abnormal movement.

In response to the demand for air defence, automatic friend-or-foe attribute assessment of high-threat or higher-threat air targets.

(2) Default rules

Parameters used by the identifier of friend or foe: The parameters of the identifier of friend or foe are divided into single target or friend or foe recognition matrix and dual target or friend or foe recognition result matrix as table 6 and table 7 follows:

Table 6. Single target identification friend or foe matrix

	No response	Civil aircraft	Friendly forces	Suspected	No recognized
result	Unknown	Neutral	Friend	Assumed	Pending

Table 7. Dual target friend or foe identification result

	No response	Civil aircraft	Friendly	Suspected	No recognized
No response	Unknown	Unknown	Unknown	Unknown	Pending
Civil aircraft	Unknown	Neutral	Neutral	Neutral	Pending
Friendly forces	Unknown	Neutral	Friend	Friend	Pending
Suspected	Unknown	Neutral	Friend	Assumed	Pending
Norecognized	Pending	Pending	Pending	Pending	Pending

Conflict between friend or foe and resolution table parameters: When the result matrix of the two friend or foe attributes conflicts, the following resolution table will be used automatically for discrimination. When the automatic recognition cannot be identified, manual intervention is required to determine the friend or foe attribute of the target. As table 8 follows:

Table 8. The friend or foe attribute automatically converted table

	Pending	Unknown	Friend	Neutral	Suspect	Hostile
Pending	Pending	Unknown	Friend	Neutral	Suspect	Hostile
Unknown	Unknown	Unknown	Friend	Neutral	Suspect	Hostile
Friend	Friend	Friend	Unknown	? ?	? ?	? ?
Neutral	Neutral	Neutral	? ?	Neutral	? ?	? ?
Suspect	Suspect	Suspect	? ?	? ?	Suspect	Hostile
Hostile	Hostile	Hostile	? ?	? ?	Hostile	Hostile
Tip:gray--consistent results; green--automatically accepted;red-- manual judgment						

The combined-rules of air target recognition mainly defines the target recognition in the actual use of air targets to enhance the effectiveness and adaptability of the recognition. In the actual application of air target recognition, especially the air target recognition under modern confrontation conditions, it is necessary to deal with the problem of dynamic changes of aerial targets in the scene.

Since the design of rules cannot exhaust all possible situations and rules, and with the increasing complexity of maritime airspace, the design of rules cannot completely and effectively solve all air target identification problems in a dynamic environment in a timely and effective manner. It is necessary to consider introducing more Externalinformation is used for support and assistance to improve the reliability of aerial target recognition.

3. Applications

3.1. Combined-rulescomposition

In the rule combination, air identification area rules, the IFF use parameter rules, and the identification warning parameter rules are used to determine the relevant areas that need target identification according to different combat tasks. Set a set of specific data parameters used, as table 9 follows:

Table 9. Parameter setting

Rule	Name	Setting basis	Main parameters
Air recognition area rules	Air defense identification warning zone	According to the ship's air defense and defense capabilities, the air defense identification warning area is set to meet the reaction time requirements of air defense operations.	Area type: custom area Area shape: sector shape, radius of 100 kilometers Position mode: Relative to the ship

Rule	Name	Setting basis	Main parameters
Identify alarm parameter rules	Abnormal target identification alarm parameters	Preset the abnormal target maneuvering (such as high-speed approach, abnormal descent, abnormal steering, etc.), and set the identification and warning parameters of abnormal targets.	[Air-High Speed Approach] Target type: air Friend or foe attribute: pending Target CPA: less than 4000m Target speed: greater than 300m/s
	Threat direction identification alarm parameters	According to the judgment of the enemy situation in advance, the possible threat direction is set, and the identification warning parameters are set according to the target characteristics.	Friend or foe attribute: pending Target area: the threat area generated in the threat direction (upper and lower limits)
Parameter rule of friend or foe evaluation	Civil target evaluation parameters	For possible civilian targets, set evaluation parameters that are automatically judged as unknown.	[Air-Civil Aviation] Applicable area: in the civil navigation channel Target type: air Friend or foe attribute: pending Target CPA: greater than 50000m Target speed: 220-275m/s Target height: 8000-10000m

3.2. scene setting

Suppose the sea formation is cruising in the waters of someisland, and Japan dispatches air force and navy surface ships to attack me. Our two frigates, own ship is in front, and the friendly ship is 2km apart in the 135-degree direction of own ship. The enemy dispatched 2 F-15 aircraft with the support of early warning aircraft. The two fighter jets flew in the direction of our ship at a distance of about 200KM from our ship at a direction of 30 degrees respectively. After releasing the ASM-2 air-to-ship missiles at appropriate positions, they flew away from the ship. As Figure 1 follows:

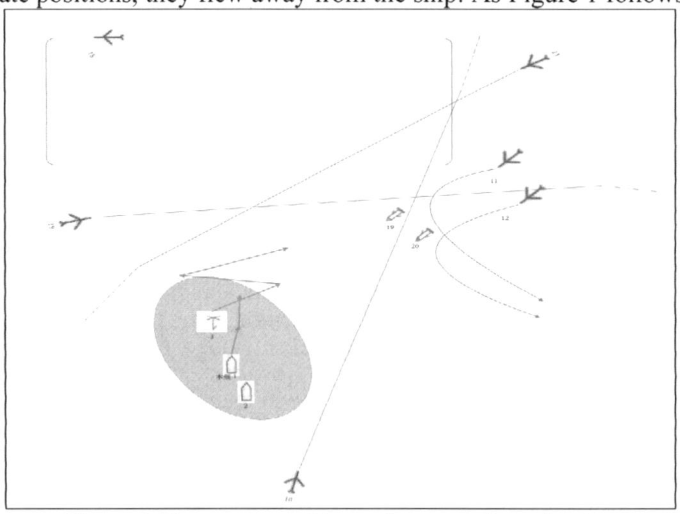

Figure 1.Target recognition rules scene imagination

3.3. Identification process

At present, this sea area is used as a public airspace; for the targets detected in the situation, according to the identification results, the corresponding icons can be automatically generated according to different attributes of the targets that confirm the attributes of the enemy and the friend. According to

ADS-B information, confirmed as a civil aircraft, and the identification result is neutral; target helicopter No. 3 in front of the ship is identified as our ship-based helicopter, and the identification result is our side; target No. 2 directly behind is identified as our ship, and the identification result is For our side; others are unrecognized target information, displayed as "unresolved" targets, and the identification rules can be used to further identify the attributes of the enemy and the friend.

1). Current situational target identification

According to the rules of target identification friend or foe, in the current situation map, the "pending" target information that appears, if the operator does not change the content of the rule parameters, the current rule pending target will be used for identification.

2). Civil aircraft identification

According to the ADS-B information, the three high-altitude targets have been identified as three civilian aircraft and marked as neutral targets. There is also a civilian aircraft flying target 16 on the civil aircraft route, and its flight speed is much faster than the other two civilian aircraft. The target continues to be tracked, and after a period of identification, it is found that the target has deviated from the civil aviation route. According to the identification warning rules, a track position warning is issued, and the target is determined in real time using the identification rules, and the position of the radiation source information can be detected from time to time. Although it is unstable, but by comparing the attributes of the civil aviation aircraft in the database, the radiation source information is not possessed by the civil aviation aircraft, so the target is manually determined to be a suspected enemy attribute.

3). High-altitude target recognition

In the high altitude, there are 2 unresolved targets No. 11 and No. 12, the flight speed reaches supersonic speed, and the IFF did not receive the mission feedback. It uses the current IFF rules to identify the target, confirms that the target is an air target, and the flight speed reaches supersonic speed. And its movement direction is close to the direction of our ship, the flight altitude is 7500 meters, and the CPA value is about 850 meters after calculation, which can be determined as a suspected enemy attribute. Combined with the current area as a disputed area, the probability of the target being the enemy is greatly increased. Therefore, according to the current rules and conditions, the system automatically identifies two aircraft targets as enemies with the attributes of friend or foe.

4). Added target recognition

When tracking two enemy aircraft, the situation gives new track information, which are similar to the current aircraft track information respectively. After detection, it is found that the correlation with the aircraft track is greatly reduced, and the speed of the aircraft is the initial speed. and grow rapidly. According to the alarm recognition rules, the target is a new potential threat target, and the movement trend is abruptly changed, and the friend or foe cannot be identified. Therefore, the system issues an IFF alarm and a movement trend alarm at the same time. After a period of detection, the radar detected that the reflection cross-sectional area of this type of target is less than 0.1 square meters, the flight speed has increased to close to 800 m/s, and its flight trajectory trend is getting closer and closer to the ship, and it is close to the ship. meters, the height drops to 100~150 meters, and it is determined as a sea-skimming missile target. Combined with the current identification rules of friend or foe, it is determined that the two targets 19 and 20 are enemy missile targets, and the attribute is the enemy.

5). Missile target recognition

Under the conditions of the identification of friend or foe and the identification and warning rules, the anti-ship missile appears in the situation as an air target. The flight direction is close to our ship. At 2000 meters, the altitude decreases, and the speed increases at this time, and the maximum rate of change reaches 50%. The speed is greater than 400 m/s. According to its motion parameters, the calculated CPA value is 200 meters and less than 300 meters. At this time, according to the identification alarm rules, issue an air-to-air self-defence identification alarm, and at the same time, according to the identification rules of friend or foe, determine that the target is a suspected enemy. When the target is 20 kilo-meters away from the ship, the radiation source signal is detected. According to the information of the radiation source, the identification result of the friend or foe is updated, and the attribute of the target is determined to be the enemy. Identification process as table 10 follows:

Table 10.Recognition process of objects in the scene

	2	3	11	12	15	16	17	18	19	20
Before	Known	Known	Unknown	Unknown	Unknown	Unknown	Unknown	Unknown	No appear	No appear
2. IFF	Friend	Friend	Noresponse	Noresponse	Noresponse	Civil	Civil	Civil	---	---
3.	Friend	Friend	Unknown	Unknown	Unknown	Neutral	Neutral	Neutral	---	---
4.	Friend	Friend	Aircraft	Aircraft	Aircraft	Aircraft	Neutral	Neutral	---	---
5.	Friend	Friend	Suspected	Suspected	Suspected	Suspected	Neutral	Neutral	Unknown	Unknown
6.	Friend	Friend	Hostile	Hostile	Hostile	hostile	Neutral	Neutral	Suspected	Suspected
7.	Friend	Friend	Hostile	Hostile	Hostile	Hostile	Neutral	Neutral	Missile	Missile
8.	Friend	Friend	Hostile	Hostile	Hostile	Hostile	Neutral	Neutral	Hostile	Hostile
After	Friend	Friend	Hostile	Hostile	Hostile	Hostile	Neutral	Neutral	Hostile	Hostile
Note: The yellow mark indicates the target whose attributes have changed during the recognition process										

3.4. identification result

By analysing the process in the identification table, the possible type of the target can be judged as military aircraft by the movement characteristics of the target in the case that the IFF can't recognize it. According to the movement trend of the target is close to our side and the CPA value is very small, the possible threat degree of the target to our side is increased. According to the characteristics of the current area is disputed area or hostile area, the probability of the target being the enemy increases and tends to be judged as the enemy target. The newly emerged supersonic target is determined as missile target by rules, which verifies the accuracy and effectiveness of the previous determination of unknown air target as enemy aircraft.

4. Conclusions and Prospect

The air target recognition method based on combined rules proposed in this paper can effectively increase the identification range and efficiency of air targets in complex sea areas through the practical application of combined rules. By designing the rule theory of air target recognition and constructing a set of effective rule combination and target recognition rule system, it can not only quickly identify pending or unknown targets, but also alarm and re-judge the results of automatic recognition error, reduce the misjudgement rate and assist the commander in air target recognition.

Design could not have exhausted all possible situations and rules, and associated with sea airspace is more and more complex, the design rules will not be able to completely solve the dynamic environment in a timely and effective manner all the air target identification problem. In order to improve the reliability of aerial target recognition, we need to consider introducing more external information to support and assist.

References

[1] MuzafferKapanoglu, Mete Alikalfa. Learning IF-THEN Priority Rules for Dynamic Job Shops Using Genetic Algorithms[J]. Robotics and Computer Integrated Manufacturing, 2011, 27(1): 47-55.
[2] K.Y. Fung, C.K. Kwong, K.W.M. Siu, K.M. Yu. A Multi-objective Genetic Algorithm Approach to Rule Mining for Affective Product Design[J]. Expert Systems with Application, 2012, 39(8):7411-7419.
[3] Yan S, Xu D, Zhang B. Graph embedding and extensions: A general framework for dimensionality reduction[J]. IEEE transaction on pattern analysis and machine intelligence, 2007, 29(1): 40-51.
[4] Ghanem B, Ahuja N. Sparse Coding of Linear Dynamical Systems with an Application to Dynamic Texture Recognition[C]. 20th International Conference on Pattern Recognition, 2010: 987-990.

[5] Belkin M, Niyogi P. Laplacianeigenmaps and spectral techniques for embedding and clustering[J]. Adv Neural Inform ProcSyst, 2002, 14: 585-591.
[6] Hu H F. Orthogonal neighborhood preserving discriminant analysis for face recognition[J]. Pattern Recognition, 2008, 41(6): 2045-2054.
[7] Kokar. M., C. Matheus, K. Baclawski. Ontology-based Situation Awareness[J]. Information Fusion, 2009, 10(1):83-98.
[8] Chia-Feng Juang. A Fuzzy System Constructed by Rule Generation and Iterative Linear SVR for Antecedent and Consequent Parameter Optimization[J]. IEEE Transactions on Fuzzy Systems, 2012, 20(2):372-384.
[9] MURPHY C K. Combining belief functions when evidence conflicts [J]. Decision Support System, 2000, 29 (1): 1- 9.
[10] H. Yu, X. Zhang, S. Wang, B. Hou. Context-based Hierarchical Unequal Merging for SAR Image Segmentation[J]. IEEE Transactions on Geoscience and Remote Sensing, 2013, 51(2): 995-1009.
[11] Xia Chen, Weidong Hu. Approach Based on Interval Type-2 Fuzzy Logic System for Emitter Identification[J]. IET Electronics Letters, 2012, 48(18): 1156-1158.

Research On Machining Product Quality Prediction Based On Machine Learning

Chen Lan[1a*], Wan Fei[1b*]

[1]School of Electrical and Electronic Engineering, Shanghai Institute of Technology, Shanghai 201499, China

[a*]2802241184@qq.com, [b*]wanf13@foxmail.com

Abstract—Data empowerment has attracted widespread attention from manufacturing companies. Enterprises control the production process by mining the data generated in the processing process, so as to achieve the purpose of improving product quality. This paper mainly uses the classification algorithm in data mining to model the data in the processing process and predict the product quality. Using the accuracy of the model prediction as the evaluation standard, the ID3, CART, SVM, KNN and Adaboost algorithms are compared. The experimental results show that the Adaboost classifier is significantly better than other algorithms in the evaluation index of accuracy, so the Adaboost classifier is selected as the final classifier.

1. Introduction

Since the proposal of "industry 4.0", various intelligent manufacturing modes have emerged rapidly, which makes the manufacturing industry face more severe challenges, such as production rate, product qualification rate and production cost. Information technologies such as data mining and cloud computing have been introduced into the manufacturing industry, not only to solve the above problems, but also to comply with the trend of the times[1]. Data mining can find rules from some unconnected data and mine meaningful data. It has been widely used in many fields such as production control and product quality, especially in the field of product quality[2]. A large amount of data will be generated in the processing of mechanical products. These data can be predicted through data mining technology, which can not only improve the qualified rate of products, but also optimize some technological processes in the production process, but also provide data support for product research and development[3].

A large number of scholars at home and abroad have done a lot of research on the application of data mining counting in industrial manufacturing. Hovhanes sadoyan et al. Predicted the behavior of manufacturing system through SVM, KNN and other algorithms[4]. Wang et al. Proposed an adaptive method of assembly quality to improve the accuracy of mechanical products under uncertain conditions[5]. Zhang Zhengang et al. Predicted the quality qualification rate of parts through random forest model and bagging model[6].

This paper classifies the data set made from the data generated in the machining process through different classification algorithms in data mining, takes the product quality as the target value and uses the accuracy as the evaluation standard, compares the algorithms such as decision tree, support vector machine (SVM), k-nearest neighbor and AdaBoost, and finally determines to use AdaBoost algorithm as the final method to solve the problem.

2.Theoretical Basis of Decision Tree Algorithm

2.1. Theoretical basis of classification algorithm

At present, the classification algorithms widely used in the field of machine learning include decision tree, support vector machine (SVM), k-nearest neighbor and AdaBoost.

Decision tree is a method of using graphics to represent the relevant steps of solving problems and the conditions and results of each step[7]. There is only one root node in a decision tree, because the root node is the top of the tree and represents the complete set of samples; Internal node refers to the node representing judgment conditions in the middle of the tree, which represents attributes. Leaf node refers to the node at the lowest end of a tree, which represents the result of selection according to different attributes[8].

The construction is a significant stage that must be experienced when generating a decision tree, because this is the process of selecting attributes as nodes. The key of decision tree learning is how to select the optimal partition attribute for construction[9]. For the binary classification problem, it is to make the divided samples trend to the same category, that is the attribute with the highest "purity". When the uncertainty grows, it will contain the more information, and the purity will be low. How to measure "purity" requires the introduction of the concept of "information entropy" in information theory. "Information entropy" is used to measure the uncertainty of information. The mathematical formula is as follows:

$$Entropy(t) = -\sum_{i=0}^{c-1} p(i \mid t)\log_2 p(i \mid t) \tag{1}$$

In formula (1), p(i|t) represents the probability that node t is classification i.

2.2. ID3 Algorithm

The core of ID3 algorithm is to calculate information gain[10]. When constructing the decision tree using ID3 algorithm, information gain will be selected as the division condition. After such division, the subset obtained has high "purity". Information gain refers to the increase of "purity" and the decrease of information entropy. In the calculation process, the normalized information entropy of each node is calculated, according to the probability of each child node in the parent node, so the mathematical formula of information gain is::

$$Gain(D,a) = Entropy(D) - \sum_{i=1}^{k} \frac{|D_i|}{|D|} Entropy(D_i) \tag{2}$$

In formula (2), D is the parent node, Di is the child node, and a in gain (D, a) is selected as the attribute of D node.

2.3. CART Algorithm

Cart (classification and regression tree) can be used as both classification tree and regression tree [12]. Unlike ID3 algorithm, which can generate binary tree or multi tree, cart algorithm can only generate binary tree. When constructing the classification tree, cart algorithm will select the attribute with the smallest Gini coefficient as the attribute division. Gini coefficient reflects the uncertainty of the sample. The smaller the Gini coefficient, the higher the purity of the sample. The mathematical formula of Gini coefficient is:

$$GINI(t) = 1 - \sum_{K} [p(C_k \mid t)]^2 \tag{3}$$

In formula (3), P (c_k|t) represents the probability that node T belongs to C_K.

2.4. Support Vector Machine

Support vector machine (SVM) is a supervised machine learning method. It has achieved good results in solving the classification and regression problems of small samples and high latitude. Suppose there are two classes of objects X and y, and a linear function is found in the high-dimensional space. Through the segmentation of the linear function, the samples of X and y are divided on both sides of the linear function, and the distance from the nearest sample point on both sides to the linear function to the linear function is maximized. The linear function is named "hyperplane", and the mathematical formula is:

$$g(x) = \omega^{x}T + b(\omega, x \in R) \tag{4}$$

In formula (1.4), ω and x are vectors in n-dimensional space, x is a function vector, ω It's a normal vector.

2.5. K-nearest neighbor algorithm

KNN (k-nearest neighbor) is a supervised classification algorithm. The working principle of KNN algorithm is to first calculate the distance between the target object and the object of defined category, and then find the K neighbors closest to the target object. The target object belongs to the category with the highest probability of occurrence among the K neighbors, and the selected neighbors are the objects that have been correctly classified.

2.6. Adaboost Algorithm

AdaBoost (adaptive boosting) is an implementation of boosting algorithm. [9]. Boosting algorithm is an integrated algorithm. It has several different weak classifiers combined into a strong classifier according to the weight. For the weak classifier, if its classification effect is better, its weight should be greater, so that the classification effect of the final strong classifier can be better. Therefore, the weight of the weak classifier is determined according to the error rate of its classification effect, which is expressed by formula:

$$\alpha_{\mathrm{i}} = \tfrac{1}{2}\log\tfrac{1-e_i}{e_i} \tag{5}$$

In formula (1.5), e_i represents the classification error rate of the I classifier.

Suppose the weak classifier is *Gi(x)*, the weight is e_i, and the formula for strong classification is *f(x)*:

$$f(x) = \sum_{i=1}^{n} \alpha_i G_i(x) \tag{6}$$

3.Training and classification

In this paper, a variety of algorithms are used to create a classification model for the data set, so a general process is defined to complete it, as shown in Figure 1.

The overall process is divided into two parts: training and testing. The training part starts with data acquisition, then preprocesses the data set, extracts and selects the features of the processed data, and then transmits these data to the algorithm model for training. The test stage also has the same data processing method as the training stage. The difference is that the original model selection is changed to model prediction, and then the test data is transmitted to the algorithm model after training for prediction.

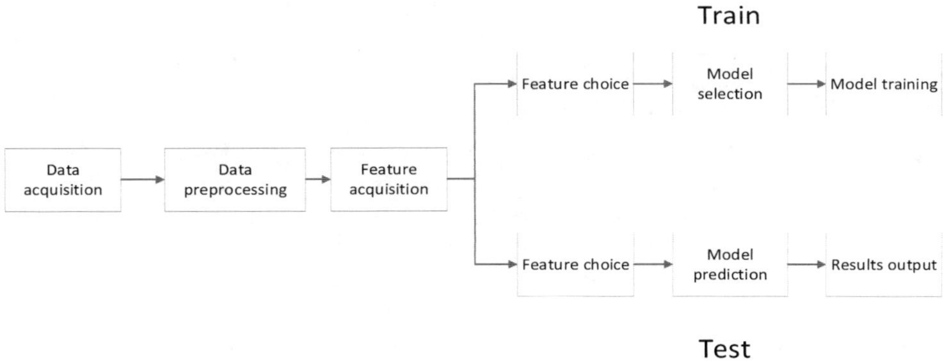

Fig.1 Schematic diagram of the overall process

3.1. Data Acquisition

The data set used in this paper comes from the measurement of various physical parameters of field processed objects through sensors, and then obtained and made.

During processing, online measurement technology is used to measure the physical parameters such as the length of the processed object, as shown in Figure 2. It can be seen that there is a chuck on the left inner wall of the machine tool. Wait for the robot to put the object into the chuck and fix it. The thimble on the right will begin to extend until it supports the processed object. Because the positions of the ejector pin and the chuck on the inner wall of the machine tool are fixed, the distance between them is also certain. At this time, as long as the extended distance of the ejector pin is collected through the displacement sensor, the length of the object can be collected indirectly.

Fig. 2 Schematic diagram of internal object fixation of machine tool

However, because the machine tool is rotating during machining, it is impossible to measure other parameters on-line, which requires off-line measurement technology to put the object into a special instrument for measuring physical parameters such as radius..

The data set comes from the measurement of various physical parameters of field processed objects through sensors, and then obtained and made. The dataset contains 1062 instances with 7 characteristics and attributes indicating whether the product is qualified or unqualified.

Table 1 shows a list of features and the number, mean, standard deviation, maximum and minimum values for each feature. Except that there is no unit in the ID column, the units in other columns are millimeters.

Table 1 Feature list

	ID	length	length_groove	radius	radius_groove	width	weight	qualified
count	1062	1062	1062	1062	1062	1062	1062	1062
mean	531.50	229.98	149.99	55.00	15.01	9.80	7.50	0.65
std	306.72	0.74	0.37	0.36	0.36	0.53	0.30	0.48
min	1.00	228.00	149.00	54.00	14.00	8.01	7.00	0.00
25%	266.25	229.40	149.72	54.72	14.75	9.59	7.23	0.00
50%	531.50	229.99	149.99	54.99	15.01	9.88	7.52	1.00
75%	796.75	230.55	150.28	55.29	15.29	10.18	7.77	1.00
max	1062.00	231.98	151.00	55.98	15.99	10.50	8.00	1.00

3.2. Data Preprocessing

This stage is very important for machine learning, because whether the data is processed or not may have a significant impact on the accuracy and speed of model training.

Because the sensor may collect null values when collecting data, and some algorithms in machine learning cannot process the missing values, which will affect the accuracy of the model, it is necessary to operate the missing values[13].

In addition to operating on null values, data transformation is also required. Because there may be data measured in different units in a data set, it is very important to translate different features to the vicinity with 0 as the center and 1 as the standard deviation. Different magnitudes of variables may affect the results, resulting in biased estimation [14]. For example, X1 = 10000, X2 = 1, X3 = 0.5, there is an order of magnitude gap between these three features. Once the value of X1 is wrong, it will directly affect the prediction of target variables. After normalization, the target variable is controlled by multiple characteristic variables with the same measurement. And if the transformation is not carried out, too many characteristic variables in the data set will increase the calculation cost. In addition, the data is too concentrated, and it is difficult to find the characteristics between the data.

In this paper, two data transformation methods are used, one is standardscaler, the other is minmaxscaler. The standardscaler standardizes the features by subtracting the mean value and scaling the data to the unit variance. The features after standardization obey the standard normal distribution. Minmaxscaler scales each feature to a given range through the maximum and minimum value. The default is [0,1]

3.3. Data Set Partition

Feature extraction is used to extract the fields that have a greater impact on the result, and then delete the fields in the dataset that have a smaller impact on the result. On the one hand, it can reduce the size of the dataset and reduce some adverse effects of some irrelevant fields on the result.

In the data set used in this paper, it can be seen that the "Id" field is only a simple number, which has no impact on the quality of the object, so it is directly removed. The remaining fields after removing "Id" and "qualified" are used as the characteristic values of the training set, and "qualified" is used as the target value of the training set.

3.4. Algorithm Application

In this paper, ID3 algorithm, cart algorithm, SVM algorithm, KNN algorithm and AdaBoost algorithm are selected to classify the data set. In the data set, 75% of the data are randomly selected as the training set of the model, and the remaining 25% of the data are used as the test set to test and evaluate the classification results of the model.

4. RESULTS ANASYSIS

4.1. Model evaluation

In this paper, the accuracy is used to measure and quantify the performance of each classifier and evaluate the model.

The accuracy of the above classifiers is displayed in the form of tables and visualized through graphics.

Table 2 Model classification accuracy

Algorithm	Accuracy
ID3	0.8233
Cart	0.8459
SVM	0.6090
KNN	0.7970
AdaBoost	0.9211

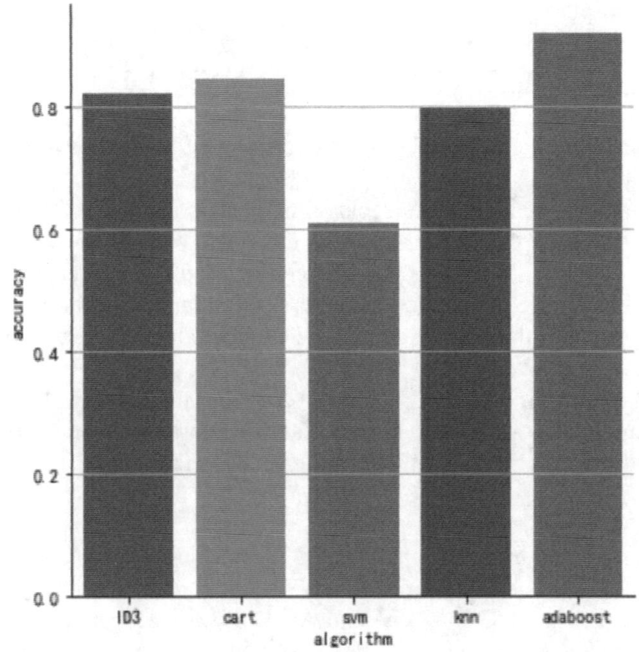

Fig.3 Model classification accuracy

As can be seen from Fig. 3, for the data set used in this paper, the classification effect of AdaBoost algorithm is the best, and the accuracy of the model reaches 92%; The classification effects of ID3 algorithm, cart algorithm and KNN algorithm are almost the same, which can reach 80%; The classification effect of SVM algorithm is poor, and the accuracy is only 60%.

4.2. Comparison between AdaBoost model and decision tree model

Theoretically, any model can be used as a weak classifier in AdaBoost, but decision tree and neural network are widely used. In the AdaBoost algorithm used in this paper, decision tree is used as a weak classifier.

Based on the original data set, taking the error rate as the evaluation index, the classification accuracy of ID3 model, cart model and AdaBoost model is compared.

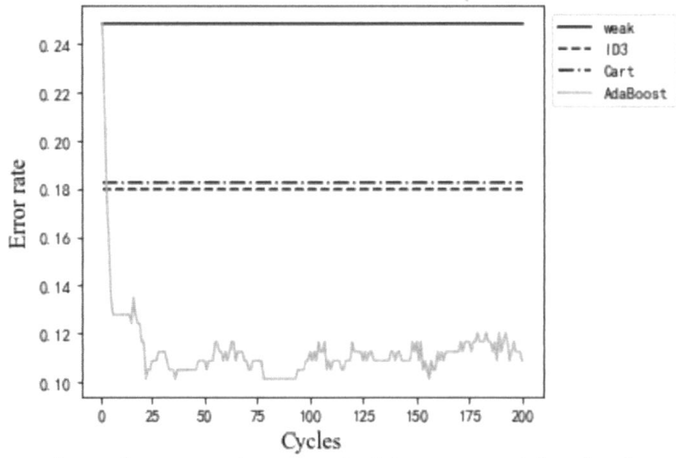

Fig.4 Comparison of error rate between decision tree model and AdaBoost model

As can be seen from the fig.4, when the decision tree weak classifier classifies separately, the error rate is the highest, which is higher than 25%. The error rate of the decision tree model classifier is significantly reduced, and the error rate is 18%. It can be seen that the error rate of AdaBoost is the same as that of the decision tree weak classifier at the beginning, but after several iterations, the error rate drops sharply. After more than 30 iterations, the change rate of the error rate tends to be flat, and the error rate is only about 10%.

Although the classification effect of a single decision tree weak classifier is not good, the classification accuracy of the AdaBoost strong classifier formed by the combination of multiple decision tree weak classifiers is better than that of the decision tree model..

5. CONCLUSION

In this paper, the physical parameters in the process of machining products are collected by sensors, and these physical parameters are made into a data set. After the data set is produced, classification algorithms such as decision tree, KNN, SVM and AdaBoost are designed to classify the data set, and the algorithm is implemented using Python. The experimental results show that the SVM algorithm has poor classification accuracy for this data set, the average accuracy of ID3, CART and KNN algorithms are all around 80%, but the average accuracy of AdaBoost algorithm reaches 90%, so Adaboost classifier is selected as the final classifier.

REFERENCES

[1] Kadlec, P., B. Gabrys, and S. Strandt, Data-Driven Soft Sensors in the Process Industry. Computers & Chemical Engineering, 2009. 33(4): p. 795-814.

[2] Ge, Z., et al., Data Mining and Analytics in the Process Industry: the Role of Machine Learning. IEEE Access, 2017. PP(99): p. 1-1.

[3] Qin, S.J., Survey on data-driven industrial process monitoring and diagnosis. Annual Reviews in Control, 2012.

[4] Sadoyan, H., A. Zakarian, and P. Mohanty, Data mining algorithm for manufacturing process control. The International Journal of Advanced Manufacturing Technology, 2006. 28(3-4): p. 342-350.

[5] Wang, X., et al., Research on assembly quality adaptive control system for complex mechanical products assembly process under uncertainty. Computers in Industry, 2015. 74: p. 43-57.

[6] Z.G.Zhang,D.Lin, Prediction of quality compliance rate of specific workpiece in data enabled discrete manufacturing. Industrial Engineering and management, 2021. 26(04): p. 28-35.

[7] Supianto, A.A., A.J. Dwitama, and M. Hafis. Decision Tree Usage for Student Graduation Classification: A Comparative Case Study in Faculty of Computer Science Brawijaya

University. in 2018 International Conference on Sustainable Information Engineering and Technology (SIET). 2018.
[8] Ibrahim, Z., RusliI D, Predicting Students' Acdemic Performance: Comparing Artificial Neural Network, Decision Tree and Linear Regression, 2007. 2007.
[9] Quadri, M.M.N. and N.V. Kalyankar. Drop Out Feature of Student Data for Academic Performance Using Decision Tree Techniques. 2010.
[10] Thakur, D., N. Markandaiah, and D.S. Raj. Re optimization of ID3 and C4.5 decision tree. in 2010 International Conference on Computer and Communication Technology (ICCCT). 2010.
[11] Yasami, Y. and S.P. Mozaffari, Anovel unsupervised classification approach fornetwork anomaly detection by k-Means clustering and ID3 decision tree learning methods. The Journal of Supercomputing, 2010. 53(1): p. 231-245.
[12] Lewis, R.J., An Introduction to Classification and Regression Tree (CART) Analysis. annual meeting of the society for academic emergency medicine.
[13] Liu, H., S. Shah, and J. Wei, On-line outlier detection and data cleaning. Computers & Chemical Engineering, 2004. 28(9): p. 1635-1647.
[14] Xu, S., et al., Data cleaning in the process industries. Reviews in Chemical Engineering, 2015.

Journal of Physics: Conference Series 2258 (2022) 012051 doi:10.1088/1742-6596/2258/1/012051

Feature Sharing Attention 3D Face Reconstruction with Unsupervised Learning from In-the-Wild Photo Collection

Xiaoxiao Yang[1a*]

[1a] School of Information Engineering, Ningxia University, Yinchuan, Ningxia, 750021, China

[a*] xiaoxiaoyang2446@stu.nxu.edu.cn,

Abstract. 3D face reconstruction is an active research topic in the field of computer vision research. It has a wide range of application scenarios. The number of 3D face annotation samples is small and the accuracy of the 3D reconstruction algorithm is insufficient. In this paper, we propose a shared attention module and a feature point constraint mechanism, and proposes a 3D face reconstruction algorithm that takes a single face image as input and uses a encoder-decoder network to predict the 3D face. The prediction view and light network introduces the attention mechanism module, which can improve the feature extraction ability of the network. The albedo and depth introduces shared features to help extract richer facial feature information. At the same time, the self-monitoring method of single image input avoids the high requirements of traditional methods for datasets. We conducted comparative experiments and ablation experiments on the BFM, and CelebA face datasets. The experimental results show that compared with Unsup3d and other representative 3D Face reconstruction algorithms, it has a certain improvement in the two evaluation indexes of Scale Invariant Depth Error (SIDE) and Mean Angle Deviation (MAD).

1. Introduction

3D face reconstruction based on monocular image has always been a hot research direction in the field of computer vision. The demand for 3D face reconstruction is increasing in the fields of 3D games, AR[1] /VR[2] and so on. The early parametric method represented by 3DMM[3] has achieved certain reconstruction effect by iteratively optimizing the reprojection error of face key points. With the rise of deep learning technology, some work directly learn 3DMM basic parameters through neural network and add more loss constraints to further improve the reconstruction accuracy[4, 5, 6], these parametric reconstruction methods are often limited by the upper limit of the expression ability of 3DMM parameter base itself, and can not obtain enough ideal reconstruction effect. At the same time , these methods are generally lack of 3D expression of facial details and high-frequency features.

Shape from X is another common 3D reconstruction modeling method, which mainly restores the specific aspect from the image such as shading[10, 11], silhouettes[14], texture[15], symmetry[16, 17] etc. Recently, ([18, 19, 20]) for unsupervised 3D shape learning, we usually learn to infer the viewpoint and shape of each image in the way of" composite analysis". Although the results of these methods are impressive, they usually assume that the shape of the object is symmetrical (symmetry assumption)[20] to prevent trivial solutions, which are difficult to extend to asymmetric objects. In addition, Wu et al[18] used an automatic encoder, which can decompose each input image into four components: depth, albedo, viewpoint, and light. which can vary accurately restore the 3D shape of human face, cat face and car from monocular images. However, in extreme facial expressions, extreme lighting conditions, and

contributions are as follows: extreme postures, the reconstruction quality is low[21]. Moreover, the reconstructed model area is prominent and the mouth model is deformed, which has been well solved.

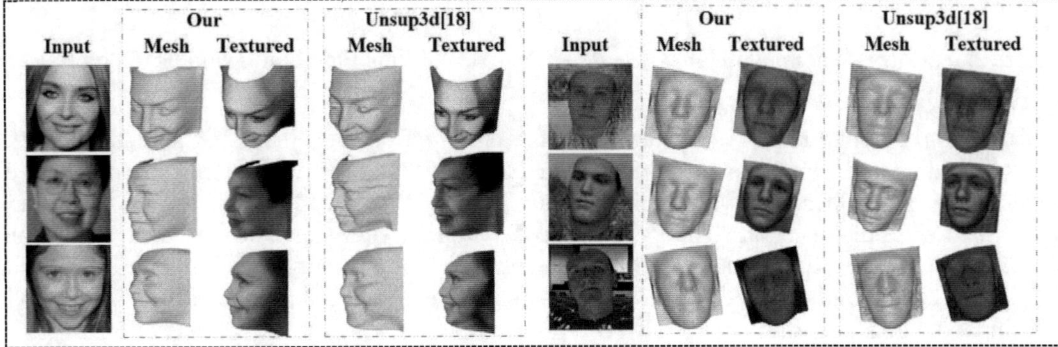

Figure 1. We propose a Progressive Cross-Stage Network with Unsupervised Learning from Images in the Wild. Comparison of qualitative results of Celeba [23] and BFM [22]. From left to right, we are testing the 3D reconstruction of ours and Unsup3d [18]. For each 3D model, we provide mesh, textured. It can clearly see that our method is more realistic in mesh and texture

In this paper, Facing the problems of existing 3D face reconstruction methods, we propose a feature sharing attention network for unsupervised 3D face reconstruction. In our method, view and light networks are introduced into the attention mechanism module to improve the feature extraction ability of the network, and shared features are introduced into albedo and depth to help extract richer facial feature information. The reconstructed loss and feature consistency loss are constrained to make the reconstructed shape more perfect, texture clearer and realistic, In general, our main contributions are as follows:

(1) We propose a feature sharing attention network to separate 3D faces from a image in the wild without using 3dmm hypothesis.

(2) By adding attention mechanism, our method can learn the correlation between channels ,the feature extraction ability of the network is improved.

(3) We introduce shared features, which can use the global feature information to extract information and effectively decompose depth and albedo.

(4) Extensive experiments on the BFM datasets show that our method can effectively reconstruct 3D face shape, which is smoother and better than the previous methods in shape reconstruction.

2. Method

In this section, we will introduce our proposed Feature Sharing Attention 3D Face Reconstruction network based on detail completion of natural images. As shown in Figure 2. With the training, the network is gradually guided to shift attention from global features to richer local details to encourage the model to learn stable details.

2.1 Unsupervised 3D Shape Learning

Overall, Given an single images of an object category, for example: human faces, cat, car ,celeba, our main goal is to learn a model Φ, input an image of an object instance and produces as output a decomposition of it into 3D shape, light, albedo and viewpoint, as show in Fig. 2. In order to mine the clues of viewpoint, light, depth and albedo, we need to model these factors explicitly. Specifically, here we follow the photo-geometric autoencoding design in [18]. For an image $I \in R^{3\times H\times W}$,we adopt a function that is designed to predict four factors (d, a, v, l), including a depth map $d \in R^{H\times W}$, an albedo image $a \in R^{3\times H\times W}$,a viewpoint $v \in R^6$, and a light direction $l \in S^2$, this function thus is implemented with four corresponding sub-networks (d, a, v, l),The four factors are recomposed to form the original image I via a rendering process Φ, which consists of lighting Λ and reprojection Π, as follows:

$$I' = \Phi(d, a, v, l) = \Pi(\Lambda(d, a, l), d, v) \quad (1)$$

Here Λ performs shading with albedo a, depth map d, and light direction l, while Π is reprojecctin.

2.2. Feature Sharing Attention Network

In this section, we introduce the architecture of the proposed Feature Sharing Attention Network in detail. According to the input image, our method consists of four parts, which are respectively used to predict depth, light, albedo and view. When predicting albedo and depth, we first share the extracted features and enter the encoder-decoder, When predicting light and view, we introduce the attention mechanism to focus on the effective information. generate normal from depth, combine light to form shading, and combine albedo to reconstruct the canonical albedo, using view, render back the same view as the input image for constraints.

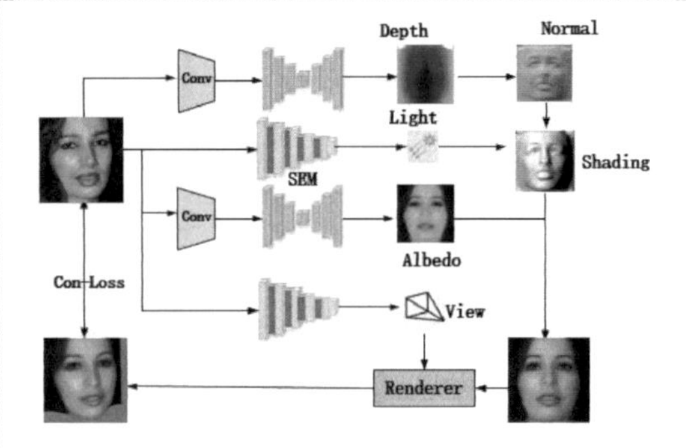

Figure 2. We introduce Feature Sharing Attention framework: According to the input image, the framewoork consists of four parts, which are respectively used to predict depth, light, albedo and view.

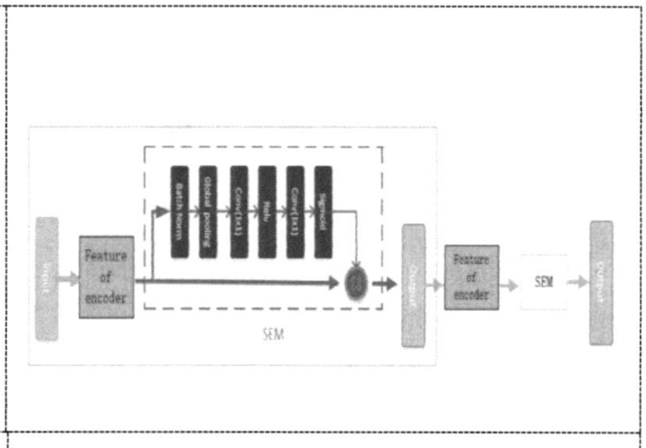

Figure 3. The architecture of the embedding attention module in light and view.

2.2.1 SEM

In our network, when the local details are unclear, the attention mechanism is added to the encoder of light and view. The attention mechanism can help the model give different weights to the input x, extract more critical and important information, and make the model make more accurate judgments without bringing more costs to the calculation and storage of the model. We use the extracted feature graph to learn the weight distribution, and then apply the learned weight with facial detail contours to the original feature graph. Finally, we add weights to different spatial regions on the spatial scale. We want to explicitly model the interdependencies between functional channels.

The original feature graph $C \times W \times H$ is global average pooling, and then a $1 \times 1 \times C$ feature graph is obtained, which has a global receptive field. The output $1 \times 1 \times C$ feature map is then processed by two fully connected neural networks, finally, a mechanism similar to the gate in the recurrent neural network is used. Weights are generated for each feature channel by parameters, in which parameters are learned to explicitly model the correlation between feature channels. The results obtained by exception are used as weights, and then weighted to the C channels by multiplication to complete the recalibration of the original features in the channel dimension, and as the input data of the next level. As shown in Figure 2.

2.3. Loss function

We use the objective functions to train our model our datasets. our loss function is defined as:

$$\mathcal{L} = \sum_{i=1}^{M}(\mathcal{L}(I', I, \Theta) + \lambda^{con}\mathcal{L}^{con}(g(I'), g(I), \Theta)) \quad (2)$$

Where $\mathcal{L}$, $\mathcal{L}^{con}$ are functions to compute perceptual loss components, $g(*)$ is a function to extract the k-th layer feature f of a VGG-16 network pre-trained on ImageNet, and $\lambda^{con} = 0.5$ is a weighting hyper-parameter. Con stands for consistency loss.

3. Experiments

In this part, we evaluate the performance of our proposed method in 3D face reconstruction task. We first introduce the datasets used in the our experiments. Then, we compare our results with the latest methods in quantitative and qualitative comparison. Finally, we provide a comprehensive ablation study.

3.1 Datasets and Metrics

BFM is a synthetic data set of 200K face images proposed by [18]. Each image is rendered using a 3D shape and texture randomly sampled from the Basel Face Model[22]. a random view, and a spherical harmonic estimated from the Celeba image[23], it also provides a real 3D depth map of the ground. We used this dataset to quantitatively evaluate our method and compare it with other baselines.

CelebA[23] is a popular facial dataset of more than 200K celebrity images. The images were captured under in-the-wild conditions. It is split into three subsets for training, validation, and testing with 162K, 20K, and 20K images.

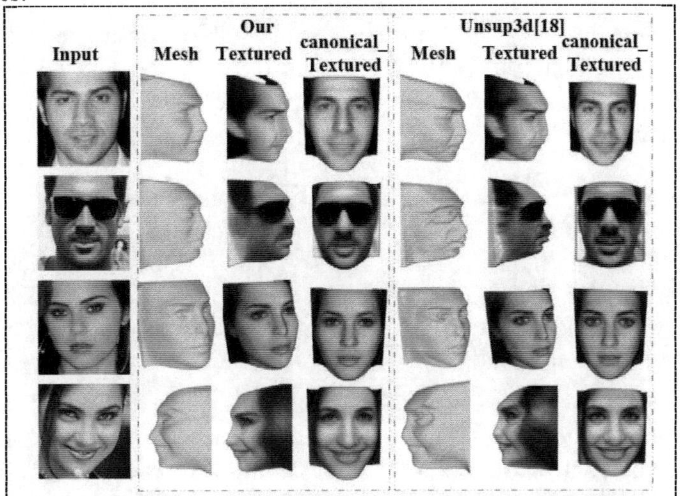

Figure 4. Qualitative results on predicted geometry. We compare our method with Unsup3d [18] For each 3D model, we provide mesh, textured, canonical albedo

3.2 Qualitative Evaluation

We qualitatively compare our method to the Unsup3d baseline on the mentioned datasets in Fig. 1 and Fig. 4. In all experiments, Our reconstruction effect can be very clear and better than the existing methods in terms of expression and texture details.. We qualitatively compared our method with the unsup3d [18] baseline. It can be seen that our method is more similar in shape and more realistic in texture.

3.3 Quantitative Evaluation

This paper mainly focuses on 3D face reconstruction, Refer to [18, 21, 24], Two evaluation indexes of Scale Invariant Depth Error (SIDE) and Mean Angle Deviation (MAD) are used to estimate the reconstruction effect of the algorithm.

Table 1. Comparison with baselines. on the BFM dataset, our reconstructions compared a fully-supervised and trivial baselines.

No	Baseline	*SIDE*(× 10^{-2}) ↓	MAD(deg.) ↓
(1)	Supervised	0.410 ± 0.103	10.78 ± 1.01
(2)	Const. null depth	2.723 ± 0.371	43.34 ± 2.25
(3)	Average g.t depth	1.990 ± 0.556	23.26 ± 2.85
(4)	Unsup3d(Wu et al., 2020)[18]	0.793 ± 0.140	16.51 ± 1.49
(5)	LeMul [24]	0.834 ± 0.140	**15.49 ± 1.50**
(6)	Gan2shape [21]	0.721 ± 0.128	15.53 ± 1.42
(7)	Ours	**0.689 ± 0.145**	15.78 ± 1.28

Scale Invariant Depth Error (SIDE) is the error between the face depth with our reconstructions and the actual face depth, which is defined as follows:

$$\mathrm{SIDE}(d^*, d') = \sqrt{\frac{\sum_{u,v} \Delta_{u,v}^2}{W \times H} - \left(\frac{\sum_{u,v} \Delta_{u,v}}{W \times H}\right)^2} \quad (3)$$

In the formula, $\Delta_{u,v} = \ln(d^*) - \ln d'$, d' is the depth value of the reconstructed face, and d^* is the depth value of the actual face.

Mean Angle Deviation (MAD) is defined as the average error between the normals with our reconstructions and the actual face.

$$\mathrm{MAD}(n^*, n') = \frac{\sum_{u,v} r(n^*, n')}{W \times H} \quad (4)$$

r is the angle between two vectors starting from the same pixel, n^* is the surface normal vector calculated by using the depth truth value of the data set, and n' is the surface normal vector calculated by using the predicted depth value, and the results are reported in Tab.1. In comparison, our two indicators decreased by 0.11 and 1.2 respectively, this means that the reconstructed 3D structure is better and the details are restored more clearly.

3.4 Ablation Study

We run ablation experiments to evaluate each network blck contribution to our result on the BFM dataset. From Unsup3d as the baseline. in Table 2, each of our proposals positively affects the MAD and SIDE numbers. We achieve the best reconstructed 3D surfaces when combining both techniques.

Table 2. Ablatioon studies on BFM dataset. sem is the module for predicting view and light, and con is the module added by depth and albedo.

No	Method	SIDE(× 10^{-2}) ↓	MAD(deg.) ↓
(1)	Unsup3d[18]	0.793 ± 0.140	16.51 ± 1.49
(2)	+sem	0.770 ± 0.142	15.76 ± 1..55
(3)	+sem+con	0.702 ± 0.153	18.59 ± 1.28
(4)	+sem+con+loss	0.689 ± 0.145	15.78 ± 1.28

4.Conclusions

In this paper, we propose a feature sharing attention network method based on the unsupervised from Images in the Wild, which can learn a 3D model from a single view unconstrained setting. The network combines a new attention mechanism. High fidelity 3D face reconstruction with a single view can be obtained. This is done without any supervision only under a single picture without supervision. Our model is superior to the most advanced unsupervised 3D face reconstruction method.

References

[1] V. Blanz and T. Vetter, "A morphable model for the synthesis of 3d faces," pp. 187–194, 1999.
[2] B. Gecer, S. Ploumpis, I. Kotsia, and S. Zafeiriou, "Ganfit: Generative adversarial network fitting for high fidelity 3d face reconstruction," pp. 1155–1164, 2019.
[3] X. Zhu, Z. Lei, X. Liu, H. Shi, and S. Z. Li, "Face alignment across large poses: A 3d solution," pp. 146–155, 2016.
[4] B. Browatzki and C. Wallraven, "3fabrec: Fast few-shot face alignment by reconstruction," pp. 6110–6120, 2020.
[5] X. Zeng, X. Peng, and Y. Qiao, "Df2net: A dense-fine-finer network for detailed 3d face reconstruction," pp. 2315–2324, 2019.
[6] B. Amberg, S. Romdhani, and T. Vetter, "Optimal step nonrigid icp algorithms for surface registration," pp. 1–8, 2007.
[7] P. Paysan, R. Knothe, B. Amberg, S. Romdhani, and T. Vetter, "A 3d face model for pose and illumination invariant face recognition," pp. 296–301, 2009.
[8] F. Liu, D. Zeng, Q. Zhao, and X. Liu, "Joint face alignment and 3d face reconstruction," pp. 545–560, 2016.
[9] F. Wu, L. Bao, Y. Chen, Y. Ling, Y. Song, S. Li, K. N. Ngan, and W. Liu, "Mvf-net: Multi-view 3d face morphable model regression," pp. 959–968, 2019.
[10] Y. Liu, A. Jourabloo, W. Ren, and X. Liu, "Dense face alignment," pp. 1619–1628, 2017.
[11] L. Tran and X. Liu, "Nonlinear 3d face morphable model," pp. 7346– 7355, 2018.
[12] L. Li and S. Wu, "Dmifnet: 3d shape reconstruction based on dynamic multi-branch information fusion," arXiv preprint arXiv:2011.10776, 2020.
[13] W. Zhu, H. Wu, Z. Chen, N. Vesdapunt, and B. Wang, "Reda: reinforced differentiable attribute for 3d face reconstruction," pp. 4958–4967, 2020.
[14] A. Jourabloo and X. Liu, "Large-pose face alignment via cnn-based dense 3d model fitting," pp. 4188–4196, 2016.
[15] A. S. Jackson, A. Bulat, V. Argyriou, and G. Tzimiropoulos, "Large pose 3d face reconstruction from a single image via direct volumetric.
[16] Y. Feng, F. Wu, X. Shao, Y. Wang, and X. Zhou, "Joint 3d face reconstruction and dense alignment with position map regression network," pp. 534–551, 2018.
[17] C. Cao, Y. Weng, S. Zhou, Y. Tong, and K. Zhou, "Facewarehouse: A 3d facial expression database for visual computing," IEEE Transactions on Visualization and Computer Graphics, vol. 20, no. 3, pp. 413–425, 2013.
[18] Wu S , Rupprecht C, Vedaldi A. Unsupervised Learning of Probably Symmetric Deformable 3D Objects from Images in the Wild[J]. 2019.
[19] Henzler P , Mitra N , Ritschel T . Escaping Plato's Cave: 3D Shape From Adversarial Rendering[C]// 2019 IEEE/CVF International Conference on Computer Vision (ICCV). IEEE, 2019.
[20] Suwajanakorn S, Kemelmachershlizerman I, Seitz S M. Total moving face reconstruction[C]// Springer International Publishing. Springer International Publishing, 2014.
[21] Pan X , Dai B , Liu Z , et al. Do 2D GANs Know 3D Shape? Unsupervised 3D shape reconstruction from 2D Image GANs[J]. 2020.
[22] Pascal Paysan, Reinhard Knothe, Brian Amberg, Sami Romdhani, and Thomas Vetter. A 3d face model for pose and illumination invariant face recognition. In Proceedings of International Conference on Advanced Video and Signal based Surveillance, 2009. 3, 5.
[23] Ziwei Liu, Ping Luo, Xiaogang Wang, and Xiaoou Tang. Deep learning face attributes in the wild. In Proceedings of the International Conference on Computer Vision, 2015. 5, 8.
[24] Ho L N, Tran A T, Phung Q, et al. Toward Realistic Single-View 3D Object Reconstruction with Unsupervised Learning from Multiple Images[C]//Proceedings of the IEEE/CVF International Conference on Computer Vision. 2021: 12600-12610

Model Reference Adaptive Speed Observer Control of Permanent Magnet Synchronous Motor Based on Single Neuron PID

Hao Liu[1a], Huixian Zhang[1b], He Zhang[1c], Gaoqi Chen[1d]

[1]Key Laboratory of Magnetic Suspension Technology and Maglev Vehicle, Ministry of Education, Southwest Jiaotong University, Chengdu 611730, China

[a]614438996@qq.com, [b]1113035744@qq.com, [c]ZhangHe@my.swjtu.edu.cn, [d]975291402@qq.com

Abstract: For the speed sensorless control of permanent magnet synchronous motor (PMSM), a design method based on model reference adaptive observer and single neuron PID speed loop was proposed. In this method, the model reference adaptive observer is used to identify the speed online, the Popov hyperstability is used to determine the reference adaptive law of the speed observer, the neural network module is used to control the q-axis current, and the LMS algorithm is selected to modify the weights of the neural network speed loop online. The simulation results show that the model reference adaptive speed observer control of PMSM based on single neuron PID is effective and robust.

1. Introduction

Permanent magnet synchronous motor has the advantages of simple structure, reliable operation, small size, light weight, low loss, high efficiency, and flexible shape and size of the motor. In recent years, with the continuous development of material technology, the continuous improvement of the performance of permanent magnet materials, as well as the mature of permanent magnet motor control technology, PMSM has been widely used in civil, aerospace, military and other fields[1]. In the field oriented vector control algorithm, in order to achieve efficient control system, the rotor speed and position information need to be obtained in advance, but the installation of traditional resolver will make the whole system become bulky, and also vulnerable to high temperature and electromagnetic interference. In this paper, the sensorless control method is proposed, which uses the voltage and current signals and the mathematical model of the motor to estimate the speed online [2].

In this paper, a model reference adaptive observer is used to identify the speed online, and the feedback speed is compared with the given speed and then sent to the single neuron PID module. LMS is used to update the weights online, and the given q-axis current is output to realize the vector control of PMSM. The simulation results show that the adaptive speed sensorless control method of PMSM based on single neuron PID has good robustness and dynamic response speed.

2. Speed sensorless design

For the three-phase surface-mounted PMSM, the voltage equation in the synchronous rotating coordinate system is [3]:

$$\begin{cases} u_d = Ri_d + L_s\dfrac{di_d}{dt} - \omega_e L_s i_q \\ u_q = Ri_q + L_s\dfrac{di_q}{dt} + \omega_e(L_s i_d + \psi_f) \end{cases} \tag{1}$$

The formula (1) is expressed by current, and some transformations are made:

$$\frac{d}{dt}\begin{bmatrix} i_d + \dfrac{\psi_f}{L_s} \\ i_q \end{bmatrix} = A\begin{bmatrix} i_d + \dfrac{\psi_f}{L_s} \\ i_q \end{bmatrix} + b\begin{bmatrix} u_d + \dfrac{R\psi_f}{L_s} \\ u_q \end{bmatrix} \tag{2}$$

In the formula:$A = \begin{bmatrix} -\frac{R}{L_s} & \omega_e \\ -\omega_e & -\frac{R}{L_s} \end{bmatrix}$;$b = \frac{1}{L_s}$.

Define the following:

$$\begin{cases} i'_d = i_d + \dfrac{\psi_f}{L_s} \\ i'_q = i_q \\ u'_d = u_d + \dfrac{R\psi_f}{L_s} \\ u'_q = u_q \end{cases} \tag{3}$$

Write formulas (3) in the form of state equation:

$$\frac{di'}{dt} = Ai' + Bu' \tag{4}$$

In the formula: $i' = \begin{bmatrix} i'_d \\ i'_q \end{bmatrix}$,$u' = \begin{bmatrix} u'_d \\ u'_q \end{bmatrix}$, $A = \begin{bmatrix} -\frac{R}{L_s} & \omega_e \\ -\omega_e & -\frac{R}{L_s} \end{bmatrix}$,$B = \begin{bmatrix} \frac{1}{L_s} & 0 \\ 0 & \frac{1}{L_s} \end{bmatrix}$.

The speed information is included in the matrix a of the formula (4), so the formula can be used as an adjustable model, the speed ω_e is used as an identification parameter, and the reference model is a permanent magnet synchronous motor. Next, we will determine the adaptive law.

By express formula (4) as estimate value, we obtain:

$$\frac{d\hat{i}'}{dt} = \hat{A}\hat{i}' + Bu' \tag{5}$$

In the formula: $\hat{i}' = \begin{bmatrix} \hat{i}'_d \\ \hat{i}'_q \end{bmatrix}$,$\hat{A} = \begin{bmatrix} -\frac{R}{L_s} & \hat{\omega}_e \\ -\hat{\omega}_e & -\frac{R}{L_s} \end{bmatrix}$.

Defining generalized error $e = i' - \hat{i}'$,subtracting formula (5) from formula (4) yields:

$$\frac{de}{dt} = Ae - J\left(\omega_e - \hat{\omega}_e\right)\hat{i}' \tag{6}$$

In the formula: $\boldsymbol{J} = \begin{bmatrix} 0 & -1 \\ 1 & 0 \end{bmatrix}$.

Formula (6) may be further abbreviated as:

$$\frac{de}{dt} = Ae - W \tag{7}$$

In the formula: $W = J(\omega_e - \widehat{\omega}_e)\hat{\imath}'$.

According to Popov hyperstability, in order to make the system asymptotically stable, we must satisfy: (1) transfer matrix $H(s) = (sI - A)^{-1}$ is a strictly positive definite matrix;(2) $\eta(0, t_1) = \int_0^{t_1} V^T W dt \geq -\gamma_0^2, \forall t_1 \geq 0, \gamma_0$ is any finite positive number [4]. When these requirements are met, $\lim_{t\to\infty} e(t) = 0$, the model reference adaptive observer tends to be stable. The adaptive law can be obtained by reversely solving the integral inequality of the condition (2), and the final result is:

$$\hat{\omega}_e = \int_0^t K_i \varepsilon_i d\tau + K_p\, \varepsilon_i + \hat{\omega}_e(0) \tag{8}$$

3. Single neuron adaptive PID

Neuron is the basic unit of neural network, which has the advantages of learning, adaptive and simple structure. The traditional PID also has the characteristics of simple structure and convenient parameter adjustment, but in the actual project, it can't modify each parameter online, making the control system can 't adjust in real time according to the operation state, and the dynamic performance is poor. So this paper presents a neural network algorithm combined with PID, the system is not only simple to control, but also has the ability of online learning and adaptive [5]. The structure of the adaptive PID controller implemented by the neural network is shown in Fig.1.

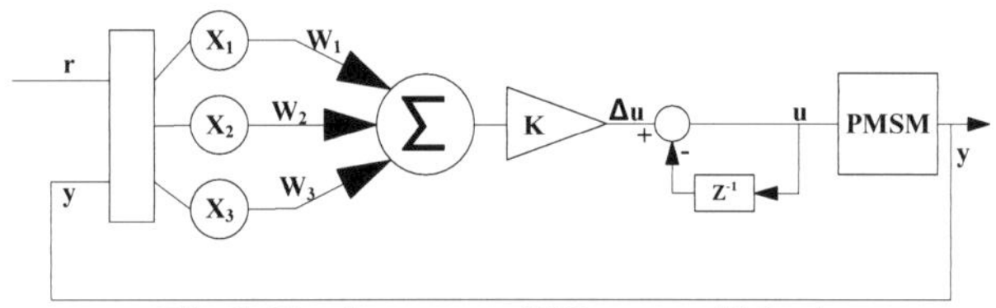

Fig.1 Single neuron PID control block

$$\begin{gathered} x_1 = r(k) - y(k) = e(k) \\ x_2 = e(k) - e(k-1) = \Delta e(k) \\ x_3 = e(k) - 2e(k-1) + e(k-2) = \Delta^2 e(k) \end{gathered} \tag{9}$$

$\omega_i (i = 1,2,3)$is the weight of $x_i (i = 1,2,3)$, $K(K > 0)$ is the proportional coefficient of the neuron [6].

The change is known as voltage:

$$\Delta u(k) = Kf(I) = K\left(\sum_{i=1}^{3} \omega_i x_i\right) \tag{10}$$

Bring formula (9) into formula (10):

$$\begin{gathered} \Delta u(k) = K\{\omega_1 e(k) + \omega_2[e(k) - e(k-1)] \\ +\omega_3[e(k) - 2e(k-1) + e(k-2)]\} \end{gathered} \tag{11}$$

In this paper, least mean square (LMS) is used to update the weights. It is assumed that *y (K)* is the actual output value of the control system and *r (K)* is the set value of the control system. Then the learning formula of the weight is as follows, $\eta_i > 0$ is learning efficiency:

$$\omega_i(k+1) = \omega_i(k) + \eta_i e(k) x_i(k) \tag{12}$$

4. Simulation result

This paper uses Matlab/Simulink simulation platform, the overall block diagram is shown in the Fig.2.

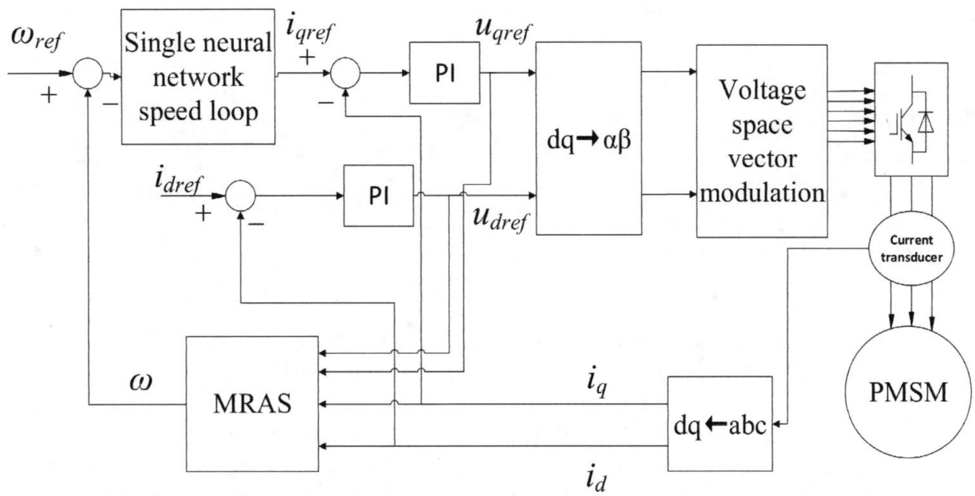

Fig.2 Control Block Diagram of Adaptive Speed Observer for PMSM

For unsupervised learning neural network, it is very important to set the parameter of learning efficiency. Since the weights are updated in real time with the input data, the initial values $\omega_i(0)$ can be selected at will. The larger the proportional coefficient *K* of the neuron is, the better the rapidity of the system is, but the overshoot will become larger, which will lead to the instability of the system; The smaller the value of *K*, the faster the system will not be enough. Therefore, it is also important to choose a reasonable value of *K*.

Tab.1 Parameters of the SPMSM

Parameter	Figure	Unit
Stator resistance	2.875	Ω
Quadrature and direct axis inductance	8.5	mH
Moment of inertia	4.8×10^{-6}	$kg\cdot m^2$
Main pole flux	0.175	Wb
Number of pole pairs	4	
Sampling Time	1×10^{-5}	*s*

After many experiments and simulations, the selected parameters are as follows:

The proportional coefficient: $K=1$; initial value of weight: $\omega_1(0)=0.3$, $\omega_2(0)=0.3$, $\omega_3(0)=0.3$; learning efficiency: $\eta_1=1\times10^{-3}$, $\eta_2=1\times10^{-3}$, $\eta_3=1\times10^{-4}$.

Fig.3 is the speed output waveform under different working condition:

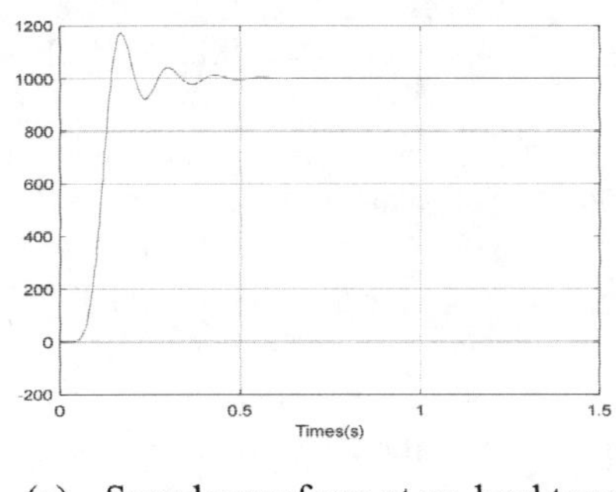

(a) Speed waveform at no-load torque

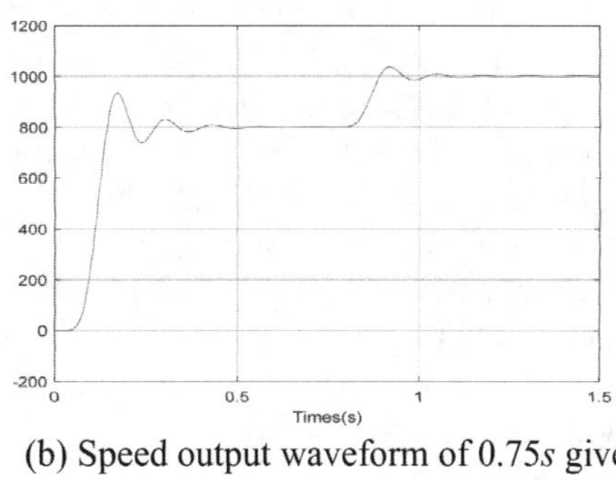

(b) Speed output waveform of 0.75*s* given speed step change

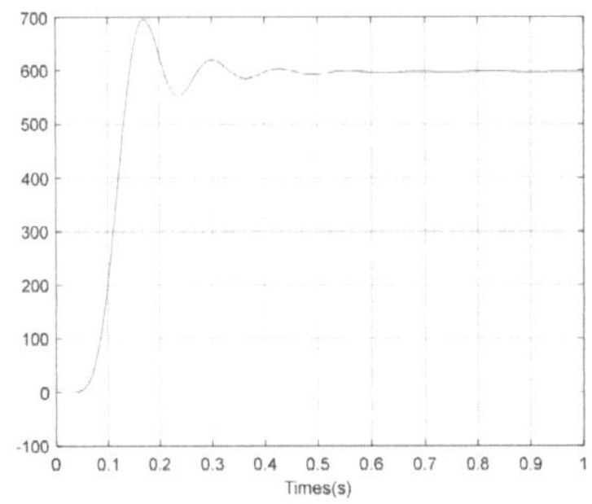

(c) 0.65s load torque becomes $5N \cdot m$ speed output waveform

Fig.3he speed output waveform under different working condition

Fig. (a) output waveform at a given speed of 1000*r/min* without load torque, it can be seen that the output waveform can quickly follow the given speed and remain stable. Fig. (b) is the speed following waveform when the initial given speed is 800r/min and the step is 1000*r/min* at 0.75 *s*. As shown in the figure, it can be seen that after the step change of the speed, the speed can quickly follow the given change within 0.25 *s* and remain stable, it is proved that the dynamic response speed of this method is fast. Fig. (c) shows that the given rotation speed is kept at 600*r/min*, and then the load torque changes from $0N \cdot m$ to $5N \cdot m$ at 0.65*s*. According to the waveform in the figure, it can be seen that the change of the load torque has little interference on the control system, indicating that the system has strong anti-interference capability.

5.Conclusion

In this paper, the speed sensorless control of PMSM is realized by combining the model reference adaptive speed observer and single neuron PID speed loop, which can not only identify the speed online in real time, but also automatically modify the parameters of the speed loop to make the system stable again when it is disturbed by external disturbance.

According to the simulation results, the model reference adaptive speed observer control strategy of PMSM based on single neuron PID can quickly reach the steady-state speed with small overshoot, and show better adaptability when the load disturbance occurs, and has stronger robustness to parameter changes and load torque disturbance.

References

[1] Chen R. Situation and Development of AC permanent magnet synchronous servo system[J]. Electric Age, 2005, (09): 104-107.

[2] Qiao Z W, Shi T N, Wang Y D, et al. Newsliding-mode Observer for Position Sensorless Control of Permanent-magnet Synchronous Motor[J]. IEEE Trans on Industrial Electronic, 2013,60(2): 710-719.

[3] Maurizio Cirrincione, Angelo Accetta, Marcello Pucci, et al. MRAS Speed Observer for High-performance Linear Induction Motor Drives Based on Linear Neural Networks[J]. IEEE Trans. on Power Electronics, 2013,28(1): 123-134.

[4] M Zerikat, S Chekroun, A Mechernene. Fuzzy-neural Net-works Controller-based Adaptation Mechanism for MRAS Sensorless Induction Motor Drives[A]. International Conference on Electrical Machines Lille[C]. 2009: 1-7.

[5] Cai C, Chu F, Wang Z, et al. Identification and control of PMSM using adaptive BP-PID neural network[C]. Dalian, China, 2013: 155-162.

[6] Xianglong Jiang, Jin Zhao, Hui Luo. "Neural network speed control of induction motor drive based on direct MRAC method," The 29th Annual Conference of the IEEE, vol 3, pp.2531—2536, Mar. 2003.

Research on the application of OpenPose in escalator safety systems

Xuehan Zhao

College of Resources, Shandong University of Science and Technology, Tai'an 271000, China

zhaoxh2001@foxmail.com

Abstract—This paper describes the application of the OpenPose based human pose detection model in escalator safety systems. The YOLO target detection is used to obtain the position of the human body in the image, and then the human limb information is extracted by OpenPose, and the human joint connection algorithm connects the identified human limb information to obtain the human pose information in the image. The experimental results show that the algorithm can effectively identify the human body posture and can provide timely alarms for dangerous postures, thus improving the safety of escalator riders.

1. Introduction

With the continuous development of urbanization, the production and use of lifts have increased significantly. Due to the advantages of strong capacity to transport passengers, uninterrupted transportation, small space occupation, and high adaptability, they are commonly used in shopping malls, metro stations, airports, and other places with a large flow of people, which brings safety problems that should not be underestimated. In the past five years, there have been 251 safety accidents caused by escalators in China, resulting in 198 deaths.[1] Therefore, it is important to establish a complete escalator safety system in order to reduce the number of safety accidents caused by them.

In recent years, several scholars have conducted research on escalator safety systems. Li Deyuan used the VGG-16 network to identify and classify dangerous human postures [2]; Chang et al. judged human motion by obtaining the distance and direction between the human head, legs, and upper body based on the morphological features of the human body [3]; Xu Honghua et al. designed an escalator monitoring and communication system through LabView and Wifi [4]; Azkune et al. used convolutional neural networks for fall posture detection. [5].

In current escalator safety management, it is mainly managed manually by placing video capture devices at the entrances and exits of escalators to obtain real-time images, which are then monitored by dedicated personnel. With the development of machine vision in recent years, the introduction of image detection-related algorithms into this monitoring system to replace manual detection and respond quickly to hazards in the video is of great significance in reducing safety incidents. This paper focuses on the application of the OpenPose algorithm to escalator safety systems, where deep convolutional neural networks can effectively correspond to various complex environments and improve the accuracy of recognition.

2. Human target detection and identification methods

To solve this problem, the Snet-Yolo algorithm [6], which introduces a convolutional neural network structure of 131, is used. YOLOv3 detects the human body through a Darknet-53 network, which has a residual [7] component (Figure 1). The problem of degradation in deep networks is solved by adding the constant mapping identity so that the performance of the network does not degrade after residual learning of the features x in the stacked structure, and overcoming the gradient disappearance by the activation function relu to speed up the training. And after three down sampling, 1*1, 3*3, and 1*1 convolutional kernels are introduced to increase the linearity of the network, and then the feature pyramid network (FPN) is used for feature fusion at different levels to enhance the detection of pedestrians of smaller size in the image and solve the problem of insensitive detection due to pedestrian occlusion or distance from the image acquisition device is solved, Figure 2 illustrates the flow block diagram for human target detection.

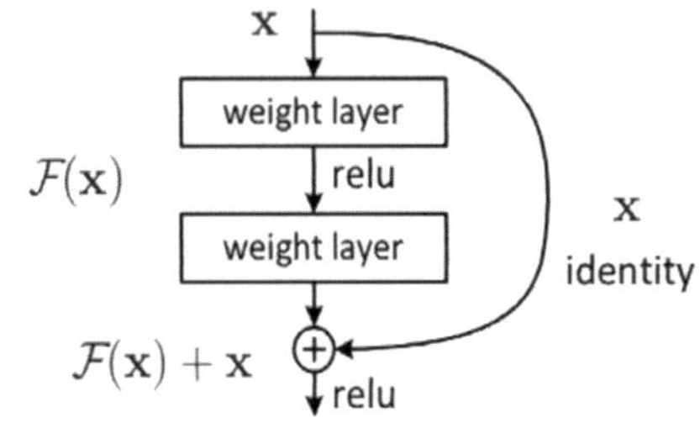

Fig.1 Residual network component

Fig.2 YOLOv3 human detection solution

2.1. *Model VGG-19*

VGG [8] is a new generation of convolutional neural networks developed by Visual Geometry Group and Googled DeepMind, whose researchers have developed a new deep convolutional neural network model, VGGNet, and used its model to win second place in the 2014 ILSVRC competition.

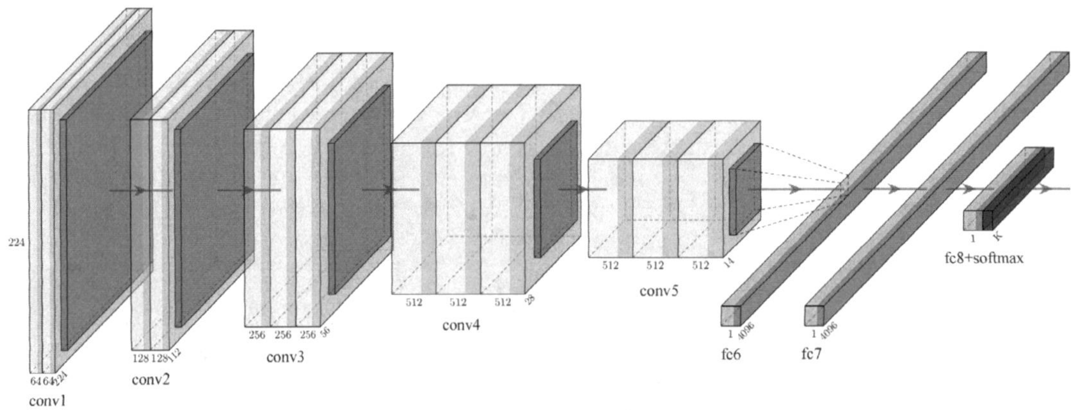

Fig.3 the basic structure of VGG-19

VGGNet-19 investigates the performance of the model as a function of the depth of the convolutional neural network, which consists of 5 convolutional layers, 3 fully connected layers and 1 output layer of SoftMax, all interconnected using pooling layers (maximum pooling layer) and all hidden layers using the activation unit of the ReLU activation function. 5 convolutional layers, 3 fully connected layers and 1 output layer of SoftMax The output layer is composed of 5 convolutional layers, 3 fully-connected layers and 1 SoftMax layer, with each layer interconnected using pooling layers (maximum pooling layers) and activation units of the ReLU activation function between each hidden layer. Compared to other widely used CNN models, VGGNet-19 is simpler to use, mainly because the convolutional layers used in this model are all Conv3, using a very small 3*3 convolutional kernel in the convolutional layer, with a minimum size of 3*3 kernels capable of obtaining pixel 8 field information; two stacked 3*3 convolutional kernels instead of 5*5 convolution, and three stacked 3*3 convolutional kernels instead of 7*7 convolution, the perceptual field size remains the same, multiple convolutional layers are stacked to be able to bring in deeper layers, which can bring in more non-linear functions, thus improving the decision function judgement and reducing the number of parameters and increasing the speed of operations. The pooling layers are all sized using 2*2 with a step size of 2, and the maximum pooling layer is used. The model has more convolutional layers and pooling layers, which can form a relatively deeper network structure, thus achieving improved performance.

2.2. The OpenPose algorithm

OpenPose is an open-source library developed by Carnegie Mellon University based on Convolutional neural network with Supervised Learning in the Caffe deep learning framework[9]. VGG-19 extracts the image features from the acquired image information through 10 layers of the convolutional neural network, and then completes the key point acquisition, key point connection, and pose recognition of a person in 6 steps.

Human key point detection is a routine task in machine vision, for human movement recognition, human behaviour analysis, etc. The task is to pinpoint the main key points of the human body, such as the top of the head, the five senses, the limbs, the neck and other major joint areas.

For traditional key node detection methods, this is usually done by a top-down pose estimation algorithm, i.e. human detection is performed first so that the individual human images are processed and then the individual images completed are subjected to individual key point detection. However, in places with a high volume of people, the number of individuals to be detected is very large, resulting in long-running times, which, combined with the possibility of problems such as mutual occlusion of people, affects the recognition accuracy of key points.

Based on these problems, OpenPose offers the Part Affinity Fields solution, which is able to store orientation and position information in the image region of each kind of limb so that a 2D vector of information exists for each of its pixels. [10]

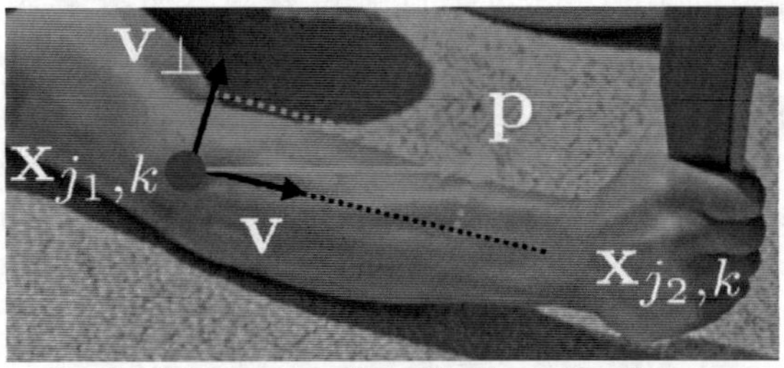

Fig.4 Diagram of key limb connections

$$L^*_{c,k}(p) = f(x) = \begin{cases} v, & if\ p\ on\ limb\ c,k \\ x, & otherwise \end{cases} \tag{1}$$

where: $x_{j1,k}$ and $x_{j2,k}$ are two adjacent limb key points.

$$v = \frac{(x_{j2,k} - x_{j1,k})}{\left\| x_{j2,k} - x_{j1,k} \right\|} \tag{2}$$

which is denoted as the unit vector of the two key points of the limb.

Determine whether point P is located on the limb c of the body at the key point k according to equation (1) and equation (2).

$$0 \leq v \cdot (p - x_{j1,k}) \leq l_{c,k} \tag{3}$$

$$0 \leq v \cdot (p - x_{j1,k}) \leq l_{c,k} \tag{4}$$

where the width of the limb is σ_l , the length of the limb is $l_{c,k} = \left\| x_{j2,k} - x_{j1,k} \right\|_2$, the vertical vector of $\vec{v}$ is $v_\perp$.

If there are multiple people in the image, PAF is the average of the PAF of each body in the image, i.e.

$$L_c^*(p) = \frac{1}{n_c(p)} \sum\nolimits_k L_{c,k}^*(p) \tag{5}$$

The number of non-zero vectors is $n_c(p)$.

Two key points j_1 and j_2 were used and the PAF labels were obtained by joining the line segments between the two points in order to determine the affinity of the two key points associated in the limb.

$$E = \int_{u=0}^{u=1} L_c(p(u)) \frac{d_{j2} - d_{j1}}{\left\| d_{j2} - d_{j1} \right\|_2} d_u \tag{6}$$

$$p(u) = (1-u)d_{j1} + ud_{j2} \tag{7}$$

The path integral of the affinity vector of the two key points j_1 and j_2 and the unit vector in the direction of the concatenation of these two points is the affinity between j_1 and j_2.

The PAF calculates the direct correlation of the key points, and finally, using the key points as vertices, the correlation between the points is transformed into a bipartite graph assignment problem. The connections between the key points are obtained using the Hungarian algorithm for optimal matching, which in turn yields a connected image of each human limb.

3. Training of human posture detection algorithms and analysis of experimental results

This paper uses a PC for the training of the fall detection algorithm.

The algorithm is implemented in Python programming on i7-9750H 2.6GHZ CPU, GTX1650 GPU, 16GB RAM, Windows 10 OS.

As TensorFlow is more convenient, readable, and flexible than the Caffe deep learning framework based on Convolutional neural network and Supervised Learning used by the open-source library, the OpenPose algorithm in this paper uses the TensorFlow deep learning framework to train The human pose detection model.

3.1. Data sets and parameter settings

The dataset uses the existing MPII Human Pose Dataset, a state-of-the-art benchmark for evaluating human joint pose estimates. The dataset consists of approximately 25,000 images of the human body, including over 40,000 people with key body points annotated. In addition, the test set has a very complete annotation, including obscured body parts, 3D torso, and head orientation of the body.

1) Batch Size

The batch size determines the number of samples fed into the training network each time. The larger the Batch Size the faster the network converges, but it will take up memory resources. According to the actual situation, Batch Size is set to 16 in this paper's training.

2) Learning Rate

The learning rate controls the overall learning rate in the system by controlling for large weights and changes in network bias during training. When the learning rate is set too small, the parameter changes are too small, which tends to result in a local optimum and requires too long a fitting time. Too large a learning rate can lead to too slow a fitting rate or too large fluctuations in the loss rate but can jump out of the local optimum. Therefore, this paper uses the exponential decay method [11] to dynamically adjust the learning rate, i.e., first using a larger learning rate to make the loss rate drop significantly and then setting a lower learning rate to enable the model to converge more smoothly.

Dynamic learning rate calculation formula:

$$Learning_{rate} = Learning_{Rate_{Base}} * Learning_{Rate_{Decay}}^{\frac{Global_{stop}}{Learning_{Rate_{Stpe}}}} \tag{8}$$

Where *Learning_Rate_Base* indicates the learning rate base, the initial value of the learning rate; *Learning_Rate_Decay* indicates the learning rate decay coefficient; *Learning_Rate_Stpe* indicates the frequency of updating the learning rate.

3) Optimizer

In the optimization algorithm, Stochastic Gradient Descent (SGD) [12] is used in this case. The descent method of stochastic gradient with momentum can over make the model converge with better accuracy.

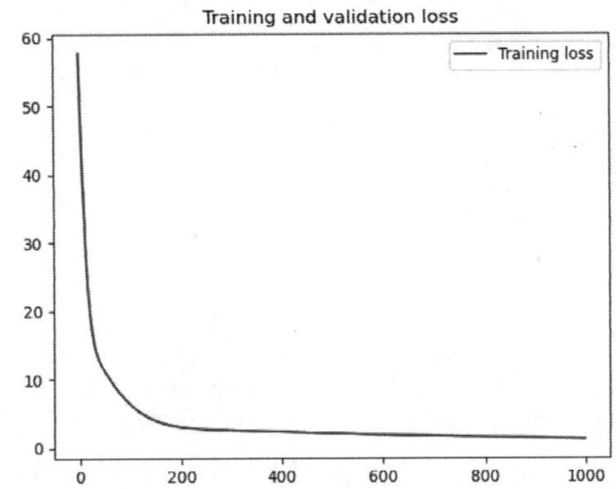

Fig.5 Loss value curve for 1000 training sessions

The training curve shown in Figure 5 shows that, with the current parameter settings, the Loss value is able to drop rapidly in the first 200 iterations, and gradually slows down until it plateaus after 200 iterations.

3.2. Experimental results

The action category is set to two labels, Stand and Fall. by capturing video information on the escalator, the images are fed into the trained model for recognition, and the posture the person is in is determined based on the position and action identified by the algorithm. The output of Figure 6 shows that the Snet-Yol model extracts the accurate coordinate position of the human body. OpenPose extracts the information of the human body joints from the acquired images, connects the joints together accurately through PAF and finally predicts the current pose of the human body.

Fig.6 Action recognition results

To test the effectiveness of the algorithm, a portion of the dataset is fed into the model as a test set for prediction, using commonly used model evaluation criteria to test how good the model is.

$$correct = \frac{TP}{TP + FP} \tag{9}$$

$$erroneous = \frac{FP}{TP + FP} \tag{10}$$

Where $correct$ is the correct rate and $erroneous$ is the error rate. TP (True Positives) is the number of correctly identified test set samples and FP (False Positives) is the number of incorrectly identified test set samples.

Table 1. Results of experimental tests

category	number of samples	Correct Number	Correct
upright	2415	2231	92.38%
trip	842	812	96.43%

The experimental results in Table 1 reflect the good overall performance of the recognition system and the high accuracy of pose recognition, with accuracy rates above 92%, indicating that the model has good robustness and can meet the demand for escalator member safety detection in practical scenarios.

4. Conclusion

For the current stage of escalator safety management which is more traditional and singular, this paper focuses on the feasibility of the OpenPose algorithm in escalator safety systems. The paper describes the implementation of Snet-Yolo for accurate recognition of human bodies, as well as the extraction of key points of human bodies by OpenPose and the connection of key points by PAF. The collected and processed dataset is trained and the new model is tested to show that the model can effectively identify whether an escalator passenger has fallen or not. The experiments show the effectiveness and feasibility of the algorithm applied to existing monitoring systems, which can greatly ensure the safety of escalator passengers, reduce the occurrence of escalator accidents and improve safety performance.

References

[1] Zhang Fei Fei, Basic situation of lift accidents and analysis of typical cases in China from 2015 to 2020, China Elevator. 2021, 32(13).(China)

[2] Li Deyuan, Research on machine vision-based safety inspection method for escalators, South China University of Technology, 2017(China)
[3] Chua J L, Lim W K.A simple vision-based fall detection technique for indoor video surveillance[J]. Signal Image and Video Processing,2015,9(3);623-633.
[4] Ni Q.W., Xu H.H.. Design of escalator monitoring communication system based on LabVIEW and Wifi [J]. Electronic Design Engineering,2018,26(12) (China)
[5] Nunez-Marcos, A, Azkune, G, Arganda-Carreras, I Vision based fall detection with Convolution Neural Networks[J]. Wireless Communications and Mobile Computing,2017(12).
[6] Gao Xinyu, Liu Xiaojian, Pedestrian detection method using Snet-Yolo framework, Information Technology and Informatization, 2021, Vol. 12. (China)
[7] Kaiming He, Xiangyu Zhang, Deep Residual Learning for Image Recognition, CVPR, 2016
[8] Simonyan K, Zisserman A, Very Deep Convolutional Networks for Large-Scale Image Recognition, Computer Science, 2014
[9] Zhe Cao, Tomas Simon, Realtime Multi-Person 2D Pose Estimation using Part Affinity Fields, Accepted as CVPR 2017 Oral.
[10] C Zhe, T Simon, SE Wei, Y Sheikh, Realtime Multi-Person 2D Pose Estimation using Part Affinity Fields[C], IEEE Conference on Computer Vision and Pattern Recognition (CVPR),2017
[11] Adams, Ralph, Testing the count rate performance of the scintillation camera by exponential attenuation: Decaying source; Multiple filters[J].Medical Physics,1988,15(3):415.
[12] Zhang T, Solving large scale linear prediction problems using stochastic gradient descent algorithms[C],International Conference on Machine Learning Omnipress 2004.

Research on Speed Control Method of Mine Hoist based on Improved Adaptive Particle Swarm

Zhang Ye[1a*], Zuo Dong-sheng[1b*],Ai Xiao-rui[1c*]

School of Electrical Engineering, Shanghai DianJi University, Shanghai 201306,China

[a*]253891469@qq.com, [b*]zuods@sdju.edu.cn, [c*]aisuperfood@qq.com

Abstract—Most of the current mine hoists use PID controllers for speed adjustment, and there are problems such as long adjustment time, excessive reliance on experience, easy overshoot and oscillation. In order to solve the above-mentioned problems, it is planned to use the adaptive particle swarm optimization based on simulated annealing (SA-APSO) algorithm to optimize the parameters of the controller, and conduct simulation experiments based on Simulink simulation. The simulation results show that the optimized control algorithm can effectively improve the control efficiency, reduce the overshoot, without vibration, and can effectively improve the speed and accuracy of the mine hoist speed control.

1. Introduction

At present, the hoist control system in coal mines in our country generally uses traditional PID (Proportion-Integral-Differential, PID) for control. Although the PID control algorithm has the advantages of simple structure and no need for detailed mathematical models, The setting of the control parameters Kp, Ki, Kd often relies on human experience, and there are disadvantages such as long adjustment time, excessive reliance on experience, easy overshoot and oscillation.

With the rapid development of computer technology and artificial intelligence technology, more and more people apply these technologies to the parameter tuning of PID control systems. Literature [1-2] applies particle swarm optimization (Particle-Swarm-Optimization, PSO) to the parameter tuning of the control system of manipulators and steam turbines. Compared with the traditional empirical tuning parameters, the system performance optimized by the PSO algorithm Significant improvement. The literature [3] realizes the adaptive control of mine hoist system through fuzzy PID controller.Although the parameter optimization schemes through these algorithms can effectively improve the dynamic performance of the system, these algorithms have a certain probability of falling into the local optimal solution when the parameters of the control system are adjusted. Control performance has an impact. In order to solve the problem of local optimal value in the process of PID parameter optimization. In this paper, the Metropolis criterion in the (Simulated Annealing, SA) algorithm is introduced into the adaptive particle swarm algorithm to avoid falling into the local optimal solution, thereby more effectively improving the speed and stability of the mine hoist speed control system sex.

2. Speed control of mine hoist based on improved adaptive particle swarm algorithm

2.1. Mathematical model of hoist speed system

Most of the current hoist speed regulation systems use frequency conversion speed regulation systems. The frequency conversion speed regulation system of mine hoists is mainly composed of motors, inverters, sensors, PLC controllers, monitoring platforms, etc. [7]. The system structure is shown in Figure 1. In the process of speed adjustment of the mine hoist, the PLC controller compares the real-time speed of the motor collected by the rotary encoder with the set speed, and then performs PID calculation on the difference between the two, and finally transmits the calculation result to the inverter. Drive the motor to run, thereby forming a closed-loop speed control.

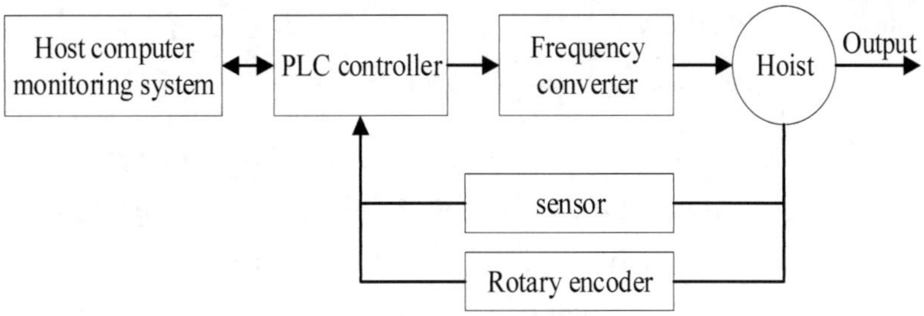

Fig.1 The composition of the speed control system of the mine hoist

The simulation three-phase asynchronous motor selected in this article is Siemens 1LE0301 three-phase asynchronous motor, the inverter model is Siemens G120C, and the acceleration time is set to 1 s. The specific parameters of the motor are shown in Table 1

Table 1 Motor specific parameter table

Rated voltage/V	Rated frequency/Hz	Motor Power /kW	Rated current/A	Rated speed r/min	Power factor	Start Time t/s
380	50	0.55	1.41	1425	0.80	0.8

the mathematical model of the asynchronous motor and the frequency converter can be approximated as a first-order inertia link. According to the above formula, the transfer function of the controlled object can be obtained as

$$G(s) = G_{MA}(s)G_{VF}(s) = \frac{142.5}{0.05s^2 + 0.45s + 1} \tag{1}$$

In the formula, G_{MA} represents the speed transfer function of the asynchronous motor; G_{VF} represents the output frequency transfer function of the inverter.

2.2 Simulated Annealing Adaptive Particle Swarm (SA-APSO) algorithm

The particle swarm optimization algorithm uses a single particle as a single individual in the bird group. Each particle has only two attributes: position and speed. Position indicates the direction of movement, and speed indicates the speed of movement. The basic idea of PSO is to find the optimal solution through collaboration and information sharing between individuals in the group [4]. The position and speed update formula is as follows

$$x_{id}^{k+1} = x_{id}^{k} + v_{id}^{k+1} \tag{2}$$

$$v_{id}^{k+1} = wv_{id}^{k} + c_1r_1(p_{id} - x_{id}^{k}) + c_2r_2(p_{gbestd} - x_{id}^{k}) \tag{3}$$

Where is x_{id}^{k} Represents the d-dimensional component of the position vector of the particle i in the k-th iteration; v_{id}^{k+1} Represents the d-th dimension component of the flight velocity vector of the particle i in the first iteration.

In order to take into account both the local search and global search capabilities, the following strategies are used to improve w、 c_1, c_2 .

Among them, the inertia factor

$$w = w_{\min} + Sigmoid(5 - 10\frac{k}{k_{\max}})(w_{\max} - w_{\min}) \tag{4}$$

Individual learning factor of particles

$$c_1 = c_{1\min} - \cos\frac{k\pi}{2k_{\max}}(c_{1\min} - c_{1\max}) \tag{5}$$

Social learning factor of particles

$$c_2 = c_{2\min} - \sin\frac{k\pi}{2k_{\max}}(c_{2\min} - c_{2\max}) \tag{6}$$

Where $i = 1,2,3...m$ Is the number of particles; $d = 1,2,3...D$ Regional dimension; p_{id} The optimal position found for particle I; p_{gbestd} The optimal position found for the particle swarm; k Is the current iteration number of the particle; r_1, r_2 Is a random number in [0,1];
$w_{\max}, w_{\min}$ Are the maximum and minimum values of the inertia factor w; $c_{1\max}, c_{1\min}$ Are the maximum and minimum values of the individual learning factor c_1 of the particle; $c_{2\max}, c_{2\min}$ Are the maximum and minimum Social learning factor of particles c_2.

The simulated annealing algorithm (Simulated Annealing, SA) is a random optimization algorithm based on Monte-Carlo iterative solution strategy. In the process of declining, the global optimal solution of the objective function is randomly searched in the solution space based on the probabilistic jump feature, that is, the local optimal solution can probabilistically jump out and eventually tend to the global optimal. Among them, Metropolis criterion is the key to SA algorithm , probability

$$p_i(k) = \begin{cases} 1 & , E_i(k) > E_b \\ e^{-\frac{E_i(k) - E_b}{T_i}} & , E_i(k) \le E_b \end{cases} \tag{7}$$

Where $E_i(k)$ ——Current solution value; T_i ——Current Temperature; E_b——Optimal value within the current population.

The Metropolis criterion can accept deteriorating solutions with a certain probability, so that the algorithm can jump out of the trap of local optimality during operation, thereby effectively avoiding falling into the local optimal solution. The optimization process of the simulated annealing adaptive particle swarm optimization algorithm is shown in figure 2

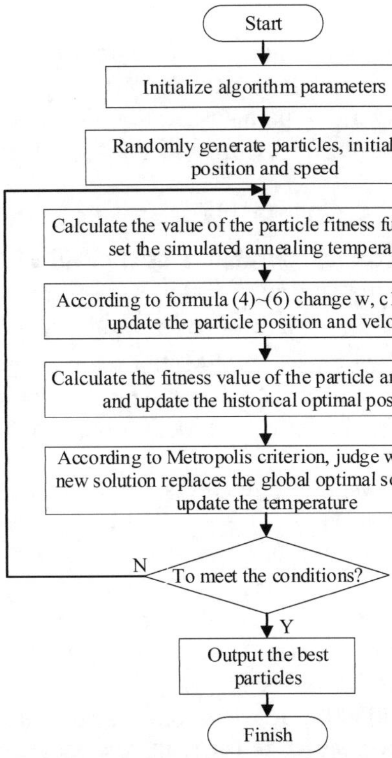

Figure 2 Simulated annealing adaptive particle swarm SA-APSO optimization process

2.3 SA-APSO algorithm optimizes fuzzy PID controller

The conventional PID control system collects the error between the given value and the actual value, and then uses the controller to control it [5], the controller outputs

$$u(t) = K_p e(t) + K_i \int_0^t e(t)dt + K_d \frac{de(t)}{dt} \tag{8}$$

Where K_p、K_i、K_d Represents the proportional, integral, and differential coefficients respectively; $e(t)$ Represents the system error.

In a conventional PID control system, the selection of proportional, integral, and differential coefficients usually requires trial and error based on experience, and once determined, online adjustment cannot be performed. Therefore, when the control system changes, the control effect may change. Poor, thus affecting the stability of the system. Therefore, the fuzzy control is introduced into the system. The fuzzy control can query the fuzzy control rule table according to the values of the deviation *e* and the deviation change rate *ec*, and then output the corresponding parameter corrections to adaptively adjust the parameters of the PID, Therefore, the SA-APSO algorithm is used to optimize the fuzzy PID controller. The main principle is to use the algorithm to find a set of optimal quantization factors and scale factors to make the system output the most stable.

The performance index of time multiplied by the absolute integral of error (ITAE) is a control system performance evaluation index with good engineering practicability and selectivity. Therefore, this paper adopts ITAE as the fitness function of evaluating system performance index, evaluation function

$$J = \int_0^{\infty} t\left|e(t)\right|dt \tag{9}$$

3. Simulation verification

According to the mine hoist speed control system model, the local mine hoist speed control simulation model is established in MATLAB/Simulink. First, the SA-APSO algorithm is established to optimize the conventional PID model, and then the algorithm parameters are set as follows: The PID parameter range is selected as [0, 100], the particle swarm size n=100, the maximum number of iterations MaxIter=50[6].

Then the parameters K_p =100, K_i =0.2561, K_d =1.124 obtained by the algorithm are brought into the fuzzy PID controller, and the SA-APSO algorithm is used again to optimize the fuzzy PID controller to obtain the best scale factor and quantization factor. The parameter settings in the fuzzy controller are as follows: input variables *e*, *ec* and output variables ΔK_p、ΔK_i、ΔK_d The fuzzy domains are all [-1,1], and the subsets are all {NB,NM,NS,ZO,PS,PM,PB}, namely {large negative, medium negative, small negative, unchanged, positive small, positive middle , positive big}, the membership function is a triangular function. The obtained control parameter optimization results are shown in Table 2.

Table 2 Parameter optimization result table of fuzzy controller

K_e	K_{ec}	ΔK_p	ΔK_i	ΔK_d
0.9944	0.6458	0.1237	0.0514	0.933

Bring the obtained parameter values into the control system, and then compare them with the optimized results of the ZN algorithm and the conventional PSO algorithm. Considering that the system will be interfered by the outside world in the actual working process, the amplitude is added to the system at 0.5 s The same pulse disturbance is used to verify the robustness of the system, as shown in Fig. 3 is the anti-jamming test result of the system under unit step input.

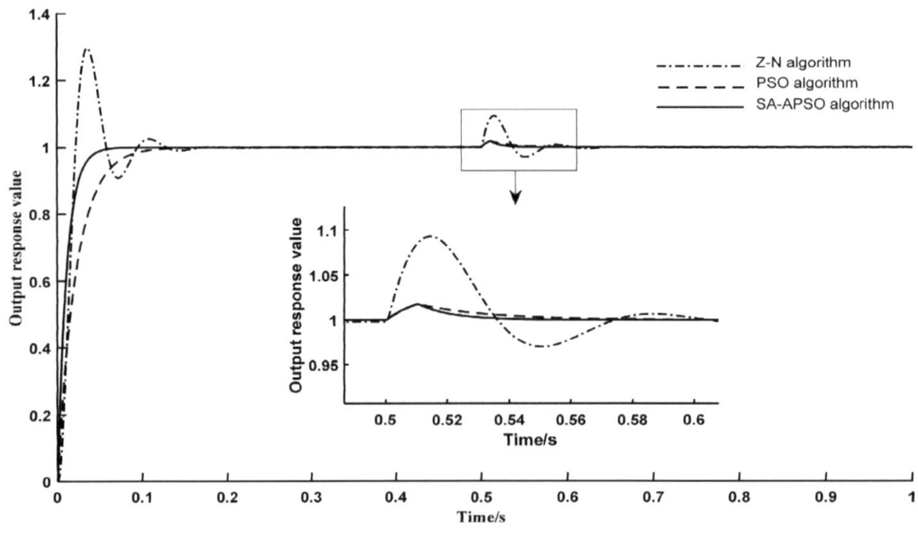

Figure 3 System robustness test results

The dynamic performance indicators of the output results of the three different methods in Figure 6 are shown in Table 3.

Table 3 Comparison table of controller output results

Control algorithm	Overshoot /%	Rise Time /s	Adjustment time /s
Z-N	29.97	0.022	0.115
PSO	0	0.049	0.089
SA-APSO	0	0.024	0.044

It can be seen from Figure 8 that the robustness of the fuzzy PID controller optimized by the SA-APSO algorithm is slightly stronger than that of the conventional PSO optimized controller, and it is obviously stronger than the control effect of the Z-N tuning method. According to Table 3, it can be seen that the controller optimized by the SA-APSO algorithm has the advantages of no overshoot and fast adjustment time.

4. Conclusion

Aiming at the problems of traditional PID control parameter tuning in mine hoist control system, this paper proposes an adaptive particle swarm optimization algorithm based on simulated annealing to optimize PID control parameters, and uses fuzzy PID control to replace conventional PID control. The simulation results show that the system parameters optimized by the algorithm can significantly improve the speed, stability and robustness of the hoist control system, The comfort level and safety and stability performance of the hoist when transporting people or things have been significantly improved, which has a certain significance for the optimization of the frequency conversion control system of the mine hoist.

References

[1]Souza Darielson A.,Batista Josias G.,dos Reis Laurinda L. N.,Júnior Antônio B. S.. PID controller with novel PSO applied to a joint of a robotic manipulator[J]. Journal of the Brazilian Society of Mechanical Sciences and Engineering,2021,43(8):

[2]Majeed Farah Ayad Abdul. Steam Turbine Governor (STG) Controller Design Based on Particle Swarm Optimization (PSO)[J]. IOP Conference Series: Materials Science and Engineering,2021,1105(1):

[3]Yan Xiang Wu,Chao Jun Zhang,Hai Bo Huo, et al. Design and Simulation of Fuzzy-PID DC Governor System Based on Mine Hoist[J]. Applied Mechanics Materials and,2013,2516(341-342):

[4]Bo Wang,Haipeng Geng,Hao Li,Wei Zheng. Particle Swarm Optimization-Based Fuzzy PID Controller for Stable Control of Active Magnetic Bearing System[J]. Journal of Physics: Conference Series,2021,1888(1):

[5]Haibo Liang,Jialing Zou,Kai Zuo, et al. An improved genetic algorithm optimization fuzzy controller applied to the wellhead back pressure control system[J]. Mechanical Systems and Signal Processing,2020,142:

[6]Mahmudur Rahman,Zhi Chao Ong,Wen Tong Chong, et al.Smart Semi-active PID-ACO control strategy for tower vibration reduction in Wind Turbines with MR damper [J].Earthquake Engineering and Engineering Vibration,2019,18(04):887-902.

Improved PD of LESO quadrotor control algorithm

Chunling Liu[a*], Jinlong Feng[b*]

College of Information Engineering, Dalian University, Dalian, Liaoning, 116622, China

[a*]liuchunling@dlu.edu.cn, [b*]1327937350@qq.com

Abstract-Aiming at the attitude instability problem caused by the quadrotor's nonlinearity, strong coupling and susceptibility to external interference, a linear extended state observer (LESO) controller with improved PD was designed. First, the incomplete differential control is formed by adding a first-order inertia system to the differential Elements of the PD algorithm. Then, LESO is used to estimate and compensate the internal and external disturbances in the control system to the original system, which can not only reduce system errors, but also enhance the anti-interference ability of the control system. The MATLAB simulation shows that the improved PD of LESO controller can quickly reach a stable state, effectively improve the stable control ability of the quadrotor's attitude, with smaller overshoot and better control effect.

1. Introduction

The four-rotor aircraft control system is a complex system with nonlinearity and strong coupling, which brings huge challenges to the control design. The classic PID design controller is widely used in the design of quadrotor control system due to its simple implementation and good anti-interference ability [1-3]. However, its ability to deal with coupling problems is poor, and its parameters cannot be adjusted adaptively [4-7].The conventional PD controller has fast optimization speed and good tracking effect, but it is particularly sensitive to high-frequency interference signals and is easy to introduce disturbances [8-10]. Aiming at the problem of attitude instability caused by the susceptibility of the quadrotor to external interference, a linear extended state observer (LESO) is designed in the feedback loop of the controller, and the improved PD control is used to enhance the dynamic characteristics of the system and improve the anti-interference ability of the system.

2. Mathematical model of quadrotor

When modeling the quadrotor, it is assumed that the airframe is a rigid body with uniform and symmetrical mass. The four rotors are rigid and will not deform, and the coordinates of the body and the center of mass coincide. M_i represents the motor of the i-th rotor. Define the relative position between the inertial coordinates E (X, Y, Z) and the body coordinates B (x, y, z) can be determined by the position coordinates. The attitude angle θ, φ, ψ of the aircraft in the ground coordinate system are pitch, roll, yaw in the airframe coordinate system. Let R(θ, φ, ψ) denote the conversion matrix between the inertial coordinate E and the body coordinate B. It can be expressed as:

$$R=\begin{bmatrix} \cos\psi\cos\varphi & \cos\psi\sin\theta\sin\varphi & \sin\psi\sin\varphi+\cos\psi\sin\theta\cos\varphi \\ \sin\psi\cos\theta & \sin\psi\sin\theta\sin\varphi & \sin\psi\sin\varphi\cos\varphi-\sin\varphi\cos\psi \\ -\sin\theta & \cos\theta\sin\varphi & \cos\theta\cos\varphi \end{bmatrix} \tag{1}$$

Non-linear motion equations are divided into translational motion and rotational motion. According to Newton's second law, the translational motion in the x, y, and z directions is expressed as : (Among them, F_i (i=1,2,3,4) is the lift of the propeller, m is the mass of the quadrotor body)

$$\begin{cases} \ddot{x}=[(\cos\varphi\sin\theta\cos\psi+\sin\varphi\sin\psi)\sum_{i=1}^{4}F_i]/m \\ \ddot{y}=[(\cos\varphi\sin\theta\sin\psi-\sin\varphi os\psi)\sum_{i=1}^{4}F_i]/m \\ \ddot{z}=(\cos\varphi\cos\theta\sum_{i=1}^{4}F_i-mg)/m \end{cases} \tag{2}$$

According to the conservation of angular momentum, the rotational motion of the quadrotor can be expressed as:(Among them I_i (i=x, y, z) is the moment of inertia of the shaft., i is the axis distance of the body. D_i (i=θ, φ, ψ) is the component of the uncertain interference in the environment decomposed to the coordinate axis)

$$\begin{cases} \ddot{\varphi}=[(I_y-I_z)\dot{\theta}\dot{\psi}+l(F_4-F_2)-D_\varphi]/I_x \\ \ddot{\theta}=[(I_z-I_x)\dot{\varphi}\dot{\psi}+l(F_3-F_1)-D_\theta]/I_y \\ \ddot{\psi}=[(I_x-I_y)\dot{\theta}\dot{\varphi}+(F_2+F_4-F_1-F_3)-D_\psi]/I_z \end{cases} \tag{3}$$

The virtual control variables U_1, U_2, U_3, and U_4 are four independent inputs, which are transformed by lift forces F_1, F_2, F_3, and F_4 respectively. The transformation formula can be expressed as:

$$\begin{bmatrix} U_1 \\ U_2 \\ U_3 \\ U_4 \end{bmatrix}=\begin{bmatrix} 1 & 1 & 1 & 1 \\ 0 & -1 & 0 & 1 \\ -1 & 0 & 1 & 0 \\ -1 & 1 & -1 & 1 \end{bmatrix}\begin{bmatrix} F_1 \\ F_2 \\ F_3 \\ F_4 \end{bmatrix} \tag{4}$$

In summary, the dynamics model of the quadrotor UAV aircraft system can be expressed as:

$$\begin{cases} \ddot{x}=[(\cos\varphi\sin\theta\cos\psi+\sin\varphi\sin\psi)U_1]/m \\ \ddot{y}=[(\cos\varphi\sin\theta\sin\psi-\sin\varphi\cos\psi)U_1]/m \\ \ddot{z}=(\cos\varphi\cos\theta U_1-mg)/m \\ \ddot{\varphi}=[(I_y-I_z)\dot{\theta}\dot{\psi}+lU_2-\mathrm{D}_\varphi]/I_x \\ \ddot{\theta}=[(I_z-I_x)\dot{\varphi}\dot{\psi}+lU_3-\mathrm{D}_\theta)]/I_y \\ \ddot{\psi}=[(I_x-I_y)\dot{\theta}\dot{\varphi}+U_4-\mathrm{D}_\psi]/I_z \end{cases} \tag{5}$$

3. Design of LESO controller for improved PD

3.1 Improved PD

The first-order inertial link $D_f=1/(1+T_f(s))$ is added to the conventional PD algorithm. The structure of the improved PD controller is shown in Figure 1:

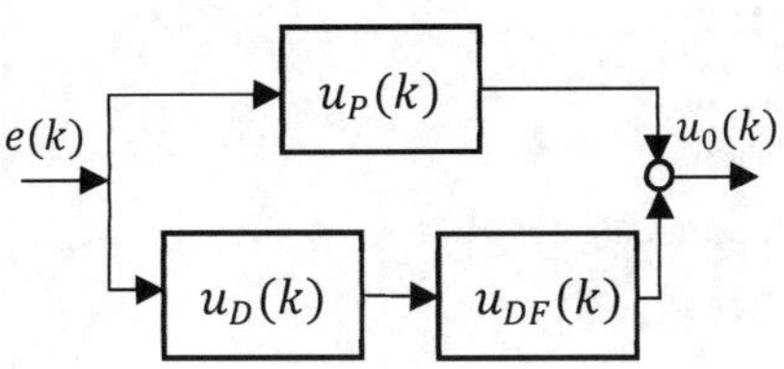

Fig.1 improved PD controller.

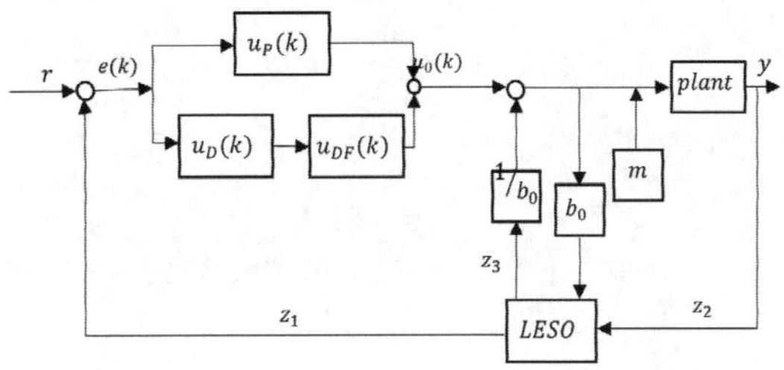

Fig.2 LESO controller of improved PD.

The control law of the improved PD controller can be expressed as: (In the formula, $u_P(k)$ is proportional control, $u_D(k)$ plus $u_{Df}(k)$ is differential control, and e(k) is the system error.)

$$u_0(k) = u_P(k) + u_D(k) + u_{Df}(k) \tag{6}$$

After adding the first-order inertia to the differential link, the differential equation form can be expressed as:

$$u_D(t) + T_f \frac{du_D(t)}{dt} = k_P T_D \frac{de(t)}{dt} \tag{7}$$

Discretize the formula (6), which can be expressed as:(T_f is the first-order inertia constant, T is the sampling period)

$$u_D(k) + T_f \frac{u_D(k) - u_D(k-1)}{T} = k_p T_D \frac{e(k) - e(k-1)}{T} \tag{8}$$

After finishing formulas (7and 8), we can get:(Where $a=T_f/(T_f+T)$, $a<1$)

$$\begin{aligned} u_0(k) &= k_p e(k) + u_D(k) + u_{Df}(k) \\ &= k_p e(k) + k_d(1-\alpha)[e(k) - e(k-1)] + \alpha u_d(k-1) \end{aligned} \tag{9}$$

The LESO controller block diagram of the improved PD is shown in Figure 2:

3.2 Linear extended state observer(LESO)

The linear expansion state observer is the core component of the linear self-anti-interference. LESO is the core component of linear self-interference. Its working principle is to expand the various disturbances in the system into a new state variable. Then the total disturbance is estimated by the system control law and output dynamics, and the total disturbance term is fed back to the original system. This not only solves the internal interference caused by the uncertain part of the quad-rotor model, but also enhances the robustness of the control system.

The design steps of LESO control algorithm based on improved PD are as follows:

(i) The nonlinear mathematical model of the quadrotor is established through Euler equations and conservation of angular momentum. The nonlinear, coupled and unknown parts of the object model are selected as the disturbance term and the dynamic linearization model is established.

(ii) According to the linearization model, the control rate of the improved PD controller is designed.

(iii) Select appropriate parameters to enable LESO to better estimate the total disturbance term and dynamically compensate it to the original system. Through the output value of the improved PD controller, the real-time control value is obtained.

4. Simulation results and analysis

Matlab2016a is used to simulate the quadrotor aircraft control system. The physical parameters of the quadrotor are shown in Table 1, and the simulation parameters are shown in Table 2:

Table 1. physical parameters of quad rotor aircraft.

sign	value	meaning
m/kg	1.2	Mass
l/m	0.2	Axis distance
g/(kg)	9.8	Gravitational acceleration
Ix/($kg \cdot m^{-2}$)	2.353E-3	x moment of inertia
Iy/($kg \cdot m^{-2}$)	2.353E-3	y moment of inertia
Iz/($kg \cdot m^{-2}$)	3.626E-3	z moment of inertia

Table 2. simulation parameters.

sign	value	meaning
T/s	0.01	Sampling period
T_f	1/27	First-order constant of inertia
b_0	10	Constant
w	50	Observer bandwidth
β_1	150	Observer parameters
β_2	7500	Observer parameters
β_3	125000	Observer parameters

The flight path of the four rotor is set from (3,2,0) to (0,0,5). During the flight along the trajectory, a disturbance is added to the roll angle to simulate the internal and external disturbances of the quadrotor during the flight loop. The disturbance is added at 3s, and the disturbance is set to 0.5*(sign(sin*(0.5t))+cos*(0.6t)+cos*(0.7t)+0.5sign(sin*(t))). The simulation is as follows:

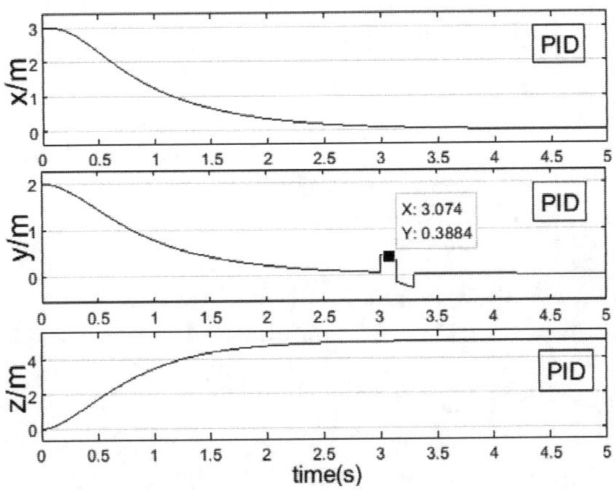

Fig.3 PID position adjustment.

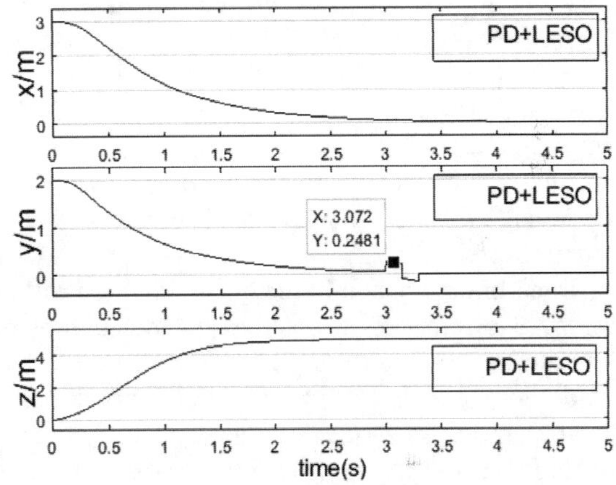

Fig.4 LESO position adjustment of improved PD

It can be seen from Figure 3 and Figure 4. After adding disturbance to the roll angle at 3s, the amplitude of the PID control roll channel position is 1.6 times that of the improved PD of LESO controller. And the former has a steeper amplitude jump. In addition, the comparison of the pitch and yaw angles of the LESO controller controlled by PID and improved PD shows that the latter has shorter adjustment time and smoother amplitude adjustment. It can be seen from the comparison that when the quadrotor is disturbed in flight. The improved PD of LESO controller has better position adjustment ability than the classic PID controller.

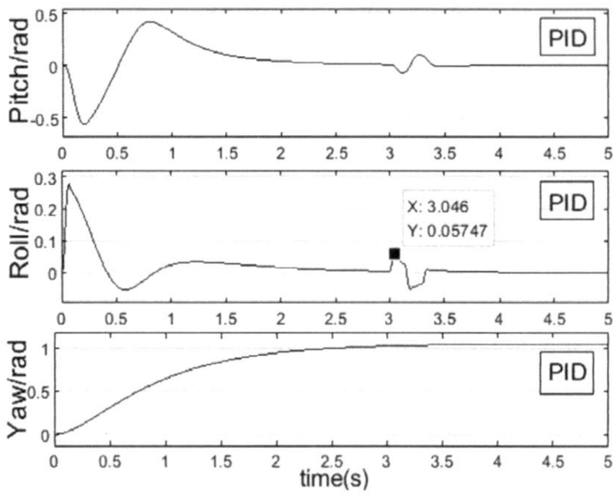

Fig.5 PID attitude adjustment.

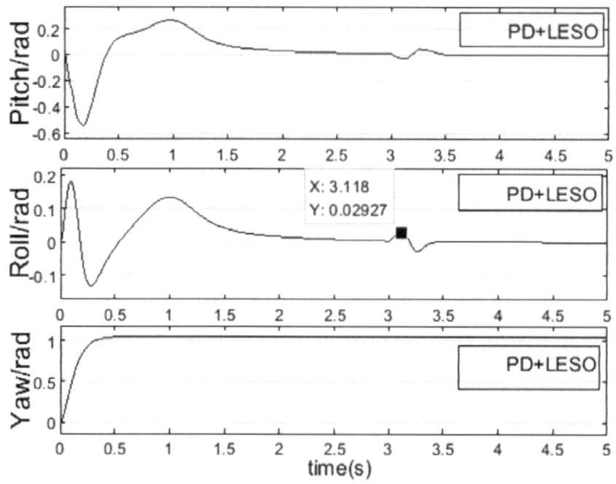

Fig.6 LESO attitude adjustment of improved PD.

From Figure 5 and Figure 6. It can be seen that after disturbance is added at 3s, the attitude adjustment amplitude of PID control roll channel is 0.057rad, and the adjustment trend is steeper after interference. Although the pitch angle adjustment is gentle, the adjustment range is large; The yaw angle adjustment time is about 2.5s. Improved PD of LESO controller roll channel attitude adjustment amplitude after interference is about half of PID controller, and the adjustment trend is gentler after interference. Moreover, the pitch angle adjustment is relatively gentle, and the adjustment amplitude is far less than that of PID control. The yaw angle adjustment time is about 0.5s, and the response time is much shorter than PID.

In addition, the amplitude of the PID control roll angle attitude is about 0.28 rad, and it is relatively steep. The pitch angle is about 0.4 rad before reaching a plateau. The amplitude of the roll angle attitude of the improved PD of LESO controller is about 0.18 rad, and it is relatively gentle. The amplitude of the pitch angle is about 0.25 rad before reaching a stable level.

It can be seen that the improved PD controller combined with LESO to compensate for disturbances. It can significantly improve the attitude angle resolution capability of each channel and reduce amplitude jumps. The simulation experiment results show that the designed controller can significantly improve the anti-interference ability of the quad-rotor rotation motion and enhance the robustness of the system.

5. Conclusion

An improved PD linear extended state observer (LESO) controller is designed. Simulation shows that the controller has a fast response time and can reduce the steady-state error of the system. In the future, the effectiveness of the controller can be verified through actual four-rotor UAV flight experiments, which will further verify its application in the field of flight control.

Acknowledgments

This work was financially supported by the Fund of Educational Department of Liaoning Province(LJKZ1184).

References

[1] WANG,H, WANG,J.H, LIU,H.X. (2020) Research on MAC-PID position control of quadrotor aircraft[J].Electronics Optics&Control,27(1):79-84.

[2] ZHAO,C.L, ZHAO,G.R, DAI,S.W, HAN,X. (2020)Flight Control of Quadrotor UAV Based on Incomplete Differential and Differential Forward PID Algorithm[J]. Computer Simulation,37(11):48-53+103.

[3] LU,X.H, ZHAN,S.W, YU,W.Q, HUANG,C.W. (2017)Flight attitude angle control and anti-chatter algorithm for UAV based on improved PID[J]. Transducer and Microsystem

Technologies, 36(03):116-119.
[4] ZHEN,K.Q, WANG,J.J. (2019) Parameter Optimization Control of Quadrotor Aircrafts Based on PD Controller[J]. Journal of Hangzhou Dianzi University (Natural Sciences),39(04):58-65.
[5] DING,S.B, XIAO,C.S, LIU,J.G, WEN,Y.Q. (2015)Modeling and Quaternion Control of X-Type Quadrotor[J]. Journal of System Simulation, 27(12):3057-3062.
[6] ZHANG,Y. (2020)Trajectory Tracking Control Method of VTOL Aircraft Based on Double Closed-loop PD Control[J]. Automation & Instrumentation, 33(04):26-29.
[7] HU,K.M, LI,Y.Z, LIU,W. (2020)Control Algorithm of the Quadrotor Aircraft Based on Kalman Filter[J].Journal of Detection&Control,42(01):50-55.
[8] JI,Y.Y, CHEN,Y.D, CHEN,M.Y.(2017) Improved Complementary Filter Based on PID Algorithm[J]. Control Engineering of China, 24(05):952-957.
[9] LI,Y.Y, WANG,J ,LI,Z.Y.(2020)A fuzzy PID control algorithm for tracking and mapping of four rotor UAV[J].Bulletin of Surveying and Mapping,(S1):1-5+12.
[10] DING,L, WANG,Z.W. (2018)A Robust Control for an Aerial Robot Quadrotor under Wind Gusts[j]. Journal of Robotics,8(1):1–8.

Non-feedback control of Lorenz chaotic system

Yang Zhao[1a*], **Shiyan Sun**[1b*], **Ping Wei**[1c]

[1]College of Ordnance Engineering, Naval University of Engineering,Wuhan 430033,China

[a*]1573380179@qq.com, [b*]540850326@qq.com, [c]597568519@qq.com

Abstract—Non-feedback chaos control of Lorenz chaotic system is realized with non-resonant parametric drive. The control input is a periodic signal whose frequency is far higher than the system characteristic frequency. The feasibility of this scheme is proved by averaging method and Lyapunov method, and the condition of control parameter is obtained. Numerical researches show that the controlled system can reach stable state rapidly even when the control signal is noisy.

1. Introduction

In 1990, Ott et al. proposed a chaos control method using state feedback for parameter perturbation, which laid a theoretical foundation for the feedback control of chaotic systems[1]. In the same year, Lima et al. proposed a non-feedback chaos control method based on parametric resonance perturbation principle, and realized chaos suppression in Duffing system[2]. Since then, the control and application of chaos have been widely concerned by researchers in various fields, and a series of research results have been obtained by using modern control theory[3-10].Since the feedback control of chaos has perfect theoretical basis and is convenient for quantitative analysis and design, most of the control strategies proposed in recent years are feedback control methods, such as adaptive control, delayed feedback control, tracking control, fuzzy control, etc. Li L. X. et al. proposed the concept of different structure synchronization to realize the tracking of various reference signals by discrete chaotic system[11]. Yang X. L. et al. gives a tracking control scheme for Rossler chaotic system, which can make a variable of the controlled system track any reference signal including chaotic signal[12].

Feedback control can make the system stable and equilibrium points or unstable periodic trajectories, and even track arbitrary reference signals, but it also has some shortcomings in some applications. Feedback control can stabilize the system at equilibrium point or unstable periodic orbit, and even track any reference signal, but there are some shortcomings in some applications. First of all, the premise of feedback control is that the state variables of the system must be physically measured, which is often difficult to achieve in engineering applications, and sometimes some state variables are unobservable. Secondly, the chaotic system is extremely sensitive to the initial value, and the state measurement process will inevitably introduce external noise, thus affecting affect the precise control of the system[13-16].

Non-feedback control does not need to measure the state variables. The control method is simple and feasible, and has unique advantages, which has been successfully applied in power systems and other fields. In recent years of research on non-feedback control of chaotic systems, the parameter resonance perturbation method and the periodic excitation method are mostly used in the control strategies. Such methods are generally applied to non-autonomous chaotic systems, and the control effect is obvious. A small control input can significantly change the system state. However, due to the lack of a strict theoretical basis, the research progress is generally slow. In addition, the bifurcation characteristics of some chaotic systems under resonant disturbance are quite complex, and are closely related to external noise. Therefore, in practical applications, the control parameters of such methods are difficult to be calculated through theoretical analysis, and their selection often relies on experience and trial, which is not conducive to popularization and application. Using non-resonant parameter excitation to realize chaos control is another non-feedback chaos control method proposed in recent years. Its feasibility has been verified by experiments, and the control parameters can be solved by strict theoretical analysis. In this paper a non-feedback control model of Lorenz chaotic system is designed based on this control principle. By applying non-resonant parameter excitation, the system transforms from chaotic state to stable state. The feasibility of the control is proved by averaging method and Lyapunov method, and the conditions for the control parameters are obtained. Numerical research shows that this method is feasible and has strong anti-interference performance.

2. Description of control systems

The mathematical model of Lorenz chaotic system is

$$\begin{cases} \dot{x} = ay - ax \\ \dot{y} = cx - xz - y \\ \dot{z} = xy - bz \end{cases} \tag{1}$$

Which is chaotic when $a = 10$, $b = 8/3$, $c = 28$. Each state variable is regarded as a harmonic signal with continuous phase and random envelope change, and the characteristic frequency of the system is defined as the average change rate of the signal phase, namely

$$\omega_0 = \lim_{T \to \infty} \frac{2\pi N(T)}{T}$$

In the above formula , $N(T)$ is the number of peaks or troughs of the signal in time T . The characteristic frequency of system (1) is $\omega_0 \approx 12\,\text{rad/s}$. Because the change of parameters a in system 1 has a great influence on the system behavior, a parametrically excited non-feedback control system can be designed as follows

$$\begin{cases} \dot{x} = ay - a\left(1 + k\cos\omega t\right)x \\ \dot{y} = cx - xz - y \\ \dot{z} = xy - bz \end{cases} \tag{2}$$

In system (2), k is the only undetermined control parameter, and excitation signal frequency ω is much higher than system characteristic frequency ω_0. Lorenz chaotic system can be stabilized at periodic state or equilibrium point by adjusting parameter k 。

3. Determination of control parameters

The averaging method is used to analyze the dynamic characteristics of the controlled system。The averaging method is an approximate analytical method for solving nonlinear vibration equations. The basic idea is that the solution of the nonlinear vibration system has two time scales of fast-varying and slow-varying. In the fast-varying period, it can be regarded as a simple harmonic vibration with constant amplitude and initial phase angle, but its amplitude and initial phase angle will vary slowly in a long time. That is, the solution of the system is a simple harmonic vibration with slow-varying envelope. The state variables in the controlled system (2) are decomposed into slow variables with the variation of system characteristic frequency ω_0 and fast variables with the variation of excitation frequency ω with external parameters. The slow variables describe the long-term trend of system evolution, while the fast variables represent the local oscillation form of system variables. Therefore, the control objective can be achieved as long as the slow-varying system is globally stable. The system variables are expressed as

$$\begin{cases} x = x_s + x_q \\ y = y_s + y_q \\ z = z_s + z_q \end{cases} \tag{3}$$

Variable subscripts q and s represent fast-varying system and slow-varying system。Suppose that the average value of the fast variable in period $T = 2\pi/\omega$ is zero, and its amplitude is much smaller than that of the slow variable. Let operator $\langle \cdot \rangle = \frac{1}{T}\int_t^{t+T} (\cdot)dt$. The formula (3) is substituted into the system (2), and each slow variable is regarded as a constant in the period T. The evolution equation of the slow variable is obtained by integrating the system variables with the averaging method.

$$\begin{cases} \dot{x}_s = ay_s - ax_s - \langle akx_q \cos \omega t \rangle \\ \dot{y}_s = cx_s - x_s z_s - y_s \\ \dot{z}_s = x_s y_s - bz_s \end{cases} \tag{4}$$

In systems (4), the higher-order small quantity with integral not zero is omitted. By comparing systems (2)and(4), the evolution equation of fast variable x_q can be obtained.

$$\dot{x}_q = ay_q - ax_q - akx_s \cos \omega t \tag{5}$$

Since the amplitude of y_q is much smaller than that of x_s, omit the ay_q in equation(5).The Solution of equation(5) is

$$x_q = \frac{-akx_s}{\omega^2 + a^2}(\omega \sin \omega t + a \cos \omega t) \tag{6}$$

Substituting equation(6) into system(4), and using the following integral relationship

$$\langle \cos^2 \omega t \rangle = 0.5, \quad \langle \cos \omega t \sin \omega t \rangle = 0$$

the slow-varying system equation is

$$\begin{cases} \dot{x}_s = ay_s - a_e x_s \\ \dot{y}_s = cx_s - x_s z_s - y_s \\ \dot{z}_s = x_s y_s - bz_s \end{cases} \tag{7}$$

The expression of a_e is

$$a_e = a - \frac{a^3 k^2}{2(\omega^2 + a^2)}$$

The Lyapunov function of system (7) is constructed as follows :

$$V = \frac{1}{2}\left(x_s^2 + y_s^2 + z_s^2\right)$$

Take the derivative of the above formula to get the following formula

$$\begin{aligned} \dot{V} &= x_s \dot{x}_s + y_s \dot{y}_s + z_s \dot{z}_s \\ &= -a_e x_s^2 - y_s^2 - bz_s^2 + (a + c)xy \\ &= -[x_s \quad y_s \quad z_s] A [x_s \quad y_s \quad z_s]^T \end{aligned}$$

The expression of matrix A is

$$A = \begin{bmatrix} a_e & -\frac{a+c}{2} & 0 \\ -\frac{a+c}{2} & 1 & 0 \\ 0 & 0 & b \end{bmatrix}$$

when $a_e \geq \frac{(a+c)^2}{4}$, Matrix A is positive definite matrix and $\dot{V} < 0$. Therefore, the system (7) is asymptotically stable at the origin. Thus, the condition that the control parameter k should satisfy is

$$k < \sqrt{\frac{(\omega^2 + a^2)(4a - a^2 - 2ac - c^2)}{2a^3}} \tag{8}$$

4. Numerical research results

When $a = 10$, $b = 8/3$, $c = 28$, the characteristic frequency of control system(2) is $\omega_0 \approx 12\,rad/s$. Take the parameter excitation frequency $\omega = 100rad/s$. According to formula(8), the condition that the control parameter should be satisfied when the controlled system is stable is $k < 84i$. The initial value of the system is $x(0) = 0$, $y(0) = 2$, $z(0) = 9$, the following is the numerical results obtained by solving system (2) using fourth - order Runge -Kutta method when k takes different values and the control input is disturbed by noise.

4.1 The controlled system is stable at the equilibrium point

Take control parameter k=84i, the control is applied when $t = 10s$. The original chaotic state of the system disappears, and the phase trajectory quickly approaches the origin. Variable x varies with time as shown in Figure 1. It can be seen that the system can be stabilized at the origin in about 5 seconds after parameter excitation control. The simulation results show that the actual control parameter value is slightly larger than the theoretical prediction value, which is because the influence of the high-order term is ignored in the derivation process. Therefore, the result of formula (8) is an approximate estimate that is easy to calculate.

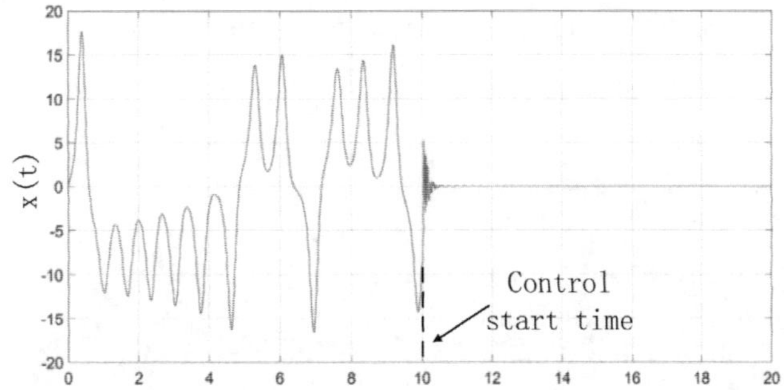

Fig.1 Variation of x over time when k=84i

4.2 The controlled system is stable in periodic orbit

Take control parameter k=70i .The chaotic state of the system disappears and stabilizes in periodic orbit after about 8s transient process, and the signal frequency contains a limited number of certain frequency components. The x-z-y phase diagram of the system is shown in Fig. 2 (the transition state has been omitted). The spectrogram of the system is shown in Fig. 3. It shows that chaos has been suppressed. Due to the large amplitude of external excitation, this orbit is not an unstable periodic orbit in chaos, but a periodic oscillation determined by the external excitation and the characteristics of Lorenz system.

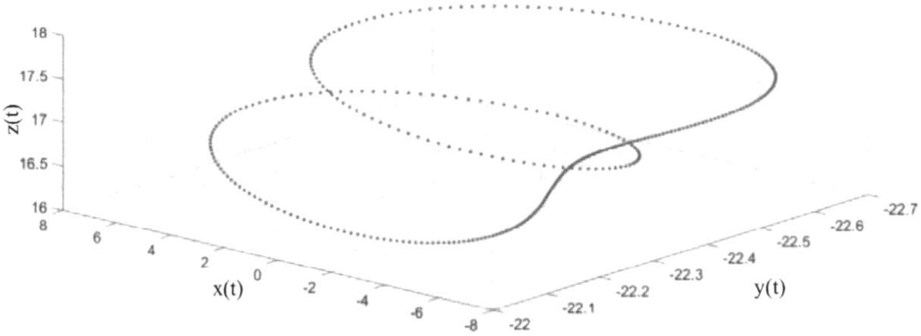

Fig.2 Phase diagram of x-y-z when $k=70i$

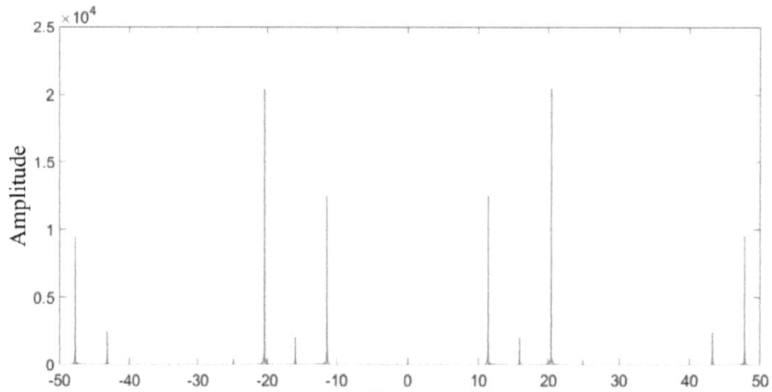

Fig.3 Spectrum

4.3 Control effect of control input disturbed by noise

In practical applications, the control input is often affected by external noise. In order to analyze the anti-interference performance of the control system, when the parameter $k=84i$, the control signal u (t) with signal-to-noise ratio of 20 dB is taken as the input, and its waveform is shown in Fig. 4.

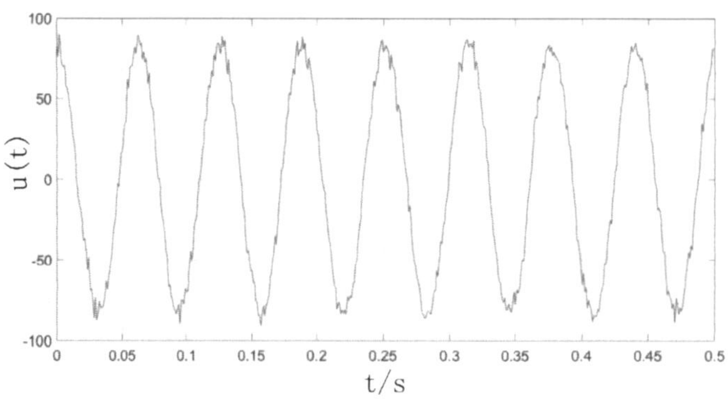

Fig. 4 Waveform of control input u(t) under noise interference

The control is applied when $t = 10s$. The system can still be quickly stabilized at the origin, where the state variable x changes over time as shown in Fig.5. Compared with the control effect without noise disturbance, the time required to stabilize the system at the origin is basically the same, and the difference is only that the fluctuation of the variable transition state is slightly larger when the control input is noisy, indicating that the anti-interference performance of the control system is good.

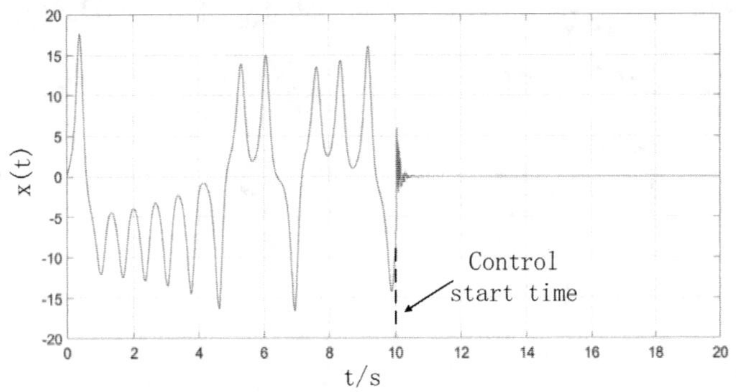

Fig. 5 Variation of x with time when the control input contains noise

5. Conclusion

Based on the principle of non-resonant parametrically excited chaos control, a non-feedback control model of Lorenz chaotic system is designed. The feasibility of the control scheme is proved by means of averaging method and Lyapunov method, and the conditions that the control parameters should meet are obtained. Numerical studies show that the theoretical analysis results of the control parameters are accurate. By selecting appropriate parameters, the system can be stabilized at the equilibrium point and periodic orbit, and has strong anti-interference performance. Compared with the parametric resonance perturbation method, this method can give strict theoretical analysis results, and the deficiency is that the disturbance to the original system is large.

References

[1] Ott E. , Grebogi C. , Yorke J. A. (1990) Controlling Chaos, Phys. Rev. Lett. ,64:1196–1199.
[2] Pecora L. M., Caroll T. L. (1990) Synchronization in Chaotic Systems, Phys. Rev. Lett., 64: 821-823.
[3] Chen G. R., Ueta T. (1997)Yet another chaotic attractor, International Journal of Bifurcation and Chaos ,9(7):1465-1466.
[4] Lu J. H., Chen G. R. (2002) A new chaotic attractor coined, International Journal of Bifurcation and Chaos, 12(3):659-661.
[5] Wang H. J., Li X. Y. (2017) Infinitely many heteroclinic orbits of a complex Lorenz system, International Journal of Bifurcation and Chaos, 27:1750110.
[6] Chen L., Wang D. S. (2007) Detection of weak square wave signals based on the chaos suppression principle with non-resonant parametric drive, Acta Phys. Sin.,56(9): 5098-5102.
[7] Chen L., Wang D. S. (2007) Adaptive tracking control of the Chen system, Acta Phys. Sin., 56(10):5661-5664.
[8] Yu H. J. (2005) Controlling chaos using time-delay nonlinear feedback, Acta Phys. Sin.,54(11):5053-5057.
[9] Wang X. Y., Shi Q. J. (2005) Tracking control and synchronization of the Rössler chaotic system, Acta Phys. Sin., 54(12):5591-5596.
[10] Zhou Y. F., Chen J.N., Xie Z. G. et al. (2004) Application of resonant parametric perturbation to the chaos control in Boost converter and its optimization, Acta Phys. Sin. 53(11):3676-3683.
[11] Zhou Y. F., Chen J. N. (2001) The Adaptive control of Chen's chaotic system, Acta Phys. Sin., 50(11):2108-2111.

[12] Li L. X., Peng H.P., Luo Y. X. et al. (2013) Problem and analysis of stability decidable theory for a class of fractional order nonlinear system, Acta Phys. Sin. 62(2):020502.
[13] Yang X. L., Xu W., Sun Z. K. (2006) Influence of harmonic and bounded noise excitations on chaotic motion of Duffing oscillator with homoclinic and heteroclinic orbits, Acta Phys. Sin. 55(4):1678-1686.
[14] Dong C. W. (2018) Periodic orbits of diffusionless Lorenz system, Acta Phys. Sin. 67: 240501.
[15] Li B. S., Ding R. Q., Li J. P. (2017) Predictability of forced Lorenz system, Acta Phys. Sin.,66:060503.
[16] Yang T., Yang Q. G.(2019) A 3D autonomous system with infinitely many chaotic attractors, International Journal of Bifurcation and Chaos 29(12):1950166.

Architecture design of battery energy storage coordinated control system based on Multi-Agent mechanism

Xuan Qiu[1a*]**, Chun-sheng Tang**[1b*]**, Wen-hao Dong**[1c]

[1] College of Automation, Nanjing University of Posts and Telecommunications, Nanjing 210023, China

[a*]qx199901@163.com, [b*]tcs2357419378@163.com, [c]dongwh123456@163.com

Abstract—Energy storage can realize rapid voltage, frequency support and energy adjustment, and is an important guarantee for promoting the efficient consumption of renewable energy and the safe and stable operation of the power system. The flexible adjustment of energy storage is not only affected by its own characteristics, but also closely related to the operation control mode of energy storage. Based on the PQ constant power and virtual synchronization control strategy of the battery energy storage system, this paper constructs the operation architecture of the battery energy storage system based on the Multi-Agent cooperative mechanism, and further gives the energy storage system Multi-Agent cooperative control system's application scenarios in active frequency modulation and reactive voltage regulation. Research shows that this architecture helps to fully tap the potential of aggregation and regulation of multi-point decentralized energy storage facilities and meet the flexible regulation needs of the power system under the access of high-permeability intermittent renewable energy.

1.Introduction

With the large-scale integration of distributed renewable energy such as photovoltaics and wind power into the grid, the intermittent, random and volatility of its output will bring negative impacts and impacts on the safe and stable operation of the grid [1-3]. Battery energy storage can achieve rapid voltage, frequency support [4-5] and energy adjustment, and is an important guarantee for promoting the efficient consumption of new energy and the safe and stable operation of the power system.

The flexible adjustment of energy storage is not only affected by its own characteristics, but also closely related to the operation control mode of energy storage. Literature [6-7] divides energy storage frequency modulation into auxiliary traditional power frequency modulation and demand-side distributed energy storage cluster frequency modulation according to operating modes. Aiming at the frequency and voltage regulation of energy storage converters, literature [8-9] proposed several improved methods for converter droop control.

Aiming at the operation control of energy storage converter and the overall system, this paper takes the Multi-Agent mechanism as the core, elaborates the operation strategy of energy storage PQ constant power control and virtual synchronization control in detail, and builds the Multi-Agent cooperative control architecture of battery energy storage system. The two-way flow of information ensures the transient active response of the local layer and the stable and reasonable distribution of the coordination layer, and provides the application scenarios of Multi-Agent collaborative control in active frequency modulation and reactive voltage regulation, which is more important for subsequent energy storage. The realization of the Agent cooperative control architecture provides a reference scheme.

2. Basic Theory and Structure of Multi-Agent System

2.1Basic concept of Multi-Agent System

Agent refers to an independent individual who can complete a specific task. In a complex environment, the Agent can perceive the surrounding environment information, and solve problems through judgment, planning, and decision-making methods based on the current state, relevant knowledge and rules, and has a certain degree of autonomy and self-adaptation.

The general Agent structure includes a perceptron module, a reasoning calculation module, a rule module and a communication module. Agent plays a role through the coordination between the four modules. The perceptron module perceives the information of the external environment while updating internal data and passing it to the inference calculation module; the inference calculation module is the core module of the Agent and is responsible for obtaining information from the other three modules, make intelligent decisions through inference and calculation, and then return the information to the other three modules; the rule module stores the rules and operation steps that determine the task execution; the function of the communication module is to set the information transfer mode between the current Agent and other Agents, transfer information, and realize the coordination work between the current Agent and other Agents.

Multi-Agent System (MAS) is composed of multiple Agents that are independent of each other in spatial locations and software networks, and have communication capabilities and reasoning learning capabilities, and are interconnected systems with more powerful processing capabilities. MAS can resolve the conflict between the constraints of the problem to be solved and the objective function by coordinating the competition and cooperation between the Agents.

2.2 Multi-Agent System architecture

The structure of MAS mainly refers to the communication and control mode between Agents, which determines the pros and cons of the entire system. Generally speaking, the structure of MAS includes: centralized structure, distributed structure and hybrid structure.

Hybrid MAS contains the excellent characteristics of centralized and distributed structures, and can be used well in complex environments, it has one or several management control Agents. The management control Agent only performs overall management and control on some Agents under certain rules, and participates in and realizes the analysis, planning and commands between these Agents. Other Agents at the same level are equal to each other, and they can make optimal decisions through competition and cooperation. Hybrid MAS is a commonly used MAS structure.

3. Battery Energy Storage Control Strategy

3.1PQ charge and discharge control

The constant power control method is also called the PQ control method, which reduces load peaks and fills load valleys by maintaining a constant power output of energy storage. The design ideas[10] are as follows:

The relationship between the three-phase voltage u and amplitude U_m of the converter output after the dq coordinate transformation is shown in formula (1),

$$\begin{bmatrix} u_d \\ u_q \end{bmatrix} = \begin{bmatrix} \cos\omega t & \cos\left(\omega t - \frac{2\pi}{3}\right) & \cos\left(\omega t + \frac{2\pi}{3}\right) \\ -\sin\omega t & \sin\left(\omega t - \frac{2\pi}{3}\right) & -\sin\left(\omega t + \frac{2\pi}{3}\right) \end{bmatrix} \begin{bmatrix} U_m \cos(\omega t) \\ U_m \cos\left(\omega t - \frac{2\pi}{3}\right) \\ U_m \cos\left(\omega t + \frac{2\pi}{3}\right) \end{bmatrix} = \begin{bmatrix} U_m \\ 0 \end{bmatrix} \tag{1}$$

It can be seen from the above formula that u_d is a constant and u_q is equal to 0.

Assuming that i_d、i_q are the d-axis and q-axis components of the converter output current in the dq rotating coordinate system, the instantaneous active and reactive power expressions of the converter in this coordinate system are shown in formula (2):

$$\begin{cases} P = u_d i_d + u_q i_q \\ Q = u_q i_d - u_q i_q \end{cases} \tag{2}$$

Assuming that P_{ref} and Q_{ref} are the given reference active and reactive power respectively, according to (1) and (2), the reference current of the converter can be obtained as:

$$\begin{cases} i_{dref} = \dfrac{P_{ref}}{u_d} \\ i_{qref} = -\dfrac{Q_{ref}}{u_d} \end{cases} \tag{3}$$

Since u_d is a constant, the active and reactive power output by the converter can be controlled by controlling the d-axis and q-axis components of the current. The active current i_d determines the active power P, and the reactive current i_q determines the reactive power Q. So as long as the reference value of d-axis and q-axis current is tracked, the tracking of active power and reactive power can be realized.

3.2Virtual synchronization control

The Virtual Synchronous Generator (VSG) control strategy uses the inverter to simulate the active frequency droop output characteristics and the reactive voltage droop output characteristics of the synchronous generator, and at the same time considers the synchronous motor stator voltage equation and rotor motion equation. The energy storage converter can simulate the external characteristics of the synchronous machine, as shown in the following formula:

$$\begin{cases} J\dfrac{d(\omega-\omega_N)}{dt} = \dfrac{1}{\omega_N}(\mathrm{P}_{ref}-P_e) - D(\omega-\omega_N) \\ \omega = \dfrac{d\theta}{dt} \end{cases} \tag{4}$$

Where: ω is the electrical angular velocity; ω_N is the synchronous electrical angular velocity; θ is the electrical angle; J is the virtual moment of inertia; D is the damping coefficient; P_{ref} is the active power reference value; P_e is the VSG output (instantaneous) active power; among which torque the relationship between and power is approximately $T_e = P_e / \omega_N \approx P_e / \omega$.

Adding traditional synchronous motor speed control to power electronic control is an important part of VSG. The resulting active power-frequency control block diagram of the virtual synchronous generator is shown in the figure below.

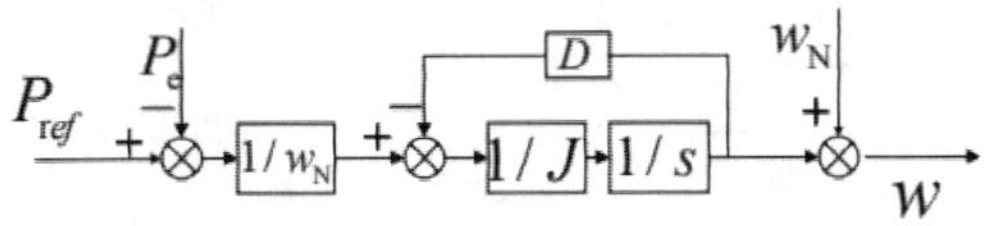

Fig.1 Active power-frequency control block diagram

The VSG reactive power-voltage control equation is:

$$\dot{E} = \left[D_q(U_n - U_o) + (Q_{ref} - Q_e)\right] / (Ks) \tag{5}$$

Where: U_n is the rated voltage amplitude; D_q is the reactive power-voltage coefficient; K is the reactive voltage integral coefficient; Q_{ref} is the reactive power reference value; Q_e is the VSG output (instantaneous) reactive power. The reactive power-voltage control block diagram of the virtual synchronous generator is shown in the figure below.

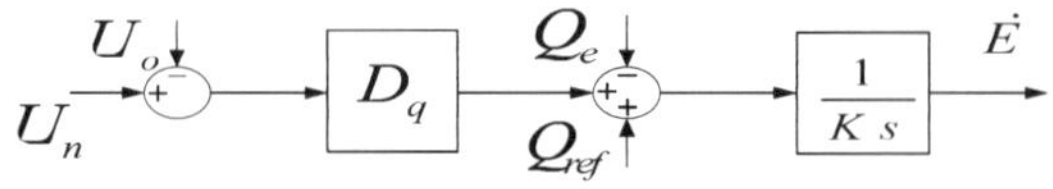

Fig.2 Reactive power-voltage control block diagram

4. Multi-Agent Mechanism for Battery Energy Storage

The Multi-Agent collaboration framework of the battery energy storage system is shown in Figure 3, the coordination Agent is connected to the executive Agent and the power grid dispatching center through dual network cables, using the IEC 61850 communication protocol. The executive Agent is mainly for the regulation of the battery and the PCS part. It can use the sensor to monitor the battery and PCS operating conditions, and control the PCS action through the built-in PQ control and virtual synchronous control mode to achieve autonomous control. Each executive Agent can be associated with each other. In addition to the ability to monitor key parameters in the area and maintain safe and stable operation, it also has the ability to coordinate with the executive Agents in the neighboring area. Coordination Agent can dispatch instructions to be reasonably consumed in each battery unit through allocation algorithms and optimization rules. Among them, the optimization rule is mainly based on the following criteria: when the energy storage unit is in a discharged state, the energy storage unit with a larger SOC should output more power, and the energy storage unit with a smaller SOC should output less power; conversely, when the energy storage unit is in the charging state, the energy storage unit with a larger SOC should absorb less power, and the energy storage unit with a smaller SOC should absorb more power.

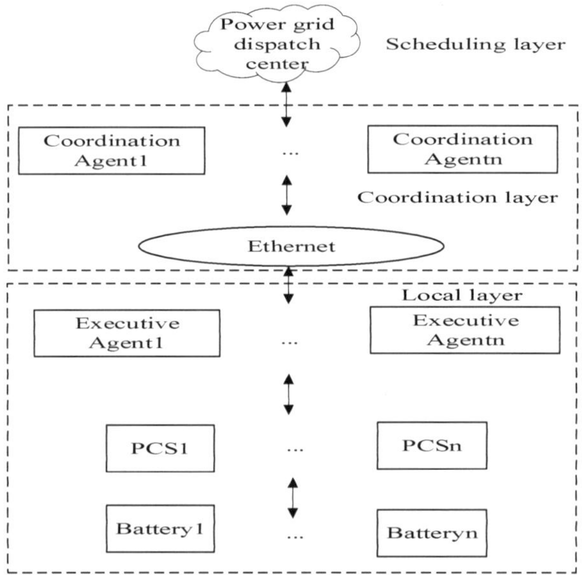

Fig.3 Multi-Agent collaborative architecture of battery energy storage system

5. Multi-Agent Collaborative Control Application Scenarios

5.1 Active power-frequency control

The active-frequency support scenario under the Multi-Agent cooperative control mechanism is shown in Figure 4. When the frequency is disturbed, the executive Agent detects the disturbance information through the perceptron, stores the relevant information in the database and uses the virtual synchronization algorithm actively responds to disturbances and provides a certain amount of inertia support. When the executive Agent cannot eliminate the disturbance or the corresponding battery SOC status has reached the upper and lower limits, the executive Agent requests assistance from the coordination Agent through the communication module and the optimization module selects the most suitable cooperative executive Agent according to the optimization rules. The power distribution module determines the active power adjustment amount of each PCS participating in frequency modulation according to the dispatching instructions of the large power grid and the information fed back by the lower-level Agents, and the database stores the relevant information of the frequency modulation process.

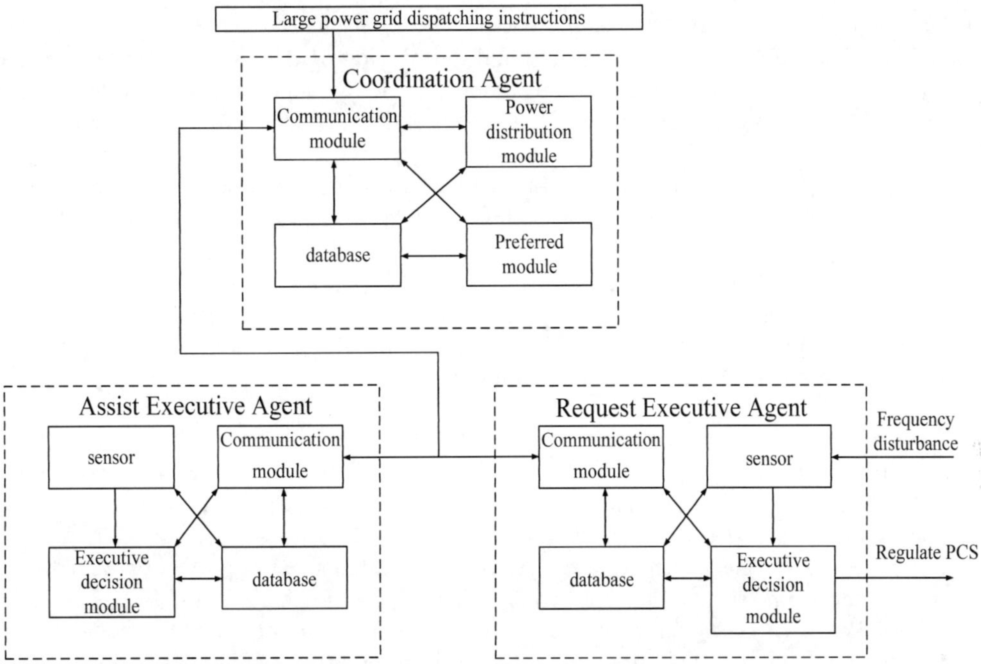

Fig.4 Active power-frequency control scenario

5.2Reactive power-voltage regulation

The Multi-Agent collaborative architecture of the battery energy storage system proposed in this paper fully considers the moderate charging and generating of energy storage to participate in the grid voltage regulation, thereby avoiding the large voltage fluctuations caused by frequent switching of capacitors/reactors. The specific regulation method is shown in Figure 5. Coordination Agent can determine the reactive power required for grid voltage regulation and the remaining reactive power margin of energy storage from the real-time voltage information fed back by the lower-level Agent. When the required reactive power output is not greater than the reactive power margin, the energy storage alone undertakes the voltage regulation task; if the required reactive power output is greater than the reactive power margin, the coordination Agent requests the grid to call the capacitor/reactor to undertake the regulation voltage task, and refined voltage regulation by energy storage auxiliary capacitor/reactor.

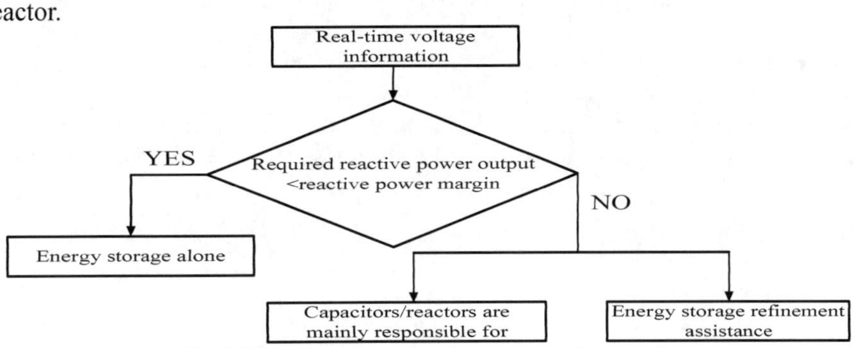

Fig.5 Reactive power-voltage control scenario

6. Conclusion

Aiming at the problem of energy storage operation control, this paper constructs a Multi-Agent mechanism-based battery energy storage system coordination control system architecture, and proposes corresponding operation control strategies and application scenarios, and draws the following conclusions:

(1) The proposed control strategy of the battery energy storage system overcomes the lack of synchronization support ability and the charging and discharging of the energy storage at a constant power under the condition of high penetration of renewable energy connected to the grid, realizes the active response to disturbances on the energy storage side and ensures the safe and stable operation of large power grids.

(2) The proposed system architecture combines upper-level instruction scheduling and lower-level information feedback through communication modules to achieve two-way circulation of information, enabling a large number of distributed energy storage facilities with scattered locations and different forms to have communication, perception and so on, give full play to the aggregation and regulation potential of distributed energy storage facilities, change the status quo that distributed energy storage is difficult to effectively participate in system regulation, and provide a reference plan for subsequent research on energy storage operation control technology.

References

[1] S. Razavia, Ehsan, R. Mohammad, S. Iavadib, et al. Impact of Distributed Generation on Protection and Voltage Regulation of Distribution Systems: A Review[J]. Renewable and Sustainable Energy Reviews, 2019, 105:157-167.

[2] Zhen-xing Li, Bin Tian, Xiang-gen Yin, et al. Active distribution network protection with distributed power sources and random loads[J]. High Voltage Technology, 2017, 43(04):1231-1238.

[3] Shou-xiang Wang, Li-ke Song, Xin Shu. Adaptive overcurrent protection scheme for active distribution network with distributed power generation and multi-load high penetration access[J]. High Voltage Technology, 2019, 45(06):1783-1794.

[4] Dong-yuan Liu. Research on Operation Control Strategy of Energy Storage System Participating in Grid Frequency Modulation[D]. Northeast Electric Power University, 2018.

[5] Zhi-peng Lv, Wan-xing Sheng, Qing-chang Zhong, et al. Virtual synchronous generator and its application in microgrid[J]. Journal of Chinese Electrical Engineering Science. 2019, 34(20): 4226-4241.

[6] Delille, Gauthier, Francois, et al. Dynamic frequency regulation using multilevel energy storge in remote area power supply system[J]. IEEE Transactions on Power Systems, 2019, 34(1):163-170.

[7] Zhao H, Hong M, Lin M, et al. Voltage and frequency regulation of microgrid with battery energy storage systems[J]. IEEE Transactions on Smart Grid, 2019, 10(1): 414-424.

[8] Tian-yuan Cao, Min-xiao Han, Wen-qiang Xie, et al. AC/DC power inverter parallel control technology research status quo analysis[J]. Journal of Electrotechnics. 2019, 34(20): 4226-4241.

[9] Tian-yang Zhang. Lead to more carbon battery parallel operation of the converter control strategy[D]. Northeast Electric Power University, 2019.

[10] Da-wei Yan, Gang Wei, Yin Hu, et al. Reliability and economy coordinated micro grid energy optimization[J]. Automation of Electric Power Systems, 2012, 36(08):18-23+51.

Research on robust control method of permanent magnet synchronous motor

Zenghui Wang[1a*], Yingfeng Zhang[1b*], Pan Sun[1c*], Jun Sun[1d*], Junhong Zhang[1e*]

[1]School of electrical engineering, Naval University of Engineering, Wuhan 430000, China

[a]*2635217103@qq.com, [b]*zyfzr1983@163.com, [c]*chinasunpan@163.com ,[d]*chinasunpan@163.com, [e]*13986151395@163.com

Abstract—During the operation of permanent magnet synchronous motor, the traditional vector control method does not match the regulator parameters with the actual parameters due to the disturbance factors such as motor parameter perturbation and load change, resulting in the decline of control performance. The research object of robust control is uncertain system, which has natural advantages for improving the robustness of control system. Firstly, this paper introduces the basic principle and implementation process of robust control, simulates and analyzes the influence of motor parameters on control performance, investigates and summarizes the improvement methods of traditional vector control, and points out the advantages and disadvantages of each scheme; Secondly, the similarities and differences of the improved methods are analyzed and compared; Finally, the development trend of permanent magnet synchronous motor control technology based on robust control is prospected.Keywords-Permanent magnet synchronous motor; robust control; linear matrix inequality

1. Introduction

Compared with traditional electrically excited motor and induction motor, permanent magnet synchronous motor (PMSM) has significant advantages such as simple structure, high power density and large torque inertia ratio. It can meet high performance requirements such as fast dynamic response, wide speed regulation range and high power factor. It has obvious advantages in high-power AC drive field, Especially in the field of advanced manufacturing, transmission system and large ship electric propulsion system

At present, PMSM widely adopts double closed-loop PI controller based on vector control in engineering. It has the characteristics of simple structure, easy implementation and certain robustness, and can better meet the control of linear system with input and single output. However, PMSM is a strongly coupled, multi input and multi output nonlinear system, and in the occasion of high-performance speed regulation, It has higher requirements for the dynamic and steady-state performance of the speed regulation system. There are uncertainties in the motor model, load changes and electromagnetic interference of the measurement module in the system. The traditional vector control may have controller parameter mismatch and unsatisfactory dynamic quality. The research object of robust control theory is uncertain system, In theory, it can solve the problems of parameter perturbation and load disturbance of permanent magnet synchronous motor speed regulation system.

The research on robust control originated from black's large gain feedback design idea for accurate systems with perturbation in 1927. The robust control theory was first proposed by Davison in 1972. It is a supplement and improvement to modern control theory and the product of the combination of modern control theory and practical engineering. Its research object is uncertain systems, The basic idea is to maintain the stability and desired performance of the system by describing the uncertainty of the controlled object model and estimating the degrees of freedom left to achieve the control goal under some specific limits

In the early 1960s, Curz and Perkins extended the sensitivity analysis idea of single input single output system to multi input multi output system, and introduced the sensitivity comparison matrix to measure the performance of closed-loop and open-loop systems. However, these early studies were mainly based on the perturbation of system small parameters, which was far from practical engineering applications, Thus, the modern robust control problem for non-small bounded perturbation uncertainty is introduced. Since the 1960s, robust control has mainly formed three representative research aspects: the frequency domain method of system transfer function, the polynomial algebra method of system characteristic polynomial, and the time domain method of system state equation.

In recent 20 years, robust control theory has made great progress, and the continuous improvement of linear matrix inequality (LMI) theory has gradually formed a complete theoretical system. Many achievements have been made in the field of motor control

2. H∞ robust control of permanent magnet synchronous motor

The mathematical model of surface permanent magnet synchronous motor (SPMSM) in d-q rotating coordinate system is:

$$\left.\begin{aligned} &\frac{\mathrm{d}\omega_{\mathrm{rm}}}{\mathrm{d}t}=\frac{3}{2}\frac{p\psi_{\mathrm{f}}}{J}i_{\mathrm{q}}-\frac{B}{J}\omega_{\mathrm{rm}}-\frac{T_L}{J},\\ &\frac{\mathrm{d}i_{\mathrm{d}}}{\mathrm{d}t}=\frac{1}{L_{\mathrm{d}}}u_{\mathrm{d}}+p\omega_{rm}i_{\mathrm{q}}-\frac{R}{L_{\mathrm{d}}}i_{\mathrm{d}},\\ &\frac{\mathrm{d}i_{\mathrm{q}}}{\mathrm{d}t}=\frac{1}{L_{\mathrm{d}}}u_{\mathrm{q}}-\frac{R}{L_{\mathrm{q}}}i_{\mathrm{q}}-p\omega_{rm}i_{\mathrm{d}}-\frac{p\psi_{\mathrm{f}}}{L_{\mathrm{q}}}\omega_{\mathrm{rm}},\\ &T_{\mathrm{e}}=1.5p\psi_{\mathrm{f}}i_{\mathrm{q}}。\end{aligned}\right\} \tag{1}$$

Where, u_d、u_q and i_d、i_q are dq-axis voltage and current respectively; R is the stator resistance; L_d=L_q is the stator inductance, ψ_f is the rotor permanent magnet flux, p is the motor pole number, T_e is the electromagnetic torque, T_L is the load torque, J is the moment of inertia, B is the viscous friction coefficient and is the mechanical angular speed of the motor.

2.1. *Standard H∞ control problem*

All For the uncertain system represented by equation (1), the standard form of H ∞ robust control problem is shown in Figure 1:

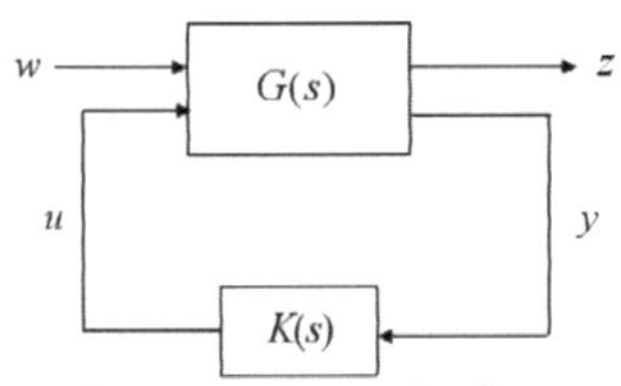

Fig.1 Schematic diagram of Standard H∞ control problem.

u is the control input, w is the external interference input, z is the regulated output, y is the measurement output, G(s) is the control object and K(s) is the controller. When the control input is 0, the closed-loop transfer function from the interference input w to the regulated output z is T(s)=$C(sI-A)^{-1}B+D$. The H∞ norm of T(s) is defined as:

$$\|T(s)\|_\infty = \sup_d \sigma_{\max}(T(j\omega))$$

That is, the peak value of the maximum singular value of the system frequency response, the H_∞ optimal control problem can be described as: finding a real rational control matrix K(s), minimizing the norm of the transfer function on the premise that K(s) stabilizes the control object G(s), that is:

$$\min_K \|T(s)\|_\infty 。$$

Assuming that the state of the system can be measured directly, a static state feedback controller in the following form is designed:u=Kx.

The closed-loop system described in equation (2) is in the following form:

$$\left.\begin{aligned} \dot{x} &= (\tilde{A} + B_u K)x + \tilde{B} d, \\ z &= (\tilde{C} + D_u K)x + \tilde{D} d \end{aligned}\right\} \quad (2)$$

The closed-loop system is required to be asymptotically stable, and the closed-loop transfer function from the disturbance input w to the regulated output z satisfies the H_∞ index.In this way, the controller K(s) is a state feedback H_∞ controller of system (2), which is transformed into LMI constraints. There are the following theorems:

Theorem : if and only if there is a symmetric positive definite matrix X and matrix W, the following LMI holds:

$$\begin{bmatrix} AX + B_u W + (A^T X + B_u W)^T & B & (CX + D_u W)^T \\ B_1^{\ T} & -\gamma I & D^T \\ CX + D_u W & D & -\gamma I \end{bmatrix} < 0 \quad (3)$$

2.2. *Standard H_∞ control problem*

The comparative simulation of robust control and PI control is carried out under the conditions of acceleration and deceleration, forward and reverse rotation, sudden loading and sudden unloading. The simulation results are as follows Fig.2-5,The robust control is scheme I and the PI control is scheme II.

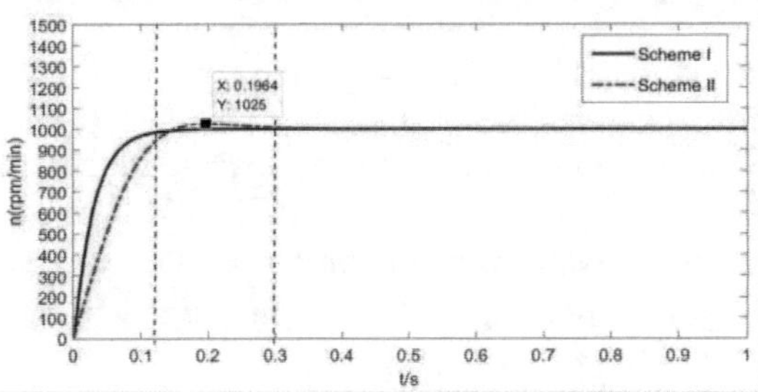

Fig.2 No load starting speed waveform.

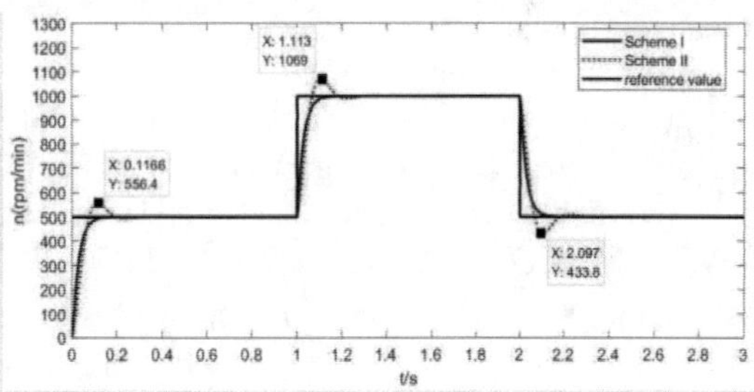

Fig.3 Acceleration and deceleration speed waveform.

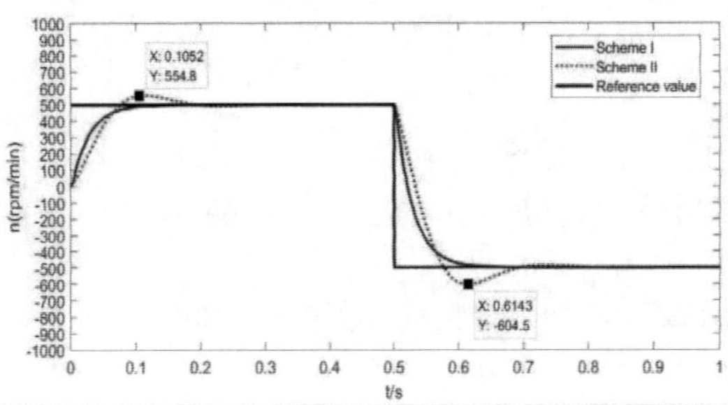

Fig.4 Forward and reverse speed waveform.

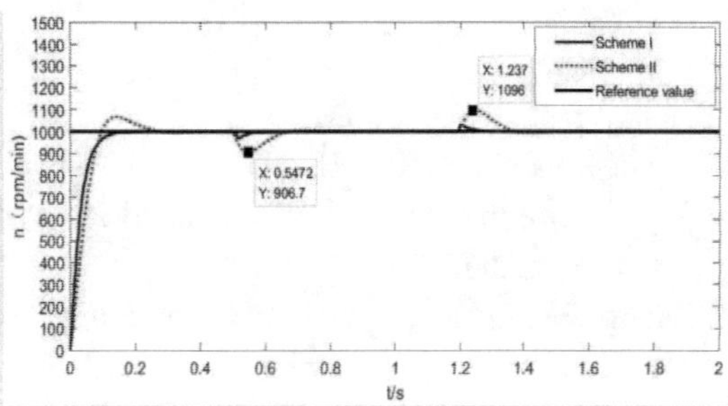

Fig.5 Sudden loading and unloading

The simulation results show that the robust control improves the robust performance of PMSM speed regulation system compared with the traditional PI control under different working conditions.

3. Other motor control applications based on robust control

The most basic robust H_∞ control only improves the robust steady-state performance of the control system. With the development of robust control theory, it has been widely used in motor control. Many scholars have made a series of achievements on this basis.

In reference [1], the mathematical model of PMSM is transformed into the standard problem of H_∞ robust control, and the appropriate weighting function is selected to obtain the Riccati equation to obtain the controller parameters, which effectively suppresses the unmodeled disturbance, but it is difficult to select the weighting function; Literature [2] designed a generalized robust PI speed controller based on convex optimization algorithm to ensure the robust steady-state performance of the system and improve the anti load disturbance and parameter perturbation ability of the speed regulation system by making the controller meet the system L1 performance index, but the dynamic regulation time of the system is long; Reference [3] designed the gain type sliding mode surface of H_∞ state feedback controller based on LMI, which ensures the robustness of speed control and improves the dynamic performance of the system, but increases the "chattering" of the system. Many other scholars have also made some achievements in permanent magnet synchronous motor speed regulation system by using robust control theory [4-10].

4. Conclusion

In control theory, robust control, as a relatively novel control method, has unique advantages and broad application prospects. For the control object of permanent magnet synchronous motor, which is multivariable, strongly coupled and has parameter uncertainty, the robust control method has natural advantages and strong engineering feasibility. With the development of control theory and computer tools, robust control has achieved many remarkable results in different applications and requirements in the field of motor control. On the basis of existing achievements, how to further solve theoretical problems and carry out engineering application is the direction of further efforts of many scholars.

References

[1] Yu Ruiwen, Zhang Hui, Tan Guojun, et al. Research on H∞ robust control strategy of permanent magnet synchronous motor [J]. Power electronics technology, 2012, 46 (02): 102-104
[2] Hou Limin, Shen Hesong, Yan Xin, et al. H∞ robust control of permanent magnetsynchronous motor speed regulation system [J]. Journal of electrotechnics, 2019, 34 (07): 1478-148
[3] VALENTA G, FORMENTINI A, PAPINI L, et al. Performance improvement of bearingless multi-sector PMSM with optimal robust position control[J]. IEEE Transactions on Power Electronics, 2018, DOI: 10. 1109 / TPEL. 2018. 2853038.
[4] Cui Jiarui, Gao Jiangfeng, Zhang Bo, et al. Sliding mode variable structure robust control of permanent magnet synchronous motor [J]. Journal of motor and control, 2016, 20 (05): 84-89.
[5] Luo Guangzhao. Research on robust control of electric actuator [D]. Xi'an: Northwest University of technology, 2003
[6] Zhang Qiaofen, Guo Haohao, Liu Yancheng, et al. Robust plug-in repetitive control for speed smoothness of cascaded-PI PMSM drive[J]. Mechanicl Systems and Signal Processing, 2021, 163.
[7] Zhang Wenjuan, Li Yu, Wu Gongping, Rao Zhimeng, et al. Robust Predictive Power Control of N*3-Phase PMSM for Flywheel Energy Storage Systems Application[J]. Energies, 2021, 14(12).
[8] Ping Zhaowu, Li Yaoyi, Song Yang, et al. Nonlinear speed tracking control of PMSM servo system: A global robust output regulation approach[J]. Control Engineering Practice,

2021, 112.

[9] Qixin Zhu, Lei Xiong, Hongli Liu, et al. Robust Two-degree-of-freedom Control Based on H∞ Method for PMSM Drive System[J]. Recent Advances in Electrical & Electronic Engineering, 2020, 13(4).

[10] Nicola CLAUDIUIONEL, Nicola MARCEL, Selişteanu DAN. Sensorless Control of PMSM Based on Backstepping-PSO-Type Controller and ESO-Type Observer Using Real-Time Hardware[J]. Electronics, 2021, 10(17).

A Change Detection Method Based on Full-scale Skip Connections and Mixed Pooling Module

Dashuai Tian[1*]

[1] The School of Information and Communication Engineering, Harbin Engineering University, Harbin, China, 150001

[*]Corresponding author's e-mail: tiandashuai@hrbeu.edu.cn

Abstract. Change detection is one of the main driving forces for the continuous development of remote sensing image processing technology. As a pixel-level detection task, change detection attaches great importance to the utilization of original spatial information as well as contextual information. In this paper, we propose a deep learning-based change detection method that combines full-scale skip connections and mixed pooling module (MPM). Firstly, full-scale skip connections make full use of the spatial information of features at different scales to improve change detection performance. In addition, we introduce MPM in this method due to the complexity of scene information in images, which can enhance the understanding of the scene with rich contextual information and boost the detection ability of strip shape targets. Finally, our experimental results on two public datasets demonstrate that the proposed method has good performance.

1. Introduction

The surface environment is changing all the time. But not all changes are good for mankind. In nature, surface changes caused by disasters such as earthquakes, mudslides, and tsunamis will have a great negative impact on human survival. If the surface changes can be monitored in time, it is of great significance for determining the source of the disaster and the estimated scope of impact. In addition to surface changes caused by natural disasters, changes in the surface caused by human activities such as urban expansion and felling of trees will also cause the deterioration of the human living environment. If we can observe these changes through a certain method, it will also be conducive to the harmonious coexistence of man and nature.

As an important means of monitoring changes on the surface, change detection has always been an significant research direction in the field of remote sensing image processing. It plays a vital role in remote sensing image analysis. The current traditional change detection methods cannot extract deep semantic information, so they do not perform well in complex scenes, and at the same time cannot meet the current industrial needs in terms of accuracy. As people's demand for change detection becomes more and more urgent, the search for automatic and high-precision change detection methods has attracted more and more attention from people in the remote sensing community. The rapid development of deep learning technology has brought the possibility of breakthroughs in various fields, and many complex needs can be met through deep learning technology. In this article, a change detection method based on deep learning is proposed, which can automatically and accurately detect changes in remote sensing images to meet current industrial needs.

To a certain extent, the change detection task can be regarded as a two-class semantic segmentation task. So many advanced methods of semantic segmentation can be used in the field of change detection. Daudt et al.[1] introduced the idea of full convolutional neural network to the change detection task, opening a new paradigm in the field of change detection. Many scholars have proposed lots of meaningful methods to improve the performance of change detection based on this idea. Although the deep learning method can extract deeper semantic information, its continuous sampling operation will lose a lot of shallow details, which can lead to inaccurate change detection results. Some scholars have alleviated this problem by using skip connections mechanism[2-5]. But as far as we know, no one has solved this problem from the perspective of using full-scale detailed information. In addition, making full use of the global information of remote sensing images is also conducive to improving the effect of change detection. The existing pyramid pooling module[6] is an example of effective use of global information. But for change detection tasks, only using global information is not conducive to detecting changes in strip shape targets such as roads. In this case, the square pooling operation will collect a lot of useless contextual information. To sum up, improving the detection effect of changing targets such as roads is also a problem that needs to be solved.

In this paper, a novel change detection network is proposed to solve the above problems. First of all, the proposed network architecture integrates the idea of full-scale skip connections[7], which can make full use of low-level detailed information and reduce the loss of spatial details. Secondly, a mixed pooling module (MPM) is introduced into the network. MPM can collect different contextual information through different types of pooling operations, and more accurately detect the changing target of strip shape such as road. Finally, we verified the effectiveness of the proposed method on two public datasets, and achieved good detection results.

2. Method

2.1. The overall architecture of network

The main body of the proposed network architecture is depicted in Figure 1(a). The bi-temporal remote sensing images to be detected are concatenated in the channel direction and then used as the input of the network.

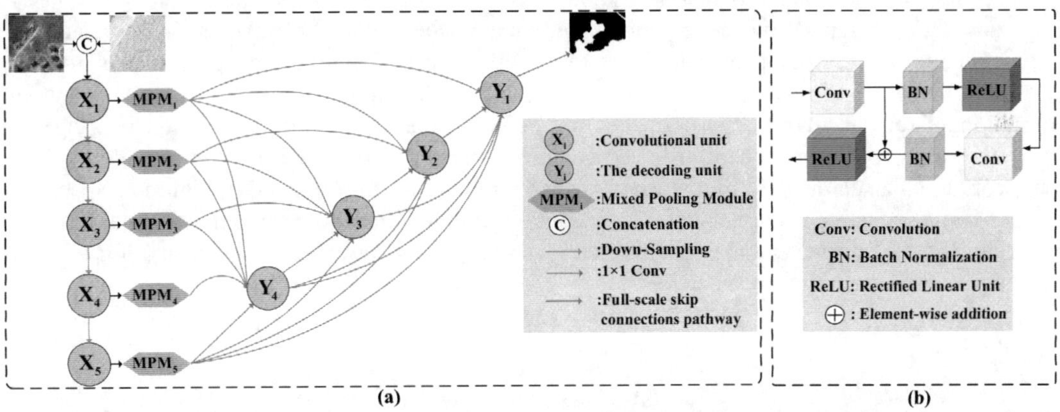

Figure 1. Illustration of the proposed network. (a)The main body of the network architecture, note that MPM in this network only differ in input. Each full-scale skip connections pathway includes a convolution operation to unify the number of channels while also unifying the spatial resolution of features using bilinear interpolation or max pooling. (b)The details of the convolution unit.

The proposed network follows the standard encoder-decoder architecture. In the encoder part, the network has five different levels of convolution units for extracting features of the input data. Figure 1(b) displays the components of each convolution unit. In the convolution unit, the short connection of the residual idea is used for feature reuse, and the feature extraction ability of the convolution unit is enhanced. Adjacent convolution units are connected through a maximum pooling operation. The number of output channels of the convolution unit is 32, 64, 128, 256, 512 from top to bottom. A mixed pooling module[8] is added at the back of each convolution unit, which can collect multiple levels of contextual information and help to improve the performance of change detection. The specific details of MPM will be introduced in detail in section 2.2. In the decoder part, only a convolutional layer with ReLU function constitutes the decoding unit. Finally, the result of the change detection is obtained through a 1×1 convolutional layer.

Taking into account the inaccurate detection problem caused by the loss of spatial details during feature extraction, we perform a full-scale feature fusion between features at different levels. This function is implemented through different full-scale skip connections pathway in the proposed network. The features used mainly include the output features of the MPM and the output features of the decoding unit. Each level of features passing through MPM contains different levels of detail information, and each feature output by the decoding unit contains rich high-level semantic information. The full fusion of these features is conducive to more accurate detection of changes in remote sensing images. Four full-scale feature fusions are performed in the proposed network and five features of different scales are fused each time. Each full-scale feature fusion needs to go through a full-scale skip connection pathway to unify the feature resolution. Specifically, when unifying the spatial resolution, one feature map among five feature maps with different resolutions is selected as the criterion, and the other four feature maps turn into the same resolution as the criterion through bilinear interpolation or max pooling operation. At the same time, each of the five obtained feature maps with the same spatial resolution is subjected to a convolution operation to unify the number of channels for the purpose of better feature fusion[7]. In this article, we choose 32 as the uniform number of channels. After that, we concatenate the five uniform resolution features to obtain a new feature map, and finally a further feature fusion is performed through the decoding unit.

2.2. Mixed pooling module

There are many changing targets of different sizes in remote sensing images, small targets such as vehicles and roads, and large targets such as mountains and rivers. Usually the changes of vehicles should accompany the road, it is not reasonable to see the changes of vehicles in the sea. Likewise, most roads should be built on the ground rather than in the water. This means that enhancing the ability of scene understanding in neural networks will help to improve the performance of change detection. For the previous work PSPNet[6], its pyramid pooling module only considers the contextual information of regular shape targets. It does not consider the contextual information of strip shape targets like roads and rivers. Therefore, we use an operation called mixed pooling module (MPM) in the network to enhance the detection ability for strip targets. MPM is depicted in Figure 2. Specifically, the input feature maps are first subjected to a convolution operation to reduce the channel dimension then generate two new feature maps, which can be used to collect contextual information of regular shape targets and strip targets respectively.

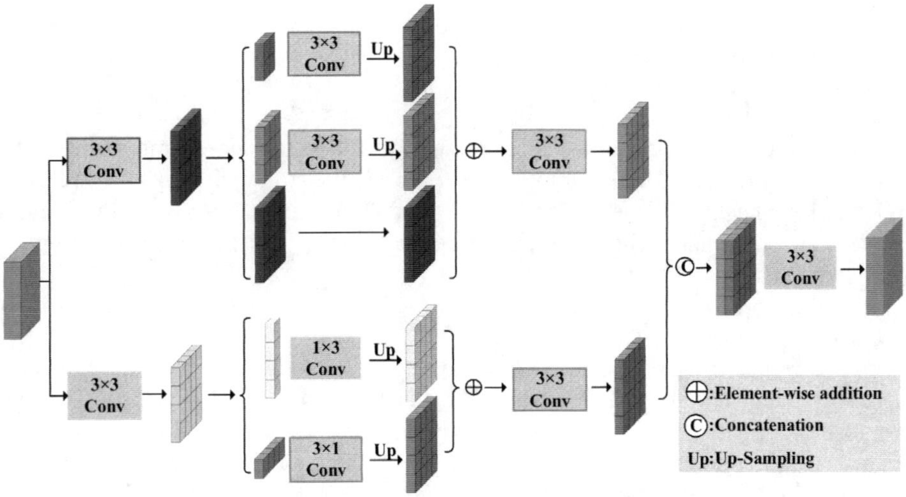

Figure 2. Illustration of the mixed pooling module.

For the branch that collects the contextual information of the regular shape targets, first we use two adaptive average pooling operations to collect contextual information of different receptive field sizes. Through the convolution operation and the up-sampling operation, the two feature maps after pooling are restored to the same resolution as the input. Finally, the two processed feature maps are added to the feature map input into this branch, and then the regular shape targets contextual information are integrated through a convolution.

For the branch that collects the contextual information of the strip shape targets, first we use the horizontal and vertical pooling operations to collect the stripe shape contextual information. The two feature maps also adopt the similar operation mode as the regular shape target branch. They are restored to the same resolution as the input branch feature map, and then the information is fused to obtain a feature map that collects the contextual information of the strip shape targets.

The feature maps output by the two branches are concatenated and then subjected to a convolution operation for information fusion, and restored to the same resolution as the original input feature map. In this way, feature maps can learn different levels of contextual information through MPM.

2.3. Loss function

In change detection task, the number of invariant pixels in the remote sensing image far exceeds the changed pixels. This usually causes a problem that the deep learning method tends to learn the features of invariant pixels, while ignoring the truly useful changing pixel features. To solve this problem, the loss function we use is a combination of dice loss function and weighted cross-entropy loss function[9]. This loss function allows the proposed network to better learn the characteristics of changed pixels and help to obtain better change detection results.

3. Experiments

3.1. Datasets and evaluation metrics

We perform experiments on the SVCD[10] and LEVIR-CD[11] datasets to verify the effectiveness of the proposed method. The SVCD dataset does not limit the types of changing objects. Buildings, vehicles, and roads are all objects that need to be detected. But the LEVIR-CD dataset is only for detecting changes in buildings. In experiments, we use the same training criteria as the dataset creators for peer evaluation. Both datasets are processed into uniform 256×256 pixel size images. Finally, the SVCD dataset used includes 16000 pairs of image tiles, and the LEVIR-CD dataset includes 10192 pairs of image files. For the SVCD dataset and the LEVIR-CD dataset, the ratios of training, validation, and test sets are 4:1:1 and 7:1:2, respectively. We use Precision, Recall and F1-Score as evaluation metrics in our experiments.

3.2. Training implementation details
All experiments are implemented on a GTX Nvidia1080Ti. The proposed method uses the Pytorch library for training. During network training, we set the number of batch size to 8 and the number of epochs to 120 when the initial learning rate to be 0.0002. We use AdamW as the optimizer in the experiment. Meanwhile, random flips and random rotations are used as means to augment the training data in the experiments.

3.3. Experimental results and comparison
We select five current state-of-the-art methods for comparison with the proposed method. These methods are CLNet[2], DifUnet++[3], HRTNet[12], DDCNN[4] and HDFNet[5]. The experimental comparison results are shown in Table 1. The experimental results of the comparative methods in Table 1 are all from their corresponding papers. We put the best metrics in bold.

To illustrate the effectiveness of MPM, we conduct an experiment without MPM. The results on both datasets demonstrate that MPM can improve the performance of the method. For the SVCD dataset, it can be seen that the proposed method significantly outperforms other methods on three evaluation metrics. For the LEVIR-CD dataset, experiment results show that our proposed method is also slightly better than other methods on the basis of evaluation metrics. In terms of the amount of parameters of the model, our proposed method is comparable to CLNet. But for the SVCD dataset, our proposed method far outperforms CLNet. In summary, our proposed method has good comprehensive performance.

Table 1. Quantitative results of the proposed method and five state-of-the-art methods on two experimental datasets.

Method	Dataset						Params(M)
	SVCD			LEVIR-CD			
	Pre (%)	Rec (%)	F1 (%)	Pre (%)	Rec (%)	F1 (%)	
CLNet[2]	94.7	89.7	92.1	89.8	90.3	90.0	8.00
DifUnet++[3]	92.15	94.63	93.37	**92.4**	87.1	89.6	-
HRTNet[12]	93.34	94.09	93.71	85.43	**91.77**	88.48	66.89
DDCNN[4]	96.71	92.32	94.46	91.85	88.69	90.24	46.68
HDFNet[5]	94.12	95.47	94.77	87.54	89.38	88.13	18.79
Ours (without MPM)	96.13	96.40	96.27	91.81	88.26	90.00	**6.75**
Ours	**96.83**	**96.99**	**96.91**	92.23	88.43	**90.29**	8.22

To further demonstrate the advancement and effectiveness of our proposed method, the visualization results of the experiments on the SVCD and LEVIR-CD datasets are shown in Figure 3.

For experiments on the SVCD dataset, the visualization results prove that the method has a good detection effect on changing objects of different scales. This is mainly due to the full-scale skip connection, which combines different levels of shallow details and high-level semantics. It can also be seen from the visualization results that the presence of MPM improves the detection performance of strip shape targets such as roads. For experiments on the LEVIR-CD dataset, the visualization results also demonstrate that our method is effective in detecting changes in buildings. Since there are few strip shape targets in buildings, the role of MPM is not obvious. But there's a small number of cases, such as the last set of results in LEVIR-CD dataset in Figure 3. MPM can solve the sticking problem during detection to a certain extent when some strip shape interfering targets exist between the two buildings.

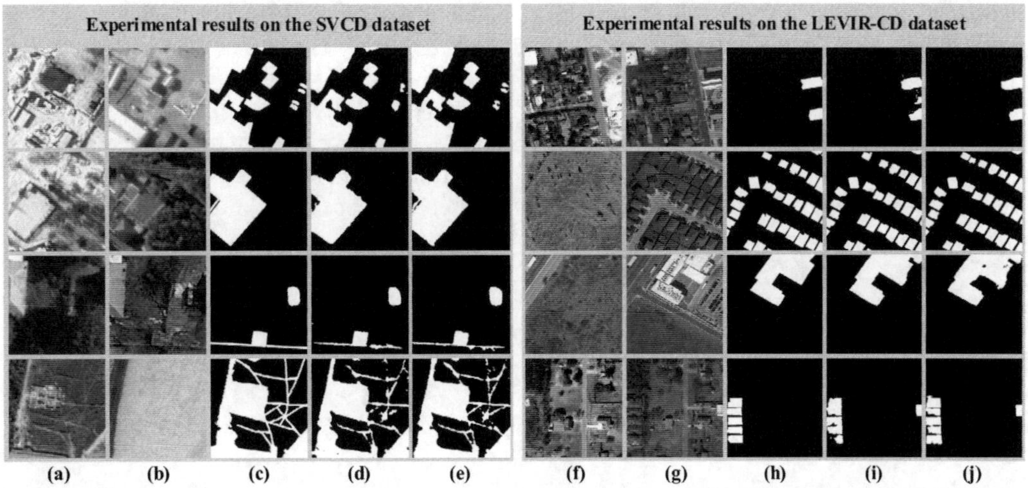

Figure 3. Visualization results of experiments on the SVCD dataset and the LEVIR-CD dataset. (a) and (f) represent the image before change. (b) and (g) represent the image after change. (c) and (h) represent the ground truth map. (d) and (i) represent the visualization results without MPM in the proposed method. (e) and (j) represent the visualization results with MPM in the proposed method.

4. Conclusion

In this paper, we propose a change detection network, which introduce full-scale skip connections to preserve enough spatial information to enhance the detection ability of objects at different scales in images. Further, strip shape targets like roads also achieve better detection results because of the mixed pooling module used. Finally, we obtain good detection results on both the SVCD dataset and the LEVIR-CD dataset.

References

[1] Daudt, R.C., Saux, B.L., Boulch, A. (2018) Fully convolutional siamese networks for change detection. In: Proceedings of IEEE International Conference on Image Processing (ICIP). Athens, Greece. pp. 4063-4067.

[2] Zheng, Z., Wan, Y., Zhang, Y., et al. (2021) CLNet: Cross-layer convolutional neural network for change detection in optical remote sensing imagery. ISPRS Journal of Photogrammetry and Remote Sensing, 175: 247-267.

[3] Zhang, X., Yue, Y., Gao, W., et al. (2021) DifUnet++: A Satellite Images Change Detection Network Based on Unet++ and Differential Pyramid. IEEE Geoscience and Remote Sensing Letters, 1-5.

[4] Peng, X., Zhong, R., Li, Z., et al. (2021) Optical remote sensing image change detection based on attention mechanism and image difference. IEEE Transactions on Geoscience and Remote Sensing, 59(9): 7296-7307.

[5] Zhang, Y., Fu, L., Li, Y., et al. (2021) HDFNet: Hierarchical Dynamic Fusion Network for Change Detection in Optical Aerial Images. Remote Sensing, 13(8): 1440-1461.

[6] Zhao, H., Shi, J., Qi, X., et al. (2017) Pyramid scene parsing network. In: Proceedings of the IEEE Conference on Computer Vision and Pattern Recognition (CVPR). Honolulu, HI, USA. pp. 2881-2890.

[7] Huang, H., et al. (2020) Unet 3+: A full-scale connected unet for medical image segmentation. In: Proceedings of International Conference on Acoustics, Speech and Signal Processing (ICASSP). Barcelona, Spain. pp. 1055-1059.

[8] Hou, Q., Zhang, L., Cheng, M.M., et al. (2020) Strip pooling: Rethinking spatial pooling for scene parsing. In: Proceedings of the IEEE Conference on Computer Vision and Pattern Recognition (CVPR). Online. pp. 4003-4012.

[9] Fang, S., Li, K., Shao, J., et al. (2021) SNUNet-CD: A Densely Connected Siamese Network for Change Detection of VHR Images. IEEE Geoscience and Remote Sensing Letters, 1-5.
[10] Lebedev, M.A., Vizilter, Y.V., Vygolov, O.V., et al. (2018) Change detection in remote sensing images using conditional adversarial networks. International Archives of the Photogrammetry, Remote Sensing & Spatial Information Sciences, 42(2): 565-571.
[11] Chen, H., Shi, Z. (2021) A spatial-temporal attention-based method and a new dataset for remote sensing image change detection. Remote Sensing, 12(10): 1662-1684.
[12] Hou, X., Bai, Y., Li, Y., et al. (2021) High-resolution triplet network with dynamic multiscale feature for change detection on satellite images. ISPRS Journal of Photogrammetry and Remote Sensing, 177: 103-115.

Journal of Physics: Conference Series 2258 (2022) 012060 doi:10.1088/1742-6596/2258/1/012060

Model-based Fault Detection of Rotor Position Sensor Fault in Permanent Magnet Synchronous Motor for Electric Vehicle

Nana Zhou[1,2a*], Lingchao Kong[2b], Jie Fan[3c], Zhenpo Wang[1d]

[1]National Engineering Laboratory for Electric Vehicles, School of Mechanical Engineering, and Collaborative Innovation Center of Electric Vehicles, Beijing Institute of Technology, Beijing 100081, China

[2] Advanced technology research institute, BIT, Jinan, 250300, China

[3] China Automotive Engineering Research Institute Co., Ltd, Chongqing 401122, China

[a*]zhounn@bit.edu.cn, [b]konglingchaobit@163.com, [c]384954890@qq.com

Abstract—Permanent magnet synchronous motors (PMSMs) are widely used in drive system of electric vehicles (EVs) because of their high efficiency, wild torque range, high power density, etc. However, rotor position, as a very critical parameter for its control strategy, often faces inaccuracies in operation. In this paper, the problem of position sensor fault detection in drive system of EV was studied. The mathematical model of EV with a specific drive system was introduced in discrete form and built in Matlab/Simulink. Based on the field oriented control strategy model, a novel fault detection scheme based on the parity space for the drive system was proposed. The proposed method considered the influence of the closed-loop control to the fault detection result. A residual generator and a residual evaluator were included. The simulation results demonstrated the effectiveness and applicability of the proposed method.

1. Introduction

The widely used motors in EV are permanent magnet synchronous motor (PMSM) which has higher energy efficiency, more excellent power density, higher torque to inertia ratio and advantageous in volume and weight comparing with other motors [1]. To maintain optimum closed loop control of PMSM, the information of rotor position is required to be accurately known. Otherwise the efficiency of PMSM drive system, the reliability and the whole system performance are also influenced and reduced [2, 3]. Therefore, how to detect the position error has strongly attracted the attention of academic researchers and industry professionals.

Nowadays, numerous studies on fault detection of motors are currently underway. A traditional solution is to add hardware redundancy to detect faults. However, for vehicle applications, weight and cost are key market factors for a vehicle manufacturer. With the development of technology, many new approaches have been proposed. The methods in the PMSM domain can be divided into two categories: high frequency signal injection method and model-based method. The first category is applied to diagnose the fault when the vehicle is not running. The principle of this method is to add a high frequency voltage (current) in stator winding and then evaluate the rotor position by the effect of the motor anisotropy on the amplitude of the correspondent stator voltage (current) [4,5]. In reality, the demodulation part is not only susceptible to the measurement error caused by the inaccuracy and limited bandwidth of the voltage sensor, but also susceptible to the imprecise extraction and calculation of the average value of the high frequency voltage pulse in the carrier cycle.

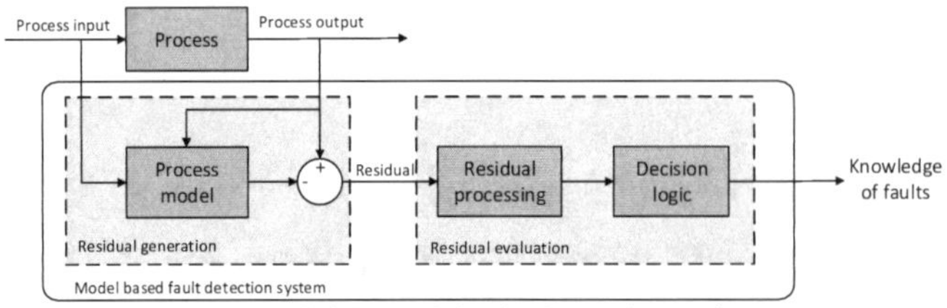

Fig.1 Model-based fault detection scheme

In other studies, model-based method is applied for the fault detection. Due to the rapid development of the computer technology and the control theory, model-based fault detection is accepted as a powerful tool to solve fault diagnose problems in technical processed [6]. The schematic diagram is shown in Fig.1. The principle of this method is to diagnose faults by the difference between the estimated position value and the actual sensor value. [7, 8]. The residual is the difference between the actual output and the model output. Model-based fault detection method can be classified into three main categories: observer-based method, parity space method and parameter identification based method [6]. The parameter identification based method was proposed to detect the unmeasurable parameter drift [9]. And there is an interconnection between the observer and parity space based residual generators established in Ref. [10,11]. Furthermore, the advantage of the parity equation method over the other two model-based methods is the complete decoupling of residual from the system control inputs and initial states. In fact, most research focus on the area of fault detection for open loop system and cannot be directly used for close loop system [12]. The drive system in EV is obviously a close loop control system. In this paper, an improved parity space based fault diagnosis method for drive system is proposed.

This paper is arranged as: In section 2, Modelling of the EV with detailed mathematical PMSM model is introduced and an improved fault detection scheme based on parity equation are introduced. Simulation results and a conclusion are given in Section 3 and Section 4, respectively.

2. Modelling and fault detection scheme

2.1. Mathematical Modelling of EV

The entire mathematical model of the EV is based on vehicle theory which is shown in Eq.1 and built in Matlab/Simulink. It could be regarded as a dynamic mathematical model of the vehicle. The parameters of the model are assumed to be known and there is no modelling error existing. For the fault detection purpose, the electric machine is also assumed to be rigidly connected to the driving axle, which means that the wheel speed is proportional to motor's speed. Therefore, the relationship between motor's speed and wheel speed is $\omega_m = \omega_{wheel}/\eta i_0$.

$$\begin{gathered} F_t = F_f + F_w + F_i + F_j \\ F_t = \frac{T_L}{r}\eta i_0, F_f = \mu mg\cos\gamma, F_w = \frac{1}{2}C_d\rho Av^2, F_i = mg\sin\gamma, F_j = m\frac{dv}{dt} \end{gathered} \tag{1}$$

where F_t is the driving force, F_f is rolling resistance, F_w is air resistance, F_i is the slope resistance, F_j is the acceleration resistance. T_L is load torque on the output side of the motor, η is the transmission efficiency, i_0 is the main deceleration ratio, r is the tire radium, μ is rolling resistance coefficient, m is the mass of the vehicle, C_d is the air resistance coefficient, ρ is the air density, A is the frontal area, v is vehicle speed, γ is the road gradient.

Fig. 2 shows the modelling block diagram of a typical PMSM drive system for EV application. The detailed operation principle of EV is shown in Ref. 13.

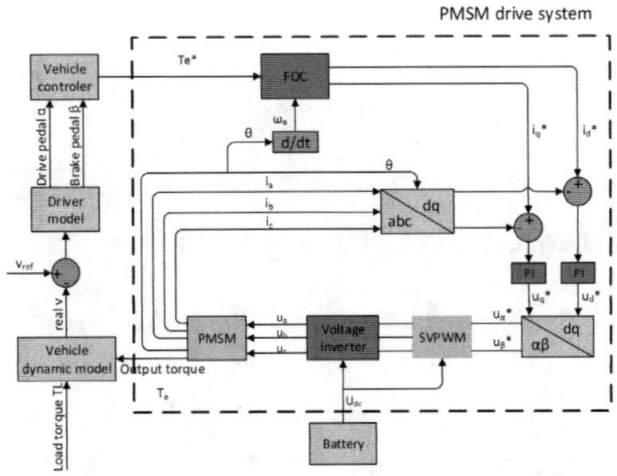

Fig. 2. Modelling block diagram of a PMSM drive system

2.2. Mathematical Modelling PMSM

PMSM is the component of interest in this paper, so this model is built in detail. After the Park's transformation and Clark's transformation, the state space form of the electrical and mechanical dynamic mathematical model of PMSM in the rotating coordinate system is given by Eq. 2.

$$\begin{cases} \frac{di_d}{dt} = -\frac{R_s}{L_d} i_d + \frac{L_q}{L_d} P\omega_m i_q + \frac{U_d}{L_d} \\ \frac{di_q}{dt} = -\frac{R_s}{L_q} i_q - \frac{L_d}{L_q} P\omega_m i_d - \frac{\lambda_m}{L_q} P\omega_m + \frac{U_q}{L_q} \\ \frac{d\omega_m}{dt} = \frac{1}{J}\left(1.5P\lambda_m i_q - b\omega_m - T_L\right) \\ \frac{d\theta}{dt} = P\omega_m \end{cases} \tag{2}$$

where U_d, U_q are phase voltages, i_d, i_q are phase currents. R, L and M represent resistance, self-inductance and mutual-inductance respectively. λ_m is peak the magnetic flux generated by the magnet in the rotor. θ is the electric angular position. ω_m is the mechanical speed. P is the number of the pole pairs, T_e is the electromagnetic torque, J is the lumped inertia of the rotating elements, b is the friction coefficient.

From Eq. 2, it could be seen that the PMSM model is a highly nonlinear system. In this paper, Taylor formula is used to approximate the real equation. The equation is expanded into a Taylor series at its operating point. The terms above the second order are ignored. Let $x=\left[i_d;i_q;\omega_m;\theta\right]$, and $u=\left[U_d;U_q\right]$. Then the system can be described by the following state space form:

$$\begin{aligned} \dot{\mathbf{x}}(t) &= \mathbf{Ax}(t)+\mathbf{Bu}(t)+\mathbf{C_0} \\ \mathbf{y}(t) &= \mathbf{Cx}(t) \end{aligned} \tag{3}$$

where $x(t)$ is the state vector; $u(t)$ is the known input vector; $y(t)$ is the measurement output vector; matrix C_0 is a constant; matrix A, B, C and C_0 are as follows:

$$\mathbf{A} = \begin{bmatrix} -\frac{R_s}{L_d} & \frac{L_q}{L_d}P\omega_{m0} & \frac{L_q P i_{q0}}{L_d} & 0 \\ -\frac{L_d}{L_q}P\omega_{m0} & -\frac{R_s}{L_q} & -\frac{\lambda_m}{L_q}P - \frac{L_d}{L_q}P i_{d0} & 0 \\ 0 & \frac{1}{J}1.5P\lambda_m & \frac{b}{J} & 0 \\ 0 & 0 & P & 0 \end{bmatrix}; \mathbf{B} = \begin{bmatrix} \frac{1}{L_d} & 0 \\ 0 & \frac{1}{L_q} \\ 0 & 0 \\ 0 & 0 \end{bmatrix}; \mathbf{C_0} = \begin{bmatrix} -\frac{L_q}{L_d}P\omega_{m0}i_{q0} \\ \frac{L_d}{L_q}P\omega_{m0}i_{d0} \\ T_L \\ 0 \end{bmatrix}; \mathbf{C} = \left[1\,1\,1\,0\right]$$

The vehicle is assumed to run at a certain speed, the other parameters of the PMSM can be obtained in Eq. 4. FOC strategy is used in PMSM control of EV, so i_d equals to zero. The electric torque T_e can be calculated by the load torque T_L which can be obtained by Eq. 1.

$$\begin{cases} \omega_{m0} = \eta i_0 v / R_\omega \\ T_{e0} = T_{L0} + b\omega_m \\ u_{d0} = -L_q P \omega_{m0} i_{q0} + R_s i_{d0} \\ u_{q0} = R_s i_{q0} + L_d P \omega_{m0} i_{d0} + P\lambda_m \omega_m \\ T_{L0} = \left(\mu mg\cos\gamma + 0.5C_d \rho A v^2 + mg\sin\gamma \right) r / (\eta i_0) \end{cases} \quad (4)$$

Considering the system disturbances $w(k)$, the system can be extended to the following model with sensor faults:

$$\begin{aligned} \mathbf{x}(k+1) &= \bar{\mathbf{A}}\mathbf{x}(k) + \bar{\mathbf{B}}\mathbf{u}(k) + \bar{\mathbf{C}}_0 + \mathbf{E}_\omega \boldsymbol{\omega}(k) \\ \mathbf{y}(k) &= \bar{\mathbf{C}}\mathbf{x}(k) + \mathbf{F}_\omega \boldsymbol{\omega}(k) + \mathbf{F}_f \mathbf{f}_s(k) \end{aligned} \quad (5)$$

where $\mathbf{x}(k) = [\mathbf{x}(t)]_{t=kT}$, $\mathbf{u}(k) = [\mathbf{u}(t)]_{t=kT}$, $\mathbf{y}(k) = [\mathbf{y}(t)]_{t=kT}$, $\bar{\mathbf{A}} = e^{AT}$, $\bar{\mathbf{B}} = \int_0^T e^{AT}\mathbf{B}dt$, $\bar{\mathbf{C}}_0 = \mathbf{C}_0 T$, $\bar{\mathbf{C}} = \mathbf{C}$ $\boldsymbol{E_w}$ and $\boldsymbol{F_w}$ are known matrices with appropriate dimensions. $\boldsymbol{f_s}(k)$ is the sensor fault, and $\boldsymbol{F_f}$ is the fault matric.

2.3. Fault detection scheme

In this paper, the fault detection steps for a closed loop control system based on the parity apace transformation are given in Fig.3.

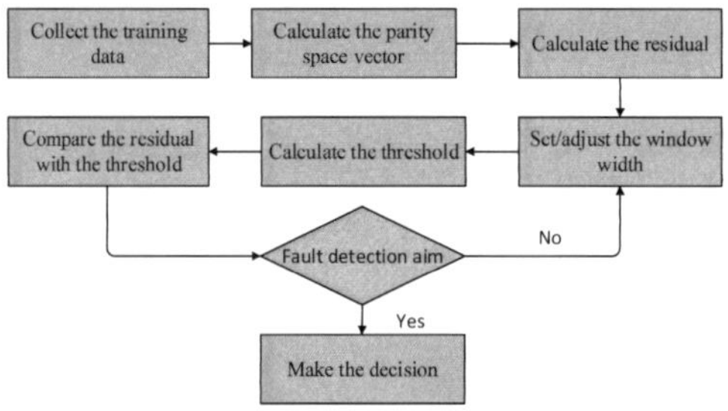

Fig. 3 The fault detection flow of the proposed method

Parity space approach is generally considered as one of the well-established model-based residual generation approaches. According to the theorem 5.9 in Ref. [10], for a given s-order parity space, an extended parity space model could be formulated as Eq.(6). The notation could be found in Ref.[10].

$$\begin{aligned} \mathbf{Y}_s(k) &= \mathbf{H}_{o,s}\mathbf{X}(k-s) + \mathbf{H}_{u,s}\mathbf{U}_s(k) + \mathbf{H}_{f,s}\mathbf{F}_s(k) \\ r(k) &= \mathbf{v}_s\left(\mathbf{Y}_s(k) - \mathbf{H}_{u,s}\mathbf{U}_s(k)\right) = \mathbf{v}_s\left(\mathbf{H}_{\omega,s}\mathbf{W}_s(k) + \mathbf{H}_{f,s}\mathbf{F}_s(k)\right) \end{aligned} \quad (6)$$

In this paper, an optimization problem is selected to make a trade-off between robustness and sensitivity when choosing a parity space vector v_s from the space $\boldsymbol{V}_s$. The optimization equation is shown in Eq. (7).

$$\min_{v_s} \frac{\sup\limits_{f=0,\omega\in l_2/\{0\}} \frac{\|\mathbf{r}\|_2}{\|\mathbf{W}\|_2}}{\sup\limits_{\omega=0,f\in l_2/\{0\}} \frac{\|\mathbf{r}\|_2}{\|\mathbf{F}\|_2}} \quad (7)$$

According to the model, the parity space vector and the residual can be obtained. Then follow the scheme to calculate the residuals and analysis the characteristics of the residuals for fault detection.

3. Test Results and Discussions

3.1. Description of specimens

The parameters of the vehicle and PMSM are presented in Tab. 1 and Tab. 2. The EV model with the detailed PMSM model is built in Matlab/Simulink based on the mathematical model.

Tab.1 Vehicle parameters

Parameters	Values (unit)	Parameters	Values (unit)
Vehicle mass	14000 (kg)	Air density	1.29 (kg/m3)
Wheel radius	0.464 (m)	Air resistance coefficient	0.55
Frontal area	6.81 (m^2)	Rolling resistance coefficient	0.0065

Tab.2 PMSM parameters

Parameters	Values (unit)	Parameters	Values (unit)
Stator resistance	0.001 (Ω)	Flux linkage	1.0714
Phase inductance	0.0015(H)	Rotor inertia of rotation	1.5625 (kg/m^2)
Pole pairs	2	Coefficient of damping friction	0.0015 (Nm.s/rad)

3.2. Simulation results and discussion

The fault detection approach proposed in section 3 is applied to detect the rotor position sensor fault occurring in the PMSM of EV. Besides, the traditional parity space approach is used as a contract. The position and angular velocity residual are both chosen as the detection basis (shown in Figs. 4-5).

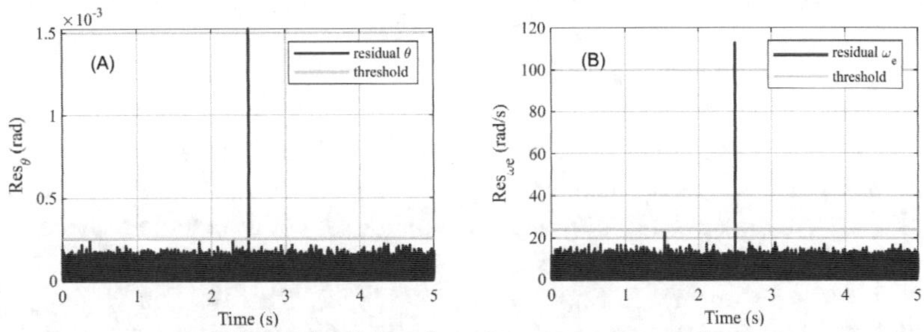

Fig. 4 Traditional parity space approach
(A) Position residual res_θ; (B) Electrical angular velocity residual $res_{\omega e}$

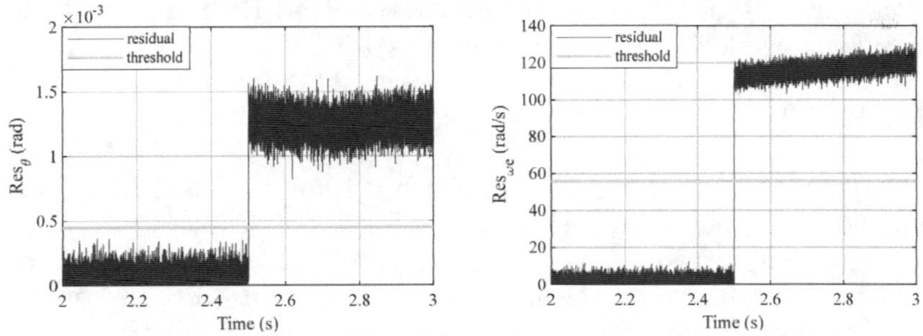

Fig. 5 Improved parity space approach for close loop control system
(A) Position residual res_θ; (B) Electrical angular velocity residual $res_{\omega e}$

From both figures, the rotor position residual is very sensitive to the fault. As the angular velocity is the derived of the rotor position, the angular velocity residual is more obvious than the position residual, which is clearly shown in the figures. The threshold is selected from the principle shown in section 2.3. It can be seen from both figures that there is a sudden change in the residual when the fault occurs. Both approaches can detect the fault. But the sudden change disappear after the fault point when using the traditional parity space approach. Most time, this abrupt change will be regarded as disturbance. This kind residual is not helpful for the fault detection. However, there is a consistent residual change when using the improved approach. From Fig.5, it can be seen that the residual evaluation values are always less than the thresholds when there are no faults occurring, while the residual evaluation values immediately exceed the threshold after the occurrence of the faults. In addition, the change trend of the angle velocity is faster than the change trend of the angular. This may be because the angular velocity is the differentiation of the angle and the sampling time is too short. In general, the improved parity equation approach is effective for detecting the rotor position sensor fault.

4. Conclusion

In this paper, an improved parity space approach is used to detect rotor position sensor fault of PMSM drive systems in electric vehicles. Firstly, an EV model with a detailed mathematical PMSM model is introduced and a linearized state equation is established using the Taylor algorithm. Secondly, the whole normal and faulty vehicle models are built in Matlab/Simulink. Then based on the traditional parity space approach, an improved approach for closed loop control system is proposed. In order to guarantee an appropriate trade-off between sensitivity and robustness against the noise, an optimization problem based on the H_∞/H_∞ index is proposed and the appropriate parity space vector is selected. Finally, the effectiveness of the proposed approach is verified by the simulation data, which clearly confirms its superiority over the traditional one.

Acknowledgments

This work is supported by the National Key R&D Program of China (Grand No. 2020YFB1600600) in part. Any opinions expressed in this paper are solely those of the authors and do not represent those of the sponsors.

References

[1] Kim, S. H. (2017). Electric motor control: DC, AC, and BLDC motors. Elsevier.
[2] Han, B., Shi, Y., Song, X., Hong, K., Mao, K. (2018). Initial rotor position detection method of SPMSM based on new high frequency voltage injection method. IEEE Transactions on Power Electronics, 34(4), 3553-3562.
[3] Zhang, Z., Hu, C., Zhang, Y., & Shen, W. (2019, June). Research on High Frequency Voltage Injection Method for PMSM. In 2019 14th IEEE Conference on Industrial Electronics and Applications (ICIEA) (pp. 266-269). IEEE.
[4] Wang, Z., Yao, B., Guo, L., Jin, X., Li, X., & Wang, H. (2020). Initial Rotor Position Detection for Permanent Magnet Synchronous Motor Based on High-Frequency Voltage Injection without Filter. World Electric Vehicle Journal, 11(4), 71.
[5] Ozturk, S. B., Kivanc, O. C., Aksoz, A., & Hegazy, O. (2020). Rotor Position Alignment of FSTPI Based PMSM Drive Using Low Frequency Signal Injection. Applied Sciences, 10(21), 7397.
[6] Frank, P.M.. (1990). Fault diagnosis in dynamic systems using analytical and knowledge-based redundancy. Automatica, 26(3), 459-474.
[7] Tian, G., Yan, Y., Jun, W., Ru, Z. Y., & Peng, Z. X. (2018, May). Rotor position estimation of sensorless PMSM based on extented Kalman filter. In 2018 IEEE International Conference on Mechatronics, Robotics and Automation (ICMRA) (pp. 12-16). IEEE.
[8] Hwang, S. H., Seo, S. W., & Jung, T. U. (2019). Initial Rotor Position Detection of Single-phase Permanent Magnet Synchronous Motor using Offset Voltage. Journal of IKEEE, 23(2), 622-627.
[9] Isermann, R.. (1993). Fault diagnosis of machines via parameter estimation and knowledge processing—tutorial paper. Automatica, 29(4), 815-835.
[10] Ding, Steven, X. (2013). model-based fault diagnosis techniques || fault identification schemes. Springer London, 10.1007/978-1-4471-4799-2(Chapter 14), 441-469.
[11] Alhelou, H. H. (2019). Fault detection and isolation in power systems using unknown input observer. Advanced condition monitoring and fault diagnosis of electric machines, 38-58.
[12] Sankavaram, C., Pattipati, B., Pattipati, K.R., Zhang, Y., Howell, M.. (2014). Fault Diagnosis in Hybrid Electric Vehicle Regenerative Braking System. IEEE (Vol.2, pp.1225-1239). IEEE.
[13] Minaker, B. P. (2019). Fundamentals of vehicle dynamics and modelling: A textbook for engineers with illustrations and examples. John Wiley & Sons.

Finite Time Speed and Position Control for Dynamics of HST with Time-Varying Delays

Runze Wang[1], **Tiantian Liang**[1*]

[1] School of Automation and Electrical Engineering, Dalian Jiaotong University, Dalian, 116028, China

[*]Corresponding author's E-mail: liangtiantian1122@163.com

Abstract. In this paper, a finite time robust controller is proposed for tracking the speed and position of high speed trains. Considering the aerodynamic flag, mechanical rolling resistance, additive resistance and wind gust, the dynamics of high speed train with time-varying delay is established in longitude. Aerodynamic flag and mechanical rolling resistance are seen as the coefficient of the state variables, wind gust and additive resistance are seen as the external disturbance, and time-varying delay in speed are considered as the train is often running in bad weather or severe working conditions. Then a delay-dependent finite time robust controller is designed. This controller not only guarantees the closed-loop error dynamics error is finite time boundness, but also is robust to the external disturbance consisting of wind gust and additive resistance in finite time. Sufficient conditions of the controller design are given, and the gain matrices of the controller are formulated as linear matrix inequality. A 5-car simulation example is given to show the effectiveness of the proposed method.

1. Introduction

At present, the railway engineering construction level of high-speed trains (HST) in China has been highly developed. Correspondingly, as the speed up of the train, running status tracking of the train becomes particularly important. Due to the complex working environment, the dynamic model of HST is becoming more complex, and the speed and position tracking is affected by many factors. Based on automatic control theory which is an important research direction for HST [1], many advanced train speed and position control methods has been proposed to ensure the safety index and comfort of trains.

Regarding the train as a whole, single particle model has been established and has been used for characterizing the motion of the train [2]. Based on this model, dynamics of HST is simplified as a single particle by ignoring the coupling characteristics and the distributed traction/braking force between every cars [3]. However, as HST adopts the distributed traction power mode, this model cannot fully describe the coupling characteristics and dynamic dispersion characteristics of the train. Thus, another model for better describing dynamics of HST to deal with tracking control emerges.

Based on the analysis of the coupling characteristics between cars, multi-particle model for dynamics of HST has been constructed and has became the main basis for the study [4]. Based on this model, controller combined with nonlinear damping has been proposed depended on multi-particle model to achieve any degree of position tracking accuracy [5]. Moreover, model predictive control(MPC) has been applied to speed and position tracking control for HST [6-7]. Further, some other researchers have been focused on the adaptive speed and position tracking control for dynamics of HST. An optimal controller has been proposed for speed and position tracking control of dynamics of HST [8], but this method only considered single particle model. For tracking control of dynamics of

HST with protection constraints, fuzzy adaptive control has been proposed based on residual nonlinearity approximation [9]. To implement adaptive speed and position tracking control for dynamics of HST based on multi-particle model, an adaptive controller has been designed based on output feedback linearization [10]. Additionally, neural networks have been introduced to improve the tracking accuracy of the adaptive algorithm. Considering the input saturation, a radius basis function network has been proposed to combine with adaptive tracking control for dynamics of HST [11]. Aiming at the speed and position control for dynamics of HST with immeasurable speed, parameter perturbation and external disturbance, a neural network observer-based output feedback trajectory tracking control method has been proposed to estimate the speed and position of the train [12]. However, these studies only considered the ideal running environment which the speed and position of dynamics of HST are real-time.

In fact, the train is often working in complex environments, and the running state of the train inevitably have time-delays due to bad weather or severe working condition. Thus, research on speed and position tracking control under state time-delays is worth noting, and researches have been reports in this field. To achieve adaptive position tracking control for dynamics of HST with time-delays [13], the piecewise constant model with unknown parameters under different train operation conditions is constructed. By designing a sampled-data tracking controller, the controller design for dynamics of HST has been transformed to the problem of stability analysis of time-varying delays system [14], but the state time-delay is ignored. Further, considering the state time-varying delays caused bad weathers, construction works, human operation, etc., a roust speed and position tracking controller is designed based on multi-particle model [15], but another factor- wind gust is not considered in this investigation.

Due to the characteristics of HST, as the speed is very quick and the relative position of each car is constraint in a very limit interval, system stability over certain time periods is more worth researching. Thus, researching on speed and position tracking control for dynamics of HST in finite time is of importance. However, the investigations about dynamics of HST with time-delays as mentioned above only considered the asymptotic stability of HST instead of the finite time stability. Based on these reasons, a robust speed and position tracking control method is proposed for dynamics of HST with state time-varying delay in finite time. Wind gust is considered as the additional factors, and is seen as part of external disturbance, which makes the research in this paper is more realistic. The delay-dependent controller is designed to track the speed and position in finite time. The finite time boundness(FTB) of the closed-loop dynamic error is analyzed, and the tracking controller design is formulated as the linear matrix inequality(LMI), which has been developed to tracking controller design for dynamics of HST in decades [16-17], so that the optimization problem is easier to solve. Further, by H-infinity analysis of the design, the robustness of the proposed controller to wind gust and the additive additive resistance is guaranteed. The contributions of this paper is summarized as follows.

Wind gust which is an important factor for dynamics of HST is concerned and is seen as the external disturbance, which is more realistic to the working conditions of the train.

State time-varying delay is considered, and a delay-dependent controller is designed. By this way, the speed and position tracking control for dynamics of HST with time-delays is implemented even under wind gust.

FTB of the closed-loop dynamic error is analyzed so that the speed and position tracking control for dynamics of HST is implemented in finite time, and the robustness of the controller to the wind gust and the external disturbance is guaranteed.

This paper is organized as follows. In section 2, considering the relative speed and relative position of each car comparing with the cruising model in equilibrium state, a relative dynamics model of HST is established based on multi-particle model. State time-delays are considered due to bad weather or severe conditions, and dynamics of HST with state time-varying delays is established. In section 3, a robust finite time controller(RFTC) for speed and position tracking is designed, its FTB is analyzed, and the robustness of the closed-loop error dynamics to wind gust and the additive resistance is guaranteed. The sufficient condition for the existence of the controller is given, and the design problem is formulate as LMI problem. In section 4, a 5-car HST model is given for simulation to show the effectiveness of the proposed method. Conclusions are drawn at the end.

2. System modeling for dynamics of HST

In this section, dynamics of HST are modeled by a cascade of cars connected with flexible couplers considering the rolling mechanical resistance, aerodynamic drag and wind gust, and the force diagram of HST is shown in Figure 1.

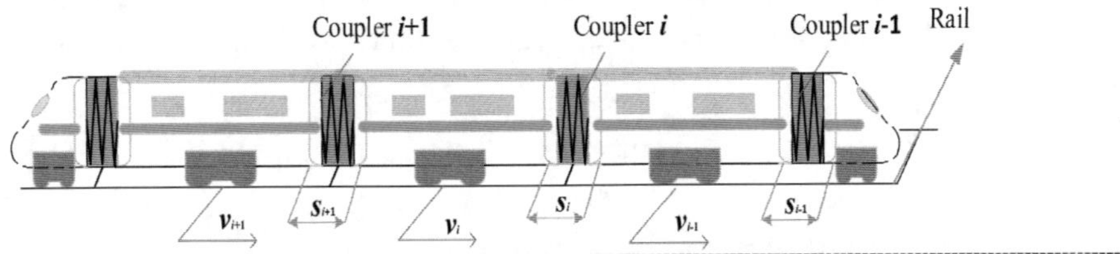

Figure 1. Force diagram of high speed train.

The couplers play an important role in the connected cars and provide the traction/braking force in longitudinal direction, and the motion of the couplers is closer to a spring model which is described as

$$f(\varepsilon) = k\varepsilon \tag{1}$$

where $k > 0$ is the stiffness coefficient. Based on the above discussion, the running resistance consists of aerodynamic drag and rolling mechanical resistance for HST is formulated as

$$R_a = l_0 + l_1 v + l_2 v^2 \tag{2}$$

where v is the car velocity, the coefficients l_0, l_1, l_2 are Davis coefficients.

Based on(1) and (2), dynamics of HST is established as

$$\begin{cases} \dot{x}_i(t) = v_i(t) - v_{i+1}(t), i = 1, \cdots, n-1 \\ m_1\dot{v}_1(t) = u_1(t) - kx_1(t) - (l_0 + l_1 v_1(t))m_1 - l_2(\sum_{i=1}^{n} m_i)v_1^2(t) + w(t) \\ m_i\dot{v}_i(t) = u_i(t) + kx_{i-1}(t) - kx_i(t) - (l_0 + l_1 v_i(t))m_i, i = 2, \cdots, n-1 \\ m_n\dot{v}_n(t) = u_n + kx_{n-1}(t) - (l_0 + l_1 v_n(t))m_n \end{cases} \tag{3}$$

where x_i is the relative position between car i and car i+1; m_i is the i-th car mass, v_i and u_i are the speed and effort of the i-th car respectively, w is the wind gust which only affects on the first car.

Assume that the equilibrium state has a desired speed $v_1 = \cdots = v_n = v_r$, and the relative position between adjacent cars are zero at the equilibrium state, i.e., $\bar{x}_1 = \bar{x}_2 = \cdots = \bar{x}_{n-1} = 0$,and $\bar{u}_1$ satisfies that

$$\bar{u}_1 = c_0 m_1 + c_1 m_1 v_r + c_2(\sum_{i=1}^{n} m_i)v_r^2, \;\; \bar{u}_i = c_0 m_i + c_1 m_i v_r, i = 2,3,\ldots,n. \tag{4}$$

Moreover, consider that the train often runs in bad weather or severe working condition so that time-varying delays exist in speed, based on (3) and (4), a dynamics model of HST is established as

$$\begin{cases} \dot{\hat{x}}_i(t) = \alpha[\hat{v}_i(t) - \hat{v}_{i+1}(t)] + \beta[\hat{v}_i(\tau) - \hat{v}_{i+1}(\tau)],\ i = 1, \cdots, n-1 \\ m_1\dot{\hat{v}}_1(t) = \hat{u}_1(t) - \alpha[k\hat{x}_1(t) + l_1 m_1 \hat{v}_1(t) + 2l_2(\sum_{i=1}^{n} m_i) v_r \hat{v}_1(t)] \\ \qquad - \beta[k\hat{x}_1(\tau) + l_1 m_1 \hat{v}_1(\tau) + 2l_2(\sum_{i=1}^{n} m_i) v_r \hat{v}_1(\tau)] + w(t) \\ m_i\dot{\hat{v}}_i(t) = \hat{u}_i(t) + \alpha[k\hat{x}_{i-1}(t) - k\hat{x}_i(t) - l_1 m_i \hat{v}_i(t)] + \beta[k\hat{x}_{i-1}(\tau) - k\hat{x}_i(\tau) - l_1 m_i \hat{v}_i(\tau)],\ i = 2, \cdots, n-1 \\ m_n\dot{\hat{v}}_n(t) = \hat{u}_n(t) + \alpha[k\hat{x}_{n-1}(t) - l_1 m_n \hat{v}_n(t)] + \beta[k\hat{x}_{n-1}(\tau) - l_1 m_n \hat{v}_n(\tau)] \end{cases} \tag{5}$$

where $\tau = t - d(t)$, $d(t)$ is time-varying and satisfies that $d(t) \le q$. $\hat{x}_i = x_i - \overline{x}_i$, $\hat{v}_i(t) = v_i - v_r$, $\hat{u}_i(t) = u_i - \overline{u}_i$.

Define $\tilde{x}(t) = [\hat{x}_1(t), \hat{x}_2(t), ..., \hat{x}_{n-1}(t), \hat{v}_1(t), \hat{v}_2(t), ..., \hat{v}_n(t)]^T$, $u(t) = [\hat{u}_1(t), \hat{u}_2(t), \cdots, \hat{u}_n(t)]^T$, considering that there exists additive resistance $\varphi(t)$, then the following systems for dynamics of HST with time-varying delays is established as

$$\dot{\tilde{x}}(t) = A\tilde{x}(t) + Bu(t) + A_d\tilde{x}(\tau) + Dw(t) + E\varphi(t) \tag{6}$$

where

$$A = \begin{bmatrix} 0 & A_{12} \\ A_{22} & A_{22} \end{bmatrix}, B = \begin{bmatrix} 0 \\ N \end{bmatrix}, N = \mathrm{diag}(m_1^{-1}, \cdots m_n^{-1}), D = \begin{bmatrix} 0_{n\times 1} \\ 1 \\ 0_{n\times 1} \end{bmatrix}, A_{11} = \begin{bmatrix} \alpha & -\alpha & 0 & 0 & ... \\ 0 & 1 & -\alpha & 0 & ... \\ 0 & ... & 0 & 1 & -\alpha \end{bmatrix}_{(n-1)\times n},$$

$$A_{12} = \begin{bmatrix} -\frac{k}{m_1}\alpha & 0 & 0 & 0 & ... \\ \frac{k}{m_2}\alpha & -\frac{k}{m_2}\alpha & 0 & 0 & ... \\ 0 & 0 & ... & \frac{k}{m_{n-1}}\alpha & -\frac{k}{m_{n-1}}\alpha \\ 0 & 0 & ... & 0 & \frac{k}{m_n}\alpha \end{bmatrix}_{n\times(n-1)}, \quad A_{21} = \begin{bmatrix} -c_1\alpha - \frac{2\alpha c_2(\sum_{i=1}^{n} m_i)v_r}{m_1} & 0 & 0 & ... \\ 0 & -c_1\alpha & 0 & ... \\ 0 & ... & 0 & -c_1\alpha \end{bmatrix}_{n\times n},$$

and A_d has the similar structure with A by replacing α by β. α and β satisfies that $\alpha + \beta = 1$, $\alpha = 1$ represents that the train is working in normal conditions, and $\alpha = 0$ means that the train is working in extreme severe condition, and $0 \le \beta \le 1$ represents the degrees of time-delays. E is the distribution matrix of $\varphi(t)$, then (6) is written as

$$\dot{\tilde{x}}(t) = A\tilde{x}(t) + Bu(t) + A_d\tilde{x}(\tau) + E_w d(t) \tag{7}$$

where $E_w = [D \quad E], d(t) = [w^T(t) \quad \varphi^T(t)]^T$ is composed of wind gust and additive resistance which is seen as the external disturbance.

3. Speed and tracking controller design for dynamics of HST

In this section, to implement the speed and position tracking control of HST in finite time, a delay dependent FTRC is designed, FTB index and robust H infinity index are analyzed to guarantee the robustness of the controller to the external disturbance $d(t)$.

The designed speed and position tracking FTRC has the structure as

$$u(t) = K\tilde{x}(t) + H\tilde{x}(\tau) \tag{8}$$

where $K \in R^{n\times(n-1)}$, $H \in R^{n\times(n-1)}$ are control parameters to be determined.

Substitute (8) into (7), then a closed-loop dynamic error equation is obtained as

$$\dot{\tilde{x}}(t) = (A + BK)\tilde{x}(t) + (A_d + BH)\tilde{x}(\tau) + E_w d(t) \tag{9}$$

Definition 1[18] Given positive scalars $c_1 > 0$, $T > 0$, a weighting matrix $U > 0$, $x(t)$ is FTB about (c_1, c_2, U, T) if the following equations hold

$$\dot{x}(t) = Ax(t) + Bw(t), x(0) = x_0 \tag{10}$$

$$\int_0^T w^T(t)w(t) < c_1 \Rightarrow x^T(t)Ux(t) < c_2, \ \forall t \in [0,T] \tag{11}$$

where c_1, c_2 are constants, T is a constant which denotes the length of finite time window.

The purpose of this paper is to design a speed and position tracking controller (9) to guarantee the following conditions

(a) For any none zeros $\|d(t)\|_2 \le c_1, d(t) \in [0,T]$, $\tilde{x}(t)$ in (9) is FTB about (c_1, c_2, U, T);

(b) $\tilde{x}(t)$ and $d(t)$ satisfy that

$$\int_0^T \|\tilde{x}(t)\|_2^2 \, dt \le \gamma^2 \int_0^T \|d(t)\|_2^2 \, dt \tag{12}$$

be satisfied at zero initial conditions.

Lemma 1 For constant matrix R>0, scalar b>0, vector function $x:[0,b] \to R^n$, the following inequality holds

$$b\int_0^b x^T(s)Rx(s)ds \ge \left(\int_0^b x(s)ds\right)^T R\left(\int_0^b x(s)ds\right) \tag{13}$$

Theorem 1 Consider the dynamic equation (9). Given a weighting matrix U, positive scalars q, σ, c_1 and δ, if there exist positive definite matrices $\bar{P}, \bar{Q}, \bar{R}$, matrices $Y_1 = \bar{P}K$, $Y_2 = \bar{P}H$ such that

$$\Xi_{i,j} < 0, i, j = 1, \cdots, 6 \tag{14}$$

where $\Xi_{i,j}$ is a symmetry matrix which satisfies that

$\Xi_{11} = A\bar{P} + BY_1 + \bar{P}A^T + Y_1^T B^T + \bar{Q} - \alpha\bar{P}$, $\Xi_{13} = A_d\bar{P} + BY_2 - q^{-1}\bar{R}$, $\Xi_{14} = E_w$, $\Xi_{15} = q^{1/2}(\bar{P}A^T + Y_1^T B^T)$, $\Xi_{16} = \bar{P}$, $\Xi_{22} = -e^{\sigma q}\bar{Q} - q^{-1}\bar{R}$, $\Xi_{23} = q^{-1}\bar{R}$, $\Xi_{33} = -2q^{-1}R$, $\Xi_{35} = q^{1/2}(\bar{P}A_d^T + Y_2^T B^T)$, $\Xi_{44} = -\gamma^2 I$, $\Xi_{45} = q^{1/2}E_w^T$, $\Xi_{55} = -2\bar{P} + \bar{R}$, $\Xi_{66} = -I$.

$$\delta c_1 < c_2 e^{-\sigma T} \tag{15}$$

then indexes (a) and (b) are satisfied simultaneously.

Proof. For the error dynamic equation (9), choose the Lyapunov-Krasovskii candidate as

$$V(t) = V_1(t) + V_2(t) + V_3(t) \tag{16}$$

where $V_1(t) = \tilde{x}^T(t)P\tilde{x}(t)$, $V_2(t) = \int_{t-q}^t e^{\sigma(t-s)}\tilde{x}^T(s)Q\tilde{x}(s)ds$, $V_3(t) = \int_{-q}^0 \int_{t+\theta}^t \dot{\tilde{x}}^T(s)R\dot{\tilde{x}}(s)dsd\theta$.

Calculating the derivative of $V(t)$ along the trajectory of (9) yields

$$\begin{aligned} \dot{V}(t) &= 2\tilde{x}^T(t)P[(A+BK)\tilde{x}(t) + (A_d + BH)\tilde{x}(\tau)] + \sigma\int_{t-q}^t e^{\sigma(t-s)}\tilde{x}(t)Q\tilde{x}^T(t)ds + \tilde{x}^T(t)Q\tilde{x}(t) \\ &- e^{\sigma q}\tilde{x}^T(t-q)Q\tilde{x}(t-q) + q\dot{\tilde{x}}^T(t)R\dot{\tilde{x}}(t) - \int_{t-q}^t \dot{\tilde{x}}^T(s)R\dot{\tilde{x}}(s)ds \end{aligned} \tag{17}$$

based on lemma 1, the following inequality is obtained as

$$-\int_{t-q}^t \dot{\tilde{x}}^T(s)R\dot{\tilde{x}}(s)ds \le -q^{-1}[\tilde{x}(\tau) - \tilde{x}(t-q)]^T R[\tilde{x}(\tau) - \tilde{x}(t-q)] - q^{-1}[\tilde{x}(t) - \tilde{x}(\tau)]^T R[\tilde{x}(t) - \tilde{x}(\tau)] \tag{18}$$

then combining (17) and (18) yields

$$\dot{V}(t) \le \varsigma^T X\varsigma + q\varsigma^T \bar{X}^T R\bar{X}\varsigma + \sigma\int_{t-q}^t e^{\sigma(t-s)}\tilde{x}^T(s)Q\tilde{x}(s)ds \tag{19}$$

where $\varsigma = \left[\tilde{x}^T(t) \ \ \tilde{x}^T(t-q) \ \ \tilde{x}^T(\tau) \ \ d^T(t)\right]^T$, $X = \left[A+BK \ \ 0 \ \ A_d + HK \ \ E_w\right]$, and $X_{i,j} < 0, i, j = 1, \cdots, 4$ is a symmetry matrix which satisfies that

$X_{11} = PA + PBK + A^T P + K^T B^T P + Q$, $X_{13} = P(A_d + HK) - q^{-1}R$, $X_{14} = PE_w$, $X_{22} = -e^{\sigma q}Q - q^{-1}R$,

$X_{23}=q^{-1}R$, $X_{33}=-2q^{-1}R$.

On the other hand, from (14) it is known that

$$\Upsilon+\bar{\Upsilon}\bar{\Upsilon}^{T}<0 \tag{20}$$

where $\bar{\Upsilon}=\begin{bmatrix}\bar{P} & 0 & 0 & 0 & 0\end{bmatrix}^{T}$, and $\Upsilon_{i,j}<0,i,j=1,\cdots,5$ is a symmetry matrix which satisfies that $\Upsilon_{11}=\Xi_{11}$, $\Upsilon_{13}=A_{d}\bar{P}+BY_{2}-q^{-1}\bar{R}$, $\Upsilon_{14}=E_{w}$, $\Upsilon_{15}=q^{1/2}(\bar{P}A^{T}+Y_{1}^{T}B^{T})$, $\Upsilon_{22}=-e^{\sigma q}\bar{Q}-q^{-1}\bar{R}$, $\Upsilon_{23}=q^{-1}\bar{R}$ $\Upsilon_{33}=-2q^{-1}\bar{R}$, $\Upsilon_{35}=q^{1/2}(\bar{P}A_{d}^{T}+Y_{2}^{T}B^{T})$, $\Upsilon_{44}=-\gamma^{2}I$, $\Upsilon_{45}=q^{1/2}E_{w}^{T}$, $\Upsilon_{55}=-2\bar{P}+\bar{R}$.

As $\bar{P}=P^{-1}>0$, then $\bar{\Upsilon}\bar{\Upsilon}^{T}>0$. Thus, from (20) it is easy to known that

$$\Upsilon<0 \tag{21}$$

Let $P=\bar{P}^{-1}$, $Q=\bar{P}^{-1}\bar{Q}\bar{P}^{-1}$, $R=\bar{P}^{-1}\bar{R}\bar{P}^{-1}$. Pre- and post-multiplying both sides of (21) by $\mathrm{diag}(\bar{P}^{-1},\bar{P}^{-1},\bar{P}^{-1},I,\bar{P}^{-1})$, then (21) is represented as

$$\Theta_{i,j}<0,i,j=1,\cdots,5 \tag{22}$$

where Θ is a symmetry matrix which satisfies that
$\Theta_{11}=X_{11}-\sigma P$, $\Theta_{13}=P(A_{d}+HK)-q^{-1}R$, $\Theta_{14}=PE_{w}$, $\Theta_{15}=q^{1/2}(A^{T}+K^{T}B^{T})P$, $\Theta_{22}=-e^{\sigma q}Q-q^{-1}R$, $\Theta_{23}=q^{-1}R$, $\Theta_{33}=-2q^{-1}R$, $\Theta_{35}=q^{1/2}(A_{d}^{T}+H^{T}B^{T})P$, $\Theta_{44}=-\gamma^{2}I$, $\Theta_{45}=q^{1/2}E_{w}^{T}P$, $\Theta_{55}=-2P+R$.

As $-PR^{-1}P\le-2P+R$, (22) implies the following inequality hold

$$\Pi_{i,j}<0,i,j=1,\cdots,5 \tag{23}$$

where Π is a symmetry matrix which satisfies that
$\Pi_{11}=X_{11}-\sigma P$, $\Pi_{13}=P(A_{d}+HK)-q^{-1}R$, $\Pi_{14}=PE_{w}$, $\Pi_{15}=q^{1/2}(A^{T}+K^{T}B^{T})P$, $\Pi_{22}=-e^{\sigma q}Q-q^{-1}R$,
$\Pi_{23}=q^{-1}R$, $\Pi_{33}=-2q^{-1}R$, $\Pi_{35}=q^{1/2}(A_{d}^{T}+H^{T}B^{T})P$, $\Pi_{44}=-\gamma^{2}I$, $\Pi_{45}=q^{1/2}E^{T}P$, $\Pi_{55}=-PR^{-1}P$
pre- and post-multiplying both sides of (23) by $\mathrm{diag}(I,I,I,P^{-1})$, then using Schur complement yields

$$\varsigma^{T}X\varsigma+q\varsigma^{T}\bar{X}^{T}R\bar{X}\varsigma-\sigma V(t)-\gamma^{2}d^{T}(t)d(t)+\sigma\int_{t-q}^{t}e^{\sigma(t-s)}\tilde{x}(t)Q\tilde{x}^{T}(t)ds+\sigma\int_{-q}^{0}\int_{t+\theta}^{t}\dot{\tilde{x}}^{T}(s)Z\dot{\tilde{x}}(s)dsd\theta<0 \tag{24}$$

Moreover, from the definition of $V_{3}(t)$ it is known that $\sigma\int_{-q}^{0}\int_{t+\theta}^{t}\dot{\tilde{x}}^{T}(s)Z\dot{\tilde{x}}(s)dsd\theta=\sigma V_{3}(t)\ge0$, then (19) and (24) implies

$$\dot{V}(t)-\sigma V(t)-\gamma^{2}d^{T}(t)d(t)\le0 \tag{25}$$

Integrate (25) from 0 to T, considering the zero initial condition and using the Granwall inequality, the following inequality is obtained as

$$V(t)<\gamma^{2}e^{\sigma t}\int_{0}^{t}d^{T}(s)d(s)ds\le\gamma^{2}c_{1}e^{\sigma T},t\in[0,T] \tag{26}$$

Additionally, as

$$\lambda_{\min}(\tilde{P})\tilde{x}^{T}(t)U\tilde{x}(t)\le\tilde{x}^{T}(t)P\tilde{x}(t)\le V(t) \tag{27}$$

where $\tilde{P}_{1}=U^{-1/2}PU^{-1/2}$. from (25)- (27), the following equation is obtained as

$$\tilde{x}^{T}(t)U\tilde{x}(t)\le(\lambda_{\min}(\tilde{P}))^{-1}e^{\sigma T}\gamma^{2}c_{1} \tag{28}$$

From (15) and (28), the following equation is obtained as

$$\tilde{x}^{T}(t)U\tilde{x}(t)<c_{2},\forall t\in\left[0,T\right] \tag{29}$$

(29) meets definition 1, so $\tilde{x}(t)$ of dynamic equation (9) is FTB.

Further, considering that $-PR^{-1}P\le-2P+R$, by pre- and post-multiplying both sides of (14) by $\mathrm{diag}(\bar{P}^{-1},\bar{P}^{-1},\bar{P}^{-1},I,\bar{P}^{-1},\bar{P}^{-1})$ yields

$$\begin{bmatrix} \Pi & \bar{\Pi}' \\ * & -PP \end{bmatrix} < 0 \tag{30}$$

where $\bar{\Pi}' = \begin{bmatrix} P & 0 & 0 & 0 & 0 \end{bmatrix}^T$, * represents the symmetric elements in a matrix.

Then pre- and post-multiplying both sides of (30) by diag(I,I,I,I,P^{-1}) yields

$$\begin{aligned} &\varsigma^T X\varsigma + q\varsigma^T \bar{X}^T R\bar{X}\varsigma - \sigma V(t) - \gamma^2 w^T(t)w(t) \\ &+\sigma\int_{t-h}^{t} e^{\sigma(t-s)}\tilde{x}(t)Q\tilde{x}^T(t)ds + \sigma\int_{-q}^{0}\int_{t+\theta}^{t}\dot{\tilde{x}}^T(s)Z\dot{\tilde{x}}(s)dsd\theta + x^T(t)x(t) < 0 \end{aligned} \tag{31}$$

from (19) and (31) it is known that

$$\dot{V}(t) < \sigma V(t) + \gamma^2 d^T(t)d(t) - \tilde{x}^T(t)\tilde{x}(t) \tag{32}$$

Integrate (32) from 0 to T, considering the initial condition and using the Granwall inequality, the following equation is obtained as

$$V(t) < \gamma^2 e^{\sigma t}\int_0^t d^T(s)d(s)ds - \int_0^t \tilde{x}^T(s)\tilde{x}(s)ds, t\in[0,T] \tag{33}$$

as $V(t)\geq 0$, (33) is equivalent to (12), index (b) is satisfied, then theorem 1 is proved.

By theorem 1, the calculation of the gain matrices K and H of the controller (8) is formulated as the LMIs (14) and (15), and the designed FTRC guarantees that $\tilde{x}$ is robust to the external disturbance $d(t)$ which composed of wind gust and additive resistance in finite time.

4. Simulation results and analysis

In this section, the model of wind gust and desired speed are given in (34), and the parameters of 5-car HST model is established to show the effectiveness of the designed FTRC [14].

$$w(t) = \begin{cases} -0.05(1+\cos(t)), & \text{if } 300 \leq t \leq 330, \\ -0.1(1+\cos(t)), & \text{if } 700 \leq t \leq 730, \\ 0, & \text{otherwise.} \end{cases}, \quad v_r = \begin{cases} 230\,\text{km/h}, & \text{if } 100\text{s} \leq t \leq 600\text{s}, \\ 220\,\text{km/h}, & \text{if } 600\text{s} \leq t \leq 800\text{s}, \end{cases} \tag{34}$$

The additive resistance $\varphi(t)$ and its disturbance is given as

$$\varphi(t) = 0.05\cos(x_4 x_5), \; E=[0,0,0,0,1,0,0,0,0]^T$$

The relative scalars are given as σ=0.8, $c_1 = 0.6$, $q = 0.2$. By theorem 1, γ and c_2 are obtained as γ=0.6, $c_2 = 0.3$. The gain K and H of FTRC are obtained as $K = Y\bar{P}^{-1}$ and $H = Y_2\bar{P}^{-1}$. To keep the paper concise, the controller gain K and H are not presented here as their dimensions are very large.

Figure 2 and Figure 3 show the control results of speed and relative displacement under the control of FTRC proposed in this paper when wind gust and additive resistance existed.

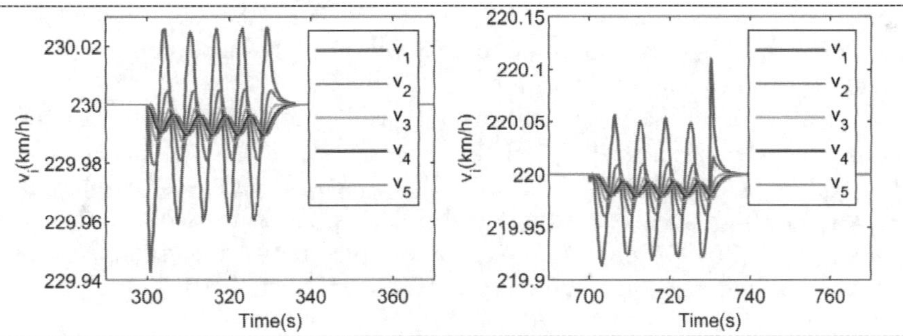

Figure 2. The velocity curves for each car of HST using FTRC

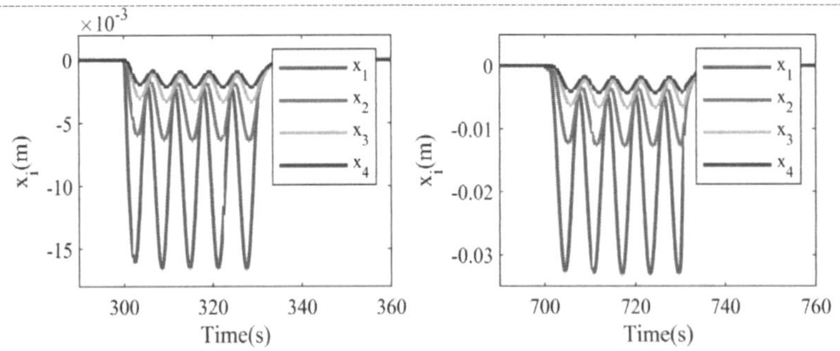

Figure 3. The relative displacements between the two neighboring cars of HST using FTRC

Figure 2 and figure 3 show that when there is no disturbance in dynamics of HST, using the propose FTRC, the speed of each car is tracked as the desired speed accurately, and the relative position between the two neighboring cars remains zero. When the disturbance appear at $300 < t < 330$ and $700 < t < 730$, the fluctuations speed of each car are reduced to the reasonable range of 0.06 km/h and 0.1 km/h respectively, and the fluctuations of relative displacements are reduced to the reasonable range of 0.015m and 0.035m. It is obvious that the proposed FTRC in this paper not only enables the desired speed to be accurately tracked, but also has good robustness performance when disturbance occurs. Simulation results show that, as the dynamics error not only is FTB but also is robust to wind gust and the additive resistance, the speed and position control is well implemented in finite time even under complex situation.

5. Conclusion

This paper proposes a FTRC for speed and position tracking of dynamics of HST considering the time-varying delay caused by bad weather or server working conditions. The finite time control for dynamic of HST is well implemented using this FTRC, because of the controller not only guarantees the closed-loop dynamics error is FTB, but also guarantee the closed-loop dynamics error satisfy the H-infinity index in finite time. By this way, the robustness of the dynamics error to wind gust and additive resistance is guaranteed in finite time. To show the effectiveness of the design, 5-car model is given in simulation. The simulation results show that, for dynamics of HST, the designed FTRC has excellent tracking ability for speed and position in finite time even the train is working in time-varying delays under severe conditions.

References

[1] Li, ZQ. (2015) Adaptive speed tracking control for high-speed EMU. Nanchang University, The Nanchang.

[2] Faieghi, M., Jalali, A., Mashhadi, SKEDM., Baleanu, D. (2018) Passivity-based cruise control of high speed trains. J Vib Control., 24: 492–504.

[3] Yang, YZ., et al. (2018) Optimal operation of high-speed trains using hybrid model predictive control. J Adv Transport., 2018: 1–16.

[4] He, J., et al. (2020) Tracking control via sliding mode for heavy-haul trains with input saturation. Meas Control-UK., 53: 1720–1729.

[5] Mao, ZH., Tao, G., Jiang, B., Yan, XG. (2018) Adaptive actuator compensation of position tracking for high-speed trains with disturbances. IEEE T Veh Technol., 67: 5706–5717.

[6] Xu, XK., et al. (2019) Adaptive model predictive control for cruise control of high-speed trains with time-varying parameters. J Adv Transport., 2019: 1–11.

[7] Li, SK., Yang, LX., Gao, ZY. (2020) Distributed optimal control for multiple high-speed train movement: An alternating direction method of multipliers. Automatica., 74: 1–8.

[8] Wu, X., Zhang, KJ., Cheng, M. (2021) Adaptive numerical approach for optimal control of a single train. J Syst SCI Complex., 32: 1053–1071.

[9] Gao, SG., et al. (2020) Fuzzy adaptive automatic train operation control with protection constraints: A residual nonlinearity approximation-based approach. ENG Appl Artif Intel., 96.
[10] Mao, ZH., Tao, G., Jiang, B., Yan, XG. (2021) Adaptive control design and evaluation for multibody high-speed train dynamic models. IEEE T Contr Syst T., 29: 1061–1074.
[11] Guo, XG., et al. (2019) Novel auxiliary saturation compensation design for neuroadaptive NTSM tracking control of HST with actuator saturation. J Franklin I., 357: 1582–1602.
[12] Liu, Y., Li, WD. (2020) Adaptive output feedback control for the trajectory tracking of high-speed trains with disturbance uncertainties on the basis of neural network observers. J Adv Transport., 2020: 1–16.
[13] Mao, ZH., Tao, G., Jiang, B., Yan, XG. (2017) Adaptive position tracking control of high-speed trains with piecewise dynamics. In:2017 American Control Conference.Seattle, WA. pp. 2453–2458.
[14] Li, SK., Yang, LX., Li, KP., Gao, ZY. (2014) Robust sampled-data cruise control scheduling of high speed train. Transport Res C-Emer., 46: 274–283.
[15] Tang, HY., Wang, QY., Feng, XY. (2018) robust stochastic control for high-speed trains with nonlinearity, parametric uncertainty, and multiple time-varying delays. IEEE T Intell Transp., 19: 1027–1037.
[16] Campos, VCD., Nguyen, AT., Palhares, RM. (2021) Adaptive gain-scheduling control for continuous-time systems with polytopic uncertainties: An LMI-based approach., Automatica., 133.
[17] Santra, S., Joby, M., Sathishkumar, M., Anthoni, SM. (2021) LMI approach-based sampled-data control for uncertain systems with actuator saturation: application to multi-machine power system. Nonlinear Dynam., 107, 967–982.
[18] Xue, WP., Mao, WJ. (2013) Asymptotic stability and finite-time stability of networked control systems :analysis and synthesis. Asian J Control., 15: 1376–1384.

Fault Diagnosis Method of Spacecraft Control Systems Based on PCA-ResNet

Xuan Wei[1*], Xiaodong Mu[1], Tao Jiang[1], Weiqiang Liu[1], Zhaoju Zeng[1]

[1]College of Operational Support, Rocket Force University of Engineering, Xi'an 710025, China

[*] Corresponding author's e-mail:287505216@qq.com

Abstract—Due to the complex space environment, spacecraft telemetry signals are accompanied by a large amount of noise, and the accuracy of fault diagnosis is low by directly using the original telemetry signals. This paper proposes a fault diagnosis method for spacecraft control systems based on principal component analysis (PCA) and residual network (ResNet). Firstly, grayscale images are generated by denoising the telemetry signal of the spacecraft control system through PCA; Secondly, the images are input into the residual network to extract deep-level features; Finally, the Softmax classifier is used for classification to realize the fault diagnosis of the spacecraft control system. The research results show that the diagnostic accuracy of the method proposed in this paper reaches 95.33%, which is higher than other diagnostic models, and the method can be used for the actual fault classification of spacecraft control systems.

1. Introduction

Our country's spacecraft undertakes the major responsibility of ensuring national defense security, developing the national economy, exploring space, and benefiting mankind. However, the space environment in which the spacecraft is located is changeable, and the performance of each component continues to decline with the increase of in-orbit operation years, which leads to an increase in the factors that cause the spacecraft to fail. The spacecraft should be diagnosed in time to understand the health of each subsystem. The situation is of practical significance to the safe and stable operation of the spacecraft.

In recent years, fault diagnosis technology has received extensive attention[1, 2], the diagnosis methods are changing with each passing day, and there are model-based, signal processing, and data-driven fault diagnosis methods[3-5]. At present, fault diagnosis methods based on model and signal processing are widely used, but most of these methods are based on the shallow phenomenon of faults, the excavation of data is not deep enough; Based on the data-driven[6, 7]method, the system failure model is established through a large amount of historical data, and the system diagnosis model is established by using the spacecraft development data, ground test data, and the telemetry data, which overcomes the defect of insufficient data utilization of the above method. As the complexity of each subsystem of the spacecraft increases, the amount of data generated increases exponentially, and the data-driven fault diagnosis method is very suitable for the spacecraft system.

Methods based on machine learning and deep learning are an important part of data-driven fault diagnosis methods, which support vector machine (SVM)[8] as a kind of machine learning method are widely used. Zhang[9] proposed a new method of diagnose based on convolutional neural network (CNN) and support vector machine (SVM). Zhao[10] proposed a fault diagnosis model based on the fusion of local mean decomposition (LMD) sample entropy and support vector machine method. However, traditional machine learning methods cannot automatically generate features, it requires extensive expert knowledge, it requires a lot of time and human resources[11, 12].

Different from traditional machine learning methods, fault diagnosis methods based on deep learning learn the deep features of data by deepening the network structure, and achieving high fault diagnosis accuracy while automatically extracting features. Wen[13]proposed a diagnosis model based on ResNet-50 to diagnose the motor bearing data. Zhang[14]designed a fault diagnosis method for wind turbine gearboxes based on channel attention residual network. Zhang[15]established the relationship between the feature transformation and the input signal, and achieved 90.00% and 90.77% diagnostic accuracy on two datasets. Wang[16] proposed a residual network fault diagnosis method combined with transfer learning, achieving a high-precision diagnosis of bearing datasets. However, the telemetry signal is different from the signal data of the bearing, the space environment where the spacecraft located is complex, and the telemetry signals received on the ground are often accompanied by a large amount of noise, using the original telemetry signals directly will lead to the low diagnostic accuracy.

In view of the above problems, this paper proposes a fault diagnosis method for spacecraft control system based on principal component analysis and residual network (PCA-ResNet). Firstly, the telemetry signal in the failure data set of the spacecraft control system is processed by PCA to realize the noise reduction of the telemetry signal, and then the signal data is converted into a grayscale image. Finally, the residual network is used to extract the features of the image, and the softmax classifier is used to classify the faults, so as to realize the fault diagnosis of the spacecraft control system.

2. Telemetry Signal-Image Processing

2.1. Principal Component Analysis

Principal Component Analysis (PCA) is mainly used in statistical data analysis, data compression and feature extraction, it is an optimal dimensionality reduction method, which can remove redundancy and remove the noise, it has good denoising performance and used in the field of fault diagnosis[17]. In this paper, the PCA method is used to select an appropriate noise reduction ratio, and the original telemetry signal is reduced in dimension and then mapped back to high-dimensional data to realize the noise reduction of the original signal. The mathematical principle is:

Assume that the spacecraft failure data set $X=[x_1,x_2,\cdots,x_n]$ has *n* samples, each sample has *k* features, and each feature corresponds to its eigenvalue.

(1) First, calculate the mean and standard deviation of each group of data, and standardize:

$$\overline{x_i}=\frac{1}{k}\sum_{j=1}^{k}x_{ij} \tag{1}$$

$$z_{ij}=\frac{x_{ij}-\overline{x_i}}{\sqrt{\delta(x_i)}} \tag{2}$$

In the formula, $\overline{x_i}$ is the feature mean of each sample, $\delta(x_i)$ is the standard deviation of each sample, $i=1,2,\cdots,n$, $j=1,2,\cdots,k$;

(2) Calculate normalization matrix M and covariance matrix *C* for *n* samples;

(3) Calculate the eigenvalues $(\lambda_1,\lambda_2,\cdots,\lambda_k)$ and eigenvectors $a_i=(a_{i1},a_{i2},\cdots,a_{ik})$ of the covariance matrix *C*, $i=1,2,\cdots,k$,and arrange the eigenvalues λ_i in descending order Sort, select the largest top *p*, and take out their corresponding p eigenvectors, we will get a set of $\{(\lambda_1,a_1),(\lambda_2,a_2),\cdots,(\lambda_p,a_p)\}$;

(4) Determine the first *P* principal components. Compute the contribution of the first *P* principal components:

$$\theta(p)=\frac{\sum_{i=1}^{p}\lambda_i}{\sum_{i=1}^{m}\lambda_i} \tag{3}$$

The larger the contribution rate, the more original information contained in the sample features. Generally, when the cumulative contribution rate is $\theta(p)\geq 85\%$, it is considered that the sample feature retains the information of the original data.

(5) Project the original data onto the selected feature vector to obtain the dimension-reduced p dimensional features. For each sample x_i, the original feature is $(x_{i1},x_{i2},\cdots,x_{ik})^T$, and the new feature after projection is $(y_{i1},y_{i2},\cdots,y_{ip})^T$, the calculation formula is as follows:

$$\begin{bmatrix} y_{i1} \\ y_{i2} \\ \vdots \\ y_{ip} \end{bmatrix} = \begin{bmatrix} a_1^T\cdot(x_{i1},x_{i2},\cdots,x_{ik})^T \\ a_2^T\cdot(x_{i1},x_{i2},\cdots,x_{ik})^T \\ \vdots \\ a_p^T\cdot(x_{i1},x_{i2},\cdots,x_{ik})^T \end{bmatrix} \tag{4}$$

where $Y=(y_{ij})_{n\times p}=(y_1,y_2,\cdots,y_p)^T$

(6) Restore the reduced dimension matrix Y to the original dimension matrix X':

$$X'=A\bullet Y \tag{5}$$

Where $A=(a_1,a_2,\cdots,a_p)$

2.2. *Signal-image conversion*

Processing the one-dimensional telemetry signal into a two-dimensional image is an effective method, it can retain the characteristics of the original signal and facilitate the residual network to perform feature extraction through the enlarged "field of view". Wen and Gong[18, 19]Normalized the maximum and minimum values of the bearing fault signal, and then convert it into a two-dimensional grayscale image. However, the telemetry signal values of the spacecraft control system are not concentrated, the value ranges of the flag bits of each component are different, and the maximum and minimum values of the telemetry signals are unstable. The normalization effect is not good, which affects the efficiency of fault diagnosis. In this paper, the telemetry signal data of the spacecraft control system has been standardized after PCA processing, which can effectively avoid the problems caused by direct data normalization.

To achieving a two-dimensional grayscale image, the processed data is rounded and converted into grayscale values, the method is as follows:

$$\varphi(i)=round(x'_i\times 255) \tag{6}$$

Where x'_i are the spacecraft telemetry signal values, $round()$ are the rounding functions, $\varphi(i), i=1,2,\cdots,k$ are the image pixel values, and the value range is [0,255].

Secondly, segment the gray scaled telemetry signal, and intercept a segments, each segment contains b data points, and all telemetry values are stacked to obtain a $a\times b$ two-dimensional matrix. At this time, the telemetry signal of the spacecraft control system is converted into grayscale with a length of a, a width of b, and $a\times b$ pixels, which implements the conversion of the telemetry signal into a two-dimensional grayscale image. The conversion method is shown in Fig.1.

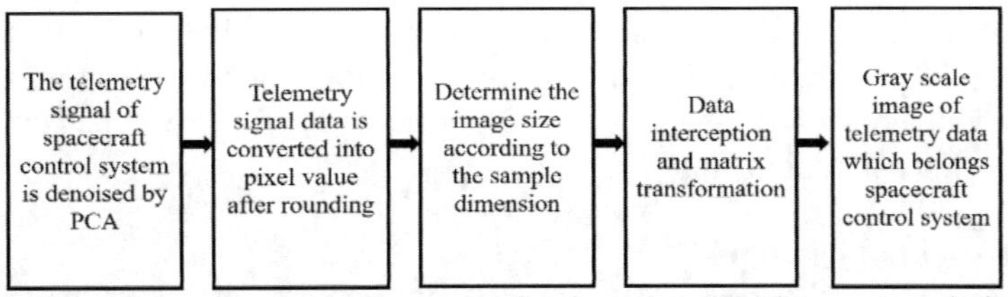

Fig.1 Conversion process of spacecraft telemetry signal to grayscale image

3. ResNet-18

Residual network-18[20]is a classic deep convolutional neural network, designed to solve the problem of network degradation caused by deepening the number of network layers, mainly composed of residual blocks. The input of the residual block is X, the output is $H(X)$, and the residual map is $F(X)$, the basic idea is to perform convolution, max pooling, and Relu activation function operations on the input X, and the output is $F(X)$, then perform convolution and activation function operations on $F(X)$, and connect with the initial input X to get the output $H(X) = F(X) + X$. The residual block can realize the identity mapping of the network, retain the original input features while increasing the number of layers, and avoid network degradation by slowing down the loss of features. Fig.2 (a) shows the structure of the residual block.

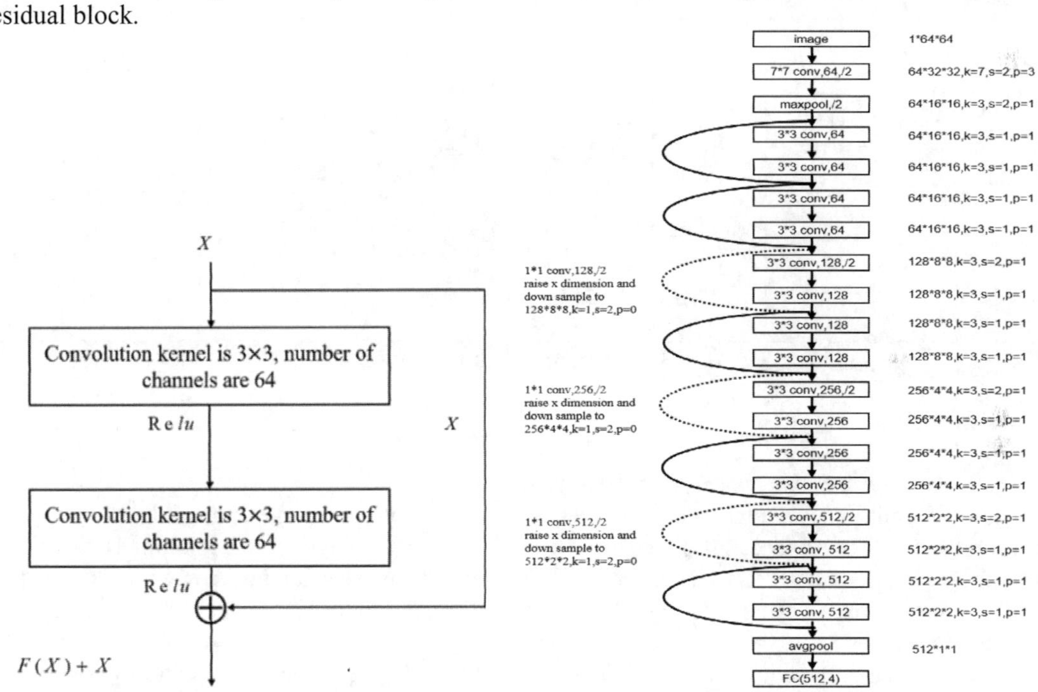

(a) The structure of residual block (b) The structure residual network-18

Fig. 2 The structure of ResNet-18

The feature extraction process of the residual network is shown in Fig.2 (b). First, the grayscale image of the spacecraft telemetry signal is used as the residual network input (the size of the input image is $1 \times 64 \times 64$, where 1, 64, and 64 respectively represent the number of channels, height, and width), through 7×7 convolution operation and maximum pool processing; Then input the feature map into the residual block whose channel number is 64, and after 3×3 convolution operation, the size of the feature map is $64 \times 16 \times 16$, because the number of output channels is 64, and the residual blocks are connected by solid lines; the feature map is taken as input, on the one hand, it is dimensioned and down sampled to a size of $128 \times 8 \times 8$, and on the other hand, when add the feature map after the 3×3 convolution operation and the after dimension increase, the image size is $128 \times 8 \times 8$. The dimension of the upper and lower network feature maps connected by the solid line in the figure has not changed and can be directly added. The dimension of the upper and lower network feature maps connected by the dotted line has doubled, so that the input and output cannot be directly superimposed, it can be add after increasing the dimension of the input x.

4. Fault diagnosis method based on PCA-ResNet

The fault diagnosis process of the spacecraft control system is divided into four steps: data preprocessing, telemetry signal-image conversion, diagnostic model training, and fault diagnosis, as shown in Fig.3.

(1) Data preprocessing. Firstly, PCA noise reduction processing is performed on the telemetry signal of the spacecraft control system fault data set and the collected spacecraft telemetry signal;

(2) Telemetry signal-image conversion. Using the signal-image conversion method from the telemetry signals of the two datasets to generate the required images of the failure set and validation set;

(3) Diagnostic model training. After inputting the fault set image into the ResNet-18, the feature vector of $512 \times 1 \times 1$ was obtained, when the feature through the fully connected layer, the Softmax classifier directly outputs the loss value and accuracy;

(4) Fault diagnosis. The validation set images are input into the model, the fault type is judged, and the diagnostic accuracy is calculated.

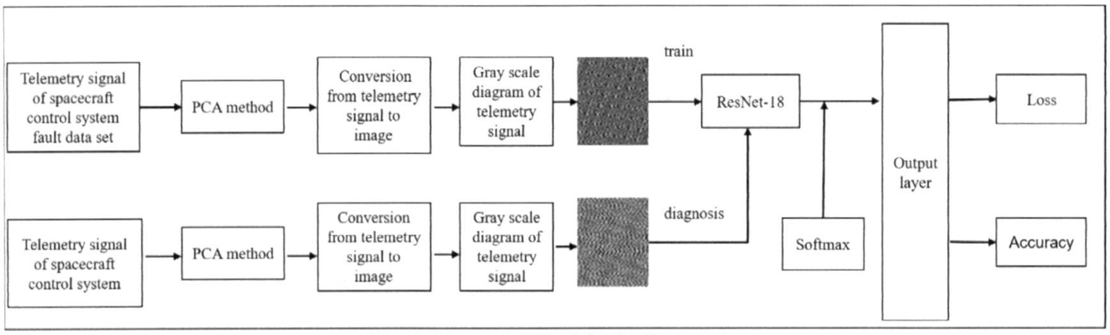

Fig.3 Flow chart of fault diagnosis

5. Case analysis

5.1. *The result of processing spacecraft telemetry signal*

The spacecraft control system fault data set and validation set used in this paper are from historical data and ground test data of a spacecraft. Fault 1 contains 32,320 sets of data, fault 2 contains 16,640 sets of data, fault 3 contains 18,240 sets of data, and fault 4 contains 23360 sets of data, each set of data contains 64 data volumes. First, all sample data are subjected to PCA noise reduction processing, and then signal image conversion is performed. Since each group of data is 64-dimensional, in order to facilitate image processing, the size of the converted image is set to 64×64, and 80% of them are randomly selected. As training set samples, 20% as a validation set samples. The converted results are shown in table 1.

Table 1. Signal-image conversion results

Fault type	Sample size	Sample length	Image size	Image number	Image number of training set	Image number of validation set	label
Fault 1	32320	64	64 × 64	505	404	101	0
Fault 2	16640	64	64 × 64	260	208	52	1
Fault 3	18240	64	64 × 64	285	228	57	2
Fault 4	23360	64	64 × 64	365	292	73	3

After the original telemetry signal is processed by PCA, a lot of noise is removed, which ensures the training effect of the residual network. Fig.4 is the converted grayscale image after the signal is denoised.

Fig.4 Grayscale image of spacecraft telemetry signal

5.2. *Experimental parameter setting*

This experiment is carried out under the framework of deep learning of Pytorch 0.4.1 (Python 3.6). The computer is configured with Intel Xeon E5-2630 v4 20 processors, 64G memory, NVIDIA GeForce GTX 1080Ti graphics card, and Adam optimizer is used in the model training process to optimize the network, configure the learning rate $\alpha = 10^{-3}$, the decay coefficient $\beta_1 = 0.9$, $\beta_2 = 0.999$, small-Value constant $\varepsilon = 10^{-8}$, batchsize is 64, epoch is 800. The validation process is the same as training, and the epoch is 100.

5.3. *Model training and verification*

In this paper, the spacecraft telemetry signal is denoised relying on the PCA method, and the one-dimensional telemetry signal is treated as a large number of numbers. The telemetry signal is mined by statistical principles to extract effective features, and the important feature information is retained according to the cumulative contribution rate of the pivot element. It can be seen from table 2 that the cumulative contribution rate reaches 95.098% of the first six pivot elements, which means the first six pivot elements contain most of the information of the telemetry signal characteristics of the spacecraft control system.

Table 2. Cumulative contribution rate of principal components of spacecraft telemetry signal

Principal component	Characteristic value	Contribution rate /%	Cumulative contribution rate /%
1	7.5039	69.631	69.631
2	1.3215	11.503	81.134
3	0.7561	6.485	87.619
4	0.4331	3.815	91.434
5	0.2649	2.427	93.861
6	0.131	1.237	95.098

In order to study whether the selection of different noise reduction ratios in the PCA method has an impact on the accuracy of fault classification, it is considered that the fault data set selected in this paper comes from the historical data of spacecraft and ground test data, and the cumulative contribution rate of the principal components is greater than 85%, which the accuracy of the entire pivot model is guaranteed. Therefore, this paper selects the PCA noise reduction ratio between 0.85 and 1. Among them, selecting "0.87" for the noise reduction ratio means to retain 87% of the main features of the original spacecraft telemetry signal, which means the first three main elements, and selecting "1" means that the data will not be denoised and the features of original spacecraft telemetry signal will be preserved. The experiments were repeated 10 times at each noise reduction ratio, and the results are shown in Fig.5.

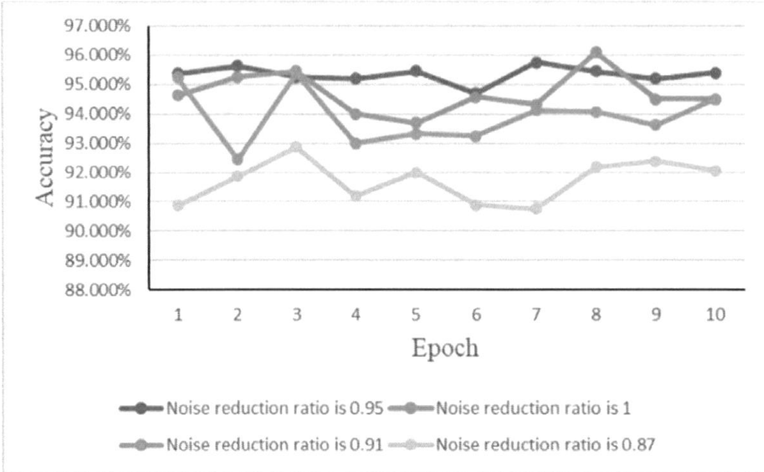

Fig.5 Accuracy under different noise reduction ratios

The statistical results of this experiment can be obtained: The average diagnostic accuracy of the validation set under different noise reduction ratios is above 90%, and the average diagnostic accuracy rate is the lowest when the noise reduction ratio is 0.87, which is 90.750%. On the other hand, when the noise reduction ratio is 1, The highest test accuracy is 96.088%, but it's average is 94.1563%. When the noise reduction ratio is 0.95, the average value of the test is 95.3325%, in several results, the mean value of which is the largest, and the robustness is good. From this table, it can be seen that elect an appropriate proportion of noise reduction which used for the spacecraft fault data set, that can improve the accuracy of diagnostic results to a certain extent.

Table 3. Statistics of diagnostic accuracy under different noise reduction ratios

Noise reduction ratio	Accuracy		
	Maximum	Minimum	Average value
0.87	92.875%	90.75%	91.706%
0.91	95.375%	92.438%	93.8912%
0.95	**95.75%**	**94.688%**	**95.3325%**
1	96.088%	93.685%	94.1563%

Fig.6 (a)shows the variation of the diagnostic accuracy of the validation set in one training with the number of iteration steps when the noise reduction ratio is 0.95. It can be concluded from the figure that with the increase of the number of iteration steps, the accuracy of diagnosis is getting higher and higher. After 70 iterations, the model converges and tends to be stable, and the diagnosis accuracy of the final validation set is 95.3325%

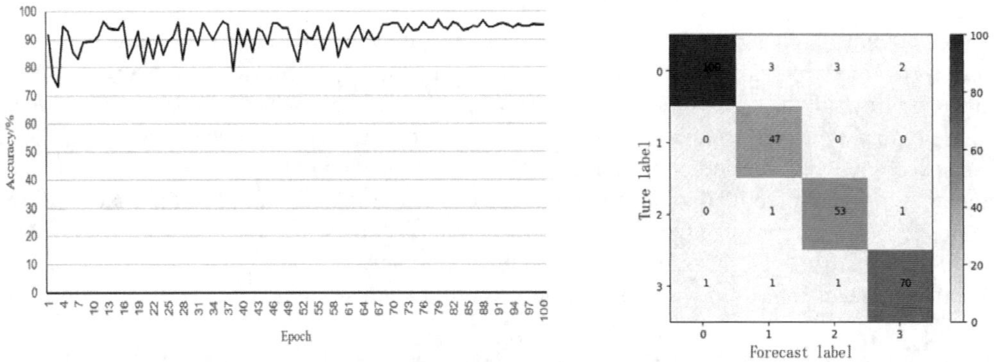

(a) Accuracy under the noise reduction ratio is 0.95 (b) Validation result confusion matrix

Fig. 6 Model training process

To observe the performance of the PCA-ResNet model on the four types of failures, a random test is performed on the validation set and the confusion matrix is plotted. Among them, the horizontal axis represents the true label of the validation set, the vertical axis represents the predicted label of the validation set, and the main diagonal is the number of correct samples predicted by the PCA-ResNet model. It can be seen from (b) that the PCA-ResNet model has high accuracy in identifying various types of faults in the spacecraft control system. The number of fault 1 samples is large, resulting in many other faults being misjudged as label 0.

5.4. *Comparative study*

In order to verify that the model proposed in this paper has a good diagnosis effect, for the same data set, SVM, LeNet-5, ResNet18 are used to compare with the model proposed in this paper. The traditional method SVM can directly diagnose the one-dimensional original telemetry signal, While the remaining three models use the transformed image as input. In order to make the comparison experiment more reliable, this paper repeats the operation for each diagnostic model 10 times, the value of epoch is 100, and the final diagnosis accuracy rate is the average of 10 times. The accuracy of each diagnostic model is shown in table 4.

Table 4. Accuracy rates of the diagnostic models

Model	epoch	Accuracy /(%)	Standard deviation
SVM+ Feature selection	100	80.17	15.57
LeNet-5	100	86.31	8.69
ResNet-18	100	91.15	5.53
PCA-ResNet	**100**	**95.33**	**4.12**

It can be seen from table 4 that:

(1) The PCA-ResNet model proposed in this paper has an accuracy rate of 95.33% for the fault diagnosis of the spacecraft control system, which is higher than that of the SVM, LeNet-5, and ResNet-18 fault diagnosis algorithms;

(2) The ResNet-18 network parameters are consistent with the residual network parameters used in this paper, but the direct use of the telemetry signal without noise reduction makes the diagnosis effect relatively poor, its overall volatility is large, and its stability is not as good as the PCA-ResNet model;

(3) The diagnostic accuracy of the three models of LeNet-5, ResNet-18, and PCA-ResNet is higher than that of the SVM algorithm, indicating that the use of the signal-image conversion method for the telemetry signal can effectively retain the characteristics of the telemetry signal and reduce the need for manual experience, and the accuracy of diagnosis is improved.

6. Conclusion

In this paper, a fault diagnosis method for spacecraft control system based on PCA-ResNet model is proposed. The telemetry signals of the fault set and the verification set are processed by the PCA method and then passed through the signal-image conversion, and the residual network is used for training, and finally, three typical types are compared. The training effect of the model can be concluded as follows:

1) The space environment where the spacecraft is located is harsh, and the PCA noise reduction of the telemetry signal can effectively reduce the interference of the external noise of the spacecraft on the fault diagnosis, thereby improving the accuracy of the fault diagnosis;

2) Using the signal-image processing method is more conducive to the deep feature extraction of the residual network, when compared with the traditional fault diagnosis algorithm, the diagnostic accuracy is improved;

3) By comparing the SVM, LeNet-5, and ResNet-18 methods, the PCA-ResNet model achieved a fault diagnosis accuracy rate of 95.33% under the condition of few spacecrafts fault samples, providing a reliable basis for accurate fault discrimination on the ground;

4) This research used PCA to process the telemetry signal and then used the residual network for deep feature extraction. Both the traditional data processing method and the deep network are used to design a fault diagnosis method for the spacecraft control system. Management provides a new way of thinking.

Acknowledgments

This work was financially finished by my teacher, and thanks to my classmate and my lover for their help

References

[1] Yin S, Xiao B, Ding S X, et al. A review on recent development of spacecraft attitude fault tolerant control system [J]. IEEE Transactions on Industrial Electronics, 2016, 63(5): 3311-3320.
[2] Wang D Y, Tu Y Y, Fu F Z. Autonomous diagnosis and reconfiguration technology of spacecraft control system[J]. Control Theory & Applications, 2019, 36(12): 1966-1978.
[3] Shen Y, Li L L, Wang Z H. A review of fault diagnosis and fault-tolerant control techniques for spacecraft[J]. Journal of Astronautics, 2020, 41(06): 647-656.
[4] Patton R J, Frank P M, Clark R N. Issues of fault diagnosis for dynamic systems [M]. Springer Science & Business Media, 2013.
[5] Frank P M. Fault diagnosis in dynamic systems using analytical and knowledge-based redundancy: a survey and some new results [J]. Automatica, 1990, 26(3): 459-474.
[6] Talebi H A, Khorasani K, Tafazoli S. A recurrent neural-network-based sensor and actuator fault detection and isolation for nonlinear systems with application to the satellite's attitude control subsystem [J]. IEEE Transactions on Neural Networks, 2008, 20(1): 45-60.
[7] Huang Y, Li S, Sun J. Mars entry fault-tolerant control via neural network and structure adaptive model inversion [J]. Advances in Space Research, 2019, 63(1): 557-571.
[8] Zhang Z S, Li L J, He Z J. Research on diagnosis method of machinery fault based on support vector machine[J]. Journal of Xi'an Jiaotong University, 2002(12): 1303-1306.
[9] Zhang C. Bearing fault diagnosis method based on CNN-SVM[J]. Modular Machine Tool & Automatic Manufacturing Technique, 2021(11): 114-116.
[10] Zhao L H, Cheng H, Li W Y, et al. Research on fault diagnosis method of plunger pump based on LMD and support vector machine [J/OL]. Machinery Design & Manufacture, 2022(01) :1-5.
[11] Jia F, Lei Y, Guo L, et al. A neural network constructed by deep learning technique and its application to intelligent fault diagnosis of machines[J]. Neurocomputing, 2018, 272: 619-628.
[12] He M, He D. Deep learning based approach for bearing fault diagnosis [J]. IEEE Transactions on Industry Applications, 2017, 53(3): 3057-3065.
[13] Wen L, Li X, Gao L. A transfer convolutional neural network for fault diagnosis based on ResNet-50[J]. Neural Computing & Applications, 2020, 32(10).

[14] Zhang K, Tang B, Deng L, et al. A hybrid attention improved ResNet based fault diagnosis method of wind turbines gearbox[J]. Measurement, 2021, 179: 109491.
[15] Zhang K, Tang B, Deng L, et al. A fault diagnosis method for wind turbines gearbox based on adaptive loss weighted meta-ResNet under noisy labels[J]. Mechanical Systems and Signal Processing, 2021, 161: 107963.
[16] Wang X, Shen C, Xia M, et al. Multi-scale deep intra-class transfer learning for bearing fault diagnosis[J]. Reliability Engineering & System Safety, 2020, 202: 107050.
[17] Wang F, Sun J, Yan D, et al. A feature extraction method for fault classification of rolling bearing based on PCA[C]//Journal of Physics: Conference Series. IOP Publishing, 2015, 628(1): 012079.
[18] Wen L, Li X, Gao L, et al. A new convolutional neural network-based data-driven fault diagnosis method[J]. IEEE Transactions on Industrial Electronics, 2017, 65(7): 5990-5998.
[19] Gong W F, Chen H, Zhang M L, et al. Intelligent diagnosis method for incipient fault of motor bearing based on deep learning[J]. Chinese Journal of Scientific Instrument, 2020,41(01):195-205.
[20] He K, Zhang X, Ren S, et al. Deep residual learning for image recognition[C]//Proceedings of the IEEE conference on computer vision and pattern recognition. 2016: 770-778.

Design of interval type-2 fuzzy fractional order PID controller based on particle swarm optimization

Hao Wu[1], Taoyan Zhao[1*]

[1] School of Information and Control Engineering, Liaoning Petrochemical University, Fushun, Liaoning, 113001, China

[*]Corresponding author's e-mail: zhaotaoyan1986@126.com

Abstract. In this paper, an interval type-2 fuzzy fractional order PID controller (IT2F-FOPIDC) based on particle swarm optimization is proposed. The proposed controller has two adjustable degree of freedom parameters, so the control range of the controller parameters becomes larger and can control the controlled object more flexibly. Aiming at the difficult problem of many controller parameters and tuning, this paper introduces particle swarm optimization algorithm to optimize the controller parameters. In order to avoid the problems of long iteration time and high computational cost caused by Karnik-Mendel reduction algorithm, the enhanced iterative with stop condition algorithm is selected for the proposed controller. Finally, the proposed controller is applied to the second order systems with time delays. Simulation results show that the proposed strategy has faster response time and smaller oscillation amplitude, and its performance is better than fuzzy controller and interval type-2 fuzzy controller.

1. Introduction

Fuzzy logic control (FLC) has been widely used in the control field because of its superior ability to deal with nonlinear, strong coupling and large time-delay systems. The establishment of interval type-2 fuzzy logic system (IT2FLS) is based on interval type-2 fuzzy set. The IT2FLS expands the dimension on the basis of traditional fuzzy system, so that a single fuzzy variable is described by two or more membership functions at different levels[1]. Therefore, its reasoning and design freedom are extended, which has potential advantages in dealing with uncertain complex systems. It has been widely used in the control field, such as manipulator[2], unicycle[3], robot[4-6], etc.

The traditional PID control is the most widely used and the most mature method in the controller system. However, its parameter selection has limitations, resulting in small stability range and low performance of the control system. Compared with traditional PID controller, fractional order PID controller has larger control range and better control effect. Pudlubny[7] proposed that fractional order controller is a milestone of fractional order calculus in the control field. The integral order calculus link of traditional PID controller is changed into fractional order calculus link, which makes the structure of the controller more flexible and the control effect better. Zarmi[8] designed a fractional order controller based on Bode diagram and analyzed it by using small gain theory and sensitivity function, which can ensure the robustness of the system when the external environment is uncertain. Thomas George[9] based state feedback gain of fractional order proportional integral derivative (FOPID) controller,the consolidation of Northern Bald ibis optimization algorithm (NOA) and Radial Basis Function Neural Network (RBFNN) controller for parameter tuning. Babaei [10] et al. used salp swarm algorithm (SSA) to tune the controller parameters of FOPID, and used the absolute error of the integral time（ITAE）of the performance index to to evaluate the time-varying time delay problem in the load frequency control (LFC) system.

The combination of T2FLC and fractional PID can greatly improve the ability of the controller to deal with the uncertainty of the system. In this paper, the advantages and disadvantages of the FOPID controller are analyzed, and the IT2F-FOPID controller is proposed by combining it with the interval type-2 fuzzy algorithm that can handle uncertain systems. The type-2 fuzzy logic system uses the EIASC algorithm, which reduces the operating cost of the system and improves the efficiency. For the parameter tuning of the controller, a particle swarm optimization algorithm with uncertainty, self-organization, evolution, and memory function is used. Other more advanced optimization algorithms are not used mainly because there are too many parameters. A simple optimization algorithm can avoid system failure, coupling, large time lag and other issues.

2. Fractional order PID controller

The basic operating operator for fractional calculus is ${}_aD_t^\alpha$, the sum a and t is the upper and lower limits of the operator, and is the order of calculus.

$$ {}_aD_t^\alpha = \begin{cases} \dfrac{d^\alpha}{dt^\alpha} & R(\alpha)>0 \\ 1 & R(\alpha)=0 \\ \int_a^t (d\tau)^{(-\alpha)} & R(\alpha)<0 \end{cases} \tag{1} $$

There are many definitions of fractional calculus, and the representative and widely used ones is GL(Grunwald-Letnikov). Definition is as follows:

$$ {}_aD_t^\alpha f(t) = \lim_{h\to 0}\frac{1}{h^\alpha}\sum_{j=0}^{[(t-\alpha)/h]}(-1)^\alpha \begin{bmatrix}\alpha \\ j\end{bmatrix} f(t-jh) \tag{2} $$

Among: h is the step size; $\begin{bmatrix}\alpha \\ j\end{bmatrix}$ is the binomial coefficient. The order of the derivative and integral terms of the traditional integer-order PID controller is extended to the fractional order field, we can get the fractional order $PI^\lambda D^\mu$ controller is as follows:

$$ C(s) = K_p + \frac{K_i}{s^\lambda} + K_d s^\mu \tag{3} $$

Figure 1 represents the value range of the differential and integral orders in the fractional controller, and they are all real numbers greater than 0. When $\lambda=1$, $\mu=1$, $C(s)$ is PID controller; when $\lambda=1$, $\mu=0$, $C(s)$ is PI controller; when $\lambda=0$, $\mu=1$, $C(s)$ is PD controller; when $\lambda=\mu=0$, $C(s)$ is P controller.

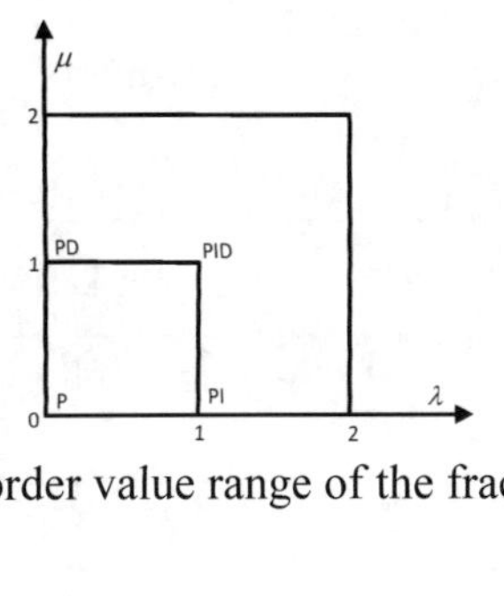

Figure 1. The order value range of the fractional controller

Since the order of fractional calculus is not an integer, it is necessary to use filters to approximate numbers to achieve fractional calculus. Improved outstaloup filter is defined as [11]:

$$s^{\alpha} \approx (\frac{d\omega_h}{b})^{\alpha}(\frac{ds^2 + b\omega_h s}{d(1-\alpha)s^2 + b\omega_h s + d\alpha})\prod_{k=1}^{N}\frac{s+\omega_k^{'}}{s+\omega_k} \tag{4}$$

In formula (4), $\omega_k^{'} = \omega_b \omega_m^{(2k-1-\alpha)/N}$, $\omega_k = \omega_b \omega_m^{(2k-1+\alpha)/N}$, $\omega_m = \sqrt{\omega_n / \omega_b}$. Under normal circumstances, the weighting parameters can be selected $b=10$, $d=9$. The fitting frequency in this application is $\omega_b = 0.001\ rad/s$, $\omega_h = 1000\ rad/s$, $N=5$.

3. Design of type-2 fuzzy fractional order controller

Fuzzy control adopts linguistic control rules, and does not need to establish an accurate mathematical model of the controlled object in the design, so that the control mechanism and strategy are easy to accept and understand, the design is convenient, and the application cost is low. The fuzzy control system has strong robustness, and the influence of disturbance and parameter changes on the control effect is greatly reduced. It is especially suitable for the control of nonlinear, time-varying and pure lag systems. Figure 2(a) and Figure 2(b) respectively represent the membership function and consequent rules of T2FLS. The membership function has a symmetrical structure. There are 25 fuzzy rules in this paper. The fuzzy rules are shown in Table 1. The structure of the IT2F-FOPID controller is shown in Figure 3, and its system output is as follows:

$$u_{IT2F-FOPID}(t) = G_{PD}U(t) + U(t)G_{PI}\frac{d^{-\lambda}}{dt^{-\lambda}} \tag{5}$$

$$U(t) = f(E(t) + \dot{E}(t)) = f(G_E e(t) + \frac{d^{\mu}e(t)}{dt^{\mu}}G_{CE}) \tag{6}$$

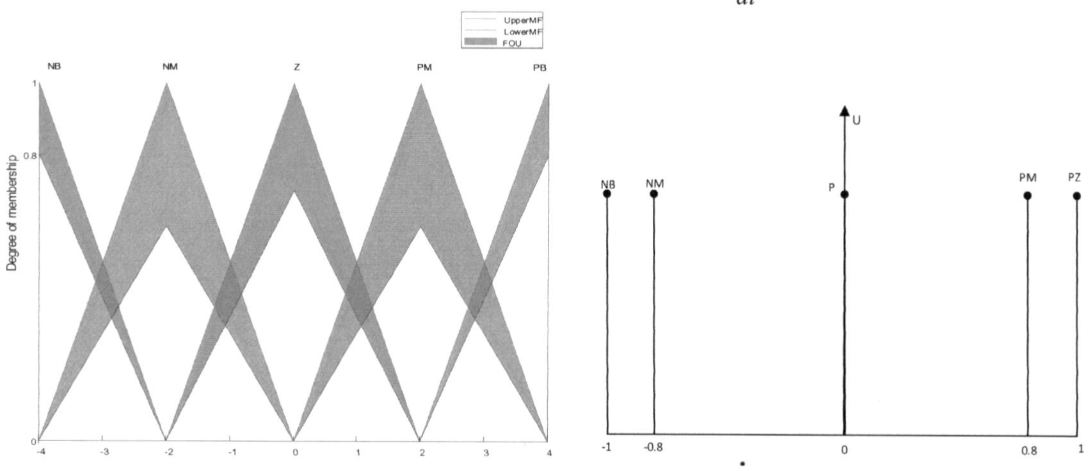

Figure 2. Membership functions of E, $\dot{E}$ and ΔU

Table 1. Type-2 fuzzy controller rule base

$E/\dot{E}$	**NB**	**NM**	**Z**	**PM**	**PB**
NB	NB	NB	NB	NM	Z
NM	NB	NB	NM	Z	PM
Z	NB	NM	Z	PM	PB
PM	NM	Z	PM	PB	PB
PB	Z	PM	PB	PB	PB

The most commonly used is the iterative Karnik-Mendel (KM) reduction algorithm in the IT2FLS, but it has disadvantages such as high computational cost, long iteration time, and poor effect of finding the left and right switching points. Therefore, the EIASC algorithm is used in this paper, which reduces the operating cost of the system, improves the operating speed of the system and the accuracy of finding the endpoints.

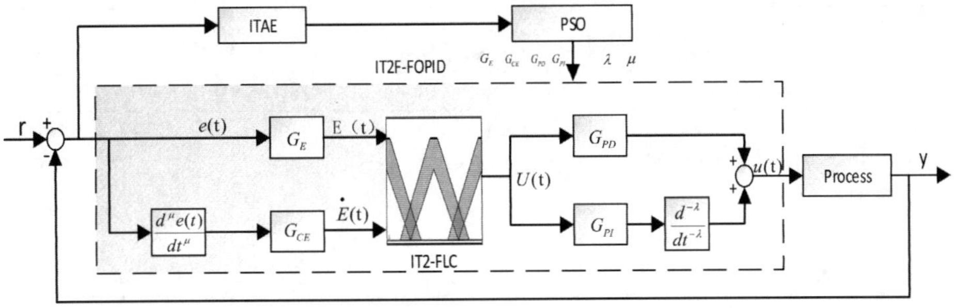

Figure 3. IT2F-FOPID controller design

According to Figure 3, under the premise of system performance indicators, the parameter domain is obtained first, and then a set of specific optimal parameters is obtained by a certain method. Aiming at the problem of controller parameter tuning used, this paper adopts the PSO optimization algorithm to take the performance index integral of time absolute error (ITAE) as the fitness value of PSO. The parameters of the optimized controller are:G_E, G_{CE}, G_{PD}, G_{PI}, λ and μ . According to the fitness value, update the speed and position of the iterative particle, update the optimal value in the local area and the global area, and judge whether the stopping condition is satisfied.The formula for ITAE is calculated as follows:

$$J_{ITAE} = \int_0^{\infty} t|e(t)|\mathrm{d}t \tag{7}$$

where $e(t)$ is the error between the expected output of the system and the actual output.

4. Simulation and discussion

Second-order linear system with hysteresis:

$$G(S) = \frac{Y(S)}{U(S)} = \frac{a_0}{b_2 S^2 + b_1 S + b_0} e^{-\tau s} \tag{8}$$

PSO parameter selection: $\omega = 1$、$c_1 = c_2 = 2$;the size of the particle swarm is 60, and the dimension is 6; the number of iterations of the population it is 200. The selection of the accused object is selected from the literature[12].

$a_0 = 1$、$b_2 = 1$、$b_1 = 2*1.125*0.45$、$b_0 = 0.45*0.45$、$\tau = 0.4$. Table 2 shows the parameter values of each controller, among which the PID controller parameters are 0.990, 0.489 and 0.907; Figure 4 is the system output; Figure 5 is the controller output; Figure 11 is the system adding a step of -0.3 in 10s Disturbance diagram; Table 3 shows the system control performance indicators for rise time, overshoot, settling time and ITAE. When adding a fractional step to the type 1 fuzzy controller, the control performance can be close to the IT2F-PID controller. From Table 3, it can be seen that the IT2F-FOPIDcontroller is slightly worse than the T1F-FOPID controller in terms of rise time, and other performance indicators are excellent. For other controllers. Figure 6 shows that when the system is disturbed, the convergence and stability of the IT2F-FOPID controller are better than other controllers.

Table 2. Controller parameter values

Controllers type	G_{PD}	G_{PI}	G_{CE}	G_E	μ	λ
T1FPID	1.524	1.056	3.127	4.219	1	1
T1F-FOPID	3.710	1.781	1.625	1.515	0.907	1.416
IT2FPID	2.514	2.015	1.545	1.631	1	1
IT2F-FOPID	3.349	1.381	1.355	1.418	0.969	1.041

Table 3. System performance indicators

	Rise Time	Overshoot	Settling Time	ITAE
PID	1.440	33.031	12.289	8.444
T1FPID	0.889	26.896	8.326	3.802
T1F-FOPID	0.743	15.029	4.900	1.766
IT2FPID	1.010	13.499	4.560	1.783
IT2F-FOPID	0.967	4.810	3.730	1.156

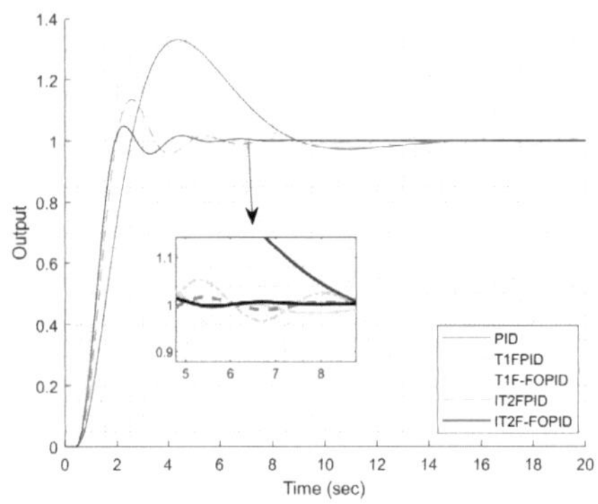

Figure 4. System output curve.

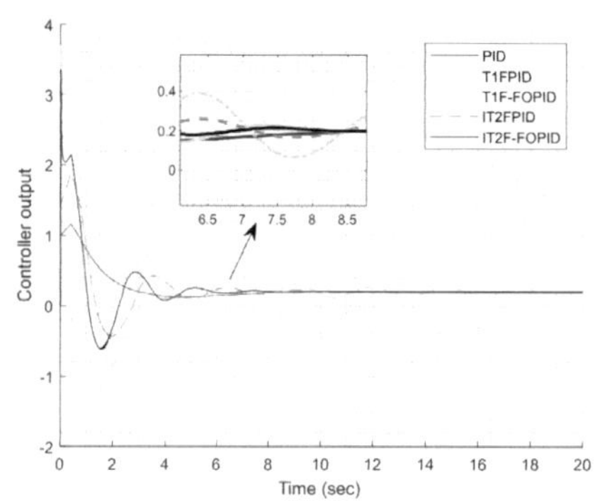

Figure 5. Controller input curve.

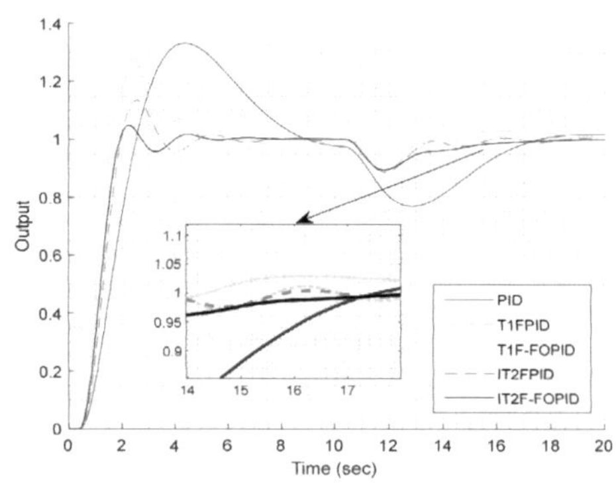

Figure 6. Output curve with disturbance.

5. Conlusion

Based on the interval type-2 fuzzy PID controller, this paper proposes the IT2F-FOPID controller. The controller has two more adjustable degrees of freedom, which makes the control effect more efficient and accurate. Another big problem of the controller is the parameter tuning problem. The PSO algorithm with low coupling degree and strong search ability is selected to improve the control structure and algorithm of the entire controller. The superiority of the proposed method is proved in the simulation. The three directions of the following research directions: 1) 2 Degrees of Freedom (2D) control, internal model control, etc. can be used to optimize the structure and feedback mechanism of the IT2F-FOPID controller to improve the control accuracy and robustness of the system. 2) The limitation of the PSO used in parameter tuning is that all particles tend to be the same, which makes the late convergence speed slow, so that the optimization cannot be continued after the optimization reaches a certain accuracy. Combine PSO with other optimization algorithms (or strategies) to form hybrid PSO algorithms, such as genetic algorithm (GA), ant colony algorithm (ACO), gray wolf algorithm (GWO), etc.3) The proposed controller is applied in real-time dynamic system, power system, UAV and other fields to verify its control performance.

Acknowledgments
This work was supported by the Program for Scientific Research Fund Project of the Educational Department of Liaoning Province of China [grant number L2019042], and the Research Initiation Fund Project of Liaoning Petrochemical University [grant number 2019XJJL-017].

References

[1] Zadeh L A. (1975) The Concept of a Linguistic Variable and Its Application to Approximate Reasoning I, II and III[J]. Information Sciences, 9(1): 43-80.
[2] Kelekci E, Kizir S. (2019) Trajectory and vibration control of a flexible joint manipulator using interval type-2 fuzzy logic[J]. ISA transactions, 94: 218-233.
[3] Chiu C H, Hung Y T. (2020) One wheel vehicle real world control based on interval type 2 fuzzy controller[J]. Mechatronics, 70(4): 102387.
[4] Ahmed S A, Petrov M G. (2015) Trajectory control of mobile robots using type-2 fuzzy-neural PID controller[J]. IFAC PapersOnLine, 48(24): 138-143.
[5] Chao F, Zhou D, Lin C M, et al. (2019) Type-2 fuzzy hybrid controller network for robotic systems[J]. IEEE Transactions on Cybernetics, 99: 1-15.
[6] Sun D, Liao Q, Stoyanov T, et al. (2019) Bilateral telerobotic system using type-2 fuzzy neural network based moving horizon estimation force observer for enhancement of environmental force compliance and human perception[J]. Automatica, 106: 358-373.
[7] PODLUBNY I. (1999) Fractional-order systems and controllers[J]. IEEE Trans on Automatic Control, 44(1): 208 -214.
[8] Monje C A, Vinagre B M, Feliu V, et al. (2008) Tuning and auto-tuning of fractional order controllers for industry applications[J]. Control Engineering Practice, 16(7): 798-812.
[9] George T, Ganesan V. (2021) Design of fractional order PID controller for higher order time delay system using NOARBFN technique[J]. Energy Systems, 1-32.
[10] Babaei F, Lashkari Z B, Safari A, et al. (2020) Salp swarm algorithm-based fractional-order PID controller for LFC systems in the presence of delayed EV aggregators[J]. IET Electrical Systems in Transportation, 10(3): 259-267.
[11] XUE D, ZHAO C, CHEN Y Q. (2006) A modified approximation method of fractional order system [C]. IEEE International Conference on Mechatronics and Automation. IEEE, 1043-1048.
[12] Shi J Z. (2020) A fractional order general type-2 fuzzy PID controller design algorithm[J]. IEEE Access, 8: 52151-52172.

Discriminator Based Resilient Multi-agent Deep Deterministic Policy Gradient Under Uncertain Faulty Agents

Jing Wang[1*], Hongyun Huang[1], Jian Hou[1]

[1] Information school, Zhejiang Sci-Tech University, Hangzhou, Zhejiang, 310016, China

*Corresponding author's e-mail: m18656897772@163.com

Abstract. In recent years, Multi-agent reinforcement learning (MARL) is widely applied in various of fields, to achieve a global goal in a centralized or distributed manner. However, during its application it is crucial to be fault-tolerance as some agents behave abnormal. In this paper, we propose a Resilient Multi-gent Deep Deterministic Policy Gradient (RMADDPG) algorithm to achieve a cooperative task in the presence of faulty agents via centralized training decentralized execution. At training stage, each normal agent observes and records information only from other normal ones, without access to the faulty ones. Meanwhile, a discriminator is generated based on the well-trained actor network to identify each faulty agent via supervised learning. Followed by executing stage, each normal agent selects its action based on local observation according to its actor network and its discriminator, so as to achieve certain system goal. Specifically, RMADDPG offers a scheme to train agents for improved resilience against arbitrary number of faulty agents. Finally, a cooperative navigation experiment is provided to validate the effectiveness of the proposed algorithm.

1. Introduction

Multi-agent systems (MAS), as a significant branch in distributed artificial intelligence, aims to achieve a global goal with local observation or interaction in a distributed manner. Recent advances in MAS in the past decades have attracted plenty of attentions and made it possible to implement various of tasks in different fields like Internet of Things [13], Blockchain [1], Wireless sensor networks et al. On the other hand, in a very closely related field of machine learning, especially reinforcement learning (RL) [14],has been proven an effective and efficient tool in optimization, control and decision making problems. In RL, each agent interacts with the environment/other agents by selecting proper actions to maximize the total rewards. It therefore seems quite natural to apply RL techniques on MAS, so-called multi-agent reinforcement learning (MARL) [4, 7, 15], so as to achieve certain cooperative or competitive goals through trial and error, and deep neural networks. [2] presents the development and evaluation of a novel system of MARL for integrated network of adaptive traffic signal controllers. MARL is utilized in [16] to build AlphaStar, in the full game of StarCraft II, through a series of online games against human players. [10] presents a general framework for intelligent agents to learn to teach in a multi-agent system, so that each agent utilizes the received advice to improve local learning. In addition, [9] introduces MARL as an automated discovery tool of turbulence models.

Most existing works up to date show solicitude for the research of ideal situation, in which each agent behavesnormal and productive [8, 11]. Unfortunately, some uncertain agents may behave way out in the left field subject to external attack or internal faults as a result of complex and time-variant running environments. In principle, a small amount of key elements in MAS may result in disastrous consequences. Thereupon, resilient MARL techniques are thus naturally presented. An absolutely loyal master agent is used in [6] to communicate with other agents to ensure resilient. A resilient consensus-based actor-critic algorithm is considered in [3], whereby the team average reward and value function are estimated and interacted by each agent and its corresponding neighbors. [17] proposes a resilient QD-learning algorithm in the presence of Byzantine agents. However, most of the existing related works require the number of faulty agents known a prior.

With regards to the issue above, we apply Multi-agent Deep Deterministic Policy Gradient (MADDPG) to train system model as well as an extra discriminator to identify the behavior trait of each agent, i.e., normal behavior or faulty behavior. At training stage of MADDPG, each faulty agent is exact known a prior for the normal ones, and thus system model under different number of normal agents is obtained. Meanwhile, the corresponding discriminator is trained via supervised learning for each normal agent. Specifically, it imports the states of all agents, and outputs the behavior trait information. By means of the discriminator and the well-trained actor network, each normal agent executes proper action to achieve cooperative task.

The rest of this paper is organized as follows. Section 2 gives the background of the paper, and the main algorithm is provided in Section 3. Section 4 conducts the experiments and the corresponding analysis. Section 5 gives a short conclusion and outlines future work.

2. Background

2.1. Markov Chain

MARL can be formulated 1 as Multi-agent Partially Observable Markov Decision Process $\mathrm{M} = \langle \mathrm{S}, \Pi, \mathrm{A}, \mathrm{P}, \mathrm{R}, \mathrm{O} \rangle$ in which state set S describes the possible configurations of all n agents, $\mathrm{A}_1,\ldots,\mathrm{A}_n$ and $\mathrm{O}_1,\ldots,\mathrm{O}_n$ are the sets of actions and observations of each agent respectively. Each agent i takes policy $\pi_{\theta_i} : \mathrm{O}_i \times \mathrm{A}_i \mapsto [0,1]$, and gets the next state through state transition function $\mathrm{P} : \mathrm{S} \times \mathrm{A}_1 \times \ldots \times \mathrm{A}_N \mapsto \mathrm{S}$. Moreover, all the agents get reward $\mathrm{S} \times \mathrm{A}_1 \times \ldots \times \mathrm{A}_N \mapsto \mathrm{R}$.

2.2. Multi-agent Deep Deterministic Policy Gradient

As a popular MARL technique, MADDPG is presented and widely used [8]. The fundamental ideology is embodied in its centralized training decentralized execution. Generally, each agent in the system consists of two actor networks and two critic networks, that all the observation information are imported to train critic network Q, while only the local information of agent i is utilized to train its own actor network μ . In addition, both actor and critic networks have a corresponding target network to stabilize the training process. During training process, tuples $(x, a_1, .., a_n, r_1, ..., r_n, x')$ are stored in the replay buffer D, in which x is the observation of all agents, and x' is a collection of observations after all agents execute their actions $a_1, .., a_n$. Then, the loss function of the critic network Q is:

$$L(\theta_i) = \mathrm{E}_{x,a,r,x'}[(Q_i^{\mu}(x, a_1, ..., a_n) - y)^2], \qquad y = r_i + \gamma Q_i^{\mu'}(x', a_1', ..., a_n')\big|_{a_j' = \mu_j'(o_j)} \tag{1}$$

and the gradient of actor network μ can be written as:

$$\nabla_{\theta_i} J(\mu_i) \approx \mathrm{E}_{x,a \sim \mathrm{D}}[\nabla_{\theta_i} \mu_i(a_i \mid o_i) \nabla_{a_i} Q_i^{\mu}(x, a_1, ..., a_n)\big|_{a_i = \mu_i(o_i)}] \tag{2}$$

2.3. Supervised Learning

Supervise learning, as a functional machine learning method, aims to train a neural network from the given labelled data. In general, each instance is composed of an input object (usually vector) and a desired output value (also known as a supervisory signal). Through data training and analysis, the supervised learning algorithm will generate an inferred feature used to map new instances. One of the well-known method is called Back Propagation (BP) neural networks [12], that adjust the weights and thresholds of the network constantly using steepest descent method via back propagation, so as to minimize mean square error of the network. The most common BP neural network includes an input layer, a hide layer and an output Layer.

3. Main Results

In this section, we propose our approach, called RMADDPG, to achieve cooperative task of n agents in the presence of $\overline{n}$ uncertain faulty agents. Our approach contains three main steps. The first step trains actor and critic networks for each agent at different scenarios. In the following, a discriminator is provided to identify the behavior trait of each agent with labelled data produced by actor networks. The last step is the training process of the discriminator's neural network.

3.1. Centralized Training

The centralized training process in this problem is similar to MADDPG algorithm, except that faulty agents are present to interfere the system, and thus a modified version is provided here. At the beginning of each episode, get system initial state x, and initialize a random process N for action exploration. Assume that $\overline{n}(<n)$ agents may behave abnormal to interrupt the system (the rest $n-\overline{n}$ agents can still accomplish the cooperative task). We define the faulty agent set as F, the normal

Algorithm 1: Train actor network with faulty agents:

for $episode = 1$ to M **do**:

Initialize a random process N for action exploration, and receive initial state x

Randomly select $\overline{n}(<n)$ agents acting as faulty ones. The faulty agent set is F, the normal agent set is G and all agent set is N

for $t = 1$ **to** $\max-episode-length$ **do:**

For each normal agent $i \in G$, replace all faulty agent information with 0 on its observation o_i to produce $\overline{o}_i$, and then select action $a_i = \mu_{\theta_i}(\overline{o}_i) + \mathrm{N}_t$

For each faulty agent $j \in F$, randomly select its action a_j

Execute actions a of all agents, and observe global reward r and next state x'

Store (x, a, r, x') in each normal agent replay buffer $\mathrm{D}_i, i \in G$, $x \leftarrow x'$

for agent $l \in N$ **do:**

Sample a random minibatch of S samples (x^k, a^k, r^k, x'^k) from D_l

Set $y^k = r_l^k + \gamma Q_l^{\mu'}(x'^k, a_i', a_j')|_{a_i'=\mu_i'(\overline{o}_i^k), a_j'=random}$

Update critic by minimizing the loss $L(\theta_l) = \frac{1}{S}\sum_k (Q_l^{\mu}(x^k, a_1^k, ..., a_n^k) - y^k)^2$

Update actor using the sampled policy gradient:

$\nabla_{\theta_l} J \approx \frac{1}{S}\sum_k \nabla_{\theta_l}\mu_l(\overline{o}_l^k)\nabla_{a_l} Q_l^{\mu}(x^k, a_i^k, a_j^k)|_{a_i=\mu_i(\overline{o}_i^k), a_j=random}$

Update target network parameters for each agent l: $\theta_l' \leftarrow \tau\theta_l + (1-\tau)\theta_l'$

agent set as G, and all agent set as $N = F \cup G$. At each step, for each normal agent $i \in G$, it obtains its own observation o_i and an induced $\overline{o}_i$ that replaces the information of all faulty agents with 0, and then selects action $a_i = \mu_{\theta_i}(\overline{o}_i) + \mathrm{N}_t$ with actor network μ. On the other aspect, each faulty agent acts with unpredictable options. In the following, by executing action a of all agents, we could get a global reward r and next state x'. Therefore, the corresponding tuple (x,a,r,x') is stored in each normal agent replay buffer $\mathrm{D}_i, i \in G$, and then $x \leftarrow x'$. When the size of replay buffer for all agents exceeds the minibatch size S, sample a random minibatch of S samples (x^k, a^k, r^k, x'^k) from replay buffer D_l for each agent $l \in N$, and update critic by minimizing this:

$$L(\theta_l) = \frac{1}{S}\sum_k (Q_l^{\mu}(x^k, a_1^k, ..., a_n^k) - y^k)^2, \qquad y^k = r_l^k + \gamma Q_l^{\mu'}(x'^k, a_i', a_j')|_{a_i' = \mu_i'(\overline{o}_i^k), a_j' = random} \quad (3)$$

In the same manner, update actor using the sampled policy gradient:

$$\nabla_{\theta_l} J \approx \frac{1}{S}\sum_k \nabla_{\theta_l}\mu_l(\overline{o}_l^k)\nabla_{a_l} Q_l^{\mu}(x^k, a_i^k, a_j^k)|_{a_i = \mu_i(\overline{o}_i^k), a_j = random} \quad (4)$$

Finally, update target network parameters for each agent l through soft update:

$$\theta_l' \leftarrow \tau\theta_l + (1-\tau)\theta_l' \quad (5)$$

3.2. Discriminator Training Set

After actor network centralized training, each agent will select its action based on its local observation. At this time, the exact faulty agent should be known a prior. Thereupon, we introduce a discriminator for each agent to identify the behavior trait of all agents based on the historical observation information. Training set, first and foremost, here is a necessary condition.

For each normal agent $i \in G$, it selects action $a_i = \mu_{\theta_i}(\overline{o}_i)$ based on local observation $\overline{o}_i$ and actor network μ_{θ_i}. For each faulty agent $j \in F$, a random action is selected so that next state is produced.Specifically, it stores the past h steps observation information o_i labelled with a vector of n -dimension, in which l denotes normal agent and 0 represents faulty one, into a training set D_i'.

Algorithm 2: Build discriminator training set:

for $episode = 1$ to M **do:**
  Initialize training set D' and step number h, and receive initial state x,
  Randomly select $\overline{n}(< n)$ agents acting as faulty ones
  for $t = 1$ **to** $\max - episode - length$ **do:**
    for agent $i \in G$ **do:**
      Select its action $a_i = \mu_{\theta_i}(\overline{o}_i)$
      if $t >= h$ **then:**
        store past h steps observation o_i, with the corresponding label in which normal agent and faulty agent indexed as 1 and 0, respectively, into training set D_i'
    For each faulty agent $j \in F$, randomly select its action a_j
    Execute actions a of all agents, and observe next state x', $x \leftarrow x'$

3.3. Discriminator Network Training

The proposed discriminator uses a neural network model, with past h steps observation information o_i as input, and outputs the behavior trait of each agent. Each agent $l \in N$ independently trains its discriminator through training set generated last step. As it is a multiple binary classification task, the Binary Cross Entropy is considered as the loss function.

Algorithm 3: **Train discriminator:**
Initialize discriminator network δ_l for each agent $l \in N$
for $epoch = 1$ to M **do**:
for $batch = 1$ to $batch - num$ **do**:
for agent $l \in N$ **do**:
Sample a random minibatch of S samples $(e^k, \hat{y}^k)$ from D_l
Set $y^k = \delta_l^k(e^k)$ Update discriminator by minimizing the loss $L(\theta) = -\frac{1}{S}\sum_k (y^k \cdot log(\hat{y}^k) + (1 - y^k) \cdot log(1 - \hat{y}^k))$

4. Experiment And Analysis

In this section, We apply our method in a cooperative navigation environment [8], in which four agents are subject to arriving at a set of given landmarks through physical actions as shown in Fig. 1. Unfortunately, some of the four agents are hacked and may behave abnormal. We denote purple and red circles as normal and faulty agents, respectively, and black points as landmarks. Therefore, Fig. 1 represents 0/1/2 possible faulty agents in the experiment. The system goal is to minimize the total running time that all normal agents reach different landmarks, subject to collision avoidance among agents.

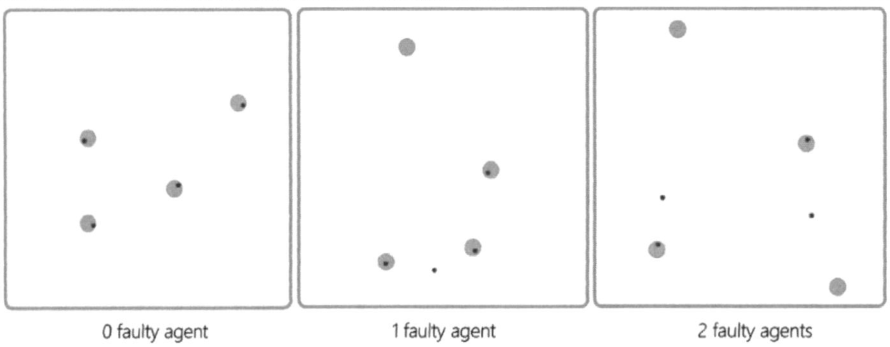

Figure 1. A cooperative navigation scenario where agents are expected to reach different landmarks within minimum time.

In the experiment, the observation information, i.e., state of each agent includes: the position and velocity of its own, the positions of the landmarks and other agents. Accordingly, the action set consists of the forces in four directions. The reward function in this experiment contains two parts. The first part evaluates the relative distance between normal agents and the corresponding landmarks. Specifically, we use Hausdorff distance [5] here to measure the distance of the agents position set P and the landmarks position set L in the metric space. That is, the Hausdorff distance $H(P,L) = max[h(P,L), h(L,P)]$, where $h(P,L) = \max_{p \in P} \min_{l \in L} \| p - l \|$, and $\|\cdot\|$ represents some underlying norm on the points of L and P (Euclidean norm in this experiment). The second part is collision avoidance with -1 per collision.

At training stage, all the faulty agents are exactly known a prior for each normal agent, and their position information is omitted in $\overline{o}_i$ for normal agent i . After training process, a discriminator is generated for each normal agent via the trained-well actor network. More specifically, the input of the discriminator includes consecutive five steps observation information of this normal agent, and then outputs a vector indicating the behavior trait of each other agent. Next, at executing stage, each normal agent selects its action based on its local observation $\overline{o}_i$ according to its actor network and its discriminator. In addition, the faulty behavior here is assumed randomly. In this experiment, both actor and critic network of RMADDPG consist of an input layer, three hidden layers and an output layer. The discriminator network consists of an input layer, a hidden layers and an output layer. The average return with respect to training episode is given as Fig. 2. From the results, the average return increases with training time, and gradually converges to -30 .

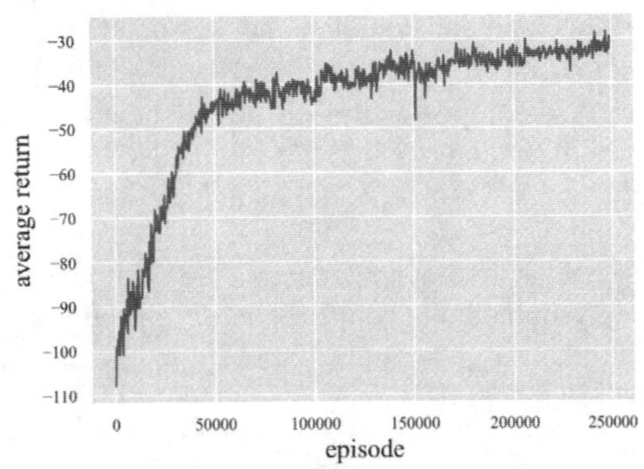

Figure 2. RMADDPG train process

In addition, we compare our approach with vanilla MADDPG with experiment results given in Fig. 3. It is obtained that when there exists no faulty agents, the average return of two approaches are almost the same size. Under other circumstances, the average return of vanilla MADDPG drops a lot, while our approach keeps pretty much the same value. The symbol $<=2$ implies uncertain faulty agents number that is no more than 2 . It again validates the effectiveness of our approach.

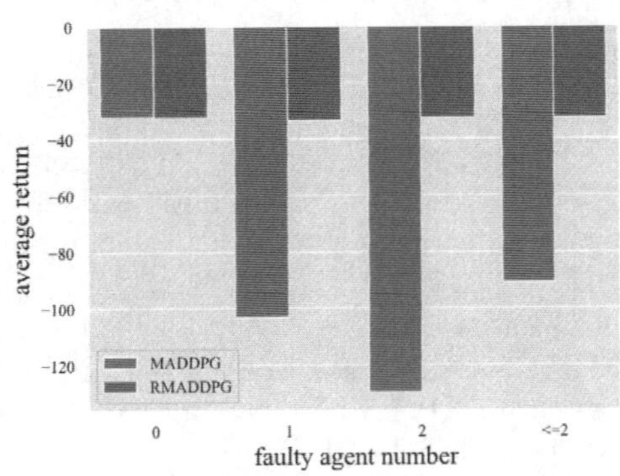

Figure 3. Comparison of our approach with vanilla MADDPG

5. Conclusion

This paper proposes a RMADDPG algorithm to achieve a certain cooperative task in the presence of uncertain faulty agents. Through training models via MADDPG with different number of normal agents, and designing a discriminator to identify the behavior trait of each agent, each normal agent will take optimal action to maximize the global return without interference from other faulty agents. The proposed scheme could achieve cooperative task against arbitrary number of faulty agents.

With random faulty behavior given at training stage, this paper shows positive results about our topic in a more ideal way when faulty agents also behave randomly at executing stage. However, due to the complex dynamic environment, the behavior trait of faulty agents is commonly uncertain and more complicated. Therefore, works with more uncertain behavior trait are expected. Besides, in the future we will extend the scenario to large-scale networks, that is a touchy issue in MARL problem.

References

[1] Berdik, D., Otoum, S., Schmidt, N., Porter, D., & Jararweh, Y. (2021). A survey on blockchain for information systems management and security. Information Processing & Management, 58(1), 102397.

[2] El-Tantawy, S., Abdulhai, B., & Abdelgawad, H. (2013). Multiagent reinforcement learning for integrated network of adaptive traffic signal controllers (MARLIN-ATSC): methodology and large-scale application on downtown Toronto. IEEE Transactions on Intelligent Transportation Systems, 14(3), 1140-1150.

[3] Figura, M., Lin, Y., Liu, J., & Gupta, V. (2021). Resilient Consensus-based Multi-agent Reinforcement Learning. arXiv:2111.06776.

[4] Foerster, J. N., Assael, Y. M., De Freitas, N., & Whiteson, S. (2016). Learning to communicate with deep multi-agent reinforcement learning. arXiv:1605.06676.

[5] Huttenlocher, D. P., Klanderman, G. A., & Rucklidge, W. J. (1993). Comparing images using the Hausdorff distance. IEEE Transactions on pattern analysis and machine intelligence, 15(9), 850-863.

[6] Lin, Y., Gade, S., Sandhu, R., & Liu, J. (2020, July). Toward Resilient Multi-Agent Actor-Critic Algorithms for Distributed Reinforcement Learning. In 2020 American Control Conference (ACC) (pp. 3953-3958). IEEE.

[7] Littman, M. L. (1994). Markov games as a framework for multi-agent reinforcement learning. In Machine learning proceedings 1994 (pp. 157-163). Morgan Kaufmann.

[8] Lowe, R., Wu, Y., Tamar, A., Harb, J., Abbeel, P., & Mordatch, I. (2017). Multi-agent actor-critic for mixed cooperative-competitive environments. arXiv:1706.02275.

[9] Novati, G., de Laroussilhe, H. L., & Koumoutsakos, P. (2021). Automating turbulence modelling by multi-agent reinforcement learning. Nature Machine Intelligence, 3(1), 87-96.

[10] Omidshafiei, S., Kim, D. K., Liu, M., Tesauro, G., Riemer, M., Amato, C., ... & How, J. P. (2019, July). Learning to teach in cooperative multiagent reinforcement learning. In Proceedings of the AAAI Conference on Artificial Intelligence (Vol. 33, No. 01, pp. 6128-6136).

[11] Rashid, T., Samvelyan, M., Schroeder, C., Farquhar, G., Foerster, J., & Whiteson, S. (2018, July). Qmix: Monotonic value function factorisation for deep multi-agent reinforcement learning. In International Conference on Machine Learning (pp. 4295-4304). PMLR.

[12] Rumelhart, D. E., Hinton, G. E., & Williams, R. J. (1985). Learning internal representations by error propagation. California Univ San Diego La Jolla Inst for Cognitive Science.

[13] Singh, R. P., Javaid, M., Haleem, A., & Suman, R. (2020). Internet of things (IoT) applications to fight against COVID-19 pandemic. Diabetes & Metabolic Syndrome: Clinical Research & Reviews, 14(4), 521-524.

[14] Sutton, R. S., & Barto, A. G. (2018). Reinforcement learning: An introduction. MIT press.

[15] Tan, M. (1993). Multi-agent reinforcement learning: Independent vs. cooperative agents. In Proceedings of the tenth international conference on machine learning (pp. 330-337).
[16] Vinyals, O., Babuschkin, I., Czarnecki, W. M., Mathieu, M., Dudzik, A., Chung, J., ... & Silver, D. (2019). Grandmaster level in StarCraft II using multi-agent reinforcement learning. Nature, 575(7782), 350-354.
[17] Xie, Y., Mou, S., & Sundaram, S. (2021). Towards Resilience for Multi-Agent QD - Learning. arXiv:2104.03153.

Journal of Physics: Conference Series 2258 (2022) 012065 doi:10.1088/1742-6596/2258/1/012065

Optimized realization of Quantum Fourier Transform for domestic DCU accelerator

Ke Ma[1a*], Lin Han[2b*],Jian-dong Shang[2c],Jing-ming Xie[2d] and Hao Zhang[1e]

[1] Information Engineering Institute, Zhengzhou University, ZhengZhou, HeNan, 450000, China

[2] National Supercomputing Zhengzhou Center, Zhengzhou University, ZhengZhou, HeNan, 450000, China

[a*]932095810qq.com,[b*]strollerlin@163.com,[c]shangjiandong@zzu.edu.cn,[d]624797981 @qq.com, [e]522954699@qq.com

Abstract. Quantum Fourier Transform (QFT) is a key part of quantum algorithms such as Shor algorithm, Grover algorithm, and quantum phase estimation algorithm. The "Songshan" supercomputer system adopts the heterogeneous architecture of domestic CPU+DCU. The simulation implementation of the quantum Fourier transform on the "Songshan" supercomputer can promote the research and optimization of related quantum algorithms. Based on the HIP heterogeneous programming model, this paper first maps the phase shift transformation R gate operation of the QFT transformation to the DCU accelerator, and initially constructs the heterogeneous implementation of QFT on the "Songshan" supercomputer; then, the Hadamard gate transformation operation originally implemented on the host side was implemented on the DCU, which greatly reduces the communication overhead caused by the quantum information data transmission between the CPU and the DCU device; finally, the quantum logic gate is analyzed, and based on the operation of the phase shift transformation R gate on the probability amplitude information in the simulation algorithm, an optimization scheme to improve the activity of the DCU thread is proposed. During the experimental verification process, the realized 28-qubit quantum Fourier transform optimization has achieved 13.440 speedup compared with the serially realized QFT simulation on the CPU. The basic quantum gate optimization method based on DCU acceleration equipment in this paper can provide a reference for the efficient simulation of other quantum algorithms on domestic heterogeneous platforms.

1. Introduction

The concept of quantum computer was first proposed by Richard Feynman in 1982 [1], which was used to simulate complex physical systems. Since the introduction of quantum computers in the 1980s to the present, they have achieved great development in the world. Although the rapid development of quantum computers has enabled theoretical physicists to carry out numerical research and calculate the basic properties of some state of matter, the current status of quantum computer research is still at an immature stage [2]. Therefore, an effective way to verify quantum algorithms is to simulate and implement quantum computing algorithms on traditional computers [3].

Quantum Fourier Transform (QFT) is similar to Discrete Fourier Transform (DFT) [4] and is a key component of many quantum algorithms, including Shor's algorithm for factorization [5] [6], Deutsch and Simon Algorithms, discrete logarithms, quantum phase estimation algorithms for estimating the eigenvalues of unitary operators, algorithms for hiding subgroups, etc. [7].

In this paper, the quantum Fourier transform algorithm is selected for simulation optimization. The quantum Fourier transform is simulated on a classical computer. This process is limited by the large-scale memory occupation and highly intensive calculations that exist in the program. Therefore, the use of supercomputers to simulate the process of quantum computing is a way to study quantum computing and assist quantum computers in performing tasks [8].

A series of single-qubit logic operations and the combination sequence of double-qubit controlled NOT gates can be combined into any quantum logic operation [9]. The purpose of this article is to decompose the logical operation of the quantum Fourier transform. On the "Songshan" supercomputer system, the algorithm is optimized based on the premise that the C language has been rewritten to HIP-C. Using HIP programming, libquantum quantum library [10] and the method and technology of quantum gate algorithm improvement, the 28-qubit QFT simulation optimization realization of the upper limit of the simulation digits in a single node is realized.

Section 2 of this article introduces domestic DCU acceleration devices and heterogeneous programming models; Section 3 introduces the principles and formulas of quantum Fourier transform; Section 4 introduces the optimization method of quantum gate algorithm improvement under DCU accelerator; Section 5 analyzes the results of the verification experiment data; Finally, summarize the full text and propose the next research direction.

2. Introduction of DCU acceleration device and heterogeneous programming model

The "Songshan" supercomputer system located in the Zhengzhou Center of National Supercomputing is equipped with the "Haiguang No. 1" DCU accelerator with a similar GPU architecture, and a large number of arithmetic operation units (Compute Unit, CU) are integrated on the accelerator. This architecture makes the advantages of DCU accelerators in terms of computing power appear when faced with many applications that are simple in logic, computationally intensive and suitable for parallel processing [11]. The DCU accelerator architecture is shown in Figure 1.

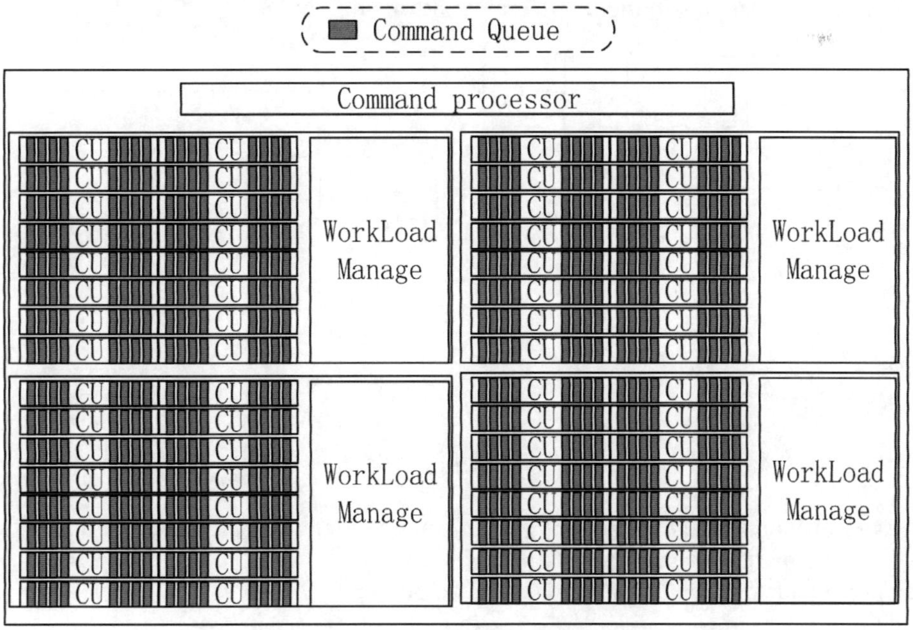

Fig.1 DCU accelerator architecture

The CPU+DCU heterogeneous programming model divides the system into host and device. The main processor (host) is the CPU, the main task is to perform logical judgment operations and serial calculations, and the programming language is C/C++; DCU is used as a co-processor (device) to perform highly intensive computing tasks [12], the code is HIP-C called the kernel function (Kernel). The host code and the device code combine to form a complete heterogeneous application[13][14].When writing kernel functions, tasks are divided according to the thread hierarchy. The thread hierarchy is thread, block, and grid. According to the scale of the problem, the number of threads is reasonably divided to achieve multi-threaded and highly parallel concurrent computing on the DCU accelerator [15].

3. Quantum Fourier transform

Quantum Fourier Transform (QFT) is a discrete Fourier transform that decomposes the original formula into a simpler product of multiple unitary matrices [16]. The standard discrete Fourier transform is defined as follows:

$$y_k \rightarrow \sum_{j=0}^{N-1} x_j e^{2\pi ijk/N} \tag{1}$$

Quantum Fourier transform is actually a discrete Fourier transform acting on Hilbert space[17]. Its input is $|0\rangle$,..., $|2^N\text{-}1\rangle$, and its role on the ground state is:

$$|j\rangle \rightarrow \frac{1}{\sqrt{2^N}} \sum_{k=0}^{2^N-1} e^{2\pi ijk/N} |k\rangle \tag{2}$$

Among them, |j> represents a quantum state source, which is converted into a linear combination of all n^2 possible quantum states [18]. When the input is $|j_1,\ldots,j_{n-1},j_n\rangle$, it means that the qubits participating in the transformation are n, and the output should be:

$$|j_1,\ldots,j_{n-1},j_n\rangle \rightarrow \frac{(|0\rangle + e^{2\pi i0.j_n}|1\rangle)(|0\rangle + e^{2\pi i0.j_{n-1}j_n}|1\rangle)\ldots(|0\rangle + e^{2\pi i0.j_1j_2\ldots j_n}|1\rangle)}{2^{n/2}} \tag{3}$$

The circuit of the quantum Fourier transform is shown in Figure 2.

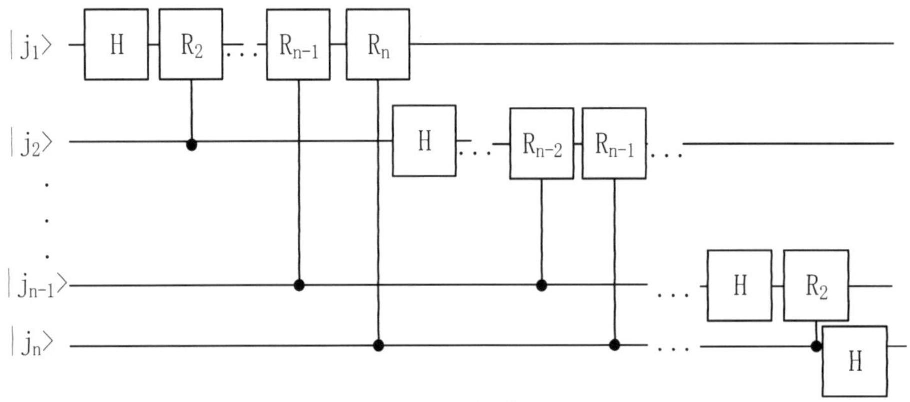

Fig.2 QFT circuit diagram

In the QFT quantum circuit diagram [19], the quantum circuit of the quantum Fourier transform is mainly composed of two quantum logic gates, Hadamard (H) gate acting on a single qubit and a controlled phase transformation R gate acting on two qubits [20].

$$H = \frac{1}{\sqrt{2}}\begin{pmatrix} 1 & 1 \\ 1 & -1 \end{pmatrix}, R = \begin{pmatrix} 1 & 0 \\ 0 & e^{2\pi i/2^k} \end{pmatrix} \tag{4}$$

4. Realization of quantum circuit optimization under DCU accelerator

4.1. Implementation of QFT simulation of CPU+DCU heterogeneous model version

The traditional quantum Fourier transform simulation program is implemented serially on the CPU, written in C/C++, and the hot spot analysis of the simulation program shows that the running time of the Hadamard gate transformation and phase shift conversion R gate accounts for 90% of the total running time of the program. Analysis of the program code found that there is no data dependency between the program execution process and other quantum registers. This program has high independence and is suitable for parallel computing on the CPU+DCU heterogeneous programming model.

According to the hardware characteristics of the DCU acceleration device, in this section, the R gate transformation operation is implemented by HIP-C, which takes a long time in the program and puts the intensive calculation part on the DCU acceleration device to run; Register initialization, probability amplitude information preparation and Hadamard transformation operations are put on the CPU side to run. The QFT simulation workflow on the CPU+DCU heterogeneous model is shown in Figure 3 [21].

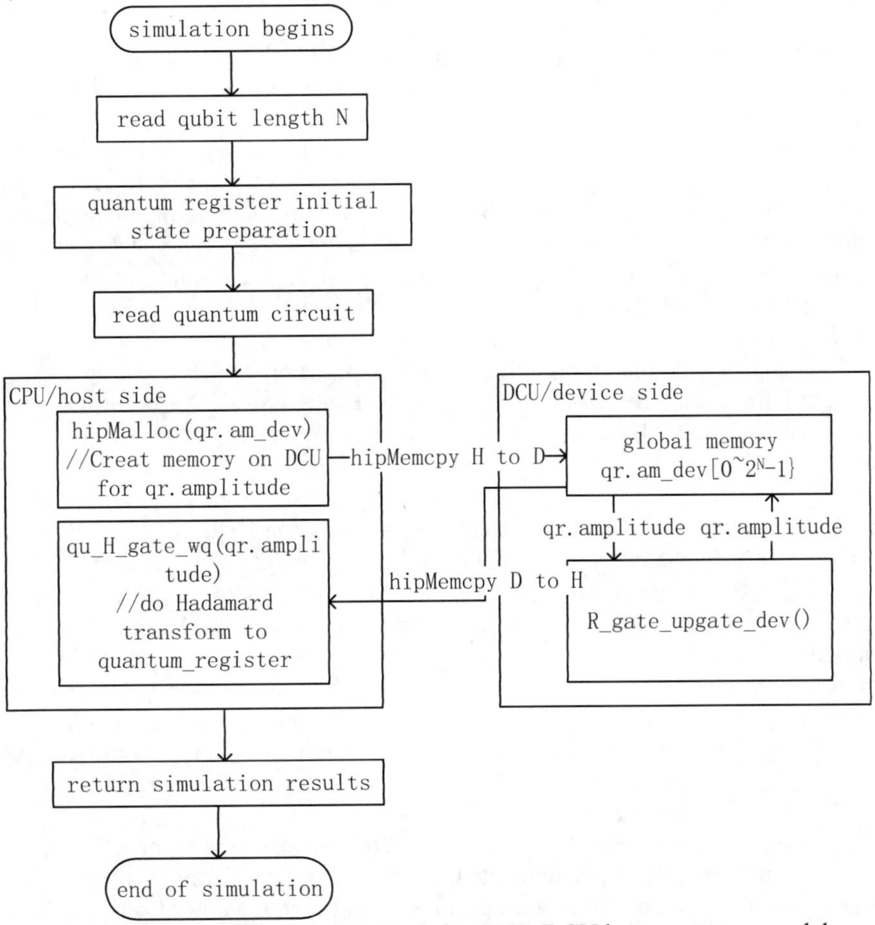

Fig.3 QFT simulation workflow on the CPU+DCU heterogeneous model

After the initial state of the quantum register is prepared, the memory space needs to be opened up to store the quantum probability amplitude information on the DCU, and the incoming probability amplitude information is divided. Then call the copy function hipMemcpy to transfer the probability amplitude information to the global memory of the DCU accelerator.

After completing the hardware configuration in the DCU accelerator, the host side starts the phase shift transformation R gate operation kernel function, and the device side code starts to participate in the operation on the DCU [22]. In the kernel function of the R gate transformation, according to the qubit and control bit information of the quantum circuit in the parameter, it is determined to start the thread with the same number of probability amplitude information corresponding to the input qubit. Through threadID and probability amplitude one-to-one correspondence, the probability amplitude information corresponding to each thread is updated by the phase update of the R gate transformation by judging the information of its own qubit and the control qubit. Realize the selection and update operation of the probability amplitude information of the phase shift transform R gate on the DCU. Compared with the cyclic traversal of the probability amplitude information in the serial simulation, the parallel implementation efficiency and time are greatly improved. After completing the phase shift transformation of the R gate on the DCU, call the hipMemcpy function again to transfer the updated probability amplitude information from the device side to the host side for Hadamard transformation. Since the Hadamard gate is not implemented on the DCU, it is necessary to continuously transmit data on the host side and the device side until all probability amplitudes are updated, and the simulation process ends.

4.2. Quantum gate optimization scheme and implementation of DCU accelerator

In the QFT implemented by the heterogeneous version simulation, the Hadamard transform (H gate) and the phase shift transform (R gate) in the quantum Fourier transform are simulated based on quantum circuits. All the initial state data prepared by the quantum register is transferred from the host side to the device side to open up the same number of threads as the quantum probability amplitude information in the quantum register. The number increases exponentially with the number of simulated qubits, resulting in an extremely large number of development threads. The phase shift operation will update the probability magnitude of the quantum register of different qubits,but after judgment and selection, a large part of the threads are not involved in the actual operation, and the actual activity of the threads is not high.

When the probability amplitude is completely updated and stored back to the quantum register, the data is transmitted from the device side back to the host side, and the global probability amplitude information starts to perform Hadamard transformation until the calculation is completed and the result is output [23], although the CPU side and the DCU side are connected by PCIe 3.0 bus , the transmission speed is relatively fast [24], but the data size of probability amplitude information increases with the number of qubits exponentially. As a result, the DCU is occupied but not involved in the calculation, which reduces the simulation efficiency of the QFT algorithm. Therefore, the simulation process of the H gate is not implemented on the DCU, and the data of the host side and the device side need to be continuously exchanged for data transmission, which causes a waste of time.

According to the above two situations, this section first completes the simulation implementation of the Hadamard gate on the DCU. Then, an optimization scheme for the R gate algorithm is proposed and implemented. The acceleration effect on the DCU is realized by optimizing and improving the two quantum logic gates, Hadamard gate and R gate, which constitute the quantum Fourier transform.

4.2.1. The realization of Hadamard gate on DCU

This section is carried out on the basis of the heterogeneous version of QFT simulation. The Hadamard gate transformation is implemented on the DCU accelerator, which can reduce the communication overhead caused by the data transmission between the device side and the host side.

The controlled quantum Hadamard gate is a very important quantum logic gate. The effect of the Hadamard gate is to transform the state of the qubit [25], and perform the conversion between the basis $\{|0\rangle , |1\rangle \}$ and $\{|+\rangle , |-\rangle \}$. After Hadamard gate transformation, its basis changes, and the measurement method follows the base to change to the X direction and performs the X direction measurement. At this time, the probability of getting 1 will be $|\alpha|^2$, that is, the value of X is 1; and the probability of -1 will be $|\beta|^2$, that is, the value of X is -1. The flow chart of Hadamard gate transformation on DCU is shown in Figure 4.

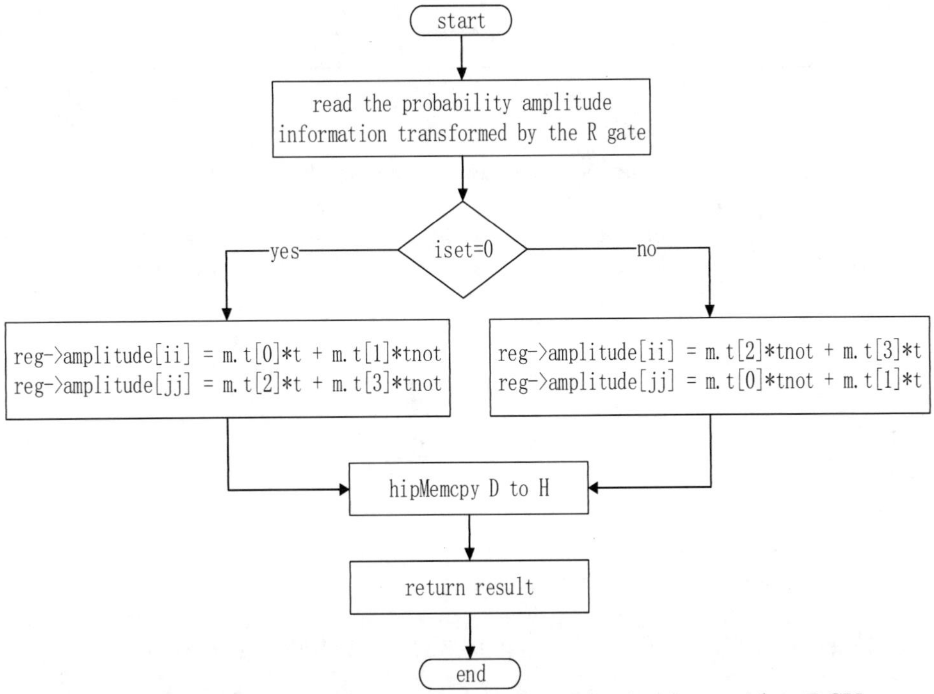

Fig.4 Flow chart of Hadamard gate transformation implemented on DCU

In the implementation process, the indexes ii and jj of reg->amplitude[ii] and reg->amplitude[jj] are all determined by the single qubit of the Hadamard transformation according to the control, that is, the target. First set a step value step to traverse the probability amplitude information after the R gate transformation; then configure the grid and thread number, start the H gate operation kernel function, determine the iset value, and perform the Hadamard gate transformation operation one by one according to the result until all the probability amplitudes information is transformed; finally, the quantum probability amplitude information after the transformation is transmitted back to the host, and the simulation ends.

4.2.2. Optimized realization of controlled phase transformation R gate

In the previous concurrent implementation, the probability amplitude information is 2^N, and 2^N threads are also allocated on the DCU side. The screening process of the phase shift transformation R gate for the effective probability amplitude is performed on the DCU, which causes too many threads to be idle. An algorithm is proposed to advance the judgment process, eliminate the probability of not participating in the operation, and put the valid data into the device side, so that the developed threads are involved in the operation, which will greatly shorten the operation time of the R gate.

The optimization scheme proposed in this paper is based on the original basis and uses binary bit operations to improve. When the number of input qubits is N, each quantum register needs to store 2^N probability amplitude information, of which the effective probability amplitude is 2^{N-2}, and the number of threads opened up is also reduced to 2^{N-2}. Filter it out and put it into the array $L[2^{N-2}-1]$, and then put it on the thread to participate in the calculation.

According to the number of input qubits, it is written as a binary representation, as shown in Figure 5.

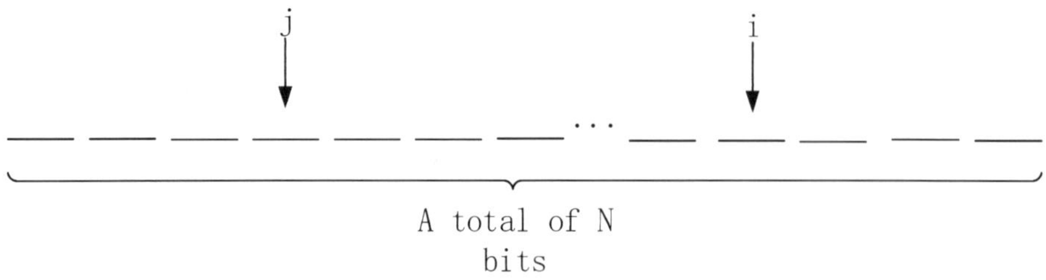

Fig.5 Schematic diagram of the number of qubits

j and i are respectively occupied binary bits, set to 1, and j>i. Find the number of bits that are not occupied between j and i : j-i-1. According to the results of j-i-1, it is divided into the following situations:

A. j-i-1=0：
   1. i=0，j=1
   2. i=N–2，j=N-1
   3. i≠0，j=i+1

B. j-i-1≠0：
   1. i=0，j=N-1
   2. i=0，j takes any value that meets the conditions
   3. i≠0，j takes any value that meets the conditions

Analyze the calculation rules, you can find L_2, L_4, L_6,..., $L_{2^{N-2}-2}$ according to the L_0 in the array $L[2^{N-2}-1]$, and then by L_0, L_2, L_4, L_6,..., $L_{2^{N-2}-2}$ plus the lowest bit that is not occupied to obtain L_1, L_3, L_5, L_7,..., $L_{2^{N-2}-1}$.

First, make sure that when N-bit qubits need to be simulated and transformed, remove the number of bits occupied by i and j, and there are N-2 vacancies. According to the found rules, L0 can be determined by L0 as:

L_2, L_4, L_6;

L_8, L_{10}, L_{12},L_{14};

L_{16}, L_{18}, L_{20},L_{22},L_{24};

…

$L_{\{1+3+...+[(N-2+1)-1]\}*2}$,……,$L_{\{1+3+...+[(N-2+1)-1]\}*2+(N-2)*2}$

Define the array a[r], the value range is 0 to i-1; the array b[r], the value range is i+1 to j-1; the array c[r], the value range is j+1 to N -1. The last set of values can be determined by L0:

$L_{\{1+3+...+[(N-2+1)-1]\}*2}=L_0+2^{cn}$

$L_{\{1+3+...+[(N-2+1)-1]\}*2+2}=L_0+2^{cn}+2^{a0}$

…

$L_{\{1+3+...+[(N-2+1)-1]\}*2+(N-2)*2}=L_0+2^{a0}+\ldots+2^{an}+2^{b0}+\ldots+2^{bn}+2^{c0}+\ldots+2^{cn}$

Its pseudo code is shown in Algorithm 1.

Algorithm 1 Filter effective probability amplitude information

```
1. int L[2^(N-2)-1],a[r],b[r],c[r] // Initialize the array
2. MAX_UNSIGNED m = (MAX_UNSIGNED) 1 << j
   MAX_UNSIGNED n = (MAX_UNSIGNED) 1 << i
3. for k=0→2^N-1 do k=k+1
4. k=m+n,then k→L[2^(N-2)-1] // Put the k value into the array L[2N-2-1]
5. if (k & 1 && k & 2)
6.   L2N-2-1= L2N-2-2+4
7.   else
8.     if(k & 1 && k & 4)
9.       L2N-2-1= L2N-2-2+2
```

10. else
11. $L_{2N-2-1}= L_{2N-2-2}+1$ // Determine the value rule for the value of the array L according to the value of k
12. if(j-i-1=0)
13. if(i=0)
14. $L_0=2^j+2^i$
$L_2=L_0+2^2$
…
$L_{(2N-2/2)-1}= L_{(2N-2/2)-2}+2^2$
$L_{2N-2/2}=L_0+2^{N-1}$
…
$L_{(2N-2/2)-1}= L_{(2N-2/2)-1}+2^{N-1}$
15. else if(i=N-2)
$L_0=2^j+2^i$
$L_0=2^j+2^i+1$
…
$L_{2N-2-1}= L_{2N-2-2}+1$
16. else
17. $L_{2N-2-1}= L_{2N-2-2}+1$
$L_{\{1+3+...+[(N-2+1)-1]\}*2}=L_0+2^{cn}$
$L_{\{1+3+...+[(N-2+1)-1]\}*2+2}=L_0+2^{cn}+2^{a0}$
…
$L_{\{1+3+...+[(N-2+1)-1]\}*2+(N-2)*2}=L_0+2^{a0}+...+2^{an}+2^{c0}+...+2^{cn}$ //Judge j-i-1=0, determine the value of the array L according to the situation
18. if(j-i-1≠0)
19. if(i=0&&j=N-1)
20. $L_0=2_j+2_i$
$L_2=L_0+2^1$
…
$L_{(2N-2/2)-1}= L_{(2N-2/2)-2}+2^1$
$L_{2N-2/2}=L_0+2^{N-2}$
…
$L_{(2N-2/2)-1}= L_{(2N-2/2)-1}+2^{N-2}$
21. else
22. $L_{2N-2-1}= L_{2N-2-2}+1$
$L_{\{1+3+...+[(N-2+1)-1]\}*2}=L_0+2^{cn}$
$L_{\{1+3+...+[(N-2+1)-1]\}*2+2}=L_0+2^{cn}+2^{a0}$
…
$L_{\{1+3+...+[(N-2+1)-1]\}*2+(N-2)*2}=L_0+2^{a0}+...+2^{an}+2^{b0}+...+2^{bn}+2^{c0}+...+2^{cn}$ //Judge j-i-1≠0, determine the value of the array L according to the situation

By rewriting the code through the above optimization scheme, it is possible to filter the effective probability amplitude information before putting the data on the device-side thread to participate in the operation, and effectively ensure that the number of threads developed will all participate in the R gate transformation.

Call hipLanchKernel to start the kernel function, so that the DCU accelerator participates in the calculation. In the implementation, the global offset of the developed threads is first calculated by hipThreadIdx and hipBlockIdx, and then the quantum probability amplitude information of the corresponding position is read from the global memory of the DCU accelerator from the calculated global offset. This paper completes the screening of the probability amplitude in advance, that is, the probability amplitude information stored in $L[2^{N-2}-1]$. After the R gate operation is over, the updated probability amplitude information is transmitted to the H gate kernel function to complete the

Hadamard transformation, until all the probability amplitudes are completed through the H gate transformation, the result is transmitted back to the host, and the simulation ends.

5. Experiment and result analysis

5.1. Experimental environment

The operating system of "Songshan" supercomputer is Centos-6.7, the CPU uses gcc/g++-7.3.1 compiler, and the DCU accelerator uses hipcc-2.9.6 compiler. The number of cores of a single "Haiguang One" CPU is 32, and the maximum number of threads that can be started at the same time is 64; the global memory of the "Haiguang No. 1" DCU accelerator is 16GB, and 60 arithmetic operation units (CU) are integrated on a single DCU chip. The ratio of CPU and DCU on a single computing node of the supercomputing platform is 1:4, and the communication between the computing nodes is maintained by the 200Gbps high-speed intercommunication network[26].

5.2. Result analysis

This paper uses the CPU+DCU heterogeneous implementation version and the quantum logic gate based on quantum circuit optimization to compare the serial version experiment to test the acceleration effect of the 28-qubit scale on the "Haiguang No. 1" DCU accelerator. Mark the four versions: serial, CPU+DCU heterogeneous on a single node, put the Hadamard gate in the DCU and improve the phase shift transform R gate as version 1, 2, 3, 4 for simulation implementation.

The 20-qubits scale version 1, 2, 3 test time-consuming comparison is shown in Table 1.

Table 1. Time-consuming comparison of the three versions of 20-qubits scale.

scale	version	time/s
20-qubits	1	8.261
	2	1.613
	3	1.332

The acceleration of version 3 compared to version 1 and 2 is shown in Table 2.

Table 2. 20-qubits scale speedup ratio of different versions.

version comparison	speedup ratio
version 3 vs. version 1	6.20
version 3 vs. version 2	1.21

According to the data in Table 1 and Table 2, the experimental results are in line with the expected results. By implementing the Hadamard gate in the DCU, a speedup of 6.20 is achieved compared to the serial implementation, which reduces data communication and improves the efficiency of the simulation implementation.

Limited by node memory, the maximum qubit simulation scale that can be carried out on a single node is 28-qubits. The 28-qubits scale version 1, 2, 3, 4 test time-consuming comparison is shown in Table 3.

Table 3. Time-consuming comparison of the four versions of 28-qubits scale.

scale	version	time/s
28-qubits	1	1097.148
	2	132.478
	3	91.968
	4	81.630

The acceleration of version 4 compared with version 1, 2, and 3 is shown in Table 4.

Table 4. 28-qubits scale speedup ratio of different versions.

version comparison	speedup ratio
version 4 vs. version 1	13.440
version 4 vs. version 2	1.623
version 4 vs. version 3	1.127

According to the data in Tables 3 and 4, the larger the scale of the analog qubit, the better the implementation effect on the DCU device., and the further improvement of the optimization scheme of the R gate algorithm achieves a speedup of 13.440 compared to the serial version. Using this optimization method effectively avoids the waste of resources caused by opening up a large number of idle threads, and improves thread activity.

The experiment realized the optimization of the 28-qubit quantum Fourier transform on the domestic DCU accelerator, and achieved a higher performance improvement. The basic quantum gate optimization implementation method for DCU acceleration equipment in this paper provides a reference for the efficient simulation of other quantum algorithms on domestic heterogeneous platforms.

6. Concluding

This paper is oriented to the domestic "Haiguang No. 1" DCU acceleration device. First, the heterogeneous version of the QFT algorithm simulation is implemented on the supercomputer platform, and then according to the quantum logic gate in the simulation implementation, an optimization and improvement plan is proposed and implemented. The experimental results verify the feasibility and effectiveness of the optimized scheme. This paper only completes the optimization of quantum Fourier transform on a single node, and implementation on multiple nodes will further improve the efficiency of simulation optimization.

References

[1] GIBNEY E. Quantum computer race intensifies as alternative technology gains steam[J]. Nature,2020,587(7834).:342-343

[2] CHO A. Google claims quantum computing milestone[J]. Science,2019,365(6460).:1364-1364

[3] LLOYD S, GARNERONE S, ZANARDI P. Quantum algorithms for topological and geometric analysis of data[J]. Nature communications, 2016, 7:10138.

[4] S. Hasanijafari,S.Parsamehr.Solving the Fourier Transform Issue Using Quantum Coherent States[J]. International Journal of Theoretical Physics,2019,58(8).

[5] Shor P . Algorithms for quantum computation: discrete logarithms and factoring[J]. In Proceedings of 35th Annual Symposium on the Foundations of Computer Science, IEEE Computer Society Press, Los Alamitos, CA, 1994:124-134.

[6] Shor, Peter W . Polynomial-Time Algorithms for Prime Factorization and Discrete Logarithms on a Quantum Computer.[J]. SIAM Review, 1999.

[7] Artyom M. Grigoryan,Sos S. Agaian. Paired quantum Fourier transform with log 2 N Hadamard gates[J]. Quantum Information Processing,2019,18(7):217

[8] Yu Z C, Li Y Z,Liu L. Summary of Quantum Computing Simulation and Optimization Methods [J/OL]. Computer Engineering:1-17[2021-11-30].https://doi.org/10.19678/j.issn.1000-3428.0061787.

[9] Miao X . Universal construction of unitary transformation of quantum computation with one- and two-body interactions[J]. physics, 2000.

[10] B. P. Lanyon, C. Hempel, D. Nigg, et al. Universal Digital Quantum Simulation with Trapped lons[J]. Science, 2011, 334(6052): 57-61.
[11] BAKHODA A, YUAN G L, FUNG W W L, et al. Analyzing CUDA workloads using a detailed gpu simulator [C]// the 2009 IEEE International Symposium on Performance Analysis of Systems and Software, 2009
[12] Sunpyo, Hong, Hyesoon, et al. An Integrated GPU Power and Performance Model[J]. Computer Architecture News, 2010.
[13] JOHN C.Professional CUDAC Programming[M].Wiley Inter Science,2014
[14] Xiao H,Guo Y H,Zhou Q L. SIFT feature matching parallel algorithm for CPU+GPU heterogeneous computing [J]. Journal of Tongji University (Natural Science Edition),2013,41(11):1732-1737.
[15] Xie J M,Hu W F,Han L.Quantum Fourier Transform Simulation Based on "Songshan" Supercomputer System [J]. Computer Science,2021,48(12):36-42.
[16] Guo G P,Chen S Y,Guo G C.Introduction to Quantum Computing and Programming [M].BeiJing：Science Press,2020：213-213
[17] Steane, A. 1998. Quantum computing. Rep. Prog. Phys. 61, 2 (Feb. 1998), 117.
[18] Liu X N, Jing L N,Wang L X. Large-scale quantum Fourier transform simulation based on Shenwei 26010 processor [J]. Computer Science,2020,47(08):93-97.
[19] HÄNER T.Steiger D S. 0.5 petabyte simulation of a 45-qubit quantun circuit [C] // International Conference for High Performance Computing.2017.
[20] LIU X,GUO H,SUN R J , et al The Characteristic Analysis and Exascale Scalability Research of Large Scale Parallel Applications on Sunway Taihulight Supercomputer[J].Journal of Computer,2018,14(10):2209-2220.
[21] Zhang P. Research on Quantum Computing Simulation Method Based on CPU+GPU Heterogeneous Cluster [D]. Nanjing University of Aeronautics and Astronautics,2016.
[22] Jing Z , Hao W , Feng W C . cuBLASTP: Fine-Grained Parallelization of Protein Sequence Search on CPU+GPU[J]. IEEE/ACM Transactions on Computational Biology & Bioinformatics, 2014, 14(4):830-843.
[23] YE C , ZHENG S G , LONG C , et al. Quantum Fourier Transform and Phase Estimation in Qudit System[J]. Communications in Theoretical Physics, 2011, 55(5):790-794.
[24] Exxact Corporation. Exxact Corporation Extends its Line of AI Development Servers to Support Latest NVIDIA A100 PCIe GPU[J]. Computer Weekly News,2020.163
[25] Simon E. Nigg. Deterministic Hadamard gate for microwave cat-state qubits in circuit QED[J]. Physical Review A,2014,89(2):
[26] Xie J M. Large-scale quantum Fourier transform simulation for "Songshan" supercomputer system [D]. HeNan: Zhengzhou University,2021.

An Improved Method of Sub-band Peak Energy Detection

Wanxiang Lou[1a*], **Qiren Fu**[1], **Kai Feng**[1]

[1]715th Institution of CSSC,Hangzhou310023,Zhejiang,China

[a*]351507369@qq.com

Abstract:In order to solve the problems that the quantity of false targets is huge and the ability of target detection is poor in the conventional method of sub-band peak energy detection(SPED),an improved method of SPED is proposed.The method screens peaks based on the levels of background noise in every sub-band and then eliminates the noise peaks. These measures reduce the quantity of false targets and improve the ability for target detection.Simulation result and experimental data analysis prove its validity and superiority.

1. Introduction

The average power spectrum of ship radiated noise is composed of continuous broadband components and narrowband components at several discrete frequencies. The corresponding detection methods are broadband detection and narrowband detection respectively.The passive sonar mainly uses beamforming to detect targets and high resolution algorithm to improve detection accuracy and azimuth resolution.

To solve the problem of large computation and lack of robustness of the high resolution algorithm, a method of sub-band Peak Energy Detection (SPED) is proposed in literature [1]-[2], by which the received signal is divided into several sub-bands, after the energy detection of each sub-band, the peak detection is carried out. Finally, the results of each sub-band are added up as the output of the detection.Based on SPED, literature [3] proposed a beam domain broadband peak energy detection method, which improves the target display effect by increasing the intensity of beam display of the target.Literature [4] combines SPED and steered minimum variance (STMV) beamforming to effectively improve azimuth resolution and display effect, thus improving broadband detection performance.

However, the conventional SPED is not satisfactory in weak target detection, especially when there is strong interference near the weak target, the weak target is often covered up.At the same time, SPED detects all the peak energy in each sub-band, which leads to too many false targets.In this paper, an improved method of SPED is proposed, which can effectively suppress the interference and improve the detection ability by screening the peak value of each sub-band and eliminating the noise.

2. Sub-band Peak Energy Detection

Assuming that the time snap received by the sonar array is $x(n)$, which is converted to the frequency snap $X(f)$ via fast Fourier transfer(FFT), where n is the time snap point and f is the FFT frequency point.The beam output obtained after frequency domain beamforming is $P(f,\theta_i)$

$$P(f,\theta_i) = w(f,\theta_i)' X(f) \tag{1}$$

Where $\theta_i (1 \le i \le I)$ is azimuth and $w(f,\theta_i)$ is driving vector.

It is assumed that the whole processing frequency band $\left[f_{\min}, f_{\max}\right]$ is evenly divided into L sub-bands, and the frequency range of the lth sub-band is $\left[f_{l\min}, f_{l\max}\right]$. Beamforming is performed for each sub-band to obtain the beam output of each sub-band,

$$P(f_l,\theta_i)=\sum_{f=f_{l\min}}^{f_{l\max}} P(f,\theta_i) \tag{2}$$

Square detection and integration of each $P(f_l,\theta_i)$ can obtain new output $P'(f_l,\theta_i)$, and then peak detection of each sub-band result is performed to extract local peak value, namely

$$P_{pks}(f_l,\theta_i)=\begin{cases} P'(f_l,\theta_i) & P'(f_l,\theta_i)>P'(f_l,\theta_{i-1})\&\&P'(f_l,\theta_i)>P'(f_l,\theta_{i+1}) \\ 0 & else \end{cases} \tag{3}$$

Finally, the output on the same azimuth is accumulated and synthesized to obtain the final result:

$$P_{subband}(\theta_i)=\sum_{l=1}^{L} P_{pks}(f_l,\theta_i) \tag{4}$$

It can be seen from Equations (2) and (3) that if there is a broadband signal in a certain azimuth, local peak values will appear in several corresponding sub-bands at that azimuth, and the signal will be strengthened after accumulation.The noise peaks appear randomly in each sub-band, and the energy is smoothed after accumulation.The two targets can be distinguished as long as they do not coincide exactly in azimuth and frequency.

The SPED not only extracts the target peak but also the noise peak when extracting the peak value. Meanwhile, the local peak value of the target is unique in each sub-band, so the number of local peak value of the noise is far more than the target peak value.With the decrease of signal-to-noise ratio, the number and energy of false alarms in sub-bands increase, and the number of false targets increases exponentially after sub-band synthesis, while the weak targets are submerged in background noise.

3. Improved SPED

In order to improve the detection ability of SPED and reduce false targets, the peak value of target should be preserved as much as possible while the peak value of noise should be reduced.In this paper, the peak value is firstly screened according to the noise energy in each sub-band, and then the peak value is removed after the screening.

This paper uses median filtering and order truncate average [5] to estimate the background noise in the beam space.The output after beamforming of each sub-band is $P'(f_l,\theta_i)$ (dimension is $L\times I$), and the background noise estimation of the beam in each sub-band is carried out.In order to eliminate the edge effect of data, select window length as $2K+1$.Since the spatial spectrum of the multi-beam system of sonar has local continuity, the edge effect can be eliminated by symmetrically extending the K point data on one side of the end point to the other side, and the expanded sub-band output data is

$$P'(f_l,\theta_{K+1}),...,P'(f_l,\theta_2),P'(f_l,\theta_1),...P'(f_l,\theta_i),P'(f_l,\theta_{i-1}),...,P'(f_l,\theta_{i-K}) \tag{5}$$

After processing, the left and right endpoints are in windows in the extended data.Formula (5) is re-expressed as

$$P_l(1),P_l(2),...,P_l(I+2K+1) \tag{6}$$

Order truncate average is performed for sub-band beam domain data from $P_l(K+1)$ to $P_l(K+I)$. The data in the window is arranged in order from smallest to largest:

$$P_l(1),P_l(2),...,P_l(2K+1) \tag{7}$$

The median of the sequence of Formula (7) is $P_l(K+1)$, and the truncate average of the order is defined as:

$$\overline{P_l} = \frac{1}{K+1}\sum_{m=1}^{k+1} P_l(K+m) \tag{8}$$

$N(f_l,\theta_i)=\overline{P_l}$ is the estimation of beam background noise. In this sub-band, peak screening is performed according to the background noise, and the results are as follows:

$$P_{pks}'(f_l,\theta_i)=\begin{cases} P_{pks}(f_l,\theta_i) & P_{pks}(f_l,\theta_i) > \alpha NP_{pks}(f_l,\theta_i) \\ 0 & else \end{cases} \tag{9}$$

Where $\alpha > 0$, and it can be adjusted according to the actual situation.

After peak screening, the non-zero value of sub-band output includes local peak value of targets and local peak value of noise.If the target signal is a wideband signal, the sub-band spectrum on the corresponding azimuth should be continuous.The local peak value of noise is randomly distributed and presents as independent points. Therefore, part of the peak value of noise can be removed according to this characteristic, namely

$$P_{pks}'(f_l,\theta_i)=\begin{cases} P_{pks}'(f_l,\theta_i) & else \\ 0 & P_{pks}'(f_{l-1},\theta_i)=0 \,\&\&\, P_{pks}'(f_{l+1},\theta_i)=0 \end{cases} \tag{10}$$

When the sub-band partition is too wide, the local peaks of the target only appear in a certain sub-band, and the process of eliminating the noise peaks will also eliminate the target peaks, thus reducing its detection performance.At the same time, when the target SNR is low and the background noise fluctuation is large, the target peak will also fluctuate in each sub-band, which may lead to the discontinuity of the sub-band spectrum.Therefore, in practical application, the sub-band width division and the selection of α directly determine the performance of the improved SPED.

4. Simulation Result

In order to verify the effectiveness of the algorithm, simulation is carried out.In the simulation, 128 element line array is used, and the target moves from 45° azimuth to -45°, and sends broadband signal.The quantity of sub-band is 100. $K=11$ $\alpha=0.9$ in the method of order truncate average.The simulation results of conventional beamforming(CBF)、the conventional SPED and the Improved SPED are as follows:

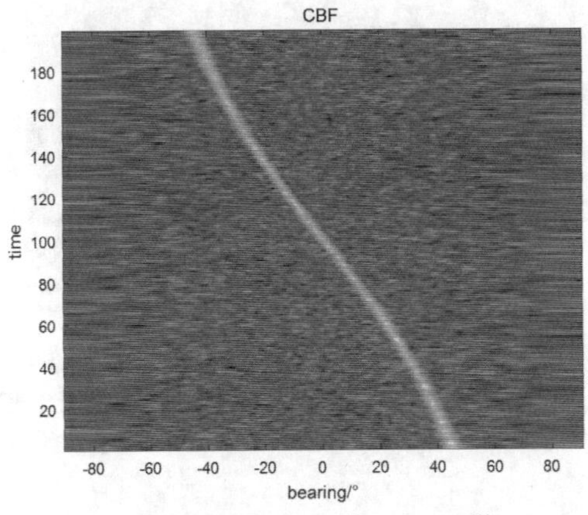

Fig.1 CBF bearing time recording

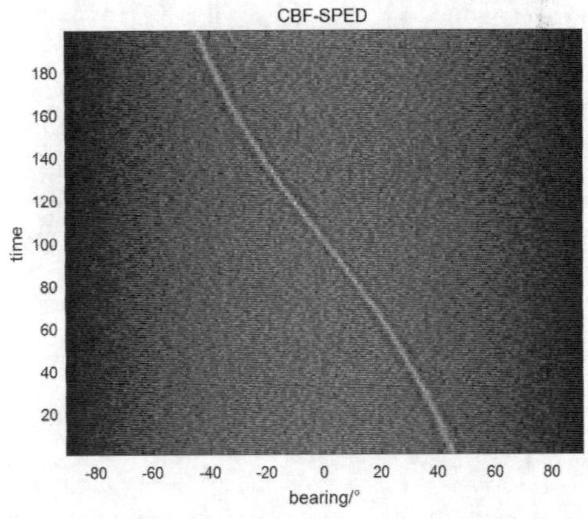

Fig.2 conventional SPED bearing time recording

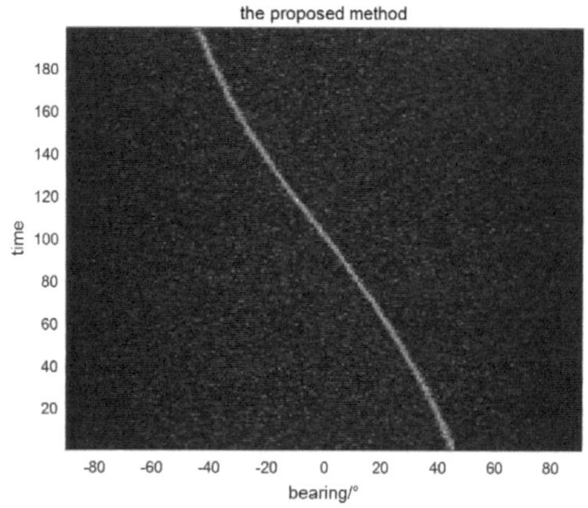

Fig.3 the proposed method bearing time recording

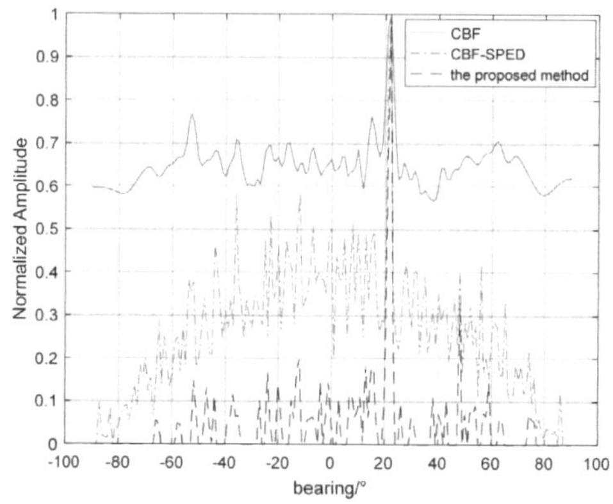

Fig.4 the beamforming results

From figure 4, it can be seen that the target peak energy to the maximum noise energy ratio of CBF、SPED and the improved SPED is 0.76、0.58 and 0.32 respectively, and the background noise energy is significantly reduced. Meanwhile, both the SPED and the improved SPED have a smaller main lobe width than CBF. Simulation results show that the improved method can effectively reduce background noise and improve the detection ability of the target.

5. Experimental Data Analysis

In order to further verify the effectiveness of the proposed method, a group of actual data results are given in this section.The experimental data come from a plane array.The target is stationary and its azimuth is 90°, which sends the broadband signal.The beamforming results and bearing time recording of the experimental data are shown below.

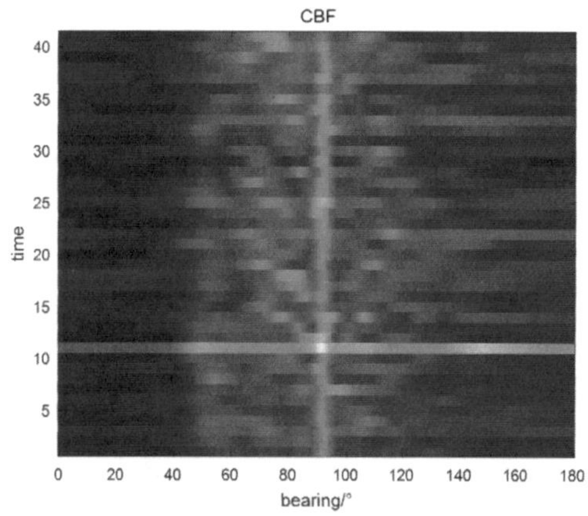

Fig.5 CBF bearing time recording

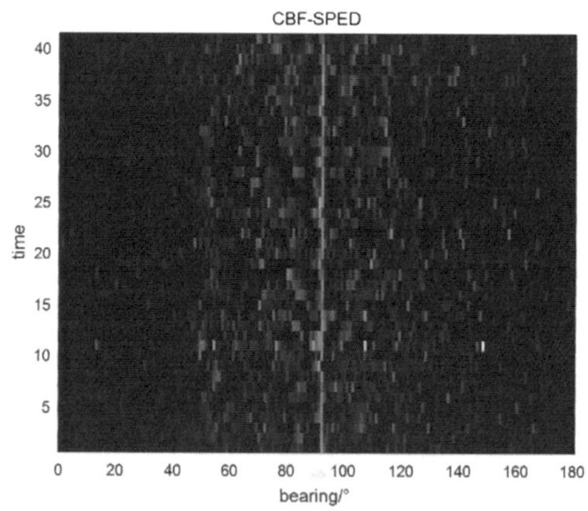

Fig.6 the conventional SPED bearing time recording

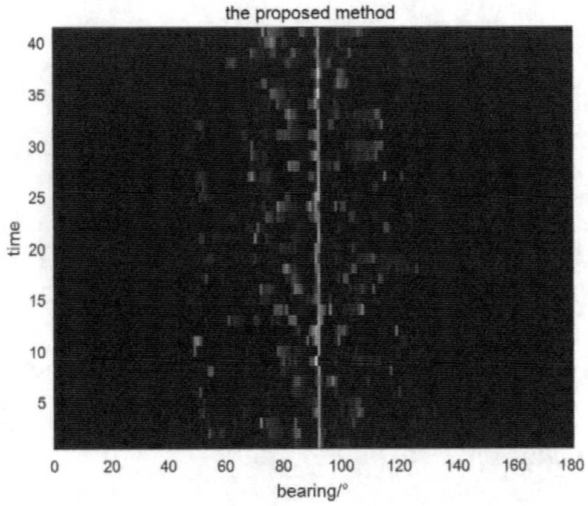

Fig.7 the proposed method bearing time recording

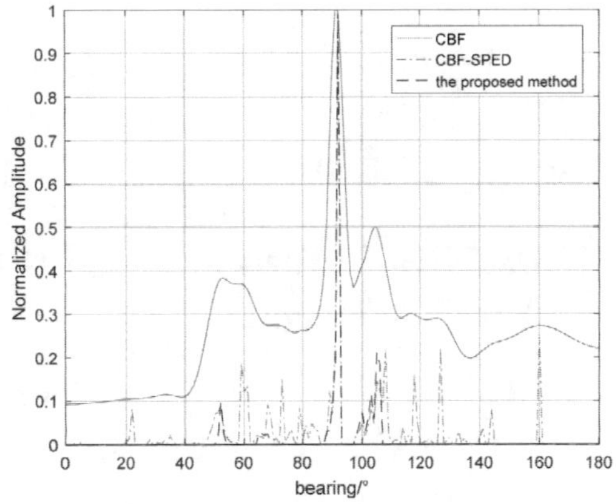

Fig.8 the beamforming results

It can be seen from FIG. 5 to FIG. 7 that the tracks of the SPED and the improved SPED are clearer than that of CBF, and the background noise of the improved SPED is significantly lower than that of the SPED. Figure 8 provides the beamforming result at a certain time. As can be seen from the figure, the target peak energy to the maximum noise peak energy ratio of CBF、 SPED and the improved SPED is 0.5,0.27 and 0.21 respectively.There are 12 false targets for SPED and 5 for the improved SPED,so the number of the false targets is significantly reduced. Experimental data analysis shows that the improved SPED noise and the quantity of false targets are significantly reduced in relative amplitude and number.

6. Conclusion

In view of the shortcomings of conventional SPED, an improved SPED method is proposed in this paper.By screening the local peak value of each sub-band and then removing some local peak value of noise, it can suppress the background noise and improve its detection ability.Simulation experiments and actual data analysis demonstrate the effectiveness of the proposed method.In practical application, background noise estimation is often inaccurate, which will affect the results of peak screening. Better peak screening methods will be improved in the follow-up work.

References

[1] Bono M. Shapo B. Sub-band Energy Detection in Passive Array Processing[R]. ADA405484, 2000.
[2] M. Bono. SEED ED vs. SPED ED Evaluation. Signal Processing Working Group, November 2,200,Applied Research Laboratories: University of Texas at Austin.
[3] Yang Chenhui, Ma Yuanliang, Yang Yixin.PeakEnergi Detection with application to passive sonar display[J].Journal of Applied Acoustics, 2003, 05 (7) : 31-35.
[4] Xiaoyong Jiang, LiyangYu,Xuanming Du. A novel efficient technique to improve performance of passive sonar broadband detection.Technical Acoustics, 2011, 30 (6) : 484-488.
[5] Li Qihu, Pan Xuebao, Yin Li.A new method of background equalization[J].Acta Acustica,2000,25(1):5-9.

Unsupervised Video Summarization Based on An Encoder-Decoder Architecture

Xin Li[1], QiLin Li[2*], Dawei Yin[3], Lijun Zhang[4], Dezhong Peng[1]

[1]Sichuan University, Chengdu, Sichuan, 610000, China

[2]Metering Center of Sichuan Electric Power Corporation, Chengdu, Sichuan, 610000, China

[3]Sichuan Zhiqian Technology Limited Company, Chengdu, Sichuan, 610000, China

[4]Chengdu Ruibei Yingte Information Technology Limited Company, Chengdu, Sichuan, 610000, China

[*]Corresponding author's e-mail: liql367x@sc.sgcc.com.cn

Abstract. The purpose of video summarization is to facilitate large-scale video browsing. Video summarization is a short and concise synopsis of original video. It usually composed of a set of representative video frames from the original video. This paper solves the problem of unsupervised video summarization by developing a Video Summarization Network (VSN) to summarize videos, which is formulated as selecting a sparse subset of video frames that best represents the input video. VSN predicts a probability for each video frame, which indicates the possibility of a frame being selected, and then takes actions to select frames according to the probability distribution to form a video summary. We designed a novel loss function which takes into account the diversity and representativeness of the generated summarization without labels or user interaction.

1. Introduction

In July 2015, YouTube revealed that it receives the video content more than 400 hours per minute, which is equivalent to uploading the video content more than 65 years per day. Consumers have increasingly participated in online video platforms and devices (such as smartphones and wearable devices) that have powerful video recording sensors and allow the captured videos to be uploaded to the Internet.

So, how can we effectively browse endless video collections and find the video content we want? The answer to this question may have been the video retrieval technology, but now I think the automatic video summarization technology is also very important. It can generate a concise summarization that cover the important parts of full-length video. In view of the excessive amount of video content on the Internet, effective video summaries can help viewers browse and browse large video collections, thereby increasing viewer engagement and content consumption.

2. Related work

Recently, the research of video summarization has made a great progress, and various methods with different characteristics have emerged.[1] Lee et al. [2] identify important objects and people by summarizing videos. Michael G et al. [3] proposed a method to predict the interest level of video frames and use these interest levels to select the key frame. Helmut G et al. [4] thought of video summarization as a subset selection problem and a multi-objective optimization sub-module function. Ejaz et al. [5] applied attention modeling techniques to extract key frames of visual saliency. These methods do not consider the temporal information of the video, but independently process each video frame. In addition, the above commonly used methods cannot train the model in an end-to-end way, which will takes additional time during testing.

Lu G. et al. [6] proposed a video summarization method with a fixed length of summarization. The fixed-size method is too rigid and lacks flexibility. M.U. Sreeja et al. [7] reported an instance-based approach. Mohamed A. et al. [8] applied a method based on reinforcement learning. In other aspects, Zhang Y. et al [9] used the key-frame to detect the human pose. Liu Y. et al [10] considered the transforming multi-concept attention.

In order to solve the above problems, we use a deep CNN to capture the feature and a LSTM to capture the long-term temporal information in the video frame. And we propose a framework based on an encoder-decoder architecture to train the network end-to-end.

3. Proposed Method

Then we will describe our network structure and optimization algorithms used.

3.1 Network Structure

The schematic diagram of the network structure we proposed is shown in Figure 1.

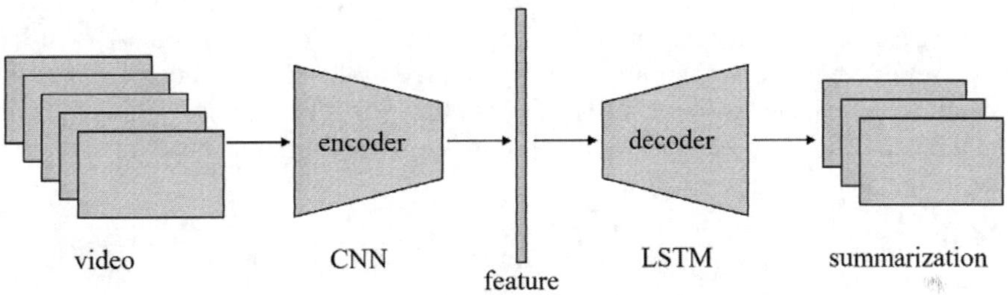

Figure 1. Network structure

Our network uses an encoder-decoder architecture which the encoder uses the CNN and the decoder uses the LSTM. Assume the length of the input video is T, which means it contains T frames $\{v_i\}_{i=1}^{T}$ from which the CNN will extract visual features $\{x_i\}_{i=1}^{T}$. Then, the LSTM uses the visual features $\{x_i\}_{i=1}^{T}$ as input and generates the corresponding hidden states $\{h_i\}_{i=1}^{T}$. Thanks to the advantages of LSTM in processing time series data, h_i not only contains past information, but also emphasizes the features around the ith frame. Finally, the sigmoid function outputs the prediction result, predicts the probability p_i of each frame being selected. Then we use this probability to sample from the Bernoulli distribution to determine the final action a_i:

$$p_i = \sigma(Wh_i), \quad (1)$$

$$a_i = Bernoulli(p_i), \quad (2)$$

where σ represents the sigmoid function, and a_i represents whether the ith frame is selected. For brevity, we ignore the bias in Eq.(1). Video summarization consists of selected frames by actions, $S = \{v_i | a_i = 1, i = 1,2,\cdots\}$.

3.2 Loss function

In general, for preserving as much temporal information in the entire video as possible, high-quality video summarization should have diversity in their content and be representative of the video.[11] To this end, we propose a novel loss function that takes the diversity and representativeness of the summarization into consideration. The loss is composed of the similarity loss and the k-medoids loss, which respectively ensure the diversity and representativeness of the summarization. The details are as follows.

We measure the degree of dissimilarity between selected frames to evaluate the generated summarization's diversity. Suppose the indices of the selected frames is $I = \{i|a_i = 1, i = 1,2,\cdots\}$, and we calculate the average value of the pairwise similarities between the selected frames as the similarity loss:

$$L_s = \frac{\sum_{i\in I}\sum_{\substack{i'\in I\\ i'\neq i}} \frac{x_i^T x_{i'}}{\|x_i\|_2\|x_{i'}\|_2}}{|I|(|I|-1)}. \tag{3}$$

The more dissimilar the selected frames are, the more diverse the summarization will be. However, Eq.(3) ignores the inherent timing structure of sequence data. In fact, we should not consider the similarity between two frames that are far apart in time. To solve this problem, we assume $f(x_i, x_{i'}) = \begin{cases} \frac{x_i^T x_{i'}}{\|x_i\|_2\|x_{i'}\|_2}, & if\ |i-i'| \le \lambda \\ 1, & if\ |i-i'| > \lambda \end{cases}$, where λ represents the distance in time between the two frames. Then we can rewrite Eq.(3):

$$L_s = \frac{\sum_{i\in I}\sum_{\substack{i'\in I\\ i'\neq i}} f(x_i, x_{i'})}{|I|(|I|-1)}. \tag{4}$$

In order to measures the degree that the summarization can represent the original video, we express the representativeness of the video summarization as the k-medoids problem.[4] We hope that the VSN chooses a set of center points to minimize the average of the sum of squared errors between each video frame and its nearest center point. With this loss function, the VSN is motivated to select frames near the medoids in the feature space, thereby making the summarization more representative. Therefore, we define the k-medoids loss as

$$L_k = \sum_{i=1}^{T} \min_{i'\in I}\|x_i - x_{i'}\|_2 . \tag{5}$$

The final loss function is composed of L_s and L_k :

$$L(S) = L_s + \alpha L_k, \tag{6}$$

where α is the weight parameter.

4. Experiments

In this section we introduce the original dataset we used to train the VSN in the experiment, as well as the evaluation metric used in this experiment, and finally gives the experimental results.

We train our model on SumMe dataset[3]. There are 25 videos in the SumMe dataset. They have various content such as festivals and sports, etc. The shortest video length is 1 minute and the longest 6 minutes. And each video is annotated by an average of 16 persons so that we can consider opinions from different people, thereby reducing the subjectivity of the video summarization.

Usually, researcher uses a subjective or an inappropriate evaluation metric to evaluate the video summarization.[12] Some researchers adopted some simple objective evaluation metrics such as fidelity and compression ratio.[13] But this is actually not a good evaluation metric for video summarization. We adopted a relatively better evaluation metric, the CUS (comparison of user summaries).[14] The CUS evaluation metric use two objective evaluation metrics which can evaluate the video summarization well. They are defined as follows:

$$CUS_A = \frac{n_{mAS}}{n_{US}}, \tag{7}$$

$$CUS_E = \frac{n_{\bar{m}AS}}{n_{US}}, \tag{8}$$

where US means the frames in user summarization. And AS means the frames from the generated summarization. The mAS and $\overline{m}AS$ stand for the matching frames and the non-matching from AS respectively. So the $n_{US}, n_{mAS}, n_{\overline{m}AS}$ mean the number of the $US, mAS, \overline{m}AS$ respectively.

As reported in Table 1, on the SumMe dataset, the VSN gets a high CUS_A and a low CUS_E. It shows that our method can have a good performance on video summarization.

Table 1. CUS_A and CUS_E of different methods on SumMe dataset.

Methods	CUS_A	CUS_E
OV [15]	0.68	0.54
DT [16]	0.59	0.31
STIMO [17]	0.73	0.56
VSUMM [18]	0.70	0.27
VSN (our)	0.80	0.26

5. Conclusion

This paper proposes a method VSN for automatic video summary generation based on an encoder-decoder architecture, where the encoder uses CNN and the decoder uses LSTM. We proposed the loss function that takes the diversity and representativeness of the summarization into consideration. From the experimental results on the SumMe dataset, the VSN we proposed has better performance compared with some other video summarization approachs.

Reference

[1] Apostolidis E, Adamantidou E, Metsai A I, et al. Video Summarization Using Deep Neural Networks: A Survey[J]. 2021.

[2] Ghosh J. Discovering important people and objects for egocentric video summarization[C] Computer Vision & Pattern Recognition. IEEE, 2012.

[3] Gygli M, Grabner H, Riemenschneider H , et al. Creating Summaries from User Videos[C] European Conference on Computer Vision. Springer International Publishing, 2014.

[4] Gygli M, Grabner H, Gool L V. Video summarization by learning submodular mixtures of objectives[C] 2015 IEEE Conference on Computer Vision and Pattern Recognition (CVPR). IEEE, 2015.

[5] Naveed Ejaz, Irfan Mehmood, Sung Wook Baik. Efficient visual attention based framework for extracting key frames from videos[J] Volume 28, Issue 1, January 2013, Pages 34-44.

[6] Lu G, Zheng J. Fixed-size video summarization over streaming data via non-monotone submodular maximization[C] MMAsia '20: ACM Multimedia Asia. ACM, 2021.

[7] Mus A, Bck B. A unified model for egocentric video summarization: an instance-based approach[J]. Computers & Electrical Engineering, 92.

[8] Aboelenien M, Salem A M. Reinforcement Learning Versus Generative Adversarial Networks for Video Summarization[C] 2021 International Mobile, Intelligent, and Ubiquitous Computing Conference (MIUCC). 2021.

[9] Zhang Y, Wang Y, Camps O, et al. Key Frame Proposal Network for Efficient Pose Estimation in Videos[C] 2020.

[10] Liu Y T, YJ Li, Wang Y. Transforming Multi-concept Attention into Video Summarization[M]. 2021.

[11] Zhou K, Qiao Y, Xiang T. Deep Reinforcement Learning for Unsupervised Video Summarization with Diversity-Representativeness Reward[J]. 2017.

[12] Kannappan S, Liu Y, Tiddeman B P. Performance Evaluation of Video Summaries Using Efficient Image Euclidean Distance[C] International Symposium on Visual Computing. Springer International Publishing, 2016.

[13] Kuanar S K, Panda R, Chowdhury A S. Video key frame extraction through dynamic Delaunay clustering with a structural constraint[J]. Journal of Visual Communication and Image Representation, 2013, 24(7):1212-1227.
[14] Ke Z, Grauman K, Fei S. Retrospective Encoders for Video Summarization[C] European Conference on Computer Vision. Springer, Cham, 2018.
[15] Daniel D, Vikrant K, David D. Video summarization by curve simplification[C] MULTIMEDIA '98: Proceedings of the sixth ACM international conference on Multimedia 1998.
[16] Padmavathi M, Yong R, Yelena Y. Keyframe-based video summarization using Delaunay clustering[J] International Journal on Digital Libraries volume 6, pages219–232 2006.
[17] Furini M, Geraci F, Montangero M, et al. Stimo: Still and Moving video storyboard for the web scenario[J]. Multimedia Tools and Applications, 2010, 46(1):47-69.
[18] Avila S, Lopes A, Luz A D, et al. VSUMM: A mechanism designed to produce static video summaries and a novel evaluation method[J]. Pattern recognition letters, 2011, 32(1):p.56-68.

Research on HPLC Experimental Platform Component Scheme

Yutu Liang[1a*], Bo Li[1b*], Xingnan Li[1c*], Xiaozhi Deng[1d*], Jiangang Lu[1e*], Zhan Shi[1f*]

[1]Power Dispatching control Center of Guangdong Power Grid Co., LTD

[a*]terminalxy@163.com,[b*]1224681971@qq.com,[c*]2404598353@qq.com,

[d*]357961393@qq.com,[e*]3257892833@qq.com,[f*] 383294769@qq.com

Abstract: Power line communication is a kind of communication mode which uses power line as transmission medium and carries on data transmission and information exchange through carrier. Power line has wide distribution, high economy, high transmission rate and good reliability. However, as a communication mode without special transmission media, there is still a lack of power line communication experimental platform in the industrial field. The communication experiment platform is composed of power line carrier field test platform, protocol consistency test platform, interconnection test platform and courts simulation platform. In this paper, the design scheme of the experimental platform is proposed from the aspects of hardware and software respectively. In terms of technology, the requirements of reliability, efficiency, enforceability and practicability are fully considered.

1. Introduction

Low-voltage power lines are connected in all directions, spread all over urban and rural areas, and directly contact the user network. They have obvious advantages such as no indoor wiring, extremely wide coverage, and plug and play connection. High-speed data after modulation is transmitted on low-voltage power lines with high cost performance. Industry experts generally believe that power line communication is one of the communication technologies with incomparable advantages in the "last 300m" solution[1]. Power line communication has natural media and cost advantages. The application of power line carrier meter reading is the most feasible technical means to realize "full coverage, full collection and full cost control", which will effectively change the situation that power users' information cannot be controlled timely, completely and accurately for a long time.

Different from wireless communication and ethernet communication, power line communication has the characteristics of short communication distance, large time variability, asymmetry and half duplex communication. As the power line is mainly aimed at transmitting electric energy, as a communication transmission medium, the low-voltage power line has the characteristics of heavy load, strong noise interference, large channel attenuation, long channel delay and bad communication environment, which restrict the transmission distance and communication reliability of the power line channel. Therefore, in the actual application of low-voltage power line communication network, technologies such as information relay, network topology control and route management are needed to extend the communication distance and improve the reliability of communication[2]. Therefore, the design of power line communication experimental platform involves many aspects.

2. Experimental platform hardware outline design

2.1. Power line carrier field test platform

The power line carrier field test platform is mainly used to collect the power line noise, impedance and other power line channel characteristic information in the field, as well as the synchronous playback attack of the power line communication in the laboratory. Low voltage broadband power line carrier communication synchronization acquisition and playback need synchronous sampling with power frequency signals to generate distorted signals synchronously with power frequency in the power line carrier test system. In order to effectively evaluate the performance of different manufacturers, it can accurately replay the attack test of test signals collected on site and reflect the operating conditions of communication system as realistically as possible in the laboratory environment. Table 1 shows the equipment composition list of the field test platform, Figure 1 shows the front display of the equipment.

Table 1. Device composition list

Hardware material	Function	Quantity
Portable suitcase	Synchronous sampling and playback of field power line noise	1

Fig.1 Front of the device

2.2. Power line carrier protocol consistency test platform

Protocol consistency test platform includes physical layer protocol test, link layer protocol test and application layer protocol test.

Physical layer protocol test: when the tested object enters the test mode, the transparent access unit at the physical layer generates test packets in different TMI modes. The tested object are parsed and reported. The transparent access unit at the physical layer compares the parsing results.

Link layer protocol test: the transparent access unit at the physical layer generates interactive packets to generate specific scenarios. Protocol interactions such as network access, network maintenance, network negotiation, routing forwarding and data interaction are performed with the tested object. Link layer checks whether the exchange packets of the object comply with the requirements of the link layer protocol.

Application layer protocol test: the virtual concentrator and electricity meter are used to test the application layer packets. The 232 interface of the communication module is used to monitor the packets to verify whether the packets meet the requirements of application layer protocols. Table 2 shows the equipment composition list of the protocol consistency test platform, Figure 2 shows the cabinet diagram of the equipment.

Table 2 Device composition list

Hardware material	Function	**Quantity**
Equipment cabinet	Shape structure carrier	2
Industrial Personal Computer	Control industrial equipment work on site	1

Display keyboard all in one machine	Human-computer interaction equipment	1
Programmable power	Station power frequency simulation	1
STA box	STA module support, provide power supply, communication, control, etc	2
CCO box	CCO module support, provide power supply, communication, control, etc	1

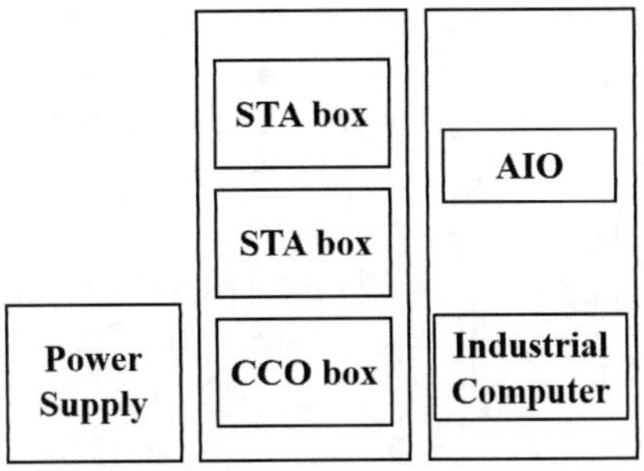

Fig.2 Equipment cabinet schematic

2.3. *Power line carrier interconnection test platform*

The interconnection test platform simulates the field environment, introduces the power frequency signal, and has the clear ability to adjust the agent relationship. Adjust the power grid topology, under different communication conditions such as crosstalk area, relay level and noise environment, the module performance is evaluated in single and mixed networking. Table 3 shows the equipment composition list of the interconnection test platform, Figure 3 shows the cabinet diagram of the equipment.

Table 3 Device composition list

Hardware material	Function	Quantity
Equipment cabinet	Shape structure carrier	2
Industrial Personal Computer	Control industrial equipment work on site	1
Display keyboard all in one machine	Human-computer interaction equipment	1
STA box	STA module support, provide power supply, communication, control, etc	3
Analog switch box	Simulation of power grid topology	3
Isolated attenuating equipment	Carrier channel simulation	3

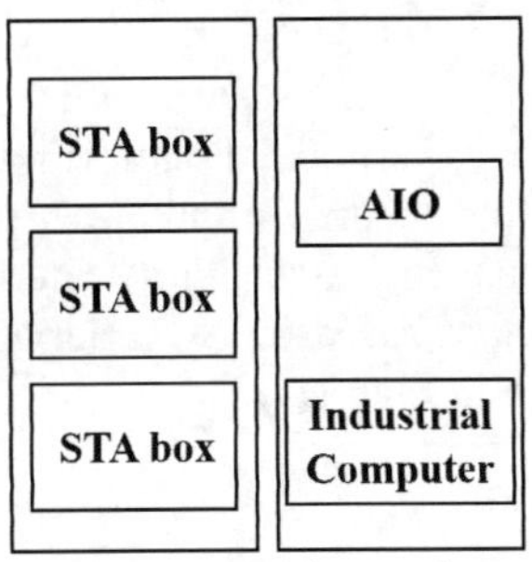

Fig.3 Equipment cabinet schematic

2.4. Power line carrier courts simulation platform
Power line carrier courts simulation platform uses physical cable, branch box, switch box and other low-voltage power equipment to build physical environment simulation device, low-voltage power line broadband carrier communication verification environment, used for testing courts identification, phase identification and other functions, while solving the problem of some differences between laboratory test and field test. The platform can simulate different cable types and different power supply topologies. Low voltage power line broadband carrier communication equipment can be flexibly connected to branch points and terminal points, and low voltage electrical equipment can be flexibly connected to signal injection points, so that the actual environment can synchronize the playback of noise signals. Table 4 shows the equipment composition list of the courts simulation platform, Figure 4 shows the cabinet diagram of the equipment.

Table 4 Device composition list

Hardware material	Function	Quantity
Equipment cabinet	Shape structure carrier	3
Industrial Personal Computer	Control industrial equipment work on site	1
Display keyboard all in one machine	Human-computer interaction equipment	1
Programmable power	Station power frequency simulation	1
STA box	STA module support, provide power supply, communication, control, etc	5
CCO box	CCO module support, provide power supply, communication, control, etc	1
Analog switch box	Simulation of power grid topology	5
Isolated attenuating equipment	Carrier channel simulation	5

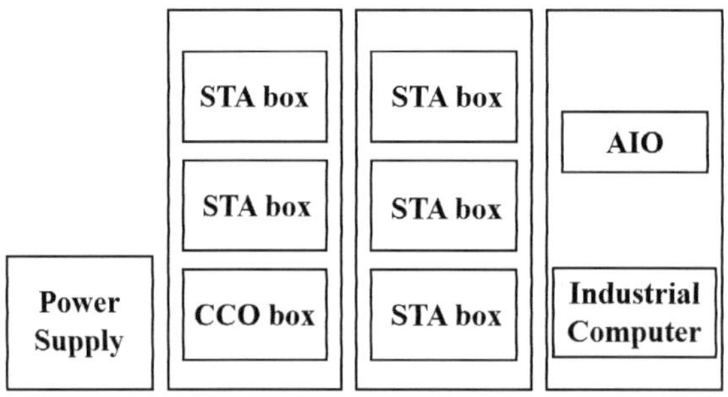

Fig.4 Equipment cabinet schematic

3. Software Technical Solution
The overall technical principles of the experimental platform revolve around "efficiency, modifiability, expansibility, multi-scheme support and security principles", which ensures that the final software has a high degree of availability, advancement, expansibility and openness. Software development adopts object-oriented technology, data storage based on SQL Server, data exchange based on TCP/IP. The system distribution structure is shown in Figure 5.

 2258 (2022) 012068
doi:10.1088/1742-6596/2258/1/012068

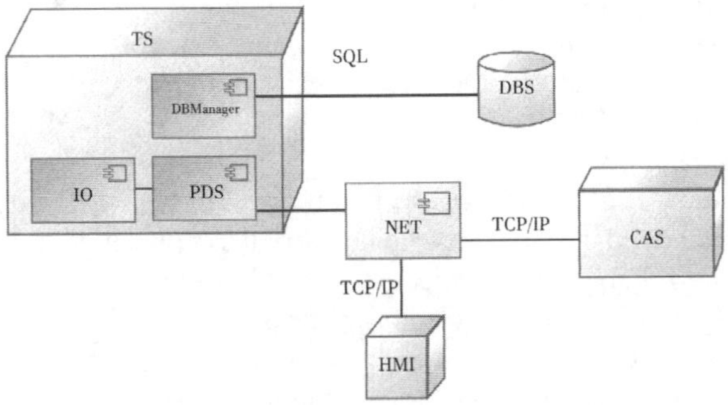

Fig.5 System distribution structure

3.1. Object oriental technique

Object oriented design is a philosophical idea that the world is composed of various objects with their own laws of motion and internal states, the interaction and communication between different objects constitute the complete real world. Object-oriented design emphasizes that the structure of the system should be directly corresponding to the structure of the real world, and the system should be constructed around objects in the real world, rather than around functions[3].

Object-oriented technology moves software development beyond procedural programming into the world of reusable programming by simplifying application development. Unlike the old way of programming, maintaining and debugging programs doesn't get more complicated as they grow in size. Object technology works at two levels:

1) At the data level, object technology can integrate many different types of information within an organization. Information from the past is no longer compatible.

2) At the program development level, object technology provides modular program constructs. At this point, the programmer develops on top of existing objects. Objects can be reused by other objects to eliminating the need to rewrite the code every time they are needed again by exploiting their procedures.

An object is a complete grouping of data that includes functions for processing data. In an object-oriented environment, the number of objects is uncountable. They might include a database record, a file, a physical resource, or even a user account. To enable developers to work with object-oriented programming languages, an object is a self-satisfying module that contains data, the structure of the data, and the ability to process the data. It is easy to redesign or extend the system because you do not need to change or decompose the entire system. In effect, these modules are discarded or modified, and new ones are added to provide enhanced functionality.

3.2. SQL Server-based data stores

Microsoft SQL Server is a comprehensive database platform that provides enterprise-level data management using integrated business intelligence (BI) tools. Microsoft SQL Server database engine provides more secure and reliable storage for relational and structured data, it can build and manage highly available and high-performance data applications for business. The SQL Server database has the following five features:

1) High-performance design that takes full advantage of Windows NT.

2) Advanced system management, support Windows graphical management tools, support local and remote system management and configuration.

3) Powerful transaction processing function, using various methods to ensure the integrity of data.

4) Support symmetric multi-processor structure, stored procedures, ODBC, and has an independent SQL language.

5) With built-in data replication function, powerful management tools.

SQL Server database compared with the Access database in the past, it is easier to save massive data, as well as distributed processing and remote connection, applicability is strong, in a large number of interactive message records, Access database has been unable to meet the functional requirements, The change to SQL Server database can store data more effectively [4].

3.3. TCP/ IP-based data exchange
TCP/IP communication protocol uses a four-layer hierarchical model structure, which is different from the OSI seven-layer model. Each layer invokes the network tasks provided by its next layer to complete the requirements. TCP is a reliable and connection-oriented protocol[5]. Figure 6 shows the TCP/IP hierarchical model structure, protocols between Application layers invoke the transport layer, the network layer and the physical data link layer step by step to realize the communication interconnection of applications at the Application layer. This paper mainly studies the communication mode, the communication connection mode and the communication message sending and receiving mode.

The IP layer receives the packets from the lower layer and sends the packets to the higher TCP layer. The IP layer also receives the packets from the TCP layer to the lower layer. The IP packet is unreliable. The IP packet contains the address of the sending host (the source address) and the address of the receiving host (the destination address). TCP is a connection-oriented communication protocol. It establishes a connection through a three-way handshake and disconnects the connection when the communication is complete. Because TCP is connection-oriented, it can only be used for end-to-end communication.

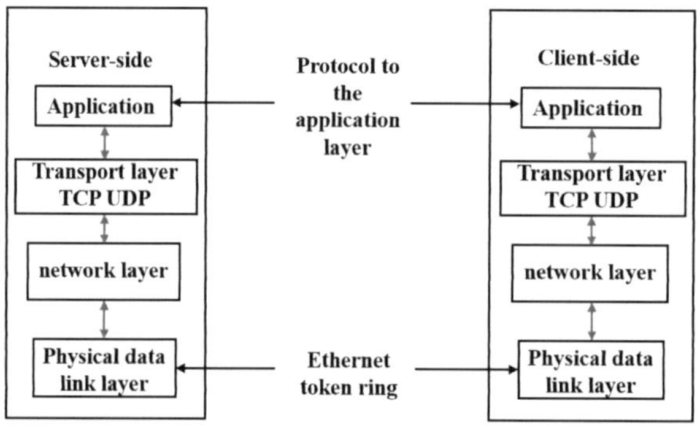

Figure 6 TCP/IP hierarchical model structure

3.4. System architecture development
Software using client/server (C/S) architecture, combined with object-oriented development methods, using C# development language, SQL SERVER database, three-tier architecture design, Visual Studio development integration tools.

The user interface (UI) layer is designed and coded using the interface designer of Visual Studio development platform, mainly used to create front-end user interface and write interface function codes.

Business layer through .NET Framework for business plug-in association, through the business operation interface layer to carry out integration and group development.

The data layer is realized by DataAdapter, which is the bridge between DataSet and data source. Figuratively speaking, the DataAdapter is a transport vehicle that transports information from the database to the DataSet, and also transports information from the DataSet to the database.

We mainly use the fill method of DataAdapter to transport database information to DataSet. The DataSet simulates a database in memory. Therefore, the information from the database should also be stored in the DataSet with the corresponding table (we usually use the real table name in the database). In other words, we tell the fill method that the target database and target data table are in memory.

The above three-tier architecture can achieve the following effects: 1. Shorten the development cycle; 2. High code reuse rate; 3. The division of labor of each layer is clear and the degree of abstraction is high, which is conducive to the modular development of the development team.

4. Conclusion

In view of the lack of power line communication experimental platform in the industrial field at present, this paper proposes an experimental platform component scheme based on object-oriented technology, SQL Server data storage and TCP/IP data exchange. Firstly, the hardware components of the experimental platform are described, and each component is briefly described. Then, the software design scheme of the experimental platform is introduced. The client program is constructed with C/S architecture, combined with network communication technology, it not only has rich UI operation functions, but also has remote communication capability. Finally, the Scheme uses Microsoft Visual Studio as the basic development platform, develops software based on. NET Framework. Overall technical scheme fully meets the efficiency, modifiability, scalability, multi-scheme support, security principles.

References

[1]. CAI wei. Research on Transmission Characteristics of Power Line Channel in Smart Distribution Network [D]. Wuhan University,2013.
[2]. Xiang Min, Wang Ping, Lv Xiaofu, Wang Shihe. Construction of low-voltage power line communication experimental teaching platform [J]. University Education,2014(10):54-55.
[3]. BOOCH G. Object-oriented Analysis and Design [M]. 3rd edition. Beijing: Publishing House of Electronics Industry,2016.
[4]. Chen Sufang. Analysis of performance optimization method of SQL Server Database [J]. Information Record Materials, 201,22(05):90-91.
[5]. Dai Wenhui. Design and implementation of high performance TCP/IP communication class library [J]. Computer knowledge and technology,2015,11(12):22-23+26.

Journal of Physics: Conference Series 2258 (2022) 012069 doi:10.1088/1742-6596/2258/1/012069

CSChecker： A binary taint-based vulnerability detection method based on static taint analysis

Wei Guo[1b], Qiang Wei[1a*], Qianqiong Wu[1], Zhimin Guo[2c]

[1] Cyberspace Security Academy, University of Information Engineering, Zhengzhou, Henan, 450001, China

[2]Electric Power Research Institute of State Grid Henan Electric Power Company,China

[*a]funnywei@163.com, [b]1037057802@qq.com,[c]zhimin.guo@163.com

Abstract. Embedded devices such as routers not only bring convenience to people's daily life, but also increase the attack surface and security risks of devices. Embedded device applications tend to be closed source and therefore cannot be searched for vulnerabilities through source code audits. Even open source applications can be insecure because they reference third-party libraries. Binary file vulnerability mining is an important means to solve this kind of problem, but it has the problems of path explosion and low efficiency. This article uses the static stain analysis with the method of combining the vulnerability characteristics, in the type of stain into classes and class assignment holes for testing. Based on function call graph, this paper uses atomic combinational optimization to detect the vulnerability of router firmware. The prototype tool -- CSChecker is implemented in D-Link, Tenda, the test was carried out on 267 firmware files of Netgear and other well-known brands, and the experimental results showed that the accuracy of CSChecker in the data set reached 92.51%, indicating that CSChecker could effectively search the injection vulnerabilities and assignment vulnerabilities of binary files.

1. Introduction

As embedded devices become popular in people's daily lives, they bring convenience but also risks such as leakage of sensitive data, especially the security of router devices. Routers, as network hubs, are able to connect most smart devices under the same premises, making them a high-risk target for attacks. Most embedded devices are closed source, which makes them more difficult to attack but also leads to their security relying mainly on a small number of people such as security testers, increasing the likelihood of vulnerabilities in the system and therefore requiring direct detection of the firmware in binary form.

Static taint analysis techniques rely on the possible presence of contaminated data streams in the target object, mainly for weblike applications[1][2] and Android[3][4] applications, as both target objects are characterised by a large number of user interactions, a large number of risk points brought about by complex components and the availability of part or all of the source code. The detection of Web-based applications is mainly for their scripting languages, such as PHP[5] and JavaScript[6]; the detection of Android applications aims at protect whether sensitive information[7] such as user privacy is stolen, for example, B-Droid[8] uses static taint analysis combined with fuzzy testing to detect privacy leaks by analysing the tested application's behaviour to detect privacy leaks, which can effectively detect the five most popular types of commercial spyware in the market.

Static detection of objects is generally low-level languages[9], and large-scale data stream detection increases the complexity of the model, leading to an increase in detection time and a decrease in accuracy. In addition, static detection technology also has some problems such as path explosion and data stream insensitivity. In this paper, we use an atomic assembly-based approach to avoid the path explosion problem and improve the detection efficiency and accuracy by detecting the disassembled C language.

The contributions of this paper are shown below

1. This paper uses the method of function atomic pooling to limit the path increase within a function, effectively avoiding the path explosion problem in the process of taint analysis.
2. This paper uses a combination of taint analysis and vulnerability features to detect tainted vulnerabilities, and optimizes the detection paths based on function call graphs by atomic combination to directly obtain reachable paths.
3. This paper designs and implements a prototype tool for binary vulnerability detection, CSChecker, for identifying injection-type and assignment-type vulnerabilities in router firmware and binary dynamic link library files, with an accuracy rate of 92.51% for vulnerabilities.

The paper is organised as follows. Section II describes the preliminaries of the paper; Section III details the specific ideas of the method used in this paper and the algorithms implemented in each section; Section IV describes the performance tests of CSChecker on a dataset and provides a comparative analysis of its experimental results; and finally discusses the limitations of the method used in CSChecker and the direction of effort for subsequent work.

As a network hub, the router usually connects all the smart components in the same scenario, making its security even more noteworthy. Similar to the two applications mentioned above, router systems also have web interfaces that interact with the user, such as login and configuration pages. In addition, the UPNP protocol, which was introduced for the sake of convenience, has been uncovered in recent years. This paper researches the percentage of vulnerability types of popular router brands such as Tenda, Netgear and D-Link from 2019 to the present on the CVE website, and the findings are shown in Table 1.

Table 1. Vulnerability type statistics

Router Name	Assignment ratio	Injected ratio
D-Link	11.54%	31.2%
Tenda	51.28%	17.95%
Netgear	13.99%	28.78%
TP-Link	15.38%	19.23%
Cicso	3.26%	43.83%

The average percentage of assignment-based vulnerabilities and injection-based vulnerabilities reached 45.68%, and they fit well with the static taint analysis-based approach proposed in this paper, so they were used as the main targets of the method.

2. Method

The method used in this paper is divided into three steps: Building the atomic pool, path network diagram optimisation, and vulnerability establishment determination.

2.1 Building the atomic pool

Using the function as the subject, the call relationship of the function, the contaminated variables within the function and the contamination of the parameters of the called function are obtained. This step is designed to perform complexity degradation on functions, treating them as atomic nodes with attributes and simplifying the complexity of constructing contamination paths.

The collected binaries are disassembled using IDA Pro. Within the function, a contamination pool is first set up, with initial contamination sources being function arguments and external inputs such as nvram_bufget, get_var, websGetVar. The internal code path of the function is traversed, variables that are directly or indirectly contaminated are collected, a list of called functions is obtained and the contamination of these function arguments is recorded. As shown in algorithm 1.

Algorithm 1. Atomic pool construction algorithm

```
Input:  FunctionArgs, InputVal                 //List of function parameters, List of external inputs
     NSList, TFH                               //List of called functions, handle of the target function object
OutPut: Result_list = []
def TaintAnalysis():
  ValPools = FunctionArgs + InputVal           //Storing contaminated variables
  while TFH.isEnd():                           //Ends the loop if it reaches the end of the function
    CodeArgs = JudgeAssignment(TFH)            //AType refers to the assignment method and is assigned the value 1
                                               //CodeArgs assignment variables and assigned variables
    if (AType == 1) and (CodeArgs[1] in ValPools) and (CodeArgs[0] not in ValPools):
                                               //If it is an assignment operation and contamination occurs
      ValPools.append(CodeArgs[0])             //Storing the assigned variable
      TFH.start()                              //Restart testing
      continue
    else:
      TFH.next()                               //If there is no assignment operation, run down
      continue
  for each_section in NSList:                  //Detects if the parameters of the next function are contaminated
    midArgs = GetArgs(TFH.each_section)        //Get the parameters of the called function
    midflag = JudgeTaint(midArgs,ValPools)     //Detects if the parameters of the called function are contaminated
    if not midflag:
      Result_list.append(each_section)
  return
```

The extracted atomic attributes are then stored using a mongoDB database.

2.2 Path network diagram optimisation

This step involves constructing a complete propagation path graph and optimising it. First, the start and end points are set; then all paths from the start point to the end point are constructed; finally, the paths are optimised using the atomic properties in the atomic pool, and unusable paths are removed.

The start point is the function that directly calls the input function. The end point is the function that calls the function for the end point that matches the vulnerability profile. The path network graph is constructed using the breadth-first algorithm to traverse all paths from the beginning to the end. The traversal algorithm is shown in Algorithm 2.

Algorithm 2. Breadth-first path network graph traversal algorithm

```
Input:  RouteList                                    //Record a list of paths
     FunctionRelation                                //Function call relationships, dictionaries
        VulFunctionList                              //List of dangerous functions, endpoints
Output: Result_list = []                             //Record all paths to the end point
def create_all_route(RouteList):
  mid_list = []
  for eRoute in RouteList:
   nSection = FunctionRelation[eRoute[-1]]           //The set of nodes next to a particular path
     mid_val = VulFunctionList ∩ nSection            //The endpoint function present in the next node set
     If mid_val:
      SaveTheRoute(Result_list,eRoute,mid_val)       //If there is a path to the end, record it
     mid_list+=AddFunction(eRoute,nSection,mid_val)  //Add the next node to the end of the path
  create_all_route(mid_list)
  return
```

The path network diagram constructed according to algorithm 2 is shown in Fig. 1.

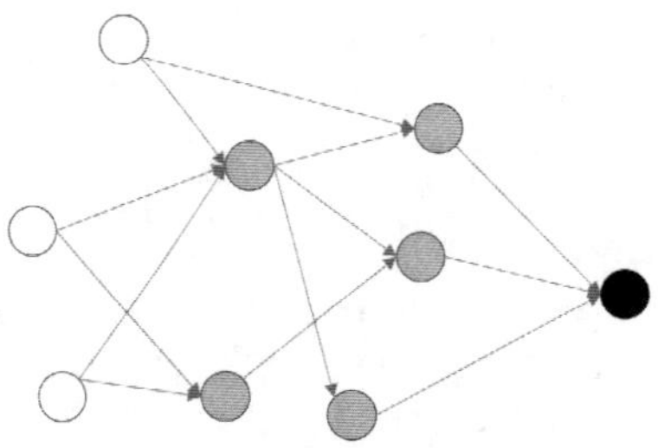

Fig. 1 Path network diagram

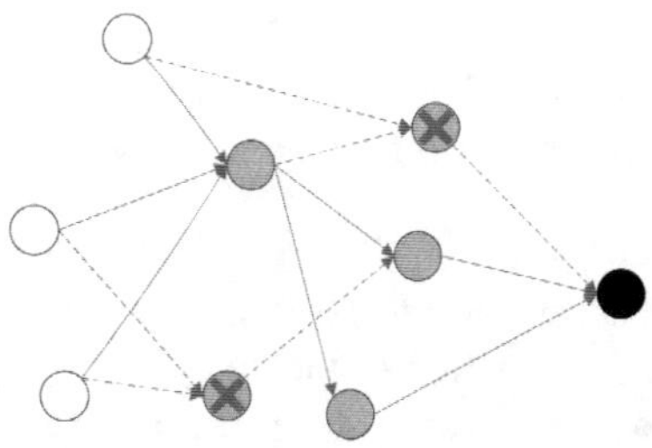

Fig. 2 Taint analysis diagram

All the nodes represented by circles in Fig. 1 are functions, with each node connected by a directed arrow, the starting point of which is the function calling function and the end point of which is the called function; the blank circle on the left is the function that can accept external input, the black solid circle on the right is the hazard function and the grey circle in the middle is the tainted propagation function.

The path network diagram contains many unusable paths, for example where the parameters of the called function are not contaminated or not contaminated at all, so that the input data cannot pass through this node and therefore such nodes are deleted. The basis for node deletion is the contamination of the called function's parameters in the atomic attributes. The optimisation process is shown in Fig. 2 and Fig. 3.

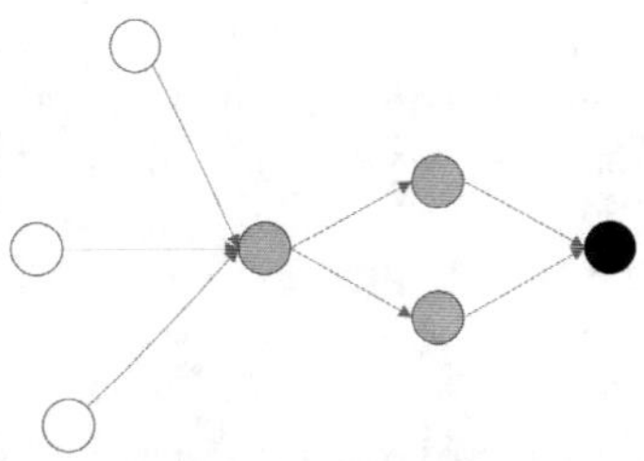

Fig. 3 Path network diagram after optimization

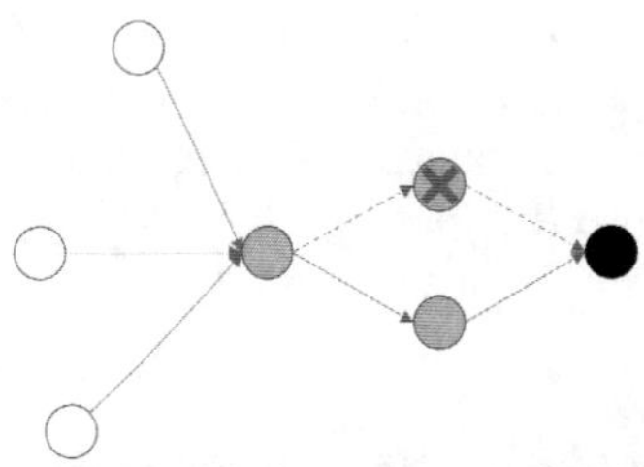

Fig. 4 Vulnerability point establishment determination diagram

2.3 Vulnerability establishment determination

So far an optimised tainted transport path has been obtained and the external external input has reached the dangerous function call function correctly. This step therefore determines whether the parameters of the dangerous function are controllable and whether the parameter characteristics are those of the injection and overflow vulnerabilities, and if so, the call point is considered to be a potential vulnerability.

Algorithm 3 Vulnerability Establishment Determination Algorithm

```
Input:  SrcArgs , DstArgs                          //Source variable, destination variable
OutPut:  True/False                                //Determine if the result is valid
def VulJudge():
  JudgeResult = JudgeTaint(SrcArgs)                //Determining whether source variable is contaminated
  VulType ==>Assignment Vulnerability
       JudgeResult1 = JudgeTaint(DstArgs)          //Determine if the destination variable points to the stack space
       if JudgeResult and JudgeResult1:
          CheckResult = CheckArgs(SrcArgs,DstArgs) //Checking the length of variables
               Set up==>return True
               Don't set up==>return False
  VulType ==>Injection Vulnerability
       JudgeResult2 = JudgeType(SrcArgs)
       JudgeResult3 = JudgeLength(SrcArgs)         //Checking variable types and length limits
       if JudgeResul2 and JudgeResult3:
            Set up==>return True
       Don't set up==>return False
  Return False
```

As shown in Fig. 4, for fragile points that are judged to be untenable, the paths for which the call relationship exists are removed, the group of call relationships that generally exist with the end of the path, and the paths that are judged to be valid are then saved to the database.

3. Experiment

This section describes the various aspects of the CSchecker performance evaluation experiments, including the construction of the dataset, the experiments and the analysis of the experimental results.

The hardware environment for this experiment is a laptop with Intel(R) Core(TM) i7-10710U CPU @ 1.10GHz 1.61 GHz, 16GB memory; the software environment is a Windows 10 system with IDA Pro 7.5 and python 3.7 installed. To facilitate data flow analysis, the FIDL framework was installed for IDA Pro. A MongoDB V4.2 database was selected for the storage of vulnerability information and data.

3.1 Dataset construction

This experiment constructs a vulnerability dataset with a data volume of 267, which consists of popular router firmware as well as dynamic link libraries, and stores the data entries in the form of (CVE number, firmware name, file name, vulnerability address) into a database.

The first step is to obtain the vulnerability information. The experiments were conducted to obtain the vulnerability information of all data set objects in the CVE website, filtering out injection and assignment vulnerabilities based on the CWE number and keywords in the CVE description information, such as 'overflow', 'command injection', etc.

The second step is to manually obtain the address of the vulnerability point. As the vulnerability information in the CVE website varies greatly, and there are many non-existent vulnerability file download links, so the experiments use a manual way to download the existence of vulnerability files, and then use IDA pro to determine whether there is a vulnerability point described in the vulnerability description information, and if it exists, store its point information, and finally stored in the database.

3.2 Performance tests

The target files are placed in the same directory and files with the same filename are named differently. Then run the program and batch the target files using IDA Python. Each target file will output individual analysis results and compare these results with the vulnerability points collected in the first step to determine if they are hits or not, thus calculating CSChecker's accuracy and redundancy rates. The relevant results are calculated as shown below.

$$HitRate = (HitPoints/TotlePoints) \times 100\% \quad (1)$$

The above equation is the formula for the accuracy rate, where HitRate indicates the accuracy rate, HitPoints indicates the number of hit vulnerability points, and TotlePoints indicates the number of all vulnerability points in the dataset.

$$ReRate = \frac{\sum_{i \in HitPoints}^{i=1} 1-(HitRate)}{HitPoints} \times 100\% \quad (2)$$

The above formula is how the redundancy rate is calculated, where ReRate, which is RedundancyRate, represents the redundancy rate, HitRate represents the accuracy rate and HitPoints represents the number of hit vulnerability points.

3.3 Analysis of experimental results

After experimenting with the above steps, the output of the CSChecker tool is as follows.

Table 2 Table of results of CSChecker experiments

Router Name	Total Vulnerabilities	Accuracy Rate	Redundancy Rate
D-Link	43	95.35%	4.43%
Tenda	26	92.31%	7.10%
Netgear	140	92.86%	6.63%
TP-Link	29	93.1%	6.24%
Cicso	29	86.21%	11.89%

As can be seen from the above table, the vulnerability accuracy of CSChecker in the five common router brands in the dataset ranges from 86.21% to 95.35%, with average accuracy of 92.51%; the hit redundancy rate does not exceed 11%. The experimental results show that CSChecker can effectively detect injection-type vulnerabilities and overflow-type vulnerabilities in router firmware.

```
v40 = a2;
strcpy(v80, v79);
memcpy(a1, a2, a3);
strncpy(v12, v32, v10);
v3 = nvram_bufget(0, "portal_manage");
...
```

Figure 5. Example diagram of direct/indirect assignment

```
valueString((int)v10, v7, v8);
v7 = get_cgdata_by_uid(v20, v28);
...
```

Figure 6. Example diagram of user-defined function pollution

Although the accuracy rate of CSChecker is high, there are still undetected vulnerability points for two main reasons. Firstly, there are still some shortcomings in the judgement of tainted variables within functions; the detection of direct and indirect assignments shown in Figure 5 is good, but there is still room for improvement in the detection of user-defined function taint shown in Figure 6. In addition to this, there is also the problem that some of the firmware functions are inclined which prevents the FIDL framework from correctly identifying functions within functions, leading to biased analysis of taint propagation.

4. Conclusion

This paper uses a combination of taint analysis and vulnerability characterisation to detect injection and assignment class vulnerabilities, and optimises the detection path using atomic attributes, as a way to detect vulnerabilities in router firmware. It was tested on 267 publicly available vulnerability points of popular brands such as D-Link and Tenda, with an accuracy of 92.51%. The experiments proved that the method is effective in detecting injection and assignment class vulnerabilities in binaries such as router firmware, and avoids problems such as path explosion.

Although the method is effective in detecting known injection and assignment vulnerabilities in router firmware, there are still areas for improvement, such as not being able to determine the full size of the memory referred to by the target variable for assignment vulnerabilities, leading to false positives. In addition, there is still room for improvement in the identification of the contamination capabilities of user-defined partial user-defined functions. These are all issues that need to be addressed in the next step of the work.

Acknowledgments

This work was financially supported by the Foundation strengthening program projects (2021JCJQZD06311).

References

[1] Maskur, A.F., and Yudistira D.W.A. (2019) Static Code Analysis Tools with the Taint Analysis Method for Detecting Web Application Vulnerability. In 2019 International Conference on Data and Software Engineering (ICoDSE), pp. 1-6.
[2] Saoji, T., Thomas H.A, and Cormac F. (2017) Using precise taint tracking for auto-sanitization. In Proceedings of the 2017 Workshop on Programming Languages and Analysis for Security, pp. 15-24.
[3] Zhang, J.B., Wang Y.Y., Qiu L.N., and Julia R. (2021) Analyzing android taint analysis tools: FlowDroid, Amandroid, and DroidSafe. In IEEE Transactions on Software Engineering.
[4] Xiong, B., Xiang G.L., Du T.L., He J.S., and Ji S.L. (2017) Static taint analysis method for intent injection vulnerability in android applications. In International Symposium on Cyberspace Safety and Security, pp. 16-31 .
[5] Yan, X.X., Wang Q.X., and Ma H.T. (2017) Path sensitive static analysis of taint-style vulnerabilities in PHP code. In 2017 IEEE 17th International Conference on Communication Technology (ICCT), pp. 1382-1386.
[6] Almashfi, N., and Lu L.J. (2019) Static Taint Analysis for JavaScript Programs. In International Conference on Tools and Methods for Program Analysis, pp. 155-167.
[7] Ferrara, P., Luca O., and Fausto S. (2018) Tailoring taint analysis to GDPR. In Annual Privacy Forum, pp. 63-76.
[8] ALmotairy, R., and Yassine D. "B-droid: A Static Taint Analysis Framework for Android Applications."
[9] Feng, C., and Zhang X. (2017) A static taint detection method for stack overflow vulnerabilities in binaries. In 2017 4th International Conference on Information Science and Control Engineering (ICISCE), pp. 110-114.

Fault Diagnosis of Civil Aircraft Electrical Power System Based on Triangular Fuzzy Number and Grey Relational Analysis

Xiaoting Jiang

Production Support Engineering Department, COMAC Shanghai Aircraft Design and Research Institute, Shanghai, 200436, China

jiangxiaoting@comac.cc

Abstract—The stable operation of the aircraft electrical power system is closely related to aircraft's function implementation and safety. Therefore, research on fault diagnosis of aircraft electrical system is important and necessary. This paper proposes a fault diagnosis method of civil aircraft power system based on triangular fuzzy number and grey relational analysis. For the purpose of obtaining complete data, it takes into account single fault symptom and historical data, and keeps the fuzzy feature from initial estimation of diagnosis scheme to the end. The example analysis shows that this method has feasibility and efficiency in practical application.

1. Introduction

Aircraft electrical power system, whose main function is supplying electric energy to all electrical equipment on board, is one of the central parts of the aircraft. It is composed of power supply system and distribution system. Power supply system includes main generators, emergency generator, second power resources, batteries, and so on. Distribution system consists of distribution equipment and controllers.

In recent years, electric drive gradually replaces traditional hydraulic and pneumatic drive, mechanical system, and aircraft accessory gearbox. The growth of airborne electrical equipment's number makes higher demand on the power supply quality [1]. Therefore, improving the system reliability and security and developing the level fault diagnosis technology have a high application value [2].

Because of the uncertain factors such as human action and environment, electrical power system's faults are intricate and repetitive. The expert system is frequently used in diagnosis, and comprehensive expert experience and technical manuals have been accumulated. Expert system is suitable for fault diagnosis of electrical power system [1]. However, when there is no obvious tendentiousness shown in estimate for fault attribute indicators by expert system, the fault scheme is hard to choose.

D-S evidence theory is often used to combine the evidence obtained by expert system, but it leads to counterintuitive results when the evidence is highly conflicting. In order to solve this problem, many scholars have researched in depth and put forward some processing methods [3]. For example, paper [1] proposes a method of using fault historical data to modify the evidence. It solves the combination problem of highly conflicting evidence. However, this method gives too much weight to the historical data, the calculation results are often consistent with the comparison results of historical data, and the impact of current fault symptoms on the calculation results is not notable enough. As the faults of civil aircraft electrical power system has the characteristics of diversity and repeatability, the historical fault data reflects the quality of finished products and assemble line, and has significance for reference. Therefore, the fault historical data should be taken as a diagnosis indicator, yet be assigned with proper weight.

The grey relational analysis is also widely used in fault diagnosis field. Based on the grey system theory, it is a method which calculates relational coefficient and relational degree between objects by comparing the geometric relations between system's statistical data [4]. Grey relational analysis has great advantages in dealing with various discrete data and multi-attribute decision-making problems. For the problem of fuzzy indicators in alternative schemes, there are some methods to calculate relational degree with triangular fuzzy number [5-6].

In this background, this paper proposes a fault diagnosis method for civil aircraft electrical power system based on triangular fuzzy number and grey relational analysis. The core idea of this method is to put historical fault data into relational degree calculation and keep the fuzzy feature throughout the calculation process. The rest of this paper is organized as follows: Section 2 reviews the concepts of triangular fuzzy number and grey relational analysis. Section 3 introduces a new fault diagnosis method based on those theories. Section 4 gives an example to show the feasibility and validity of the method. Section 5 concludes all the work done.

2. Review of triangular fuzzy number and grey relational degree

2.1. The definition and operation rulers of triangular fuzzy number

If the membership function of fuzzy number A is

$$\mu_A(x)=\begin{cases}0, & x<u\\ \dfrac{x-u}{v-u}, & u\le x\le v\\ \dfrac{w-x}{w-v}, & v\le x\le w\\ 0, & x>w\end{cases} \tag{1}$$

then fuzzy number A is called regular triangular fuzzy number, which is recorded as $A=(u,v,w;1)$. In the following, triangular fuzzy numbers are written as $A=(u,v,w)$.

Let fuzzy numbers $A_1=(u_1,v_1,w_1)$, $A_2=(u_2,v_2,w_2)$, the operation rules between two fuzzy numbers are:

- Addition: $A_1+A_2=(u_1+u_2,v_1+v_2,w_1+w_2)$;
- Multiplication: $A_1\times A_2\approx(u_1u_2,v_1v_2,w_1w_2)$;
- Division: $A_1/A_2\approx(u_1/w_2,v_1/v_2,w_1/u_1)$;
- The distance between two fuzzy numbers:

$$d(A_1,A_2)=\left[\frac{(u_1-u_2)^2-(v_1-v_2)^2-(w_1-w_2)^2}{3}\right]^{\frac{1}{2}} \tag{2}$$

Other operation rulers can be derived from above rulers [5].

Rank fuzzy numbers by their means and variances. If one fuzzy number has a higher mean and a lower variance, it is considered to be ranked forward.

For triangular fuzzy number $A_i=(u_i,v_i,w_i)$, $i=1,2,\cdots,m$, the formulas of its mean $\overline{x}(A_i)$ and variance $\sigma^2(A_i)$ are

$$\overline{x}(A_i)=\frac{1}{4}(u_i+2v_i+w_i) \tag{3}$$

$$\sigma^2(A_i)=\frac{1}{80}\left[(u_i-v_i)^2+2(u_i-w_i)^2+2(v_i-w_i)^2\right] \tag{4}$$

After means and variances have been worked out, the rulers of comparing fuzzy numbers are [6]:

if $\overline{x}(A_1)>\overline{x}(A_2)$, $A_1>A_2$;

if $\overline{x}(A_1)=\overline{x}(A_2)$ and $\sigma^2(A_1)>\sigma^2(A_2)$, $A_2>A_1$.

2.2. Grey relational degree of triangular ordered fuzzy number

According to grey relational analysis, if the comparison sequence is $C_i(j)=\{C_i(1),C_i(2),\cdots,C_i(m)\}$ and the reference sequence is $C_0(j)=\{C0(1),C_0(2),\cdots,C_0(m)\}$, the relational coefficient between the comparison sequence C_i and the reference sequence C_0 at the point j is

$$\xi_i(j)=\frac{\min\limits_i\min\limits_j|C_0(j)-C_i(j)|+\rho\max\limits_i\max\limits_j|C_0(j)-C_i(j)|}{|C_0(j)-C_i(j)|+\rho\max\limits_i\max\limits_j|C_0(j)-C_i(j)|},i=1,2,\cdots,n;j=1,2,\cdots,m \tag{5}$$

When the elements are fuzzy numbers, the comparison sequence is $C_i(j)=\{M_i(u_j,v_j,w_j)\}$, and the reference sequence is $C_0(j)=\{M_0(u_j,v_j,w_j)\}$, the relational coefficient between the comparison sequence C_i and the reference sequence C_0 at the point j can be expressed as

$$\xi_i(j)=\frac{\min\limits_i\min\limits_j|M_0(u_j,v_j,w_j)-M_i(u_j,v_j,w_j)|+\rho\max\limits_i\max\limits_j|M_0(u_j,v_j,w_j)-M_i(u_j,v_j,w_j)|}{|M_0(u_j,v_j,w_j)-M_i(u_j,v_j,w_j)|+\rho\max\limits_i\max\limits_j|M_0(u_j,v_j,w_j)-M_i(u_j,v_j,w_j)|}, \tag{6}$$

$i=1,2,\cdots,n;j=1,2,\cdots,m$

where ρ is the distinguishing coefficient。The smaller ρ is，the greater the distinguishing ability is. $\rho\in(0,1)$, commonly, it takes 0.5 [6].

3. Fault Diagnosis Decision Based on Triangular Fuzzy Number Grey Relational analysis

In this section, a fault diagnosis decision method based on triangular fuzzy number and grey correlation analysis is proposed. The steps are given below:

(a) Evaluation and historical data calculation

For a fault, there are several diagnosis schemes corresponding standard fault patterns respectively. According to data from sensors and fault symptoms, expert system evaluates the characteristic indicators for the schemes. Some or all indicators are evaluated in fuzzy language.

Calculate the historical failure rate of each fault pattern and add them into the indicator eigenvector.

(b) Data Processing

Transform the evaluation of fuzzy expression into triangular fuzzy numbers. Each diagnosis scheme's indicator eigenvector constructs a comparison sequence $C_i=\{M_i(u_j,v_j,w_j)|i=1,2,\cdots,n;j=1,2,\cdots,m\}$, here, n represents the number of diagnosis schemes, and m represents the number of indicators for the diagnosis schemes.

(c) Construction of the reference sequence

Based on the attribute type, determine the reference sequence $C_0(j)=\left\{M_0\left(u_j,v_j,w_j\right)\middle|j=1,2,\cdots,m\right\}$. When the grey relational analysis is used for comprehensive evaluation, the evaluation standard is the optimal value of each indicator. That is, when the indicator belongs to benefit type, the maximum value in each diagnosis scheme is taken; while the indicator belongs to a cost type, the minimum value in each diagnosis scheme is taken [6].

(d) Normalization

It is difficult to compare indicators directly as their different dimensions and types. Therefore, the indicators of the comparison and reference sequences should be reformed to be dimensionless.

$$C_i^{'}(j)=C_i^{'}\left\{M_i^{'}\left(u_j,v_j,w_j\right)\right\}=\frac{C_i(j)}{C_j}=\frac{M_i\left(u_j,v_j,w_j\right)}{C_j} \tag{7}$$

here, $C_j=\frac{1}{n+1}\sum_{i=0}^{n}M_i\left(u_j,v_j,w_j\right)$, $i=0,1,\cdots,n$; $j=1,2,\cdots,m$; the triangular fuzzy number $M_i^{'}\left(u_j,v_j,w_j\right)$ is the element of the comparison and reference sequences after normalization.

(e) Calculation of overall fuzzy relational degree

The fuzzy relational degree between each comparison sequences and the reference sequence is obtained according to equation (6) and equation (8).

$$\gamma_i=\sum_{j=1}^{m}\delta_j\xi_i\left(u_j,v_j,w_j\right),i=1,2,\cdots n;j=1,2,\cdots,m \tag{8}$$

here, γ_i is the fuzz relational coefficient of the indicator j, δ_j is the attribute weight of the indicator j.

Finally, the scheme with the highest relational degree is the priority troubleshooting scheme.

4. Illustrative Example

When a fault occurs in electrical power system, in the light of the general process and experience, the fault is generally diagnosed from three aspects: line problem (Target A), finished parts (Target B) and crosslinking system (Target C).

According to the line conduction report, CMS information, EICAS information, fault symptoms of finished parts, troubleshooting workload, the expert system makes an evaluation from five aspects: relativity from control logic, from circuit diagram, from CMS message, from function realization and workload.

In the fault history database, the same faults occurred 11 times: the line (Target A) caused faults 6 times, while the finished parts (Target B) has failed once, and cross-linking system (Target C) caused faults 4 times.

The evaluation of the expert system and historical data are shown in Table 1.

Table 1. Alternative Evaluation Form.

Attribute indicator	Scheme 1 (Target A)	Scheme 2 (Target B)	Scheme 3 (Target C)
Relativity from control logic	MR	R	VR
Relativity from circuit diagram	R	VR	MR
Relativity from CMS message	N	R	R
Relativity from function implementation	VR	VR	R
Workload	S	VB	B
Historical failure rate	0.55	0.09	0.36

Table 2. Linguistic variables and their corresponding triangular fuzzy numbers

Linguistic variable		
Relativity	Workload	Triangular fuzzy number
Very irrelevant(VNR)	Very small(VS)	(0,0,0.1)
Irrelevant (NR)	Small(S)	(0,0.1,0.3)
Relatively irrelevant (MNR)	Relatively small (MS)	(0.1,0.3,0.5)
Neutral(N)	Middle（M）	(0.3,0.5,0.7)
Relatively relevant(MR)	Relatively big（RB）	(0.5,0.7,0.9)
Relevant(R)	Big(B)	(0.7,0.9,1)
Very relevant(VR)	Very big(VB)	(0.9,1,1)

According to Table 2, convert the schemes' evaluation into triangular fuzzy numbers, and construct comparison sequences:

C_1={(0.5,0. 7,0.9), (0.7,0.9,1), (0.3,0.5,0.7), (0.9,1,1), (0,0.1,0.3), 0.55};

C_2={(0.7,0.9,1), (0.9,1,1), (0.7,0.9,1), (0.9,1,1), (0.9,1,1), 0.09};

C_3={(0.9,1,1), (0.5,0.7,0.9), (0.7,0.9,1), (0.7,0.9,1), (0.7,0.9,1), 0.36}.

Take the optimal value of each characteristic indicator to construct a reference sequence:

C_0={(0.9,1,1), (0.9,1,1), (0.9,1,1), (0.9,1,1), (0,0,0.1), 1}.

The comparison sequences and reference sequence are normalized to the following results:

C_1'={(0.513,0.778,1.2), (0.718,1,1.333), (0.324,0.606,1.077),(0.9,1.026,1.176),(0,0.2,0.75),1.1};

C_2'={(0.718,1,1.333), (0.923,1.111,1.333),(0.757,1.091,1.538),(0.9,1.026,1.176),(1.5,2,2.5),0.18};

C_3'={(0.923,1.111,1.333), (0.513,0.778,1.2), (0,757,1.091,1.538), (0.7,0.923,1.176), (1.167,1.8,2.5), 0.72};

C_0'={(0.923,1.111,1.333),(0.923,1.111,1.333),(0.973,1.212,1.538),(0.9,1.026,1.176),(0,0,0.25),2}.

According to equation (6), the fuzzy relational coefficients of schemes' indicates are obtained, as shown in Table 3.

Table 3. Fuzzy relational coefficients of alternative indicators.

Attribute indicator	Scheme 1 (Target A)	Scheme 2 (Target B)	Scheme 3 (Target C)
Relativity from control logic	(0.542,0.857,1.077)	(0.6,0.947,1.309)	(0.6,1,1.667)
Relativity from circuit diagram	(0.6,0.947,1.309)	(0.6,1,1.667)	(0.542,0.857,1.077)
Relativity from CMS message	(0.438,0.767,0.894)	(0.6,0.943,1.294)	(0.6,0.943,1.294)
Relativity from function implementation	(0.6,1,1.667)	(0.6,1,1.667)	(0.6,0.951,01.316)
Workload	(0.429,0.909,1.667)	(0.214,0. 5,0.556)	(0.214,0.526,0.652)
Historical failure rate	(0.349,0.690,0.758)	(0.224,0.524,0.486)	(0.296, 0.61, 0.616)

The weight vector of the characteristic indicator in the diagnosis model is {1, 1, 1, 0.3, 0.5}. Processing the relational coefficients in Table 3 according to equation (8), the overall fuzzy relational degree between each comparison sequence and the reference sequence is obtained:

$$\gamma_1 = \sum_{j=1}^{6} \xi_1\left(u_j, v_j, w_j\right) = (0.414, 0.698, 0.971),$$

$$\gamma_2 = \sum_{j=1}^{6} \xi_2\left(u_j, v_j, w_j\right) = (0.431, 0.717, 1.058),$$

$$\gamma_3 = \sum_{j=1}^{6} \xi_3\left(u_j, v_j, w_j\right) = (0.426, 0.702, 0.976).$$

The means of the fuzzy relational degrees are 0.695, 0.731 and 0.702 respectively. According to the triangular fuzzy number ranking rule, the ranking of the three fault patterns is $\gamma_2 > \gamma_3 > \gamma_1$. Therefore, among the three schemes, scheme 2 is better than the other two schemes.

5. Conclusion

In this paper, a multi-attribute decision-making method based on triangular fuzzy number and grey relational analysis is employed to the fault diagnosis of civil aircraft electrical power system.

For the purpose of obtaining complete data, the method takes not only single fault symptom and subjective evaluation from expert system but also objective historical fault data, and calculates the overall fuzzy relational degree between each scheme and the ideal scheme, so as to find out the optimal troubleshooting scheme. When the decision-maker's reliance on each indicator changes, the weight of each indicator can be adjusted and reevaluated easily. For solving the problem of fuzzy expression of expert system, the linguistic variable is transformed into triangular fuzzy number. The fuzzy feature is kept through the whole process, so the result is expressed accurately.

This method provides a theoretical basis for finding the order of diagnosis schemes, improving system reliability, and speeding up the fault diagnosis of civil aircraft electrical power system at production and flight test stages.

References

[1] Wu, Y. (2017) Fault Diagnosis of Civil Aircraft Electrical Power System Eased on Evidence Theory. Foreign Electronic Measurement Technology, 8: 5–10.

[2] LU, F., Huang, J., Kong, X. (2010) Fault Diagnosis for Sensors and Components of Aero-Engine based on Grey analysis. In: 2010 IEEE International Conference on Intelligent Computing and Intelligent Systems, ICIS 2010. Guilin. pp. 623-627.

[3] Tang, C., Jiang, W., Chen, Y., Wu, C., Kong, F. (2015) Conflicting evidence combination based on a new measure method of uncertainty. Systems Engineering-Theory & Practice, 9: 2394-2400.

[4] Zhou, L., Wang, Y., Wang, F., Yan, C., Bi, J. (2017) A Transformer Fault Diagnosis Method based on Grey Relational Analysis and Integrated Weight Determination. In: 35th IEEE Electrical Insulation Conference, EIC 2017. Baltimore. pp. 491-494.

[5] Fu, Y. (2008) The TOPSIS method of multiple attribute decision making problem with Triangular-fuzzy-valued weight. In: 2008 International Workshop on Modelling, Simulation and Optimization, WMSO 2008. Hong Kong. pp. 11-14.

[6] Feng, X., Yang, H., Xie L., Yang J (2019) Research on Multi-level Comprehensive Evaluation Model Based on Triangular Fuzzy Number Grey Relational Theory. Mechanical Engineer, 9: 168–178.

Research on LQR controller of control torque gyro self-balancing vehicle based on QPSO optimization

Zezhou Luo[1a] and Lin Xu[1b*]

[1] School of Automotive Engineering, Wuhan University of Technology, Wuhan, Hubei, 430000, China

[a]zezhouluo@whut.edu.cn, [b*]xulin508@whut.edu.cn

Abstract. The LQR control method based on quantum particle swarm optimization (QPSO) is used to solve the balance problem of the control torque gyro self-balancing vehicle. Firstly, the system dynamics model was established by Lagrange method, and the fitness function of mixed indexes was designed, and the system was simulated numerically. Finally, compared with the results of double closed-loop PID control and LQR control using empirical method, the performance of each algorithm in vehicle body balance control is analyzed. The results show that the LQR control algorithm based on PSO optimization can quickly adjust the roll Angle from the initial Angle to the balance position, and has a faster response time, smaller overshoot, and a stronger ability to resist external interference for the balance control of the torque gyro balancing vehicle.

1. Introduction

Bicycle is a typical unstable system, similar to the inverted pendulum system, the system has the characteristics of multivariable, nonlinear, strong coupling, high order and parameter uncertainty, which is an important research object in the field of automatic control. The control moment gyroscopic self-balancing vehicle is a new type of bicycle which relies on the gyroscopic moment generated by CMG precession to keep the body stable. It is necessary to design a controller to keep its roll stability, and adjust the CMG precession Angle to zero in time while keeping the bicycle's roll Angle to prevent the system instability caused by exceeding the maximum precession Angle[1-2]. At present, the control method of torque gyro bicycle is dominated by traditional control method. Lam et al. used PD controller to stabilize the bicycle roll Angle, which had good robustness and efficiency, but they could not overcome the coupling effect in the model and failed to effectively control the precession Angle zero return[3]. Zhao Mingguo et al. used pole assignment method to design state feedback controller to achieve the stability of the control system, but the design of pole position and number is highly dependent on experience[4]. Hsieh et al. adopt fuzzy sliding mode controller, which has the advantages of fast response and strong anti-interference ability[5]. Compared with the traditional control methods, LQR controller, as an optimal control algorithm based on state feedback, has the advantages of high stability and simple design.

In this paper, LQR controller is designed to control torque gyro balancing vehicle as the research object. Aiming at the problem that it is difficult to select Q matrix parameters and achieve optimal performance, QPSO algorithm is used to search the optimal Q matrix weighting coefficient that makes the system stable, and the global optimal solution of the matrix is obtained. The simulation results are compared with those obtained by PID and empirical LQR method, and the effectiveness of QPSO optimized LQR controller in controlling torque gyro balancing vehicle is proved.

2. The Dynamics Model of the control moment gyroscopic self-balancing bicycle

The overall structure of the bicycle is divided into several parts by function: control moment gyro device, motion drive part, attitude sensor and embedded control system. The two flywheels of the control moment gyro are arranged front-to-back, and the self-rotating motor drives the two gyro flywheels to reverse rotate at the same speed. The feeding motor controls the gyro flywheels to reverse rotate synchronously in the feeding direction through gear transmission, and the feeding force generated by the control moment gyro is used to offset the gravitational force brought by the body tilting sideways, so that the body attitude is balanced[6]. The control process is as follows: when the body deviates from the balance position, the microprocessor of the embedded system reads the angle value of the body lateral tilt deviation collected by the attitude sensor, and after processing by the control system, outputs the control signal to control the feeding motor to produce the feeding motion relative to the body and output the feeding torque to return the body to the balance position.

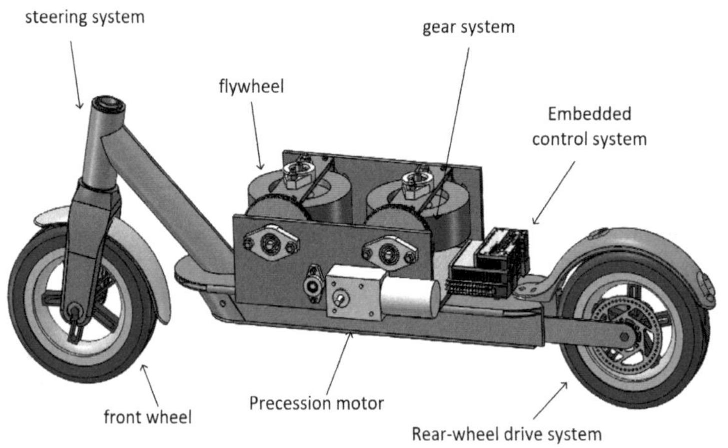

Figure 1. Control torque gyro self-balancing vehicle structure

In this paper, the zero-speed balance of bicycle is studied, and the coupling relationship between handlebar steering and body roll Angle is not considered. The following reasonable assumptions are made for the system:

(1) The bicycle has no forward speed, that is, the speed in the X direction is zero.

(2) The handlebar steering is not considered and the body and gyro frame are rigid bodies.

(3) The tire is rigid body, and the contact deformation between the tire and the ground is not considered, and the tire and the ground are point contact.

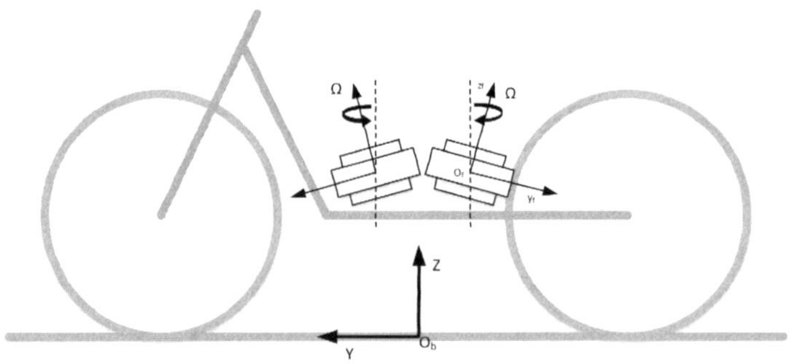

Figure 2. System reference frame

Lagrange method was used to establish the dynamics model of the system, and according to the Lagrange equation:

$$\frac{d}{dt}(\frac{\partial L}{\partial \dot{q}_j})-\frac{\partial L}{\partial q_j}=Q_j(j=1,2,\cdots,k) \tag{1}$$

The body roll Angle and gyro flywheel precession Angle were selected as generalized coordinates, and the system dynamics equation was obtained as follows:

$$[2(I_{fy}+I_{gy})\cos^2\gamma+2(m_f h_f^2+m_g h_g^2)+m_b h_b^2+I_{by}+2(I_{fz}+I_{gz})\sin^2\gamma]\ddot{\theta}+$$

$$2(I_{fz}+I_{gz}-I_{fy}-I_{gy})\dot{\gamma}\sin 2\gamma\dot{\theta}-(2m_f h_f+2m_g h_g+m_b h_b)g\sin\theta=2I_{fz}\Omega\dot{\gamma}\cos\gamma \tag{2}$$

$$2(I_{fx}+I_{gx})\ddot{\gamma}+(I_{fy}+I_{gy}-I_{fz}-I_{gz})\dot{\theta}^2\sin 2\gamma+2I_{fz}\Omega\dot{\theta}\cos\gamma=2(T_m-B_m\dot{\gamma}) \tag{3}$$

Dc motor is selected as the actuator to control the precession of torque gyro, and the motor equation is as follows:

$$T_m=K_m i \tag{4}$$

$$U=L\frac{di}{dt}+Ri+Ke\dot{\gamma} \tag{5}$$

Table 1. system parameter table

Meaning of parameters	Parameter symbol
Body mass, Height of mass of the body	m_b, h_b
Mass of flywheel, Height of mass of flywheel	m_f, h_f
Mass of the inner frame, Height of mass of inner frame	m_g, h_g
Body inertia around x-axis	I_{bx}
Flywheel rotational inertia around x-axis, y-axis, z-axis	I_{fx} ,I_{fy} ,I_{fz}
Inner frame rotational inertia around x-axis, y-axis, z-axis	I_{gx} ,I_{gy} ,I_{gz}
Vehicle roll angle	θ
Precession angle of flywheel	γ
Angular velocity of flywheel	ω
Gravity acceleration	g
Torque constant of motor	Km
Voltage constant of motor	Ke
Resistance of motor	R
Inductance of motor	L

3. LQR controller based on PSO optimization

3.1. LQR controller

Optimal control is to search the control algorithm that can minimize the objective function by optimizing the performance index. LQR control refers to the optimal control problem of a linear system in which the corresponding index function is the integral of the quadratic function of the state or control variable. The performance indexes are as follows:

$$J=\int_0^\infty(x^T Qx+u^T Ru)dt \tag{6}$$

In the formula above, $u^* = -R^{-1}B^T Px = -Kx$, And P is the solution of riccati matrix equation, matrix Q and R are mutually limited, the value of Q matrix is related to the anti-interference ability of the system, with the increase of the value of Q matrix, the anti-interference ability of the system is enhanced, and the system adjustment time is shortened. Once we have determined the Q and R matrices, the state feedback matrices are uniquely determined[7]. The quality of LQR control performance is largely determined by Q matrix, which is the weighting coefficient. However, in the actual design of LQR controller, the value of each weighting coefficient often depends on the experience of engineers or a large number of experiments to determine, which is highly subjective, which not only consumes a lot of time and energy, but also difficult to achieve the optimal performance. In this paper, quantum particle swarm optimization algorithm is used to optimize the Q matrix parameters of LQR controller, so as to solve the optimal problem caused by the selection of weighting coefficient.

3.2. Quantum particle swarm optimization algorithm

Quantum particle swarm optimization (QPSO) is a new PSO algorithm proposed from the perspective of quantum mechanics. The algorithm considers that particles have quantum behaviour, and each particle has a DELTA potential well centered on optimal point as it moves in the search space. Since particles in the quantum space have completely different properties to meet the aggregation state, there is no definite trajectory when moving, which enables the particles to explore in the whole feasible solution space to find the global optimal solution. Therefore, the global search ability of QPSO algorithm is far superior to the classical PSO algorithm[8-9]. Given that the population size of the particle is N, the current iteration number is T, the particle moves in the D-dimensional space, and the potential well of the particle in the d-dimension is PID (t), the position update equation of the i-th particle in the D-dimension can be described by the following formula at the t+1 iteration:

$$x_{id}(t+1) = p_{id}(t) \pm \beta(t) \cdot |C_d(t) - x_{id}(t)| \ln \frac{1}{u_{id}(t)} \tag{7}$$

$$p_{id}(t) = \varphi(t) P_{id}(t) + (1 - \varphi(t)) G_d(t) \tag{8}$$

$$C(t) = (C_1(t), C_2(t), \cdots, C_D(t)) = (\frac{1}{N}\sum_{i=1}^{N} P_{i1}(t), \frac{1}{N}\sum_{i=1}^{N} P_{i2}(t), \cdots, \frac{1}{N}\sum_{i=1}^{N} P_{iD}(t)) \tag{9}$$

where: I = 1,2..., N. Pi (t) is the current optimal position of the particle; G (t) and C (t) are the global and average optimal positions of the population. φ (t) and u (t) are uniformly distributed random numbers in the interval [0,1]. β (t) is the contraction and expansion factor.

3.3. Optimization steps of LQR controller parameters based on QPSO

(1) Using quantum particle swarm optimization to generate particle swarm (initialization);

(2) Assign the particles in the particle swarm to the LQR controller parameters in turn, run Simulink, transfer the output to workspace, and calculate the corresponding objective function value of the group of parameters;

(3) The objective function value obtained from (2) is transferred to the QPSO program as the fitness of the particle, and the fitness value of the new and old particles is compared to update the individual optimal and global optimal position;

(4) Judge the fitness value to check whether it meets the requirements of iteration times or performance indicators. If it meets the requirements, exit the algorithm directly to obtain the optimal parameters; If not, return to (2) and continue the loop.

The optimized program can obtain the optimal solution of Q matrix and the value of the optimal gain matrix K.

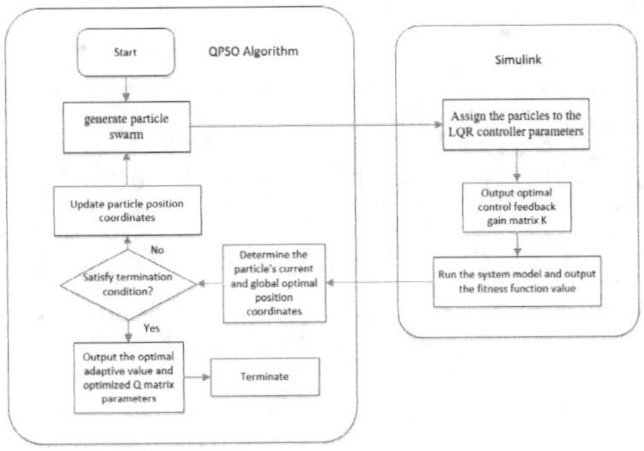

Figure 3. Optimization process of LQR algorithm based on QPSO

3.4. Mixed index fitness function

As can be seen from the above QPSO algorithm, it is necessary to establish fitness function to evaluate the merits and demerits of the global particle position, and the particle carries out the next iterative evolution according to the fitness function value of the global optimal particle, and finally converges to the optimal fitness function value. Combined with the actual situation in this paper, the system should accurately return to the equilibrium position with a small swing in as short a time as possible after deviating from the equilibrium position by disturbance, so the response time and overshoot of the system are also taken as important reference indicators. At the same time, considering that the precession motor will lose the ability of outputting torque after its rotation exceeds 90 degrees, it is necessary to make the precession back to zero. The fitness function of the precession Angle is selected to reflect the ITAE error index of the stability and rapidity of the control system:

$$f_{ITAE} = \int_0^t t|e(t)|dt = \int_0^t t|\gamma(t)|dt \tag{10}$$

Due to the large difference in the values of each indicator, the fitness function of the mixed indicator can be obtained after normalization:

$$f = \lambda_1 \ln(f_{ITAE}) + \lambda_2(t_s / T_s) + \lambda_3(y_{max} / Y_{max}) \tag{11}$$

where t_s is the system response time, y_{max} is the overshoot of the system response, Ts and Y_{max} are the maximum response time and overshoot selected based on experience respectively, and λ1, λ2, and λ3 are the weighting coefficient.

4. Simulation and result analysis

4.1. Simulation and parameter settings

Matlab/Simulink was used for simulation, and the system model was built as shown in the figure4.

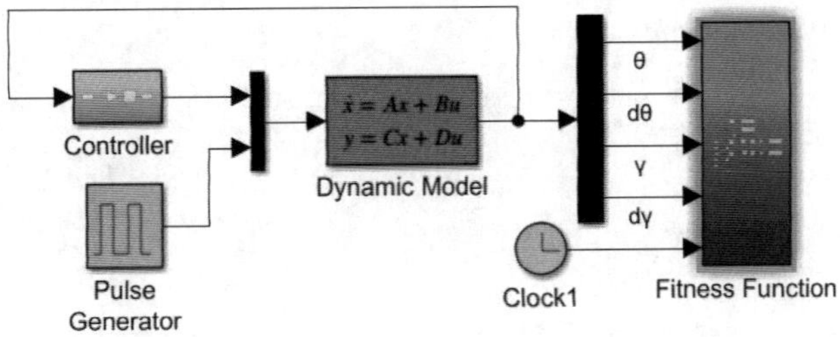

Figure 4. System simulation block diagram

The Dynamic Model module is the system dynamics Model. Fitness Function module is the system Fitness Function value, which will be transmitted to workspace for the next algorithm iteration. Controller module is LQR Controller based on the Fitness Function optimization last time; External interference is replaced by Pulse signals generated by the Pulse Generator. The system parameters are measured by the virtual physical prototype, and the specific parameter values are shown in the table2:

Table 2. System parameter value table

Symbols	Values	Symbols	Values
m_b / kg	28.95	$I_{gx} /(kg{\cdot}cm^2)$	17.29
m_f / kg	5.65	$I_{gy} /(kg{\cdot}cm^2)$	66.05
m_g / kg	1.47	$I_{gz} /(kg{\cdot}cm^2)$	54.91
h_b / m	0.42	ω /(rad/s)	463.5
h_f / m	0.22	$g/(m/s^2)$	9.81
hg / m	0.22	K_m /(N·m/A)	1.59
$I_{bx} /(kg{\cdot}cm^2)$	878.01	K_e /(V·s/rad)	1.59
$I_{fx} /(kg{\cdot}cm^2)$	87.06	R /(Ohm)	1.71
$I_{fy} /(kg{\cdot}cm^2)$	87.06	L /(mH)	0.41
$I_{fz} /(kg{\cdot}cm^2)$	159.04		

In the parameter setting of QPSO algorithm, the maximum population is 120, the number of iterations is 100, and the optimization range of q-matrix weight parameters is [1,600]. Weighting coefficient λ1=0.1, λ2=0.4, λ3=0.5. The contract-expansion factor β can effectively improve the convergence speed by adopting the linear value method, and the parameter updating method is as follows[10]:

$$\beta=1-(K-k/K)*0.5 \tag{12}$$

where, K is the maximum iteration number and K is the current iteration number. Formula 12 makes the value of β decrease linearly from 1 at the first iteration to 0.5 at the end of the iteration.

4.2. result analysis

The simulation time is 10s, the body roll Angle Theta is set as 0.174 rad at the beginning, and the body roll Angle velocity, flywheel precession Angle and flywheel precession Angle velocity are all zero. Q matrix parameters and fitness function values of QPSO algorithm in 100 iterations are shown in the figure. The algorithm achieves convergence after 60 iterations, and the parameter selected by empirical method is:

$$Q = diag(300,10,250,25) \tag{13}$$

And the parameter finally optimized is:

$$Q = diag(459.1,2.3,397.6,4.1) \tag{14}$$

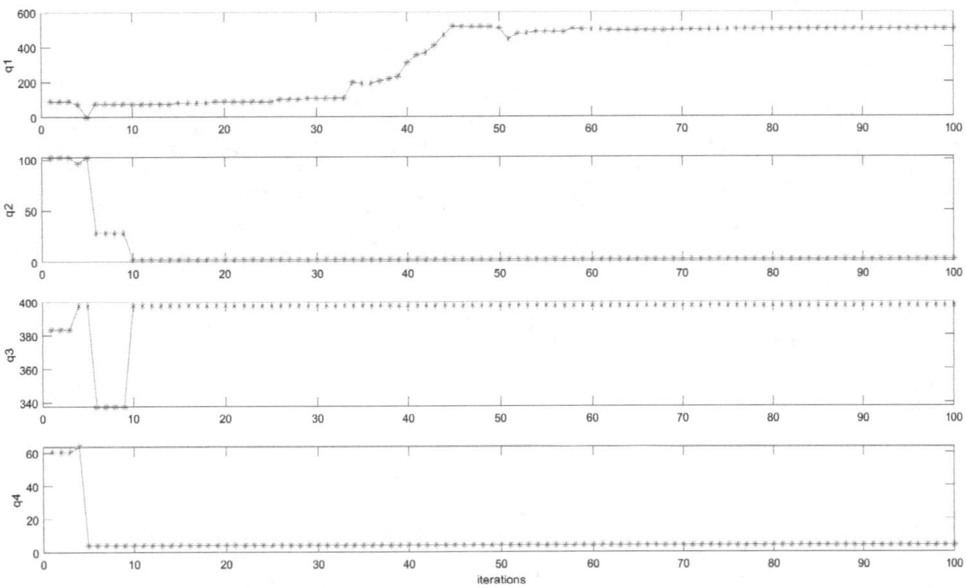

Figure 5. Q matrix parameter optimization curve

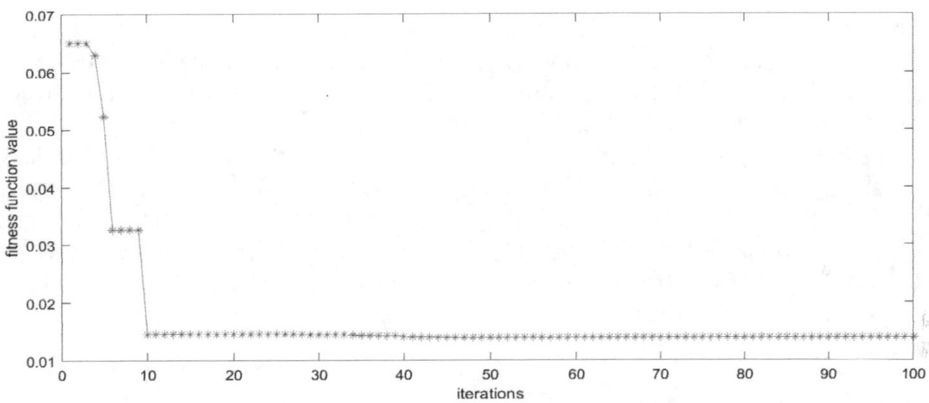

Figure 6. Variation curve of the fitness function

The double closed-loop PID control algorithm and the LQR control algorithm based on empirically selected parameters were used to simulate the system, and an interference with amplitude of 1Nm and width of 0.5s was applied at 5s (after the system was stabilized). The simulation results are shown in the figure.

As can be seen from the figure:

(1) Three different control strategies can make the body roll Angle stable to the equilibrium position and make the flywheel precession Angle return to zero. The system using double closed-loop PID control (system 1) has the body roll Angle stabilization time of 2.228s and overshoot of 0.038rad, and the flywheel precession Angle stabilization time of 2.121s and overshoot of 0.112rad. For the system controlled by empirical method LQR (system 2), the body roll Angle stabilization time is 1.821s and overshoot is 0.029rad, while the flywheel precession Angle stabilization time is 2.19s and overshoot is 0.087rad. The LQR controller system (system 3) optimized by QPSO algorithm has the body roll Angle stability time of 1.229s and overshoot of 0.021rad, and the flywheel precession Angle stability time of 2.082s and overshoot of 0.075rad.

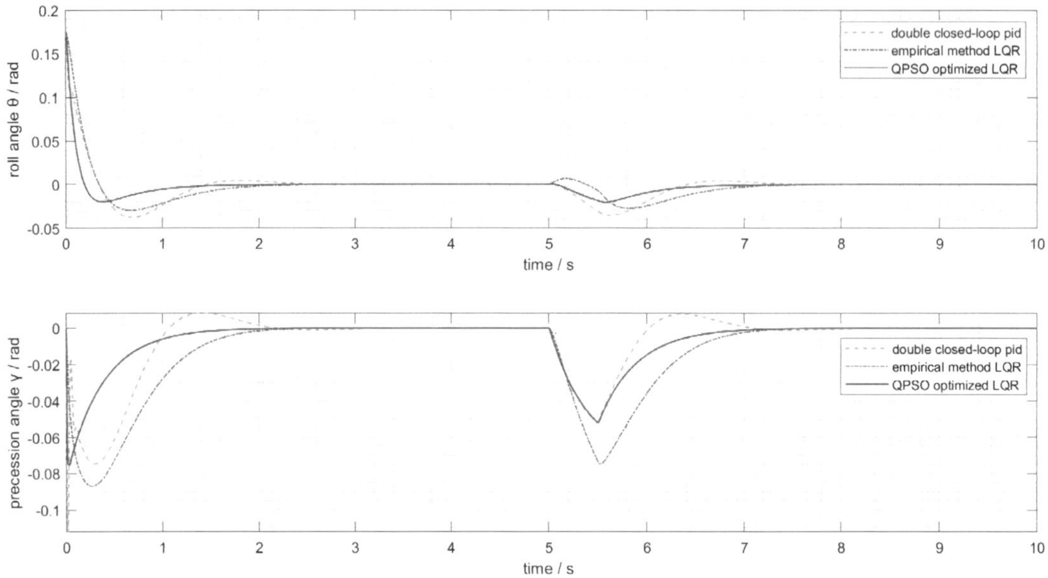

Figure 7. Simulation results of body roll Angle and flywheel precession Angle

(2) When the system is disturbed in the tenth second, the maximum tilt Angle of system 1 body deviates by 0.025rad, and the recovery time is 2.294s; For system 2, the maximum deviation of body roll Angle is 0.035rad and the recovery time is 2.535s. System 3 has a maximum tilt deviation of 0.027rad and a recovery time of 2.428s.

According to the experimental results, compared with the dual closed-loop PID control system and the empirical LQR control system, the LQR controller based on QPSO optimization is easier to make the system reach the equilibrium state, and the system convergence speed is faster. The peak value of roll Angle response decreased by 47.35% and 31.37% respectively. The convergence time is shortened by 44.84% and 32.51%. The robustness is also stronger under external disturbance, and the peak value of roll Angle response is reduced by 28.57% and 7.41% respectively. Recovery times are reduced by 9.51%, 5.52%.

5. Conclusion

In this paper, the LQR control algorithm based on QPSO is used for the control torque gyro self-balancing vehicle to achieve the balance control of the vehicle, the control is fast, accurate and stable. Compared with the double closed-loop PID control and the experience-based LQR control algorithm, it not only has smaller overshoot and shorter adjustment time, but also has stronger ability to resist external interference.

References

[1] Wenyu Liang, Huixing Zhou, Rongmin Cao, Kai Zhao, Libing Jiang, Huaqiang Wen. Manned self-balancing two-wheel control system research review [J]. Journal of control engineering, 2010,17(S2):139-144+190.DOI:10.14107/j.cnki.kzgc.2010.s2.053.

[2] Yamakita M, Utano A. Automatic control of bicycles with a balancer[C]// IEEE/ASME International Conference on Advanced Intelligent Mechatronics. IEEE, 2005.

[3] Lam.Gyroscopic stabilization of a kid-size bicycle[C]// IEEE International Conference on Cybernetics & Intelligent Systems. IEEE, 2011.

[4] He J, Zhao M. Control System Design of Self-balanced Bicycles by Control Moment Gyroscope[J]. Lecture Notes in Electrical Engineering, 2015, 338:205-214.

[5] Tamayo-Leon S, Pulido-Guerrero S, Coral-Enriquez H. Self-Stabilization of a riderless bicycle with a control moment gyroscope via model-based active disturbance rejection control[C]// 2017 IEEE 3rd Colombian Conference on Automatic Control (CCAC). IEEE, 2017.
[6] Jiguang Yue, Hongzhi He, Chenhao Wu, Lei Zhao, Ming Lu, Lin Lai. Composite control of single frame control torque gyro system based on neural network algorithm [J]. Journal of Tongji University (Natural Science),2021,49(08):1184-1191.
[7] Tingfang Zhang, Lei Wu, Xiaojian Wu. Research on Active Suspension Control Based on Road Grade Clustering Identification and LQR Weight Dynamic Optimization [J]. Noise and vibration control,2021,41(04):1-6+12.
[8] Li Xu, Qingsong Cao, Dingjun Zhang. Fractional Order Control Strategy of Active Suspension Based on Quantum Particle Swarm Optimization algorithm [J]. Vibration and Shock,2021,40(16):227-233.DOI:10.13465/j.cnki.jvs.2021.16.029.
[9] Jian Yu, Ping GUO. Quantum Particle Swarm Optimization Algorithm based on MATLAB and its application [J]. Computer and digital engineering,2007(12):38-39+160.
[10] Guobang Ban, Yutao Xu. Research on Peak Cutting and Valley Filling Strategy of DC Power Distribution Center Based on Improved Quantum Particle Swarm Optimization algorithm [J]. Power grids and clean energy,2021,37(12):86-95.

Journal of Physics: Conference Series 2258 (2022) 012072 doi:10.1088/1742-6596/2258/1/012072

Complete Coverage Path Planning of Underwater Wall-climbing Cleaning Robot

Peng Wang[1], Qian Li[2,*], Lei Zhang[3], Yulong Xu[4]

[1,2,3,4]Institute of Automation, Qilu University of Technology(Shandong Academy of Sciences), Shandong Provincial Key Laboratory of Robot and Manufacturing Automation Technology, Jinan, China, 250014

[*]Corresponding author's e-mail: liqian1@sdas.org

Abstract. Aiming at the special working environment of the underwater wall-climbing cleaning robot and the low efficiency and poor universality of the complete coverage path planning algorithm in this environment, a complete coverage path planning algorithm based on enclosing window is proposed. Firstly, the robot's floating system is briefly introduced. Then, the grid map method is used to model the environment, and the traversal priority order of the single-layer enclosing window is determined according to different grid maps. Finally, when the robot falls into the dead corner, the diffusible water-ripple enclosing window search algorithm is used to find the escape grid, then the robot escapes from dead corner and continues to traverse until all traversable grids are traversed. The simulation results show that the method in this paper can not only achieve 100% full coverage and 0% repetition rate, but also reduce path length. It has higher planning efficiency in underwater environments, and has better applications.

1. Introduction

The loss of underwater parts of hull, dam, gate, etc. due to the corrosion of underwater attachments is immeasurable. The use of underwater wall-climbing robots for cleaning operations is becoming more and more common[1], and path planning is a necessary guarantee to realize its independent operation. Path planning is generally divided into two types: one is point-to-point path planning, which requires the robot to find a path from the start point to the end point in the working area, so that the robot can smoothly reach the end point from the start point according to this path and will not encounter obstacles during travel[2-3]. The other is complete coverage path planning, which requires the robot to find a path that can traverse every feasible point in the environment in the workspace, so that the robot can traverse any passable point in the working environment at the end of operation[4]. For underwater wall-climbing cleaning robots, the path planning is mainly complete coverage path planning.

Balch[5] uses a random traversal algorithm to let the robot walk in a straight line. When encountering obstacles, it will rotate a random angle and then continue to drive. This algorithm is too simple and has great limitations. It will repeatedly clean the same area and some areas are not cleaned; Qian et al.[6] proposed a bio-inspired traversal path planning method based on inner spiral search for the high repeat coverage of traditional bio-inspired neural network traversal path planning and the sub-regional path is not optimal. This method completes the underwater map environment modeling and path planning through the biological excitation neural network algorithm in the case of unknown underwater environment information; Zhu et al.[7] proposed an improved Glasius Bio-inspired Neural Network (GBNN) algorithm for the shortcomings of the biologically inspired neural network algorithm in robot path planning, and applied it to the path planning of autonomous underwater vehicles (AUV).

In conclusion, although the complete coverage path planning algorithm has achieved fruitful results, each method can not be applied to all occasions because of its own advantages and disadvantages. According to the particularity of the working environment of underwater wall-climbing cleaning robot, a complete coverage path planning algorithm based on enclosing window is proposed in this paper. Firstly, the horizontal and vertical distances of the traversed area are compared to determine the priority of the traversal order of the single-layer enclosing window, and then the area is traversed according to the determined priority order. When it falls into a dead corner, it escapes the dead corner to the nearest non-traversed grid through the diffusible water-ripple enclosing window search algorithm, and continues to complete the traversal. The simulation results show that the algorithm can achieve the full coverage of the traversal area and 0% repetition rate, shorten the path length of the robot, and is especially suitable for the working environment of the underwater wall-climbing cleaning robot.

2. Working environment modeling of underwater wall-climbing cleaning robot

2.1. Robot's floating system

The underwater wall-climbing cleaning robot is a robot equipped with cleaning equipment that moves on the hull, dam, gate and other underwater parts to achieve the function of cleaning operations. The cleaning objects are harmful substances attached or growing on the surface, such as dirt, shells, algae and tube worms. Aiming at the underwater working environment, tasks and objects of the robot, and meeting the needs of underwater flexible movement and cleaning and full coverage operations, an underwater robot with both floating and wall-climbing functions is designed.

The four horizontal propellers are distributed in a horizontal diamond shape in the robot. By controlling the combined force generated by the work of different propellers, the robot can move in any direction in the horizontal direction. The three vertical propellers are distributed in a triangle on the upper part of the robot, which can control the robot to perform ascent and pitch motions. The thrust generated by it can ensure that the robot is close to the working surface and can also make the robot leave the working surface.

The floating system greatly increases the motion flexibility of the underwater robot, and such propellers layout also gives the robot more choices in path planning. For example, the robot can move in any direction on the horizontal plane; when the robot encounters an obstacle that cannot be crawled through, it can directly float up and then cross the obstacle.

2.2. Environment modeling based on grid map

Environmental modeling is to abstract the data and information that can represent the environment, and use mathematical methods to describe the spatial relationship between objects and them. The existing methods of constructing environment maps include: feature map method, topological map method and grid map method. Through the comparative analysis of the above methods and combined with the working environment of underwater wall climbing robot, the grid map method which is easy to implement and the most widely used in current research and application is finally selected to construct environment map. Using grid map method for environment modeling is to divide the robot's working surface environment map into several grids of the same size, the grid map method has a simple modeling process, a small amount of calculation, and can more intuitively show the path traversal situation.

3. Complete coverage path planning based on grid map

Because the underwater working environment is complex and changeable, the robot must understand the working environment in advance, that is, detect the position and environment through its own sensors, and build an environment model map based on the information fed back by the sensors, so as to realize its own positioning and complete coverage path planning. For the sake of safety, the obstacles are expanded to a certain extent.

3.1. Priority order of single-layer enclosing window

Taking the center point of a grid as the center and the side length as two grids to make a square, all grids intersecting the square are called the single-layer enclosing window of the grid. How to combine the environmental information of single-layer enclosing window with the specific requirements of path planning to design a reasonable priority order is the key to the complete coverage path planning. We specify the traversal priority order of single-layer enclosing window according to different working areas. We call the east-west direction of the working area as the horizontal direction and the north-south direction as the vertical direction. The traversal algorithm proposed in this paper adopts different priority traversal order to reduce the number of turns of the robot when the horizontal and vertical distances of the grid map are different. Taking the starting point as the upper left corner of the grid map as an example, determine the following rules:

- Rule 1: For areas where the horizontal distance is greater than the vertical distance, the traversal priority of the single-layer enclosing window is: up, left, right, down, upper left, upper right, bottom left, bottom right, as shown in Figure 1(a).
- Rule 2: For areas where the vertical distance is greater than the horizontal distance, the traversal priority of the single-layer enclosing window is: up, left, down, right, upper left, bottom left, upper right, bottom right, as shown in Figure 1(b).

(a) Rule 1 traversal priority (b) Rule 2 traversal priority

Figure 1. Priority order of single-layer enclosing window

Adopting different traversal priorities for grid maps with different horizontal and vertical distances will greatly reduce the number of turns of the robot during the traversal process, thereby increasing the efficiency of the robot to complete the full traversal of the area and improving the overall performance of the algorithm. As shown in Figure 2, in a map with a horizontal distance greater than a vertical distance, such as a 10×20 map, the number of turns for rule 1 traversal priority is 18, and the number of turns for rule 2 traversal priority is 38.

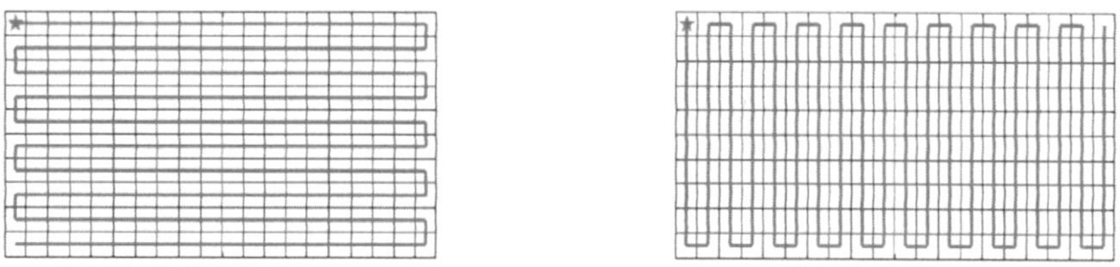

(a) Rule 1 traverse (b) Rule 2 traverse

Figure 2. Comparison of different traversal priority order

3.2. Dead corner situation

In the traversal planning, the dead corner situation is the local environment that the robot needs to pay special attention to. The ability to quickly and efficiently escape the dead corner can reflect the effectiveness and superiority of the algorithm. The dead corner situation means that when the robot executes the traversal task and reaches a certain place on the grid map, the grids around the robot are all obstacles, boundaries or traversed grids.

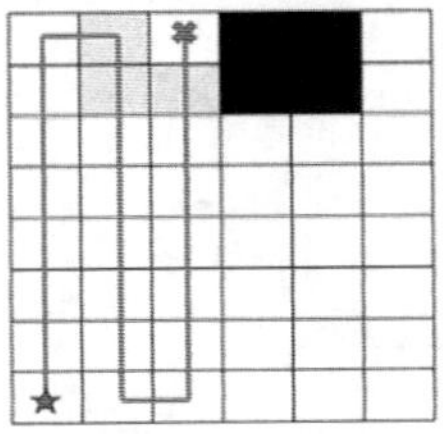

Figure 3. Dead corner situation

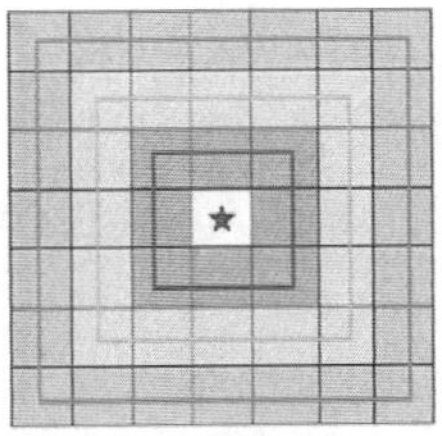

Figure 4. Diffusible water-ripple enclosing window

As shown in Figure 3, after the robot has fallen into a dead corner, it can no longer use the single-layer enclosing window traversal method for local path planning. The underwater working environment is different from the land plane and belongs to a three-dimensional interval. Combined with the robot's floating system, it can be assumed that when the robot is stuck in a dead corner, the robot does not need to find a path to bypass obstacles but can directly cross it with the help of the propellers. This can not only effectively reduce the path length of the full traversal of the robot, but also achieve the goal of zero repetition rate. For the underwater wall-climbing cleaning robot, this not only reduces the power consumption and working time of the robot, but also makes the cleaned area no longer be cleaned repeatedly, and improves the working efficiency of the underwater wall climbing cleaning robot. Based on this consideration, this paper proposes a diffusible water-ripple enclosing window search algorithm to find the next most suitable grid (escape grid) to escape the dead corner.

As shown in Figure 4, the five-pointed star is located at the midpoint of the central grid, with the five-pointed star as the center, the grid intersected by the square with the length of two grids as the side length is the first layer enclosing window (blue grid); the grid intersected by a square with the length of 4 grids as the side length is called the second layer surrounding window (yellow window); the grid intersected by a square with the length of 6 grids as the side length is called the third layer surrounding window (green window), and so on.

When the robot falls into a dead corner, the specific steps of using the diffusible water-ripple enclosing window search algorithm to find the escape grid are as follows: firstly, according to the single-layer enclosing window traversal method in the first layer enclosing window to find escape grid. At this time, the process of "flying over" by the robot through the propellers is not involved. If the escape path cannot be found, find the escape grid in the second layer enclosing window, that is, select the traversable grid (not obstacles, boundaries, and traversed grids) in the second layer enclosing window according to the priority order and calculate the Euclidean distance between the center point of each traversable grid and the center point of the central grid, then select the grid with the smallest Euclidean distance as the escape grid. If there is no escape grid in the second layer enclosing window, continue to select the escape grid in the third layer enclosing window according to the above steps, and so on until the escape grid is found or all grids have been traversed.

The priority order of the diffusible water-ripple enclosing window search algorithm mentioned above is as follows: firstly search the grids in the horizontal and vertical directions, specifically up, left, right, and down; then search the second quadrant grid counterclockwise, search the first quadrant grid clockwise, the search the third quadrant grid counterclockwise, search the fourth quadrant grid clockwise.

In summary, the flow chart of the complete coverage path planning algorithm based on the enclosing window applied to the underwater wall-climbing cleaning robot is shown in Figure 5.

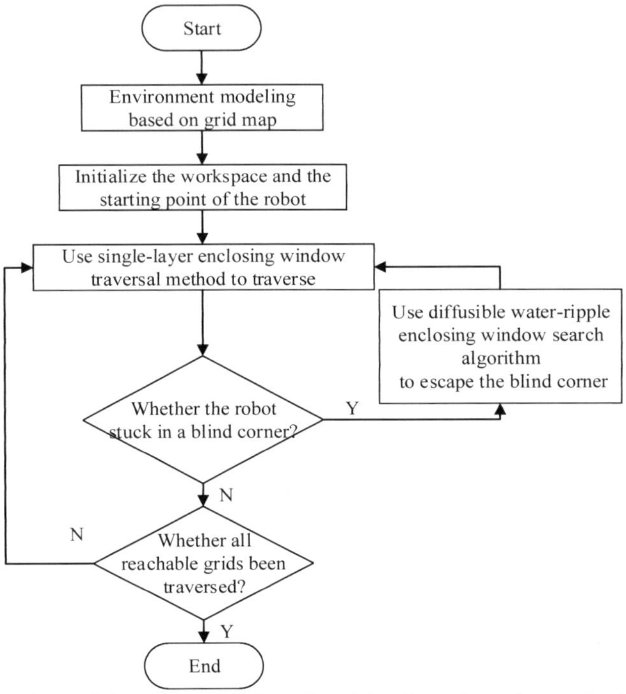

Figure 5. Flow chart of the path planning algorithm based on the enclosing window

4. Simulation analysis

In order to verify the effectiveness of the complete coverage path planning in the paper, Matlab was used for experimental simulation. The grid map is shown in Figure 6(a), which consists of 20×20 grids, among which the black areas are obstacles, and the number of grids that can be traversed is 350. Taking the upper left corner of the grid as the starting point, the complete coverage path planning algorithm based on the enclosing window proposed in this paper is used to perform the full traversal of the grid map.

Figure 6(b) shows the traversal result of the algorithm. The green path is the path for the robot to escape from the dead corner by using the diffusible water-ripple enclosing window search algorithm when the robot is trapped in the dead corner. As shown in Figure 6(c), points B and C use the first layer enclosing window to escape the dead corner, points A and D uses the second layer enclosing window to escape the dead corner, and point E uses the fourth layer enclosing window to escape the dead corner. In the same environment, the bow-shaped forward traversal method based on the improved A star is used for simulation, and the result is shown in Figure 6(d).

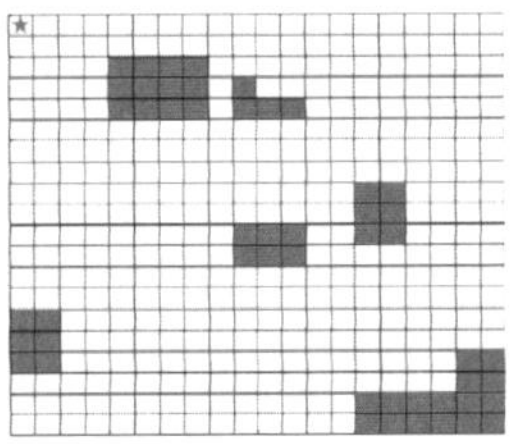

(a) Experimental simulation grid map

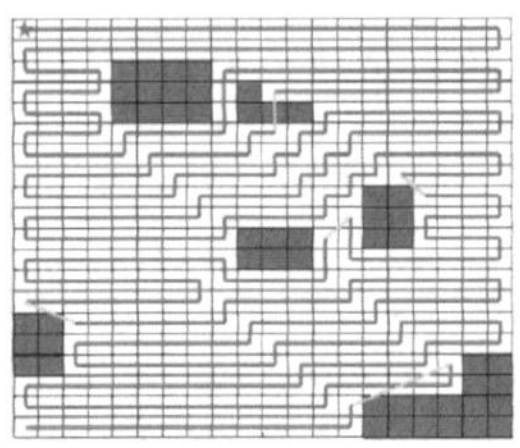

(b) Simulation traversal path diagram

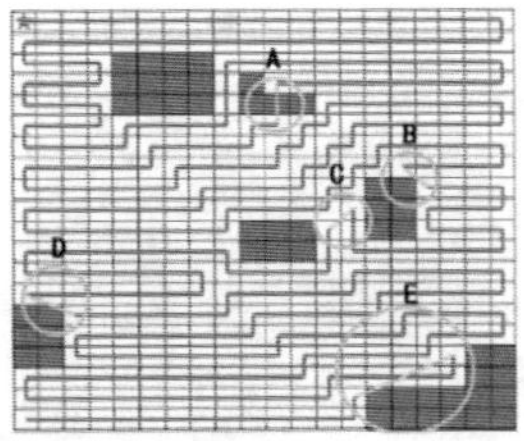

(c) Escape from dead corner

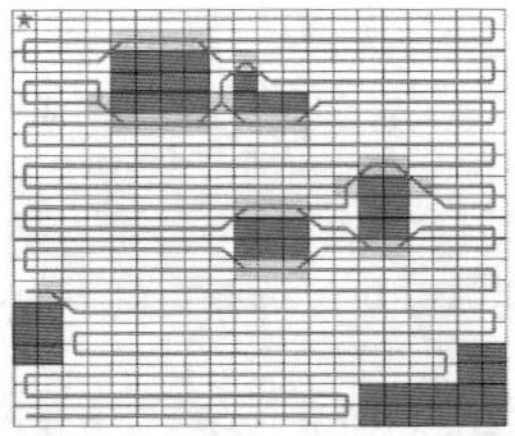

(d) The result of the bow-shaped forward traversal method based on the improved A star

Figure 6. Simulation results

When using the complete coverage path planning algorithm based on the enclosing window to traverse, we assume that the extra path length of the two processes of leaving the working surface and reaching the target point when the robot crosses the obstacle is 0.5. There are 3 situations in which obstacles are crossed in this simulation experiment, that is, the path length of the algorithm is added 1.5. It can be seen from Table 1 that although the two algorithms can achieve full coverage of the grid map, there is a big difference in the two indicators of repetition rate and path length. The repetition rate of the improved bow-shaped forward traversal method based on A star is 9.43%, and the path length is 390.11. The complete coverage path planning algorithm based on the enclosing window not only achieves the full coverage of the raster map but also has a repetition rate of 0%, and the path length is 357.04, which is much lower than the path length of the improved bow-shaped forward traversal method based on the A star.

Table 1. Performance evaluation index of simulation results

Complete Coverage Path Planning Algorithm	Coverage Rate/%	Repetition Rate/%	Path length
The bow-shaped forward traversal method based on the improved A star	100	9.43	390.11
Complete coverage path planning algorithm based on the enclosing window	100	0	357.04

5. Conclusion

Aiming at the special working environment of the underwater wall-climbing cleaning robot and the low efficiency and poor universality of the complete traversal path planning algorithm in this environment, a complete coverage path planning algorithm based on enclosing window is proposed. Firstly, this paper gives a brief introduction to the robot's floating system layout and explains the particularity of the underwater robot's working environment. Then, it is determined that the grid map method is used for environmental modeling and the priority of the single-layer enclosing window is determined according to different grid maps. Finally, the diffusible water-ripple enclosing window search algorithm is explained. It can be seen from the simulation results that the method in this paper can not only achieve full coverage of the grid map and 0% repetition rate, but also reduce the path length, and has a higher planning efficiency in an underwater environment. This paper assumes the additional path length when the robot crosses the obstacle or traverses the grid in the simulation experiment. In the next step, the algorithm will be tested by the real object to verify the algorithm.

Acknowledgments

Supported by the Industry-Study-Research Collaborative Innovation Fund of Shandong Academy of Sciences(2020-CXY05), the Pilot Project for the Integration of Science, Education and Industry of the Qilu University of Technology(2020KJC-GH05) and the Key Research and Development Plan in Shandong Province(2019GHZ006).

References

[1] Sun Lixin, Yang Xingyu, Yang Gaiwen, et al. Research on Hull Cleaning Robot Based on Negative Pressure Adsorption [J]. Machine Tool & Hydraulics, 2019, 47(17): 56-59.

[2] Connell D, Manh La H. Extended rapidly exploring random tree–based dynamic path planning and replanning for mobile robots[J]. International Journal of Advanced Robotic Systems, 2018, 15(3): 1-15.

[3] Li Caihong, Song Yong, Wang Fengying, Wang Zhiqiang, Li Yibin. A Bounded Strategy of the Mobile Robot Coverage Path Planning Based on Lorenz Chaotic System[J]. International Journal of Advanced Robotic Systems, 2016, 13(3): 1-9.

[4] D. Drake, S. Koziol and E. Chabot, Mobile Robot Path Planning With a Moving Goal[J], IEEE Access, 2018, 6: 12800-12814.

[5] Balch T. The Case for Randomized Search, workshop on Sensors and Motion. IN: IEEE International Conference on Robotics and Automation, San Francisco, 2000:213-215.

[6] Qian Jinwei, Dai Xiaoqiang, Gao Hongbo, Zhu Yanshuan. Path Planning Algorithm for Biological Incentive Traversal Based on Internal Spiral Search[J]. Computer Simulation, 2021,38(09):339-343+394.

[7] Zhu Daqi, Tian Chen, Sun Bing, Luo Chaomin. Complete Coverage Path Planning of Autonomous Underwater Vehicle Based on GBNN Algorithm[J]. Journal of Intelligent & Robotic Systems, 2019, 94(1): 237-249.

Underwater target passive detection method based on Hough transform track-before-detect

Xuemin Wang[1a*], Renwei Lu[1b*], Wenhai Li [1c]

[1] School of Aviation Operations Service, Naval Aviation University, Yantai 264001, China

[a*]18842981610@163.com, [b*]cheefoo2020@163.com, [c]ermalu188@163.com

Abstract—Aiming at the low detectability and concealment detection requirements of underwater targets such as underwater unmanned vehicles (UUV), an underwater target passive detection method based on Hough transform track-before-detect was proposed. In order to deal with the approach reconnaissance of UUV, the passive detection model of UUV was constructed; In order to obtain the target position information required for track-before-detect, the cross-positioning method of multiple passive sonar buoys was adopted to realize the data pre-processing and noise reduction before concealment detection; In order to improve the detection performance of passive sonar in complex underwater acoustic environment, Hough transform method and point trace correlation were used to realize the demand of low detectability of underwater targets. Simulation results show that this method can still maintain high detection probability under the conditions of low signal-to-noise ratio, multi-target and complex noise.

1. Introduction

With its advantages of small volume, low radiation noise and flexibility, UUV is deeply promoting the transformation and upgrading of intelligent warfare at sea. In military affairs, approach reconnaissance and key road sealing are the main combat styles of UUV. How to counter the threat of UUV? Finding the enemy first is the key to success in naval operations. When checking the sea area where there may be underwater targets, passive sonar buoys can detect and drive away incoming UUV with the help of the anti-submarine aircraft platform and prior information.

Underwater target detection technology can be roughly divided into acoustic spectrum detection and non-acoustic spectrum detection. Sound spectrum detection is divided into line spectrum detection and continuous spectrum detection. For example, the reference [1] proposed a passive detection method of underwater targets based on the correlation measurement between the received signals of three sub-arrays. The target was found by searching for correlation peaks on the two-dimensional delay plane[2], the detection performance had been improved to a certain extent. But in the case of low signal-to-noise ratio, the detection performance was affected by the accuracy of delay estimation. The reference [3] proposed a signal detection method based on adaptive step-size intermittent chaotic oscillator using line spectrum detection technology. However, it was necessary to know the range of line spectrum in advance, resulting in poor ability to detect low signal-to-noise ratio signals.

Non linear spectrum detection can be divided into matched filtering, time-frequency analysis, track-before-detect(TBD) and so on. TBD is an effective technology to detect targets with low signal-to-noise ratio[4]. It has been widely studied and applied in radar, infrared and other fields[4-10]. Common algorithms include particle filter track-before-detect (PF-TBD) and Hough transform track-before-detect (HT-TBD). With the advantages of simple algorithm, high efficiency and strong adaptability, HT-TBD is favored by researchers. For example, the reference [12] performed Hough detection on the time-frequency diagram of the sonar received signal to realize the detection of weak underwater targets; the reference [13] used the underwater unmanned platform to detect the low detectable underwater target autonomously by detecting the sonar azimuth history map.

When detecting underwater targets, active detection of sonobuoys can obtain rich target information, but it is easy to expose; the passive detection of sonobuoy has good concealment, but it cannot directly measure the position information of the target. In order to solve the above problems, this paper adopted the HT-TBD algorithm to detect the UUV with the help of the passive sonar buoy array. Firstly, according to the typical combat style of UUV, the passive detection model of UUV was constructed; Secondly, the adaptive correlation cross location method was used to realize the data preprocessing before HT-TBD detection; Finally, Hought transform and point trace correlation were carried out on the preprocessed data to realize the covert detection of underwater low detectable targets.

2. Passive detection model of UUV

For UUV approaching for reconnaissance, they were usually deployed near the base or the channel, and their direction of attack was known in advance. In these areas, targeted sonar buoy interception arrays can be set up to build warning zones and block their combat operations. Fig.1(a) showed the schematic diagram of the detection target of the linear buoy array.

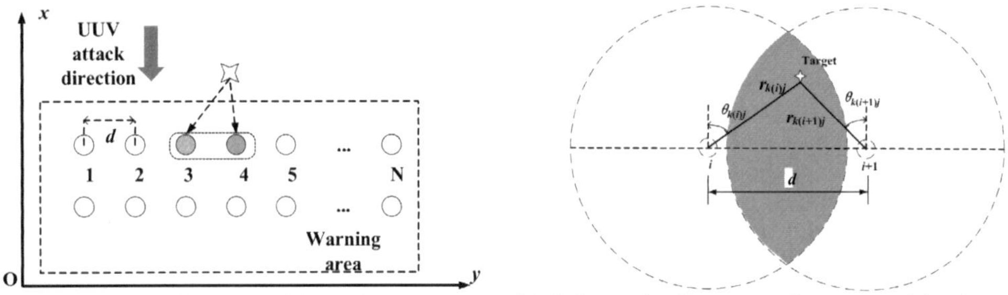

(a)Schematic diagram of detection target (b) Schematic diagram of cross positioning

Fig.1 Schematic diagram of cross positioning of passive sonar buoys

Usually, the single-line buoy interception array is taken as an example for analysis. Fig.1(b) showed the schematic diagram of cross positioning of any two adjacent passive sonar buoys. It was assumed that the single-line buoy array consists of N equidistant passive sonobuoys that were collinear. In the Cartesian coordinate system xOy, the position coordinates of the sonobuoy were $(x_{si}, y_{si})(i = 1,2,\cdots,N)$, and the distance between two adjacent buoys was d.

$$z_{ki} = \{(\theta^m_{k(i)j}, a^m_{k(i)j}) | j = 1,2,\cdots,n_j; m = 1,2,\cdots,n_m\} \quad (1)$$

where, $\theta^m_{k(i)j}$ was the azimuth angle in the received data of the sonobuoy, $a^m_{k(i)j}$ was the energy information in the received data of the sonobuoy, n_j was the number of measurement points, and n_m was the number of targets. The measurement errors of the azimuth angle $\theta^m_{k(i)j} \in (-\frac{\pi}{2}, \frac{\pi}{2})$ were independent of each other and obey a Gaussian distribution with zero mean and variance $\sigma^2_{\theta_i}$.

3. Cross-localization method for adaptive association

The detection range of a single passive sonar buoy is usually limited. According to the underwater motion characteristics of UUV, the adaptive correlation method was adopted to automatically switch to the passive sonobuoy with the best detection effect, so as to realize the detection continuity and robustness. Within the action range of the passive sonobuoy array, two passive sonobuoys for detection were determined by detecting the maximum signal strength obtained by each buoy.

At the kth measurement time, according to the receiving azimuth angle $\theta_{k(i)j}(i = 1,2,\cdots,N)$ of N passive sonobuoys its energy information $a_{k(i)j}$ can be expressed as

$$a_{k(i)j} = \begin{cases} s_{k(i)}(\theta_{k(i)j}) + \delta_{k(i)}(\theta_{k(i)j}) & target\ exit \\ \delta_{k(i)}(\theta_{k(i)j}) & no\ exit \end{cases} \quad (2)$$

Using the maximum value method, the buoy i^* corresponding to the maximum energy value was obtained, and it was used as the first detection buoy.

$$b_{k(i^*)j} = \max\left\{a_{k(i)j} \middle| i = 1,2,\cdots,N\right\} \quad (3)$$

The larger of $b_{k(i^*-1)j}$ and $b_{k(i^*+1)j}$ was selected as the second detection buoy. The two sonobuoys used for detection in the buoy array cross-locate the underwater target[14,15] and the position coordinates$(x_{k(i)j}, y_{k(i)j})$ of the target in the Cartesian coordinate system were obtained. Therefore, the target measurement data for Hough transform detection was represented as $(x_{k(i)j}, y_{k(i)j}, a_{k(i)j})$.

It can also be seen from Fig.1(b) that after the cross-location processing, the influence of most of the background noise of the marine environment was eliminated, and the clutter was only retained in the overlapping area detected by adjacent passive sonobuoys.

4. Method for detecting UUV by HT-TBD

After the pre-detection preprocessing of the cross-location was completed, the HT-TBD was used to detect the UUV. The basic idea of this detection method [16]was as follows: a data matrix was constructed based on the passive sonobuoy measurement data after cross-location processing, and the parameter matrix was established; the measurement data was mapped to the parameter space by the Hough transform; in the parameter space resolution unit, non-coherent energy accumulation and threshold detection were done; the threshold resolution unit was inversely mapped to the data space by Hough transform, and the initial starting track was obtained; through track merging and false track removal, the initial track was associated with point tracks, and a high-confidence initial track corresponding to the number of targets was obtained. If it existed, underwater weak targets were detected; if it did not exist, the above detection process was repeated. The flow chart of HT-TBD detection was shown in Fig.2

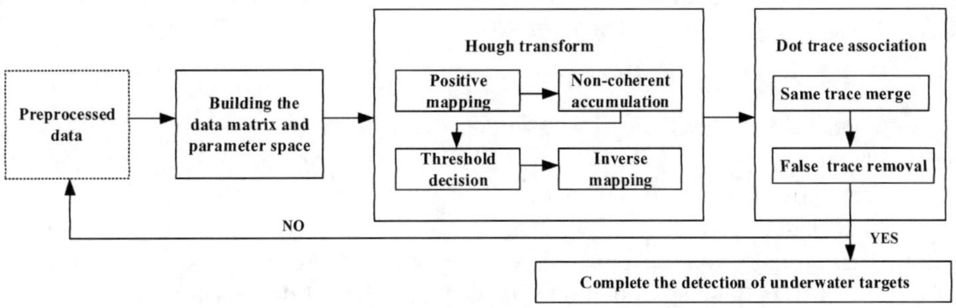

Fig.2 HT-TBD detection flow chart

4.1. Constructing the data matrix and parameter space

At time k, the measurement data received by the i-th passive sonobuoy was $z_{k(i)j} = (x_{k(i)j}, y_{k(i)j}, a_{k(i)j})(j = 1,2,\cdots,n_l)$, where $n_l(n_l < n_j)$ was the number of measurement points after cross-location processing. The corresponding data matrix B_k was

$$B_k = \begin{bmatrix} x_{k(i)1} & x_{k(i)2} & \cdots & x_{k(i)n_l} \\ y_{k(i)1} & y_{k(i)2} & \cdots & y_{k(i)n_l} \end{bmatrix} \quad (4)$$

At time k, the corresponding data energy matrix E_k was

$$E_k = \left[a_{k(i)1}\ a_{k(i)2}\ \cdots\ a_{k(i)n_l}\right] \tag{5}$$

Considering the influence of the angle measurement error, in order to effectively detect the underwater target information in the measurement data, the parameter space was discretely divided into $N_\theta \times N_\rho$ cells. The center point of each cell was

$$\begin{cases} \theta_l = \left(l - \frac{1}{2}\right)\Delta\theta & l = 1,2,\cdots,N_\theta \\ \rho_m = \left(m - \frac{1}{2}\right)\Delta\rho & m = 1,2,\cdots,N_\rho \end{cases} \tag{6}$$

where, the size of the cell was $\Delta\theta = \frac{\pi}{N_\theta}$, $\Delta\rho = \frac{r_{max}}{N_\rho}$, and r_{max} was usually twice the measurement range of the passive sonobuoy.

The parameter space accumulation matrix was established and initialized. The parameter space energy accumulation matrix was used to store energy accumulation data. The accumulation matrix shall be initialized before storage.

4.2. Hough transform

(1) Positive mapping

According to Hough transform positive mapping formula $\rho = x\cos\theta + y\sin\theta$, at time k, the positive mapping matrix H corresponding to data matrix B_k was

$$H_k = \begin{bmatrix} \cos\theta_{k(i)1} & \cos\theta_{k(i)2} & \cdots & \cos\theta_{k(i)n_l} \\ \sin\theta_{k(i)1} & \sin\theta_{k(i)2} & \cdots & \sin\theta_{k(i)n_l} \end{bmatrix}^T \tag{7}$$

By positive mapping, the matrix C_k of the parameter space was

$$C_k = H_k B_k = \begin{bmatrix} r_{1,\theta_{k(i)1}} & r_{2,\theta_{k(i)1}} & \cdots & r_{n_l,\theta_{k(i)1}} \\ r_{1,\theta_{k(i)2}} & r_{2,\theta_{k(i)2}} & \cdots & r_{n_l,\theta_{k(i)2}} \\ \vdots & \vdots & \ddots & \vdots \\ r_{1,\theta_{k(i)n_l}} & r_{2,\theta_{k(i)n_l}} & \cdots & r_{n_l,\theta_{k(i)n_l}} \end{bmatrix} \tag{8}$$

(2)Non-coherent accumulation.

At time k, the parameter elements covered by the curve were accumulated by incoherent energy and stored in the corresponding accumulation matrix.

$$E(\theta_l, \rho_m) += E(\theta_l, \rho_m) \tag{9}$$

Where, '+='indicates accumulation operation.

The accumulation was repeated. When $k > K$, the accumulation operation was ended. The reference space point accumulation histogram and energy accumulation histogram were obtained.

(3)Threshold decision.

Incoherent energy accumulation threshold δ_E was set to detect the threshold of the accumulated value of the parameter spatial resolution unit. The threshold detection decision condition was

$$E(\theta_l, \rho_m) > \delta_E \tag{10}$$

When the accumulated value of the parameter unit satisfies equation (10), the corresponding parameter unit (θ_l^*, ρ_m^*) was obtained.

Parameter unit (ρ_m^*, θ_n^*) was inversely mapped by Hough transform to find the collinear point trace set in the corresponding data space, and then the initial detection track was obtained.

4.3. Point trace correlation

In the plot set Z_K, the initial detection tracks with the same plot were merged, and the repeated initial detection tracks are deleted. Based on the combination, the false initial track was deleted with the help of the a priori constraints of the speed and angle of the UUV [16], and a one-to-one corresponding detection track with high reliability was obtained, that is, it is considered that the underwater target was effectively detected, so as to realize the detection of the UUV.

5. Simulation and analysis

5.1 Parameter setting

It was assumed that the underwater vehicle moves in a straight line at a uniform speed in a two-dimensional plane. The starting position of the underwater target was (6km, 6km), the heading was 60°, and the average speed was 5m/s. The first passive buoy sonar deployment location was (4.8km, 6.25km), the number of sonobuoys was 2, the deployment direction was 90°, the action distance was 5km, the distance between adjacent buoys was 1.2 times the action distance, and the angle measurement error was 0.5 °, the cross-location error was 100m, and the error type was zero-mean Gaussian noise. The number of observations was 5 times, and the observation period was20s. The Hough transform used 360 grids in the parameter space, and the non-coherent detection threshold was 0.8 times the maximum energy accumulation in the reference space.

The detection performance of the algorithm was simulated based on the measured radiation noise data of underwater targets. According to the literature [17], in order to ensure effective detection probability and efficient operation time, the number of parameter space accumulation should be at least 5 times. In this paper, the research was carried out 5 times, repeated 1000 times, and the average value was used as the final detection performance index.

5.2 Influence of underwater acoustic environment complexity on detection performance

Noise density is one of the main factors to measure the complexity of underwater acoustic environment. In order to verify the effectiveness of the algorithm in this paper, the simulation verification was carried out when the noise density is 100 and the signal-to-noise ratio is 5dB. First, the cross-location pre-processing was performed on the sonar observation data, and the distribution of noise and target point traces was shown in Fig.3; Then, the points in the data space in Fig.3(b) were Hough transformed to the parameter space, and the parameter space was discretized and non-coherent energy accumulation was carried out respectively. The obtained energy accumulation histogram of the parameter space was shown in Fig.4(a).

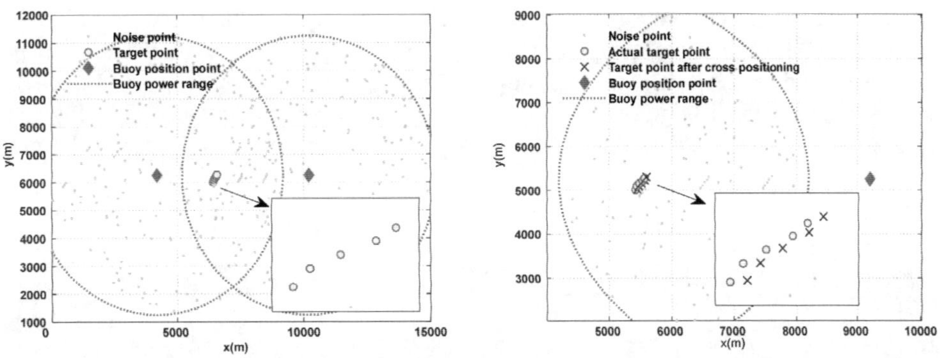

(a)Trace distribution before cross positioning (b)Trace distribution after cross positioning
Fig.3 Distribution of noise points and target point traces before and after cross-positioning

In order to effectively eliminate false tracks and clutter points in the track, and reduce the number of repeated tracks, the point track association processing was performed on the point track set that satisfies Fig.4(a). The specific method was shown in Section 4.3, and the processed result was shown in Fig.4(b). Backtracking was performed on the dot traces satisfying Fig.4(b), and the HT-TBD detection track was obtained, as shown in Fig. 4(c).

In order to further verify the effectiveness of the algorithm in this paper, simulation verification is carried out when the noise density was 200 and the signal-to-noise ratio was 5dB. Fig.5(a) showed the distribution of noise and target point traces, the detection result was shown in Fig.5(b), and the detection track was shown in Fig.5(c). It can be seen from the simulation results in Fig. 5 that after increasing the noise density, the method in this paper can still effectively detect the underwater target track.

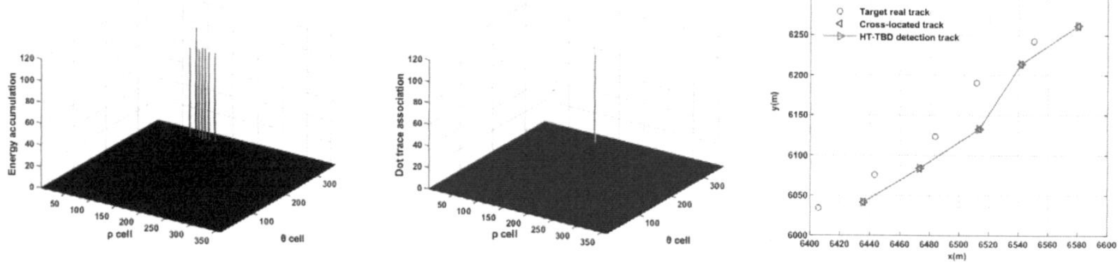

(a) Energy accumulation histogram (b) dot trace correlation processing (c) Detection track map

Fig.4 Single-target HT-TBD detection track diagram

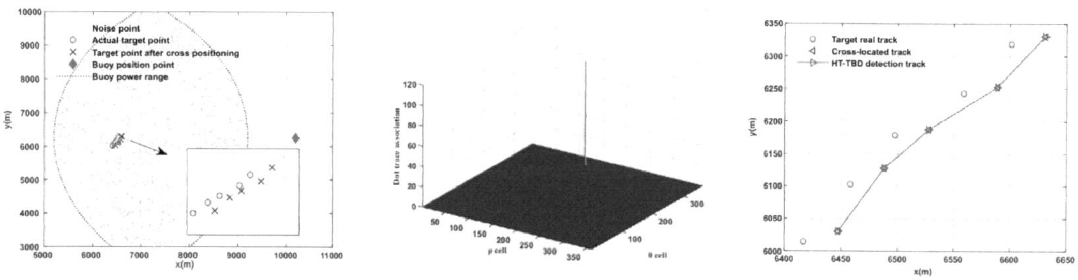

(a)Dot trace distribution (b) Detection histogram (c) Detection track map

Fig.5 Schematic diagram of HT-TBD detection at high noise density

5.3 Influence of multiple targets on detection performance

When the noise density was 100, the signal-to-noise ratio was 5dB, and the number of targets was 3, the simulation verification was carried out. The initial positions of the three targets were (6 km, 6 km), (6 km, 7 km), (6 km, 7 km), and the navigation was 60°. Fig.6(a) showed the distribution of noise and target point traces after cross-location processing, the detection result was shown in Fig.6(b), and the detection track was shown in Fig.6(c). It can be seen from the simulation results in Fig.6 that after increasing the number of targets, the method in this paper can still effectively detect the underwater target track.

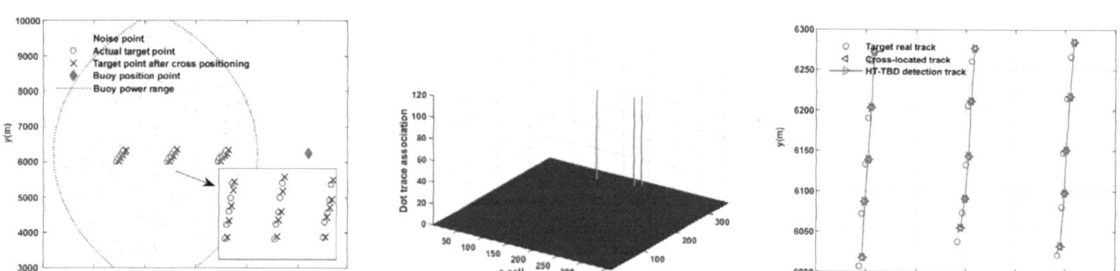

(a)Dot trace distribution (b) Detection histogram (c) Detection track map

Fig. 6 Schematic diagram of multi-target HT-TBD detection

5.4 Method validity analysis

In order to further verify the performance of the algorithm, when the signal-to-noise ratio was equal to 5 dB, 10 dB, 15 dB, and 20 dB, and the rest of the parameters were set in Section 5.1, the time-frequency method WVD [18] and the method in this paper（HT-TBD）were both carried out 1000 times of Monte Carlo simulation experiments. The comparison of the detection performance of the two algorithms was shown in Table 1.

Table 1. Comparison of the detection performance of the two methods.

Signal-to-noise ratio(dB)	Detection probability(%)		average running time(s)		Detection probability error range(%)	
	WVD	HT-TBD	WVD	HT-TBD	WVD	HT-TBD
5	67	74	25.7	37.2	±6.2	±4.4
10	72	79	23.6	23.8	±3.4	±1.8
15	78	80	23.3	23.3	±1.8	±0.7
20	80	81	22.1	22.2	±1.6	±0.6

The following results can be obtained from Table 1:

(1) The detection probability of both algorithms increased with the increase of signal-to-noise ratio, but the detection probability of HT-TBD was better than that of WVD, especially when the signal-to-noise ratio was low. For example, when the SNR was 5dB, the detection probability of HT-TBD was 10% higher than that of WVD.

The average running time of both algorithms decreased as the signal-to-noise ratio increases. Under the condition of lower signal-to-noise ratio, the real-time performance of the WVD algorithm was better, but when the signal-to-noise ratio exceeded 10dB, the average running time of the two algorithms was basically the same.

The detection probability error range of the two algorithms decreases with the increase of signal-to-noise ratio, but the detection error range of HT-TBD algorithm is less than that of WVD algorithm. Therefore, in general, HT-TBD algorithm is better than WVD algorithm.

6. Conclusion

Based on the results and discussions presented above, the conclusions are obtained as below:

(1) This paper presented a method of cross location and HT TBD detection based on passive sonar buoy array. It can still maintain high detection probability under the conditions of low signal-to-noise ratio, multi-target and complex noise.

(2) Compared with the existing time-frequency detection methods, the detection performance was better. When the signal-to-noise ratio was 5dB, the detection probability was increased by 10% compared with WVD method. The detection probability error range decreased with the increase of signal-to-noise ratio, and was better than WVD method. When the signal-to-noise ratio exceeded 10dB, the detection speed was consistent with the WVD method.

Reference

[1] Jiang K Y., Yao Z X., Yin J X. (2012) A passive positioning method for underwater targets based on Three-element-array. J . Acta Armamentarii., 33: 1107-1111.

[2] Song X J. (2004) Research on Passive Ranging Sonar of Digital Noise Target. D . Harbin Engineering University, Harbin . pp.15-21.

[3] Cong C., Li X K., Song Y. (2014) A method of detecting line spectrum of ship-radiated noise using a new intermittent chaotic oscillator. J . ActaPhysica Sinica., 63: 168-179.

[4] Yang W., Fu Y W., Pan X G., et al. (2014) Track-before-detect technique for dim targets: An overview. J . Acta Electronica Sinica., 42: 1786-1793.

[5] Carlson B D., Evans E D., Wilson S L. (1994) Search Radar Detection and Track with the Hough Transform, Part I: System Concept. J . IEEE Transactions on Aerospace and Electronic

Systems., 30: 102-108 .
[6] Carlson B D., Evans E D., Wilson S L. (1994) Search radar detection and track with the Hough transform, Part II: Detection statistics. J . IEEE Transactions on Aerospace and Electronic Systems., 30: 109-115.
[7] Carlson B D., Evans E D., Wilson S L. (1994) Search radar detection and track with the Hough transform. Part III: Detection performance with binary integration. J . IEEE Transactions on Aerospace and Electronic Systems., 30: 116-125.
[8] Liu Z X., Xie W X., Yang J. (2004) Fuzzy hough transform track initiation algorithm for an asynchronous passive sensor system. J . Systems Engineering and Electronics., 11: 1545-1548,1591.
[9] Wang J. (2019) Track before detect preprocessing for active sonar. J . Technical Acoustics., 5: 133-134.
[10] Chen H J. (2021) Subspace projection based track-before-detect scheme for small moving target in complex underwater environment. J . Journal of Electronics & Information Technology., 3: 826-833.
[11] Liu H M., Yang X. (2021) Research on the Passive Sonar Target Tracking Algorithm before Detection Based on Particle Filter. J . Acoustics and Electronic Engineering., 143: 17-21.
[12] Barbarossa S. (1995) Analysis of multicomponent LFM signals by a combined Wigner-Hough transform. J . IEEE T rans on Signal Processing., 43: 1511-1515.
[13] Li Z G., Li S Q.,Wen S L. (2017) Automatic detection of an underwater target based on UUV. J .Journal of Harbin Engineering University., 38: 104-109.
[14] Tian T. (2010) Sonar Technology (Second Edition). Harbin Engineering University Press, Harbin.
[15] Bai J.,Wang G H., Wang N., et al. (2009) Study on optimum cut angles in bearing-only location systems. J . Acta Aeronautica et Astronautica Sinica., 30: 298-304.
[16] Wang G H., Kong M., He Y. (2005) Hough transform and its application in information processing. Weapon Industry Press, Beijing.
[17] Li L., Wang G H., Yu H B., et al. (2017) A pre-detection tracking algorithm for hypersonic targets in adjacent space. J . Acta Astronautics., 38 : 420-427.
[18] BOASHASH B, O' SHEA P. (1988) Time frequency analysis applied to signature of underwater acoustic signals. In: ICASSP-88, International Conference on Acoustics, Speech, and Signal Processing. New York. pp. 2817-2820.

Journal of Physics: Conference Series 2258 (2022) 012074 doi:10.1088/1742-6596/2258/1/012074

Design of Gust Load Alleviation Control Based on UD-PSO for Civil Aircraft

Xiaoyu Qu*, Tao Li

Shanghai Aircraft Design and Research Institute, Commercial Aircraft Corporation of China, Ltd, Shanghai,201210, China

*Corresponding author's e-mail: chusay@126.com

Abstract. Civil aircraft will produce undesirable normal overload under the disturbance of vertical turbulence or gusts, which will lead to deterioration of the ride quality of passenger. To solve this problem, a design scheme of gust load alleviation (GLA) control based on UD-PSO optimization is proposed. Firstly, design a classical gust load alleviation control strategy in the control system. The GLA controller will automatically generate the control command, which is calculated by the sensor feedback signal of turbulence or gust. That is, the aileron and spoiler are deflected in the same direction to counteract the undesired normal overload; at the same time, the elevators are deflected to ensured the longitudinal stability of the aircraft. Then, selects a suitable reference model and a reasonable objective function to achieve the desired ride quality performance index. Finally, the UD-PSO algorithm is used to automatically optimize the controller parameters. The simulation results show that the load alleviation control strategy optimized based on UD-PSO can effectively reduce the gust interference, improve the ride comfort, and achieve a good control performance, which shows the effectiveness of the proposed method.

1. Introduction

During flight phases, civil aircraft often produce undesirable aerodynamic forces and moments due to the influence of gusts or turbulence, which brings additional overload to the aircraft. As a result, it reduces the flight quality of the aircraft, which not only makes the aircraft difficult to control, but also makes the pilots and passengers feel extremely uncomfortable. On the other hand, it will also make the aircraft structure bear greater dynamic load, and the body is prone to structural fatigue damage. This is an urgent problem for aircraft design, especially for civil aircraft design.

Using load alleviation control (LAC) or gust load alleviation (GLA) [1] in active control technology, when encountering gusts or turbulence disturbances, the controller automatically commands the response The control surface reduces the interference of gusts or turbulence and the change of overload, and improves the comfort of passengers[2]. In addition, during take-off, landing, and go-around flight phases that require high trajectory control, load alleviation control can improve flight trajectory accuracy, reduce human operation errors caused by gust interference, and improve the stability and safety of the aircraft[3].

To solve the problem that the gust disturbance reduces the ride quality during flight of civil aircraft, a load alleviation control system design based on UD-PSO optimization is proposed in this paper, and uses active control technology to reduce the additional overload caused by gust disturbance. The effectiveness of this method is verified by mathematical modeling and simulation with Matlab/Simulink.

2. Problem Formulation

2.1. Gust disturbance

In the natural atmospheric environment, due to the influence of factors such as friction, vortex and geographical environment, wind disturbance does not appear in a pure form, but is often accompanied by turbulence. The appearance and formation of atmospheric turbulence are also related to many factors, such as terrain induction, heat, moisture, wind shear and so on.

During the flight of a civil aircraft, the wind speed measured by the airborne sensor fluctuates around a certain value, which reflects the general movement of the air, called wind; and the fluctuation represents the local air flow, called turbulence.

Flow, its moving speed is represented by V_w. The turbulence velocity is a random variable, and its average value μ_w is zero. The mean square error of turbulence is defined as:

$$\sigma_w^2 = \lim_{T\to\infty} \frac{1}{2T} \int_{-T}^{T} V_w^2 dt \tag{1}$$

where the σ_w is the standard deviation of the turbulent flow velocity, also known as the root mean square value.

The mean square deviation or standard deviation of the turbulence velocity is closely related to the intensity of the turbulence field. Table 1 is based on the maximum turbulence velocity V_{w_max} as the turbulence intensity classification.

Table.1 Classification of gust intensity

Maximum turbulence velocity (m/s)	intensity level
2 ~ 6	weak
6 ~ 12	medium
12 ~ 16	strong
16 ~ 20	Extremely strong

2.2. Gust model

In terms of mathematical definition, turbulence is defined as a uniform, stable, ergodic and isotropic random process superimposed on the constant wind. The statistical characteristics of the random process do not change with time. In this paper, only vertical gust disturbances are considered, and the Dryden model is used in the gust model[4].

The power spectrum function of the vertical wind speed of Dryden model is as follows:

$$\Phi_{w_g}(\Omega) = \sigma_w^2 \frac{L_w}{\pi} \frac{1+3(L_w\Omega)^2}{[1+(L_w\Omega)^2]^2} \tag{2}$$

where $\Omega = 2\pi/\lambda$ is the spatial frequency, λ is the sinusoidal wavelength; σ_w is the standard deviation of the gust component; L_w is the scale of the turbulence; w is the gust component.

Because of $\omega = V_0\Omega$, Eq. (2) is converted into time variable from space variable, as shown in Eq. (3):

$$\Phi_{w_g}(\Omega) = \frac{\sigma_w^2 L_w}{\pi V_0} \frac{1+3(\frac{L_w}{V_0})\omega^2}{[1+(\frac{L_w}{V_0})^2\omega^2]^2} \tag{3}$$

In order to facilitate the simulation analysis, a shaping filter can be introduced to generate the required colored noise power spectrum function. Suppose a random process x(t) is generated by white noise r(t) of unit intensity passing through a filter with a transfer function of G(s), and the spectral density function of x(t) is:

$$j_x(\omega) = G^*(j\omega)G(j\omega) = |G(j\omega)|^2 \tag{4}$$

Decomposing the spectral density shown in Eq. (4), the transfer function of the shaping filter for a given spectral density function can be obtained:

$$G_w(s) = \sqrt{K_w}/(s+\beta_w)/(s+\lambda_w)^2 \tag{5}$$

where $K_w = 3V_0\sigma_w^2/L_w\pi, \beta_w = V_0/\sqrt{3}L_w, \lambda_w = V_0/L_w$.

In order to facilitate the simulation analysis, $G_w(s)$ is simplified to the first-order function:

$$G_w(s) = \sqrt{K_w}/(s+\lambda_{w1}) \tag{6}$$

2.3. The impact of gust on aircraft movement

The atmospheric turbulence described in section 2.1 is usually based on the geodetic coordinate system and does not change with time. However, when analysing the influence of vertical turbulence on the aircraft, in the body coordinate system, the turbulence not only changes with time, but also changes with the coordinates, that is, at a certain moment, the turbulence speeds experienced by different positions of the wings are different. Even at the same location, the turbulence velocity experienced at different moments is different. At this time, this change over time is caused by the continuous forward movement of the aircraft. Therefore, when studying the influence of vertical turbulence on the aircraft, the turbulence gradient distribution on the Z axis of the aircraft must be considered. Usually the distribution is linear, that is:

$$\omega_g(x,y) = \omega_g(0) + \partial\omega_g/\partial x\Big|_0^x + \partial\omega_g/\partial y\Big|_0^y \tag{7}$$

The influence of gusts or turbulence on aircraft motion is due to the change of the relative air flow during aircraft balance, resulting in additional aerodynamic changes[4-6]. The influence caused by aerodynamic change can be expressed by the equivalent change of state quantity in aircraft state equation. In the vertical direction, it is generally considered that the turbulence velocity ω_g can cause the disturbance of angle of attack as follows:

$$\triangle\alpha_g = -\omega_g/V_0 \tag{8}$$

When the aircraft is regarded as a particle, the influence of gradient on the aircraft can be ignored. In the vertical direction, the aerodynamic change caused by turbulence can be completely equivalent to $\triangle\alpha_g$. The disturbance of angle of attack increment $\triangle\alpha_g$ is introduced into the linear motion equation of aircraft, and the influence of turbulence on aircraft motion is transformed into angle of attack $\triangle\alpha_g$. To sum up, the motion equation of the aircraft under the action of vertical turbulence is:

$$\begin{bmatrix}\triangle\dot{V}\\ \triangle\dot{\alpha}\\ \triangle\dot{\beta}\\ \triangle\dot{\theta}\end{bmatrix} = \begin{bmatrix} -X_v & -X_\alpha & 0 & -X_\theta \\ -Z_v & -Z_\alpha & 1 & 0 \\ -\left(M_v - M_{\dot{\alpha}} Z_v\right) & -(M_\alpha - M_{\dot{\alpha}}Z_\alpha) & -\left(M_q + M_{\dot{\alpha}}\right) & 0 \\ 0 & 0 & 1 & 0 \end{bmatrix} \cdot \begin{bmatrix}\triangle\bar{V} + \triangle V_g \\ \triangle\alpha + \triangle\alpha_g \\ \triangle q \\ \triangle\theta\end{bmatrix} +$$
$$\begin{bmatrix} 0 \\ Z_{\delta_e} \\ -\left(M_{\delta_e} - M_{\dot{\alpha}}z_{\delta_e}\right) \\ 0 \end{bmatrix} \cdot\triangle\delta_e = A_L\begin{bmatrix}\triangle\bar{V}\\ \triangle\alpha\\ \triangle q\\ \triangle\theta\end{bmatrix} + B_L\cdot\triangle\delta_e + \begin{bmatrix} -X_v & -X_\alpha \\ -Z_v & -Z_\alpha \\ -(M_v - M_{\dot{\alpha}}Z_v) & -(M_\alpha - M_{\dot{\alpha}}Z_\alpha) \\ 0 & 0 \end{bmatrix}\begin{bmatrix}\triangle V_g \\ \triangle\alpha_g\end{bmatrix} \tag{9}$$

3. Design of load alleviation controller based on UD-PSO

3.1. Uniform design particle swarm optimization

The classical particle swarm optimization algorithm[7] can converge quickly in the initial stage of optimization, but in the later stage of optimization, the algorithm may fall into local optimization and cannot calculate the global optimal solution. In order to enhance the global search ability, the particle swarm optimization algorithm is improved. The concentration concept in uniform design[8] is introduced to promote low concentration particles and suppress high concentration particles, so as to maintain the diversity of particles and overcome the defect that the classical particle swarm optimization method is easy to fall into local optimization.

Combining uniform design (UD) with classical particle swarm optimization algorithm, first select a group of random particles for initialization, then divide the area searched by particle swarm optimization algorithm through uniform design, calculate the effect value and regression analysis, and change the current optimal position of particle swarm to fall into the effective area selected based on uniform design, It not only ensures the effectiveness of the current optimal value variation, but also changes the direction and trend of the whole particle swarm search, increases the gap between particles, guides the continuous evolution of particle swarm, and finally achieves the purpose of global optimization.

In the classical particle swarm optimization algorithm, Eq. (10) and (11) update the speed and position of particles:

$$V_i(n+1) = \lambda_1 \cdot Rand_1 \cdot (P_{best} - X_n) + \lambda_2 \cdot Rand_2(G_{best} - X_n) + \omega \cdot V_i(n) \quad (10)$$

$$X_i(n+1) = X_i(n) + V_i(n+1) \quad (11)$$

where V_i is the velocity of the i-th particle, X_i is the position of the i-th particle, ω is a weighted number, Rand1 and Rand2 take random numbers between 0 and 1, λ_1 and λ_2 are Learning factor; G_{best} is the position of the optimal solution of the particle group; P_{best} is the position of the optimal solution of the particle itself.

The particle swarm search area is divided into n levels uniformly distributed, and the search table selects the appropriate column corresponding to the representative area. After the current global optimal position mutation is performed, it must be accompanied by a recalculation of the inertia weight ω in the particle swarm. Increasing the value of ω can increase the search ability of the particle swarm. Before mutation, this paper chooses to use a linear decreasing strategy, as shown in Eq. (12) for calculation.

$$\omega = \omega_{\max} - \frac{\omega_{\max} - \omega_{\min}}{T} d \quad (12)$$

where d is the current iteration algebra, and T is the total iteration algebra.

Suppose that after n iterations, the global optimal position mutates. At this time, as the inertia weight becomes smaller, the search ability of the particle swarm decreases. In order to improve the search ability of the mutated particle swarm, the mutated particle swarm is regarded as a new one. The total iteration algebra is T=T-n, the iteration algebra t starts counting from 1 again, and the inertia weight calculation equation becomes as follows:

$$\omega = \omega_{\max} - \frac{\omega_{\max} - \omega_{\min}}{T - n} d \quad (13)$$

The inertia weight ω has been appropriately improved, which satisfies the better global search in the early stage after the mutation, and the accurate local search in the later stage[9].

3.2. Design of load alleviation controller based on UD-PSO

3.2.1. Design of load alleviation controller

The flight control system of modern civil aircraft usually controls the attitude and track of the aircraft through the pilot's manipulation of the elevator, aileron, rudder and multi-function spoiler. The purpose of load alleviation control is the aircraft can be actively controlled. When encountering gust disturbances, the designed controller automatically controls the appropriate control surface to produce aerodynamic changes equal to the disturbance in the opposite direction, which can offset the change of overload and structural load increasing caused by the influence of gusts, and improve riding quality.

According to the characteristics of the civil aircraft's aerodynamics and various control surfaces, the structure of the vertical load alleviation control system designed in this paper is shown in Figure 1. Control surfaces includes elevator, aileron and two pairs of outside multifunctional spoilers. The feedback signal is the normal overload of the aircraft center of gravity, and the signal will feed back to the aileron and multi-function spoiler commands through the washout filter to improve the ride quality.

Normally, the normal overload feedback will increase the damping of the aircraft and reduce the natural frequency of the short period, delaying the response of the aircraft to the pilot input. For this reason, the load alleviation control must be combined with the aircraft's longitudinal stabilization control, and the comprehensive design of the pitch axis direction must be introduced. At this time, the pitch rate signal is introduced, and then feed back to the elevator command through the low-pass filter and washout filter, so as to improve the control characteristics of the aircraft and ensure the flight quality. Each feedback channel of load alleviation control uses a washout filter to eliminate the influence of the aircraft on the steady-state acceleration and pitch rate when turning.

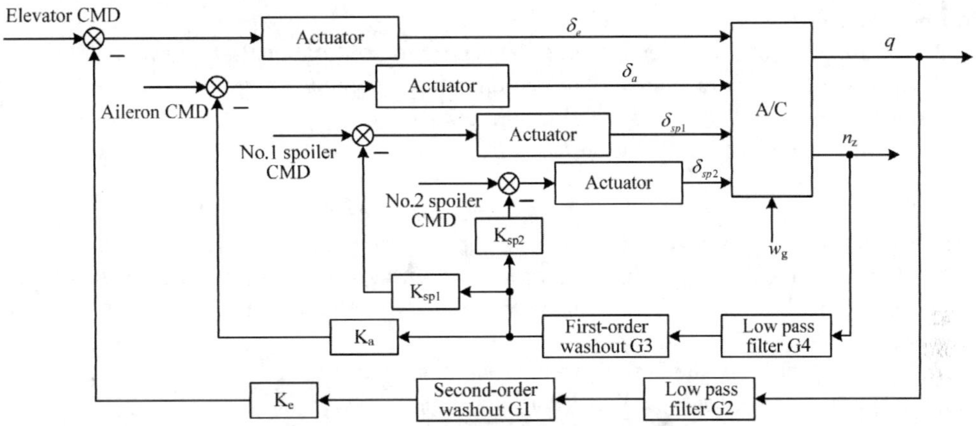

Figure 1. Load alleviation control design

As shown in Figure 1, the commands of the elevator, aileron spoiler No.1 and No.2 multi-function spoiler are shown in Eq. (14) respectively.

$$\begin{aligned} \delta_e &= R_e - G1 \cdot G2 \cdot K_e \cdot q \\ \delta_a &= R_a - G3 \cdot G4 \cdot K_a \cdot N_z \\ \delta_{sp1} &= R_{sp1} - G3 \cdot G4 \cdot K_{sp1} \cdot q \\ \delta_{sp2} &= R_{sp2} - G3 \cdot G4 \cdot K_{sp2} \cdot q \end{aligned} \quad (14)$$

3.2.2. Parameter optimization based on UD-PSO algorithm

When the framework of the control system is determined, the quality of its control effect depends on the definition of the parameters in the controller. Since there is no obvious mapping relationship between the gain parameters, filter parameters and the expected response of the system in the load alleviation controller, choosing a suitable objective function is a difficult point to be solved by using the UD-PSO algorithm to optimize the controller parameters. This paper chooses the optimization strategy of load alleviation controller based on the reference model, and its structure is shown in Figure 2.

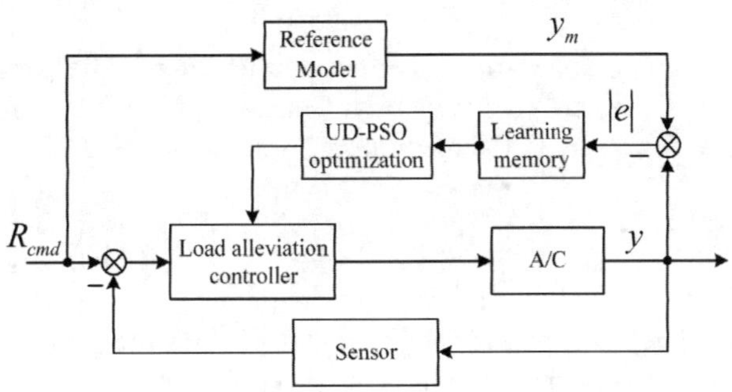

Figure 2. Load alleviation controller based on UD-PSO

In Figure 2, R_{cmd} is the pilot command, y is the output state response of the aircraft, y_m is the reference model, |e| is the absolute value of the difference between y_m and y, and |e| is integrated and used as the objective function. Therefore, the UD-PSO algorithm program will automatically adjust the controller parameters according to the deviation between the actual output y and the reference output y_m until the deviation meets the performance index requirements. At this time, the controller parameters are the optimized results.

The design process of load alleviation controller based on UD-PSO is as follows:

Step1: Determine the load alleviation control plan and structure.

Step2: Choose a suitable reference model according to the requirements of ride quality on performance indicators.

Step3: Choose a reasonable objective function according to flight quality requirements.

Step4: Clarify the population size N, the random velocity $V_i$and position X_i of a single particle, and select reasonable learning factors λ_1,λ_2 and weight ω.

Step5: Compare the optimal solution G_{best} of the group with the optimal solution P_{best} of the particle itself and update the value.

Step6: When the global optimal position of the particle does not change for x consecutive iterations, calculate the mutation probability p. If rand<p is satisfied, the current global optimal position will be mutated and fall into the corresponding area of uniform design according to the probability, and at the same time according to Eq. (11)~(13) recalculate the inertia weight ω, and update the particle velocity and position;

Step7: Determine whether the optimization process is over: if the maximum number of iterations is reached, the optimization ends; otherwise, return to Step5.

3.3. Ride quality evaluation

One purposes of load alleviation control is to make passengers on the aircraft feel comfortable in a gusty environment, so it is also called ride quality control. The control effect of the two definition is same, which both use active control technology to reduce the additional overload caused by gust interference. The control effect is also evaluated by measuring the degree of restraining the additional overload change under the gust interference.

Ride comfort index is an evaluation index for the riding quality of civil aircraft. The US MIL-F-9490D document has the following definitions:

$$J_{RD} = kC_{L\alpha}/(W/S) \tag{15}$$

where (W/S) is the load on the wing, $C_{L\alpha}$ is the derivative of wing lift to angle of attack α, k is the proportional constant. When the comfort index is less than 0.1, the passengers feel comfortable or the workload of the crew is relaxed; When the comfort index is greater than 0.28, in order to reduce the impact of gust disturbance on aircraft motion, the pilot must control the aircraft to change the flight state and track of the aircraft.

For the large civil aircraft, criterion C is selected as the riding quality evaluation index in this paper, and the evaluation is as follows.

$$C = 1.15 + 6.8a_z \tag{16}$$

where $a_z$is the root-mean-square value of the normal acceleration at the passenger's place caused by the gust. The smaller index C, the better ride quality. Generally, if $C<3$, the ride quality is comfortable or the crew's workload is light; $3<C<4$, it is medium; $C\geq4$, it means the ride quality is uncomfortable.

4. Simulation verification

Based on the principle of linear small disturbance, taking a typical aircraft as an example, given the flight state: H = 6000m, Ma = 0.55, perform trim linearization to obtain the linear equation of the aircraft at this state point. The wingspan of the aircraft is 50.5m. The gust chooses the Dryden model and defines L_w = 533.4m. The specific value of σ_w is given by the curve in the relevant line standard. In the simulation, the gust disturbance selects the medium turbulence form (maximum speed 12m/s) .

The control surface model should consider the characteristics, position limiting and velocity limiting of the actuator. The deflection range of the elevator δ_e is -29°～15°, and the deflection range of the aileron is -25°～10°. The 0° position of the multi-function spoiler is pre-biased at 25°, which can produce positive and negative lift changes, So that the effect of the spoiler deflection range is ±25°. The deflection rate of the aileron and elevator shall not be less than 100deg/s, and the deflection rate of the spoiler shall not be less than 200deg/s.

Figure 3 is the normal overload response without GLA function under gust disturbance, Figure 4 is the normal overload response with GLA function; Figure 5 shows the riding quality contrast curve between without GLA function, and with GLA function.

Figure 3 and Figure 5 show that under gust disturbance, the normal overload N_z of the control system without GLA function is 0.425g, corresponding to the ride quality index C_{max}=4.21, and the ride quality is already uncomfortable. Figure 4 and Figure 5 show that the control system with GLA function, in the event of a gust disturbance, uses the rapid deflection of the aileron and multi-function spoiler to offset the effect of the gust, and improves the short cycle by the deflection of the elevator. The dynamic characteristics of the aircraft enhance the stability of the aircraft. In the end, the maximum normal overload N_z is only 0.259g, corresponding to the ride quality index C_{max}=2.81, the ride quality is comfortable, and the control effect is remarkable.

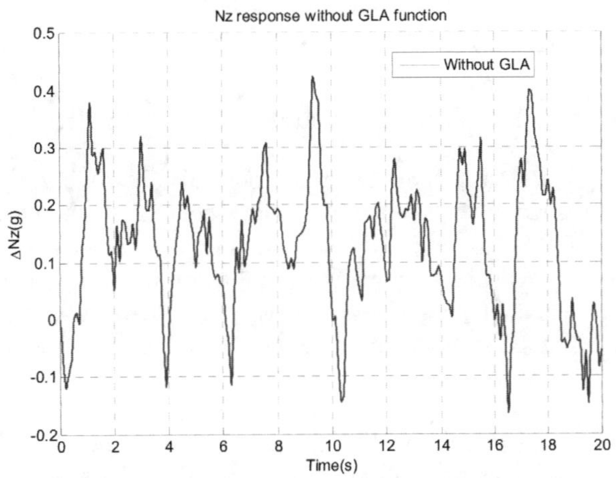

Figure 3. Nz response without GLA function

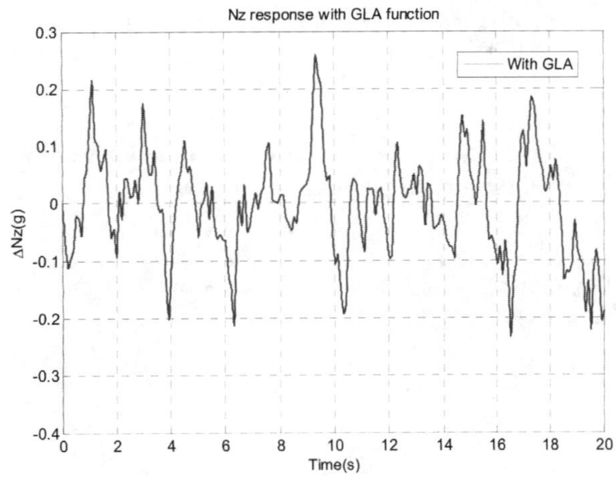

Figure 4. Nz response with GLA function

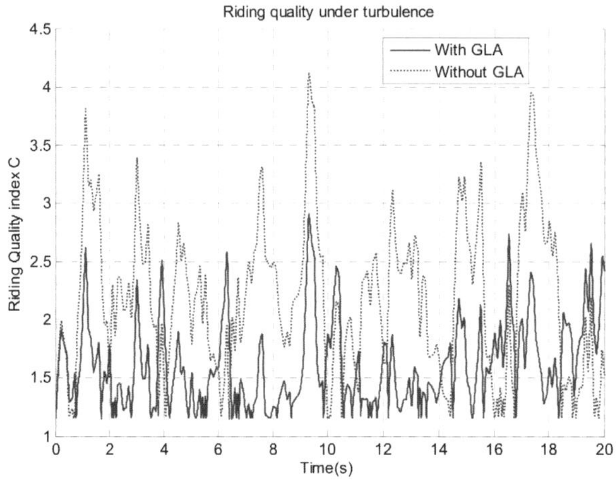

Figure 5. Riding quality under turbulence (with GLA and without GLA)

Table.2 Gust load alleviation control effect

GLA Function	Maximum normal load root mean square	The ride quality index C	Ride quality	Alleviation rate
With	0.259g	2.81	Comfortable	33.3%
Without	0.425g	4.21	Uncomfortable	

Table 2 shows that under the gust disturbance, the control effect without load alleviation control design is not good, and it already feels uncomfortable under medium gusts. The load alleviation control system based on UD-PSO optimization designed in this paper can effectively alleviate the undesired load increment under gust disturbance, the alleviation rate can reach 33.3%, the ride quality under gusts has been improved and the comfortable index is maintained throughout the whole process. This shows the effectiveness of the load alleviation control system based on UD-PSO designed.

5. Conclusions

To solve the problem of passenger's ride quality decreased when large civil aircraft flying in vertical gust turbulence, a load alleviation control system based on UD-PSO optimization designed method is proposed in this paper. The simulation results show that, under the medium turbulence interference, the application of the design method proposed in this paper can significantly reduce the dynamic disturbance under gusts, effectively ensure the comfort of passengers, improve the ride quality, and achieve good control effects. This design method is suitable for engineering applications and can provide a reference for the design and parameter optimization of the active control system of civil aircraft. In addition, this paper only studies a load alleviation control method for vertical gust disturbances, the next study will be focused on the crosswinds load alleviation control method to improve the ride quality during the entire flight phase.

Acknowledgments

This study was supported by the scientific research project of Science and Technology Commission of Shanghai Municipality (No. 18XD1423400).

References

[1] Qing X.X. (2020) Research on gust load alleviation technology of civil aircraft based on feedforward control [D]. TianJin: Civil Aviation University of China, pp. 3-18.

[2] Zhang W.G, Li A.J, Li G.W, Liu X.X. (2009) Modern Flight Control System Design[M]. Xi'an: NWPU Press.

[3] Moulin B, Karpel M. (2007) Gust Loads Alleviation Using Special Control Surfaces[J]. Journal of Aircraft, 44(1): 17-25.
[4] Real T.R. (1993) Digital Simulation of Atmospheric Turbulence for Dryden and von Karman Models[J]. Journal of Guidance, Control, and Dynamics. Vol. 16, No. 1, January-February.
[5] M. G. SKARPETIS.v(1995) Disturbance Rejection for an Aircraft Flying in Atmospheric Disturbances[J]. Dynamics and Control.
[6] Xu D. (2018) Dynamic Gust Loads Analysis for Flexible Aircraft with Nonlinear Control Effects[D]. TianJin: Civil Aviation University of China, pp. 7-24.
[7] Kennedy J, Eberhart RC. (1995) Particle swarm optimization[C]. Proceeding of IEEE International Conference on Neural Networks, Perth, Australia, pp.1942-1948.
[8] Fang K.T, Ma C.X. (2001) Orthogonal and uniform experimental design[M]. Beijing: Science Press, pp.89-105.
[9] Liu B.N, Zhang W.G, Li G.W, Liu X.X. (2011) Improved particle swarm optimization and its application in the flight control system [J]. Flight Dynamics.

Identification of voltage sag source based on Improved GRA

Ma Li[1a*], **Pengyuan Dong**[1b*]

[1] School of Electrical and Control Engineering, Xi'an University of Science and Technology, Xi'an 710054, China

[a*]710849937@qq.com, [b*]19206204047@stu.xust.edu.cn

Abstract—Accurate identification of fault source causing voltage sag is of great significance to the governance and further research of voltage sag. In this paper, a voltage sag source identification method based on improved grey correlation analysis (GRA) is proposed. In view of the deficiency that the traditional GRA only considers the similarity, the proposed method considers the geometric area and slope of sequence into the calculation of correlation coefficient, so that the improved GRA has attributes of both sequence similarity and data closeness. Eight features are extracted from six common types of sag waveforms to form standard and test sequence; The improved GRA is used to calculate the correlation degree between the test and standard sequences, then the identification of sag source can be realized. Simulation model is built in MATLAB. Results show that this proposed method can effectively identify the voltage sag source. Compared with the traditional GRA, this method has higher identification accuracy and stronger identification ability.

1. Introduction

Voltage sag is a kind of power quality phenomenon caused by short-duration voltage disturbance [1], which can be caused by either grid side or user side. The causes from grid side include short-circuit fault, lightning strike, switch operation, switching of transformer and capacitor bank, etc. And the main cause from user side is the start of impact loads [2]. When voltage sag occurs, sensitive equipment is easily affected, resulting in huge economic losses. The purpose of voltage sag source identification is to find out the specific type of fault causing voltage sag, which is helpful to the mitigation of voltage sag and the determination of accident responsibility.

In recent years, scholars have done a lot of research on identification of voltage sag source. Various signal processing methods are used to extract the different features of sag waveform, and then the identification is realized mainly by statistical methods or machine learning. In [3], S-transform and multi-dimensional fractal constitute the feature indices of voltage sag signal. Support vector machine (SVM) is used to realize the classification and identification of voltage sag disturbance sources. In [4], Hillbert-Huang transform (HHT) is used to extract features, and decision tree is constructed to adaptively determine the threshold value of features to identify four types of sag sources including lightning strike. In [5], Mahalanobis distance is introduced and the distance distinguish analysis is applied to sag source identification. In [6], the traditional GRA and entropy weight method are combined to identify sag source. GRA is an effective method to analyze uncertain system with small amounts of data and information, which is widely used in various fields, such as GIS quality evaluation, short-term load forecasting and transformer condition evaluation [7,8,9,10]. However, the traditional analysis model based on Deng's correlation degree only considers the sequence similarity, which is not perfectly suitable in the field of sag source identification [11].

In view of the defects of traditional GRA, this paper proposes an improved GRA by considering the geometric area and slope of two discrete sequences into the calculation of correlation coefficient. Then, six common types of sag sources are simulated in MATLAB, eight kinds of features are extracted to form the test sequence and standard sequence. By using the improved GRA, the maximum correlation degree is selected, and the sag identification is realized. Confusion matrix and statistical evaluation indices are introduced to evaluate the identification result. The analysis shows that the proposed method has the advantages of higher identification accuracy and stronger identification ability for single voltage sag source.

2. Improved GRA

2.1. Traditional GRA and similarity correlation coefficient

As an important part of grey system theory, the principle of GRA is selecting different evaluation indices to construct standard sequence and test sequence, calculating the correlation degree which is also the similarity level of geometric shapes between the two sequences; if the correlation degree is the largest, then the corresponding test sequence is optimal [10]. Expressions of the correlation coefficient and Deng's correlation degree are shown in (1) and (2).

$$\xi_1(k)=\frac{\min\limits_{k}\left|x(k)-y(k)\right|+\varepsilon\max\limits_{k}\left|x(k)-y(k)\right|}{\left|x(k)-y(k)\right|+\varepsilon\max\limits_{k}\left|x(k)-y(k)\right|} \tag{1}$$

$$r_1=\sum_{k=1}^{n}\frac{1}{n}\xi_1(k) \tag{2}$$

In (1), ε is a factor that directly affects the resolution of GRA, which is between [0,1]. Neither too large or too small ε can accurately reflect the correlation between sequences, so it is usually set 0.5 [12]. $\xi_1(k)$ is correlation coefficient, it is also called similarity correlation coefficient in this paper. In (2), n is sequence length; r_1 is correlation degree, its value is between (0,1]. For a fixed ε, the larger the r_1, the stronger the correlation between the two sequences [11].

In fact, there are obvious defects using the above correlation degree to analyze and make decisions. Because traditional GRA essentially reflects the similarity of two sequences only from the aspect of the absolute difference $\left|x(k)-y(k)\right|$, the closeness of sequence data is ignored. This closeness belongs to the error range for simulation [13]. In addition, there are two test sequences with different trends, but their correlation degree with the standard sequence may be the same. Both defects seriously affect the analysis and decision-making. Therefore, it is necessary to improve the existing GRA.

2.2. Traditional GRA and similarity correlation coefficient

For test sequence $x(k)$ and standard sequence $y(k)$, the closeness correlation coefficient $\xi_2(k)$ is defined in (3), where a is the envelope area of the two sequences at [k, k+1]. Δt is sampling time.

$$\xi_2(k)=e^{-|a|} \tag{3}$$

$$a=\frac{\left|x(k)-y(k)\right|+\left|x(k+1)-y(k+1)\right|}{2}\Delta t \tag{4}$$

Slope correlation coefficient $\xi_3(k)$ is defined in (5), where s is the absolute slope difference.

$$\xi_3(k)=e^{-|s|} \tag{5}$$

$$s=\left|\frac{x(k+1)-x(k)}{\Delta T}-\frac{y(k+1)-y(k)}{\Delta T}\right| \tag{6}$$

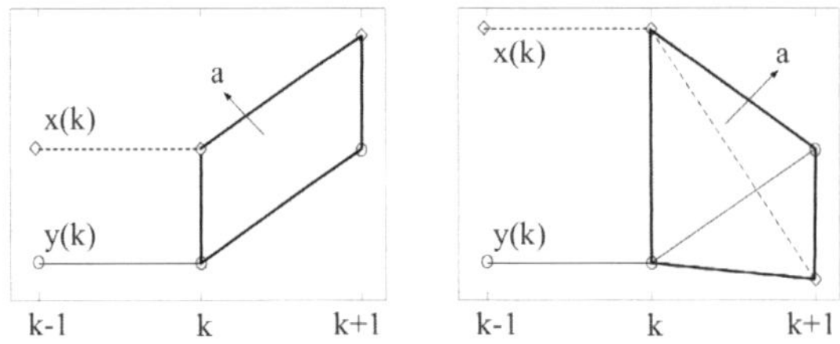

(a) Positive correlation (b) Negative correlation

Fig.1 Envelope area

To sum up, the improved correlation degree r is shown in (7). $\alpha_1, \alpha_2, \alpha_3$ are weight of similarity, closeness and slope correlation coefficient respectively. The sum of them is 1. The specific value of weight can be set according to the actual sequence data and application scenarios. In this paper, they are all equally set $1/3$ for convenience. However, more proper weight makes the analysis model more suitable for data samples with different characteristics, which deserves attention.

$$r = \sum_{k=1}^{n} \frac{\alpha_1 \xi_1(k) + \alpha_2 \xi_2(k) + \alpha_3 \xi_3(k)}{n} \tag{7}$$

3. Identification of voltage sag source and evaluation indices

3.1. *Identification of voltage sag source*

Definition of voltage sag as per IEEE Std 1159 is a decrease in root mean square (RMS) voltage between 0.1 and 0.9 p.u. and a time duration from 0.5 cycles to 1 min [1]. Voltage sag can occur by various events, such as short-circuit fault, induction motor start, transformer operation, capacitor bank switching, lightning strike and switch operation in the transmission and distribution system [14]. Six types of voltage sag sources studied in this paper include single-phase-to-ground short-circuit(C1), phase-to-phase short-circuit(C2), double-phase-to-ground short-circuit(C3), three-phase short-circuit(C4), saturated transformer start(C5) and induction motor start(C6). Typical waves of these six types of voltages sag are shown in Fig.2.

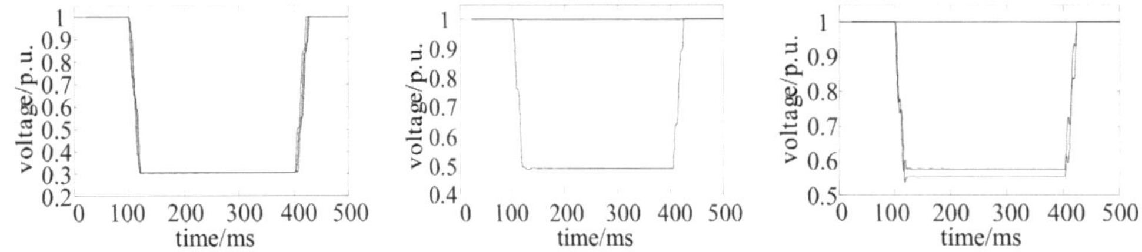

(a) Three-phase short-circuit (b) Single-phase-to-ground Short-circuit (c) Phase-phase short-circuit

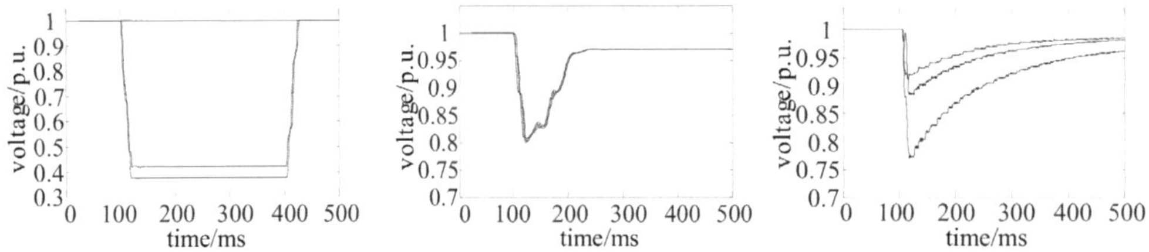

(d) Double phase-to-ground Short-circuit (e) Induction motor start (f) Saturated transformer start

Fig. 2 Six common types of sag waveforms

Eight features selected in [6] are used to reflect the waveform difference in this paper. They are Mean Value, Variance, RMS Value, Skewness, Kurtosis, Crest Factor, Form Factor and Logarithmic Energy.

The steps of voltage sag source identification are as below:

a. Obtain sag waveforms;
b. Standardize voltage value into per unit value;
c. Extract multiple features from known types of voltage sag waveforms;
d. Form test sequence x, standard sequence y_{Ci} (Ci is the type of sag source; n is the number of features);
e. Calculate correlation degree between x and y_{Ci} using (7);
f. Sort the correlation degree.

$$x=[x(1) \quad x(2) \quad \cdots \quad x(n)] \tag{8}$$

$$y_{Ci}=[y_{Ci}(1) \quad y_{Ci}(2) \quad \cdots \quad y_{Ci}(n)] \tag{9}$$

x belongs to Ci-type which the correlation degree between x and y_{Ci} is the largest.

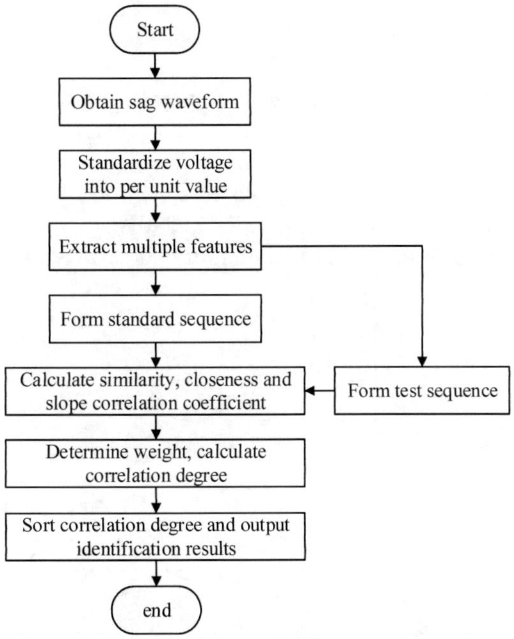

Fig.3. Flowchart of voltage sag source identification

3.2. Evaluation index for identification performance

Confusion matrix is introduced to evaluate the identification performance, as shown in Tab.1. d_{ij} is the number of $Ci\text{-}type$ sag source identified as $Cj-type$. $i,j=1,\cdots,6$.

Tab.1 Confusion matrix of voltage sag source identification

Actual	Identification result					
	C1	C2	C3	C4	C5	C6
C1	d_{11}	d_{12}	d_{13}	d_{14}	d_{15}	d_{16}
C2	d_{21}	d_{22}	d_{23}	d_{24}	d_{25}	d_{26}
C3	d_{31}	d_{32}	d_{33}	d_{34}	d_{35}	d_{36}
C4	d_{41}	d_{42}	d_{43}	d_{44}	d_{45}	d_{46}
C5	d_{51}	d_{52}	d_{53}	d_{54}	d_{55}	d_{56}
C6	d_{61}	d_{62}	d_{63}	d_{64}	d_{65}	d_{66}

$$P_{accuracy} = \frac{d_{11} + d_{22} + d_{33} + d_{44} + d_{55} + d_{66}}{\sum_{i=1}^{6}\sum_{j=1}^{6} d_{ij}} \tag{10}$$

$P_{accuracy}$ indicates the proportion of correctly identified test sequences in all test sequences. The larger the value is, the higher identification accuracy is.

$$P_{sensitive}^{Ci} = \frac{d_{ii}}{\sum_{j=1}^{6} d_{ij}} \tag{11}$$

$P_{sensitive}^{Ci}$ indicates the proportion of sequences identified as $Ci\text{-}type$ in all $Ci\text{-}type$ sequences. It is used to indicate the identification ability to $Ci\text{-}type$ sequence. The larger the value is, the stronger the identification ability is.

4. Simulation and analysis

4.1. Simulation model

In order to verify the proposed method in this paper, a voltage sag simulation model is built in MATLAB with the reference of [5] and [6], as shown in Fig.4. In this simulation model, the power supply voltage is 10kV, 50Hz. K1 and K2 are three-phase switches. F means the short-circuit fault. L is a 500kVA resistive inductive load, and M is a 150HP squirrel-cage asynchronous motor. The connection mode and transformation ratio parameters of transformer T1, T2 and T3 are shown in Fig.4, saturation of T1 is considered. Fig.5 is the visualization figures of simulation model in MATLAB.

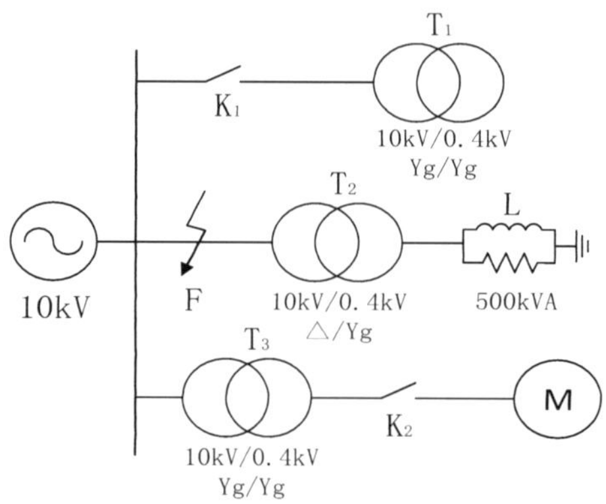

Fig.4 Voltage sag simulation model

Change the short-circuit fault type and short-circuit point location, the nominal power of the transformer and the motor to simulate different sag sources and collect 100 sag waveforms of each type. Extract features from these waveforms and obtain a total of 600 data samples. Divide the data samples in the ratio of 3:2 to form standard and test samples. Use the proposed method to identify the type of test samples.

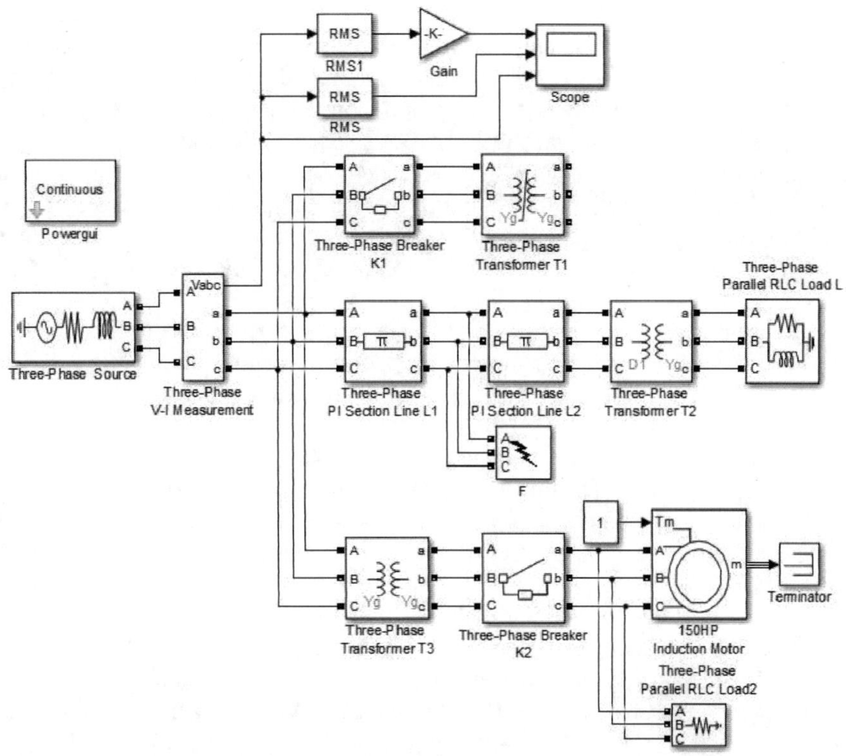

Fig.5 Voltage sag simulation model in MATLAB

4.2. *Result and analysis*

Identification results are shown in Tab.2. Based on traditional GRA, most of sag sources can be identified, but misidentification exists to all types of sag sources. Based on improved GRA, C3-type and C5-type can be accurately identified. And the misidentification number of each type is less.

From Tab.3, the overall identification accuracy of improved GRA is higher than that of traditional GRA. The identification ability to C2-type and C6-type maintains the same based on both methods. For other types, the identification ability based on improved GRA is stronger, which gets to 100% when it comes to C3-type and C5-type. While it is no more than 95% based on traditional GRA.

Tab.2 Confusion matrix of identification result

Actual	Identification result (GRA/Improved GRA)					
	C1	C2	C3	C4	C5	C6
C1	35/38	0/2	5/0	0/0	0/0	0/0
C2	0/0	38/38	0/0	0/2	2/0	0/0
C3	2/0	0/0	38/40	0/0	0/0	0/0
C4	0/0	5/2	0/0	35/38	0/0	0/0
C5	0/0	0/0	0/0	0/0	36/40	4/0
C6	0/0	4/4	0/0	0/0	0/0	36/36

Tab.3 Identification result evaluation

Method	$P_{accuracy}$ / %	$P^{Ci}_{sensitive}$ / %					
		i=1	i=2	i=3	i=4	i=5	i=6
GRA	90.83	87.5	95	95	87.5	90	90
Improved GRA	95.83	95	95	100	95	100	90

5. Conclusion

Traditional analysis model based on Deng's correlation degree only considers the sequence similarity, which is not perfectly suitable in the field of voltage sag source identification. In this paper, an improved GRA is proposed, which takes the geometric area and slope of sequence into account to calculate the correlation coefficient, hence it has attributes of both sequence similarity and data closeness. Six common sag sources are simulated in MATLAB and identified by the proposed method. Identification results are evaluated. The results show that the improved GRA has the advantages of higher accuracy and stronger ability for single voltage sag source identification. How to apply the proposed method to complex sag sources identification will be the direction of future work.

References

[1] IEEE Std 1159^{TM}-2019, IEEE Recommended Practice for Monitoring Electric Power Quality.
[2] He Juan, Liu Xiaobo, et al. Cluster analysis of factors affecting voltage sag[J]. The Journal of New Industrialization, 2020, 48(19): 9-16.
[3] Yang Xiu, Zhang Tongyao, et al. Voltage sag source feature identification with S transform and multidimensional fractal [J]. Power System Technology, 2021, 45(2): 672-679.
[4] Li Xin, Wu Lihui, Fan Xinming, et al. Voltage sag sources identification method based on Hilbert-Huang transform and decision tree[J]. Science Technology and Engineering, 2019, 19(34): 198-205.
[5] Wang Ying, Wang Huan, et al. Identification method of voltage sag source based on distance discriminant analysis[J]. Power System Protection and Control, 2020, 48(19): 9-16.
[6] Wang Ying, Wang Huan, Wang Xin. A method of voltage sag source identification based on improved grey relational analysis[J]. Electrical Measurement & Instrumentation, 2020, 57(15): 1-7.
[7] Hao Yanpeng, Zhang Lei, et al. GIS quality evaluation based on improved grey relational analysis[J]. Electric Power Automation Equipment, 2017, 37(7): 161-164.
[8] Wu Yun, Lei Jianwen, et al. Short-term load forecasting based on improved grey relational analysis and neural network optimized by bat algorithm[J]. Automation of Electric Power System, 2018, 42(20): 67-72.
[9] Liu Jun, Zhao Liqun, et al. State evaluation method for power transformer based on the TOPSIS and grey relational analysis[J]. Journal of Electric Power Science and Technology, 2019, 34(4): 63-68.
[10] Hu Yuwei, Ma Ping, et al. A comprehensive consistency test method based on improved grey relational analysis for simulation results[J]. Transactions of Beijing Institute of Technology, 2013, 33(7): 711-715.
[11] Meng Xiaocheng, Che Renfei, Gao Shi. Outage rate estimation for transmission lines based on improved gray relational analysis model[J]. Proceedings of the CSU-EPSA, 2019, 31(11): 1-6.
[12] Wang Linjun, Luo Bin, et al. Influence factors of wind load coefficient of dish solar concentrator based on fuzzy grey correlation analysis[J]. Acta Energiae Solaris Sinica, 2021, 42(5): 328-333.
[13] Wu Jing, Wu Xiaoyan, Chen Yongxing. Validation of simulation models based on improved grey relational analysis[J]. Systems Engineering and Electronics, 2010, 32(8): 1677-1679.
[14] Bollen M H J. Algorithms for characterizing measured three-phase unbalanced voltage dips[J]. IEEE Transactions on Power Delivery, 2003, 18(3): 937-944.

Research on Process Control Strategy of Asphalt Tank Inventory Based on Dynamic Programming Algorithm in Post Epidemic Era

Dong Pang

Department of Business Administration, Dalian Polytechnic University, Dalian 116033, China

857054235@qq.com

Abstract—Based on the background of post epidemic era, starting with the traditional industrial inventory process control, this paper discusses the dynamic inventory process control strategy under the asphalt tank inventory mode, and analyzes the intelligent inventory mode under the trend of intelligent logistics through the dynamic programming method, so as to provide reference for the development direction of intelligent inventory of unconventional storage materials such as chemical industry and energy. At the same time, with the recurrence of short-term epidemic and the dilemma of low return to work and production rate, through the integration and interconnection of mobile communication, intelligent analysis, digital management and other technologies with all links, the strategies of inventory dynamic control, inventory risk control, improving inventory efficiency and optimizing inventory cost are put forward.

1. Introduction

The interaction between the transformation of the digital age and the post epidemic era makes the digital transformation of traditional industrial production enterprises more and more urgent. In the face of the dilemma between the rigor of epidemic control and the low rate of resumption of work and production in stages, the natural physical barrier of the digital economy ensures that during the epidemic prevention and control period, work is stopped without production, and humans and machines work together. In the intelligent industrial production scenario, the industrial production process covers the integration and interconnection of mobile communication, intelligent analysis, digital management and other technologies and various links, which greatly simplifies the industrial production process, improves production efficiency, optimizes production strategies and costs, and integrates industrial. The concept of production is taken to a new level. [1]. In traditional industrial production, the inventory of asphalt raw materials is affected by the construction period demand, and the storage container is a tank-type storage tank. During the heating process of storage, there will be corresponding fuel loss and increase the inventory cost. Therefore, it is necessary to effectively control the inventory process of asphalt raw materials in different production periods, not only to avoid inventory waste caused by excessive tank inventory pressure, but also to ensure sufficient production supply. Under the background of the post-epidemic era, cost management and risk planning combined with intelligent data analysis are the core and key processes of inventory process control.

2. Related Work

2.1 Classical Control Theory

The classical control theory was initially formed in the 1950s, and the frequency response method and the root locus method were applied to study sampling control systems and simple nonlinear control systems [2]. One input variable, one output variable and a univariate steady system whose parameters do not change with time are taken as the research object, the Laplace transform is used as the mathematical basis, and the transfer function between the input and output is used as the mathematical model of the system [3]. Classical control theory mainly solves relatively simple control system analysis and design problems, but the research object is mainly limited to single-variable, steady-state systems.

2.2 Modern Control Theory

Modern control theory was initially formed in the 1960s and developed on the basis of classical control theory. The research objects include single-variable systems, multi-variable systems, stationary systems and time-varying systems, based on the time domain [3]. Although the research on modern control theory seems to be stagnant, it has been widely penetrated and applied to various branches of industries, produced in large quantities and quickly implemented [4].

2.3 Inventory Cost Control

Inventory cost refers to the cost of goods stored in the warehouse, including purchase fee, order fee and storage fee [5]. Inventory costs are generally mainly composed of the following three costs: inventory holding costs, inventory procurement costs and inventory out-of-stock costs.

a. Inventory holding cost: It refers to the cost related to the quantity of inventory, which is composed of many different parts, that is, the related cost incurred by the enterprise holding the inventory.

b. Production preparation cost: Refers to the cost of issuing purchase orders to external suppliers or to internal production preparation costs.

c. Inventory shortage cost: Refers to the losses caused by the interruption of inventory supply, including the loss of work stoppage caused by the interruption of material supply, the loss of delayed delivery caused by the shortage of finished product inventory, and the loss of lost sales opportunities.

3. Inventory Process and Dynamic Programming

Configure the asphalt production material ratio according to the current total production demand, store the production raw materials in the metering tank and the batching insulation tank, conduct semi-finished product testing, limit the inventory of production raw materials and semi-finished products, and control the process variables such as temperature, humidity, viscosity and so on. The oil furnace equipment is driven into the asphalt storage tank at high temperature to complete the tank storage. When there are no raw materials and finished products in the warehouse, the storage tank maintains a low-consumption operation state; when there are raw materials and finished products in the warehouse, the high temperature is heated to the standard warehousing temperature. Asphalt inventory process control is shown in Figure. 1 below:

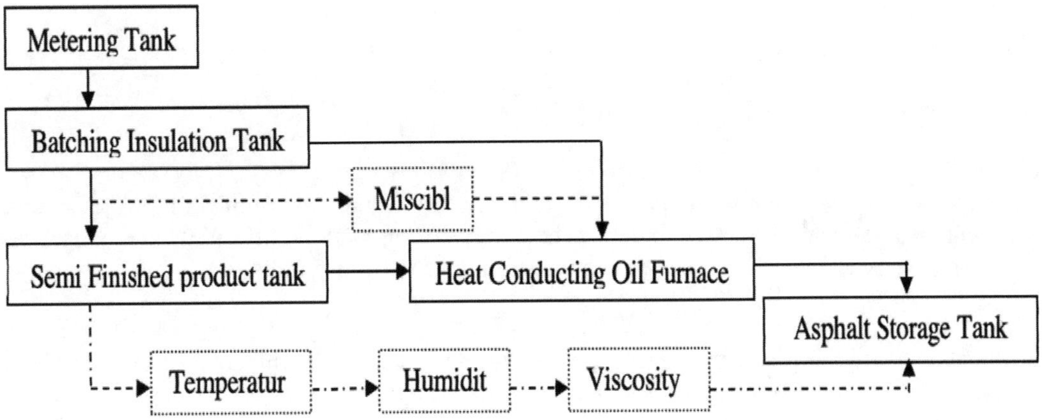

Figure.1 Flow Chart of Asphalt Inventory

3.1 Model Description

Let H denote the storage capacity limit. Assuming that the construction period is divided into stages, k is the stage number, which means the ordinal number of the week; x_k is the decision variable, which represents the amount of asphalt raw material purchased in each stage; s_k is the state variable, representing the inventory of asphalt raw materials (before the asphalt raw materials are released from the warehouse at this stage, and after the asphalt raw materials are stored in the previous stage); d_k indicates the demand for asphalt raw material production at each stage; α_k indicates the consumption of each ton of asphalt raw materials stored in each stage Unit heat storage cost.

Suppose the state variable s_k represents the beginning of the stage k, which is the asphalt raw material at this stage is not out of the warehouse and is before production; the previous stage asphalt raw material has been purchased and put into the warehouse.

The decision variable x_k represents the amount of asphalt raw materials purchased in the stage k. State transition equation:

$$s_{k+1} = s_k + x_k - d_k\,, \quad (k = 0,\ 1 \dots n) \tag{1}$$

$$d_k \le s_k \le H \tag{2}$$

$$D_k(s_k) = \{s_k; s_k \ge 0, d_{k+1} \le s_k + x_k - d_k \le H\} \tag{3}$$

Let the optimal value function $f_k(s_k)$ denote the minimum inventory cost from the stage k to the last stage when the stock quantity from the stage k is s_k. Therefore, under the above constraints, the recurrence relation of dynamic programming can be listed as:

$$f_k(s_k) = \min_{x_k \in D_k(s_k)} [\alpha_k x_k + f_{k+1}(s_k + x_k - d_k)],\ k = 0,1,\dots,n \tag{4}$$

$$f_{n+1}(s_{n+1}) = 0 \tag{5}$$

3.2 Model Solving

In the reverse equation of the dynamic programming equation established above, the stage k (k=0, 1,..., n) is expressed as stage n, and the inverse method is used to reverse inversely from the last stage k=n. The recurrence process is as follows:

When stage k=n, due to the condition of the target problem, it is assumed that the ending inventory is 0, which is 0. As of the last stage, raw materials no longer need to be purchased and put into storage.

$$x_n = 0 \tag{6}$$

$$f_n(s_n) = 0 \tag{7}$$

$$s_n = d_n \tag{8}$$

When stage k=n-1， find the optimal solution x_{n-1}^*

$$s_n = s_{n-1} + x_{n-1} - d_{n-1} \tag{9}$$

$$f_{n-1}(s_{n-1}) = \min_{x_{n-1} \in D_{n-1}(s_{n-1})} (\alpha_{n-1} x_{n-1}) \tag{10}$$

x_0^* is the value of the optimal solution at the initial stage of the goal problem at stage 0. Reverse substituting the relational expression to derive the corresponding optimal decision value of each stage, and the objective function value of is the optimal solution of the target problem.

3.3 Model Test

A chemical raw material production enterprise needs to formulate a monthly asphalt inventory plan to control the inventory cost to the greatest extent. It is suggested that the inventory capacity is h = 500, the initial inventory is 78, and the ending inventory is 0. The actual production demand and inventory cost data are shown in Table 1:

Table.1 Production Demand and Unit Inventory Cost

Production Stage	Production Demand (ton)	Unit Inventory Cost (yuan/ton)
0	—	1.26
1	312	0.67
2	330	0.63
3	390	0.54
4	297	—

Suppose the decision variable x_k represents the quantity of asphalt raw materials stored in stage k.

$$s_{k+1} = s_k + x_k - d_k\,, \quad (k = 0, 1, 2, 3, 4) \tag{11}$$

$$D_k(s_k) = \{s_k; s_k \geq 0, d_{k+1} \leq s_k + x_k - d_k \leq H\} \tag{12}$$

$$f_k(s_k) = \min_{x_k \in D_k(s_k)} [\alpha_k x_k + f_{k+1}(s_k + x_k - d_k)],\ k = 0,1,2,3,4 \tag{13}$$

when stage k = 4, 3, 2, 1, 0, Reverse substituting the relational expression to derive the corresponding optimal decision value of each stage , and the objective function value of is the optimal solution of the target problem.

$$\begin{aligned} f_0(s_0) &= \min_{x_0 \in D_0(s_0)} [\alpha_0 x_0 + f_1(s_0 + x_0 - d_0)] \\ &= \min_{x_0}[1.26x_0 + 836.22 - 0.67(s_0 + x_0)] \\ &= \min_{x_0}[0.59x_0 - 0.67s_0 + 836.22] \end{aligned} \tag{14}$$

$$312\text{-}s_0 \leq x_0 \leq 500\text{-}s_0 \tag{15}$$

According to the above recursive calculation sequence, the optimal decision value of each stage is obtained as follows: $x_0^* = 234$, $x_1^* = 312$, $x_2^* = 330$, $x_3^* = 390$.

4. Inventory process control strategy

4.1.Dynamic Inventory Control

The core of inventory process control is to manage the dynamic changes of inventory. It is necessary to enable effective change management processes in time according to changes in demand, and to provide real-time and accurate inventory data for inventory forecasting. According to demand forecast, production data and inventory data, process variables such as temperature variables, humidity variables, and viscosity in the process of asphalt production and inventory are controlled to maximize the security of finished product inventory. Combined with the comprehensive conditions of contribution to the industrial production process, production speed, inventory efficiency, data deviation, etc., the safety stock value is calculated to avoid high inventory expectations caused by uncertain factors, and effectively realize the dynamic management of inventory.

4.2.Inventory Risk Control

To ensure normal production turnover, set safety stocks and the procurement lead time of required raw materials to drive industrial production and supply. At the same time, adjustments can be made at any time according to changes in demand to avoid excess inventory and loss of inventory costs. Although the existence of excess inventory is a reasonable phenomenon, it is necessary to control it within a reasonable range, pay attention to key data such as turnover inventory, safety stock, and excess inventory to avoid the risk of inventory retention and eventually form risk inventory.

5. Conclusion and significance

In the traditional industrial production industry, the optimization and stable control of the production process and inventory are the prerequisites to ensure that enterprises reduce operating costs, improve production efficiency, and build core competitiveness. The purpose of this research is to take tank storage asphalt as the research object, control the inventory process by applying dynamic programming algorithm, explore its stability in the process of high temperature thermal storage, and control the variables in the inventory process to realize real-time inventory dynamic control and control. Inventory risk control ensures the maximum security of inventory, which in turn helps enterprises reduce inventory cost losses. The disadvantage of this research is that it focuses more on the theoretical level, and the model establishment is relatively simple. The future research direction can focus on the algorithm control of the complex inventory process and provide a deeper design idea for the automation system of industrial production.

References

[1] Qin, W., Shiyan, W. (2020) The Motivation and Prospects of ASEAN Countries Implementing the 'Industry 4.0' Strategy. Asian-Pacific Economic. pp. 36-43+150.

[2] Brogan, L. (1995) Classical Control Theory Applied to Potentially Chaotic Systems. Computer Applications in Engineering Education. pp. 111-120.

[3] Chuanbo, W., Min, L. (2006) Comparative Study of Modern Control Theory and Classical Control Theory. Machinery Management Development. pp. 6-8.

[4] Ramirez, A., Giral, A., Holguin, M. (2021) Physics and Differential Geometry in Modern Control Theory Applied to A Synchronous Generator. Journal Of Physics: Conference Series, 2046(1): 012025.

[5] Phillips, E., White, R. (1981) Minimizing Inventory Cost. Interfaces. pp. 42-47.

Rapid Reduction Dimensionality Transfer Alignment Method Based on CKF for SINS of Ship-borne Aircraft with GNSS Assistance

Yongjun Wang[1a*], Jingshuo Xu[2b] and Fengjuan Gao[2c]

[1] Yantai Institute of Science and Technology, Yantai, Shandong, 265600, China.

[2] Qingdao Campus of Naval Aviation University, Qingdao, Shandong, 266041, China.

[a]wang94095@sina.com,[b] xujingshuo@sina.com, [c]27638397@qq.com

[*]Corresponding author's e-mail: wang94095@sina.com

Abstract. Aiming at the problem of rapid transfer alignment of inertial navigation system in shipborne environment with large azimuth misalignment angle, a dimension reduction fast transfer alignment scheme of shipborne aircraft inertial navigation system based on CKF algorithm is designed. Firstly, the nonlinear attitude error equation, velocity error equation and position error equation of the Global Navigation Satellite System (GNSS) assisted strapdown inertial navigation system in the case of large azimuth misalignment are analyzed. Secondly, the nonlinear dimensionality reduction rapid transfer alignment state equation and measurement equation are derived. Finally, the CKF algorithm is simulated and verified, and compared with EKF algorithm and UKF algorithm. The simulation results show that the reduced dimension fast transfer alignment method of shipborne aircraft inertial navigation system based on CKF algorithm can complete the estimation of attitude error angle in 80s, with horizontal estimation error of 1 ' and azimuth error of 4', and the estimation accuracy is equivalent to EKF and UKF, with less calculation and better rapidity.

1. Introduction

The layout of the parking lot on the ship deck determines to a great extent the problem of large azimuth angle between the frame coordinates of carrier-borne aircraft and ship coordinate system. Theoretically, according to the precise coordinates of the docking station relative to the mothership center, inertial navigation data from the mothership center can be converted to the virtual inertial navigation system of the docking station coordinate system through coordinate transformation, so that the main inertial navigation system (virtual inertial navigation system) and the sub-inertial navigation system (carrier-borne aircraft inertial navigation system) can meet the small angle condition, which can use Kalman filter to complete the estimate of the misalignment angle. However, large azimuth misalignment angle may exist in the carrier-borne aircraft body coordinate system and the aircraft parking coordinate system due to the deformation of the hull and the aircraft parking position error.

To solve the problem of large azimuth misalignment, nonlinear filtering methods such as EKF [1] and UKF [2-3] are widely used at home and abroad. However, there are problems of complicated Jacobian matrix calculation and large amount of deterministic sampling calculation respectively. A filtering method with precision equal to or higher than UKF and computation less than UKF is sought, which can effectively shorten the alignment time, reduce computation and improve the alignment speed. CKF [4-6] filtering algorithm based on Cubature transform has been widely used in the fields of state estimation and target location due to its high filtering accuracy and no need to calculate Jacobian matrix. Based on this, in the case of large azimuth misalignment angle, CKF algorithm is proposed to be used in the study of carrier-borne aircraft strapdown inertial navigation system, aiming to achieve rapid transfer alignment of carrier-borne aircraft inertial navigation system, and compared with the traditional EKF algorithm and UKF algorithm, providing reference for engineering applications.

2. Establish system model

2.1. Nonlinear error model of SINS with large azimuth misalignment angle

The related frames are defined as follows:

(1) Carrier frame (b): Three axes point to "right- front –above" along the carrier.

(2) Navigation frame (n): Three axes point to the east-north-up of the geographic frame.

Because the euler attitude error angle is not small, it is rotational noncommutative. In other words, there is rotation exchange error between the ideal navigation coordinate system n and the calculated navigation coordinate system n′.

According to the relation between quaternion and direction cosine matrix and assuming that horizontal attitude angle is small, the coordinate transformation matrix from n system to n′ system is

$$\boldsymbol{C}_n^{n'} = \begin{bmatrix} \cos\phi_z & \sin\phi_z & -\phi_y \\ -\sin\phi_z & \cos\phi_z & \phi_x \\ \phi_y\cos\phi_z + \phi_x\sin\phi_z & \phi_y\sin\phi_z - \phi_x\cos\phi_z & 1 \end{bmatrix} \tag{1}$$

The differential equation of attitude error angle under the condition of large azimuth misalignment angle is

$$\dot{\boldsymbol{\phi}} = \boldsymbol{C}_{\phi_\psi}^{-1}\boldsymbol{\omega}_{nn'}^{n'} = \begin{bmatrix} 1 & 0 & -\phi_y \\ 0 & 1 & \phi_x \\ \phi_y & 0 & 1 \end{bmatrix}^{-1} \boldsymbol{\omega}_{nn'}^{n'} \tag{2}$$

The nonlinear attitude error equation and nonlinear velocity error equation [3] of the strapdown inertial navigation system are respectively:

$$\dot{\boldsymbol{\phi}} = \boldsymbol{C}_{\phi_\psi}^{-1}[(\boldsymbol{I} - \boldsymbol{C}_n^{n'})\hat{\boldsymbol{\omega}}_{in}^n + \delta\boldsymbol{\omega}_{in}^n - \boldsymbol{C}_b^{n'}\boldsymbol{\varepsilon}_b^b - \boldsymbol{C}_b^{n'}\boldsymbol{\varepsilon}_w^b] \tag{3}$$

$$\delta\dot{\boldsymbol{V}}^n = (\boldsymbol{I} - \boldsymbol{C}_n^{n'})\boldsymbol{C}_b^{n'}\hat{\boldsymbol{f}}^b - \left(2\hat{\boldsymbol{\omega}}_{ie}^n + \hat{\boldsymbol{\omega}}_{en}^n\right)\times\delta\hat{\boldsymbol{V}}^n - \left(2\delta\boldsymbol{\omega}_{ie}^n + \delta\boldsymbol{\omega}_{en}^n\right)\times\hat{\boldsymbol{V}}^n + \boldsymbol{C}_b^{n'}\nabla_b^b + \boldsymbol{C}_b^{n'}\nabla_w^b \tag{4}$$

2.2. Position error model of SINS

The Position error equation [7] of the strapdown inertial navigation system is

$$\begin{cases} \delta\dot{L} = \dfrac{1}{R_M + h}\delta v_N - \dfrac{v_N}{\left(R_M + h\right)^2}\delta h \\ \delta\dot{\lambda} = \dfrac{\sec L}{R_N + h}\delta v_E + \dfrac{v_N \sec L \tan L}{R_N + h}\delta L - \dfrac{v_E \sec L}{\left(R_N + h\right)^2}\delta h \\ \delta\dot{h} = \delta v_U \end{cases} \tag{5}$$

2.3. Simplified dimension reduction for nonlinear rapid alignment state models

Considering that the alignment time is short, the estimation of gyro bias and accelerometer bias cannot be completed in a short time. In addition, the values of gyro bias and accelerometer bias are small, which has little impact on alignment, and the core of alignment is attitude misalignment angle estimation. In order to reduce the amount of calculation and improve the speed of operation, it is not necessary to include them in the state when constructing the mathematical model.

Define

$$X = \begin{bmatrix} \varphi^{n,\mathrm{T}} & \delta V^{n,\mathrm{T}} & \delta P^{\mathrm{T}} \end{bmatrix}^{\mathrm{T}} \tag{6}$$

According to equations (3) to (5) , the state space model of nonlinear dimension reduction rapid alignment system can be written as:

$$\dot{X} = f(X) + W \tag{7}$$

Where

$$f(X) = \begin{bmatrix} C_{\phi_\psi}^{-1}[(I - C_n^{n'})\hat{\omega}_{in}^n + \delta\omega_{in}^n - C_b^{n'}\varepsilon_w^b] \\ (I - C_n^{n'})C_b^{n'}\hat{f}^b - (2\hat{\omega}_{ie}^n + \hat{\omega}_{en}^n)\times\delta\hat{V}^n - (2\delta\omega_{ie}^n + \delta\omega_{en}^n)\times\hat{V}^n + C_b^{n'}\nabla_w^b \\ \dfrac{1}{R_N + h}\delta V_N - \dfrac{V_N}{(R_N + h)^2}\delta h \\ \dfrac{\sec L}{R_M + h}\delta V_E + \dfrac{V_E \sec L \tan L}{(R_M + h)^2}\delta L - \dfrac{V_E \sec L}{(R_M + h)^2}\delta h \\ \delta V_U \end{bmatrix} \tag{8}$$

and $W = \begin{bmatrix} C_b^{n'}\varepsilon_w^b & C_b^{n'}\nabla_w^b & 0_{6\times1} \end{bmatrix}^T$ is system noise.

2.4. Establish measurement equation

Using satellite navigation information such as GNSS for inertial navigation alignment of shipborne aircraft is one of the common methods. This paper intends to use GNSS as the reference information of auxiliary alignment.

The difference between the speed and position information output by the shipborne aircraft inertial navigation system and that of the GNSS system is selected as the measured value, so

$$Z = \begin{bmatrix} V_{SINS}^n - V_{GNSS}^n \\ P_{SINS} - P_{GNSS} \end{bmatrix} = \begin{bmatrix} \delta V_{SINS}^n - V_w \\ \delta P_{SINS} - P_w \end{bmatrix} = HX + V \tag{9}$$

Where $H = [0_{6\times3} \quad I_{6\times3}]$, $V = [-V_w \quad -P_w]^T$ is measurement noise, V_w and P_w are the velocity measurement noise and position measurement noise of the GNSS , respectively.

3. Simulation and Results

3.1. Simulation

The parameter setting of the swing model is the same as that in the literature [8].

CKF filtering algorithm [9] is used as nonlinear filtering method. In order to facilitate comparison, EKF and UKF filtering methods are also used in the simulation.

Set the initial horizontal misalignment angle as 1° and the azimuth misalignment angle as 30°. The standard deviation of GNSS speed measurement noise and position measurement noise are 0.1m/s and 10m respectively.

The motion trajectory of the mothership is as follows: the ship sail in a straight line with an acceleration of 1 m / s^2 in 0 to 20s, sail at constant speed in 20s to 40s, sail in a straight line with uniform deceleration of -1 m / s^2 in 40s to 60s, and sail in a straight line with an acceleration of 1 m / s^2 in 60s to 90s.

The filtering period is 100ms. Twenty transfer alignment simulations were carried out, and the curve was drawn after calculating the standard deviation of each error quantity, as shown in Fig.1 to Fig.3.

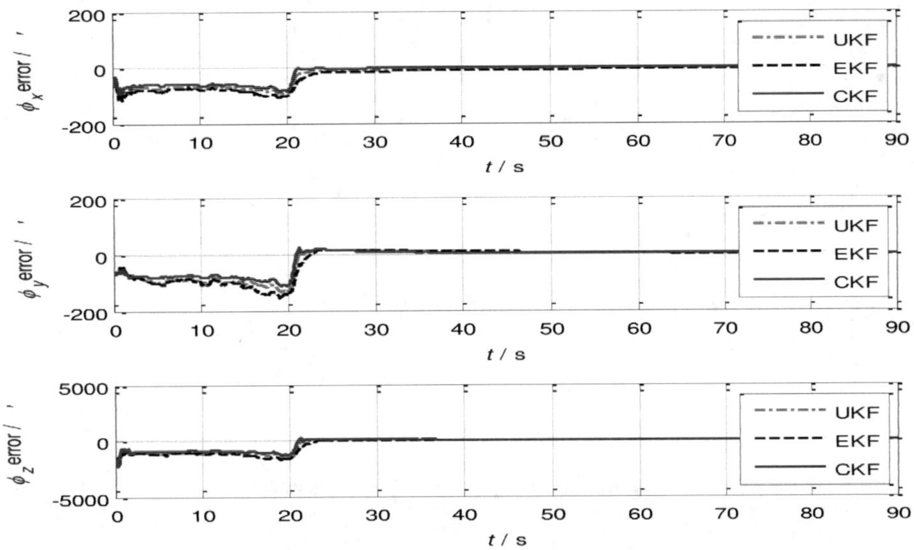

Fig.1 Misalignment angle error with GNSS assistance

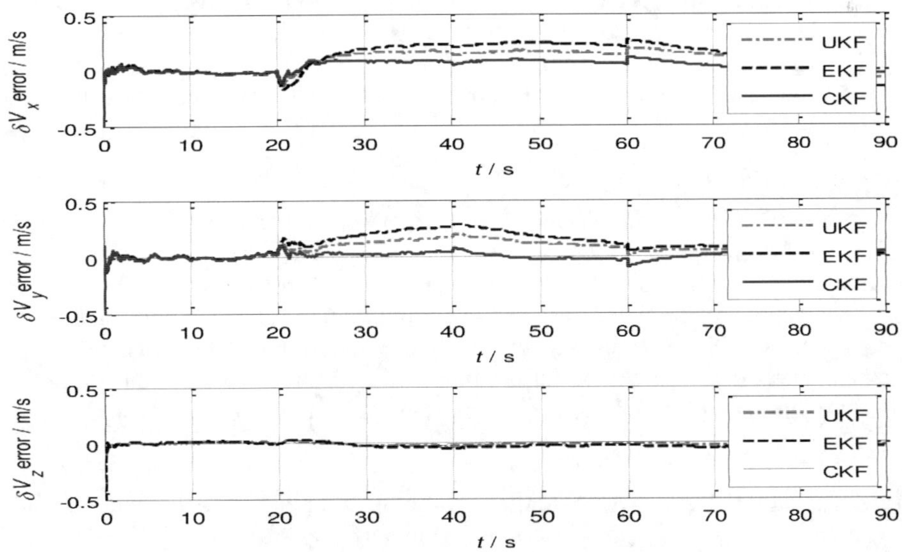

Fig.2 Velocity error with GNSS assistance

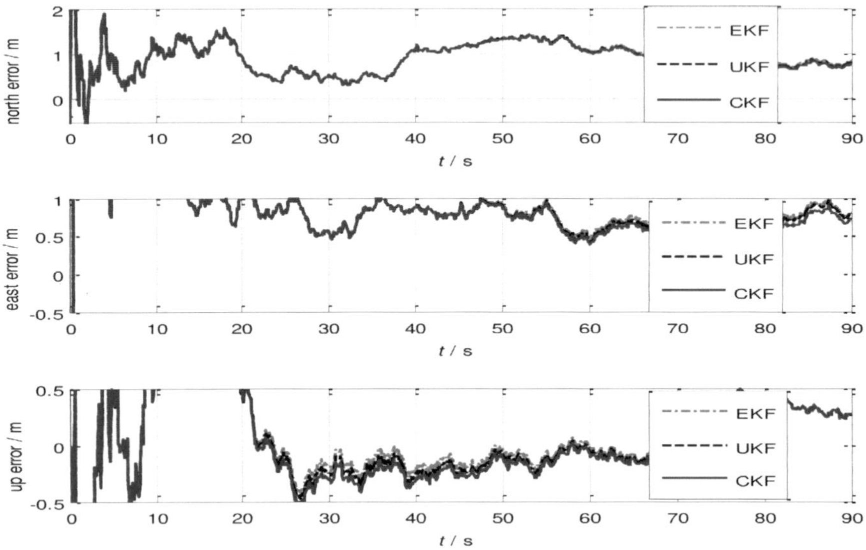

Fig.3 Position error with GNSS assistance

3.2. Results

According to Fig.1~ Fig.3, the following conclusions can be drawn:

(1) The CKF filtering can be used to estimate the attitude error angle under the condition of large azimuth misalignment angle. The time azimuth error begins to converge to within 30 ' in 30s, reaches the steady state in 80s, the horizontal estimation error reaches 1' and the azimuth error reaches 4 '.

(2) The CKF has the same accuracy as the EKF and the UKF, but the CKF does not need to calculate complex Jacobian matrix, and the amount of calculation is less than the UKF.

(3) The CKF filtering can complete the estimation of velocity error in 80s, with an estimation accuracy of 0.05m/s, and the estimation of position error within 30s, with an estimation accuracy of 1m.

4. Conclusions

Aiming at the problem of large azimuth misalignment angle alignment of shipborne aircraft inertial navigation system in shipborne environment, the alignment scheme based on CKF can effectively estimate the attitude error angle in a short time, realize the rapid transfer alignment of shipborne aircraft inertial navigation system in the case of large azimuth misalignment angle, and the estimation accuracy is equivalent to UKF with less calculation. The simulation results verify the feasibility and correctness of this method, which can be further modified and improved by combining with the sea test in the next step, and provide reference for engineering application.

References

[1] J Reid, M Tucker, R Dayan. An extended Kalman filter for the estimation of transfer alignment errors to an airborne vehicle[C]. AIAA Guidance & Control Conference, 1980, pp: 54-61.

[2] Julier S J, Uhlmann J K. Unscented filtering and nonlinear estimation[J]. Proceedings of the IEEE, 2004, 92(3): 401-422.

[3] YAN G.M., YAN W.S., XU D.M., Application of simplified UKF in SINS initial alignment for large misalignment angles.JOURNAL OF CHINESE INERTIAL TECHNOLOGY. 2008, 16(3).253-264. (in Chinese)
[4] Arasaratnam I., Haykin S.,Hurd T.R. Cubature Kalman filters. IEEE Trans on Automatic Control, 2009,54(6):1254-1269.
[5] Arasaratnam I., Haykin S.,Hurd T.R. Cubature Kalman Filtering for Continuous-Discrete Systems: Theory and Simulations. IEEE Trans on Signal Processing, 2010,58(10):4977-4993.
[6] DANG L.J., CHEN B.D., Cubature Kalman Filter Under Minimum Error Entropy With Fiducial Points for INS/GPS Integration.IEEE/CAA Journal of Automatica Sinica. 2022,9(03)
[7] QIN Y.Y. (2014) Inertial Navigation (2nd Edition). Science Press.Beijing. (in Chinese)
[8] XU J.S, WANG Y.J. Coarse Alignment Technique Based on SVD Method for SINS on Swing Base.2018 IEEE 3rd Advanced Infromation Technology,Electronic and Automation Control Conference,Chongqing,China.2278-2281.
[9] SUN F., TANG L.J., Estimation Precision Comparison of Cubature Kalman filter and Unscented Kalman Filter.Control and Decision.2013,28(2):303-308.

Experimental research on occlusion face detection method based on attention mechanism

Kexin Zhang [1a], Qing Zhu[2b] and Weiran Li[1*]

[1] School of Software Engineering, Beijing University of Technology, Beijing 100020, China

[2] School of Software Engineering, Beijing University of Technology, Beijing 100020, China

[a] zkx@emails.bjut.edu.cn, [b] ccgszq@bjut.edu.cn, [*]liweiran@bjut.edu.cn

Abstract. Face occlusion is a common problem in real life. Aiming at the problem that the face detection effect is poor when there is occlusion in the natural environment, this paper proposes the RetinaNet occlusion face detection network based on the fusion image enhancement network and the attention mechanism. Traditional RetinaNet, as a typical single-stage detector, does not explicitly give the process of extracting candidate regions and directly obtains the results of object detection. Compared with the two-stage detector, the processing speed is faster, but the accuracy will also be reduced. In this paper, the attention mechanism and image enhancement strategy are introduced to improve the detection accuracy while ensuring the detection speed.The experimental results on the MAFA dataset show that the RetinaNet model based on the attention mechanism proposed in this paper achieves 89.7% accuracy in detecting occluded faces. The feasibility and effectiveness of the model are verified.

1. Introduction

In recent years, with the rapid development of computer vision, target detection is a basic and important research direction. Face is the most important biological feature of human body. Face detection is an indispensable pre-work in face recognition[1]. It is also the basic step of computer vision tasks such as face correction, face verification and the basic premise of face related applications. Nowadays, the performance of face detection task has been greatly improved and can be applied in reality. However, face detection for occlusion caused by masks, glasses and other objects or body parts is still difficult[2]. Occlusion face detection algorithm often affects the detection performance because of input image destruction in practical scene. Therefore, how to achieve accurate face detection for people whose faces are occluded in real scenes is a very important research direction.

Traditional face detection algorithms often require manual feature extraction, which is vulnerable to external factors and unsatisfactory for face detection in occluded faces and complex scenes. Derived from the manually designed face detector, the robust features learned by the detector based on deep learning through the deep convolutional neural network greatly improve the detection performance, and are now widely used in face detection and recognition tasks. The target detection algorithm based on deep learning can be divided into two categories from the perspective of ideas. One is the two-stage target detector based on candidate box generation and classification, and the other is the one-stage target detector based on regression. The two-stage detector divides the whole detection process into two parts: generating candidate boxes and identifying objects in the box. There are four main tasks:

feature extraction, generating candidate boxes, classification and location regression. Rich features and high accuracy can be obtained, but the detection speed is relatively slow. The common two-stage detectors are R-CNN and Faster R-CNN[3]; the single-stage detector does not need to generate the candidate box, but only needs to complete three tasks: feature extraction, classification and location regression. The detection speed is faster, and it is suitable for use on mobile terminals and other platforms. Common single-stage detectors are RetinaNet[4], YOLOv3[5], etc.

Aiming at the problems that each backbone network is susceptible to background noise interference in the process of feature extraction, deep convolution is prone to lose spatial detail information and does not consider the multi-scale attribute of the target, attention mechanism is introduced into deep learning. The neural network model based on attention mechanism has been widely used in computer vision fields such as face detection. Therefore, based on the single-stage target detector RetinaNet, this paper combines the attention mechanism and increases the pre-network of image enhancement to study the occlusion face detection method based on the attention mechanism. The input image is enhanced, and the face feature information of the unoccluded part is strengthened. The context information is well preserved, and the information of the visible area is highlighted. The accurate and rapid detection of occluded faces in real scenes is realized.

2. Original RetinaNet detector

RetinaNet is a classic one-stage detector. Different from the traditional single-stage detection model, RetinaNet has improved both feature pyramid (Feature Pyramid Networks, hereinafter referred to as FPN)[6] and focus loss (Focal Loss, hereinafter referred to as FL)[7], and achieved dense and small-scale object detection. Very satisfied with the effect. However, it does not perform well when detecting on a dataset of occluded faces.

2.1. FPN

In the application of traditional computer vision, FPN is commonly used to solve the problem of objects with different sizes in the image to be detected. An image is first subsampled to form a series of pyramids with lower resolution and smaller size images; then hand-designed features are extracted from each layer in the pyramid to detect objects.

With the development of deep learning, hand-designed features are no longer used, but more efficient Convolutional Neural Network (CNN) is used. The output size of the feature map in the CNN architecture is reduced after each successive convolution operation, forming a pyramid structure. The feature pyramid network combines low-resolution semantic enhancement and high-resolution semantically weak features with rich semantics at all levels.

RetinaNet combines FPN and adds classification and regression sub-networks for model detection. The model architecture is divided into four parts, as shown in Figure 1.

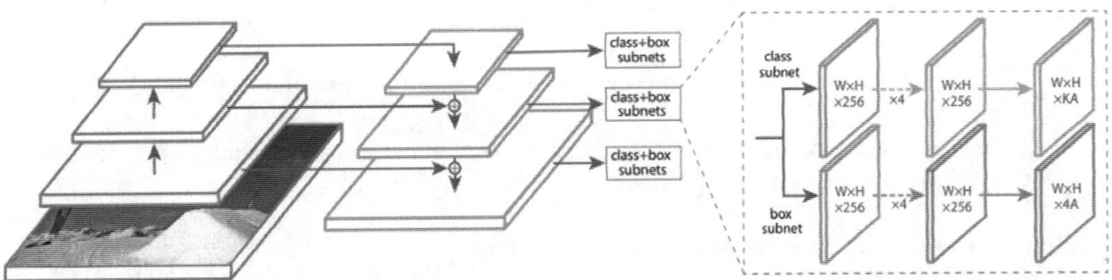

Figure 1. RetinaNet Model Architecture.

- Bottom-up path-self-convolutional backbone network: responsible for computing convolutional feature maps of different sizes on the entire input image;

- Top-down paths and lateral connections: top-down paths upsample coarser feature maps at higher pyramid levels; lateral connections merge bottom-up layers with top-down layers with the same size; so the network effectively builds rich multi-scale features from a single resolution input image;
- Classification subnet: predicts the probability of object existence at each spatial location for each anchor box and object class;
- Regression Subnet: Regresses the bounding box offset from each object's anchor box.

2.2. Focal Loss

FL is an enhancement to the cross-entropy loss to deal with the class imbalance problem of single-stage detection models. Due to the dense sampling of anchor boxes, the single-stage detection model suffers from severe foreground-background imbalance. In RetinaNet, there can be thousands of anchor boxes on each pyramid level, but the vast majority are background classes, and only a small fraction will be assigned to real objects. This loss function reduces the weight of a large number of simple negative samples during training.

The binary cross-entropy loss function is:

$$L = -y\log y' - (1-y)\log(1-y') = \begin{cases} -\log y', & y=1 \\ -\log(1-y'), & y=0 \end{cases}$$

Where y' is the output of the activation function, and the value range is between 0-1. It can be seen that the ordinary cross-entropy loss function for positive samples, the larger the output probability, the smaller the loss; for negative samples, the smaller the output probability, the smaller the loss. The loss function at this time is slow in the iterative process of a large number of simple samples and may not be optimized to the optimal solution. Based on this, factors are added on the original basis for improvement, and FL is obtained:

$$L_{ft} = \begin{cases} -(1-y')^{\gamma}\log y', & y=1 \\ -y'^{\gamma}\log(1-y'), & y=0 \end{cases}$$

Where $\gamma>0$ reduces the loss of easy-to-classify samples and pays more attention to difficult and misclassified samples.

In addition, a balance factor α is added, which can be used to balance the unbalanced proportion of positive and negative samples.

2.3. Summary

Although RetinaNet has been widely used in face detection applications, there are still the following problems:

- RetinaNet is easily affected by the decline in the quality of the input image, thereby reducing the recall rate;
- It is easy to be interfered by image background and face occlusion, resulting in false recall;
- Multi-scale information of input image is not fully considered.

3. RetinaNet Network Based on Attention Mechanism

Aiming at the problems of RetinaNet, this paper proposes an occlusion face detection algorithm based on attention mechanism[8]. In order to enhance the quality of the input image, the pre-general recognizable image enhancement network URIE is used. At the same time, the branch learning attention map is added, and the point after exponential operation of the attention map is multiplied to the original feature map, so as to strengthen the visible region information, Strengthen the visible area information, improve the recall rate and reduce false recall.

3.1. Image enhancement pre-network

In view of the possible distortion of input images in practical applications, this paper conducts experiments on the front URIE[9] network before occlusion face detection, and improves the detection performance by enhancing the input data. The URIE network is designed based on Selective Enhancement Module (SEM), and the architecture is shown in Figure 2.

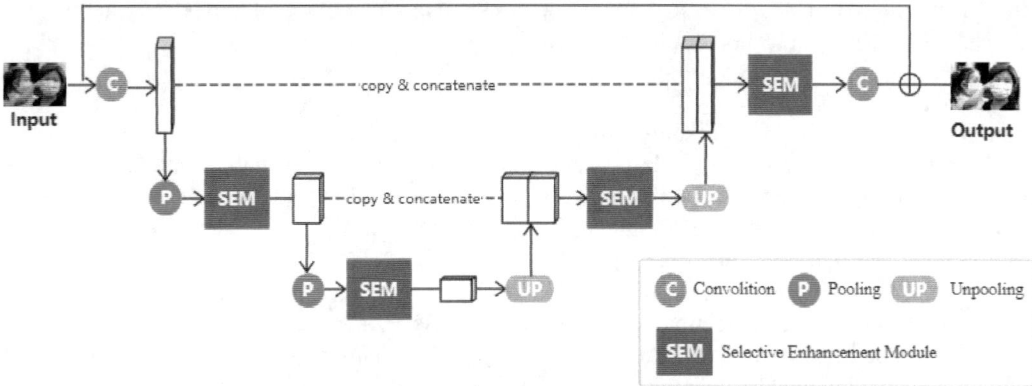

Figure 2. URIE Architecture.

The detailed structure of SEM is shown in Figure 3.

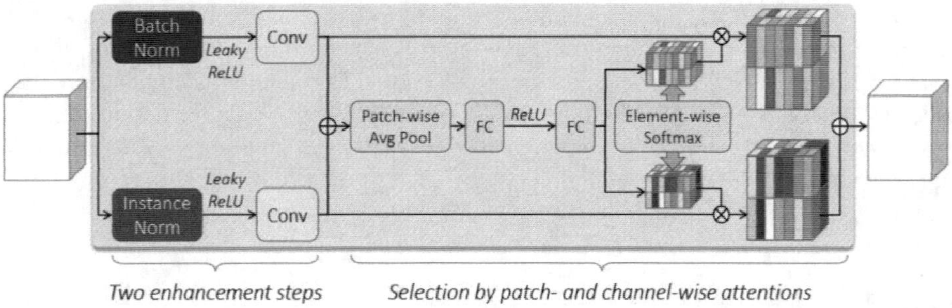

Figure 3. Detailed structure of SEM.

Each SEM provides two different image enhancement steps. Therefore, by cascading SEM, the network will have a multiplier enhancement process. In addition, each SEM has a focus mechanism that can assign greater attention weight to its enhancement steps, which enables URIE to dynamically select useful enhancement processes according to potential types. In the training process, URIE is coupled with the pre-trained image classification network on the original ImageNet dataset. Then, it is trained in an end-to-end manner to improve the performance of the classifier in distortion processing. Using ImageNet-c as training data, the original ImageNet image is compressed by 19 types of distortion. The training goal of the training strategy is not to output the perceptual quality of the image, but to restore the image so that the occlusion face detection algorithm can better process the image.

In URIE, the image enhancement process is represented as a series of composite layers. As a basic component of the network, SEM provides a separate step (convolution layer) for image enhancement, so a series of steps form an enhancement process. Specifically, a single point model has two different enhancement steps and can estimate which of them is more suitable for processing input distortion.

The two enhancement steps in SEM have independent convolution layers, which are related to instantiation specification (IN) and batch processing standardization (BN), respectively. These two different normalization operations are used to learn a pair of different and complementary enhancement functions. IN is applied to neural type conversion, and the image style is normalized while maintaining the main content of the image by normalizing the statistics of each image. When the distortion type is regarding as an image style, the enhancement step can adjust the distortion effect of a single image. On the other hand, the enhancement step with BN maintains another distortion information that is not considered, and is trained by considering different distortion images simultaneously in each small batch. Two outputs of these enhancement steps are fed to the attention module, which determines which of them is more useful for processing input image distortions for each channel and for each local color block entered.

3.2. Attention mechanism

Human attention mechanism is obtained from intuition, using limited attention resources to quickly select high-value information from large amounts of information. Attention mechanism in deep learning draws lessons from human attention thinking mode, and is widely used in various types of deep learning tasks such as Nature Language Processing (NLP)[10], image classification and speech recognition, and has achieved remarkable results.

The essence of attention mechanism is to locate the information of interest and suppress useless information. Occlusion face detection is a very difficult task. For such tasks, it is necessary to ensure that there are enough receptive fields to provide context information. Attention mechanism can help the network learn the visible region better. According to different principles, attention models are divided into spatial attention model, channel attention model, spatial and channel mixed attention model.

3.3. Implementation of occlusion face detection model based on attention mechanism

Based on sections 3.1 and 3.2, the image enhancement module is added before RetinaNet to enhance the input image quality; by integrating the attention mechanism, the data of human faces in MAFA data are analyzed, and a new anchors strategy is redesigned to mine the deep, shallow feature space, multi-scale and other attributes of the image, enhance the visible region information of the unoccluded part, and realize the occlusion face detection model based on attention mechanism. The detailed structure of the model is shown in Figure 4.

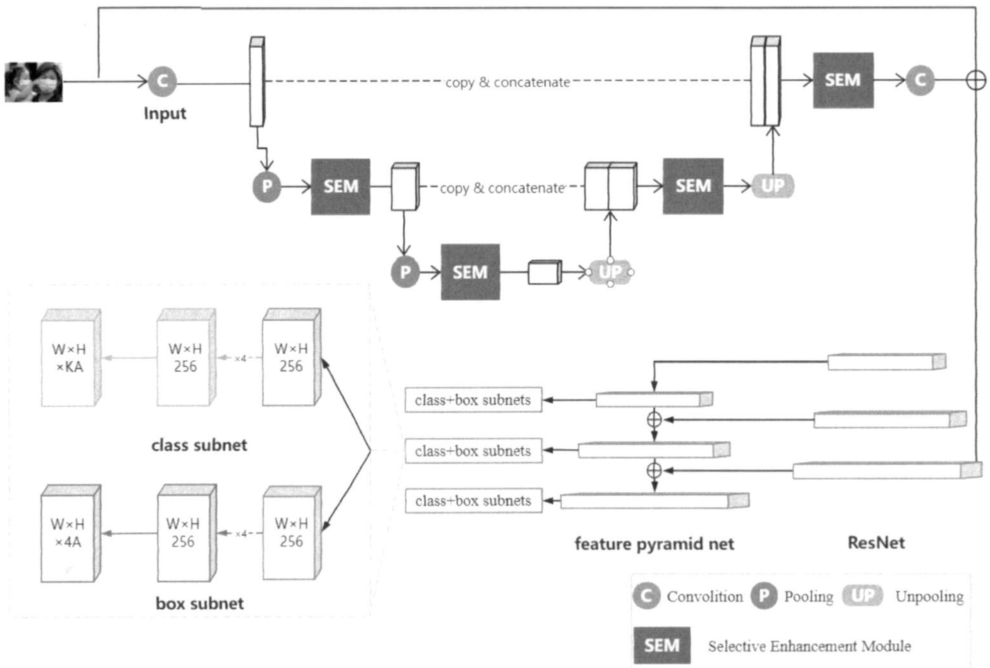

Figure 4. Overall structure of the model.

4. Test Results and Discussions

4.1. Description of specimens

The experimental environment is configured as follows: Google Colab server, memory 12GB, NVIDIA-SMI 460.32.03 graphics card. All programs are designed based on Pytorch framework and implemented by Python programming language.

4.2. Evaluating indicator

In this paper, average precision (AP) and mean average precision (mAP) are selected as the evaluation indexes of this occluded face detection algorithm. AP calculation formula is:

$$v_{AP} = \int_0^1 P(R)dR$$

The mAP calculation formula is:

$$v_{mAP} = \frac{1}{N}\sum_{i=1}^{n} AP_i$$

Among them, N is the total number of categories, Pi is the precision rate of category i; P(R) is the curve composed of precision rate (P) and recall rate (R), and the calculation formula is:

$$\begin{cases} P = T_p / (T_p + F_p), \\ R = T_p / (T_p + F_N), \end{cases}$$

Where Tp is the number of positive classes identified as positive classes; Fp is the number of negative classes identified as positive classes; FN is the number of unidentified positive classes marked as negative classes.

4.3. Parametric analysis

The important parameters that need to be considered when setting the attention network in this occlusion face detection model are the aspect ratio, size and density of anchors. By setting reasonable anchors, each anchor has a larger receptive field and implicitly learns context information.

4.3.1. anchors size. Firstly, the pixel distribution of face size in MAFA dataset is analyzed, as shown in Figure 5.

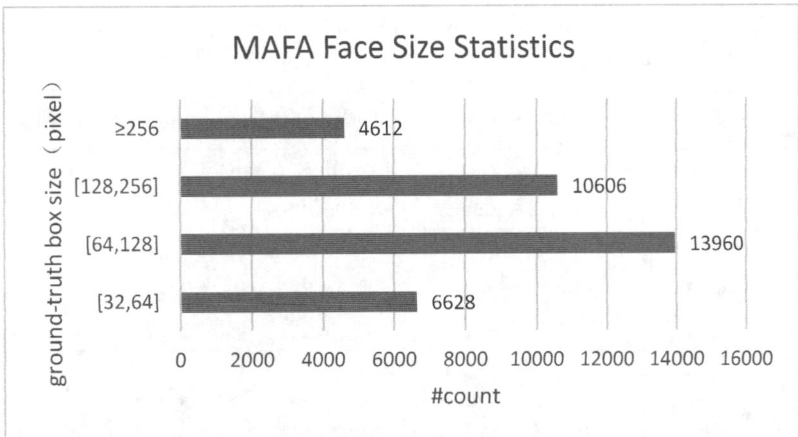

Figure 5. Pixel interval of face size in MAFA dataset.

Calculate the proportion of face size in all data sets:

Table 1. Proportion analysis of face size in all datasets.

Face size (pixel)	Proportion (total 35806)
32-64	19
64-128	39
128-256	30
≥256	12

Face size is concentrated in 32-256 pixels, accounting for 88%. Considering the low resolution of small size images, if it is included in the training data, noise may be introduced, but the detection performance will be reduced. In this experiment, whether the anchors size contains [32,64] pixels is tested under the condition of laying a anchor in each layer. The experimental results of different anchors size are shown in Table 2:

Table 2. Analysis of experimental results of different anchor size.

Anchor scale (pixel)	Result of survey		
	weak	medium	heavy
32-256	75.5	74.1	72.8
64-512	74.9	72.4	70.2

Because [32,64] pixel face accounts for 19% of the total data set, it will reduce the amount of training data without considering this part of the data, and eventually lead to a slight decline in the detection results. Considering the experimental data comprehensively, this study sets the anchors size to be laid from 32 * 32 pixels. In addition, considering the different orientations of faces in MAFA dataset, the aspect ratio of anchors is set to 1:1 and 1:1.5, mainly considering the front face and side face. Face, aspect ratio is 1:1; lateral face, aspect ratio is 1:1.5.

4.3.2. anchors laying density. In order to allocate each ground truth to the appropriate anchor as much as possible, this study adopts the method of laying multiple anchors in each layer. The density of each layer of anchor was tested, and the experimental results are shown in Table 3:

Table 3. Analysis of experimental results of different anchors densities.

Anchors density (pixel)	Result of survey		
	weak	medium	heavy
2	78.5	77.5	72.1
3	78.2	77.6	71.9
4	76.8	75.5	69.7

The experimental results show that there is little difference between the detection results of two anchors and three anchors in each layer, but when more anchors are laid, the false positive situation increases, and the performance decreases. Therefore, the method of laying two anchors at each layer is finally adopted to step by 2 ^ (− 1/2).

The final anchors are laid as shown in Table 4:

Table 4. Aanchor laying results.

Detection layer	Anchor scales	Radio
P3	32-45	1:1； 1:1.5
P4	64-90	1:1； 1:1.5
P5	128-181	1:1； 1:1.5
P6	256-362	1:1； 1:1.5

Among them, P3, P4, P5 and P6 are the layers of attention network, respectively. From P3 to P6, they represent more and more deep information. Different layers have different ground - truth face sizes. P3 and P4 focus on smaller faces, P6 focus on larger faces, and there is no much correlation between attention maps at all levels. The attention map is exponentially amplified and added to the feature, which not only highlights the face region features, but also retains more context information.

4.4. Experimental results and comparative analysis

This paper proposes a method to improve the quality of input image by adding a pre-network and fuse the attention mechanism to improve RetinaNet to complete occlusion face detection. In order to highlight the effectiveness of the proposed algorithm, the detection results of traditional HH, HPM, MT, LLE-CNNs and FAN algorithms based on MAFA dataset are compared. The average accuracy (AP) of each algorithm is shown in Figure 6.

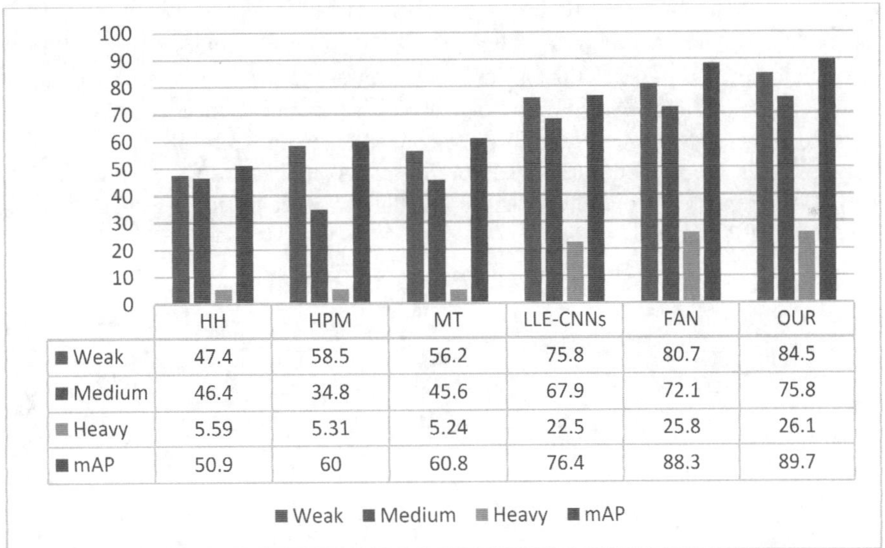

Figure 6. Average precision(%) on the testing set of MAFA.

5. Conclusion

In view of the difficulty in detecting occluded faces, this paper proposes a RetinaNet network based on attention mechanism, improves the traditional RetinaNet detector, pre-positions the image enhancement network that can improve the recognition performance, and integrates the attention mechanism. While improving the quality of the input image, the visible region information is highlighted by exponential amplification of the attention map. Different sizes of anchors are laid to fully detect different sizes of faces. The experimental results on MAFA and WilderFace face datasets show that compared with the existing face detection methods based on deep learning, the improved RetinaNet detection network can better and quickly detect occluded faces, which has a good application prospect in real scenes.

The effect of this method in detecting large-scale changed faces and highly occluded faces is still not ideal, and further exploration is needed. The basic network model proposed in this paper is a large one. Next, we will consider how to miniaturize the model.

References

[1] Yunqian, F. (2021) Research on Face Recognition Algorithm Based on Convolutional Nerve.J. Journal of Physics: Conference Series, 1966(1).

[2] Chen, L., Pang, Y. Context-Sensitive Multi-Scale Face Detection. Laser & Optoelectronics Progress, 56(4). p.041003.

[3] Jiang, H., Learned-Miller, E. (2017) Face detection with the faster R-CNN. In 2017 12th IEEE international conference on automatic face & gesture recognition (FG 2017). pp. 650-657.

[4] Zhang, T., Zhang, L. (2021) Multiscale Feature Fusion-Based Object Detection Algorithm. Laser & Optoelectronics Progress, 58(2). p.0215003.

[5] Cao, C., Yuan, J. Mask-wearing detection method based on YOLO-mask. Laser & Optoelectronics Progress, 58(8). p.0810019.

[6] Lin, T.Y., Dollár, P., Girshick, R., He, K., Hariharan, B., Belongie, S. (2017) Feature pyramid networks for object detection. In Proceedings of the IEEE conference on computer vision and pattern recognition. pp. 2117-2125.

[7] Charoenphakdee, N., Vongkulbhisal, J., Chairatanakul, N., Sugiyama, M. (2021) On focal loss for class-posterior probability estimation: A theoretical perspective. In Proceedings of the IEEE/CVF Conference on Computer Vision and Pattern Recognition. pp. 5202-5211.

[8] Wickens, C. (2021) Attention: theory, principles, models and applications. International Journal of Human–Computer Interaction, 37(5). pp.403-417.

[9] Son, T., Kang, J., Kim, N., Cho, S., Kwak, S. (2020) URIE: Universal Image Enhancement for Visual Recognition in the Wild. In European Conference on Computer Vision. pp. 749-765.

[10] Hu, D. (2019) September. An introductory survey on attention mechanisms in NLP problems. In Proceedings of SAI Intelligent Systems Conference. pp. 432-448.

[11] S., Ge, J., Li, Q. Ye, Z. Luo. (2017) Detecting Masked Faces in the Wild with LLE-CNNs. IEEE Conference on Computer Vision and Pattern Recognition (CVPR). pp. 426-434, doi: 10.1109/CVPR.2017.53.

Journal of Physics: Conference Series 2258 (2022) 012079 doi:10.1088/1742-6596/2258/1/012079

Research on electrical automatic control system based on PLC

Jingxiao Feng*, Xijuan Wang

School of physics and electronic information, Luoyang Normal University,

Luoyang, 471000, China

*Email: 1850508590@qq.com

Abstract: At present, automatic control technology is widely used in the field of drying. Although the existing processing technology can basically ensure the quality of drying, there are some problems, such as low production efficiency, low thermal efficiency, uneven hot air cloth, inaccurate temperature control and so on. The application of electromagnetic technology can improve the efficiency of drying and processing. It has remarkable energy-saving effect. In the drying process, aiming at the technical problems such as incomplete volatilization of water and difficult stability of temperature in rapid drying, a drying system based on PLC control is designed in this paper. Using PLC as the main controller of the control system, through the reasonable control of the changes of temperature and humidity in the drying process, a more efficient processing process can be realized. The drying quality can be improved, and the process flow of mass processing and production can be realized. Through the experimental analysis of the drying process, it is proved that the control accuracy of the research content in this paper is more accurate, which not only improves the automation level in the drying process, but also reduces the labor cost in the process, and has high technical application value.

1 Introduction

Drying requires high temperature and wind speed. Most of the traditional temperature control methods are manual control. Coal is added to the hot blast stove manually, and the hot blast temperature is controlled by the amount of coal added. This method will lead to excessive temperature fluctuation in the process of coal adding. Obviously, this method cannot be used for drying. Using some intelligent means in the drying process can not only improve the processing efficiency, but also improve the drying quality [1]. In order to improve the drying quality and production efficiency, the electromagnetic heating coupled hot air, heat pipe waste heat recovery and fluidized bed drying technology are introduced into the drying. The control system is debugged to realize the automatic control of drying. Verify the temperature regulation effect to realize the accurate control of temperature in the drying process [2].

Therefore, it is necessary to design a set of hardware control system, which is used to analyze the requirements for the control system in the drying process, select the hardware according to the analysis results, and allocate the I / O ports required by the hardware after selecting the appropriate hardware. Then draw the hardware electrical schematic diagram and wiring diagram, make the electrical cabinet corresponding to the wiring diagram, and install and connect the hardware. Design the hardware operation program for debugging, and carry out the next test after debugging [3]. The hardware of drying control system mainly takes PLC controller as the core. The reason why it is selected as the controller is that it has good stability and can avoid faults in the test process. Compared with other controllers, PLC has relatively simple structure, more function expansion modules, flexibility and convenience, relatively low maintenance cost and high efficiency [4].

The input signals involved in this design mainly include temperature value and humidity value, which are analog signals. The Siemens PLC used can only receive digital signals, so analog-to-digital conversion is required. The output signal is the speed of the blower, and the blower controller used can only receive analog signals. Therefore, the expansion module needs to be used for conversion. The overall design of the control system is shown in Figure 1.

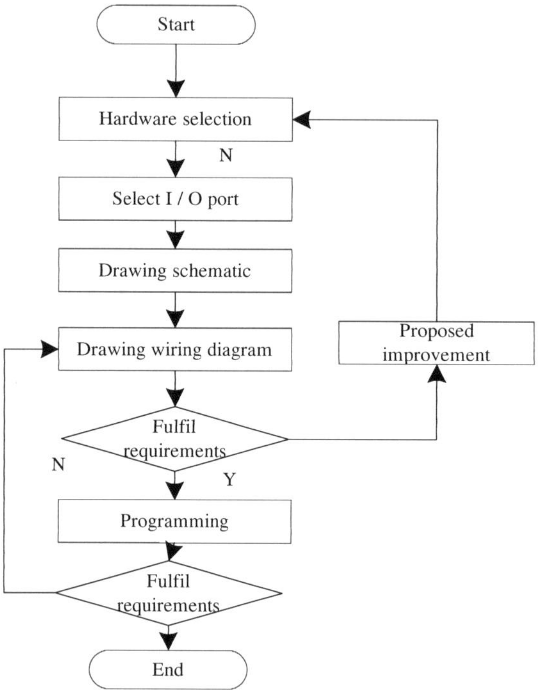

Figure 1 Overall design flow chart of drying control system

The PLC automatic control system is built. The electrical components are selected, and the Siemens CPU226 PLC is selected as the system controller to realize the automatic control of drying and reduce too much manual interference [5]. The communication between PC and PLC is realized; the I / O port of PLC is allocated and the electrical wiring diagram in line with the control scheme is drawn; the electrical cabinet is designed and the hardware installation is completed to realize the accurate control of temperature in the drying process. There are still many parameters that can be optimized for this test-bed. The structure and function of the control system can also be optimized. The accuracy of temperature control can also be studied to control the temperature in a more accurate range

2 Device selection of drying control system

2.1 Selection of PLC controller

At present, there are many kinds of PLC on the market. The power supply, I / O port, CPU, etc. are centrally installed in one casing, which is called integral structure PLC. Several slots are installed on the bus board, and each module can be inserted into the slot for use. This kind of PLC is called modular structure PLC. The difference between small and medium-sized PLCs lies in their I / O number and storage capacity. In practical production and application, sufficient I / O ports need to be reserved according to the demand, and the storage capacity should also be considered [6]. This design mainly selects the PLC from the number of I / O ports, memory capacity and scanning speed: from the number of I / O ports, this design not only needs to reserve a certain number of I / O ports for switching value, but also needs to consider that a certain number of I / O ports should be reserved for analog input and output modules. In terms of memory capacity, the approximate range of capacity is 10 times of digital quantity points, 100 times of analog quantity points, and 25% margin; in terms of scanning speed, PLC usually has hysteresis. And the scanning speed is about 1-2 scanning cycles. In the scanning process, the working sequence of PLC is reading input, executing user program, processing communication request, CPU self diagnosis and rewriting output [7].

Simens S7-200 PLC is selected in this design. The specific model of the PLC is CPU226AC / DC / relay. The pins are shown in Figure 2

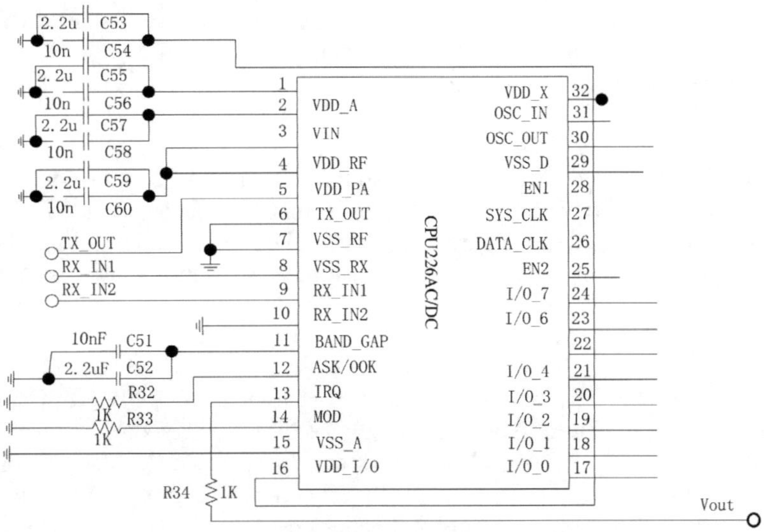

Figure 2 PLC pins

There are 24 input ports and 16 relay output ports inside. Cables can also be used to connect 8 analog inputs and 4 analog outputs. Its memory has a storage capacity of 16384B.

2.2 Selection of data acquisition card

This paper needs to explore the influence of different temperature and wind speed on drying. A large number of tests need to be done to find the law. A large number of data will be generated in the test process, and a conclusion will be drawn through the research and analysis of these data. Therefore, according to the requirements of the drying control system, the data acquisition card needs to be selected for selection. The main functions need to include the ability to realize multi-channel sampling and specify the required number of channels. The sampling accuracy is good. And it has large enough range, sufficient input and output ports, etc. At present, the common data acquisition cards include Altai multi-functional data acquisition card, hengkai data acquisition card, hengruifeng USB series data acquisition card, NI data acquisition card, etc. The USB-6215 data acquisition card produced by American NI company is a multi-functional DAQ device with isolation, powerful function, support of external digital trigger function and high sampling accuracy. At the same time, it can be directly connected with the bolt terminal, which reduces the setting time [8].

This acquisition card mainly collects the output data of temperature sensor and humidity sensor during the drying test, and uses AI0 and AI1 channels to detect the analog data of temperature and humidity parameters respectively so as to obtain their data as well as provide data support for the drying and baking effect test.

2.3 Selection of other hardware

The control system adopts the method of multi-point sampling for the temperature in the drying box. PT100 platinum thermistor temperature sensor is selected. The drying process requires high-precision temperature value. This sensor has high measurement accuracy and stable performance, and can meet the test requirements. The basic working principle is that the internal resistance value of the sensor will change with the change of temperature. The specific temperature can be measured by measuring the change of resistance [9]. PR-3002-WS temperature and humidity transmitter is selected to realize multi-point sampling. The extended metal waterproof probe of this transmitter is used as the temperature measuring element to mainly sample the temperature and humidity in the drying box during drying. The basic principle is: the collected temperature and the humidity signal are converted into linear voltage signal output after circuit processing (including steady-state filtering, operational amplification, etc.). The specific temperature value and humidity value are obtained after linear calculation. The measurement accuracy is high, which meets the requirements of this design.

The temperature transmitter selected for the control system is SBWZ-2280, which is used as the signal conversion element of the thermal resistance temperature sensor. The PT100 thermal resistance sensor is used as the signal input. After passing through the transmitter, the signal is converted into a voltage output signal of 0-10V. The two-wire transmission mode is adopted, which has the characteristics of wide ambient temperature range, high precision, stable and reliable operation and low energy consumption. The basic error of this transmitter is ± 0.2%, which meets the requirements of this design [10].

3. Overall scheme

The model of AC contactor selected for the control system is CJX2-0910. The function of AC contactor in the system is to control the start and stop of blower. The control coil of AC contactor is connected with the control end of PLC. The pull in and break of coil contact is controlled by PLC. This AC contactor adopts silver alloy contact, which can carry large current. And it has high sensitivity and strong safety, and meets the requirements of this design.

The model of the solid-state relay selected for the control system is FDR 3-d48100z. The solid-state relay is mainly used to control the start and stop of the heater. Because the test has high requirements for temperature control and the time required for each round of test is long, the selection requirements of the solid-state relay are high. The control current range of this solid-state relay is 3-36A DC. The maximum through current can reach 100A, which can avoid damaging the device due to excessive instantaneous current and has the advantages of long service life. The shell is flame-proof, strong toughness, high temperature resistance, no deformation during long-time operation, and fast heat dissipation. This solid state relay can meet the requirements of this test.

LSA-H3p40YB voltage regulating module is selected for the control system. This module is mainly used to adjust the blower speed, which has the advantages of convenience, accuracy and efficiency. In addition, this module supports PWM wave, analog quantity and potentiometer regulation. Its control end is connected with the analog output module, which is controlled by PLC. The specific parameters obtained by consulting the data are shown in Table 1.

Table 1 Parameter table of voltage regulating module

setting	parameters
control voltage	0-5V，0-10V
Control current	4-20mA，0-10mA
Control potentiometer	10K
Rated voltage	220V/380V
Minimum load response current	<10ms
turn-off delay	<10ms

The Siemens model SITOPPS207 power module is selected for the control system. The function of this module is to provide 24V DC power supply for the system with strong safety performance. Its 24V DC output interface is 2 pairs and the rated current is 2.5A, which can meet the requirements of this test.

Table 2 shows the hardware configuration statistics of the control system. The I / O address allocation is shown in Table 3.

Table 2 Main hardware configuration of drying control system

Names	Types
Programmable logic controller	S7-200CPU226CN
Analog input module	EM231CN
Analog output module	EM232CN
touch screen	SMART700IEV3
DC power module	SITOPPS207
Thermal Resistive Sensors	PT100
Temperature and humidity transmitter	PR-3002-WS
AC contactor	CJX2-0910
Three phase solid state relay	FDR3-D48100Z
data acquisition card	USB-6215
Blower	EM180B-3
Voltage regulating module	LSA-H3P40YB
Wind speed sensor	GM-816

Table 3 Signal address allocation table of drying control system

Terminal address	Connected device	Description
I0.0	Start button	Device start
I0.1	Stop button	Device stop
Q0.0	Solid state relay	Start and stop of heater
Q0.4	AC contactor	Start and stop of blower
AIW0	Temperature transmitter1	Temperature acquisition
AIW2	Temperature transmitter2	Temperature acquisition
AIW4	humidity transmitter	humidity acquisition
AQW0	Voltage regulating module	blower speed

The hardware is wired and installed according to the electrical wiring diagram, and the internal structure of the completed PLC electrical cabinet is shown in Figure 3.

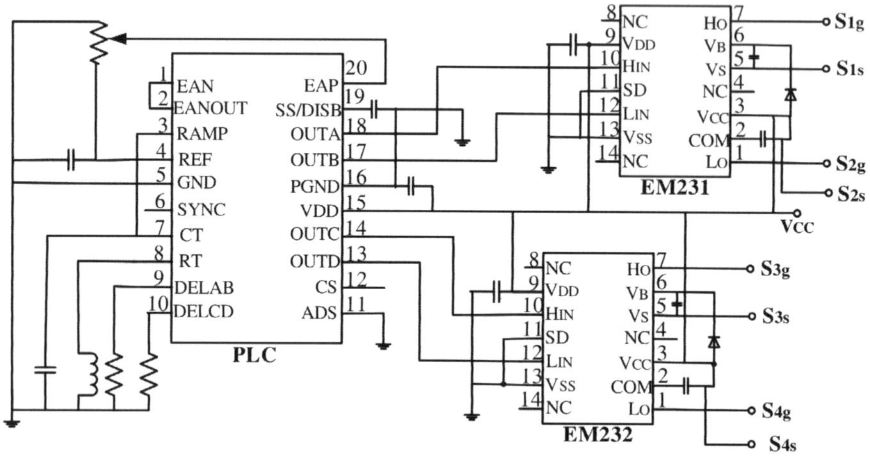

Figure 3 Hardware and electrical wiring diagram of drying control system

4 Effect test verification

In the process of drying, the requirements for temperature are high. Too high, too low or too much temperature fluctuation will affect the drying quality. Therefore, it is necessary to ensure that the temperature in the drying box is in a stable state after reaching the preset value. The preset temperature range of this test is 45 ℃ - 70 ℃. Considering the accuracy requirements of actual production temperature control, the test sets the temperature gradient as a gradient every 5 ℃ to verify the temperature PID effect. The preset temperature output characteristic curves of 55 ℃ and 70 ℃ are shown in Figure 4.

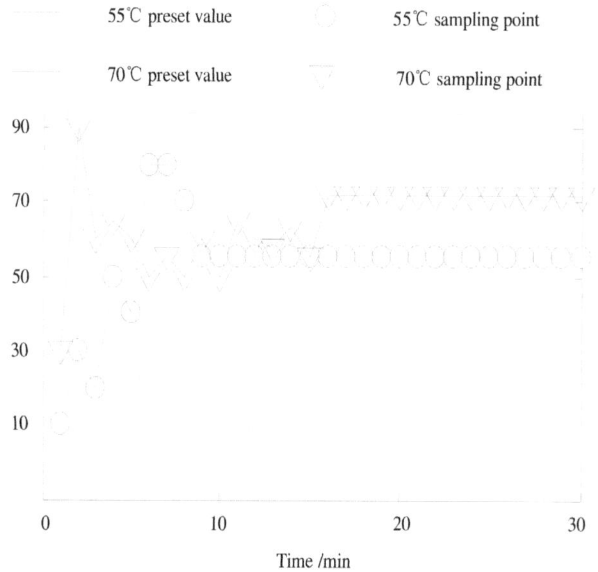

Figure 4 temperature curve

In Figure 4, the temperature effect in the drying box is remarkable. The temperature in the drying box will experience a peak when it rises from the room temperature to the preset value. When the temperature reaches the peak, it gradually tends to be stable. The temperature change range in the whole regulation process is relatively gentle. The test data under different preset temperatures are shown in Table 4.

Table 4 experimental data

Preset value /℃	Highest value /℃	Minimum value /℃	average value /℃
45	48.796	43.257	45.719
50	54.617	47.65	50.521
55	58.579	53.137	55.329
60	63.759	57.306	60.302
65	68.54	62.243	65.194
70	73.941	67.95	70.84

In Table 4, when the preset temperature values are different, the difference between the maximum and minimum temperature and the preset temperature in the whole regulation process is within ± 5 ℃. The difference between the average value and the preset temperature is small, indicating that the temperature control method can provide a relatively stable drying environment for drying.

5. Conclusion

Aiming at the problems existing in the drying process, this paper designs a set of hardware control system. The paper introduces the hardware design principle of the drying control system in detail, and draws the overall design flow chart of the drying control system. PLC is selected as the controller of this paper. According to the requirements of the control system, the hardware of the control system is counted. The I / O address is allocated by combining with the drying test. The hardware circuit wiring diagram is drawn, and finally the design and wiring of the electrical cabinet are completed.

There are still many parameters that can be optimized for this test-bed. The structure and function of the control system can also be optimized. The accuracy of temperature control can also be studied, which can control the temperature in a more accurate range.

References

[1]Wei Wang, Yiling, Li Wang. Research on Grain Drying Control System based on Microcontroller[J]. World Scientific Research Journal, 2020, 6(11).

[2]Amir Ebrahimifakar, David Yuill. Inverse estimation of thermophysical properties and initial moisture content of cereal grains during deep-bed grain drying[J]. Biosystems Engineering, 2020, 196.

[3]Kulothungan , Kamalakannan , Thirugnanam. Agriculture robot for irrigation and automation[J]. Bulletin of Pure & Applied Sciences-Geology. 2018, 37 (1): 110-115.

[4]Raikwar, Laxmikant1arun kumar, Sreenivasulu Rao. Hardware-in-the-Loop test automation of embedded systems for agricultural tractors.[J].Measurement.2019:271-280.

[5]Fu Li , Fu Xw , Zhang Yx. Design on automatic rolling system for agricultural greenhouse[J]. Aip Conference Proceedings. 2018:020031.

[6]Bouketir. An automatic irrigation system for water optimization in the Algerian agricultural sector [J]. Agricultural Science and Technology. 2019, 11 (2):133-137.

[7]Dai An, Zhou Xg, Liu Xd. Design and Simulation of a Genetically Optimized Fuzzy Immune PID Controller for a Novel Grain Dryer[J].IEEE Access.2017:14981-14990.

[8] An Taoyu. Analysis of the Common Problems and Coping of Siemens PLC in Factory Application [J]. Science and Technology Innovation Dress, 2020, 17 (03): 69-70.

[9] Du Wei. Application of PLC anti-interference technology in industrial control system [J]. Industrial Technology Innovation, 2020, 2 (34): 41-42.

[10] An Feipeng, Niu Xinjie, Bai Fail. Research on anti-interference technology of automated control system [J].145 + 148.

AUTHOR INDEX